Springer Series in Plasma Science and Technology

Plasma Science and Technology covers all fundamental and applied aspects of what is referred to as the "fourth state of matter." Bringing together contributions from physics, the space sciences, engineering and the applied sciences, the topics covered range from the fundamental properties of plasma to its broad spectrum of applications in industry, energy technologies and healthcare.

Contributions to the book series on all aspects of plasma research and technology development are welcome. Particular emphasis in applications will be on high-temperature plasma phenomena, which are relevant to energy generation, and on low-temperature plasmas, which are used as a tool for industrial applications. This cross-disciplinary approach offers graduate-level readers as well as researchers and professionals in academia and industry vital new ideas and techniques for plasma applications.

More information about this series at http://www.springer.com/series/15614

Werner Ebeling · Vladimir E. Fortov
Vladimir Filinov

Quantum Statistics of Dense Gases and Nonideal Plasmas

Springer

Werner Ebeling
Institut für Physik
Humboldt Universität Berlin
Berlin
Germany

Vladimir Filinov
Joint Institute for High Temperatures
Russian Academy of Sciences
Moscow
Russia

Vladimir E. Fortov
Russian Academy of Sciences
Moscow
Russia

ISSN 2511-2007 ISSN 2511-2015 (electronic)
Springer Series in Plasma Science and Technology
ISBN 978-3-319-66636-5 ISBN 978-3-319-66637-2 (eBook)
DOI 10.1007/978-3-319-66637-2

Library of Congress Control Number: 2017952008

Printed on acid-free paper

This Springer imprint is published by Springer Nature
The registered company is Springer International Publishing AG
The registered company address is: Gewerbestrasse 11, 6330 Cham, Switzerland

Preface

The physics of dense gases and nonideal plasmas, along with the physics of matter with high energy density, form together a growing field. The physics of dense gases is connected with the name of the pioneer Johannes Diderik van der Waals from the Netherlands (Nobel Prize 1910). The foundations for the theory of nonideal plasmas were laid 50 years later by another pioneer from the Netherlands, Peter Debye (Nobel Prize 1936), with the introduction of the screening concept. So it should be no surprise that the quantum statistics of dense gases also began in the Netherlands with the work of Uhlenbeck and Beth in 1936/1937. Important contributions to the physics of real gases are due to subsequent work by Mayer, Fuchs, Kirkwood, Bogolyubov, Yvon, Born, and Green. This development culminated in the van der Waals Centennial Conference in Amsterdam 1973 and international conferences on statistical physics organized by the IUPAP.

The physics of nonideal plasmas, otherwise known as strongly coupled plasmas, began with the work of Debye and Hückel, Onsager, Falkenhagen, Bjerrum, Eggert, and Saha in the 1920s and culminated in the 1960s. The first conference on "Strongly Coupled Plasmas" was held in 1977 in Orleans la Source, France, and was followed by a series of conferences under this name. A parallel series called "Physics of Nonideal Plasmas" started in 1980 in a village near Rostock, and then continued biennially, with the latest in Almaty, Kazakhstan, in 2015. The physics of extreme states of matter is still in *status nascendi.* The increasing interest in dense gases and plasmas is connected with the fact that more than 99% of the visible Universe is evidently in this state, and a large part of this exists at extremely high energy density. We are only just beginning to explore the world outside the narrow window of the little solid state planet on which we live. In order to understand our place in the Universe, we have to extend research to dense gases and plasmas including exotic states.

In this view, we feel like part of a large international collaboration which includes many researchers, universities, and very big research organizations such as CERN in Geneva, DESY in Hamburg, UNILAC and SIS in Darmstadt, ITEP in Moscow, SPS, LHC, and so on. A second group of big international collaborations like COBE and subsequent projects is concentrating on the study of astrophysical objects. This is also at present at the forefront of international research. It is based on new possibilities for observing distant objects from satellites and rockets. Many exciting insights have come recently from this field.

However, this book is definitely not about big colliders, telescopes, and space missions. Our aim is to explore the basic physics and in particular the quantum statistical thermodynamics and kinetics of states of matter, starting from dense gases and nonideal plasmas and ending with matter at high energy densities. We are convinced that thermodynamics and quantum statistics are still the main foundation on which even the most advanced research is being built. In spite of the development of so many modern concepts, the seemingly old concepts of thermodynamics and transport are still the basis for the whole field.

The body of the present book is based on lectures at universities and presentations at seminars and international conferences. Furthermore, the book draws upon many original publications and in particular on a long collaboration between the present authors which started in the 1970s. We mention in particular:

- The courses of lectures on quantum statistics and plasma physics given by Werner Ebeling at the University of Rostock between 1970 and 1979, at the University of Paris VI in 1977, at the Humboldt University in Berlin, between 1980 and 2001, and in guest lectures in Minneapolis 1986, Moscow 2003, and Krakow 2005. The latest full lecture cycle was given at Humboldt University in Berlin by Werner Ebeling in collaboration with Thorsten Pöschel and several former coworkers, including Dieter Beule, Andreas Förster, Lutz Molgedey, Jens Ortner, Waldemar Richert, and Ilya Valuev, and several former students and aspirants, including Jörn Dunkel, Hendrik Hache, Stefan Hilbert, Dirk Holste, Ines Leike, Ulf Leonhardt, Burkhard Militzer, Thomas Pohl, Saltanat Sadykova, Friedemann Schautz, Michael Spahn, Mario Steinberg, and others.
- Lecture courses given by Vladimir Fortov at the Moscow Institute of Physics and Technology on extreme states of matter on earth and in the cosmos (Fortov 2008, 2009, 2011).
- Presentations of the present authors at a series of international conferences, such as the IUPAP conferences under the headings "Statistical Physics" (STATPHYS), "Strongly Coupled Coulomb Systems" (SCCS), and "Physics of Nonideal Plasmas" (PNP).
- The long-standing collaboration between the authors and their colleagues in Berlin, Rostock, Moscow, and other research centers around the world, resulting in many shared articles and several shared books since the 1980s.

Following the personal interests of these authors, which go along the same lines in the tradition of quantum statistical thermodynamics in Berlin, Rostock, and Moscow, we concentrate on the development of fundamentals and applications to gases and plasmas, including dense nonideal and exotic gases and plasmas. Most of the existing textbooks and monographs on quantum statistics have some bias toward condensed matter and solid states.

This book is written at an intermediate level and addressed to students and young scientists at an advanced level. We include a few results obtained only recently. In general, the book should be accessible to students of the higher semesters, doctorands, and young researchers in the field. We have tried to be self-contained, repeating, and explaining the most relevant tools.

During their careers, the present authors, or at least some of them, have actually met several of the pioneers of the quantum statistics of dense gases and plasmas, such as Alexander A. Abrikosov, Nikolay N. Bogolyubov, Alexander S. Davydov, Hans Falkenhagen, Michael Fisher, Vitali L. Ginzburg, Günter Kelbg, Yuri L. Klimontovich, Rolf Landauer, Joel Lebowitz, Joseph E. Mayer, Ruslan L. Stratonovich, Alexander A. Vedenov, Yakov B. Zeldovich, and others. These people influenced our views through their advice and personal discussions, and so we must express our gratitude to them. In particular, we are grateful to Günter Kelbg, Yuri L. Klimontovich, and Yakov B. Zeldovich. Furthermore, we are very grateful for a long and very fruitful collaboration with many colleagues in the field, including David Blaschke (Wroclaw), Michael Bonitz (Kiel), Alexander Chetverikov (Saratov), Dietmar Ebert (Berlin), Viktor A. Gryaznov (Chernogolovka), Holger Fehske (Greifswald), Yuri B. Ivanov (Moscow), Wolf D. Kraeft (Greifswald), Dietrich Kremp (Rostock), Pavel Levashov (Moscow), Genri Norman (Moscow), Gerd Röpke (Rostock), Ronald Redmer (Rostock), Heidi Reinholz (Rostock), Manfred Schlanges (Greifswald), Boris Sharkov (Moscow), Werner Stolzmann (Kiel), Sergey Trigger (Berlin), and Manuel G. Velarde (Madrid). We should also mention that the last two chapters were written in close collaboration with A.S. Larkin, who provided the main contributions to the results. Particularly sincere thanks go to Thorsten Pöschel, who actively participated in the lectures at Humboldt University in Berlin and wrote lecture notes describing the basic tools of quantum statistics, which will appear separately.

In conclusion, let us express the wish that the present book might contribute to the general education of the present generation of physicists in the field of dense fluids. Two of the authors are theoreticians, while the other (V.E.F.) took part in many pioneering experiments in the field, including experiments with extreme pressures generated by shocks or laser beams which helped, e.g., in experimentally confirming plasma phase transitions. V.E.F. has also been in charge of major international research projects and experiments. When running these and other big research projects in our field, we have reached the conclusion that physicists of the younger generation need more knowledge and ability to solve problems involving dense gases, nonideal plasmas, and extreme states of matter. We hope the book will help them to understand the problems and methods of a rather new and fascinating field.

Berlin, Germany — Werner Ebeling
Moscow, Russia — Vladimir E. Fortov
Moscow, Russia — Vladimir Filinov
May 2017

Contents

Chapter 1
Physics of Dense Gases, Nonideal Plasmas, and High Energy Density Matter

Here we summarize the most important results in this field of physics, which is growing due to the dominant role of these forms of matter in the cosmos. We describe the progress made in physical studies and the statistical theory of dense gases and nonideal plasmas, including their historical roots in the work of van der Waals, Debye, Saha, Planck, Einstein, and others. We present the basic tools required for the quantum statistical description of nonideal fluid systems, including analytical methods and computer simulations, and we discuss studies of plasma-like matter with high energy density.

1.1 Strongly Coupled Fluid Matter: A New Field of Physics

In 1873, in his dissertation presented at the University of Leiden, the Dutch physicist Johannes Diderik van der Waals developed a new model of dense gases and fluids. This work opened the way to understanding matter in nonideal states (van der Waals 1873) (see Fig. 1.1). By *nonideal states* we mean here states of matter that are not described by the classical ideal gas law, i.e., the relation between pressure, density, and temperature $p = nk_{\mathrm{B}}T$. Fifty years later, Peter Debye (1884–1966), another eminent Dutch scientist, founded the science of nonideal Coulomb systems with a lecture given in 1923 to the Nederlandsch Natuuren Congres (see Fig. 1.2). The enormous progress made in the science of nonideal gases and nonideal plasmas in the 100 years after the initiation by van der Waals and his school and 50 years after the work of Debye and his school was summarized in 1973 at the van der Waals Centennial Physics Conference in Amsterdam.

Many physicists believe that this field is of less importance than the physics of condensed matter, since solid state electronics has such a big impact on our everyday life. But this is certainly a form of ignorance on their part, since less than one percent of matter in the Universe is in a condensed state. More than 99 percent of our world

W. Ebeling et al., *Quantum Statistics of Dense Gases and Nonideal Plasmas*, Springer Series in Plasma Science and Technology, DOI 10.1007/978-3-319-66637-2_1

Fig. 1.1 J. D. van der Waals (1837–1923), Nobel Prize 1910. Photo from Nobelprize.org

Fig. 1.2 Peter Debye (1884–1966), Nobel Prize 1936, and his coworkers (Hans Falkenhagen, left, Kasimir Fajans, second from the right, and Lars Onsager, right). Photo from Falkenhagen's archive

is in a gaseous, plasma, or extreme fluid state. Our planet is surrounded by a gas atmosphere and we use gases in many technological devices. Space around the Earth and interstellar matter is mostly in a gaseous or plasma state. Furthermore, we should not forget the matter in the Sun and the many stars, from the giants to the white dwarfs and exotic stars such as neutron stars and others, as well as many less known objects like 'dark matter'.

In the last few decades, condensed states of matter have dominated physicists' education, but we should bear in mind that, on the scale of the Universe, the solid state is a relatively rare form of matter. This rather special state is based on bound states of electrons and nuclei, forming atoms and molecules. Bound states exist only in a small region of density and temperature. Such conditions came into existence a few billion years ago on our planet and made the evolution of life and technology possible (see, e.g., Feistel et al. 1989, 2011). Life is based on the existence of atoms and molecules as bound states of electrons and nuclei. However, in most parts of our Universe, e.g., in stars, atoms do not exist, since the densities and temperatures are too high. Most matter is in extreme plasma states under very high pressure, like the matter in white dwarfs and neutron stars. The physics of gases and plasmas at high pressures and temperatures plays an important role in our understanding of the structure and evolution of astrophysical objects: neutron and 'strange' quark stars, black holes, pulsars, supernovas, magnetars, giant planets, and exoplanets. In the future, this new physics may also be relevant for technological development (Fortov 2011, 2013).

In the last few decades there has been extensive work on dense plasmas, with applications ranging from inertial confinement fusion, Z-pinch experiments, X-ray Thompson scattering, and exploding wire experiments to describing the astrophysics of white dwarfs and the interiors of giant planets (Fortov 2011, 2013).

In the present book we consider mostly dense forms of hydrogen, helium, and other noble gases and plasmas, as well as alkali plasmas. A topic of special interest to us is chemical and phase equilibria. Central issues are the influence of strong coupling on the equation of state and transport properties. Experimental methods are not discussed in detail in the present book. A minimum of information is given about important experiments and references (for more detail, see Fortov 2009, 2011, 2013).

1.2 Physics of Dense Classical Fluids

1.2.1 Van der Waals Equation of State and Interactions

The theory of gases developed in the dissertation presented by van der Waals in 1873, and this may be considered as the starting point for the modern theory of nonideal fluids and phase transitions. Van der Waals' approach was based on a simple physical model of interactions between particles which takes into account short range repulsive and long range attractive forces.

The model equation for the pressure is

$$p = \frac{k_B T}{v - b} - \frac{a}{v^2}, \qquad v = \frac{V}{N} = \frac{1}{n}, \tag{1.1}$$

where V is the volume, N the particle number, n the density, and T the temperature. The van der Waals model predicts that, below a critical temperature T_c, there will coexistence two phases which differ from each other by the density of molecules. In connection with the development of more rigorous theories, it became clear that van der Waals' approach is restricted to relatively weak attractive forces which either decay with the distance faster than $1/r^3$ or satisfy the so-called Kac conditions. Therefore the conditions for applicability to Coulomb forces are not given.

Prototype Models of Interaction

There are hundreds of models describing the interactions between molecules, atoms, or elementary particles (Hirschfelder et al. 1954). There are forces of attraction like van der Waals forces, chemical forces, etc., and repulsive forces due to Coulomb repulsion between charges and the Pauli principle. The typical shape is a decaying function with a minimum. Such systems are called prototype models when they play a special role, either because they possess some universality for classes of real substances or because they allow exact solutions. We restrict our study mainly to three-dimensional problems.

The simplest model of interactions is the hard-sphere model. The interaction between two hard spheres with diameter d is described by the hard core potential:

$$U_{\mathrm{HC}}(r) = \begin{cases} \infty & \text{if} r < d, \\ 0 & \text{if} r \geq d. \end{cases} \tag{1.2}$$

Including a range of attractions yields the piecewise constant square-well potential. Many results exist also for a prototype model with softer repulsion (Hansen et al. 1976):

$$U_{\mathrm{SR}}(r) = \varepsilon \frac{a^n}{r^n}, \qquad n = 6, 9, 12, \ldots . \tag{1.3}$$

These potential models are more or less empirical. The Coulomb potential is the special case with $n = 1$. Calculations based on quantum-mechanical first order perturbation theory yield an exponential repulsion based on the forces due to the Pauli exclusion for overlapping atomic core wave functions. In the simplest approximation, this leads to an exponential repulsive law:

$$U_{\mathrm{E}}(r) = U_0 \exp\big[- b(r - \sigma)\big] . \tag{1.4}$$

This potential contains the hard-core potential in the limit $b \to \infty$. Another interaction potential based on quantum mechanical calculations is the Morse potential:

$$U_{\rm M}(r) = D\Big\{\exp\big[-2b(r-\sigma)\big] - 2\exp\big[-b(r-\sigma)\big]\Big\}\,. \tag{1.5}$$

Here the positive term describes the repulsive forces due to the Pauli exclusion for overlapping atomic core wave functions. The negative term models qualitatively the attraction due to induced quantum-mechanical dipole–dipole forces.

The Toda model is an exponential potential with an additional (nonphysical) linear attraction, leading to a minimum of the potential at $r = \sigma$. Beside the depth of the potential $-D$, further important physical information is contained in the frequency of oscillations around the minimum $m\omega_0^2 = 2Db^2$ and in the stiffness of the potential which is proportional to the parameter b. The Toda potential is very useful since it allows fully analytical calculations. Figure 1.3 shows that the Toda potential and the Morse potential agree nicely at smaller distances, up to the minimum and a bit beyond. Note that a good fit of the Toda and Morse potentials is obtained near the minimum if the Toda parameters are related to the Morse parameters by the relations $a_{\rm T} = (2/3)bD$, $b_{\rm T} = 3b$ (Chetverikov et al. 2011).

A prototype model closely related to the exponential and the Morse model of interactions is the Yukawa model:

$$V_{ab}^{\rm Y}(r) = g_{ab}\frac{\exp(-\eta r)}{r}\,. \tag{1.6}$$

This potential was developed by Hideki Yukawa in 1935 in order to describe strong forces in elementary particle physics, mediated by the exchange of massive particles. The Yukawa potential now plays a paradigmatic role in statistical physics, since it has found applications in many fields (Fortov 2013). We will use this potential as a standard model for a gas with weak interactions and in particular for a quantum gas with weak interactions. A specially important property of the Yukawa potential is the existence of a Fourier transform, which is defined by

$$\tilde{V}_{ab}^{\rm Y}(\mathbf{t}) = \int_V \mathrm{d}\mathbf{r}\, V_{ab}^{\rm Y}(r)\exp(\mathrm{i}\mathbf{t}\cdot\mathbf{r}) = \frac{4\pi g_{ab}}{t^2+\eta^2}\,. \tag{1.7}$$

The Yukawa potential contains the Coulomb potential in the limit $\eta \to 0$. The interaction between two charges e_a and e_b is described (in rational units) by the potential

$$V_{ab}(r) = \frac{e_a e_b}{\varepsilon_{\rm r} r}\,. \tag{1.8}$$

In most cases we will assume without comment that the charges are in a vacuum $\varepsilon_{\rm r} = 1$.

The Coulomb potential is long-range. It was Joseph E. Mayer (1904–1983) who first noticed that the Fourier transform of the Coulomb potential is of primary importance for solving problems of screening and cluster theory (Mayer 1950). Strictly speaking, the integral over the volume is divergent for Coulomb interactions when the volume is infinite. This is an important problem and the Coulomb potential in its

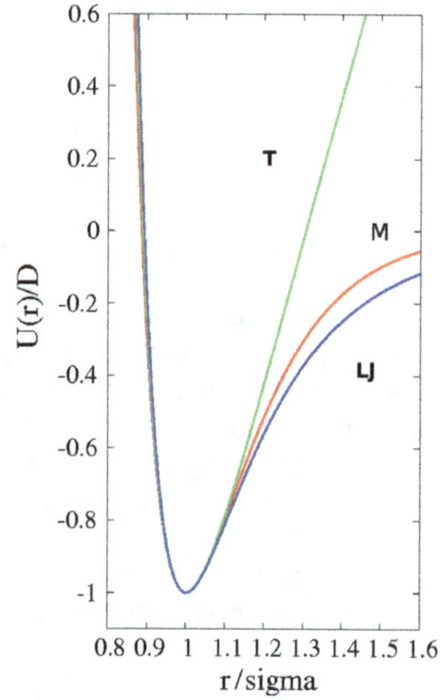

Fig. 1.3 Interaction between the atoms: The depth, frequency, stiffness, and other details about the interaction of atoms may be fitted by different potentials. Here we represent the Toda potential (*upper curve*), the Morse potential (*middle curve*), and the (r^{-12}, r^{-6}) Lennard-Jones potential (*lower curve*), suitably scaled around the minimum to have identical values at the minimum and identical values of the second derivative (frequency) and the third derivative (stiffness)

original form, needs some regularization. The great pioneer of the statistical theory of Coulomb systems, Joseph Mayer, proposed to introduce the Coulomb potential as the limit of a Yukawa potential for small $\eta \to 0$. This leads to a well defined limit for the Fourier transform.

One of the most popular models in the theory of gases and liquids is the Lennard-Jones (6–12)-potential which is semi-empirical with respect to the repulsive part, but well founded by quantum calculations for the attractive part. We scale here in such a way that the minimum $(-D)$ is at $r = \sigma$:

$$U_L(r) = D\left(\frac{\sigma^{12}}{r^{12}} - 2\frac{\sigma^6}{r^6}\right) . \tag{1.9}$$

A more general form is the Lennard-Jones (n-m)-potential. Characteristics of the potential are the energy at the minimum D, the distance r_0 where the potential energy goes from positive to negative values (crosses zero), and the location of the minimum at $r = \sigma$. Some typical data are given in Table 1.1. Note that there are many other potential models on the market, each adapted to certain practical applications (Hirschfelder et al. 1954).

We consider in this book two classes of these prototype models with quite different methods:

1. Potentials with a hard core like the hard sphere, Morse, and Lennard-Jones potentials. Here the classical limit is well defined and normally quantum effects including degeneracy give only small corrections. A special case are gases at

Table 1.1 Lennard-Jones potential. Energy at the minimum D and location of the minimum σ for various molecules

Substance	D [eV]	σ [Å]
H_2	4.5	0.75
O_2	5.1	1.20
C_2	5.6	1.31
Cl_2	2.5	1.98

extremely small temperatures, where quantum effects are strong. However, these special systems are not the main focus of the book.

2. Potentials with no singularity at $r = 0$ or a weak singularity like r^{-1} and a relatively long tail at $r \to \infty$.

The main assumption is, however, that the Fourier transform is well defined. Examples are the Coulomb potential and the Yukawa potential. For this class of systems, quantum effects are strong as a rule. In particular, for the Coulomb system, the ground state is determined completely by quantum effects. Therefore perturbation theories around the classical limit make no sense in this case and quantum-statistical tools are essential from the very beginning. On the other hand, expansions with respect to interactions including weak correlations, but without any restriction due to degeneracy, are sometimes appropriate.

1.2.2 Statistical Theory of Dense Classical Gases

The van der Waals theory has a long prehistory which began in the seventeenth century (Simonyi 1990). Robert Boyle (1627–1691) experimented with gases and found the first gas law (Boyle's law), which says that at a constant temperature T, the volume V of a given mass of gas is inversely proportional to the pressure ($p = C/V$, where here and in the following C denotes an appropriate constant). The second perfect gas law says that, at constant volume, the absolute pressure is proportional to the absolute temperature ($p = CT$). One application is the hydrogen thermometer. Standard temperature and pressure are defined as 273.15 K (0 degrees Celsius) and 101.325 kPa (760 mm Hg).

To change the state of a gas, heat is either added or taken away from it. If the state of a gas is altered without a change in heat, we speak about an adiabatic change. If a compressed gas expands adiabatically, cooling occurs. Since atoms and molecules interact by attractive forces, energy is required as the gas expands to overcome the intermolecular forces. A gas cools as it expands, and if it is rapidly compressed, its temperature rises. Dalton's law of partial pressures says that, in a mixture of gases, the pressure each gas exerts is the same as if it alone occupied the volume. Avogadro's law states that equal volumes of gases at the same temperature and pressure contain equal numbers of molecules. A mole is the amount of a substance containing the number of particles 6.022×10^{23}. Combining the perfect gas laws and Avogadro's law, we get the universal ideal gas law due to Gay-Lussac (1778–1850):

$$p = \nu RT/V \ , \qquad \text{or} \quad p = nk_\mathrm{B}T \ , \qquad n = N/V \ , \tag{1.10}$$

where ν is the number of moles, N is the corresponding number of molecules, R is Avogadro's constant, and k_B is Boltzmann's constant. The energy density of an ideal gas is given by

$$\rho_\mathrm{E}(T) = c_v nT \ . \tag{1.11}$$

Here, c_v is the specific heat, which for normal gases is $c_v = fk_\mathrm{B}/2$, with f the number of degrees of freedom.

An adiabatic process is one without transfer of heat or matter between a system and its surroundings. The equation of state for such processes reads

$$p = Cn^\gamma \ , \qquad \gamma = \frac{2+f}{f} \ , \tag{1.12}$$

with $\gamma = 5/3$ for normal gases and C an appropriate constant.

In a first approach, the definition of temperature scales may be fixed by using the universality of the ideal gas law and fixing the triple point of water by definition to $T_\mathrm{c} = 273.16\,\mathrm{K}$. The triple point of water is that unique temperature at which pure ice, pure water, and pure water vapor can coexist at equilibrium. The triple point is important since there is only one pressure at which all three phases can be in equilibrium with each other.

Soon after van der Waals' work, the method of virial expansions for non-ideal gases was developed. This expansion, which expresses the pressure of a many-particle system in equilibrium as a power series in the density, was introduced in 1901 by Heike Kamerlingh Onnes (1853–1926, Nobel Prize 1913), is a natural generalization of the ideal gas law. Kamerling Onnes represented the pressure of a gas with density n and temperature T as a power series in the density n ($\beta = 1/k_\mathrm{B}T$):

$$\beta p = -\left(\frac{\partial(\beta F)}{\partial V}\right)_{T,N} = n\left[1 - nB_2 - 2n^2B_3(T) - \cdots\right] . \tag{1.13}$$

The corresponding expansion for the free energy is

$$F = F_\mathrm{id} - k_\mathrm{B}TV\left[n^2B_2(T) + n^3B_3(T) + \cdots\right] . \tag{1.14}$$

An elementary way to get the virial functions including interaction effects starts from the binary correlation functions $g(r)$. By definition, in a fluid of density n, the number of particles in a shell of thickness $\mathrm{d}r$ at distance r from a given center particle is

$$g(r)(4\pi nr^2)\mathrm{d}r \ . \tag{1.15}$$

In a first classical approximation, the binary correlations are given by a Boltzmann factor $g(r) \sim \exp(-\beta U(r)$ and provide the first order mean potential energy

$$g(1,2) = \exp\big[-\beta V(1,2)\big]\,,$$
$$U = \big\langle U(r)\big\rangle = \frac{1}{2}\int d\mathbf{r}_1 d\mathbf{r}_2 U(\mathbf{r}_1,\mathbf{r}_2)\exp\big[-\beta U(\mathbf{r}_1,\mathbf{r}_2)\big]\,. \tag{1.16}$$

Since the internal energy U and the free energy F are connected by the thermodynamic relation $U = \partial(\beta F)/\partial\beta$, for the second virial coefficient, we find by integration

$$B_2(T) = \frac{1}{2}\int d\mathbf{r}\Big\{\exp\big[-\beta V(\mathbf{r})\big] - 1\Big\}\,. \tag{1.17}$$

For the higher order virial coefficients we may derive explicit expressions by using the cluster expansion methods of statistical mechanics worked out by Joseph Mayer (1904–1983), Klaus Fuchs (1911–1988), and others (Hirschfelder et al. 1954; Hill 1956; Friedman 1962; Barker and Henderson 1967).

A survey of the state of the art in the theory of nonideal gases was given in 1973 at the Van der Waals Centennial Conference. The program demonstrated that most prominent scientists in the field like de Boer, Lebowitz, Langer, Widom and Wilson honored van der Waals were working to develop this further (see Fig. 1.4). In particular, we mention reports on the statistical foundations of the van der Waals equation by Klein and Lebowitz and the theory of fluid phase transitions by Langer, Levelt-Sengers, Widom, and Wilson.

1.3 Quantum Physics of Strongly Coupled Gases

1.3.1 Correlations in Bose–Einstein and Fermi–Dirac Gases

The notion of *ideal gas* is not uniquely defined. Often one understands as 'ideal' a gas without interactions, i.e., having an additive Hamiltonian. However, this point of view overlooks the fact that coupling between the particles is more essential than the formal aspect of additivity. At high densities, Fermi–Dirac gases and Bose–Einstein gases are strongly coupled gases due to the Pauli principle and the corresponding exchange effects. So here we shall take an *ideal gas* to be just the usual classical ideal gas consisting of independent particles. Since coupling due to exchange effects plays an important role in dense gases, we first recall the theory of Bose–Einstein and Fermi–Dirac gases. Then we go on to consider the quantum statistics of real gases by including interaction forces, as was first done by Uhlenbeck and Beth in 1936–1937. As we pointed out above, Bose–Einstein and Fermi–Dirac gases are strictly speaking not ideal in the classical sense, since they show rather strong correlations with increasing degeneracy.

Van der Waals Centennial Conference on Statistical Mechanics

PROGRAM

Sunday, 26 August

Registration Desk	: 14.00 - 17.00	Registration
Lounges	: 17.00 - 18.00	!nformal get together

Monday, 27 August

Main Hall

10.00	Official Opening	
10.30	J. de Boer	: Opening address
11.30	M. Klein	: Historical origins of the Van der Waals equation
14.00	J. L. Lebowitz	: Exact derivation of the Van der Waals equation
15.30	J. S. Langer	: Metastable states
16.30	J. M. H. Levelt Sengers	: Equation of state and thermodynamic correlations near the critical point

Rijksmuseum

20.30 Official receotion

Tuesday, 28 August

Blue Room

9.00	B. Widom	: The critical point and scaling theory
10.30	K. Wilson	: The renormalization group and the critical point
11.30	A. V. Voronel	: High resolution specific heat measurements

14.00 Section meetings

Blue Room	: Critical phenomena
I - II	: Equilibrium properties of liquids
III	: Equation of state, equilibrium properties of gases
IV	: Correlation functions and transport

Fig. 1.4 Johannes D. van der Waals died in 1923, in the year Peter Debye presented the first theory of nonideal charged particle systems. Fifty years later, in 1973, the van der Waals Centennial was held in Amsterdam. Facsimile of the first page of the program

The first quantum statistics of a gas without interactions was in fact already developed in 1900 by Max Planck (1858–1947). Unfortunately, he started with a quite complicated case, the photon gas. This is due to the fact that his colleagues in Berlin had collected a lot of experimental data on radiation which needed theoretical interpretation. The photon gas is a relativistic system of particles with zero mass, i.e., it is a quite special case of a gas. The generalization of Planck's theory was given by Einstein in 1924 for particles with rest mass and integer spins, and for particles with half-integer spins by Fermi in 1926. Einstein's work was based on the happy occasion when the Bengali physicist Satyendra Nath Bose (1894–1974) sent his article *Planck's law and the hypothesis on light quanta* to Einstein. Einstein translated it into German and sent the article to Zeitschrift für Physik (1924), appending several remarks that essentially generalized Bose's approach. He immediately saw the power of Bose's method and the way to apply it to gases (Einstein 1924; Kirsten and Körber 1975; Ginzburg 2001).

Bose had in fact proposed a new method for counting the probabilities of the macrostates. In Bose's interpretation, the radiation field looks like a gas of photons. Einstein applied Bose's new counting method to derive the quantum statistics of monatomic ideal gases. Only eight days after the official date of receipt of Bose's paper, he presented his new results to a session of the Prussian Academy held on 10 July in Berlin. What Einstein appreciated above all was Bose's new method for counting probabilities based on the indistinguishability of identical particles (see Ebeling and Hoffmann 1991, 2014). Einstein successfully applied Bose's new method to derive the quantum statistics of monatomic ideal gases in three papers all printed in the "Sitzungsberichte" of the Prussian Academy of Science in Berlin.

The work in Bose and Einstein's papers was heavily criticized by Ehrenfest, Planck, and other colleagues, since neither Einstein nor Bose gave any deeper foundation to justify the new way of counting. Nernst and Schrödinger were among the first colleagues to support Einstein's new views. It was definitely Einstein who understood the connection between the new statistics and the indistinguishability of identical particles. The new view was a genuine revolution in physics, something that was first clearly understood by Planck and Ehrenfest, who expressed serious protests against Einstein's new views. The reason for these protests was in fact that these scientists immediately understood that the independence of the quantum gas particles was lost and that Einstein's method introduced as yet unknown quantum correlations between noninteracting particles.

The first physicist who understood the general physical principle behind the new mysterious correlations found by Einstein was Wolfgang Pauli, who formulated his exclusion principle in 1925. The first application of the new quantum statistics to the electron gas was given in 1926 by Enrico Fermi (1901–1954) and in the same year by Paul Dirac (1902–1984). Fermi–Dirac statistics applies to identical particles with half-integer spin in a system in thermodynamic equilibrium. The particles in this system have negligible mutual interactions. Note that very important applications of the new statistics were soon given by Fowler and Sommerfeld, who were treating electron plasmas with the new Fermi–Dirac statistics.

Today, particles that obey the exclusion principle, such as particles with spin $s = 1/2$, are called fermions, and particles with integer spins which are like the atoms in Einstein's theory are called bosons. Like Bose–Einstein gases, dense Fermi–Dirac gases also behave at low temperatures in a completely different way to the standard classical ideal gas. In order to demonstrate this, we consider the energy density, defined as the energy per unit of volume $\rho_{\mathrm{E}} = E/V$. For Boltzmann gases, the energy density and the pressure increase linearly with density n and temperature

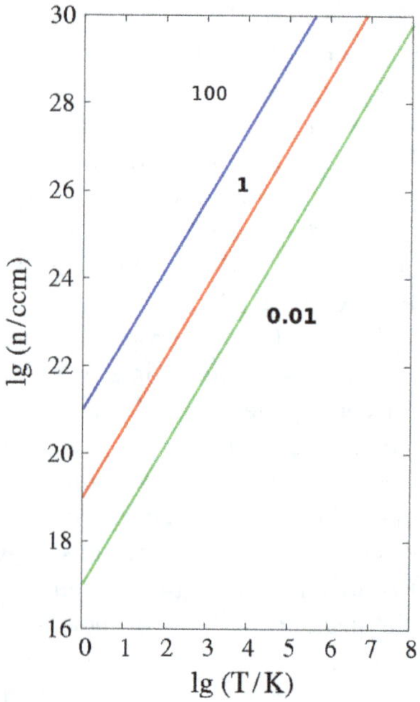

Fig. 1.5 Density–temperature plane for an electron gas on a log-scale, including the lines $n\Lambda^3 = 100, 1, 0.01$ separating strongly degenerate (*above upper line*) from moderately degenerate (*between lines*) from non-degenerate (*below lower line*)

T. For degenerate Fermi gases, the temperature dependence is rather weak, but the energy density and the pressure increase strongly with the density:

$$\rho_{\rm E} = Cn^{4/3} , \qquad p = Cn^{4/3} . \tag{1.18}$$

As we will show later, due to the high pressures and energy densities in dense systems, all bound states will be destroyed at higher densities. Very dense fermion systems behave like Fermi–Dirac gases. The transition, which is rather sharp, occurs at a density where the thermal de Broglie wavelength begins to overlap (see Fig. 1.5).

We note that the behavior of a quantum gas changes at the transition from non-degenerate gas to degenerate gas, and the masses play a big role here. Typical Fermi systems consist of light electrons which soon reach degeneracy and heavy particles like nuclei which need extreme densities to become degenerate. For Bose gases, the changes at the degeneracy line may be even more dramatic. This was already observed by Einstein, who predicted a new condensation phenomenon in a second talk at the academy, given on 8 January 1925. As a consequence of the theory, he described the phenomenon now known as Bose–Einstein condensation. We note as a matter of fact that the prediction of a condensation phenomenon belongs clearly only to Einstein (Ebeling and Hoffmann 2014). The negative reaction to the idea of Einstein condensation changed only later, when Uhlenbeck, Bogolyubov, and others succeeded in including interactions in the theory, and finally, 70 years later, when experimentalists succeeded in reaching low enough temperatures.

The Einstein condition for condensation is this: in a cube with length equal to 10 thermal de Broglie wavelengths should be more than 2 612 gas particles. Here the thermal de Broglie wavelength is the wavelength corresponding to the thermal momenta at temperature T. Erwin Schrödinger found this role of the De Broglie matter waves very interesting and inspiring, and exchanged letters with Einstein which were important for his formulation of wave mechanics. But Schrödinger could not believe that such conditions could be reached for real gases. The experimental verification of Einstein's prediction of a condensation of atomic gases at low temperatures was confirmed only 70 years after the prediction, in 1995. At the International Conference of Laser Spectroscopy on the island of Capri, Eric Cornell reported experiments at the University of Boulder which confirmed Einstein's prediction, as did parallel experiments by Wolfgang Ketterle's group at MIT. Nowadays, many groups around the world work in that field, and several Nobel Prizes have been attributed. Note that the effects we have discussed here, from the destruction of bound states in fermion gases to the condensation phenomena in boson systems, are clearly strong correlation effects.

1.3.2 Quantum Statistics of Interacting Gases

One of the difficulties in confirming experimentally the early theories of quantum gases was that the particles in real gases or plasmas are always interacting. The models by Einstein and Fermi, based on additive Hamiltonians, were just an abstraction. So the need to include at least weak interactions as a perturbation was seen immediately. For weakly degenerate systems, i.e., when the gas is still below the degeneracy line, in order to go from classical statistics to the quantum case only a few changes are needed in the classical theory. To see this, we study the mean potential energy, which in quantum statistics is expressed by an expectation value and given by a trace, viz.,

$$U = \langle V \rangle = \frac{1}{2}\mathrm{Tr}\big[U(12)\hat{\rho}_2(1,2)\big]\,, \tag{1.19}$$

where $\hat{\rho}_2$ is a two-particle density operator. An easy way to proceed is by using the coordinate representations. The pair probability given classically by a Boltzmann factor is to be replaced by its quantum-statistical counterpart, the Slater sum of pairs. In this way, as shown by Beth and Uhlenbeck, the classical expressions remain valid if we simply replace the Boltzmann factor by binary Slater sums (Uhlenbeck and Beth 1936). The second virial coefficient expressed in terms of the binary Slater function or Slater sum is

$$B_2(T) = \frac{1}{2}\int \mathrm{d}\mathbf{r}\big[S_{ab}(\mathbf{r}) - 1\big]\,, \tag{1.20}$$

where **r** is the displacement vector between the two particles. In the classical case, the Slater sum is identical to the Boltzmann factor. The pioneers Uhlenbeck and Beth represented the Slater sum in terms of the wave functions, including bound and scattering states. They thus expressed the second virial coefficient in terms of the energy spectrum and the density of states $n_l(k)$. The density of states may be expressed in terms of the scattering phase shifts or the Jost functions, using the relations between scattering phase shifts and the Jost functions of scattering theory (Uhlenbeck and Beth 1936; Kraeft et al. 1986; Kremp et al. 2005; Ebeling et al. 1976; Blaschke et al. 2014).

The extension of the statistical theory to strongly degenerate quantum gases can be attributed to several scientific schools. We mention, for example, Alexei A. Abrikosov, Nikolai N. Bogolyubov, Richard Feynman, Vitaly L. Ginzburg, Ryogo Kubo, Walter Kohn, Lev D. Landau, Elliot Montroll, and Julian Schwinger (Abrikosov et al. 1962; Bogolyubov and Bogolyubov 1992; Feynman 1972; Martin and Schwinger 1959; Montroll and Ward 1958). Among them, Nikolai N. Bogolyubov holds a special place in our view. Coming originally from the Kiev school of nonlinear mechanics and mathematics, he turned to statistical physics in the 1940s and 1950s, in particular to the method of distribution functions, where he developed with others the method of the BBGKY hierarchy to derive the kinetic equations. He then turned to the theory of superfluidity and superconductivity, formulated the microscopic theory of superfluidity, and made other essential contributions. Later he worked on quantum field theory, introduced the Bogolyubov transformation, and formulated and proved several theorems now named after him.

1.4 Ionic Fluids and Dense Low-Temperature Plasmas

1.4.1 Coulomb Forces and Debye–Hückel–Wigner Theories

The law of interaction between charged particles was formulated by Charles Augustin de Coulomb around 1785. The force between two charges e_a and e_b is as a function of the distance r in the radial direction (in Gaussian units)

$$F = -\frac{e_a e_b}{\varepsilon_\mathrm{r} r^2} \,. \tag{1.21}$$

This force is repulsive for equal charges and attractive for opposite charges. We assume here that the two particles belong to species a and b and that ε_r is the relative dielectric constant of the imbedding medium. The corresponding potential was introduced above. The potential at distance r from a charge e and the related field $\mathbf{E}(r)$ decay as $1/r$. In the following, we work in the Gaussian system of units. In the denominator, ε_r is the relative dielectric constant of the medium. In most cases we will assume without comment that the plasma is imbedded in a vacuum, so that

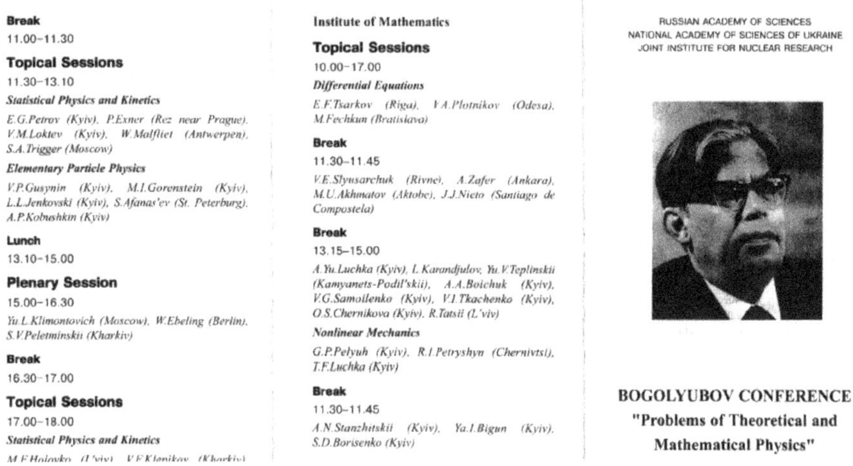

Break
11.00–11.30
Topical Sessions
11.30–13.10
Statistical Physics and Kinetics
E.G.Petrov (Kyiv), P.Exner (Rez near Prague), V.M.Loktev (Kyiv), W.Malfliet (Antwerpen), S.A.Trigger (Moscow)
Elementary Particle Physics
V.P.Gusynin (Kyiv), M.I.Gorenstein (Kyiv), L.L.Jenkovski (Kyiv), S.Afanas'ev (St. Peterburg), A.P.Kobushkin (Kyiv)
Lunch
13.10–15.00
Plenary Session
15.00–16.30
Yu.L.Klimontovich (Moscow), W.Ebeling (Berlin), S.V.Peletminskii (Kharkiv)
Break
16.30–17.00
Topical Sessions
17.00–18.00
Statistical Physics and Kinetics

Institute of Mathematics
Topical Sessions
10.00–17.00
Differential Equations
E.F.Tsarkov (Riga), V.A.Plotnikov (Odesa), M.Fechkan (Bratislava)
Break
11.30–11.45
V.E.Slyusarchuk (Rivne), A.Zafer (Ankara), M.U.Akhmatov (Aktobe), J.J.Nieto (Santiago de Compostela)
Break
13.15–15.00
A.Yu.Luchka (Kyiv), L.Karandjulov, Yu.V.Teplinskii (Kamyanets-Podil'skii), A.A.Boichuk (Kyiv), V.G.Samoilenko (Kyiv), V.I.Tkachenko (Kyiv), O.S.Chernikova (Kyiv), R.Tatsii (L'viv)
Nonlinear Mechanics
G.P.Pelyuh (Kyiv), R.I.Petryshyn (Chernivtsi), T.F.Luchka (Kyiv)
Break
11.30–11.45
A.N.Stanzhitskii (Kyiv), Ya.I.Bigun (Kyiv), S.D.Borisenko (Kyiv)

RUSSIAN ACADEMY OF SCIENCES
NATIONAL ACADEMY OF SCIENCES OF UKRAINE
JOINT INSTITUTE FOR NUCLEAR RESEARCH

BOGOLYUBOV CONFERENCE
"Problems of Theoretical and Mathematical Physics"

Fig. 1.6 Poster for a conference in 1999 in the Ukrainian capital Kiev, where Nikolai N. Bogolyubov (1909–1992) was educated and begin his career

$\varepsilon_r = 1$. The electric field is source-free, except at the locations of the point charges, which are the sources of the field. This leads to the Poisson equation for the potential, which is the basic tool for the Debye–Hückel theory Fig. 1.6.

The idea due to Milner in 1912 is that free charges in an electroneutral ensemble of charges are always surrounded by opposite charges which screen the Coulomb fields. Milner's theoretical treatment, which used mainly graphical tools, was quite complicated. In 1923, Peter Debye presented a simpler theory based on Poisson's law and showed that the Milner effect is responsible for an exponential screening of the fields. This regularization is needed since the integral over the naked Coulomb potential is divergent, whence direct application of the method of virial expansions fails. All the standard virial coefficients known from the statistical theory of gases diverge for Coulomb potentials. As a consequence, it is known from the work of Milner, Debye, and Debye–Hückel (1923) that the form of the thermodynamic functions for gases or solutions containing particles satisfying Coulomb's law, like plasmas and electrolytic solutions, show essential deviations from those of typical gases. Density expansions of thermodynamic functions fail and completely different series expansions are needed. In particular, the pressure or its analogue in solutions, the osmotic pressure, cannot be expanded in Taylor series with respect to density. In the statistically well founded approach due to Debye and Hückel (Debye 1923; Debye and Hückel 1923; Kelbg 1963, 1972), the modern screening concept was developed. Further important contributions to the classical statistical theory of screening effects were made by Bogolyubov, Mayer, Meeron, Zubarev, Yukhnovsky, Kelbg, and Friedman (Friedman 1962; Bogolyubov and Bogolyubov 1992).

Let us now briefly discuss the concept of Debye screening based on the Poisson equation and some extensions, such as the Bogolyubov theory based on integral equations. As pointed out above, statistical integrals like the virial coefficients do not exist for Coulomb potentials, so the standard expressions for the statistical theory of gases are not applicable. In a real plasma, long-range potentials are screened. Around a negative charge we will find a region where positive charges are dominant, and around a positive charge we will find a region where negative charges prevail. As known from electrostatics, a sheet of charges of opposite sign screens the central charge, since the force lines end at the counter charges. In his lectures at Rostock University in the 1950s, Professor Hans Falkenhagen compared the situation with a dance party. At a party, we will find with higher probability a boy near a girl, and near a boy, we expect to see on average one girl, in spite of all the dynamical changes in the party. In plasmas and electrolytes, we refer to this effect, based on attractive and repulsive forces between partners, as *screening*. The Coulomb forces are screened by these effects and converted into short-range interactions. Due to screening, the interactions do not decay according to power laws, but instead according to exponential laws. Technically the long-range divergencies may be removed by introducing exponentially screened potentials instead of the long-range Coulomb potentials:

$$V_{ab} = \frac{e_a e_b}{\varepsilon_\mathrm{r} r} \longrightarrow V_{ab}^\mathrm{D} = \frac{e_a e_b}{\varepsilon_\mathrm{r} r} \exp\left(- \frac{r}{r_\mathrm{D}} \right) . \tag{1.22}$$

Here r_D is some screening length. Debye used a Poisson–Boltzmann equation to find his screened potential (Falkenhagen 1971). According to this idea, the effective potential $\Phi(r)$ at distance r satisfies the Poisson–Boltzmann equation ($\beta = 1/k_\mathrm{B}T$)

$$\Delta\Phi(\mathbf{r}) = -4\pi e n \Big\{ e \exp\big[- e\beta\Phi(\mathbf{r})\big] - e \exp\big[+ e\beta\Phi(\mathbf{r})\big]\Big\} \approx \kappa^2 \Phi(\mathbf{r}) , \tag{1.23}$$

where the last step follows from linearization and leads to the Debye equation with $\kappa^2 = r_\mathrm{D}^{-2} = 8\pi\beta n e^2/\varepsilon_\mathrm{r}$.

Bogolyubov proposed an alternative procedure based on the solution of an equivalent integral equation which he derived from the so-called BBGKY hierarchy (Bogolyubov 2005–2009). According to Debye, the mean Coulomb energy is given by the potential energy of a charge against its screening cloud. For a two-component system, this is $e^2\kappa/\varepsilon_\mathrm{r}$. Summing over all charges, we still have to divide by two in order to avoid double counting. For the total Coulomb energy, we then get (Figs. 1.7 and 1.8)

$$U_\mathrm{c} = \frac{1}{2}(N_+ + N_-)\frac{e^2}{r_\mathrm{D}} = -V\frac{\kappa^3}{8\pi} . \tag{1.24}$$

In fact, Debye and Hückel developed the modern screening concept. The extension to non-equilibrium phenomena was given by Lars Onsager (1903–1976, Nobel Prize 1968) and Hans Falkenhagen (1895–1971). The statistical foundation of this theory on the basis of the concept of molecular distribution functions was laid later by Bogolyubov. Further important contributions to the classical statistical theory of

Fig. 1.7 Erich Hückel (1896–1980), who developed the concept of screening and the statistical thermodynamics of electrolytes with Peter Debye in 1923/24. Courtesy of K. Beneke (1999)

Fig. 1.8 Eugene Wigner visiting as young scientist Berlin and other scientific centers (by courtesy of Thomas Hüsing)

screening effects were made later by Mayer (1950), Meeron, Zubarev, Yukhnovsky, and Friedman (1962).

Wigner Lattices of Coulomb Systems

At higher densities the behavior of nondegenerate Coulomb systems is determined by the formation of lattices. This was first shown in the 1930s by the young Hungarian physicist Eugene Wigner (1902–1995). Wigner enrolled at the Technische Hochschule in Berlin in 1921 (today the Technische Universität), where he studied chemical engineering, like his fellow student from Budapest grammar school, Johann von Neumann (1903–1957). They also attended the famous Wednesday afternoon colloquia of the German Physical Society. These colloquia featured such luminaries as Max Planck, Max von Laue, Rudolf Ladenburg, Werner Heisenberg, Walther Nernst, Wolfgang Pauli, and Albert Einstein. No doubt, these colloquia influenced Wigner and von Neumann very much, as well as other young physicists. Wigner met here another physicist from Hungary, Leo Szilard, who at once became Wigner's closest friend. Later Wigner worked at the Kaiser Wilhelm Institute for Physical Chemistry and Electrochemistry (now the Fritz Haber Institute), where he met Michael Polanyi. Polanyi supervised Wigner's dissertation *Formation and Decay of Molecules*. After a stay in Budapest he returned to Berlin in 1926 and worked with Karl Weissenberg and later with Richard Becker at the Technical University. Note that, from 1928 to 1933, his old friend Johann von Neumann was the youngest *Privatdozent* at the Berlin University Unter den Linden, and wrote here his groundbreaking book on the mathematical foundation of quantum mechanics and quantum statistics (von Neumann 1932).

It was under these circumstances that Wigner explored quantum mechanics and prepared to become one of the most brilliant scientists working on the analytical behavior of plasma thermodynamic functions. He was the first to understand that, at high densities, the analytical behavior of Coulomb systems changes and approaches a regime where the Coulomb energy is determined by lattice effects and increases more slowly than in the Debye theory. Wigner first studied a system of electrons with a positive smeared out background in the form of a bcc lattice, nowadays called a Wigner lattice, and estimated the energy to be (Wigner 1934)

$$-\frac{U_{\mathrm{c}}}{N} \approx 0.9\frac{e^2}{\varepsilon_{\mathrm{r}} r_0} \sim n^{1/3}\,, \qquad \frac{V}{N} = \frac{4\pi}{3} r_0^3\,, \tag{1.25}$$

where r_0 is the mean distance between the charges. This estimate is based on the assumption that the radial distribution function has a strong maximum at a distance r_0 that is typical for lattices and lattice-like structures. We expect a scaling of the specific Coulomb energy as $r_0^{-1} \sim n^{1/3}$ to be quite typical for most high density structures. We shall leave the explanation of how Wigner-type laws may be derived statistically to Chaps. 3–6.

1.4.2 Ionization and Association Equilibria

Typical phenomena in dense systems of charged particles are association and bound state effects connected with chemical equilibria and changes in the composition depending on the thermodynamic conditions. Historically, the case of plasmas was studied first. Plasmas are gases consisting essentially of charged particles, mainly electrons and ions, created by the ionization of atoms at high density or high temperature. The development of the quantum statistics of ionization and plasmas around

1920 was intimately connected with the development of chemical thermodynamics, and also with the new scientific discipline of astrophysics. Although astronomy is as ancient as recorded history itself, it was for a long time considered to be quite separate from physics.

This changed only at the end of the nineteenth century, when Bunsen and Kirchhoff discovered that, by decomposing light from the Sun, a multitude of spectral lines were observed. Laboratory experiments showed that the same lines could be observed in the spectra of gases, specific lines corresponding to well defined chemical elements. In this way it was proved that the chemical elements found in the Sun (mainly hydrogen) were also found on Earth. The element helium was first discovered in the spectrum of the Sun, and only later on Earth, hence its name. As a consequence a new branch of science developed in order to understand the astronomical and experimental observations in spectroscopy.

In order to bring some systematics into the observations, a theory of abundances was required. The theoretical basis for this was established by the work of the German physical chemist John Eggert (1891–1973) and the Indian physicist Meghnad Saha (1893–1956). The work of Eggert and Saha is based on Rutherford and Bohr's new model of atoms. In 1913 Niels Bohr developed a dynamical model based on Rutherford's ideas, which describes the electron orbits and the corresponding energies in a semi-classical approximation. According to the Bohr theory, there are stationary radii of electron orbits and the corresponding energies are

$$a_s = s^2 a_\mathrm{B}\ , \quad a_\mathrm{B} = \frac{\hbar^2}{\mu e^2}\ , \quad E_s = -\frac{\mu e^4}{2\hbar^2 s^2}\ , \quad \mu = \frac{m_\mathrm{e} m_+}{m_\mathrm{e} + m_+}\ , \tag{1.26}$$

where μ is the reduced mass. We denote the so-called main quantum number by s, in order not to confuse with the notation n for the density. According to the Bohr model, there are infinitely many levels close to the limit $s \to \infty$.

Let us give a short derivation of the Eggert–Saha equation for a binary Coulomb system with n_+ positive ions (cations) and n_- negative charges (electrons) per cubic centimeter, where $n_+ = n_-$. The density of free atoms is n_a. The total density is $n_0 = n_+ + n_\mathrm{a} = n_- + n_\mathrm{a}$. In the following we will use the plasma notation, i.e., we call the negative charges 'electrons' and the positive charges simply 'ions'. In the spirit of Nernst's and Eggert's ideas, we assume a chemical equilibrium between electrons, ions, and atoms, viz.,

$$e^- + i^+ \longrightarrow a_0\ , \qquad \mu_\mathrm{e} + \mu_\mathrm{i} = \mu_\mathrm{a}\ , \tag{1.27}$$

where the $\mu_k = \partial F/\partial N_k$ are the chemical potentials for the corresponding species, with F the free energy. For the atoms, we assume (I - ionization energy)

$$\mu_\mathrm{a} = I + k_\mathrm{B}T \ln n_\mathrm{a} + k_\mathrm{B}T \ln \lambda_\mathrm{a}^3\ , \tag{1.28}$$

and for the charges

$$\mu_{\rm e} = k_{\rm B}T \ln n + k_{\rm B}T \ln \Lambda_{\rm e}^3 \,, \quad \mu_{\rm i} = k_{\rm B}T \ln n + k_{\rm B}T \ln \Lambda_{\rm i}^3 \,. \tag{1.29}$$

The last terms are the constants first found by Sackur and Tetrode, which contain Planck's constant h (Badino 2009):

$$\Lambda_k = h/\sqrt{2\pi m_k k_{\rm B} T} \tag{1.30}$$

The condition of chemical equilibrium implies the Eggert–Saha formula

$$\frac{n_{\rm a}}{n_{\rm e} n_{\rm i}} = K(T) = \Lambda^3 \exp\left(\frac{I}{k_{\rm B}T}\right), \quad \Lambda = \frac{\Lambda_{\rm e}\Lambda_{\rm i}}{\Lambda_{\rm a}} \,. \tag{1.31}$$

A plasma is a neutral gaseous system with free electrons and ions. However, plasmas also exist in the solid state. For example the conduction electrons in metals and semiconductors are a special form of plasma.

The behavior of a plasma depends on the density and temperature. Degenerate and non-degenerate plasmas behave in a quite different way. In the logarithmic density–temperature plane (see Fig. 1.5), the line $n_{\rm e}\Lambda_{\rm e}^3 = 1$ is the border between the degenerate and non-degenerate regions. The length $\Lambda_{\rm e}$ is the de Broglie wavelength of electrons with thermal momentum, and $n_{\rm e}$ the density of free electrons. Free electrons are formed in chemical equilibrium with bound electrons. This depends crucially on the binding energies. As a rule of thumb, we may consider bound states to be relevant if $I/k_{\rm B}T > 10$, i.e., if $T < I/10k_{\rm B}$, where I is the ionization energy. For example, in the case of hydrogen, $I = |E_{10}|$ and $I/k_{\rm B} \sim 156\,000\,{\rm K}$. Measuring T in electronvolt, ($1\,{\rm eV} \approx 10^4\,{\rm K}$), for hydrogen with $I = 13.6\,{\rm eV}$, we find that, according to our rule, bound states are relevant for $T < 15\,000\,{\rm K}$.

Bjerrum's Theory of Ionic Association

In spite of the fact that ionization equilibria in electrolytes were considered much earlier by Arrhenius and others, the connection with Coulomb forces was not studied until 1926, by the Danish physicist Niels Bjerrum (1879–1956). Bjerrum studied the strength of the forces between positive and negative charges in an electrolyte and defined a characteristic length, the Bjerrum length $\ell_{\rm B}$. This is the separation of charges at which the electrostatic interaction is comparable in magnitude with the thermal energy scale $k_{\rm B}T$, where as usual $k_{\rm B}$ is the Boltzmann constant and T the absolute temperature in kelvin. In rational units, the Bjerrum length is given by $\ell_{\rm B} = e^2/\varepsilon_{\rm r} k_{\rm B} T$, where $\varepsilon_{\rm r}$ denotes the relative dielectric constant of the medium. This length scale arises naturally in discussions of electrostatic, electrodynamic, and electrokinetic phenomena in electrolytes and plasmas.

As for plasmas we consider the chemical equilibrium between associates (ion pairs) with concentration n_0, and free positive or negative charges with concentrations n_+ or n_-. The assumed mass action law reads

$$\frac{n_0}{n_+ n_-} = K(T) \,. \tag{1.32}$$

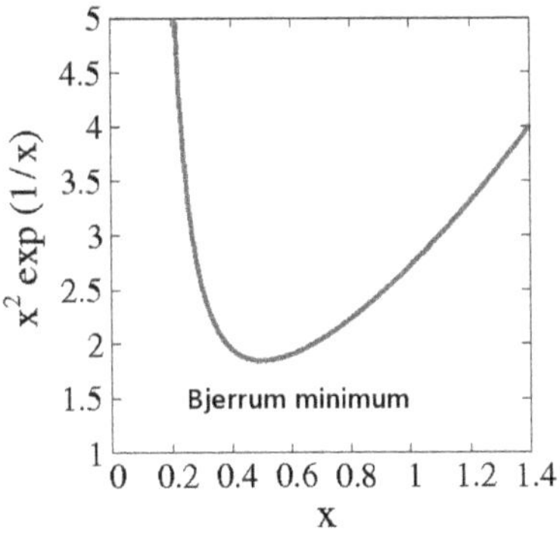

Fig. 1.9 Bjerrum's best choice for the upper limit of the integral defining the classical mass action constant corresponds to the minimum of the integrand. We will apply a similar procedure to quantum plasmas

Bjerrum studied ions with a hard core a, assuming $\ell_B > 2a$, which means relatively strong Coulomb forces. For the mass action constant, Bjerrum assumed the following integral (Falkenhagen et al. 1971):

$$K(T) = \int_a^R \mathrm{d}r \,(4\pi r^2) \exp(\beta e^2/\varepsilon_\mathrm{r}) \ . \tag{1.33}$$

For the moment, for the upper limit, we still have a free choice. Bjerrum looked at the r-dependence of the expression under the integral and found that there is a best choice corresponding to a minimum of the sub-integral function with respect to r (see Fig. 1.9). So in fact Bjerrum's choice for the upper limit was $e^2/2\varepsilon_\mathrm{r} k_\mathrm{B} T$. This choice is only arbitrary to some extent. With regard to this problem, we may mention an often cited statement by Lars Onsager. In a discussion at the electrolyte conference held in Montpellier in 1968, he made the following remark about Bjerrum's choice of the mass action constant for ion association (reported by Hans Falkenhagen; see also Hemmer et al. 1996, p. 249):

> Bjerrum's choice is good but we could vary it within reason. In a complete theory this would not matter; what we remove from one side of the ledger would be entered elsewhere with the same effect.

The problem of defining the mass action constant for electrolytic association is completely equivalent to the problem of the choice of a plasma partition function and the mass action constant. Applying Onsager's theorem to our problem, we come to the conclusion that there is indeed some freedom in defining the partition function, but in a complete theory this would not matter because "what we remove from one side of the ledger would be entered elsewhere with the same effect".

In correct theories, the different choices of partition functions may be used to some extent equivalently. Following Onsager, we would prefer to apply the term 'right' or 'wrong', not to the bound state partition function, but only to the expression for the complete thermodynamic functions. In the end, only the thermodynamic quantities are measurable quantities.

1.5 Quantum Statistics of Coulomb Systems

1.5.1 Quantum Interactions, Screening, and Regularization

The problem with Coulomb systems is that in a classical theory there are several divergencies, which are due to the peculiarities at $r = 0$ and at $r = \infty$. We already noticed that the most critical *basic problems of Coulomb systems* are the screening problem and the bound state problem. These problems are more or less due to the unique properties of the Coulomb potential and its specific character. The first is connected with its long-range nature and the second with the divergence at $r = 0$. Strictly speaking, a classical statistics of many-component Coulomb point charges does not exist, since the divergence at the origin can only be lifted in a quantum treatment. The development of a systematic statistical theory for systems with Coulomb interactions involves the solution of problems which are in part connected with the divergencies.

The Coulomb potential energy between the two charges e_a and e_b is long-range and has to be screened, and it is singular at short distances. In the quantum statistics developed by Günter Kelbg, it was shown that the short distance divergence is removed by quantum effects, e.g., the Heisenberg uncertainty (Kelbg 1963). The expression derived by Kelbg contains hypergeometric functions and will be introduced later (see Chap. 3). Beside his mathematically strict solution, Kelbg also proposed a simple approximate smooth potential with the correct asymptotic at $r = 0$ (Kelbg 1963, 1972):

$$\varphi_{ab}(r) = \frac{e_a e_b}{r}\left[1 - \exp(-\alpha_0 r)\right], \qquad \alpha_0 = \frac{\sqrt{\pi m_{ab} k_B T}}{\hbar}. \tag{1.34}$$

This form of regularized potential was first used by Kramers and also Hellmann in the early quantum theory. At large radii, the Kelbg potential has the same behavior as the Coulomb potential and needs screening.

We will start with a discussion of the screening problem using the method developed by Bogoliubov and then continue with a discussion of the divergence of the Bohr atom studied by Herzfeld and Planck. The screening theory of Debye and Hückel cannot be applied directly to the quantum-statistical problem since the Poisson equation is no longer valid. To find this effective potential, Debye used an extended Poisson equation as shown above. Instead, in 1946, Nikolai N. Bogoliubov (1909–1992) proposed an integral equation for the relation between the potential and the screened potential $\varphi_{ab}^{\mathrm{D}}$:

$$\varphi_{ab}^{\mathrm{D}}(1,2) = \varphi_{ab}(1,2) + \sum_c n_c \int \mathbf{d}\mathbf{r}_3\, \varphi_{ac}(1,3)\varphi_{bc}^{\mathrm{D}}(2,3). \tag{1.35}$$

The radial distribution and the correlation function g_{ab} are connected to the screened potential by

$$F_{ab} = 1 + g_{ab} , \qquad g_{ab} = -\beta \varphi_{ab}^{D} . \tag{1.36}$$

In this way we see a connection between the Bogoliubov integral equation and the Ornstein–Zernike equation introduced above. According to Debye and Bogolyubov, the mean Coulomb energy is given by

$$U_{\mathrm{c}} = \frac{V}{2} \sum_{ab} n_a n_b \int d\mathbf{r} \frac{e_a e_b}{r} F_{ab}(r) . \tag{1.37}$$

Carrying out the integration gives

$$U_{\mathrm{c}} = -V k_{\mathrm{B}} T \frac{\kappa^3}{8\pi} \frac{1}{\sqrt{1 + (\sqrt{\pi}/2)(\kappa\lambda)}} , \qquad \lambda = \frac{\hbar}{2\pi \mu k_{\mathrm{B}} T} . \tag{1.38}$$

This means that the Coulomb energy density per particle increases as in the Debye theory, i.e., as the square root of the density, but with quantum corrections. We note that averaging the Kelbg potential instead of the Coulomb energy (as, e.g., in appendix of Kraeft et al. 2015) has no physical meaning. The reason is that the Kelbg potential is not a physical energy, but is already the result of a quantum averaging. In fact, this averaging of the Kelbg potential may be responsible for deviations from the correct results (Alastuey et al. 2015).

The free energy may be obtained by Debye's charging procedure, integrating with respect to e^2, which leads to a prefactor $3/2$. Note that, strictly speaking, there is also a contribution linear in e^2. Carrying out the summation first, the linear term results in a zero contribution due to electroneutrality. However, carrying out the integration first, we find a sum of infinities which is indefinite. We may conclude that on a classical level the problem of Coulomb divergencies cannot be decided on a purely mathematical level. In spite of this several authors conclude that such terms have no justification. We will come back to this problem in later chapters. A different procedure for deriving the free energy due to Mayer and Macke involves the summation of chain or ring diagrams of Coulomb lines. For the classical case, using the charging procedure for a pure classical Coulomb system, we get the Debye free energy.

The quantum theory of screening was developed by Macke, Gell-Man, Brueckner, Montroll, Ward, Vedenov, Larkin, DeWitt, Kelbg, and others. The most important results in the present context were obtained in different research centers. Vedenov and Larkin (1959) in Moscow obtained the first quantum corrections to Debye's results using field theoretical methods. Another center was founded by Kelbg in Rostock, for early results see, e.g., (Kelbg 1963; Ebeling et al. 1976; Kraeft et al. 1986). A first *Conference on Strongly Coupled Plasmas* was organized 1977 in Orléans by Kalman and Carini (1978). The Rostock school first used the method of effective potentials to include quantum corrections and later more refined methods with Green functions and other methods (Kelbg 1964; Ebeling et al. 1976).

A different approach based on the Montroll–Ward method was developed by DeWitt. For the main results, see (Montroll and Ward 1958; De Witt 1962; De Witt et al. 1995). The school of Nevil Mott (1905–1996, Nobel Prize 1977) studied the influence of interactions on bound states in a plasma. Another important approach based on the Feynman–Kac methods was developed by Alastuey and Perez (1992), and an approach based on the effective field methods of quantum field theory by Brown and Yaffe (2001). All these quite different methods lead at lower densities to essentially the same results (Alastuey et al. 2015).

In the region of strong correlations, i.e., *strongly coupled gases and plasmas*, which is located in the so-called *corner of correlations*, the physical properties of plasmas may change drastically in comparison to ideal or weakly non-ideal plasmas. In particular, the ionization phenomena in plasmas may depend strongly on non-ideality effects and the continuum is influenced by density and temperature. With increasing non-ideality the levels merge into a continuum and the discrete states disappear (Mott effects).

One idea which played an essential role in the development of the theory of correlated plasmas involves effective potentials or pseudo-potentials. This idea was first introduced in quantum chemistry by Hellman and Gombas, in solid state physics by Heine and others, and in statistical physics by Morita (1959). Since 1962, Kelbg and coworkers at Rostock University have developed this method into a powerful tool in plasma physics (Kelbg 1963, 1972). The idea of Günter Kelbg was to replace the Coulomb potential by an effective potential which is finite at zero distance due to quantum effects (see Chaps. 4 and 5). Kelbg's theory was originally based on quantum perturbation theory for the diagonal matrix elements (Slater sums) of the density matrix, and he succeeded in obtaining exact expressions for the first order in e^2.

His theory was developed by a group at Rostock University(including Ahlbehrendt, Ebeling, Hetzheim, Hoffmann, Kraeft, Kremp, and Schmitz, later joined by Röpke, Schlanges, Redmer, Blaschke, Bonitz, and others). The Rostock school of quantum statistics formed by Günter Kelbg concentrated on analytical calculations of thermodynamic functions based on effective potentials (Kraeft et al. 1986). Several authors and in particular Norman and coworkers started Monte Carlo calculations using effective potentials (Zamalin et al. 1977; Fehske et al. 2008). Deutsch (1977) introduced useful approximations and Hansen and Mc Donald (1976) gave first applications to molecular dynamics. Further, Kleinert (1995) developed a variational approach.

Let us discuss now the problems connected with stronger coupling, i.e., large deviations from non-ideality. Strongly correlated Coulomb systems have been in the focus of recent investigations in many fields, including dense plasmas in space and in the laboratory (Kraeft et al. 1986; Ebeling et al. 1991; Kremp et al. 2005), in electron–hole plasmas in semiconductors, and charged particles confined in traps or storage rings. In these systems the Coulomb interaction energy U is on average higher than the mean kinetic energy K, i.e., the coupling parameter $\Gamma = |\langle U \rangle| / \langle K \rangle > 1$. In particular, Coulomb and Wigner crystallization, which may occur when $\Gamma \gg 1$ (typically of the order of 100), has attracted much attention in recent years. Coulomb

Fig. 1.10 The first Conference on Strongly Coupled Plasmas 1977 in Orléans: Forrest Rogers (*left*) and Hugh De Witt (*right*) with one of the authors. See also (Whitley et al. 2015). Photo from author's archive

crystals have been observed in ultracold trapped ions, e.g., in dusty plasmas, and in storage rings (Figs. 1.10 and 1.11).

Moreover, there exist many strongly correlated Coulomb systems where quantum effects are important. Examples are dense astrophysical plasmas in the interior of giant planets or white dwarf stars as well as electron–hole plasmas in condensed matter, and few-particle electron or exciton clusters in mesoscopic quantum dots (see also Chaps. 4, 9, and 10). The formation of Coulomb bound states such as atoms and molecules or excitons and bi-excitons, and also Coulomb liquids and electron–hole droplets, are examples of the wide variety of correlation phenomena that exist in these systems (see Chaps. 9 and 10 and Kremp et al. 2005). At the same time, recombination of electrons and positive charges to form neutral bound complexes greatly reduces the Coulomb coupling and thus acts against formation of Coulomb crystals in two-component charged particle systems.

Regularization of the Partition Function

Bound states of pairs of oppositely charged Coulomb particles were first described by the Bohr model and this was a real revolution in the understanding of the structure of atoms. In 1916, the Austrian physicist Karl Herzfeld detected a serious problem with the Bohr model, namely that the statistical partition function is divergent. This is a fundamental problem of Coulomb systems which will be discussed here and at various points in this book. Eggert and Saha ignored the open problems with the partition function, and succeeded in finding the solution to one of the most urgent

Fig. 1.11 Günter Kelbg with his coworkers Wolf-Dietrich Kraeft (*right*), Dietrich Kremp, and a guest from Krakow, Andrej Fulinski (*left*), standing in front of Rostock University. Author's archive

problems in astrophysics at the time, viz., the ionization phenomenon (Eggert 1919; Saha 1920).

In their derivation, Eggert and Saha used the correct theoretical tools, such as Sackur–Tetrode's formula for the chemical constant, to calculate the chemical potential of electrons. On the other hand they clearly missed the significance of the partition function in the chemical potential of the atoms and ignored the fact that the atomic partition function is divergent for Coulomb atoms. The problem of the divergence of the partition function of the Bohr atom was already noticed by Bohr, and was then studied in detail by Herzfeld, and later by many others, including Planck. For an atom with energy levels E_{sl} depending on the main quantum numbers s and l, the atomic partition function is defined by

$$\sigma(T) = \sum_{s,l} \exp(-\beta E_{sl}) \ . \tag{1.39}$$

For hydrogen plasmas we have $E_{sl} = E_s = -I/s^2$. The terms in the atomic partition function for hydrogen diverge as the square of the main quantum number s^2 (see Fig. 1.12). Although as mentioned the problem of the divergence of the atomic partition function was seen already by Bohr and in particular by Herzfeld, it took a long time before the mathematical background for a serious treatment of this problem

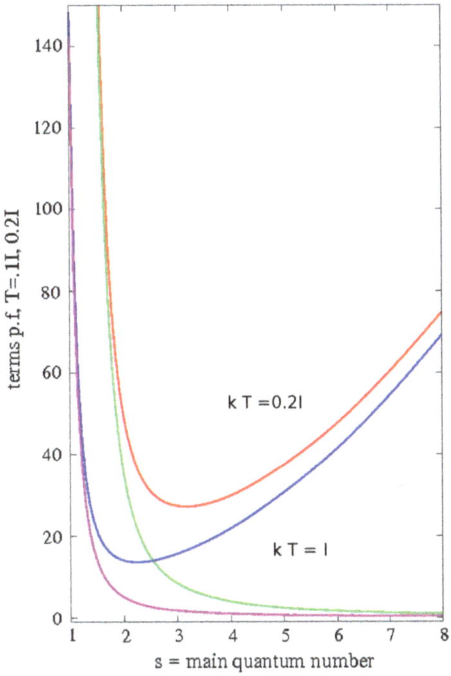

Fig. 1.12 Terms in the atomic partition function defined by (1.39) for two temperatures corresponding to $\beta I = 5$ (*red and green curves*) and $\beta I = 10$ (*blue and violet curves*). Up to terms of order $E_s^{\max} \approx k_B T$, the terms in the usual partition function decrease monotonically, then begin to increase as s^2, leading to the Herzfeld divergence problem. For comparison, we show the Planck–Brillouin–Larkin regularized partition function with the terms of a convergent series

became available. A detailed discussion of this issue is one of the major tasks of the present book, and we will come back to it soon. However, those who urgently needed to solve the ionization problems in astrophysics and plasma physics had no time to wait for strict solutions. As pointed out above, for the most urgent calculations, the problem was simply ignored and the partition function was estimated by the first term only, or possibly by the sum of the few first terms.

One should admit that there were also some good mathematical reasons for such a procedure. Looking at the terms in Fig. 1.12, we see a strong similarity with the so-called asymptotic series in mathematics, where the terms first decrease and then increase again. In such cases mathematical theories suggest cutting off the series at the smallest terms. In our case this is the term with $|E_s| \approx k_B T$. A more physical argument was, that terms with lower energies than the thermal energy $|E_s| \approx k_B T$ are not sufficiently stable and may not contribute (Riewe and Rompe 1938).

The first reasonable solution of the divergence problem of the Coulomb partition function was given by Planck (1924). Planck divided the atomic partition function Σ into three parts, viz.,

$$\Sigma = \Sigma_1 + \Sigma_2 + \Sigma_3 \, , \tag{1.40}$$

where Σ_1 is the contribution of the scattering states, Σ_2 is the contribution of the high bound states near the series limit, and Σ_3 is the contribution of the lowest bound states. According to Planck's estimate, the boundary between the lower and higher bound states is given by some appropriate choice $-\epsilon'$, located between an upper and a lower limit:

$$e^2/d < \epsilon' < k_B T \ .$$

In other words, the boundary is an energy between the thermal energy and the Wigner energy. In the next step, Planck estimated the sum of the first two contributions using a semiclassical estimate and found it to be equal to the sum over states for free particles, viz.,

$$\Sigma_1 + \Sigma_2 \approx V/\Lambda^3 \ .$$

In other words, the interaction parts (which are both divergent, but of opposite sign) compensate each other. Due to this compensation, we have

$$\Sigma = V/\Lambda^3 + \Sigma_3 \ . \tag{1.41}$$

Since the free contribution is not counted as an internal atomic contribution, the sum is only carried out over $E_{sl} < -\epsilon'$. In the more general case where the energy levels depend on the density, which will be discussed later, we need an appropriate density and temperature-dependent choice of the upper border $\epsilon'(n, T)$. At low density, $\epsilon'(n, T)$ should be near the terms in the partition function where the minimum of the terms in the sum is located.

In Fig. 1.12, we compare the terms in the atomic partition function with the regularized formula. Up to terms of order $E_s^{\max} \approx k_B T$, the terms decrease monotonically and then start to increase as s^2. Notice the structural similarity with Fig. 1.9. For comparison, we show the regularized Planck–Brillouin–Larkin partition function. A disadvantage with the Planck procedure, or other generalizations working with discrete upper limits of the s summations depending on T and/or n, is the possibility of discontinuities, when levels cross with changing T and/or n limit. A procedure which avoids this problem was worked out by Brillouin (1930) in one of the first books on the new science of quantum statistics.

Much later, in the 1960s, a more rigorous basis was worked out in a two-page letter by Larkin (1960) and in extended works by one of the present authors in collaboration with Kremp and Kraeft (Ebeling 1967, 1968, 1974; Ebeling, et al. 1976). We do not find this procedure explicitly in Planck's paper, but it was clearly presented by Brillouin (with reference to Planck) and much later in Larkin's letter, evidently without knowledge of the earlier work. The regularization is based on a Taylor expansion of the partition function for hydrogen levels, subtracting the first divergent terms:

$$\sigma(T) = \sum_{s=1}^{\infty} s^2 \Big[\exp(-\beta E_s) - 1 + \beta E_s \Big] \ . \tag{1.42}$$

This convergent (regularized) form of the atomic partition function given by (1.42) is nowadays often called the Planck–Brillouin–Larkin partition function. The convergence is due to the fact that, in the series, the divergent contributions up to $O(e^4)$ were omitted. The practical equivalence of the two procedures, at least in the low density region, can be seen in Fig. 1.12. The deeper physical justification for the procedure in (1.42) is based on the fact that these contributions are influenced by screening effects and have to be treated separately. An alternative way to express this is the observation that there are compensation effects between contributions of the terms just below and just above the series limit. Due to these and other physical effects, the sum of contributions to any thermodynamic quantity is always finite.

The important work of Planck on the partition function published in 1924 and the subsequent work by Leon Brillouin (1889–1969) (Brillouin 1930) was largely ignored for about 30 years. Then the formula (1.42) appeared in a letter by Larkin (1960) as a byproduct of the more rigorous field-theoretical statistical thermodynamics of Vedenov and Larkin (Vedenov and Larkin 1959; Larkin 1960). The whole complexity of the problem and the connection with the Eggert–Saha equation became clear only in later work by the Rostock school (Ebeling 1967, 1968, 1974) and was then summarized in several books (Ebeling et al. 1976, 1991; Kraeft et al. 1986; Kremp et al. 2005).

Note that, at high density, the bound states may depend on density and temperature and may possibly even disappear completely. These effects were studied first in early papers by Inglis and Teller (1939), dedicated to the merging of levels due to the Stark effect. Nevil Mott (1905–1996) investigated the level shifts due to screening and also showed that this effect plays an essential role in semiconductors (Mott 1961).

The more advanced theory of the Mott effects, i.e., the disappearance of energy levels due to screening effects is based on the solution of the so-called Bethe–Salpeter equations (Kremp et al. 2005; Redmer et al. 2010). In the region where Mott effects begin to play a role, the partition function has to be replaced by an effective partition function which depends on the effective energy levels (see Chaps. 3–6).

1.5.2 Coulomb Phase Transitions

In 1943, Landau and Zeldovich discussed new possibilities for phase transitions in Coulomb systems connected with metal–insulator transitions in metal fluids in the vicinity of critical points. A first systematic study of Coulomb transitions in plasmas was given by Norman and Starostin (1968). The first order transition predicted by these authors was called the plasma phase transition (PPT). The PPT was predicted as a possible result of the competition between effective Coulomb attraction and quantum repulsion in the partially ionized dense plasma.

The qualitative picture was similar to the van der Waals model, where the phase transition is a result of the long-range attraction between neutral molecules and their short-range repulsion. We remember that the two phases in the van der Waals model differ from each other by the density of molecules. The two phases in PPT have different number densities of charged particles and different degrees of ionization. Atoms, which are present in the two coexistent phases, were treated originally as

Fig. 1.13 Michael Fisher in the discussion at the Conference on Strongly Coupled Plasmas in Budapest 2011. From SCCS conference photos

an ideal gas, later including interactions with neutrals (Norman and Starostin 1968, 1970). Since the estimates of this transition lead to very high pressures, densities, and temperatures, the location of the PPT in plasmas was for a long time not clearly identified experimentally (Weir et al. 1996; Ternovoi et al. 1999; Fortov 2011, 2013). Only recent experimental work seems to detect this transition clearly (Fortov et al. 2007; Knudson et al. 2015).

The main theoretical approaches will be explained in Chaps. 3–6. In Chap. 3, we begin to explain the properties and the conditions for a PPT on the basis of very simple models, essentially a combination of the Debye–Hückel and the van der Waals theory. We will show that simple models already show a separate Coulomb transition, which is as shown in particular by Michael Fisher (see Fig. 1.13).

Phase transitions in classical Coulomb systems have received more attention. The first results on phase transitions in ionic systems go back to the late 1960s. In 1970, Voronzov-Veliaminov found a coexistence line and a critical point in the course of Monte Carlo studies of charged hard spheres imbedded in a dielectric medium (Voronzov-Veliaminov 1970). A first analytical estimate of the critical point and the coexistence line for electrolytes based on the Debye–Hückel theory was given in a short note by one of the authors (Ebeling 1971), predicting the critical temperature for ionic solutions at

$$T_{\rm cr} \approx \frac{e^2}{16\varepsilon_{\rm r} k_{\rm B} a} \,. \qquad (1.43)$$

A systematic theory of classical Coulomb phase transitions including a comparison with numerical and experimental data was given later in the pioneering work of Michael Fisher and coworkers (Fisher and Levin 1993). These authors also made an extensive investigation of the state of the art in this field. Summarizing these results for classical systems we may say that theory and experiment (Monte Carlo data as well as measurements on electrolytes) are in quite good agreement. There is no longer any doubt that a Coulomb phase transition exists in classical systems.

We will not repeat all the arguments here, but refer to the careful investigations of Fisher et al. (1993). The classical Coulomb transition is due to a balance between a hard-core repulsion and a Coulomb attraction. We will show here that the PPT is a balance between quantum repulsion between point charges and Coulomb attractions. In this respect a PPT is a kind of quantum variant of the classical Coulomb transition. While the existence of a Coulomb transition in ionic systems is now well confirmed, there are still open problems about the PPT in plasmas. From the theoretical point of view, this is connected with the difficulties involved in deriving an accurate equation of state for non-ideal quantum plasmas. From the experimental point of view the difficulties are connected with the very high pressures at which the PPT could occur. However, as we will point out, there are now several experimental and numerical data which point to the existence of a PPT in real plasmas (Fortov et al. 2007; Knudson et al. 2001, 2015).

In order to demonstrate how the expressions for plasma thermodynamic functions, in particular the pressure, may become unstable for higher densities, let us consider a model. We shall describe the so-called quantum Debye–Hückel approximation (QDHA) or Lambda approximation (Ebeling et al. 1976) in which quantum effects are expressed in terms of just one characteristic length, the thermal de Broglie wavelength Λ, in combination with Debye–Hückel type approximations. As we will show in Chaps. 4 and 6, the Lambda approximation is in good agreement with the exact quantum-statistical results for small densities and $T \approx 10^4$ K (Ebeling et al. 1976; Kraeft et al. 1986). This approximation requires the plasma to be non-degenerate, i.e., $n^* \Lambda^3 \ll 1$, where n^* denotes the density of free particles. Furthermore, only a small fraction of the atoms should be bound in molecules $\beta_2 \ll 0.5$.

In this way the classical and the quantum case might be treated in the present rough approximation by the same procedure. The easiest way to check for stability of the system is to investigate the sign of the derivative of the pressure. The region of instability is

$$\frac{\partial(\beta P)}{\partial n} < 0 \,. \tag{1.44}$$

Introducing here the Debye–Hückel-type expression with a temperature-dependent effective radius $a(T)$, we may show that instability occurs at temperatures below the critical value (Ebeling et al. 1976; Fisher and Levin 1993)

$$T \leq T_{\mathrm{cr}} \approx \frac{e^2}{16\varepsilon_0 k_{\mathrm{B}} a(T)} \,. \tag{1.45}$$

In the case of classical ionic solutions, the parameter a is to be identified with the diameter of ions. In the case of dense quantum plasmas, we have to introduce the quantum effective diameter $a(T) = \Lambda/8$. In this way we have shown that the PPT is the analogue of the known classical Coulomb transition in ionic solutions. At least in the simplified theory given above, the classical transition in ionic solutions and the quantum PPT in dense plasmas are fully analogous. In the classical case we get a critical value of the Bjerrum parameter $b_{\mathrm{cr}} = e^2/\varepsilon_{\mathrm{r}} k_{\mathrm{B}} T_{\mathrm{cr}} a \sim 16$. In the quantum case we find the corresponding estimate for the critical temperature to be

$$T_{cr} \approx 12\,570\,\mathrm{K}\,. \tag{1.46}$$

In the case of electrolytic solutions or semiconductors the critical temperature is much lower due to the appearance of the dielectric constant in the denominator. Another effect which decreases T_{cr} is the finite size of the counterions in a plasma. Including the finite ion size and other nonideality effects may lead to a significant decrease in the critical temperature to values far below 10 000 K (Redmer and Röpke 2010).

1.6 Development of Computer Simulation Methods

1.6.1 The Metropolis Algorithm

Dense gases and plasmas are typical many-particle systems and need special approaches, in particular numerical methods. The first successful algorithm for treating the difficult many-particle problem which is of basic importance for plasma physics is the Metropolis algorithm which was developed in 1953 by the Los Alamos group. This algorithm belongs to a class of stochastic algorithms which are called Monte Carlo algorithms. It was named after Nicholas Metropolis, a member of the Los Alamos group. He was one of the authors of the first paper, in 1953, along with Arianna W. Rosenbluth, Marshall N. Rosenbluth, Augusta H. Teller, and Edward Teller, entitled *Equation of State Calculations by Fast Computing Machines*.

Here the authors developed an algorithm for treating plasma in the canonical ensemble. There is controversy over the credit for discovery of the algorithm. Edward Teller states in his memoirs that the five authors of the 1953 paper worked together for "days and nights". Marshall Rosenbluth, in an oral history recorded much later, gives credit to Teller for posing the original problem, himself for solving it, and Arianna Rosenbluth (his wife) for programming the computer.

Some time later, in 1970, W.K. Hastings extended the method to the more general case. Brush, Sahlin, Teller, De Witt, Hansen, Pollock, Zamalin, Norman, Binder, and others developed the method further (Brush et al. 1966; Zamalin et al. 1977; Binder 1979; Bonitz and Semkat 2006; Fehske et al. 2008). The Metropolis–Hastings algorithm works by generating a sequence of sample values in such a way that, as more and more sample values are produced, the distribution of values approximates the desired distribution $P(x)$. These sample values are produced iteratively, making the sequence of samples into a Markov chain.

In statistics and statistical physics, the Metropolis–Hastings algorithm is based on generating Markov chains to obtain a sequence of random samples from a probability distribution for which direct sampling is difficult. This sequence can be used to approximate the distribution, i.e., to generate a histogram, or to compute an integral, such as an expected value. The Metropolis–Hastings and other Monte Carlo algorithms are generally used for sampling from multi-dimensional distributions,

Fig. 1.14 Edward Teller as a young scientist visiting Berlin. Courtesy of Thomas Hüsing

especially when the number of dimensions is high. For single-dimensional distributions, other methods are usually available (e.g., adaptive rejection sampling), which can directly return independent samples from the distribution and are free from the problem of auto-correlated samples that is inherent in these methods. For details see the last few chapters of this book (Figs. 1.14 and 1.15).

The Metropolis–Hastings algorithm, which will be explained in more detail in Chap. 9, can draw samples from any probability distribution $P(x)$, provided one can compute the value of a function $f(x)$ which is proportional to the density of $P(x)$. If we believe in the memoirs of Rosenbluth, then neither Metropolis nor Teller participated in any way to developing the algorithm. Many authors now consider Rosenbluth's account of events as the most likely version of the way the new algorithm was invented, but its name is for ever connected with the name Metropolis.

Later in the 1950s and 1960s, Marshall Rosenbluth made many important contributions to the burgeoning discipline of plasma physics and laid the foundation for the solution of many important problems in the field, particularly the theory of plasma instabilities. So for sure, Marshall Rosenbluth is one of the most important pioneers in our field.

1.6.2 Monte Carlo and Molecular Dynamics Simulations

The Metropolis algorithm is cited in *Computing in Science and Engineering* as being among the top ten algorithms having the "greatest influence on the development and practice of science and engineering in the 20th century". There are is now many applications to dense gases and plasmas (Zamalin et al. 1977; Binder 1979; Kalman

Fig. 1.15 Marshall Rosenbluth developed algorithms and solved important problems of plasma instabilities. By courtesy of Thomas Hüsing

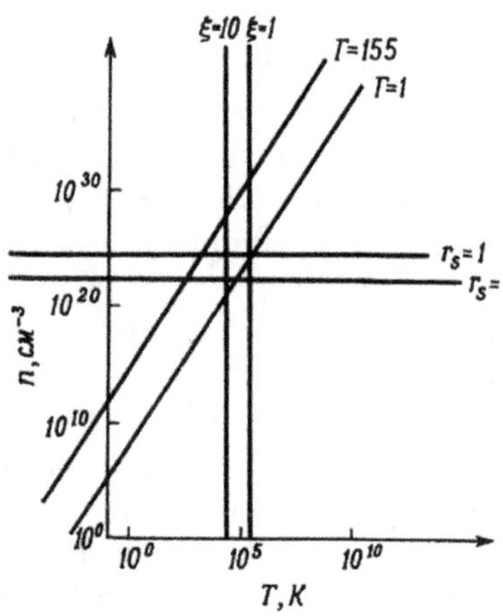

Fig. 1.16 The region of stronger correlations in a one-component plasma starts in the corridors $\Gamma > 1$ and $r_s > 1$ located in the left corner of the density–temperature plane

et al. 1998; Militzer and Pollock 2000; Bonitz and Semkat 2006; Fehske et al. 2008). Let us come back to the problem of simulating dense plasmas. An important paper for the development of plasma physics was the treatment of a one-component plasma by Monte-Carlo methods in the 1960s by Brush, Sahlin, and Teller. The corresponding region of strong correlations is shown in Fig. 1.16. Brush et al. (1966) observed a transition in a 32 particle classical Monte Carlo simulation at $\Gamma > 125$. Hansen and Pollock and Hansen followed with improved calculations, and found $\Gamma > 155$. Using the empirical Lindemann ratio melting criterion, a transition at $\Gamma_{\mathrm{m}} \sim 172$ was predicted. Later studies have converged on the estimate of $\Gamma_{\mathrm{m}} \sim 178$.

In 1976, Hugh De Witt presented rather precise Monte Carlo simulations for one-component plasmas which were later extended to many-component plasmas.

Some of these results will be discussed in more detail in later chapters. Another important class of algorithms for many-particle problems was developed only a few years later by Alder and Wainwright, in late 1950s, and independently by Rahman in the 1960s. Alder and Wainwright simulated perfectly elastic collisions between hard spheres. In 1964, Rahman simulated liquid argon using a Lennard-Jones potential. The transport properties calculated in this way, such as the coefficient of self-diffusion, compare well with experimental data. (see also Berne et al. 1998; Ceperley 1995; Magro et al. 1996; Bonitz and Kraeft 2005).

1.7 Transport Theory of Nonideal Gases and Plasmas

1.7.1 Extension of Boltzmann's Theory to Dense Gases

The development of the kinetic theory was the first essential step to an understanding of transport in gases. This theory was originally based on a simplified particle description of a gas. August Krönig (1822–1879) and Rudolf Clausius (1822–1888) in Berlin developed a simple theory based on the idea that the atoms are small elastic balls moving stochastically. From this idea many gross properties of the gas could be derived (for a historical review, see Rompe et al. 1987; Ebeling and Hoffmann 1991). A few years later James Clerk Maxwell (1831–1879) and Ludwig Boltzmann (1844–1906) developed this theory into a central tool of nineteenth century physics. At the end of nineteenth century, van der Waals was able to introduce interactions other than hard core into these concepts. In this way, a generalization of the theory to a real gas of moderate density and higher temperature became possible.

The kinetic theory was a big step in the understanding of nonequilibrium processes in gases. The pioneer of the kinetic theory of gases was Ludwig Boltzmann (1844–1906). His kinetic equation is strictly speaking valid only for ideal gases. The first big step in extending the Boltzmann transport theory to dense gases was made by David Enskog (1884–1947) in Sweden (Enskog 1917) and in further work by Chapman and Cowling (1991). The Boltzmann-type kinetic equation for the distribution function $f(\mathbf{r}, \mathbf{v}, t)$ reads

$$\frac{\partial f}{\partial t} + \mathbf{v}\cdot\frac{\partial f}{\partial \mathbf{r}} + \frac{1}{m}\mathbf{K}(\mathbf{r})\cdot\frac{\partial f}{\partial \mathbf{v}} = \left.\frac{\partial f}{\partial t}\right|_{\text{coll}} . \tag{1.47}$$

Here the left-hand side describes Liouville-type mechanical flows in the phase space and the right-hand side jumps in the phase space due to collisions between the particles. The kinetic theory of Boltzmann is a theory which includes collisions. However, as already mentioned, other nonideal gas effects are not taken into account. This can be seen from the corresponding transport equations. We will explain this point in Chap. 7. The first transport theory including the correlations due to exchange effects was developed by Uehling and Uhlenbeck (1933).

1.7.2 Kinetic Theory of Dense Plasmas

As already noted, a plasma has a characteristic length, the Debye length $r_D = \sqrt{k_B T / 4\pi n_e e^2}$ (for electron plasmas on a uniform background), which is not just the radius of the screening cloud, but characterizes more generally the scale of all charge non-uniformities in a plasma. Due to the existence of the Coulomb field, larger non-uniformities lead to strong restoring forces. The existence of fields which try to avoid the separation of charges leads also to characteristic oscillations.

In order to estimate this frequency, let us imagine that we move the electrons in a uniform plasma with density n_e by the amount ξ in the x direction. Newton's equation of motion is then

$$m_e \frac{d^2\xi}{dt^2} = -e\,\delta E \,, \tag{1.48}$$

where δE is the extra field generated by the distortion, which is given by Poisson's equation

$$\delta E = 4\pi n_e e \xi \,. \tag{1.49}$$

This leads to an oscillatory equation for ξ, viz.,

$$\frac{d^2}{dt^2} + \frac{4\pi e^2 n_e}{m_e} = 0 \,, \qquad \omega_{pe} = \sqrt{4\pi e^2 n_e / m_e} \,. \tag{1.50}$$

Here there appears a characteristic frequency, the plasma frequency of electrons ω_{pe}. We may conclude that, for all kinetic phenomena in a plasma, there exists a characteristic time $\tau_{pe} \sim 1/\omega_{pe}$ defined by oscillatory electronic motions. Correspondingly, we may also define characteristic times for oscillations of the heavier plasma components, the ions.

The kinetic theory of plasmas started by adapting the classical kinetic equation of Boltzmann to the specific conditions. The Boltzmann-type kinetic equation for the distribution function $f(\mathbf{r}, \mathbf{v}, t)$ of the particles of an electron plasma in an electromagnetic field reads

$$\frac{\partial f}{\partial t} + \mathbf{v} \cdot \frac{\partial f}{\partial \mathbf{r}} + \frac{e}{m} \left(\mathbf{E} + \frac{1}{c} \mathbf{v} \times \mathbf{H} \right) \cdot \frac{\partial f}{\partial \mathbf{v}} = \left. \frac{\partial f}{\partial t} \right|_{coll} \,. \tag{1.51}$$

As above, the left-hand side describes Liouville-type mechanical changes and the right-hand side the influence of collisions. Here the Maxwell equations for the fields have to be added.

In order to estimate the role of plasma waves, we derive a kinetic equation for the distortion of the distribution δf and the field δE:

$$\frac{\partial \delta f}{\partial t} + \mathbf{v} \cdot \frac{\partial \delta f}{\partial \mathbf{r}} + \frac{e}{m} \mathbf{E} \cdot \frac{\partial \delta f}{\partial \mathbf{v}} = -\Delta f \,. \tag{1.52}$$

Here we have neglected second order perturbations and the magnetic field, and estimated the role of collisions by a small damping term Δf. For a wave-like perturbation

$$\delta f(r, p, t) = C_f \exp(\mathrm{i}\omega t - \mathrm{i}\mathbf{k} \cdot \mathbf{r}) , \quad \delta E(r, p, t) = C_E \exp(\mathrm{i}\omega t - \mathrm{i}\mathbf{k} \cdot \mathbf{r}) , \quad (1.53)$$

the kinetic equation implies

$$\mathrm{i}C_f \big(\omega - \mathbf{k} \cdot \mathbf{v} + \mathrm{i}\Delta\big) = -e\mathbf{E} \cdot \frac{\partial f}{\partial \mathbf{p}} . \quad (1.54)$$

The Poisson equation assumes the form

$$k^2 \delta E = 4\pi \int \mathrm{d}\mathbf{p}\, \delta f; \qquad \delta \mathbf{E} = -4\pi e \mathrm{i} \frac{\mathbf{k}}{k^2} \int \mathrm{d}\mathbf{p}\, \delta f . \quad (1.55)$$

Defining the dielectric function of the medium by

$$D(\omega, \mathbf{k}) = \epsilon(\omega, \mathbf{k}) E(\omega, \mathbf{k}) , \quad (1.56)$$

we get for the dielectric function of the plasma the expression

$$\epsilon(\omega, \mathbf{k}) = 1 + \frac{4\pi e^2}{k^2} \int \frac{\mathbf{k} \cdot \partial f / \partial \mathbf{p}}{\omega - \mathbf{k} \cdot \mathbf{v} + \mathrm{i}\Delta} \mathrm{d}\mathbf{p} . \quad (1.57)$$

This complex function determines most of the kinetic properties of plasmas, as known from many studies and as we shall explain later in more detail (Pines 1961; Silin and Rukhadse 1961; Balescu 1963; Klimontovich 1964, 1982; Pines and Nozieres 1966).

The dielectric function also plays a determining role in the collision term which appears in the kinetic equations of dense plasmas (Klimontovich 1964). The direct application of the Boltzmann theory leads to divergencies connected with the long range of the Coulomb forces. Lev D. Landau (1908–1968) showed in 1937 how the Boltzmann collision theory can be reformulated for Coulomb forces and how the divergencies can be avoided. Anatoly A. Vlasov (1908–1975) then discussed in detail how the electromagnetic fields which modify the left-hand side of the Boltzmann equation influence the kinetic theory. Note that Vlasov neglected the collision integral on the right-hand side. However, he studied the influence of the fields using Maxwell's equations and found solutions of the resulting nonlinear equations. In particular, he investigated the coupling to the Poisson equation connecting the electric field with the density (Figs. 1.17 and 1.18).

The Vlasov equation, which has the same structure as the Liouville equation, the 'collisionless' Boltzmann equation, is fundamental to plasma physics. Vlasov showed that this equation, which takes into account the collective interactions in the plasma, explains many plasma effects such as plasma waves. The Vlasov equation describles reversible phenomena in the plasma, while damping phenomena were

Fig. 1.17 Yuri Klimontovich (1924–2002), pioneer of kinetic plasma theory, visiting Rostock University in the 1970s. Author's archive

Fig. 1.18 Radu Balescu (1932–2006), pioneer of plasma kinetic theory, at a Conference on Plasma Theory in Kiev in the 1970s. Author's archive

observed in particle beams interacting with plasmas. These effects are based on the energy exchange between an electromagnetic wave with phase velocity v_{ph} and particles in the plasma with velocity approximately equal to v_{ph}, which can interact strongly with the wave. The theoretical interpretation of these effects was given by Landau, so this effect is called Landau damping. According to Landau's picture, particles with velocities slightly less than v_{ph} will be accelerated by the electric field of the wave to move with the wave phase velocity. The Vlasov–Landau theory studies the deviations of the distribution functions from a Maxwellian distribution.

Another highlight in the development of the kinetic theory of plasmas was the development of a more rigorous kinetic theory of plasmas including collisions in the work of Bogoliubov, Klimontovich, Ssilin, Rukhadse, Balescu, and others. The collision term obtained in the Lenard–Balescu theory reads

$$\left.\frac{\partial f}{\partial t}\right|_{\mathrm{coll}} = 2e^4 \frac{\partial}{\partial \mathbf{p}} \cdot \int d\mathbf{k}\, d\mathbf{p}' \frac{\mathbf{k}\mathbf{k}}{k^4} \frac{\delta(\mathbf{k}\cdot\mathbf{v} - \mathbf{k}\cdot\mathbf{v}')}{|\epsilon(\mathbf{k}\cdot\mathbf{v}, \mathbf{k})|^2} \cdot \left[\frac{\partial f(\mathbf{p})}{\partial \mathbf{p}} f(\mathbf{p}') - f(\mathbf{p}) \frac{\partial f(\mathbf{p}')}{\partial \mathbf{p}'}\right] . \tag{1.58}$$

The first reviews of the theory and its applications were given in the monographs by Spitzer (1961); Balescu (1963); Klimontovich (1964), and Ichimaru (1992). (see also Landau and Lifshits 1979; Lifshits and Pitaevskii 1979; Magro et al. 1996; Zubarev et al. 1996; Ginzburg 2001; Bonitz and Kraeft 2005; Haken et al. 2016; Bonitz 2016).

1.8 Dense Gases and Plasmas in the Laboratory and in the Sun

1.8.1 Studies of Ionization Phenomena

Now that we have spent some time considering the theoretical aspects, let as pay a little attention to some experimental aspects and applications. Plasma is one of the four fundamental states of matter, the others being solid, liquid, and gas. The idea that there are four basic states of matter goes back to the Greek philosopher Empedocles (about 450 BC). He established four ultimate elements which make all the structures in the world: fire (plasma), air (gas), water (liquid), earth (solid). Empedocles called these four elements 'roots'. The state referred to as 'fire' in Empedocles system is the plasma state, which is one of the two targets of this book. The contemporary physical investigation of the plasma state began around 1816 with the investigations of Faraday and was continued around 1879 by Sir William Crookes. Note that Crookes called this state 'radiant matter' or 'cathode rays'. After more detailed investigation, in particular by the British physicist Sir J.J. Thomson in 1897, the term 'plasma' was coined by Irving Langmuir in 1928.

The plasma state is rather special and has properties unlike those of the other states. A plasma is by definition a neutral system with free electrons and ions. Like gases, plasmas do not have a definite shape or definite volume unless enclosed in a container. We consider here mostly gaseous plasmas. Plasmas are by far the most common phase of ordinary matter in the universe, both by mass and by volume. Essentially, nearly all of the visible light from space comes from stars, which are plasmas with temperatures such that they radiate strongly at visible wavelengths. However, plasmas also exist as subsystems in the solid state. For example, the conduction electrons in metals and in semiconductors may be considered as a special form of plasmas.

The behavior of plasmas depends on density and temperature. Degenerate and non-degenerate plasmas behave in a quite different way. In the logarithmic density–temperature plane (see Fig. 1.5), the line $n_e \Lambda_e^3 = 1$ is the boundary between the degenerate and non-degenerate regions. Here $\Lambda_e = h/\sqrt{2\pi m_e k_B T}$ denotes the de Broglie wavelength of electrons with thermal momentum and n_e the density of free electrons. On Earth plasmas appear in many forms, such as discharges, flames, electrical sparks, lightnings, the ionosphere, the magnetosphere, the chromosphere, and so on.

There is a series of international conferences concerned with the properties of plasmas. In order to give just one example, we mention the *International Conferences on Phenomena in Ionized Gases* (e.g., in Berlin 1977; in Swansea 1986; in Pisa 1991). The name of this conference series reflects one of the basic phenomena, namely *ionization*. In order to create a plasma state, one needs ionization processes. A plasma can be created by heating a gas or subjecting it to an electromagnetic field, applied with a laser or microwave generator. The presence of a significant number of charge carriers makes plasmas electrically conductive so that they respond strongly to electromagnetic fields. As stated in the preface and introduction, plasmas constitute the most abundant form of ordinary matter in the Universe, not taking into account other still hypothetical forms such as dark matter. We will now show that the theoretical understanding of ionization processes around 1920 was a major step forward which simultaneously opened the way to astrophysical applications.

Under standard conditions on Earth, electrons are not free but bound in the shells of atoms. Free electrons are formed in equilibrium with bound electrons, e.g., the core electrons of atoms. This equilibrium depends crucially on the binding energies. Our main interest is in hydrogen, helium, and other noble gas plasmas. As already pointed out, the ionization theory of plasmas is closely connected with questions of astrophysics. Let us have a brief look at the problems examined by the two pioneers Eggert and Saha.

John Eggert, born 1891 in Berlin, was a student of chemistry at the *Technische Hochschule Berlin* and worked for some time as Nernst's assistant. In this way, in his studies of ionization he was well prepared to use the results on chemical equilibrium obtained in Nernst's group and the results on the entropy of gases obtained by Sackur and Tetrode. Eggert studied ionization as a chemical equilibrium between electrons, ions, and atoms, and used the Sackur–Tetrode results for calculating the chemical potential of the particles. He studied the chemical equilibrium between electrons, ions, and atoms and transformed the relation for the chemical potentials into the Eggert–Saha formula. When he tried to use these equations to study the multiple ionization of iron atoms in the interior of stars, Eggert did not succeed. As far as we can see, his approach was entirely correct, but the application to phenomena in the interior of stars was too ambitious (Fig. 1.19).

Knowing the work of Eggert, the young Indian physicist Saha was more successful with applications. At first Saha realized the significance of the ionization potential I of atoms. By applying the results from thermodynamics and Bohr's theory to stellar matter, Saha derived a formula by which the degree of ionization in a very hot stellar gas could be expressed in terms of its temperature and electron pressure. For quantitative astronomy, the work of Eggert and Saha was a real breakthrough.

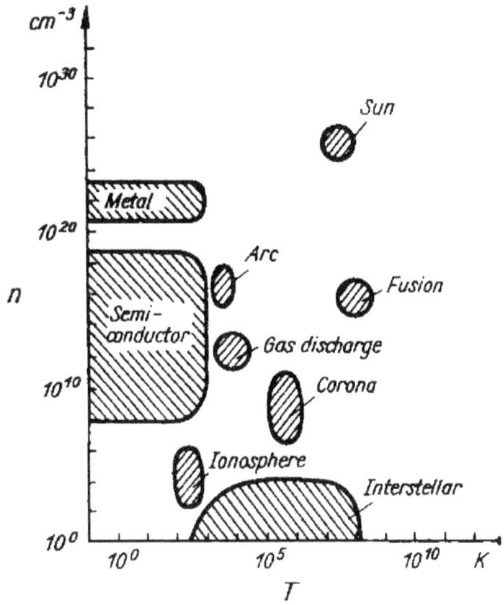

Fig. 1.19 The location in the density–temperature plane of several real plasma systems on Earth and in the cosmos

Fig. 1.20 Megnat Saha (1893–1956) in Berlin in around 1922. According to Eddington, his work was a landmark in the history of astronomy. Courtesy of Thomas Hüsing

Saha's greatest contribution is the generalization of Eggert's ionization theory and the correct application to stellar atmospheres. His ionization theory was first given in the paper *On ionization in the solar chromosphere*, published in the Philosophical Magazine in 1920. Saha traveled to Europe in the 1920s, visiting centers in England and Germany (see Fig. 1.20).

The work of Eggert and Saha showed that the analysis of stellar spectra can provide much information about the state of a stellar atmosphere, not only as regards the chemical composition, but also as regards the temperature. The impetus given to plasma physics and astrophysics by the work of Eggert and Saha cannot be overestimated, as nearly all later progress in this field was influenced by it. So this work may be considered as the starting point of the close relation between plasma physics and astrophysics which is nowadays extended to the physics of quark matter.

During recent decades, a considerable amount of experimental and theoretical effort has been applied to describe the behavior of alkali metals in the liquid and plasma states produced when they are heated (Hensel et al. 1985; Bonzel et al. 1989; Sizyuk et al. 2006). Of particular interest are Li+, Na+, K+, Rb+, Cs+, and Be2+ plasmas. Of special interest has been the liquid–vapor critical point (Hensel et al. 1985). In recent years, hot alkali plasmas have also been widely employed in many technical projects. For instance, lithium plasmas are planned to be used in inertial confinement fusion devices, in solar power plants, for electrochemical energy storage, in magnetohydrodynamic power generators, and in many other applications. Recent advances in the field of extreme ultraviolet lithography have revealed that laser-produced Li plasmas are source candidates for next-generation microelectronics (Sizyuk et al. 2006). For this and other reasons, the investigation of static and dynamic properties of alkali plasmas are certainly of great interest.

Also of considerable interest are semiconductor systems, since the free charges in a semiconductor form a kind of plasma gas. In solid state materials, the conducting band and the valence band are separated by a nearly empty gap. Near to the conducting or valency band, one finds energy states which are due to impurities (doping). The Fermi edge of semiconductors is within the band. We have to distinguish between donor and acceptor semiconductors. Both consist of a host lattice with some imbedded impurity atoms. *Donors* are impurity atoms which have one excess electron in comparison with the atoms of the host lattice. For example, we may consider the semiconductor germanium doped with arsenic atoms which have one electron more in the outer shell as a Ge lattice. This electron is only loosely bound and may escape from the donor, moving freely in the lattice and behaving in a similar way to the free electron in a plasma. In particular, it may form hydrogen-like states.

1.8.2 Generation of Fluids with High Energy Densities

So far we have mainly been discussing gases and plasmas at low and standard pressures. For the generation of high pressures, pulsed methods have been developed, based on the compression and heating of matter in intense shock waves and waves of adiabatic expansion. To generate shock impulses in the megabar pressure range, cylindrical and spherical condensed high explosive charges, laser and corpuscular beams, high velocity impacts, and soft X-rays are widely used. The experimental data obtained by shock wave techniques and physical models of strongly coupled plasmas at high pressures, temperatures, and energy densities provide a lot of new insights, e.g., about the metallization and dielectrization of highly compressed matter, high temperature thermodynamics and phase transitions, and deformation of the energy spectrum of compressed ions and atoms. Wide-range semi-empirical equations of state and states and models of plasma transport properties are constructed on the basis of the experimental data. These semi-empirical models have been used for multidimensional numerical simulation of pulsed high-energy processes. As a result several new technologies based on high pressures have been developed (Fortov 2009, 2011).

The ultimate objective of experiments in high-energy-density physics consists in the generation of extreme material parameters, whose values are at the frontiers of modern experimental capabilities. Plasma states with peak pressures of hundreds or thousands of megabars, temperatures up to 10 billion degrees, and energy densities of 10^9 J/cm^3, which is comparable to the energy density of nuclear matter, have already become the subject of laboratory investigations (Fortov 2009, 2011). According to the ideas developed to date, to implement a controlled thermonuclear reaction with inertial plasma confinement requires an energy of several megajoules to be delivered to a spherical target in 10^{-9} s to generate at its center a deuterium–tritium plasma with extremely high parameter values $T \sim 10^8$ K, $\rho \sim 200$ g/cm^3, $p \sim 150$–200 Gbar, which is close to the conditions at the center of the Sun. The corresponding output laser power should exceed the total power of all terrestrial electrical power plants by several orders of magnitude. These conditions, which are required for the ignition of a controlled thermonuclear reaction, are quite extraordinary by terrestrial standards, but are quite typical for the great bulk of matter in the universe, compressed as it is by gravitational forces in the interior of stars and other astrophysical objects. In this case, in the physics of high energy densities, there arise several absorbing problems, and progress in this and allied fields of knowledge depends on their solution.

The ignition of thermonuclear reactions with inertial plasma confinement under controlled conditions is the principal pragmatic objective of research in high-energy density physics. Here, lasers are in the lead, although electrodynamic techniques (Z-pinches) and heavy-ion schemes are making rapid progress (Fortov 2009, 2011, 2013). The operation of such targets is basically close to supernova explosions, allowing the vast wealth of experimental results and sophisticated computer codes for the calculation of fusion micro-targets to be employed in astrophysics.

Of fundamental significance is the study of the equation of state of matter and plasma composition in a broad domain of the phase diagram, including the conditions inherent in giant planets, exo-planets, dwarfs, giants, and neutron stars, and the determination of quasi-classicality bounds, e.g., the Thomas–Fermi model, in thermodynamics. Of considerable interest are the properties of degenerate compressed plasmas, their thermodynamics, equilibrium, kinetic, and transport properties in the region of strong non-ideality and in the presence of intense magnetic fields, as well as the properties of quark–gluon plasmas and the existence and properties of their hypothesised phase transition (Blaschke and Sedrakian 2006).

The construction and verification of the structure and evolutionary models of planets and exo-planets calls for the derivation of reliable experimental data on matter under similar laboratory conditions. For Jupiter and other planets it is vital to ascertain or disprove the existence of a hard core and determine the dimensions of the domain occupied by metallic hydrogen and the metallization bound for H_2 and $H_2 + He$. Of fundamental importance is the analysis of Jupiter's energetics with the inclusion of phase layering of the mixtures He–H, C–O, etc., as well as the study of the origin and dynamics of its magnetic field. Similar problems are also encountered in studies of giant planets. In order to explore these states, a big role is played by shock-wave experiments, which enable the metallization bounds to be determined and the occurrence of a plasma phase transition to be ascertained. The intriguing question of a phase transition of the first kind in a strongly non-ideal plasma has remained open for almost 80 years (see the first data on the observation of this transition in deuterium discussed in (Fortov 2011, 2013).

The more distant future will see studies of the conditions for pycno-nuclear reactions ('cold' fusion). We remember that gaseous fusion reactions are divided into two main classes: thermonuclear and pycnonuclear. In a gas of low density and high temperature the Coulomb-barrier penetration probability is only slightly affected by electron screening, and the main contribution to the reaction probability comes from the relatively few fast-moving nuclei at the energy of the Gamow peak. These are temperature-sensitive thermonuclear reactions. As the density increases and the temperature decreases, the potential barriers are depressed by electron screening and the Gamow peak is increased in height and displaced towards a lower energy. At high densities and low temperatures, when the potential barriers are depressed by a larger amount, the Gamow peak is shifted across the origin and only its tail-end remains. The reaction probability is now strongly density sensitive, and the main contribution comes from the relatively abundant slow-moving nuclei. These reactions are no longer thermonuclear and are described as 'pycnonuclear'.

The methods of laboratory energy cumulation now being developed make it possible to obtain relativistic jets and intense collisional, collision-less, and magnetohydrodynamic shock waves, much like those observed in astrophysical objects (Fortov 2009, 2011). Under discussion are projects involving the production of radiation-dominated hot plasmas, like the conditions in black holes and accretion disks of neutron stars, as well as the stability of these regimes (see the experiments involving nuclear explosions, lasers, and Z-pinches Fortov 2009, 2011). Interesting suggestions have been put forward concerning the generation of radiative collapsing magnetohydrodynamic and collision-less shocks, fast particles, relativistic jets, and their focusing (Fortov 2009, 2011).

The high-energy-density techniques under development open up the possibility of generating ultrahigh magnetic fields ($B > 1$ GGs) and investigating their effect on the physical properties of matter. The fields obtained in laser-produced plasmas already range into hundreds of mega-gauss. The list may easily be lengthened and is perhaps limited only by our imagination. At present it is hard to tell, even to within decades, which if any of these fascinating problems of high-energy-density physics will be solved at existing and projected laboratory facilities.

Laser-Generated High Energy Plasmas

The rapid progress of laser technology has opened up the possibility of generating ultrashort laser pulses in the nano–pico–femto–atto second range and bringing existing and projected laser complexes into the petawatt–zetawatt power range. This way it has become possible to span a wide range of power densities up to the highest values achievable today, namely, 10^{22}–10^{23} W/cm^2, which will continue to rise as time goes by. The action of these enormous intensities on targets leads to diverse new physical effects in hot plasmas, such as multi-photon ionization, self-focusing and filamentation of different types, the generation of super-intense electric and magnetic fields, electron and ion acceleration to relativistic velocities, nuclear reactions caused by these fast particles, relativistic plasma transparency, nonlinear modulation and multiple generation of harmonics, ponderomotive effects in hydrodynamics, and many other effects, which are the subject of intensive research today.

With the invention of lasers it became possible to record numerous nonlinear optical effects, such as the deformation of intra-atomic and molecular fields by laser radiation, stimulated Raman, Brillouin, and Thomson scattering, multi-photon ionization, and several other nonlinear effects related to bound electrons. Since the advent of the first laser, raising the peak power of laser radiation has been, and still is, one of the principal goals of quantum electronics. The very notion of high peak power is permanently changing, and today it means a power of no less than 1 PW (10^{15} W). The rapid progress in output laser power in the 1960s and 1970s was based on the principles of Q-switching and mode locking, which enabled the laser pulse duration to be reduced from microseconds to picoseconds over 40 years (Fortov 2009, 2011). Further advancement in this direction has been limited by the large dimensions and cost of lasers and the need to operate at the limit of radiation resistance of the optical elements. The present-day 'renaissance' in laser physics is related to the invention of the chirped optical pulse technique in 1985, which opened up the way for multi-terawatt, petawatt, and even exawatt laser systems to raise the intensities on target to 10^{22} W/cm^2, with the theoretical limit around 3×10^{23} W/cm^2 (Fortov 2009, 2011).

The existing and projected petawatt lasers are divided into three types according to the amplifying medium (Fortov 2009, 2011):

- neodymium glass,
- sapphire,
- KDP and DKDP parametric crystal amplifiers.

In all three laser types the energy (in the form of population inversion) is stored in neodymium ions in glass. In the first case this energy is directly converted into the energy of a chirped pulse, which subsequently undergoes compression. In the second and third cases the stored energy is converted into the energy of a narrow-band nanosecond pulse, which is converted into the second harmonic to serve a pump for the amplifiers of chirped pulses. This pump either produces population inversion in a sapphire crystal or decays parametrically into two chirped pulses in a nonlinear crystal (Fortov 2009, 2011).

Unlike neodymium-glass lasers, sapphire lasers provide broadband amplification, which makes it possible to compress pulses down to very short time intervals of the order of 10 to 20 fs. The present advanced crystal growth technology allows one to obtain an aperture of sapphire crystals which does not exceed 10 cm. This is important for attempts to cross the petawatt threshold. The small aperture means one can limit the chirped-pulse energy due to optical breakdown and self-focusing. Today, tens of petawatt lasers are either already in operation in present-day laboratories (see Fortov

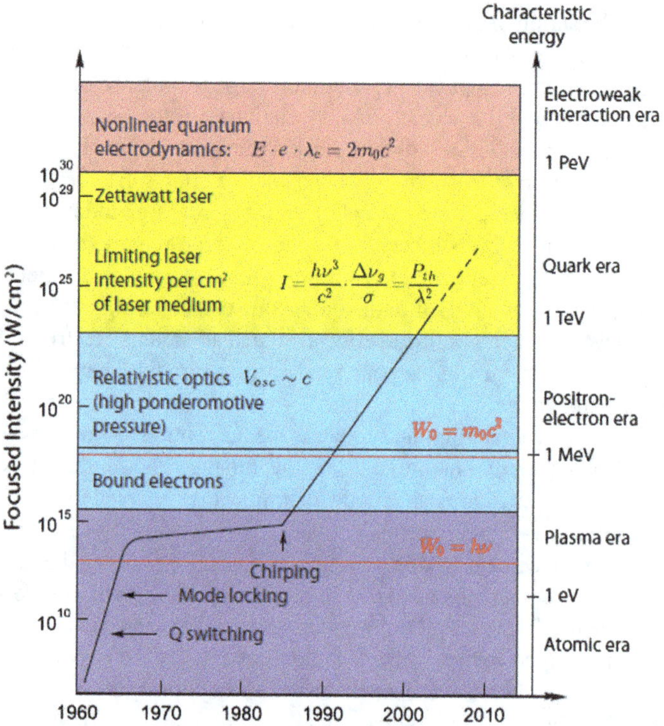

Fig. 1.21 Evolution of the laser focused intensity in watt per cm^2 over the years (Fortov 2011)

2009, 2011) or will soon be developed. Present-day cutting-edge technologies give hope that ultrahigh intensities of the order of 10^{28} W/cm^2 will be obtained in the future (see Fig. 1.21).

Significant progress has also been achieved by application of fast Thomson scattering in experiments using laser-driven shock waves. This method was employed in particular for constructing the equations of state of shock-compressed high-density plasmas. This way more information was obtained from shock-wave experiments in comparison with the ordinary technique of high-power shock waves (Fortov 2009, 2011). High-power pulsed lasers, which were developed for controlled thermonuclear fusion and nuclear substance simulation, provide a unique possibility to generate light pulses with energies ranging into the megajoules and powers of several hundred terawatts. Irradiating a condensed medium by such pulses produces record local energy densities, which no other method is able to provide under laboratory conditions. This circumstance makes high-power laser systems a highly promising tool for investigating substances in a plasma state with extremely high energy densities.

Measuring the parameters of a compressed substance is a special problem (Fortov 2009, 2011). The employment of high-power lasers for the generation of shock waves makes it possible to substantially broaden the attainable pressure range.

It is evident that in laser experiments, too, the classical approach reliant on the use of plane quasistationary shock waves holds the greatest promise for executing quantitative measurements. To date, the majority of experiments in the laser generation of intense shock waves have been carried out in precisely this traditional formulation, which imposes certain restrictions on the target and laser pulse parameters. In this formulation it is currently possible in metals to generate laser-driven shock waves with behind-front pressures of tens of megabars and even gigabars. These values are several times higher than those attainable with the conventional technique of condensed explosives and light-gas 'guns'. They are comparable only to the pressures achieved in the near zone of underground nuclear explosions.

Immediately after the advent of lasers their unique properties were employed to advantage for generating intense shock waves and obtaining extreme substance states with their aid. Laser methods may by helpful not only in studying highly compressed substances with above-solid densities. Using the expansion of a material heated by a laser-driven shock wave or nonthermal electrons, a broad spectrum of states can be obtained in an isentropic unloading wave. These include the domain of strongly nonideal Boltzmann plasma, the neighborhood of the high temperature boiling curve, and the metal–dielectric transition domain.

A problem of special interest connected with the physics of clusters (Krainov and Smirnov 2002, 2007) is the attack of molecular clusters, say H-clusters, with strong laser beams. The interaction of an ultrahigh-power laser pulse of femtosecond duration with cluster targets allows one to obtain plasmas with unique properties (Zweiback et al. 2000; Krainov et al. 2007). On the one hand, at cluster locations, a plasma forms with a density far greater than the critical value. It consists of multiply charged ions and electrons with kilo-electronvolt energies. On the other hand, owing to the smallness of the average density of the medium and individual cluster size, the electromagnetic wave propagates freely and interacts with all atoms of the target, i.e., it freely penetrates the domains with above-critical electron density and is not reflected, unlike the case with solid-state targets. Furthermore, the heating of conduction electrons in the case of metallic clusters (or primary ionized electrons in the case of rare-gas atomic clusters), on the one hand, and the absence of the possibility of fast heat removal as in ordinary plasmas, on the other, allow one to attain a substantially higher excitation of the electron subsystem in comparison with isolated atoms and molecules. As this takes place, atomic ions are barely heated. After the rapid primary multiple ionization during the course of the main part of the laser pulse, the cluster ion substance is an ideal plasma consisting of electrons and multiply charged atomic ions. These plasmas may have quite unusual properties, e.g., non-Maxwellian velocity distributions (Ebeling et al. 2009).

Under strong excitation, the clusters transform to completely ionized plasma, which initially occupies the region occupied by the cluster and freely expands into the ambient space later on. To excite the cluster matter, the time of energy deposition must therefore be short in comparison with the plasma expansion time. For a characteristic cluster size of about 10 nm (10^4 atoms per cluster) and an ion expansion velocity of 10^7 to 10^8 cm/s, the ion flight time amounts to 10^{-13} to 10^{-12} s, which calls for femtosecond excitation times of cluster excitation and plasma production. Therefore, in a cluster target, it is possible to attain a high specific absorption of laser energy and, accordingly, a higher specific substance excitation in comparison with other targets. The characteristic charge of the resulting plasma exceeds the average charge of the plasma produced in the interaction of laser radiation with solid or gaseous targets. The fraction of laser pulse energy converted to X-ray radiation with the use of a cluster target amounts to about 10 to 15 percent and is of non-thermal nature.

These results in the high-efficiency generation of fast electrons and multiply charged ions with kinetic energies range from tens of keV to hundreds of MeV. The simultaneous use of laser pulses with a high temporal contrast ratio of 10^8 to 10^{10} and submicrometer-sized clusters provides a high laser–cluster interaction efficiency and makes it possible to generate different high-intensity photon beams of soft and hard X-ray (up to gamma-ray) radiation, as well as intense high-energy electron and ion fluxes. Investigations into the laser radiation–cluster interaction lead to new information regarding the fundamental properties of matter under extreme conditions (Fortov 2011), while providing the possibility of using new approaches to the solution of several applied problems such as initiation of nuclear reactions, electron and heavy particle acceleration, and production of a bright X-ray radiation source for medical, biological, and lithographic purposes (Fortov 2011, 2013).

1.9 Relativistic Plasmas and Matter with Extreme Energy Density

1.9.1 Relativistic, Subhadronic and Quark–Gluon Plasmas

Dense gases and plasmas at extremely high pressures and temperatures where relativistic effects become essential play an important role for understanding the structure and evolution of astrophysical objects: neutron and 'strange' quark stars, black holes, pulsars, supernovas, magnetars, giant planets, and exoplanets. Furthermore, this field may be relevant for the development of new technologies.

So far we have mainly been considering the statistics of particles with nonzero rest mass m at velocities far below the velocity of light. However, as we have known since Planck, Bose, and Einstein, photon gases behave in a completely different way from normal gases and plasmas. As a first sign of this specific behavior, we remember the Stefan–Boltzmann law which states that the relation between energy density, pressure, and temperature of a photon gas is given by

$$\rho_{\mathrm{E}}(T) = AT^4 , \qquad p = \frac{1}{3}\rho_{\mathrm{E}}(T) = \frac{A}{3}T^4 . \tag{1.59}$$

This means that the pressure and energy density increase very fast, with the fourth power of the temperature. Another fundamental relation connects the temperature

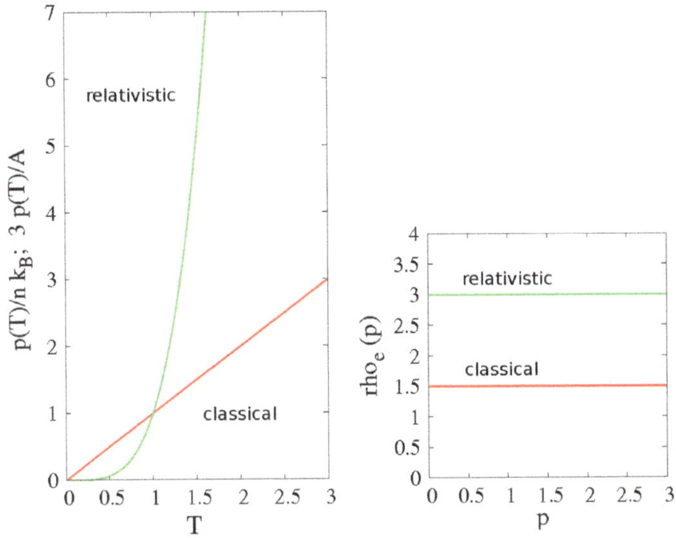

Fig. 1.22 Comparison of the pressure (*left*) and the energy density (*right*) of non-relativistic plasmas, according to ideal gas laws and relativistic plasmas obeying the Stefan–Boltzmann law

and mass density ρ_M for adiabatic processes of photon gases:

$$\rho_M(T) = \text{const.} \times T^3 \ . \tag{1.60}$$

The class of systems which follow these or similar laws is quite broad. Besides the well-known photon gas, it also includes a large part of the plasmas in our Universe. In the following we will refer to this class as *relativistic plasmas*.

In the standard plasma as discussed so far, photons are always present and the equilibrium between matter and radiation plays an essential role. However, the energy density per unit volume is dominated by matter with rest mass and density n and more or less obeys the rules for a Maxwell gas:

$$\rho_E(T) = \frac{3}{2} n k_B T \ , \qquad p = n k_B T = \frac{2}{3} \rho_E(T) \ . \tag{1.61}$$

Figure 1.22 compares the energy density of Maxwell gases at different densities with the universal expression for the energy density of relativistic gases.

As already mentioned, the photon-dominated plasmas are not the only examples of relativistic plasmas in our world. Since the 1960s, it has become more and more clear that under specific conditions subhadronic matter may form relativistic plasmas (Yagi et al. 2005; Satz 2012, 2013). At first several apparently unrelated phenomenological ideas were put forward in the middle of the 1960s, to bring some order into the zoo of known hadrons and leptons. The most important was the idea of Gell-Man and Zweig regarding the basic constituents of nuclear matter, the quarks, which marked the beginning of sub-hadronic physics. These authors showed a possible way to systematize and arrange the extensive and seemingly chaotic zoo of hadrons into some acceptable order by making some simple assumptions.

The basic hypothesis is that all hadrons are built of a minimal composition of more elementary underlying particles, the massive u, d, and s quarks and the corresponding antiquarks combined in the proper way. These constituent quarks play a basic role in quantum chromodynamics (QCD). The model turned out to be surprisingly fruitful for qualitative understanding, and sometimes even for describing the soft hadronic interaction in a quantitative way in a number of experiments. The details of all these exciting developments cannot be explained in this book. We mention only that the renormalized Yang–Mills theory was considered one of the biggest achievements of physics in the last century, and 't Hooft and Veltman were awarded the Nobel Prize in Physics in 1999 for this work. The work of many authors from Brout to Veltman established a mechanism known today as 'broken symmetry' in the quantum theory of fields, the theory that describes all known basic interactions of nature except gravity (Ebert 1989). This mechanism implies the existence of an auxiliary particle, a scalar boson, known today as the Higgs boson, also connected with a recent Nobel Prize in Physics.

These developments in theoretical physics supplied a decisive element in the construction of the *standard model of elementary particles*. The standard model represents so far the most complete theory we have to understand the basic interactions of nature. The broken symmetry mechanism, which is an essential part of the standard model, shows how very short range interactions between elementary particles, such as the weak nuclear force responsible for radioactive decay, and long range interactions, such as electromagnetic fields, can have a common origin. This picture was worked out by Glashow, Salam, and Weinberg, who constructed a theory unifying electromagnetic and weak interactions, which led to their being awarded the Nobel Prize in 1979.

This field of theoretical physics is now called quantum chromodynamics (QCD). This is the theory of strong interactions, a fundamental force describing the interactions between quarks and gluons which make up hadrons such as the proton, neutron, and pion. QCD is a type of quantum field theory called a non-Abelian gauge theory with symmetry group SU(3). The QCD analog of electric charge is called 'color'. The role the photons play for the electromagnetic force is played by particles called gluons. They are the force carriers of the theory (Figs. 1.23, 1.24, 1.25, 1.26 and 1.27).

We do not want to go into any detail and mention only two peculiar properties of QCD:

- Confinement, which means that the force between quarks does not decay as the particles are separated. When we separate one quark from other quarks, the energy in the gluon field is enough to create another quark pair. Therefore they are forever bound into hadrons like the proton and the neutron.

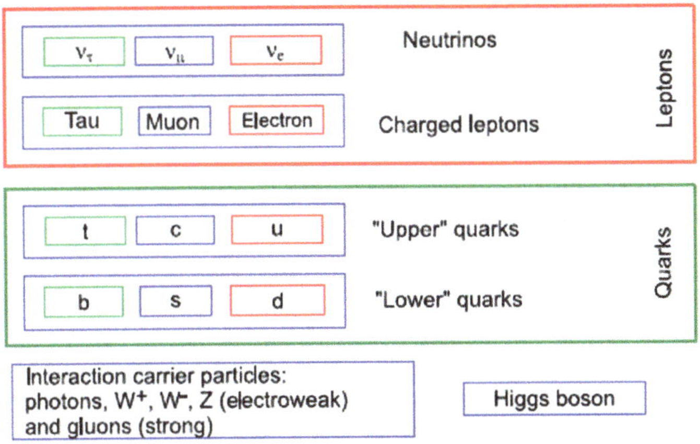

Fig. 1.23 Schematic view of the standard model (Fortov 2009, 2011)

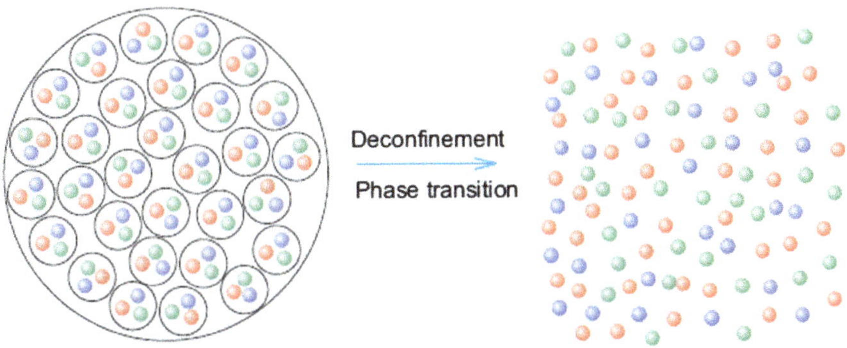

Fig. 1.24 Symbolic transition from confined quarks to a deconfined phase

- Asymptotic freedom, which means that in very high-energy states, quarks and gluons interact very weakly, creating a quark–gluon plasma. This prediction of QCD was first discovered in the early 1970s by Politzer, Wilczek, and David Gross. For this work they were awarded the 2004 Nobel Prize in Physics.

This theory is certainly not a closed chapter of theoretical physics and definitely not part of the present book. However, we cannot overlook the fact that in the mean time a huge body of experimental evidence for the predictions of the standard model and QCD has been gathered over the years. Certainly, the physics of matter with very high energy densities has introduced important new concepts into plasma physics. Therefore we will devote the last two chapters of this book to the quantum statistics of relativistic plasmas with high energy densities. What is relevant in the present context is so far only the fact that relativistic plasmas exist in nature and may be created even under laboratory conditions.

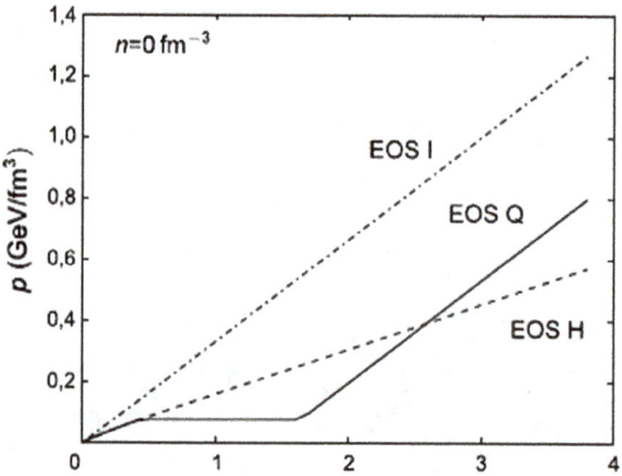

Fig. 1.25 Sketch showing the difference between the equation of state (EOS) for an ideal relativistic gas as represented in Fig. 1.22 and the EOS of a quark–gluon plasma which is like the transition from an ideal gas to a van der Waals gas (Fortov 2011)

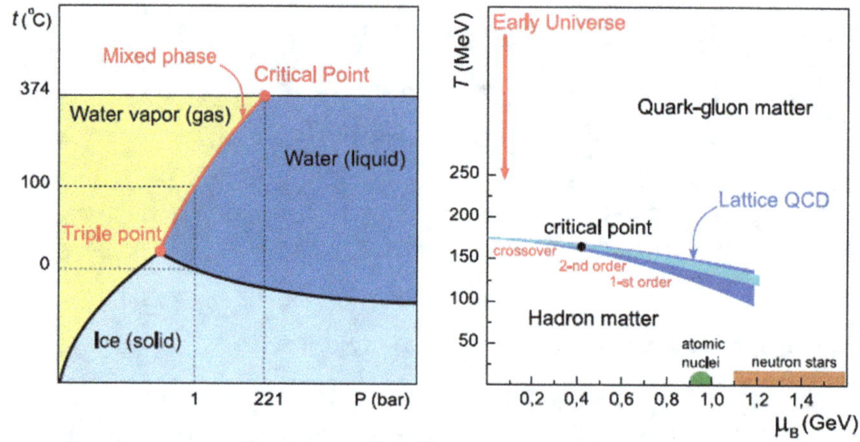

Fig. 1.26 Analogies between the phase transitions in water and in a QGP (Fortov 2011)

Quark–Gluon Plasmas

A quark–gluon plasma (QGP) or quark soup is a phase of quantum chromodynamics (QCD) which is hypothesized to exist at extremely high temperature, high density, or both high temperature and high density. This phase is thought to consist of asymptotically free quarks and gluons, which are several of the basic building blocks of matter (Satz 2012, 2013). As stated above, it is believed that, up to a few milliseconds after the Big Bang, the Universe was in a quark–gluon plasma state.

In normal matter quarks are confined, while in the QGP, quarks are deconfined. Under special conditions, quarks can be freed of their strong attraction. As discussed above, this may happen in the laboratory but also in many situations in the cosmos,

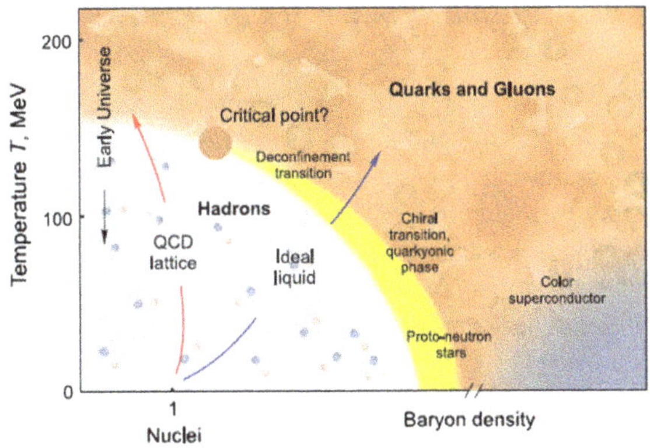

Fig. 1.27 Phase diagram of quarks, gluons, and hadrons (Fortov 2009, 2011)

where extremely high energy densities are normal. In classical QCD, quarks are the fermion components of mesons and baryons, while gluons are considered to be the boson components of such particles. Gluons are the force carriers, or bosons, of the QCD color force, while the quarks by themselves are their fermionic matter counterparts.

Note that the concept of gauge theories going back to Hermann Weyl and Vladimir Fock played a central role in unification (Ebert 1989; Rothe 2005). The new theory of elementary particles led to quantum chromodynamics and the model called the standard model (Satz 2012, 2013). In the years after 1967, the full standard model of elementary particle theory was developed through the work of many contributors as Weinberg, Salam, Glashow, Higgs, and others. As a result there was a unification of several field theories. In particular the weak and electromagnetic interactions were made consistent with the theory of strong interactions between quarks. This part of physics is exciting but not yet finished. Parts of the new theory deal with electroweak interactions and neutrinos. The theory of electrodynamics and the theory of weak interactions has been tested and found to be extremely accurate.

Another area of great interest for our considerations is finite temperature field theory, a branch that attempts to understand particle physics under high temperature conditions. These investigations are crucial for understanding the early evolution of our universe and a new generation of observations (Blaschke and Sedrakian 2006; Satz 2012, 2013). The cross-over temperature from the normal hadronic to the QGP phase is about 175 MeV. This cross-over may actually be a second order phase transition, e.g., belonging to the universality class of the three-dimensional Ising model, as some theorists expect (Satz 2012, 2013). The phenomena involved correspond to an energy density less than 1 GeV/fm^3. For relativistic matter, the equation of state, i.e., the relation between the energy density and the pressure, has been found through lattice computations, perturbation theory, and string theory. This is still a matter of active research (Yagi et al. 2005; Blaschke et al. 2006; Satz 2012, 2013).

1.9.2 Plasmas Generated by Relativistic Particle Beams

The highest energy densities attainable under terrestrial conditions are generated in relativistic heavy-ion collisions. Accelerators required for this purpose operate in several laboratories throughout the world and are well known as the principal experimental tool in nuclear physics, elementary particle physics, quantum chromodynamics, and super-dense nuclear matter physics research, i.e., in the areas which have always been at the forefront of the natural sciences (Fortov 2009, 2011). Most of the experimental achievements have been made by the big international collaborations. The discovery at CERN in 1983 of the massive bosons predicted by the standard model led to the Nobel prize being awarded to the experimenters Rubbia and van der Meer. The discovery of the Higgs boson is also connected with one of the latest Nobel prizes. The phase transition temperature between confinement and freedom has been measured by the ALICE experiment to be around 160 MeV. Below this temperature, confinement is dominant, while above it, asymptotic freedom becomes dominant.

There is a demand for constant advancement in the domain of higher energies and higher phase densities of accelerated particle beams. Investigations in the realm of high energy physics and relativistic nuclear physics have led to the discovery that the main laws controlling the motion and interaction of elementary particles are highly unusual and quite simple. They are based on symmetry and the principle that everything not contradicting the symmetry is not only allowed, but may and must actually occur. This is fully realized in gravitational, electromagnetic, weak, and strong particle interactions.

Accelerator science and technology have come a long way from the first 1.2 MeV proton cyclotron invented by E. Lawrence in 1932 to the Large Hadron Collider (LHC) just beginning operation at CERN, with energies of 7 million MeV in either of the counter-propagating beams, which is 7000 times the proton rest energy. In the center-of-mass system this corresponds to a proton collision energy in the TeV range. During the last few decades, the world has seen the construction of thousands of accelerators of different types. These are giant electrical facilities incorporating cutting-edge engineering ideas and exhibiting a high degree of reliability. Today, accelerators seem to hold all the records in high-energy-density physics.

Experiments at CERN's Super Proton Synchrotron (SPS) first tried to create the QGP in the 1980s and 1990s. The results led CERN to announce indirect evidence for a "new state of matter" in 2000. Current experiments at the Brookhaven National Laboratory's Relativistic Heavy Ion Collider (RHIC) on Long Island (NY, USA) and at CERN's recent Large Hadron Collider near Geneva (Switzerland) are continuing this effort, by colliding relativistically accelerated gold (at RHIC) or lead (at LHC) with each other or with protons. Although the results have yet to be independently verified as of February 2010, scientists at Brookhaven RHIC have tentatively claimed to have created a quark–gluon plasma with an approximate temperature of 4 trillion (4×10^{12}) degrees Celsius. Three experiments running on CERN's Large Hadron Collider (LHC), on the spectrometers ALICE, ATLAS, and CMS, continue to study properties of the QGP. Starting in November 2010, CERN began colliding lead ions for the ALICE experiment, looking to create a QGP. A record breaking temperature was set by ALICE (a large ion collider experiment) at CERN, in the range of 5.5 trillion (5.5×10^{12}) degrees Celsius. Of course, these numbers will already have become old hat! For the latest achievements, one has to consult today's newspapers and recent laboratory reports.

However, many questions remain open in this field, and further experiments and more theoretical work must be carried out. In any case, there is now a large body of evidence that sub-hadronic matter does exist, and that it needs further exploration, also by quantum-statistical methods.

1.10 Dense Gases and Plasmas in Astrophysics

1.10.1 High Energy Densities in Astrophysical Systems

As we have seen in previous sections, the origin of astrophysics was from the very beginning closely connected with plasma physics. As a first important example, we discussed the Eggert–Saha equation for ionization equilibria. This close connection still exists and is very fruitful for the development of both sides. The contemporary understanding of the structure of astrophysical objects and their evolution, which takes place under the action of gravitational forces and thermonuclear energy release, is in particular closely connected with high-energy-density physics and our knowledge of sub-nuclear plasmas (Yagi et al. 2005; Blaschke and Sedrakian 2006).

It is hard to imagine a domain of scientific activity that underwent greater changes over the past century than cosmology. Modern scientists have radically revised their viewpoint on the structure of the surrounding world. The scale and diversity of these phenomena are astounding, and comprehending them calls for the latest ideas and data from the cutting edge of modern physics. In this case, the observations of astrophysical objects are carried out over a broad range of the electromagnetic spectrum, from hard X-rays to waves with meter-long wavelength, providing unique information about the physical processes at ultrahigh energies unattainable by modern accelerators (Fig. 1.28).

One of the pioneers in this particular field was the Russian physicist Yakov B. Zeldovich. He is remembered for his remark that "the universe is an accelerator for the poor", which does not involve costs in the billions of euros for the construction of experimental facilities. The Universe consists of ordinary matter, photons, relict radiation, hidden mass, and 'vacuum-like' matter which manifests itself as a nonzero cosmological constant. By ordinary matter we mean primarily protons, electrons, and neutrons. Hydrogen is the dominant element. There is also helium and a small amount of lithium. Heavy atoms are present in a very small amount in the Universe. The number of protons in our Universe of size 10^{28} cm is of the order of the Eddington–Dirac number $N \approx 10^{80}$. The mean density of heavy matter in the Universe is

Fig. 1.28 Yakov B. Zeldovich, during a lecture in the Einstein house in Potsdam in the 1980s. Author's archive

$\rho \sim 10^{-31}\,\mathrm{g\,cm^{-3}}$. This matter exists in the form of stars, comets, interstellar gas, meteorites, and cosmic rays.

According to modern views, about 74 percent of the mass-energy is due to the anti-gravitating 'dark energy'. The gravitating mass ('dark', or hidden mass) accounts for about a quarter of the average density of the Universe. About 22 percent is due to the 'dark' matter. The ordinary baryon substance represented in Mendeleev's periodic table accounts in this view for only about 4 percent. This 4 percent of the matter is contained in stars, planets, and the interstellar medium. The interstellar medium accounts for 4/5 of the mass of baryon matter, and only 0.5 percent of the average density of the mass and energy in the Universe. It is precisely in the stars that irreversible thermonuclear transformations take place, producing heavy elements, and generating exotic forms of matter inaccessible to us, such as neutronic matter, quark–gluon plasma, etc. And while our information about the first hundreds of thousands of years of the evolution of the Universe is provided by relict radiation, we assess the history of the next few billion years by stellar observations.

The range of matter parameter variations in the universe is extremely broad from empty space and the very thin intergalactic gas with a density of $10^{-30}\,\mathrm{g/cm^3}$ to extremely high neutron star densities of 10^{14} to $10^{17}\,\mathrm{g/cm^3}$. The temperature of the intergalactic gas with a density of $n \sim 10^{-4}$ to $10^{-3}\,\mathrm{cm^{-3}}$ ranges from 10^7 to 10^8 K. However, it may range up to a billion degrees when heated by shock waves (arising from the shedding of stellar shells, stellar collisions and explosions, the collisions of gas clouds, etc.). Inside neutron stars, it varies from 10^8 to 10^{11} K. About 99 percent of visible matter is heated to temperatures exceeding 10^5 K. The central elements for reconstructing the structure and evolution of astrophysical objects are the physical properties, and in particular the equation of state, of strongly compressed and heated matter, the composition of the plasma, its optical properties, and its transfer coefficients, including also the viscosity, thermal conductivity, diffusion, electrical conductivity, stopping power for particles, etc. These data and the corresponding theory are required in a wide range of the parameters of state. We hope that certain chapters of this book will be useful in this respect and will supplement earlier approaches by providing the necessary data (Ebeling et al. 1984, 1991; Ebeling et al. 2007).

Under extreme conditions, i.e., extreme high pressure and temperatures, ordinary matter undergoes a transition to a series of exotic states of matter. These states are of great interest to astrophysicists, because of the scenarios of the history of our Universe, to be discussed briefly in the following, and further because these high-pressure conditions are believed to exist inside stars that have used up their nuclear fusion 'fuel', such as white dwarfs and neutron stars.

1.10.2 Relativistic Plasmas in Our Universe

At any given time our knowledge of history is limited by a time horizon. At the present time, this horizon is somewhere between 10 and 20 billion years ago. This means that it does not make much sense for us to ask what there was before. Any history has to start with some moment which is given by records, with at least a few pieces of data that are not purely speculative. In our case, the first observed piece of data is the redshift, interpreted as due to an expansion of our cosmos and first observed by Edwin P. Hubble in 1929. The second important piece of data is the so-called black body radiation, which was first deduced theoretically in 1946 from an adiabatic expansion of our Universe, on the basis of the Einstein–Friedman model, by George Gamov, Ralph Alpher, and Hans Bethe. This black-body radiation was observed by Arno Penzias and Robert Wilson nearly 20 years later, in 1964. The third kind of data includes the relative abundance of protons and neutrons, which is 3 : 1, and the estimated abundances of the elements, and in particular the ratio between hydrogen and helium in the universe.

These three observations and others with them are interpreted today as being connected with some singular event which happened more than 10 billion years ago (Green 2004; Linde 2005; Yagi et al. 2005; Hoyng 2006). So let us take the Big Bang event as a model assumption, similar to Pandora's opening of a certain box in Ancient Greece. According to mythology, Pandora opened this box and many sins escaped, leaving only hope behind them. In our case, Pandora's box released a relativistic, optically transparent quantum gas of extremely high density and high temperature which started to expand. Modern researchers do not believe in Greek mythology, but strangely enough they believe in the thermodynamics of gases and plasma, and they believe in the Friedman–Gamov model.

So let us assume, following the standard picture, that the relativistic quantum gas satisfied the relation between temperature T and density r which is valid for the ideal relativistic adiabatically expanding plasma, as given above:

$$T \sim \rho^{1/3} . \tag{1.62}$$

The model further assumes that the number density of particles $n(t)$ and the temperature changed with some scaling $R(t)$ of distances in the early Universe according to

$$n(t) \sim R(t)^{-3} , \qquad T(t) \sim R(t)^{-1} . \tag{1.63}$$

In words, we expect the temperature, which is assumed to be very high at the start (the opening of Pandora's box), to fall as the reciprocal of the scaling distance $R(t)$. For the initial stage of some quantum gas of massless particles, the solution of Einstein's general relativistic equations for an expanding relativistic gas cosmos found by Friedman yields the following time dependence (Greene 2004; Hoyng 2006):

$$R(t) \sim \sqrt{t}, \qquad \rho(t) \sim t^{-2}, \qquad T(t) \sim t^{-1/2} . \tag{1.64}$$

Introducing here some known facts, such as the fact that nowadays, after more than 10 billion years, the radiation has a temperature of about 3 K, we can estimate that the temperature has decreased since the Big Bang approximately according to the rule of thumb

$$T(t)\,[\mathrm{K}] \sim \frac{10^{10}}{\sqrt{t\,[\mathrm{s}]}} . \tag{1.65}$$

Of course this is a very rough estimate based on several serious assumptions, in particular:

- The whole process is assumed to be adiabatic in the thermodynamic sense.
- The matter in the Universe is ultra-relativistic and radiation-dominated.

A brief sketch of the scenario of events after the Big Bang may be given in the form of a short story consisting of 12 epochs (Feistel et al. 1989, 2011). We consider here only the first eight epochs. These concern the expansion of plasma matter, which was very hot and dense at the 'beginning' and cooled down during the later adiabatic expansion:

Epoch 1. Physical Vacuum and Space-time Field Foam

Not much is known about this early epoch, sometimes called the Planck era, which ends with the formation of what we know as space and time at the so-called Planck time

$$t \sim 10^{-43}\,\mathrm{s} .$$

In the Higgs field hypothesis, all elementary interaction forces were still unified and their carrier bosons were massless, similar to photons today.

Epoch 2. Mining the Vacuum

This is the epoch when the 'primary soup' formed from the vacuum, i.e., a fluid form of matter with high density and a very high temperature, viz.,

$$T \sim 10^{32}\,\mathrm{K} .$$

At the beginning of this epoch, the Universe was expanding extremely rapidly. This expansion was called 'inflation'. In the currently widely accepted inflation theory, it is assumed that the early Universe expanded in diameter by a huge factor in just 10^{-35} s, much faster than light (Greene 2004). Driven by temporarily repulsive gravity, this process can be thought of as a sudden phase transition of an under-cooled Universe, similar to the explosive growth of a supercritical nucleus, during which the fundamental symmetries between elementary particles were broken, the Universe became flattened, and initial quantum fluctuations were frozen in to give rise to the presently visible lumpy structure of galaxies and their clusters.

Perhaps the most convincing observational evidence for this theory is the fact that the angular correlation of temperature fluctuations of the cosmic background radiation as measured by the COBE and WMAP satellites has a complicated shape that is perfectly consistent with related theoretical predictions carried out for acoustic oscillations of a dense quantum gas. Thus, the present background radiation appears as a frozen-in image of the dense universe prior to the inflation event. The primary soup which was left at the end of the second epoch, at time $t \sim 10^{-33}$ s, consisted of quarks, antiquarks, leptons, photons, and other particles. At the end of the second epoch, the soup had cooled down to

$$T \sim 10^{28}\,\mathrm{K}\,.$$

These are the starting conditions for the third epoch.

Epoch 3. Quark–Gluon Plasma

In this epoch, the universe was a kind of quark–gluon plasma of high density, perhaps more similar to a 'soup' than a gas. The high temperature supported a state wherein the constituents of atomic nucleons—quarks and gluons—existed unbound. The situation was in some respects similar to what is investigated in recent big accelerator experiments (see Fig. 1.29). Therefore we can express the hope that these studies will contribute to the understanding of the early Universe. At the end of this third epoch, the soup had cooled down to

$$T \sim 10^{15}\,\mathrm{K} \approx 1\,000\,\mathrm{GeV}\,.$$

The corresponding time is about $t \sim 10^{-12}$ s. At the characteristic energy of this epoch, which was on the order of 1 000 GeV, the electroweak symmetry was broken, according to the Weinberg–Salam theory. This change in symmetry would have led to a change in the composition of matter due to quark annihilation, and this in turn would lead to the next epoch.

Epoch 4. Quark Annihilation

At the beginning of the next time interval $10^{-11} < t < 10^{-6}$ s, quarks were still dominant in the Universe. However, near to the end of this epoch, the temperature went down to

$$T \sim 10^{13}\,\mathrm{K} \approx 1\,\mathrm{GeV}\,. \tag{1.66}$$

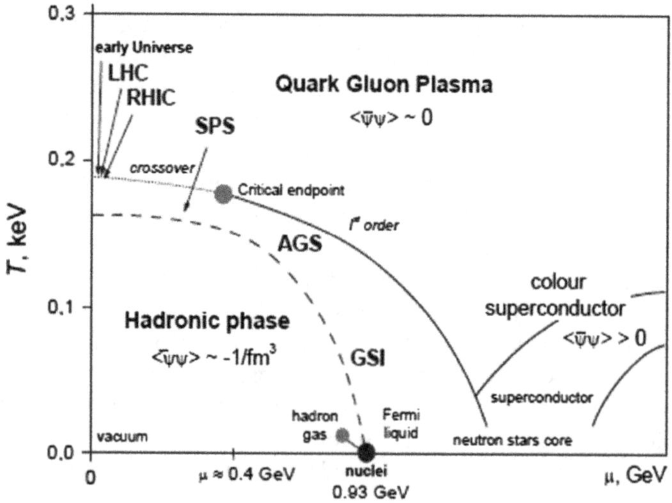

Fig. 1.29 Phase diagram in the temperature versus chemical potential plane, identifying situations which were present in the early Universe and in present day accelerators (Fortov 2009, 2011)

This corresponds to a particle energy at which the annihilation of quarks becomes possible, including the reactions between quarks and antiquarks, viz.,

$$q + \tilde{q} \to 2\gamma \,, \qquad q + \tilde{q} \to e + \tilde{e} \,. \tag{1.67}$$

In this epoch nearly all quarks were annihilated, except for a small number of surplus quarks.

Epoch 5. Formation of a Nucleon–Lepton–Photon Plasma

The time interval $10^{-6} < t < 10^{-3}$ s is the epoch of nucleon formation. Due to attractive chromodynamic forces between quarks, the remaining quarks could form nucleons, either protons or neutrons:

$$u + u + d \to p \,, \qquad u + d + d \to n \,. \tag{1.68}$$

Besides nucleons, in this fifth epoch the metagalaxy was filled with electrons, positrons, photons, and neutrinos. Due to further cooling going as $T \sim 1/\sqrt{t}$, the temperature was about 30 MeV at the end of the fifth epoch.

Epoch 6. Neutrino Decoupling

We consider now the time interval $10^{-3} < t < 1$ s. Due to further expansion and decreasing temperatures, the mass density and the temperature approached the values

$$\rho \sim 108 \, \text{g/cm}^3 \,, \qquad T \sim 10^{10} \, \text{K} \,. \tag{1.69}$$

Under such conditions, neutrinos would have decoupled from the other particles. As we know from many experiments, in the dense matter on the Earth, neutrinos have very long mean free paths and may travel over great distances. This already began in the sixth epoch, and was the source of a sea of neutrinos which now fill our Universe.

Epoch 7. Breaking the Neutron–Proton Symmetry

The age of the universe is now about one second and the temperature around 1 MeV, corresponding to $T \sim 10^{10}$ K. So far the number of protons and neutrons was nearly equal. However, from now on, due to temperatures below 1 MeV, a certain proportion of the neutrons would have changed into protons, which have a lower mass. Finally, the relative abundances were 75 percent for protons and 25 percent for neutrons. These are the abundances of protons and neutrons which we observe today, and this is one of the correct predictions of the Big Bang model.

Epoch 8. Formation of Helium and Other Nuclei and the Fixing of Element Abundances

In our Universe the standard plasmas in which H and He are dominant and other nuclei are present with low abundances were formed within just two hours, i.e., basically in the time interval $10^2 < t < 10^4$ s, temperatures would have approached

$$T \sim 10^9\,\mathrm{K} \sim 100\,\mathrm{eV}\,, \tag{1.70}$$

and fusion reactions like

$$2p + n \to \mathrm{He}^3, \qquad \mathrm{D} + \mathrm{D} \to \mathrm{He}^4 \tag{1.71}$$

would have become possible. The abundances of the chemical elements became fixed due to the insufficient plasma temperature for other nuclear fusion reactions. Note that the heavier elements beyond helium, which we find now on Earth, were formed only in later epochs in which self-organization started on Earth (Feistel et al. 1989, 2011).

Our story started with relativistic plasmas and ended with the epochs 7 and 8, in which the standard plasmas and gases were formed. Many questions still remain open. A few hot questions about the origin and structure of our Universe were explained recently in a lecture by V. Mukhanov at a conference in Berlin (Mukhanov 2015). One of them is the problem about the origin of the observed non-uniformities and the long correlations observed in our Universe. On one hand the we observe hierarchies of structures as stars, galaxies, and metagalaxies and on the other hand we observe a nearly constant composition and the same mechanisms acting everywhere. Evidently, in the distant past there were strong correlations between parts of the Universe which are now far apart and seemingly uncorrelated.

One of the pioneers who first understood the deep problems in the model of the expanding Universe was Yakov B. Zeldovich (1914–1987). Among the most serious problems, in part still open, is the horizon problem. This is the problem of determining why the Universe appears statistically homogeneous and isotropic in accordance with the cosmological principle. In contrast to a rather uniform gas contained in a certain volume, the situation is quite different in the Big Bang model without inflation. The gravitational expansion does not give the early universe enough time to equilibrate. In a Big Bang with only matter and radiation as in the standard model, two widely separated regions of the observable universe cannot have equilibrated because they move apart from each other faster than the speed of light and have never come into causal contact. Because separated regions had no interaction, it is difficult to explain why they have the same temperature and mean density, and why they are structurally so similar.

One of the approaches to solve these problems is the inflationary hypothesis, developed in the 1980s by physicists Alan Guth, Andrej Linde, Vjacheslav Mukhanov, Alexej Starobinsky, and others. This hypothesis tries to explain the origin of the large-scale structure of the cosmos by quantum fluctuations. According to this assumption, quantum fluctuations in the microscopic inflationary region were magnified to cosmic size and became the seeds for the growth of structure in the Universe. Following the inflationary period, the Universe continued to expand, but at a less accelerated rate (Mukhanov and Winitzki 2007; Mukhanov 2015).

These problems are not the topic of the present book, which is devoted to the basic physics of gases and plasmas, in the present-day state of our Universe, or rather, in the state which is nowadays accessible to observation. Here we find dense gases, strongly coupled plasmas, and extreme states of high energy density in many places in the observable cosmos. For hot topics and questions connected with new astrophysical insights, for example connected with superdense QCD matter and compact stars, the reader should turn to the specialized literature (Greene 2004; Blaschke et al. 2006, 2014; Mukhanov and Winitzki 2007; Fortov 2011, 2015; Satz 2012, 2013). In the present book, we shall not follow the evolutionary path from high-energy density plasmas to normal matter, but will go the opposite way from normal gases and plasmas to relativistic high energy density matter.

References

A.A. Abrikosov, L. Gorkov, I.E. Dzyaloshinskii, *Methods of Quantum Field Theory in Statistical Physics (in Russian)* (Moscow, 1962)

A. Alastuey, A. Perez, Virial expansion ot the equation of state of a quantum plasma, Europhys. Lett. **20**, 19–24 (1992)

A. Alastuey, V. Ballenegger, W. Ebeling, Comment on direct linear term in the EOS of plasmas, Phys. Rev. E **92**, 047101 (2015)

M. Badino, The odd couple: Boltzmann, Planck and applications of statistics to physics 1900–1913. Ann. Phys. (Berlin) **18**, 81–101 (2009)

R. Balescu, *Statistical Mechanics of Charged Particles* (Wiley, London, 1963)

J.A. Barker, D. Henderson, Perturbation theory and EOS for fluids. J. Chem. Phys. **47**, 2856 (1967)

K. Beneke, Erich Hückel und seine Arbeiten zur theoretischen Chemie, Mitt. Kolloid-Gesellschaft 274–304 (1999)

B.J. Berne, G. Ciccotti, D.F. Coker (eds.), *Classical and Quantum Dynamics of Condensed Phase Simulation* (World Scientific, Singapore, 1998)

K. Binder (ed.), *Monte Carlo Methods in Statistical Physics* (Springer, Berlin, 1979), p. 1986
D. Blaschke, D. Sedrakian, *Superdense QCD Matter and Compact Stars* (Springer, Berlin, 2006)
D. Blaschke, M. Buballa, A. Dubinin, G. Röpke, D. Zablocki, Generalized Beth–Uhlenbeck approach to mesons and diquarks in hot dense quark matter. Ann. Phys. **348**, 228–255 (2014)
N.N. Bogolyubov, *Collected Papers*, vol. 1–12 (Fizmatlit, Moscow, 2005–2009)
N.N. Bogolyubov, N.N. Bogolyubov Jr., *Introduction to Quantum Statistical Mechanics* (Gordon and Breach, New York, 1992)
M. Bonitz, *Quantum Kinetic Theory* (B.G. Teubner, Stuttgart, 1998); 2nd edition (Springer, Berlin, 2016)
M. Bonitz, W.D. Kraeft (eds.), Kinetic theory of nonideal plasmas. J. Phys.: Conf. Ser. **11** (2005)
M. Bonitz, D. Semkat (eds.), *Introduction to Computational Methods for Many Body Systems* (Rinton Press, Princeton, 2006)
H.P. Bonzel, A.M. Bradshaw, G. Ertl (eds.), *Physics and Chemistry of Alkali Metal Adsorption* (Elsevier, Amsterdam, 1989)
L. Brillouin, Les statistiques quantiques et leurs applications, Paris 1930; German translation, Berlin (1931)
L.S. Brown, L.G. Yaffe, Effective field theory of highly ionized plasmas. Phys. Rep. **340**, 1–164 (2001)
S.G. Brush, H.L. Sahlin, E. Teller, Monte Carlo study of a one-component plasma I. J. Chem. Phys. **45**, 2102 (1966)
D.M. Ceperley, Path integrals in the theory of condensed Helium. Rev. Mod. Phys. **65**, 279–356 (1995)
S. Chapman, T.G. Cowling, *The Mathematical Theory of Non-uniform Gases: An Account of the Kinetic Theory of Viscosity, Thermal Conduction and Diffusion in Gases*, 3rd edn. (Cambridge Mathematical Library, Cambridge, 1991)
A.P. Chetverikov, W. Ebeling, M.G. Velarde. Local electron distributions and diffusion in anharmonic lattices mediated by thermally excited solitons. Eur. Phys. J. B **70**, 217–227 (2011)
P. Debye, Handelingen Nederlandsch Natuuren Congres (1923)
P. Debye, E. Hückel, Theorie der Elektrolyte I. Physik Z. **24**, 185 (1923)
C. Deutsch, Nodal expansion in a real matter plasma. Phys. Lett. A **60**, 317–318 (1977)
H.E. De Witt, Evaluation of the quantum-mechanical ring sum with Boltzmann statistics. J. Math. Phys. **3**, 1216–1228 (1962)
H.E. De Witt, M. Schlanges, A.Y. Sakakura, W.D. Kraeft, Low density expansion of the equation of state for a quantum electron gas. Phys. Lett. A **197**, 326 (1995)
W. Ebeling, Statistical thermodynamics of bound states in plasmas (in Dt.). Ann. Physik (Leipzig) **19**, 104–112 (1967)
W. Ebeling, Derivation of free energy of quantum plasmas from exact scattering shifts (in Dt.). Ann. Physik **22**, 33–39, 383–391 (1968)
W. Ebeling, Equation of state and Saha equation of partially ionized plasmas. Physica **38**, 378 (1968); **73**, 573–584 (1974)
W. Ebeling, On the possibility of diffusion instabilities in weak electrolytes. Z. Phys. Chem. **247**, 340 (1971); Quantum statistics of ionization. Phys. Stat. Sol. (b) **46**, 243–255 (1971)
W. Ebeling, D. Hoffmann, The Berlin school of thermodynamics founded by Helmholtz and Clausius. Eur. J. Phys. **12**, 1–9 (1991)
W. Ebeling, D. Hoffmann, Eine Vorlage Einsteins in der Preuischen Akademie der Wissenschaften 1924, Leibniz Online, EbelingHoffmann.pdf (2014)
W. Ebeling, W.D. Kraeft, D. Kremp, *Theory of Bound States and Ionisation Equilibrium in Plasmas and Solids* (Akademie-Verlag, Berlin, 1976). Extended Russ. translation Mir, Moscow (1979)
W. Ebeling, V.F. Fortov, YuL Klimontovich, et al. (eds.), *Transport Properties of Dense Plasmas* (Birkhäuser, Boston, 1984)
W. Ebeling, A. Förster, V.E. Fortov, V.K. Gryaznov, AYa. Polishchuk, *Thermophysical Properties of Hot Dense Plasmas* (Teubner, Stuttgart, 1991)

W. Ebeling, A. Förster, V. Fortov, V. Gryaznov, A. Polishchuk, *Thermophysical Properties of Hot Dense Plasmas* (Teubner, Stuttgart, 1991) (Russ. transl. Moskva, Ishevsk, 2007)
W. Ebeling, M.Yu. Romanovsly, I.M. Sokolov, Velocity distributions and kinetic equations including Levy power law tails. Contr. Plasma Phys, **49**, 704–712 (2009)
D. Ebert, *Eichtheorien: Grundlage der Elementarteilchenphysik* (VCH-Verlag, Weinheim, 1989)
J. Eggert, Über den Dissoziationzustand der Fixsterngase. Physikalische Zeitschrift **20**, 570–574 (1919)
A. Einstein, Quantentheorie des einatomigen idealen gases. Sitzungsber. Preuss. Akad. Wiss. Phys.-math. Kl. **22**, 261–267 (1924); **23**, 3–14 (1925)
D. Enskog, Kinetische Theorie der Vorgänge in mässig verdünnten Gasen (författare) (Almqvist and Wiksell, Uppsala, 1917)
H. Falkenhagen, W. Ebeling, Equilibrium properties of ionized dilute electrolytes, in *Ionic Interactions*, ed. by S. Petrucci (Academic Press, New York, 1971)
R. Feistel, W. Ebeling, *Evolution of Complex Systems* (Kluwer, Dordrecht, 1989)
R. Feistel, W. Ebeling, *Physics of Self-Organization and Evolution* (Wiley-VCH, Weinheim, 2011)
H. Fehske, R. Schneider, A. Weibe (eds.), *Computational Many-Particle Physics* (Springer, Berlin, 2008)
R.P. Feynman, *Statistical Mechanics* (Benjamin, Reading Mass, 1972)
M.E. Fisher, Y. Levin, Criticality in ionic fluids: Debye–Hückel theory, Bjerrum, and beyond. Phys. Rev. Lett. **71**, 2138 (1993)
H. Friedman, *Ionic Solution Theory* (Wiley Interscience, New York, 1962)
V. Fortov, M. Mochalov et al., Phys. Rev. Lett. **99**, 185001 (2007)
V.E. Fortov, *Extreme States of Matter (Russ)* (FizMatGis, Moskva, 2009). New edition 2015
V.E. Fortov, *Equation of State of Matter* (FizMatGis, Moskva, 2013). (in Russian)
V.E. Fortov, *Extreme States of Matter: On Earth and in the Cosmos*, The Frontiers Collection (Springer, Berlin, 2011); *Extreme States of Matter, High Energy Density Physics* (Springer, Berlin, 2016)
V.L. Ginzburg, *The Physics of a Lifetime: Reflections on the Problems and Personalities of 20th Century Physics* (Springer, Berlin, 2001)
B. Greene, *The Fabric of the Cosmos* (Knopf, New York, 2004)
H. Haken, P. Plath, W. Ebeling, YuM Romanovsky, *Beiträge zur Geschichte der Synergetik* (Springer, Berlin, 2016)
J.P. Hansen, I.R. McDonald, *Theory of Simple Liquids* (Academic Press, New York, 1976); J.N. Henn, J.C. Plefka, *Scattering Amplitudes in Gauge Theories*, Lecture Notes in Physics (Springer, Berlin, 2014)
P.C. Hemmer, H. Holden, S. Kjelstrup Ratkje (eds.), *The Collected Works of Lars Onsager* (World Scientific, Singapore, 1996)
F. Hensel, S. Juengst, F. Noll, R. Winter, in *Localisation and Metal Insulator Transitions*, ed. by D. Adler, H. Fritsche (Plenum, New York, 1985)
T.L. Hill, *Statistical Mechanics, Principles and Selected Applications* (McGraw Hill, New York, 1956)
J.O. Hirschfelder, C.F. Curtis, R.B. Bird, *Molecular Theory of Gases and Liquids* (Wiley, New York, 1954)
P. Hoyng, *Relativistic Astrophysics and Cosmology* (Springer, Dordrecht, 2006)
S. Ichimaru, *Statistical Plasma Physics* (Addison-Wesley, Redwood, 1992)
D.R. Inglis, E. Teller, Astrophysics **90**, 430 (1939)
G. Kalman (ed.), *Strongly Coupled Coulomb Systems* (Pergamon Press, Oxford, 1998)
G. Kelbg, Quantenstatistik der Plasmen. Annalen der Physik **12**, 354 (1963)
G. Kelbg, Einige Methoden der statistischen Thermodynamik hochionisierter Plasmen, *Ergebnisse der Plasmaphysik aund Gaselektronik*, vol. 3 (Akademie-Verlag, Berlin, 1972)
C. Kirsten, H.-G. Körber (eds.), *Physiker über Physiker* (Akademie-Verlag, Berlin, 1975)
H. Kleinert, Path integrals in quantum mechaniscs, statistics, polymer physics and financial markets, World Scientific, Singapore (1995)

Yu.L. Klimontovich, *Statistical Theory of Nonequilibrium Processes in Plasmas* (Izdat MGU, Moscow, 1964); English translation (Pergamon, Oxford, 1967)
Yu.L. Klimontovich, *Statistical Physics (Russ.)* (Nauka, Moscow, 1982); English translation (Harwood, New York, 1986)
W.D. Kraeft, D. Kremp, W. Ebeling, G. Röpke, *Quantum Statistics of Charged Particle Systems* (Akademie-Verlag and Pergamon Press, Berlin and New York, 1986)
W.D. Kraeft, D. Kremp, G. Röpke, Direct linear term in the equation of state of plasmas. Phys. Rev. E **91**, 013108 (2015)
M.D. Knudson, M.P. Desjarlais, A. Becker, R.W. Lemke, K.R. Cochrane, M.E. Savage, D.E. Bliss, T.R. Mattsson, R. Redmer, Science **348**(6242), 1455–1460 (2015)
G. Kalman, P. Carini (eds.), *Strongly Coupled Plasma* (Plenum Press, New York, 1978)
V.P. Krainov, M.B. Smirnov, Cluster beams in the super-intense femtosecond laser pulse. Phys. Rep. **370**, 237–331 (2002)
V.P. Krainov, M.B. Smirnov, B.M. Smirnov, Femtosecond excitation of cluster beams. Phys. Usp. **50**, 907–931 (2007)
D. Kremp, M. Schlanges, W.D. Kraeft, *Quantum Statistics of Nonideal Plasmas* (Springer, Berlin, 2005)
A.I. Larkin, Thermodynamic functions of a low-temperature plasma. JETP **11**, 1363–1364 (1960)
J.D. Landau, E.M. Lifshits,*Statistical Physics (Part I)*, Nauka Moska 1976, German translation Berlin 1979, English translation Statistical Physics, Part 1, vol. 5, 3rd edn. (Butterworth–Heinemann, London, 1980)
E.M. Lifshitz, L.P. Pitaevskii, *Physical Kinetics, Course of Theoretical Physics*, vol. X (Publishing Science, Moscow, 1979)
A. Linde, *What Energy Drives the Universe* (Stanford University, Stanford, 2005)
W.R. Magro, D.M. Ceperley, C. Pierleoni, B. Bernu, Phys. Rev. Lett. **76**, 1240 (1996)
P.C. Martin, J. Schwinger, Theory of many-particle systems. Phys. Rev. **115**, 1342–1432 (1959)
J.E. Mayer, Theory of ionic solutions. J. Chem. Phys. **18**, 1426–1436 (1950)
B. Militzer, R. Pollock, Variational density matrix method for warm condensed matter and dense hydrogen. Phys. Rev. E **61**, 3470–382 (2000)
E. Montroll, J. Ward, Quantum statistics of interacting particles. Phys. Fluids **1**, 55 (1958)
T. Morita, Equation of state of high temperature plasma. Prog. Theor. Phys. (Kyoto) **22**, 757–774 (1959)
N. Mott, The transition to the metallic state. Phil. Mag. **6**, 287–309 (1961)
V. Mukhanov, S. Winitzki, *Introduction to Quantum Effects in Gravity* (NOOK Book, 2007)
V. Mukhanov, *Quantum Universe* (Plenary lecture DPG, Berlin, 2015)
J. von Neumann, *Mathematische Grundlagen der Quantenmechanik* (Springer, Berlin, 1932)
G.E. Norman, A. Starostin, High Temp. **6**, 394 (1968); **8**, 381 (1970)
D. Pines, *The Many Body Problem* (Benjamin, New York, 1961)
D. Pines, P. Nozieres, *The Theory of Quantum Liquids* (Benjamin, New York, 1966); Russ. Translation Mir, Moskva (1967)
M. Planck, Zur Quantenstatistik des Bohrschen Atommodells. Annalen der Physik **75**, 673–684 (1924)
R. Redmer, G. Röpke, Progress in the theory of dense strongly coupled plasmas. Contr. Plasma Phys. **50**, 970–985 (2010)
R. Redmer, B. Holst, F. Hensel (eds.), *Metal to Non-Metal Transitions* (Springer, Berlin, 2010)
K. Riewe, R. Rompe, Ann. Phys. (Leipzig) **17**, 126 (1938)
R. Rompe, H. Treder, W. Ebeling, *Zur grossen Berliner Physik* (Teubner, Leipzig, 1987)
H.J. Rothe, *Lattice Gauge Theories* (World Scientific, Singapore, 2005)
M.N. Saha, Phil. Mag. **40**, 472 (1920); Z. Phys. **6**, 40 (1921)
H. Satz, *Extreme States of Matter in Strong Interaction Physics: An Introduction* (Springer, Berlin, 2012)
H. Satz, *Ultimate Horizons: Probing the Limits of the Universe* (Springer, Berlin, 2013)
K. Simonyi, *Kulturgeschichte der Physik* (Urania Verlag and Harri Deutsch, Frankfurt, 1990)

V. Sizyuk, A. Hassanein, T. Sizyuk, J. Appl. Phys. **100**, 103106 (2006)
L. Spitzer, *Physics of Fully Ionized Plasmas* (Wiley, New York, 1961)
V.Ya. Ternovoi, A.S. Filimonov, V.E. Fortov, S.V. Kvitov, D.D. Nikolaev, A.A. Pyalling, Phys. B **265**, 6 (1999)
G.E. Uhlenbeck, E. Beth, The quantum theory of the non-ideal gas I. Deviations from the classical theory. Physica **3**, 729–745 (1936)
E.A. Uehling, G.E. Uhlenbeck, Phys. Rev. **43**, 552 (1933)
J.D. van der Waals, Over de Continuiteit van den Gas- en Vloeistoftoestand (Continuity of fluid and gaseous state). Dissertation, University of Leiden (1873)
A.A. Vedenov, A.I. Larkin, Equation of state of plasmas (in Russ.). Zh. eksp. teor. Fiz. **36**, 1133–1142 (1959); Sov. Phys. JETP **9**, 806–821 (1959)
J. Von Neumann, Mathematische Grundlagen der Quantenmechanik (1932)
P.N. Voronzov-Veliaminov, A.M. Eliashevich, V.P. Morgenstern, V.P. Chassovskich, Teplophys. Vysokh. Temp. **8**, 277 (1970); **14**, 199 (1976)
S.T. Weir, A.C. Mitchell, W.J. Nellis, Phys. Rev. Lett. **76**, 860 (1996)
H. Whitley, A. Alastuey, J.A. Gaffney, R. Cauble, W.D. Kraeft, M. Bonitz, A tribute to the pioneers of the strongly coupled plasmas: Hugh E. De Witt, Bernard Jancovici, Forest A. Rogers. Contr. Plasma Phys. **55**, 102–115 (2015)
K. Yagi, T. Hatsuda, Y. Miake, *Quark-Gluon Plasma: From Big Bang to Little Bang* (Cambridge University Press, Cambridge, 2005)
E.P. Wigner, Interaction of electrons in metals. Phys. Rev. **46**, 1002–1011 (1934)
V.M. Zamalin, G.E. Norman, V.S. Filinov, *Monte Carlo Method in Statistical Mechanics (in Russ.)* (Nauka, Moscow, 1977)
D. Zubarev, V. Morozov, G. Röpke, *Statistical Mechanics of Nonequilibrium Processes* (Akademie, Berlin, 1996)
J. Zweiback, T.E. Cowan, T. Dimitre et al., Phys. Rev. Lett. **85**, 3640 (2000)

Chapter 2
Strong Correlations and Equation of State of Dense Gases

We summarize the physics and the statistical theory of strongly coupled gases. Starting from the binary correlation functions, we develop systematic expansions with respect to the density, the so-called virial expansions, and also fugacity expansions. Further, we discuss solutions of integral equations (PY and HNC) for dense systems. Quantum effects are treated by the methods of Slater and Beth–Uhlenbeck. Finally, we discuss strongly correlated Fermi–Dirac and Bose–Einstein gases, and Yukawa fluids.

2.1 Classical Molecular Distribution Functions and Density Expansions

2.1.1 Distribution Functions and Ornstein–Zernike Relations

From the classical kinetic theory of gases, we know the equation of state of ideal gases $\beta p = n$ (see Chap. 1). For real gases, the interaction forces lead to corrections to the ideal gas equation of state. We mention the classical theory of van der Waals and the systematic expansions with respect to the density, the so-called virial expansions with respect to density (Hirschfelder et al. 1954). We also mention fugacity expansions, solutions of integral equations and perturbation theories (Barker and Henderson 1967; Croxston 2009; Friedman 1962; Henderson 2009; Ross 1986).

The easiest way to get thermodynamic functions including interaction effects starts from the binary correlation functions. We already pointed out in Chap. 1 that the basic tool of fluid physics is connected with the concept of molecular distribution functions. We define pair distribution functions $g(r)$ for a fluid of average density

W. Ebeling et al., *Quantum Statistics of Dense Gases and Nonideal Plasmas*, Springer Series in Plasma Science and Technology, DOI 10.1007/978-3-319-66637-2_2

n by writing the number of particles in a shell of thickness $\mathrm{d}r$ at distance r from a selected particle in the form

$$ng(r)(4\pi r^2)\mathrm{d}r\ . \tag{2.1}$$

Unfortunately, this notation is not unique. Like the French physicist Yvon a decade earlier, Bogoliubov and shortly afterwards also Born, Green, and Kirkwood considered the pair distribution function as the special case of a hierarchy of molecular distribution functions F_s with $s = 1, 2, 3, \ldots$, for $s = 2$ (in Bogoliubov's notation). Hence, Bogoliubov denoted the pair distribution by $F_2(r)$. We will use both notations parallel depending on the context.

For two particles of species a and b at locations 1 and 2, the pair distribution is denoted by $F_{ab}(1, 2)$. The pair function converges for uniform systems to unity at large distances where there are no longer any correlations. In the radially symmetric case, we define the correlation function by

$$h(r) = g(r) - 1 = F_2(r) - 1\ . \tag{2.2}$$

When there is no radial symmetry, we use a different notation for the function, describing the *total correlation* between molecules 1 and 2 by $h(\mathbf{r}_{12}) = g(\mathbf{r}_{12}) - 1$, which is a measure of the influence of molecule 1 on molecule 2 at a distance $\mathbf{r}_{12}$, with $g(\mathbf{r}_{12})$ the so-called radial distribution function.

An important branch of fluid physics starts with an integral equation introduced by Leonard Ornstein and Frits Zernike in 1914. In statistical mechanics, the Ornstein–Zernike equation is an integral equation defining the so-called *direct correlation function*. It is in fact just another tool for describing the correlation between two molecules, which Ornstein and Zernike proposed to split into two contributions, a direct and an indirect part. The direct contribution is denoted by $c(r_{12})$. The indirect contribution is due to the influence of molecule 1 on a third molecule, labeled 3, which in turn affects molecule 2, directly and indirectly. This indirect effect is weighted by the density and averaged over all possible positions of the particles 3 in the neighborhood.

This decomposition is formulated by

$$h(r_{12}) = c(r_{12}) + \rho \int \mathrm{d}\mathbf{r}_3\, c(r_{13})h(r_{23})\ . \tag{2.3}$$

In the following, we will introduce several integral equations based on additional closure relations between $h(r_{12})$ and $c(r_{12})$. These will be useful for calculating the properties of dense fluids. The mean potential energy is related to the pair correlation function by

$$U_{\text{pot}} = \langle U \rangle = \frac{1}{2}n^2 \int \mathrm{d}\mathbf{1}\,\mathrm{d}\mathbf{2}\, U(12)F_2(12)\ . \tag{2.4}$$

Another simple relation connecting thermodynamics with the pair function is the *virial formula for the pressure*, valid for simple fluids with radially symmetric potentials:

$$\beta p = n + \frac{2\pi}{3} n^2 \int_0^\infty \mathrm{d}r\, r^3 \frac{\mathrm{d}U(r)}{\mathrm{d}r} F_2(r) \,. \tag{2.5}$$

In the simplest possible way, we may approximate the pair function by a Boltzmann factor, whence the mean potential energy becomes

$$U_{\text{pot}} = \langle V \rangle = \frac{1}{2} n^2 \int \mathrm{d}1\, \mathrm{d}2\, U(12) \exp\big[- \beta V(1,2) \big] \,. \tag{2.6}$$

Since the internal energy U and the free energy F are connected by the thermodynamic relation $U = \partial(\beta F)/\partial\beta$, the second virial coefficient is given by integration as

$$B_2^{\text{cl}} = \frac{1}{2} \int \mathrm{d}\mathbf{r} \Big\{ \exp\big[- \beta U(\mathbf{r}) \big] - 1 \Big\} \,. \tag{2.7}$$

Bogoliubov–Born–Green–Kirkwood–Yvon Hierarchy

In the 1930s and 1940s, a number of authors, namely, Bogoliubov, Born, Green, Kirkwood, and Yvon, developed a very fruitful method, based on a hierarchy of equations for molecular distribution functions called the BBGKY hierarchy. The authors are not listed chronologically here, but alphabetically. As far as we know, the French physicist Yvon was the first to use this method. We follow here mainly the rather clear and exhaustive work published in 1946 by N.N. Bogoliubov, since it was given in both a classical and a quantum-statistical version (see Bogoliubov 2005–2009).

We start from the classical N-particle density in configuration space, viz.,

$$\rho_N(q_1, \ldots, q_N) = \frac{1}{Q_{N,\text{int}}} \exp\big[- \beta U_N(q_1, \ldots, q_N) \big] \,. \tag{2.8}$$

We define a series of distribution functions by

$$\begin{aligned} F_s(1, \ldots, s) &= V^s \rho_s(1, \ldots, s) \\ &= V^s \int \mathrm{d}(s+1), \ldots, \mathrm{d}N \rho_N(1, \ldots, N) \,. \end{aligned} \tag{2.9}$$

For a gas mixture of species $a, b, c, \ldots$, the pair distributions F_{ab} and the mean potential energy are given in the classical case by

$$\begin{aligned} F_{ab}(1,2) &= \exp\big[- \beta U_{ab}(1,2) \big] \,, \\ U_{\text{pot}} &= \frac{1}{2} \sum_{ab} n_a n_b \int \mathrm{d}1\, \mathrm{d}2\, U_{ab}(12) \exp(-\beta U_{ab}) \,. \end{aligned}$$

For the higher virial coefficients, we may derive corresponding expressions (Bogoliubov 2005–2009; Hirschfelder et al. 1954). The pair correlation functions are only a special case of the set of molecular distribution functions.

2.1.2 *Virial Expansions*

Since the internal energy U and the free energy F are connected by the thermodynamic relation $U = \partial(\beta F)/\partial\beta$, the free energy is given by integration as

$$F(T, V, N) = F_{\text{id}} + F_{\text{int}} = Nk_BT\big[\ln(n\Lambda^3) - 1\big] \\ - k_BTV\big[n^2B_2^{\text{cl}}(T) + n^3B_3^{\text{cl}}(T) + \cdots\big], \quad (2.10)$$

and for the second virial coefficient

$$B_2^{\text{cl}} = \frac{1}{2}\int \mathbf{dr}\left\{\exp\big[-\beta V(\mathbf{r})\big] - 1\right\}. \quad (2.11)$$

By differentiating with respect to the volume V, i.e., $p = -\partial F/\partial V$, we obtain the equation of state

$$\beta p = -\beta\left(\frac{\partial F}{\partial V}\right)_{T,N} = n\Big[1 - nB_2^{\text{cl}}(T) - 2n^2B_3^{\text{cl}}(T) - \ldots\Big]. \quad (2.12)$$

2.2 Integral Equation Methods and Prototype Hard Sphere Systems

2.2.1 *Percus–Yevick and Hypernetted-Chain Equations*

The hard-sphere model for molecular interactions is a rather crude approximation, but it is quite useful as a limiting case (Croxston 2009). The interactions are defined by

$$U(r) = \begin{cases} 0, & r > d, \\ \infty, & r < d. \end{cases} \quad (2.13)$$

We now formulate the PY equation for this system, starting from the Ornstein–Zernike equation. Defining the displacement vector between two molecules to be $\mathbf{r}_{ij} := \mathbf{r}_i - \mathbf{r}_j$, for $i, j = 1, 2, 3, \ldots$, the OZ equation can be rewritten using a convolution:

$$\begin{aligned} h(\mathbf{r}_{12}) &= c(\mathbf{r}_{12}) + \rho\int \mathbf{dr}_3\, c(\mathbf{r}_{12} - \mathbf{r}_{32})h(-\mathbf{r}_{32}) \\ &= c(\mathbf{r}_{12}) + \rho\int \mathbf{dr}_{32}\, c(\mathbf{r}_{12} - \mathbf{r}_{32})h(\mathbf{r}_{32}) \\ &= c(\mathbf{r}_{12}) + \rho(c * h)(\mathbf{r}_{12}), \end{aligned} \quad (2.14)$$

where $\rho = N/V$ is the number density of molecules. If we then denote the Fourier transforms of $h(\mathbf{r})$ and $c(\mathbf{r})$ by $\hat{H}(\mathbf{k})$ and $\hat{C}(\mathbf{k})$, respectively, and use the convolution theorem, we obtain

$$\hat{H}(\mathbf{k}) = \hat{C}(\mathbf{k}) + \rho\hat{H}(\mathbf{k})\hat{C}(\mathbf{k}) \,, \tag{2.15}$$

which yields

$$\hat{C}(\mathbf{k}) = \frac{\hat{H}(\mathbf{k})}{1+\rho\hat{H}(\mathbf{k})} \,, \qquad \hat{H}(\mathbf{k}) = \frac{\hat{C}(\mathbf{k})}{1-\rho\hat{C}(\mathbf{k})} \,. \tag{2.16}$$

One needs to solve for both $h(r)$ and $c(r)$ (or, equivalently, their Fourier transforms). This requires an additional equation, known as a closure relation. The Ornstein–Zernike equation can be seen formally as a definition of the direct correlation function $c(r)$ in terms of the usual correlation function. We may approximate $c(r)$ by

$$c(r) = \mathrm{e}^{-\beta w(r)} - \mathrm{e}^{-\beta[w(r)-u(r)]} \,. \tag{2.17}$$

Introducing the function $y(r) = \mathrm{e}^{\beta u(r)} g(r)$ into the approximation for $c(r)$, one obtains

$$c(r) = g(r) - y(r) = \mathrm{e}^{-\beta u} y(r) - y(r) = f(r)y(r) \,. \tag{2.18}$$

This is the essence of the Percus–Yevick approximation. If we substitute this result into the Ornstein–Zernike equation, we obtain the Percus–Yevick equation:

$$y(r_{12}) = 1 + \rho \int f(r_{13})y(r_{13})h(r_{23})\mathrm{d}\mathbf{r_3} \,. \tag{2.19}$$

The approximation was made by Percus and Yevick 1958. For hard spheres, the PY equation has an analytical solution (Henderson 2009).

The hypernetted-chain equation follows from a closure relation to solve the Ornstein–Zernike equation which relates the direct correlation function to the total correlation function. It is commonly used in fluid theory to obtain, e.g., expressions for the radial distribution function. It is given by

$$\begin{aligned} \ln y(r_{12}) &= \ln g(r_{12}) + \beta u(r_{12}) \\ &= \rho \int \big[h(r_{13}) - \ln g(r_{13}) - \beta u(r_{13})\big] h(r_{23})\mathrm{d}\mathbf{r_3} \,. \end{aligned} \tag{2.20}$$

In the following step we approximate $c(r)$ by w using (2.17). By expanding the indirect part of g(r) in the above equation and introducing the function $y(r) = \mathrm{e}^{\beta u(r)} g(r)$, we can approximate $c(r)$ by writing

$$\begin{aligned} c(r) &= \mathrm{e}^{-\beta w(r)} - 1 + \beta\big[w(r) - u(r)\big] = g(r) - 1 - \ln y(r) \\ &= f(r)y(r) + \big[y(r) - 1 - \ln y(r)\big] \,, \qquad f(r) = \mathrm{e}^{-\beta u(r)} - 1 \,. \end{aligned} \tag{2.21}$$

This is the essence of the hypernetted-chain equation. Equivalently, we may write

$$h(r) - c(r) = g(r) - 1 - c(r) = \ln y(r) \ . \tag{2.22}$$

If we substitute this result into the Ornstein–Zernike equation we obtain the hypernetted-chain equation:

$$\begin{aligned} \ln y(r_{12}) &= \ln g(r_{12}) + \beta u(r_{12}) \\ &= \rho \int \big[h(r_{13}) - \ln g(r_{13}) - \beta u(r_{13})\big] h(r_{23}) \mathrm{d}\mathbf{r_3} \ . \end{aligned} \tag{2.23}$$

2.2.2 *Hard Core Fluids and Fluid Mixtures*

All the electrons in atoms and molecules are subject to the Pauli principle, which stipulates that the same atomic state cannot be occupied by two electrons. As a consequence, other electrons are simply excluded and this appears like a hard core. In the zeroth approximation, atoms and molecules are like hard spheres with well defined radius. Consequently, in the zeroth approximation, dense fluids are like a mixture of hard spheres. For the hard sphere fluid, the correlation functions are known a priori in special distance regions, i.e., for physical reasons we have

$$F_2(r) = 0 \ , \qquad h(r) = -1 \ , \qquad \text{if} \quad r < d \ . \tag{2.24}$$

With this expression for the binary distribution function, we find the classical second virial coefficient for hard-core potentials. Correspondingly, the EOS for hard sphere gases is

$$\beta p = n \left(1 + \frac{2}{3} \pi n a^3 + \cdots \right) \ .$$

The Percus–Yevick approximation is obtained by completing this exact relation by the approximate relation

$$c(r) = 0 \ , \qquad \text{if} \quad r > d \ . \tag{2.25}$$

We note also the exact relation for the pressure which follows from (2.5)

$$\beta p = n + \frac{2\pi}{3} n^2 d^3 F_2(d+) \ . \tag{2.26}$$

In the history of the theory of fluids, much effort went into calculating the higher virial coefficients in the series

$$\beta p = n + \sum_{k=2}^{\infty} B_k n^k \ . \tag{2.27}$$

Hutchinson and Rushbrook calculated seven coefficients using the PY equation. Their result can be written in the form (see Croxston 2009)

$$B_k = \frac{8}{4^k}(k-4)\left(\frac{2\pi}{3}d^3\right)^{k-1} . \tag{2.28}$$

Summing these expressions, we find for the pressure and contact value

$$\beta p = n\frac{1+2\eta+3\eta^2}{(1-\eta)^2} , \qquad F_2(d+) = \frac{1+\eta/2}{(1-\eta)^2} . \tag{2.29}$$

In this approximation, the pressure may be integrated to give the Helmholtz free energy (Henderson 2009)

$$\beta F = \beta F_{\text{id}} + 2\ln(1-\eta) + \frac{6\eta}{1-\eta} . \tag{2.30}$$

Note that exact solutions of the PY equations for hard spheres were also derived. The essential approximation is due to Wertheim, who assumed for the direct correlation function of hard spheres (Croxston 2009)

$$c(r) = \begin{cases} 0 , & \text{if } \; r > d , \\ -\alpha - \beta(r/d) - \gamma(r/d)^2 - \delta(r/d)^3 , & \text{if } \; r < d . \end{cases} \tag{2.31}$$

Explicit solutions based on these assumptions were given by Thiele and Wertheim using Laplace transforms, and by Baxter using Fourier transforms (Henderson 2009).

There exists also a very useful semi-empirical formula which is compatible with the PY solution, but fits the computer simulation data even better. This is the Carnahan–Starling approximation (Ross 1996):

$$\beta F = \beta F_{\text{id}} + \frac{4\eta - 3\eta^2}{(1-\eta)^2} , \qquad p = nk_{\text{B}}T\frac{1+\eta+\eta^2+\eta^3}{(1-\eta)^3} . \tag{2.32}$$

Mixture Rules

Most real fluids are mixtures of several components, and the number of possible combinations is very large. This means that simple rules for calculating the properties of mixtures from those of the pure components are of importance for applications. The linear mixing model (Ross 1996) is a semi-empirical model for calculating thermodynamic functions of a binary system. The assumption is that the pressures p_i, the internal energies ε_i, and the free energies per particle of the two components are known. Defining the mole fractions of the components by

$$x = \frac{N_1}{N_1+N_2} , \qquad (1-x) = \frac{N_2}{N_1+N_2} , \tag{2.33}$$

the *linear mixing rule* assumes that the pressure and specific energy have mixing that is linear in the composition x, i.e.,

$$p = xp_1 + (1-x)p_2 \,, \qquad E = xE_1 + (1-x)E_2 \,. \tag{2.34}$$

For the free energy per particle, we assume the following functional of the composition x :

$$\phi(n, T; x) = x\phi_{\text{pl}} + (1-x)\phi_{\text{nl}} + 2x \log \frac{2x}{1+x} + (1-x)\log\frac{1-x}{1+x} \,. \tag{2.35}$$

Here the last term represents a kind of mixing entropy. The linear mixing functional has a rather simple structure. When the composition is free and chosen by the system by adapting the free energy, we find the actual composition by looking for the minimum corresponding to the composition denoted by γ.

A special mixture of great interest for later studies of partially ionized plasma is the mixture of neutral spheres of density n_0 with diameter d and packing parameter $\eta = \pi d^3 n_0/6$ and point charges with density n_1. These point charges may be considered as a kind of classical neutral 'electrons' with radius zero. We assume that the point charges behave like a classical ideal gas, but they are not allowed to penetrate into the interior of the neutral spheres:

$$\beta F = \beta F_{\text{id}} + Vn_0 \frac{4\eta - 3\eta^2}{(1-\eta)^2} - n_1 \ln(1-\eta) \,. \tag{2.36}$$

The last contribution to the free energy is called the *excluded volume contribution*. It plays an important role for dense plasmas. The presence of point particles in a hard-sphere system greatly increases the free energy, and many point particles may even destroy the hard-sphere subsystem (Ebeling et al. 1976, 2012).

2.3 Quantum Effects

2.3.1 *Bose–Einstein and Fermi–Dirac Gases*

Virial expansions were originally developed for the classical case, but the same expansion with modified coefficients works for the quantum case. The extension of this classical formula to the quantum case was first given by Uhlenbeck and Beth (1936). Note that, in the classical case, the first term in the expansion is the ideal Boltzmann contribution, but in the quantum case, one has the Fermi–Dirac or Bose–Einstein expressions given in Chap. 3. We see that interactions are not the only source of deviations from the ideal term, i.e., deviations from the simple law $\beta p = n$. Quantum effects also contribute to the virial coefficients. The classical ideal gas law $\beta p = n$ is not valid for quantum gases. We have to distinguish between Bose gases and Fermi gases. For a *Fermi–Dirac gas*:

$$\beta p = (2s + 1)\Lambda^{-3} f_{5/2}(z) \,. \tag{2.37}$$

For a *Bose–Einstein gas*:

$$\beta p = \frac{2s + 1}{V} \ln (1 - z) + (2s + 1)\Lambda_{-3} g_{5/2}(z) \,. \tag{2.38}$$

Here z is the fugacity, related to the density by the relation $n = z\partial(\beta p)/\partial z$. For small densities, we may expand. In the first approximation, we find

$$\beta p = n \left[1 \pm \frac{1}{2^{5/2}} \left(\frac{n\Lambda^3}{2s + 1} \right) + O\left(n^2\right) \right] , \tag{2.39}$$

where the upper sign is valid for fermions and the lower sign for bosons. The parameter in the expansion is $n\Lambda^3$. We define $n\Lambda^3 = 1$ as the degeneracy line. On a log-scale we get the line $\log n = (3/2) \log T + \text{const.}$

2.3.2 Density Expansions Including Interaction Effects

The easiest way to get the virial functions including interaction effects starts from the binary correlation functions F_{ab}. In the first classical approximation, the binary correlations are given by a Boltzmann factor and lead to the first order mean potential energy given above. Furthermore, for the second virial coefficient, we find

$$B_2^{\text{cl}} = \frac{1}{2} \int d\mathbf{r} \Big\{ \exp\left[- \beta V(\mathbf{r}) \right] - 1 \Big\} \,. \tag{2.40}$$

For the higher virial coefficients we may derive corresponding expressions (Bogoliubov 2005–2009; Hirschfelder et al. 1954).

In the quantum case, a few changes are needed. The mean potential energy is given by a trace

$$U = \langle V \rangle = \frac{1}{2} \sum_{ab} \text{Tr}\big[V_{ab}(12) F_{ab}(1, 2) \big] \,. \tag{2.41}$$

An easy way to proceed is to use the coordinate representations. The pair probability given classically by a Boltzmann factor is replaced by its quantum-statistical counterpart, the Slater sum of pairs introduced in Chap. 4. In this way, as shown by Beth and Uhlenbeck, the classical expressions remain valid if we just replace the Boltzmann factor by binary Slater sums (Beth and Uhlenbeck 1937; Uhlenbeck and Beth 1936):

$$\exp\big[-\beta V(\mathbf{r})\big] \longrightarrow S_2(\mathbf{r}) = C\sum_n \exp\big[-\beta E_n\big]\phi_n(\mathbf{r})\phi_n^*(\mathbf{r})\phi_n(\mathbf{r})\,. \qquad (2.42)$$

Beth and Uhlenbeck (1937) derived other useful representations. In particular, they studied the role of bound and scattering states and introduced phase shifts.

General Density Expansions

For a more general derivation of virial expansions for gases, we start from the free energy of an N-particle system:

$$F(T,V,N) = -k_\mathrm{B}T\ln Q_N(T,V)\,,$$
$$Q_N(T,V) = \mathrm{Trace}_{(1,\ldots,N)}\Big[\exp(-\beta\hat{H}_N)\Big]\,. \qquad (2.43)$$

Using representations in terms of Slater functions (or Slater sums), the quantum-statistical problem may be mapped to a classical form. We use the decomposition

$$Q_N(T,V) = Q_{N,\mathrm{id}}\cdot Q_{N,\mathrm{int}}\,, \qquad Q_{N,\mathrm{id}} = \frac{(2s+1)^N\;V^N}{N!\;\Lambda^{3N}}\,, \qquad (2.44)$$

where the ideal part is defined by classical Boltzmann statistics. For the free energy, using Stirling's formula $\ln N! = (N\ln N - N)$, we obtain the representation

$$F(T,V,N) = F_\mathrm{id} - k_\mathrm{B}T\ln Q_{N,\mathrm{int}} = Nk_\mathrm{B}T\left[\ln\left(\frac{n\Lambda^3}{2s+1}\right)-1\right]$$
$$-k_\mathrm{B}T\ln\left[\frac{N!\Lambda^{3N}}{(2s+1)^N V^N}\right]\sum_{\sigma_1,\ldots,\sigma_N}\int \mathrm{d}\mathbf{r}_1\ldots\mathrm{d}\mathbf{r}_N\big\langle\ldots|\exp(-\beta\hat{H}_N)|\ldots\big\rangle\,.$$

This representation of the free energy already has a classical form and may be treated, e.g., by the well known technique of Mayer expansions (Friedman 1962). Following this route, as proposed by Slater (1939) and Morita (1959), we now introduce the so-called Slater sums or Slater functions.

The N-particle Slater sums are defined by

$$S^{(N)}(\mathbf{r}_1,\ldots,\mathbf{r}_N) = \frac{N!\Lambda^{3N}}{(2s+1)^N}\sum_{\sigma_1,\ldots,\sigma_N}\big\langle\ldots|\exp(-\beta\hat{H}_N)|\ldots\big\rangle\,. \qquad (2.45)$$

At small densities we may use a pair approximation

$$S^{(N)}(\mathbf{r}_1,\ldots,\mathbf{r}_N) = \prod_{i<j}^{N} S^{(2)}(\mathbf{r}_i,\mathbf{r}_j)\,, \qquad (2.46)$$

where, by definition,

$$S^{(2)}(\mathbf{r}_i, \mathbf{r}_j) = \frac{2!\lambda^6}{(2s+1)^2} \sum_{\sigma_1\sigma_2} \left\langle \mathbf{r}_1\mathbf{r}_2 \right| \exp(-\beta \hat{H}_2) \left| \mathbf{r}_1\mathbf{r}_2 \right\rangle . \tag{2.47}$$

The quantum statistical partition function corresponds directly to the classical form, so the quantum-statistical second virial coefficient is represented by

$$B_2^{qm}(T) = \frac{1}{2V} \int d\mathbf{r}_1 d\mathbf{r}_2 \left[S^{(2)}(\mathbf{r}_i, \mathbf{r}_j) - 1 \right] . \tag{2.48}$$

For a gas mixture, the expansions for the free energy and the EOS read

$$\begin{aligned} F(T, V, N) &= F_{\text{id}} - k_B T V \left[\sum_{ab} n_a n_b B_{ab}(T) + \sum_{abc} n_a n_b n_c B_{abc}(T) + \cdots \right], \\ \beta p = -\beta \left(\frac{\partial F}{\partial V} \right)_{T,N} &= \sum_a n_a \left(1 - \sum_b n_b B_{ab} - 2 \sum_{bc} n_b n_c B_{abc} - \cdots \right), \end{aligned} \tag{2.49}$$

where the virial coefficients are expressed by 2-particle and 3-particle Slater functions, etc.

2.4 Pair Correlations and Beth–Uhlenbeck Method

2.4.1 Slater Sums for Pairs and Second Virial Coefficient

In order to calculate the binary Slater sum for two particles of different species a and b and spins s_a and s_b, we write

$$\begin{aligned} S_{ab}^{(2)}(\mathbf{r}_1, \mathbf{r}_2) = (1 + \delta_{ab})! \, & \frac{\Lambda_a^3 \Lambda_b^3}{(2s_a + 1)(2s_b + 1)} \\ & \times \sum_{\sigma_1, \sigma_2} \langle \mathbf{r}_1\mathbf{r}_2\sigma_1\sigma_2 | \exp(-\beta \hat{H}_{ab}) | \mathbf{r}_1\mathbf{r}_2\sigma_1\sigma_2 \rangle . \end{aligned} \tag{2.50}$$

Using the properties of energy eigenfunctions and the definitions

$$\lambda_{ab} = \frac{\hbar}{\sqrt{2 m_{ab} k_B T}} , \qquad m_{ab} = \frac{m_a m_b}{m_a + m_b} , \tag{2.51}$$

we get the final result

$$S_{ab}^{(2)}(r) = 2\pi^{1/2}\lambda_{ab}^3 \left\{ \sum_{l=0}^{\infty}(2l+1)\left[\sum_{s=1}^{l-1} \exp(-\beta E_{sl})\,|R_{s\ell}(r)|^2\right] + \int \mathrm{d}k \exp\left(-\frac{\beta\hbar^2 k^2}{2m_{ab}}\right)|R_{k\ell}|^2 \frac{\mathrm{d}n_l(k)}{\mathrm{d}k}\right\} . \tag{2.52}$$

Here we have introduced the density of states $\mathrm{d}n_\ell(k)/\mathrm{d}k$, which gives the number of states with quantum number ℓ in the interval $\mathrm{d}k$ (Beth and Uhlenbeck 1937; Huang 1987; Uhlenbeck and Beth 1936, 2001). Taking symmetry into account, we find the Slater sum with exchange effects:

$$S_{ab}^{(2)}(\mathbf{r}) = (1+\delta_{ab})!\, 2\sqrt{\pi}\lambda_{ab}^3 \sum_{l=0}^{\infty}(2l+1)\left[1 \pm \delta_{ab}\frac{(-1)^l}{(2s_a+1)}\right] \times \left[\sum_{s=l+1}^{\infty} \mathrm{e}^{-\beta E_{sl}}|R_{sl}(r)|^2 \int_0^\infty \mathrm{d}k\, \mathrm{e}^{-\lambda_{ab}^2 k^2}\frac{\mathrm{d}n_l(k)}{\mathrm{d}k}\right]|R_k l(r)|^2 .$$

Second Virial Coefficient

The second virial coefficient can be expressed by the binary Slater function or Slater sum, as shown in the last section. Using the normalization of the wave functions, we get (Beth and Uhlenbeck 1937; Uhlenbeck and Beth 1936)

$$B_{ab}(T) = (1+\delta_{ab})!\, 4\pi^{3/2}\lambda_{ab}^3 \sum_{l=0}^{\infty}(2l+1)\left[1 \pm \frac{(-1)^l}{(2s_a+1)}\right] \times \left\{\sum_{s=l+1}^{\infty} \mathrm{e}^{-\beta E_{sl}} + \int_0^\infty \mathrm{d}k\, \mathrm{e}^{-\lambda_{ab}^2 k^2}\left[\frac{\mathrm{d}n_l(k)}{\mathrm{d}k} - \frac{\mathrm{d}n_l(k)^0}{\mathrm{d}k}\right]\right\} . \tag{2.53}$$

Once again, we have the density of states of pairs $n_\ell(k)$ and the corresponding density $n_\ell(k)^0$ for the noninteracting system, which corresponds to the -1 in the virial formula.

A different and sometimes more convenient way of writing the second virial coefficient may be obtained from a relation between scattering phase shifts and Jost functions (Ebeling et al. 1976; Kraeft et al. 1969; Kremp et al. 1971). We can express the second virial coefficient in terms of the two-particle trace and then use the resolvent representation

$$\begin{aligned} B_{ab}(T) &= \text{const.} \times \mathrm{Tr}\Big[\exp(-\beta H_{ab}) - \exp(-\beta H_{ab}^0)\Big] \\ &= \text{const.} \times \frac{1}{2\pi \mathrm{i}}\int_c \exp(-\beta z)\mathrm{Tr}\left(\frac{1}{H_{ab}-z} - \frac{1}{H_{ab}^0 - z}\right) . \end{aligned} \tag{2.54}$$

In this way we arrive at a representation of the second virial coefficient by a complex integral along a path enclosing the real axis:

$$B_{ab} = \frac{4\pi^{3/2}(1+\delta_{ab})\lambda_{ab}^3}{(2s_a+1)(2s_b+1)} \frac{1}{2\pi i} \int_c \exp(-\beta z) F(z) dz \,,$$
$$F(z) = \mathrm{Tr}\left(\frac{1}{H_{ab}-z} - \frac{1}{H_{ab}^0 - z}\right) \,. \tag{2.55}$$

After some transformations, we arrive at the following complex representation in terms of Jost functions $D_\ell(z)$. To our knowledge this was first obtained in independent work by Kraeft et al. (1969) and Kano and Mishima (1969). The Jost functions $D_\ell(z)$ are analytical functions, well known from scattering theory, with poles at the bound states and a branch cut at the positive real axis defined by the scattering phase shifts. These functions, lie other scattering quantities, are exactly known, e.g., for Coulomb systems, and will appear in later calculations. When the attractive well of the interatomic forces is deep enough for bound states to form, the contribution from scattering states may be neglected, at least at low temperatures. Specific problems connected with bound state contributions will be treated later.

2.4.2 Beth–Uhlenbeck Representation for Real Gases

In many cases the attractive van der Waals forces between atoms or molecules in a gas are so weak that the formation of bound states is not possible. As known from quantum mechanics, only deep wells may form bound states. A common example for weak attractive forces are the noble gases. In this case the second virial coefficient contains only scattering contributions. At temperatures below 100 K, classical calculations do not work and we need quantum-statistical calculations based on the density of states (Kilpatrick et al. 1954; Kraeft et al. 1969, 1986; Costa et al. 2013).

In order to find the density of states, we have to analyze the asymptotic of the radial wave functions. For large r, we have

$$\lim_{r\to\infty} R_{sl}(r) = \frac{\sqrt{2/\pi}}{r} \sin\left[kr - \frac{\pi}{2}l + \delta_l(k)\right] \,, \tag{2.56}$$

where $\delta_l(k)$ are the scattering phase shifts. Enclosing the system in a sphere with radius R, assuming $R_{sl}(r \geq R) = 0$, the following relation holds:

$$kR - \frac{\pi}{2}l + \delta_l(k) = n_l(k)\pi \,, \qquad n_l(k) = 0, \pm 1, \pm 2, \dots \,. \tag{2.57}$$

Using the same relation for free particles, we find

$$n_l(k) - n_l^0(k) = \frac{1}{\pi}\delta_l(k) \,. \tag{2.58}$$

To calculate the second virial coefficient, we need the phase, and we usually have to do numerical calculations (Kilpatrick et al. 1954; Costa et al. 2013). For a few cases such as a hard-sphere, square-well potentials, and Coulomb potentials, exact analytical expressions are available (Beth and Uhlenbeck 1937; Kraeft et al. 1970). The numerical calculation of the phase shifts $\delta_\ell(k)$ requires a precise numerical integration of the the radial Schrödinger equation for the case of large distances. Comparing the result with the asymptotic solution

$$\psi_\ell(r) = A \sin\left(kr - \frac{l\pi}{2} + \delta_\ell\right) , \tag{2.59}$$

we find the phase shifts as a function of k and l. For example, for He–He interactions the phase shifts with $\ell = 0, 1, 2, 3, 4, \ldots,$ have been calculated by Kilpatrick et al. (1954). After calculating the phase shifts for the given interaction potential in analytical or numerical form, the virial coefficient may be calculated as a function of temperature (Costa et al. 2013).

At higher temperatures the properties of gases are only weakly influenced by the spins. This is different at lower temperatures, say $T < 100$ K, where the spins of the atoms (or molecules) forming the gas play an essential role. Note that H atoms with spin $s = 0$ and He-4 atoms with spin $s = 1$ are bosons, in contrast to He-3 atoms, which have $s = 1/2$ and are therefore fermions. This leads in particular to essential differences between the properties of the fermionic gas He-3 and the bosonic gas He-4. In an oft-cited paper by Kilpatrick et al. (1954), the second virial coefficients of He-3 and He-4 were calculated at closely spaced temperatures over the range 0.3–60 K using the Lennard-Jones 12-6 potential with parameters given by de Boer and Michels and based on a calculation of the phase shifts. Some results are shown in Fig. 2.1.

A more recent calculation of the virial expansions for He-4 in the region $3 < T < 100$ K was given, e.g., by Costa et al. (2013). According to this study, the ideal quantum term contributes approximately 11 percent at 3 K and less than 1 percent at 100 K. The He-4 diatomic molecule possesses a very weak bound state which contributes less than 0.1 percent at 3 K and much less at higher temperatures. Therefore this contribution may in general be neglected, even in comparison with the ideal Bose contribution.

For most rare gases, such as hard-sphere gases, the virial expansion is converges well, since the virial terms decay quickly. In particular, this means we need conditions like $nB_2(T) \ll 1$, $n^2B_3(T) \ll 1$, and so on. For example, for a gas of spheres with diameter d, this gives the condition $nd^3 \ll 1$ or $d \ll n^{-1/3}$. In other words the effective diameter of particles d should be small in comparison to the average distance between particles in the gas. For most rare gases, this is a weak condition which is often satisfied. We have shown that He-4 is well described by the cluster representation, even down to a few degrees kelvin. In all these cases, the second virial coefficient is not too large and is dominated by the scattering contributions. However, there are physical conditions in which the contribution of the second virial coefficient is very large in comparison to the ideal contributions, and even increases

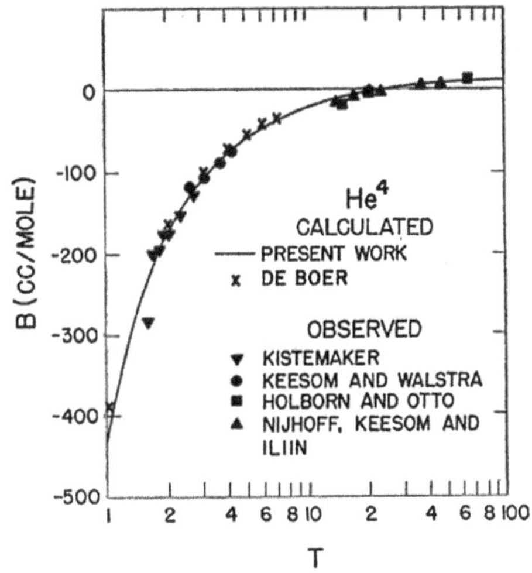

Fig. 2.1 Temperature dependence of the second virial coefficient of helium-4 in the range 1–100 K (Kilpatrick et al. 1954) in comparison with some experimental points

exponentially with the reciprocal temperature $\beta = 1/k_{\mathrm{B}}T$. In this case we have to search for other representations of the pressure.

2.5 Representations in the Grand Canonical Ensemble

2.5.1 *Fugacity Expansions*

From the theory of ideal quantum gases, we we know that the pressure may be represented in terms of the density, but also in terms of expansions with respect to the fugacity z. We shall now derive such fugacity expansions for arbitrary interacting gases. Following the pioneering work of Terrel L. Hill 1960, we will show that fugacity expansions are appropriate tools for gases with deep bound states, showing association and reaction effects (see also Ebeling 1974; Friedman and Ebeling 1979). In order to check the power of these expansions, which have been the appropriate tool since Hill's work, we recall the basic formulae for the grand canonical ensemble.

In this ensemble the basic independent quantity is the chemical potential or the fugacity $z = \mathrm{e}^{\beta\mu}$, rather than the density. The density operator is given by

$$\hat{\rho} = \Xi^{-1} \exp\left(\beta\hat{N}\mu - \beta\hat{H}_N\right) , \tag{2.60}$$

and the grand canonical partition function is

$$\Xi = \sum_{N=0}^{\infty} \mathrm{e}^{\beta\mu N} \mathrm{Tr}_{\mathcal{H}_N} \mathrm{e}^{-\beta\hat{H}_N} . \tag{2.61}$$

This is a kind of generalized trace, including an N summation, corresponding to a so-called Fock space.

The thermodynamic functions follow from

$$pV = -k_B T \ln \Xi \,, \qquad \langle N \rangle = \frac{\partial}{\partial \mu}(pV) = \beta z \frac{\partial}{\partial z}(pV) \,. \tag{2.62}$$

These formulae provide an expansion of the pressure and the densities with respect to the fugacities. Just for convenience we introduce a new quasi-density $\tilde{z} = \Lambda^{-3} z$, where Λ is again the thermal wavelength. Note that $\tilde{z}$ has the dimension of a density and converges for the classical case to the density n. (In the future we shall sometimes omit the tilde if the meaning is clear from the context.)

The virial function follows by expansion with respect to the quasi-density $\tilde{z}$, which is related to the former density expansion. The virial coefficients follow by comparison of coefficients on both sides of the series expansions. Using the expansion of the log-function, we find

$$\begin{aligned}\beta p = \frac{1}{V}\sum_{j=1}^{\infty} b_j \tilde{z}^j &= \frac{1}{V}\left(\tilde{Q}_1 \tilde{z} + \frac{1}{2!}\tilde{Q}_2 \tilde{z}^2 + \cdots\right) \\ &- \frac{1}{2V}\left(\tilde{Q}_1 \tilde{z} + \frac{1}{2!}\tilde{Q}_2 \tilde{z}^2 + \cdots\right)^2 + \frac{1}{3V}\left(\tilde{Q}_1 \tilde{z} + \frac{1}{2!}\tilde{Q}_2 \tilde{z}^2 + \cdots\right)^3 .\end{aligned} \tag{2.63}$$

Comparing the coefficients of equal powers $\tilde{z}^k$, we get

$$\begin{aligned} b_1 &= \frac{1}{V}\tilde{Q}_1 \,, \qquad b_2 = \frac{1}{2!\,V}\left(\tilde{Q}_2 - \tilde{Q}_1^2\right) , \\ b_3 &= \frac{1}{3!\,V}\left(\tilde{Q}_3 - 3\tilde{Q}_1\tilde{Q}_2 + 2\tilde{Q}_1^3\right) . \end{aligned} \tag{2.64}$$

The general relations follow from Thiele's semi-invariants or equivalent methods of cumulant expansions. For the first terms in the series, we get

$$\beta p V = b_1 \tilde{z} + b_2 \tilde{z}^2 + b_3 \tilde{z}^3 + \cdots \,, \tag{2.65}$$

$$n = \tilde{z}\frac{\partial(\beta p)}{\partial \tilde{z}} = b_1 \tilde{z} + 2b_2 \tilde{z}^2 + 3b_1 \tilde{z}^3 + \cdots \,. \tag{2.66}$$

In principle, this virial expansion is equivalent to the density expansion. However, as we will show, the convergence may differ significantly from case to case. Explicitly, we find

$$b_1(T) = \frac{1}{V}\tilde{Q}_1 = \frac{\Lambda^3}{V}\mathrm{Tr}_1\left(\mathrm{e}^{-\beta \hat{H}_1}\right) = 1 \,, \tag{2.67}$$

$$b_2(T) = \frac{1}{2V}\left(\tilde{Q}_2 - \tilde{Q}_1^2\right) = \frac{1}{V}\Lambda^6 \mathrm{Tr}_{(12)}\left[\mathrm{e}^{-\beta \hat{H}_2} - \mathrm{e}^{-\beta \hat{H}_2^{(0)}}\right] . \tag{2.68}$$

Comparing with Sect. 5.2, it follows that

$$b_2(T) = B_2(T) . \tag{2.69}$$

For higher orders, the relations are more complicated (Ebeling 1974, 1976).

2.5.2 *Fugacity Expansions and the Chemical Picture*

From the theory of ideal quantum gases we already know that the pressure may be represented in terms of the density or in terms of expansions with respect to the fugacity z. In the last section we showed that we have these two possibilities for any gas. Let us look first at a hard core system.

Figure 2.2 compares the density expansion for the compressibility coefficient as a function of the dimensionless parameter $y = \pi n d^3/6$ with the Carnahan–Starling expression, which fits well with numerical MC and MD data for hard spheres. The agreement between a first order (second virial approximation) density expansion, viz.,

$$p/nk_BT = 1 + \frac{2}{3}(\pi n d^3) + \cdots , \tag{2.70}$$

and the Carnahan–Starling approximation is rather good.

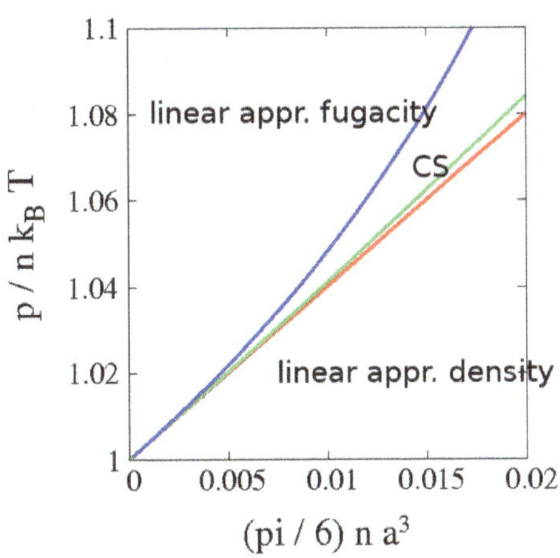

Fig. 2.2 Pressure in relation to ideal pressure for a classical gas of hard spheres. The result of a virial expansion with respect to density as a function of the characteristic hard sphere parameter $(\pi/6)na^3$ is compared with the result of a fugacity expansion. Both expansions are cut off after the second virial coefficients. The results are compared with the Carnahan–Starling formula which is nearly exact. The Carnahan–Starling curve which is in between is in rather good agreement with the expansion linear in the density

Figure 2.2 also compares, in the second virial approximation, the fugacity expansion (upper curve) with the Carnahan–Starling formula, which is close to the exact EOS for hard sphere systems. As shown already in earlier work (Friedman and Ebeling 1979), we see that, up to $\pi na^3/6 \sim 0.03$, a density expansion cut off after linear terms in the density is just a bit below the exact curve and describes the real hard-sphere gas quite well. On the other hand, the corresponding fugacity expansion provides too high a pressure and deviates significantly from the exact Carnahan–Starling curve.

Our conclusion is that, for systems with dominating repulsive forces, such as the hard sphere system, the density expansion is much 'better' than the fugacity expansion. Here 'better' only refers to fast convergence, since in principle the two expansions are equivalent.

In a second example, we consider systems with deep bound states, like a gas of hydrogen atoms which are able to form the bound state H_2 with a binding energy of about 4.7 eV. For systems with deep bound states, we have

$$B_2(T) \approx 8\pi^{3/2}\lambda^3 \sum_{sl} \exp(-E_{sl}/k_{\mathrm{B}}T) \,. \tag{2.71}$$

When the ground state energy E_{10} has modulus larger than $k_{\mathrm{B}}T$, the exponential function may give a very large contribution. For example, a typical binding energy is 0.1 eV, expressed in 10^3 kelvin. Then temperatures below room temperature may generate such large values of $B_2(T)$ that, in the virial series, the pressure becomes negative and the expansion is no longer physically meaningful.

Hydrogen Gas and the Chemical Picture

We now analyze a hydrogen gas as a simple model case of a gas where the fugacity expansion including only the second virial coefficient yields excellent results. We thus write

$$\beta p = \tilde{z} + B_2(T)\tilde{z}^2 \,, \qquad n = \tilde{z} + 2B_2(T)\tilde{z}^2 \,. \tag{2.72}$$

We may consider (2.72) as an implicit EOS and exclude the fugacity by solving the quadratic equation $2B_2(T)\tilde{z}^2 + \tilde{z} - n = 0$. Taking the relevant solution, we thus get the pressure

$$\beta p = \frac{1}{2}n + \frac{1}{2}\tilde{z} = \frac{n}{2} + \frac{1}{8B_2(T)}\left(\sqrt{1 + 8nB_2} - 1\right) \,. \tag{2.73}$$

This equation contains only the second virial coefficient. However, in comparison with the density expansion containing only second virial contributions, the new expression for the pressure can never assume negative values, even if $B_2(T)$ assumes very large values. This result suggests that fugacity expansions may be more appropriate for gases with deep bound states than density expansions (Hill 1960; Ebeling 1974, 1976; Friedman and Ebeling 1979).

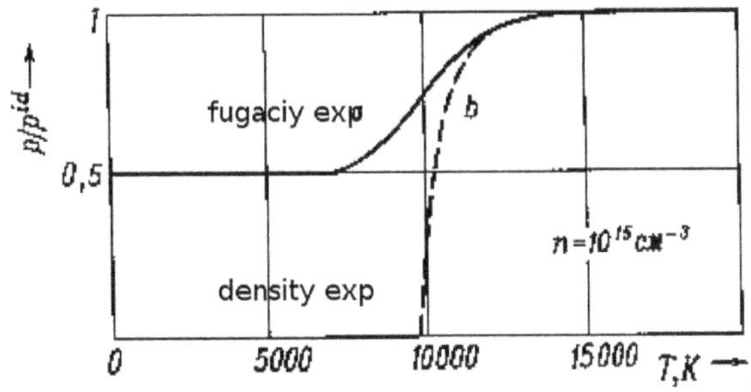

Fig. 2.3 Pressure of hydrogen gas as a function of temperature, calculated from the fugacity expansion (curve a) and the density expansion (curve b). At lower temperatures the density expansion fails due to the exponential increase from bound state contributions

Let us study us an example an atomic hydrogen gas in the region where molecule formation plays a role. Assuming a bound state energy of molecules (dissociation energy) of $D = 4.7$ eV, we get for the second virial coefficient of the gas

$$B_{\mathrm{H}}^{(2)}(T) \approx \frac{h^3}{\sqrt{2\pi m_{\mathrm{H}} k_{\mathrm{B}} T}^3} \exp\left(\frac{4.7\ \mathrm{eV}}{T\ [\mathrm{eV}]}\right) . \tag{2.74}$$

Figure 2.3 shows the pressure as a function of temperature for a hydrogen gas. Continuous lines show the fugacity expansion and dashed lines the density expansion, including only the second virial coefficient. The curves we obtained show that the fugacity representation describes the physics of a hydrogen gas correctly. There are two limiting cases:

- At low temperatures with $p \to n k_{\mathrm{B}} T/2$ for $T \to 0$, the 'effective particle density' is just 50% of the density at high temperatures. This is the region where molecules are formed. The number of hydrogen molecules H_2 per cm^3 is half of the total density of atoms.
- At high temperatures there are no molecules present and the 'effective particle density' is equal to the total density of atoms.

We see that the fugacity expansion gives a correct physical description of the equilibrium between atoms and molecules in the gas.

So far we have described the system without using any explicit chemical notation, just using physical terms like virial coefficients. So how would a chemist describe the system we are studying here? In a first approximation, the chemist would consider it as a mixture of atoms with density n_{A} and molecules with density n_{M}. In the chemical picture, due to the assumption of ideality, we would write for the pressure

$$\beta p = n_{\mathrm{A}} + n_{\mathrm{M}} , \qquad n_{\mathrm{M}}/n_{\mathrm{A}}^2 = K(T) , \tag{2.75}$$

where the second relation is the mass action law, with a mass action constant $K(T)$ which is proportional to an exponential function of E_0, the binding energy of formation of a molecule from two atoms. Since the total density of the reacting mixture is given by $n = n_{\mathrm{A}} + 2n_{\mathrm{M}}$, we get the equation of state

$$\beta p = n_A + n_A^2 K(T) , \qquad n = n_A + 2n_A^2 K(T) . \tag{2.76}$$

A comparison with the fugacity expansion shows the full equivalence between the physical description given by (2.72) and the chemical description given by the foregoing equations. The two pictures are identical if we use the notation $\tilde{z} = n_A$, $K(T) = B_2(T)$. This means that we may interpret $\tilde{z}$ as the density of free atoms and $B_2(T) \sim \tilde{z}^2$ as the density of free molecules.

This leads us to the insights that for systems with attractive forces and two deep particle bound states, under the condition $\exp(-E_{10}/k_B T) \gg 1$, the method of density expansions is badly convergent and may even lead to negative pressures. On the other hand, fugacity expansions give an excellent description from low to high temperatures. Of course, we may have in reality both strongly repulsive forces with hard sphere core and attractive forces leading to deep bound states in the same system. The way out here is to introduce a description which lies somewhere between fugacity expansions and density expansions. Such 'mixed expansions' may be based on a more complicated quasi-particle picture (Hill 1960; Ebeling 1974). A systematic investigation of the thermodynmaic formulation of chemical association in statistical mechanics has been given by Fisher and Zuckerman (1998). Several aspects of the relation between the physical and chemical description of plasmas will be investigated in more detail in Chap. 5. In particular, the class of systems with Coulomb interactions which are in part repulsive and in part attractive requires more sophisticated expansions situated somehow between density and fugacity expansions (Ebeling 1974, 1976; Friedman and Ebeling 1979; Kremp et al. 2005).

2.6 Strong Exchange Correlations in Fermi–Dirac Gases

2.6.1 Pair Correlations and Thermodynamics

In classical physics non-ideality means, that the EOS is different from $\beta p = n$. In quantum gases, the situation is more complicated. We already know that Fermi or Bose gases without interactions may develop strong correlations due to exchange effects. For this reason we must look primarily at the correlations between the particles which reflect the degree of coupling.

By definition, we speak about correlations in a fluid if the spatial correlation function $g(1, 2)$ differs from unity. Recall that the pair distribution function of a fluid was defined by $g(1, 2) = 1 + h(1, 2)$. Fermi–Dirac gases show rather strong correlations at high densities due to the Pauli principle. In this sense, they are not 'ideal' in the classical meaning of independent particles, but are correlated. In the quantum case, $g(1, 2)$ is the quantum statistical pair distribution function and $h(1, 2)$ the corresponding correlation function of two noninteracting Fermi or Bose particles (Pines and Nozieres 1966; Kremp et al. 2005).

Here we start by investigating Fermi particles (electrons). Then, in the low density limit, we have

$$h(1,2) = -\frac{1}{2}\exp(-r^2/\lambda^2)\,, \qquad \lambda^2 = \frac{\hbar^2}{mk_BT}\,. \tag{2.77}$$

In the opposite case of high density (strong degeneracy), the correlation function of noninteracting electrons is given by (Pines and Nozieres 1966)

$$h(r) = -\frac{9}{2}\left[\frac{\sin(p_Fr) - p_Fr\cos(p_Fr)}{p_F^3r^3}\right]^2, \qquad p_F^{\frac{1}{}} = .521a_Br_s\,. \tag{2.78}$$

Due to the correlations, the EOS is not linear in the density, but increases very fast at higher densities. Our task is now to discuss this function in the full density range. In order to solve this problem we have to express the pressure and other thermodynamic functions explicitly in terms of the densities.

So far we have calculated the thermodynamic functions in the standard way, working in the grand canonical ensemble and considering the fugacities as independent input parameters (see e.g. Huang 2001, Kremp 2005). Only then did we express the pressure in terms of the density for low densities. On the other hand we must consider that the densities are the primary input parameters under normal experimental conditions. Fugacities (chemical potentials) are quantities which are difficult to control directly in experiments in most cases. In order to compare theory with experiment, we need as a rule to represent the thermodynamic functions as an explicit function of the density over a wide range. The implicit representation of the EOS by two equations for p and n depending on the fugacity is not convenient, and in many representations not well convergent. So we have to solve the so-called inversion problem.

In principle it should be enough to get one thermodynamic potential like the free energy $F(T, V, N)$, for example, as a density function and obtain the other potentials in standard way by differentiation from Maxwell relations. In spite of the fact that one thermodynamic potential is enough, because all others could be derived from it, we find it more convenient, to start from two or three explicit thermodynamic functions. This procedure allows us to obtain all other functions just by arithmetic operations such as addition or subtraction, noting that differentiation can introduce inaccuracies.

We thus study the Gibbs potential G_e, free energy F_e, and internal energy U_e, and derive the pressure p_e and the entropy S_e by simple arithmetic relations:

$$p_eV = F_e - G_e\,, \qquad TS_e = U_e - F_e\,.$$

In the previous chapter, in the framework of the grand canonical ensemble, we expressed the pressure and the thermodynamic functions in terms of the activity z and the chemical potential μ. With the abbreviations $z = \exp(\alpha)$ and $\alpha = \mu/k_BT$, for the density of a Fermi–Dirac gas, we find (Huang 1987; Landau and Lifshits 1980, 2001)

$$\frac{n\Lambda^3}{2s+1} = \frac{4}{\sqrt{\pi}} \int_0^\infty \mathrm{d}x \frac{x^2 z}{\mathrm{e}^{-x^2} + z} \equiv f_{3/2}(z)\,. \tag{2.79}$$

The function $f_{3/2}(z)$ is analogous to the g-function for a Bose gas and may also be represented by expansions in z (in the region of convergence):

$$f_\lambda(z) = \sum_{l=1}^{\infty} (-1)^{l+1} \frac{z^l}{l^\lambda}\,. \tag{2.80}$$

Besides this definition of Fermi functions, we will use also another definition (DeWitt 1961; Kraeft et al. 1986):

$$I_\nu(y) = \frac{1}{\Gamma(\nu+1)} \int_0^\infty \mathrm{d}x \frac{x^\nu}{\exp(x-y)+1}\,. \tag{2.81}$$

There is an extensive mathematical literature on the properties of this class of functions (see, e.g., Wasserman et al. 1970, Zimmermann 1987, Lether 2000, Fukushima 2015). In the first approximation, for the corrections to the Boltzmann expressions for the chemical potential, we find the series

$$\frac{\mu}{k_\mathrm{B}T} = \ln z = \ln \frac{n\Lambda^3}{2s+1} + \frac{1}{2\sqrt{2}} \frac{n\Lambda^3}{2s+1} + \cdots\,. \tag{2.82}$$

Similar expressions follow for the pressure, viz.,

$$\begin{aligned} p = k_\mathrm{B}T \left(\frac{2s+1}{\Lambda^3}\right) f_{5/2}(z) &= \frac{k_\mathrm{B}T(2s+1)}{\Lambda^3} \left(z - \frac{z^2}{2^{2/5}} + \cdots\right) \\ \approx nk_\mathrm{B}T \left[1 + \frac{1}{4\sqrt{2}} \left(\frac{n\Lambda^3}{2s+1}\right)\right]\,. \end{aligned} \tag{2.83}$$

In general, for applications, we must express the pressure and the chemical potential in terms of densities and temperatures. The task of excluding the activities and expressing them in terms of densities is called the *inversion problem*.

As mentioned above, we find it most convenient to start with the chemical potential of the electrons (which is the Gibbs potential per electron) and the free energy per electron, both measured in $k_\mathrm{B}T$ (or possibly in Rydberg) as primary quantities:

$$\alpha_\mathrm{e} = \frac{\beta G_\mathrm{e}}{N_\mathrm{e}}, \qquad \phi_\mathrm{e} = \frac{\beta F_\mathrm{e}}{N_\mathrm{e}}\,. \tag{2.84}$$

The main contribution to these functions comes from the ideal terms which were already considered in detail on a rather elementary basis. The extension to the region of weak correlations requires greater accuracy and more precise studies of the series with respect to the degeneracy parameter. Note that there is a long list of mathematical studies of the Fermi integrals (see, e.g.,Wasserman et al. 1970, Lether 2000). The problem of inversion form the variable α to n is non-trivial (Zimmermann 1987). Note that the known limiting cases have to be respected:

$$\alpha_e = \ln y\ , \quad \text{if } y \ll 1\ , \quad \alpha \longrightarrow \left(\frac{3\sqrt{\pi}}{4}\right)^{2/3} y^{2/3}\ , \quad \text{if } y \gg 1\ . \tag{2.85}$$

We use a convenient formula proposed by Zimmermann (1988), consisting of two different analytical expressions for low and high densities, and given in dimensionless units $y = n_e\Lambda_e^3/2$ which are smoothly interpolated:

$$\begin{aligned}\varphi_e^{FD}(y) &= w_1(y-5.5)\Big[\ln(y) - 1 + 0.1768y - 0.0165y^2 + 0.000031y^3\Big] \\ &\quad + w_2(y-5.5)\Big(0.7254y^{2/3} - 2.040y^{-2/3} - 0.45y^{-2}\Big)\ ,\end{aligned} \tag{2.86}$$

$$\begin{aligned}\alpha_e^{FD} &= w_1(y-2.5)\Big[\ln(y) + .3536y - 0.00198y^2 + 0.000124y^3\Big] \\ &\quad + w_2(y-2.5)\Big(1.209y^{2/3} - .6803y^{-2/3} + 0.45y^{-2}\Big)\ ,\end{aligned} \tag{2.87}$$

where the interpolating weight factors are defined by

$$w_1(x) = \left[1+\exp(+bx)\right]^{-1}\ , \qquad w_2(x) = \left[1+\exp(-bx)\right]^{-1}\ , \tag{2.88}$$

with $b \approx 10$. These expressions turn out to be very convenient for practical calculations of the thermodynamic functions, including the equation of state:

$$\pi = \beta n p^{FD} = \varphi_e^{FD}(y) - \alpha_e^{FD}(y)\ . \tag{2.89}$$

Finally, we note once again that the ideal properties of a Fermi gas change in the presence of an external magnetic field B, since the particles, e.g., electrons have a finite Larmor radius $\omega_c = eB/m_e$. The field effect may be estimated by the replacement (Steinberg et al. 1998)

$$n_e\Lambda_e^3 \longrightarrow n_e\Lambda_e^3\frac{\tanh x_e}{x_e}\ , \qquad x_e = \frac{\hbar\omega_c}{2k_BT}\ . \tag{2.90}$$

Note that this effect leads to an effective decrease in the degeneracy of light charged particles such as electrons.

2.6.2 Hartree–Fock Contributions

In order to demonstrate the role of physical interactions we consider the prototype system referred to as a 'Yukawa gas' and closely related models. Here only weak interactions will be taken into account and in a linear approximation, i.e., in the Hartree–Fock approximation. Our basic assumption about the potential is that it has a Fourier transform. The Coulomb case is excluded so far, because of the specific singularities.

As a standard example of a gas with weak interactions with Fourier transform, we consider a quantum gas with Yukawa or related exponential interactions:

$$V_{ab}^{\mathrm{Y}}(r,\eta) = g_{ab}\frac{\exp(-\eta r)}{r}\,, \qquad V^{\mathrm{E}} = -\frac{\partial V^{\mathrm{Y}}(r,\eta)}{\partial \eta}\,. \tag{2.91}$$

The first potential was developed by Hideki Yukawa in 1935 to describe strong forces in elementary particle physics, mediated by the exchange of massive particles. This potential now plays a paradigmatic role in statistical physics, since it has found applications in many fields (Fortov et al. 2005; Langin et al. 2016; Ott et al. 2014). The Coulomb case appears here as a limit of very small η, so the limit $\eta \to 0$ will be mostly excluded. As already discussed in Chap. 1, the Yukawa potential has a singularity at $r = 0$ which may lead to difficulties at higher orders of perturbation expansions. A regularization at $r = 0$ is possible by adding a term of the same form as the one used by Kramers in electrolyte theory, Heller in quantum chemistry, and Kelbg, Deutsch, and others in the quantum statistics of plasmas:

$$V_{ab}^{\mathrm{KD}}(r) = g_{ab}\left[\frac{\exp(-\eta r)}{r} - \frac{\exp(-\alpha r)}{r}\right]\,. \tag{2.92}$$

At small distances, the so-called Kelbg–Deutsch potential has a finite value. We assume in the following that $\alpha \gg \eta$ has a sufficiently large value and serves only to avoid divergencies in higher orders of perturbation theory due to singularities at $e = 0$.

There are different ways to calculate the Hartree–Fock approximation (HFA). The first and simplest is to begin by calculating the correlation functions of the free Fermi gas in coordinate space, and then obtain the thermodynamic functions by averaging the interaction energy

$$\langle U_{\mathrm{int}}\rangle = \frac{1}{2}N\Big\langle V^{\mathrm{Y}}(1,2)\big[1 + h^{\mathrm{FD}}(1,2)\big]\Big\rangle\,, \tag{2.93}$$

where $h^{\mathrm{FD}}(1,2)$ is the quantum statistical pair correlation function of an ideal Fermi–Dirac system. In the low density limit, we have the standard result for fermions given in the last section. Averaging the Yukawa potential just with the first term '1' for the pair distribution function gives the Hartree contribution

$$E^{\mathrm{H}} = F^{\mathrm{H}} = 2\pi k_{\mathrm{B}} T V g \lambda^2 \frac{1}{6\eta^2} . \tag{2.94}$$

For the Coulomb limit $\eta = 0$, the Hartree term is formally infinite, but this term cancels with the contribution from the background of counter charges (Kremp et al. 2005). Note that this term does not appear at all in regularized models (Pines and Nozieres 1967; Bobrov et al. 2015). The second term coming from averaging with the term $\exp(-r^2/\lambda^2)$ is the so-called Hartree–Fock term. This term is convergent and, in the Coulomb case $\eta = 0$, after elementary integration, yields the known Hartree–Fock contribution to the interaction energy and free energy:

$$E^{\mathrm{HF}} = F^{\mathrm{HF}} = -V n g \pi \int_0^\infty \mathrm{d}r r \exp(-r^2/\lambda^2) = -\frac{V}{2} \pi n g \lambda^2 . \tag{2.95}$$

Including the contribution from $\eta > 0$, we get

$$E^{\mathrm{HF}} = F^{\mathrm{HF}} = -\pi n g \left\{ -\frac{\lambda^2}{4} + \frac{\sqrt{\pi}}{8} \exp(\eta^2 \lambda^2/4) \left[1 - \Phi(\eta\lambda/2) \right] \right\} . \tag{2.96}$$

In the opposite case of high density (strong degeneracy), the correlation function of noninteracting electrons is given by (Pines and Nozieres 1966)

$$h^{\mathrm{FD}}(r) = -\frac{9}{2} \left[\frac{\sin(p_{\mathrm{F}} r) - p_{\mathrm{F}} r \cos(p_{\mathrm{F}} r)}{p_{\mathrm{F}}^3 r^3} \right]^2 , \qquad p_{\mathrm{F}}^{\frac{1}{}} = .521 a_{\mathrm{B}} r_s . \tag{2.97}$$

By integration, for the Coulomb case, we obtain the ground state energy

$$E^{\mathrm{HF}} = -N_{\mathrm{e}} \frac{3}{5} \frac{p_{\mathrm{F}}^2}{2m} . \tag{2.98}$$

In order to get the full expression for the Hartree–Fock contributions for any degree of degeneracy, representations in the momentum space are more convenient (Kraeft et al. 1986; Kremp et al. 2005):

$$\langle U_{\mathrm{c}} \rangle = V \int \frac{\mathrm{d}\mathbf{p}_1 \, \mathrm{d}\mathbf{p}_2}{(2\pi)^6} V^{\mathrm{Y}}(\mathbf{p}_1 \mathbf{p}_2) f(\mathbf{p}_1) f(\mathbf{p}_2) . \tag{2.99}$$

These analytic expressions depend on the full Fermi functions f taken as functions of the momenta. The result of this integration is known only for the Coulomb limit $\eta = 0$. After some transformations of the integral, we derive (De Witt 1996; Kraeft et al. 1986; Wasserman et al. 1970)

$$\langle U_{\mathrm{c}} \rangle = 2V \frac{g}{\Lambda^4} \int_{-\infty}^{\alpha} \mathrm{d}y \, I_{-1/2}^2(y) . \tag{2.100}$$

The Hartree–Fock contribution to the free energy follows by charging (integration with respect to e^2), and by derivation with respect to α, we find a particularly simple expression for the chemical potentials at any degeneracy:

$$\alpha^{\mathrm{HF}} = -\frac{g}{k_{\mathrm{B}} T \Lambda} I_{-1/2}(\alpha) . \tag{2.101}$$

Note that density and ideal chemical potential are connected by the implicit equations

$$\frac{1}{2} n_{\mathrm{e}} \Lambda_{\mathrm{e}}^3 = I_{1/2}(\alpha_{\mathrm{e}}) , \qquad \alpha = \beta \mu_{\mathrm{e}}^{\mathrm{FD}} . \tag{2.102}$$

The Hartree–Fock approximations given here were derived as the first corrections to ideality in the grand canonical ensemble. If we are interested in the corrections in an ensemble with given density and temperature, we have to introduce the canonical ideal chemical potentials from the Fermi expressions, which have to be inverted (Kraeft et al. 1986). The task of excluding the activities and expressing them in terms of the densities is another *inversion problem*, similar to the one for the ideal Fermi gas. As pointed out above, for the ideal case, we find it most convenient to take as primary quantities the Gibbs potential per electron, which is the chemical potential of the electrons, and the free energy per particle measured in $k_{\mathrm{B}} T$:

$$\alpha^{\mathrm{HF}} = \frac{\beta G^{\mathrm{HF}}}{N} , \qquad \varphi = \frac{\beta F^{\mathrm{HF}}}{N} . \tag{2.103}$$

After obtaining these two thermodynamic functions in the Hartree–Fock approximation, all others, e.g., the pressure, follow just by algebraic operations.

First we find α^{HF} as a function of the dimensionless electron density $y = n\Lambda^3/2$, where the limiting cases of small or large density have to be respected. As for the Fermi–Dirac case, we use the method of piecewise representations and interpolations developed for the ideal Fermi contributions by Zimmermann (1987):

$$\varphi^{\mathrm{HF}}(y) = -\frac{g}{k_{\mathrm{B}} T \Lambda_{\mathrm{e}}} \Big[w_1 (y - 1.8)\big(0.5y - 0.3540y^2 + 0.1825y^3\big) \tag{2.104}$$
$$+ w_2 (y - 1.8)\big(0.9307y^{1/3} - 0.6976y^{-1} + 0.7635y^{-7/3}\big)\Big] ,$$

$$\alpha^{\mathrm{HF}}(y) = -\frac{g}{k_{\mathrm{B}} T \Lambda_{\mathrm{e}}} \Big[w_1 (y - 0.8)\big(y - 1.0612y^2 + 0.7299y^3\big) \tag{2.105}$$
$$+ w_2 (y - 0.8)\big(1.241y^{1/3} - .6981y^{-1} + 1.018y^{-7/3}\big)\Big] .$$

Here the interpolating weight factors are the same as in the previous section. By construction of the interpolating procedure these expressions are compatible with the known approximations for low or high degeneracy. For alternative representations

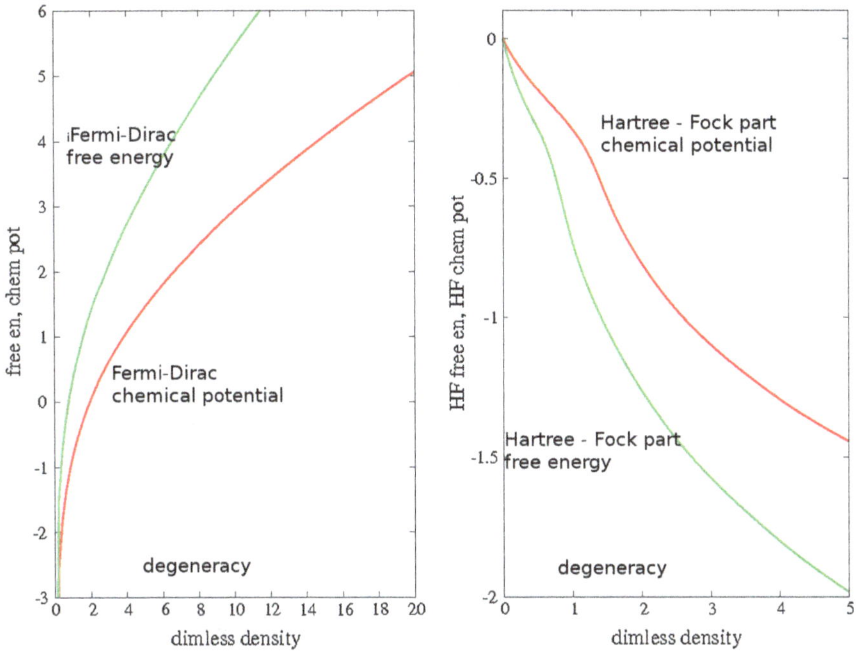

Fig. 2.4 Free energy per electron (*green*) and chemical potential (*red*) of an ideal electron gas in the Fermi–Dirac model (*left*) and the (additional) Hartree–Fock contribution (*right*). The EOS is found as the difference between the two curves. The density is given in the dimensionless units $y = n_e \Lambda_e^3/2$. Deviations from the thermodynamics of a classical gas increase significantly with the density

including the higher order terms and covering the full range, see Perrot and Dharma-Wardana (1984) and Kraeft et al. (1986).

The dependence of the Hartree–Fock functions on the density is shown in Fig. 2.4 (right panel). We observe here a smooth transition between the two regions of the expansions, in the region of intermediate degeneracy. For the intermediate region $E_F < T < 12E_F$, a parametrization of the Hartree–Fock contributions was given by Perrot and Dharma-Wardana (1984). This approximation, which is considered as rather appropriate, will be given explicitly in Chap. 6, along with other, simpler but less precise approximations. The expressions for the Hartree–Fock term are also useful for calculations of the Hartree–Fock contribution to the equation of state

$$\beta p^{\mathrm{HF}}(y) = \alpha_e^{\mathrm{HF}}(y) - \varphi_e^{\mathrm{HF}}(y) . \tag{2.106}$$

The Hartree–Fock contributions increase strongly with the density, changing their analytical shape at certain value of the degeneracy parameter, where non-degenerate behavior goes more or less abruptly over into degenerate behavior.

In certain regions of the density and temperature, roughly speaking the region with $r_s > 1$ and $\Gamma > 1$, the Hartree–Fock contribution (shown in Fig. 2.4 right panel) is not small, and gives a substantial correction to the thermodynamic properties of the ideal Fermi gas. As mentioned, the Hartree–Fock contribution increases strongly with the density and changes its analytical shape to a $n^{1/3}$ law at the transition to strong degeneracy. We have found here full expressions for a Yukawa gas only in the nondegenerate case, and closed expressions for all degrees of degeneracy only for small η, which is the Coulomb limit.

The region where Hartree–Fock approximations work is a relatively large part of the density–temperature plane, covering a region near the boundary of the ideal region. Here correlations due to interactions are still very weak, but nevertheless strongly influence the behavior. A characteristic property of Fermi particles is that they occupy (in the isotropic case) all energy states in momentum space inside the Fermi sphere with the Fermi momentum p_F as radius. We note that the first careful studies of the Green's functions for electrons and other Fermi particles at low T were presented in key papers by Gellman and Brueckner 1957, Galitskii et al. 1958a, Galitskii (1958), Montroll and Ward (1958), and Vedenov and Larkin (1959). It has been shown in particular that the pole of the Green's functions defines the energy spectrum of the Fermi quasi-particles, information that is critical for the low temperature thermodynamics. The microscopic derivation of the Fermi liquid theory using the diagrammatic methods of quantum field theory for very low T was carried out by Landau in 1959 and in 1960. Applications to the theory of superconductivity were developed by Gorkov in 1958. For a summary, see the monograph by Abrikosov, Gorkov and Dzyaloshinsky (1965). Several results for applications to plasmas will be given in Chaps. 5 and 6.

2.7 Quantum Statistics of Prototype Yukawa Gases

2.7.1 Perturbation Theory for Pair Density Operators

In the previous section we considered Hartree–Fock approximations (HFA) for a quantum gas with interactions having a Fourier transform. On this basis, we studied the region of weak correlations for any degree of degeneracy. The HFA is valid in the border region near to the region of ideal Fermi or Bose gases. Here the interactions are weak and may be taken into account in a linear approximation with respect to the interaction parameter, i.e., in the HFA. We shall now study Yukawa gases as a particular example of interacting quantum gases, still in the region of weak correlations and low density. Then we will study higher order approximations in the interaction parameter, including the role of bound states and assuming weak degeneracy.

We consider the Yukawa gas as a prototype of interactions having a Fourier transform, which is here rather simple $4\pi g_{ab}/(t^2 + \eta^2)$, but may be replaced by other interactions of same type. The Coulomb case appears as a limiting case and there are many other interesting applications (Fortov et al. 2005). We will then go beyond the HFA and include large values of the interaction parameter. In some sense the prototype Yukawa gas is just opposite to the prototype hard-sphere gas. As before, we use the virial expansion (Beth and Uhlenbeck 1937) and develop a systematic perturbation theory with respect to the interaction parameter g_{ab} for the Fermi and the Bose cases, including strong interactions. Generalizing the previous result, we include mixtures of gases with arbitrary spin, but restrict the calculations to weak degeneracy here.

Since the Slater function of a pair is the diagonal element of the pair density operator, we can take advantage of the general expansion technique for density operators. Using g for the interaction parameter and following Abrikosov et al. (1965) we get the expansion

$$\rho_2(\beta, g) = \exp(-\beta H_0) \sum_{n=0}^{\infty} (-1)^n g^n \int_0^{\beta} \mathrm{d}\tau_n V^{\mathrm{Y}}(\tau_n) \cdots \int_0^{\tau_2} \mathrm{d}\tau_1 V^{\mathrm{Y}}(\tau_1) ,$$

with

$$V^{\mathrm{Y}}(\tau) = \exp(\tau H_0) V^{\mathrm{Y}} \exp(-\tau H_0) . \tag{2.107}$$

We may now proceed to the region of weak correlations by systematic expansions of the Slater sums (Slater functions) with respect to the interaction parameter, and calculate the contributions term by term (Kelbg and Hoffmann 1964).

We denote the expansion of the binary Slater function with respect to the g-parameter by

$$S_{ab} = S_{ab}^{(0)} + S_{ab}^{(1)} + S_{ab}^{(2)} + \cdots .$$

It is useful to introduce representations of the Slater sums by plane waves. This scheme was also used in the original papers by Uhlenbeck and Beth (Uhlenbeck and Beth 1936, 1937). Following here the early work of the Kelbg school, we shall systematically use this representation by plane wave eigenfunctions (see Ebeling et al. 1967):

$$S_{ab}(r) = \mathrm{A} \int \mathrm{d}\mathbf{k} \exp(-\mathbf{k} \cdot \mathbf{r}) \exp(-\beta \hat{H}_{ab}) \exp(+\mathrm{i}\mathbf{k} \cdot \mathbf{r}) , \tag{2.108}$$

$$\hat{H}_{ab} = -\frac{\hbar}{2m_{ab}} \Delta + V_{ab} , \tag{2.109}$$

where m_{ab} is the relative mass. The Fourier representation in the distance coordinate space reads

$$\tilde{S}_{ab}^{(n)}(\mathbf{t}) = \int \mathrm{d}\mathbf{r}\, S_{ab}^{(n)}(\mathbf{r}) \exp(\mathrm{i}\mathbf{t} \cdot \mathbf{r}) . \tag{2.110}$$

In order to find the virial coefficient, we only need an integral over the Slater sum, and it is sufficient to get the Fourier transforms at the zero point $t = 0$. To calculate the Fourier transforms of the Slater function, we define

$$v_{ab}(r) = \exp(-i\mathbf{k}\cdot\mathbf{r})\exp(-\beta\hat{H}_{ab})\exp(i\mathbf{k}\cdot\mathbf{r}) ,$$

and the Fourier transform

$$v_{ab}(\mathbf{k},\mathbf{t}) = \int d\mathbf{r}\, \exp(i\mathbf{t}\cdot\mathbf{r})v_{ab}(\mathbf{k},\mathbf{r}) . \tag{2.111}$$

With these definitions we can use our general perturbation scheme by transforming to the functions v_{ab}. For the first two iterations, we find

$$v_{ab}^{(1)}(\mathbf{k},\mathbf{t}) = -\beta\int_0^\beta d\beta'\tilde{V}_{ab}(\mathbf{t})\exp\left[-\lambda_{ab}^2(\mathbf{t}^2 - 2\mathbf{k}\cdot\mathbf{t})\left(1-\frac{\beta'}{\beta}\right)\right] , \tag{2.112}$$

$$v_{ab}^{(2)}(\mathbf{k},\mathbf{t}) = \frac{1}{8\pi^3}\int_0^\beta d\beta'\int_0^{\beta'} d\beta''\int d\mathbf{t}'\tilde{V}_{ab}(\mathbf{t}-\mathbf{t}')\tilde{V}_{ab}(\mathbf{t}') \tag{2.113}$$
$$\exp\left\{-\lambda_{ab}^2\left[(\mathbf{t}^2 - 2\mathbf{k}\cdot\mathbf{t})\left(1-\frac{\beta'}{\beta}\right) + (\mathbf{t}'^2 - 2\mathbf{k}\cdot\mathbf{t}')\left(\frac{\beta'}{\beta}-\frac{\beta''}{\beta}\right)\right]\right\} .$$

On the basis of these approximations for the Bloch functions, we find the free energy up to second order in the coupling g. In the same way, we may go to any order.

2.7.2 *Perturbation Expansion for the Free Energy*

Introducing the first two iterations of the functions v_{ab} given before into the formula for the Slater functions, the corresponding Fourier transforms can be written

$$\tilde{S}_{ab}(\mathbf{t}) = \frac{\lambda_{ab}^3}{\pi^{3/2}}\int d\mathbf{k}\exp(-\lambda^2\mathbf{k}^2) \tag{2.114}$$
$$\times\left[v_{ab}(\mathbf{k},\mathbf{t},\beta) + \frac{(-1)^{2s_a}}{(2s_a+1)}\delta_{ab}v_{ab}(\mathbf{k},\mathbf{t}+2\mathbf{k},\beta)\right] .$$

In Chaps. 4–6, we will use similar expressions for plasmas. For Yukawa gases, we get the first (linear) approximation of the Slater function in the Fourier representation:

$$\tilde{S}_{ab}^{(1)}(\mathbf{t})) = -\beta\frac{4\pi g_{ab}}{t^2+\eta^2}\exp\left(-\frac{1}{2}\lambda_{ab}^2t^2\right)M\left(\frac{1}{2},\frac{3}{2};-\frac{1}{4}\lambda_{ab}^2t^2\right) \tag{2.115}$$
$$+\frac{(-1)^{2s_a}}{(2s_a+1)}4\pi\delta_{ab}\beta g_{ab}\int d\alpha\int\frac{d\mathbf{k}}{\eta^2+(2\mathbf{k}+\mathbf{t})^2}\exp\left\{-\lambda_{ab}^2\left[k^2+\alpha\mathbf{t}\cdot(\mathbf{t}+2\mathbf{k})\right]\right\} ,$$

where M is the Kummer function (or confluent hypergeometric function). We get the second virial function and the free energy as deviations from the Boltzmann term in the form of an expansion with respect to the interaction parameter:

$$F = F^{\mathrm{B}} - \frac{1}{2} V \sum_a n_a \sum_b n_b \cdot \sum_{p \geq 1} \left[\tilde{S}_{ab}^{(p)}(\mathbf{t} = 0) \right] + \cdots . \tag{2.116}$$

Summarizing these calculations, we find the free energy by carrying out the integrations. For the pressure we get

$$\beta p = \sum_a n_a + \sum_a n_a \sum_b n_b \cdot \sum_{p \geq 1} \left[\tilde{S}_{ab}^{(p)}(\mathbf{t} = 0) \right] + \cdots . \tag{2.117}$$

For the exponential potential, we do not have any divergence problems when carrying out the integrals in the series, neither at $r \to 0$, nor at $r \to \infty$. For Yukawa potentials, there are no divergencies up to second order. However, beginning at the third order, we have to include regularization, e.g., a finite α parameter, in order to avoid divergencies in the classical limit at $r = 0$.

In the classical limit $\hbar = 0$, all integrations are elementary. A well known approximation for the excess pressure and the corresponding sound velocity used, e.g., in the theory of the Bose gas by Bogoliubov, is

$$\beta p_{\mathrm{ex}} = +n^2 \tilde{V}(\mathbf{t} = 0) , \qquad c_v = \sqrt{n \tilde{V}(\mathbf{t} = 0)} . \tag{2.118}$$

For the free energy of the classical Yukawa gas, we get

$$F = F^{\mathrm{B}} - V k_{\mathrm{B}} T \sum_a n_a \sum_b n_b \left(-4\pi \frac{\beta g_{ab}}{\eta^2} + 2\pi \frac{\beta^2 g_{ab}^2}{\eta} \right) . \tag{2.119}$$

Including quantum effects leads to more complicated integrals. For the free energy, the result for the first order which provides the Fock and the Hartree–Fock contributions of a mixture with arbitrary spins is

$$\begin{aligned} F^{\mathrm{HF}} &= 2\pi k_{\mathrm{B}} T V \sum_{ab} \left\{ g_{ab} \lambda_{ab}^2 \frac{1}{6\eta^2} \right. \qquad (2.120) \\ &\left. + \frac{\delta_{ab}}{2(2s_a + 1)} \pi g \lambda_{ab}^2 \left[1 - \frac{1}{2} \eta \lambda_{ab} \exp\left(\frac{1}{4} \eta^2 \lambda_{ab}^2 \right) \left[1 - \Phi(\eta \lambda_{ab}/2) \right] \right] \right\} , \end{aligned}$$

$$\Phi(x) = \frac{2}{\sqrt{\pi}} \int_0^x \mathrm{d}t \exp(-t^2/2) . \tag{2.121}$$

For the case $g \to e_a e_b$, $\eta \to 0$, (2.121) reduces to a relation which will be used for plasmas. In the second order of perturbation theory, we find

$$\tilde{S}^{(2)}_{ab}(\mathbf{t}=0) = \frac{\pi}{\eta}\beta^2 g^2_{ab} - \frac{\pi\sqrt{\pi}}{2}\lambda_{ab}\beta^2 g^2_{ab} + \frac{\pi}{\eta}\beta^2 g^2_{ab} f_2(\zeta) \,. \tag{2.122}$$

The first term is purely classical and the second is a quantum contribution. The third is a mixed contribution, given by an integral f_2 that cannot be solved. An estimate may be found by a median value approximation, viz.,

$$f_2(y) \approx \left(1 + \frac{1}{2}y^2\right)\left[\exp(y^2) erfc(y/2) - 1\right] , \tag{2.123}$$

where

$$erfc(z) = \frac{2}{\sqrt{\pi}} \int_z^\infty \mathrm{d}t \exp(-t^2) \,.$$

Putting together these results we arrive at the following expressions for the free energy of Yukawa gases containing the first two orders in the coupling g_{ab} in the direct and the exchange contributions. The 'virial functions' Q for the direct Heisenberg term and E for the exchange term are in principle infinite series in the coupling g:

$$F = F^{\mathrm{id}} - V k_{\mathrm{B}} T \sum_{ab} n_a n_b 2\pi \lambda^3_{ab} \left[Q(\gamma_{ab}, \zeta) \pm \frac{(-1)^{2s_a}}{(2s_a+1)} E(\gamma_{ab}, \zeta) \right] . \tag{2.124}$$

The coupling parameter γ and the decay parameter ζ are defined by

$$\gamma_{ab} = \beta \frac{g_{ab}}{\lambda_{ab}} \,, \qquad \zeta = \eta \lambda_{ab} \,. \tag{2.125}$$

For the Yukawa case, the first expansion terms of the virial functions read

$$Q(\gamma, \zeta) = \frac{2\gamma}{\zeta^2} + \frac{1}{2}\pi^{3/2}\gamma^2 + \frac{\gamma^2}{\zeta}[1 + f_2(\zeta)] \,, \tag{2.126}$$

$$E(\gamma, \zeta) = \frac{\sqrt{\pi}}{4} + \frac{\gamma}{2}\left[1 - \frac{\zeta}{2} erfc(\zeta/2)\right] . \tag{2.127}$$

The other limit $\eta \to 0$ does not exist, but we will show in Chap. 6 that, for small η, a known result for Coulomb plasmas [we mean the second term in (2.122)] arises, which was first obtained for Coulomb systems (Ebeling et al. 1967, 1976).

2.8 Analytical Properties of Thermodynamic Functions of Yukawa Systems

2.8.1 Bound States and Analytical Properties

This section investigates the general structure of the thermodynamic functions of fermion systems with interactions possessing a Fourier transform. Extending the perturbation theories now to arbitrary order, we will also be able to include bound state situations where the coupling is strong and where bound states may play an important role. The first model to be treated is a one-component Yukawa gas. In contrast to the previous section, we use here the notation λ for the strength of the interaction. This is the main parameter determining the analytical properties of the gas. The one-component Yukawa gas plasma (OCP) is used nowadays as an idealized model in which particles interact through a Coulomb potential going as r^{-1} with an additional exponential decay with length $1/\eta$. The model is considered as paradigmatic and is used to approximately describe a wide range of physical systems, especially strongly coupled plasmas of charged particles on a background, such as dusty plasmas (Fortov et al. 2005; Langin et al. 2016; Ott et al. 2014). Note that all physical properties of Yukawa OCPs are expected to be universal in η when expressed in appropriate scaled units (Langin et al. 2016). Most of the existing calculations of thermodynamic functions are purely classical, and indeed for many applications, such as dusty plasmas, quantum effects are not relevant. Nevertheless, we will continue here with the quantum-statistical approach used in the previous section, in order to close an existing gap for Yukawa systems.

This section is mainly dedicated to two-component electro-neutral gaseous systems and is heavily based on the analytic properties of the system with respect to the λ-parameter. The first terms of Taylor expansions in λ were obtained in the previous section. We will show here that a two-component symmetrical Yukawa system is paradigmatic for a whole class of systems with bound states, such as hydrogen plasmas.

We begin with the analytical properties of one-component systems with respect to the interaction parameter λ and then go to a model of a 'two-component plasma' in which the two components have opposite 'charges'. The basic assumption is that the partition function of the system, namely,

$$Tr\Big\{\exp\big[-\beta H_0 - \beta\lambda V(r)\big]\Big\}\,, \tag{2.128}$$

is analytic with respect to the interaction parameter. In spite of the fact that we do not have exact theorems giving the conditions of validity, we are convinced that the class of potentials for which analyticity holds is sufficiently big. At least for the second virial coefficient, which is a partition function for $N = 2$, a proof of analyticity exists for special potentials (Ebeling et al. 1976; Kremp et al. 2005, 1971). This proof is relevant to integrable interaction potentials with the property

$$\int_0^\infty \mathrm{d}r\, r|V(r)| < \infty\,, \qquad \int_0^\infty \mathrm{d}r\, r^2|V(r)| < \infty\,. \tag{2.129}$$

This condition is in particular satisfied by our Yukawa-type interaction potentials.

We assume that the interaction strength λ may have positive or negative values, and study first one-component systems, changing λ from positive to negative values, with special interest in the bound state effects for $\lambda < 0$. Note that the bound state energies of Yukawa systems are not analytic functions of λ. They have been studied in great detail analytically as well as numerically (Rogers et al. 1970; Bessis et al. 1975; Kilimann and Ebeling 1990; Lehmann and Ebeling 1991). For small η, the potential may be approximated by an expansion in η:

$$V(r) = \lambda\frac{1}{r} - \lambda\eta + \frac{1}{2}\lambda\eta^2 r + \cdots\,. \tag{2.130}$$

A linear perturbation theory taking into account the negative sign of the bound state energies leads to the simple estimate

$$E_{sl} = \begin{cases} -\dfrac{\lambda^2\mu}{\hbar^2 s^2} + |\lambda|\eta\,, & \text{if} \quad s^2 < \dfrac{|\lambda|\mu}{\hbar^2\eta}\,, \\ 0\,, & \text{otherwise}\,. \end{cases}$$

In the first approximation, the shifts relative to the Coulomb levels are positive and level-independent. At the point where they cross zero, the bound states disappear. Note that details of the bound states have been carefully investigated numerically and by perturbation methods (Rogers et al. 1970; Kilimann and Ebeling 1990; Lehmann et al. 1996).

A useful formula for approximating the s-states has been given by Dutt and Mukherjee (1981):

$$E_{s0} = -\frac{\lambda^2\mu}{\hbar^2}\left(\frac{1}{s} - \frac{s\eta}{2}\right)^2 + \eta\Big[0.61\exp(-s^2\eta) - 0.11\Big]\,. \tag{2.131}$$

A special and quite relevant property is that the number of bound states is always finite for finite η. For the ground state and the excited states, the shifts may be estimated by perturbation expansion with respect to η (Kilimann and Ebeling 1990). According to Mott's estimate (Mott 1961), the ground state disappears for $\eta^{-1} > 2a_\mathrm{B}$, where a_B is the Bohr radius. Rogers et al. (1970) found $\eta a_\mathrm{B} = 1.19$ numerically for the critical value.

Variational methods are also quite effective using test wave functions (Rogers et al. 1970) of the form

$$\psi(r,\alpha) = (\pi\alpha^3)^{-1/2}\exp(-r/\alpha)\,. \tag{2.132}$$

With simple analytical calculations one may obtain results which agree sufficiently well with Rogers et al. (1970) (see Rogers et al. 1970). Introducing the dimensionless parameter $X = e^2\eta$, several authors have shown that, at certain value X_{sl}, the level with quantum numbers s, l disappears and merges with the continuum (Rogers et al. 1970), Bessis et al. (1975). Following Rogers et al. (1970) we get, for example

$$\begin{aligned}
&X_{10} = 1.19\,,\\
&X_{20} = 0.3103\,,\quad X_{21} = 0.2202\,,\\
&X_{30} = 0.1394\,,\quad X_{31} = 0.1127\,,\quad X_{32} = 0.0913\,,\\
&X_{40} = 0.0788\,,\quad X_{41} = 0.0679\,,\quad X_{42} = 0.0581\,,\quad X_{43} = 0.0498\,.
\end{aligned}$$

Note that several other approximate representations of the levels are known (Kilimann and Ebeling 1990; Rogers et al. 1970). For example Kilimann and Ebeling (1990) use a fourth order polynomial in X to represent the levels. Here we use a more convenient (less precise) approximation by means of a second order polynomial with $a_{\mathrm{B}} = \hbar^2/\mu|\lambda|$ and $X = \eta a_{\mathrm{B}}$:

$$E_{sl}(X) = -\frac{\lambda^2}{2\hbar^2 s^2}\left[1 - 2s^2 X + (2s^2/X_{sl})X^2 - X^2/X_{sl}^2\right], \tag{2.133}$$

with $E_{sl} = 0$ if $X > X_{sl}$. For later applications it may be useful to realize that $X \sim |\lambda|$ and that the X_{sl} are just numbers, which are tabulated.

Let us assume from now on that the bound state levels are known with sufficient accuracy. We will study first one-component systems and then symmetric pseudo-plasmas, bearing in mind the postulated analyticity with respect to λ. Note that we define $B_2(T)$ here in such a way that Bose and Fermi correlations are included, not in the ideal part, but in the second virial coefficient. The two main ways to split the second virial coefficient into two parts are:

- Decomposition into a symmetric and an antisymmetric contribution:

$$B_2(T,\lambda) = B_2^{\mathrm{sy}}(T,\lambda) + B_2^{\mathrm{as}}(T,\lambda)\,. \tag{2.134}$$

with

$$\begin{aligned}
B_2^{\mathrm{sy}}\Big[T,\lambda) &= \frac{1}{2}(B_2(T,\lambda) + B_2(T,-\lambda)\Big]\,,\\
B_2^{\mathrm{as}}\Big[T,\lambda) &= \frac{1}{2}(B_2(T,\lambda) - B_2(T,-\lambda)\Big]\,.
\end{aligned} \tag{2.135}$$

Under the assumption that $B_2(\lambda)$ is an analytic function with the expansion

$$B_2(\lambda) = a_1\lambda + a_2\lambda^2 + a_3\lambda^3 + \cdots\,, \tag{2.136}$$

the even part is given by the sum of the even terms

$$B_2^{\mathrm{sy}}(\lambda) = a_2\lambda^2 + a_4\lambda^4 + \cdots\,. \tag{2.137}$$

The method of decomposition of second virial coefficients into even and odd parts is well known from classical statistical thermodynamics (Falkenhagen 1971). It has been shown that, for strong interactions, the asymptotic is determined only by the symmetric part. This important property will play a role in the following decomposition.

- The second virial coefficient is split into a bound state and a scattering state contribution (Beth and Uhlenbeck 1937):

$$B_2(T,\lambda) = B_2^{\mathrm{bs}} + B_2^{\mathrm{sc}} . \tag{2.138}$$

This decomposition is sometimes complicated. The two constituents are in general nonanalytic functions. From the whole structure of the virial coefficient, it is evident that the bound state part determines the asymptotic for strong coupling. In the Coulomb case, the two methods of decomposition are identical. This is connected with the property $E_s \sim \lambda^2$ mentioned above. For a Yukawa system, the bound states are less simple. In our approximation for the bound states, we get

$$B_2^{\mathrm{bs}}(T;\lambda) = \frac{2}{\sqrt{\pi}}\Lambda^{3/2}\exp(-\eta\beta|\lambda|)\sum_{k=1}^{\infty}\frac{1}{\Gamma(k+1)}\left(\frac{\beta\lambda^2}{\hbar^2}\right)^k \zeta^*(2k-2,\eta) ,$$

where we have introduced a modified zeta-function. For $\eta \to 0$, this is the original Riemann zeta function. We see that for Yukawa potentials the bound state part and the symmetrical part are not equal, but they agree for the Coulomb case $\eta \to 0$. For finite values of η, the symmetric part and the bound state part agree only with respect to the low temperature asymptotic. In the Coulomb case, the two representations are identical. This case will be important in the following.

2.8.2 Exact Virial Coefficient and Thermodynamic Functions

We first investigate the one-component Yukawa gas and write for the second virial coefficient

$$B_2(T;\lambda) = \frac{2}{\sqrt{\pi}}\Lambda^{3/2}\sum_{\ell=0}^{\infty}(2\ell+1)\left[1 \pm \frac{(-1)^\ell}{2s+1}\right] B_\ell(T,\lambda) , \tag{2.139}$$

with $\Lambda = h/\sqrt{2\mu k_B T}$ (μ-reduced mass) and

$$B_l = B_l^{\mathrm{bs}} + B_l^{\mathrm{sc}} , \qquad B_l^{\mathrm{bs}} = \textstyle\sum_{s>l}\left[\exp(-\beta E_{sl}) - 1\right] ,$$
$$B_l^{\mathrm{sc}} = \frac{\beta}{\pi}\int_0^\infty \mathrm{d}E\,\delta_l(E)\exp(-\beta E) . \tag{2.140}$$

Here the (-1) under the sum stems from a partial integration and use of Levinson's theorem (Kraeft et al. 1969; Kremp and Kraeft 1972; Kremp et al. 1971, 2005). With this definition, we define the bound state part in such a way that free particles do not have a bound state part. In the Coulomb limit, this function of λ has the analytical form

$$B_l(\lambda) = \Theta(-\lambda) f^{\mathrm{b}}(|\lambda|) + \mathrm{sign}\,(\lambda) f^{\mathrm{f}}(|\lambda|) \,, \tag{2.141}$$

and correspondingly the l-terms have the expansions

$$B_l(\lambda) = \tilde{a}_{1l}\lambda + \tilde{a}_{2l}\lambda^2 + \tilde{a}_{3l}\lambda^3 + \cdots \,. \tag{2.142}$$

If the system possesses bound states $\lambda < 0$, it follows that that the bound state contribution is even in the interaction parameter λ:

$$f^{\mathrm{b}}(|\lambda|) = 2a_2\lambda^2 + 2a_4\lambda^4 + \cdots \,. \tag{2.143}$$

This is the special case we mentioned above. At lower temperatures, the bound states give the dominant contribution to the total virial coefficient, i.e., we have the asymptotic

$$B_2 \sim B_2^{\mathrm{bs}} = 2a_2\lambda^2 + 2a_4\lambda^4 + \cdots \,. \tag{2.144}$$

In other words, the low temperature behavior of a gas with bound states is dominated by the even coefficients. We will see that, under certain conditions, the asymptotic is even exact. This result is typical and of great value for Coulomb systems. Note that, for Coulomb systems with $\lambda \sim e^2$, the bound state energies are of order $\lambda^2 \sim e^4$. Accordingly, the bound state contributions to the virial functions are even in λ^2.

The analytical properties of the second virial coefficient as a function of the interaction parameter have rather deep consequences. In order to understand some of these consequences, let us study the pseudo-plasma consisting of two Yukawa gases with opposite 'charges' defined by (2.145). We assume a mixture of two types of 'charges' with opposite sign and the same concentration. This defines a kind of binary plasma model which has 50% positive interactions $\lambda > 0$ of charges with the same sign and 50% negative interactions $\lambda < 0$ of charges with opposite sign. We refer to this artificial system, which is a caricature of a Coulomb plasma, as a Yukawa gas plasma:

$$V_{\mathrm{ee}} = V_{\mathrm{ii}} = \frac{e^2}{r}\exp(-\eta r) \,, \quad V_{\mathrm{i.e.}} = -\frac{e^2}{r}\exp(-\eta r) \,. \tag{2.145}$$

Up to the second virial coefficient, the free energy is given by

$$\delta F = -k_{\mathrm{B}} T V \left(n_{\mathrm{e}}^2 B_{\mathrm{ee}} + 2n_{\mathrm{e}} n_{\mathrm{i}} B_{\mathrm{ei}} + n_{\mathrm{i}}^2 B_{\mathrm{ii}}\right) . \tag{2.146}$$

Introducing $\lambda = e^2$, we expand with respect to this parameter and get

$$\delta F = -k_B T V \Big[n_e^2(a_1 e^2 + a_2 e^4 + \cdots) \\ + 2 n_e n_i(-a_1 e^2 + a_2 e^4 + \cdots) + n_i^2 n(a_1 e^2 + a_2 e^4 + \cdots) \Big] . \tag{2.147}$$

With $n_e = n_i = n_0$, it follows that

$$\delta F = -k_B T V 4 n_0^2 \big(a_2 e^4 + a_4 e^8 + \cdots \big) = -V k_B T 2 n_0^2 B_2^{bs} . \tag{2.148}$$

After adding the prefactors and the l-summations we get explicit

$$\delta F = -k_B T V 2 n_0^2 \frac{\Lambda^{3/2}}{\sqrt{\pi}} \sum_{l=0}^{\infty} (2l+1) \sum_{s>l} \big[\exp(-\beta E_{sl}) - 1 \big] . \tag{2.149}$$

This result is exact. No approximations were made, except that we omitted the exchange terms.

Let us now estimate the exchange contributions to the second virial coefficient. This contribution exists only for pairs of identical species, i.e., for positive coupling parameters corresponding to repulsion. Note that there is no compensation with respect to exchange, since exchange contributions are missing for the $\pm$ interactions. We write this symmetry-dependent term in a form which we borrow from known results for Coulomb systems (Kraeft et al. 1986):

$$B_{ab}^S = \delta_{ab} \frac{(-1)^{2s_a} \delta_{ab}}{2(2s_a + 1)} E(\tilde{\xi}_{ab}; \eta) , \qquad \tilde{\xi}_{ab} = -\frac{g_{ab}}{\lambda_{ab}} . \tag{2.150}$$

For Coulomb systems, this function may be calculated using the exact scattering phases. This will be shown in Chap. 4. The result is a function which decays quickly with increasing (negative) coupling. For Yukawa plasmas, we know only the linear term

$$E(g, \eta) = E(\tilde{\xi}) = \frac{\sqrt{\pi}}{4} + \frac{\tilde{\xi}}{2} + \cdots \approx 0.1768 \exp\big(- 2.828 |\tilde{\xi}| \big) . \tag{2.151}$$

This surprising result tells us that the exchange effects disappear exponentially with increasing coupling, and may be neglected in a certain temperature range where $|\tilde{\xi}| \gg 1$.

To sum up we may say that the direct terms are strongly determined by compensation effects and that exchange terms are (presumably) exponentially small. This result is not influenced by drastic approximations. However, we should stress that we have assumed here without comment that the masses of all particles are equal, i.e., all thermal wavelengths are equal, otherwise the cancelation would not be complete and our final result would not be exact. (Note also that we should not confuse the thermal wavelength Λ with the interaction parameter λ.) This surprising exact result tells us the following things about our model plasma:

- The free energy is determined by the bound state contributions alone.
- All contributions from scattering states cancel out completely.
- Only the even powers g^4, g^8, ..., contribute.
- Exchange effects give exponentially small contributions.

In spite of the fact that these statements cannot be directly transferred to the more complicated Coulomb systems, we will see later that the basic features remain valid for the Coulomb case (Ebeling et al. 1976).

The result we got for the free energy may be obtained in an equivalent way for the pressure in the grand canonical ensemble for 'electrons and positrons' interacting via Yukawa forces. As shown by the above considerations, a regularized partition function arises, which is similar to the PBL partition function discussed in Chap. 1. This choice is not an arbitrary invention, but deeply connected with the analyticity of the functions with respect to the interaction parameter. In later chapters, we will see that the choice of the Brillouin–Planck–Larkin partition for Coulomb systems has some deep roots in the analytical properties of thermodynamical and partition functions.

2.9 Strongly Correlated Bose Gases at Low Temperatures

2.9.1 Noninteracting Bose Gases

Here we use a well known method to find expressions for the pressure and the density of noninteracting Bose gases (Huang 1987; Landau and Lifshits 1980, 2001). We consider the total system as a conjunction of subsystems in quantum state k. The energy E, total particle number N, and pressure p are additive:

$$E = \sum_k E_k \,, \qquad N = \sum_k N_k \,, \qquad p = \sum_k p_k \,. \tag{2.152}$$

Note that here the summation index k runs over all the microscopic states and not only over the different energy states. Here the spin states are to be counted as extra quantum states. The partial pressures p_k of subsystems are like partial pressures in a mixture of ideal gases and are therefore additive according to (2.152). We apply the laws of grand canonical ensembles to the subsystem k, where N_k is the particle number in state k. For the partial pressure p_k and the mean particle number $\bar{N}_k$ of the subsystem in state k, we thus get the following expressions (Huang 1987; Landau and Lifshits 1980, 2001):

$$p_k = -\frac{k_B T}{V} \ln(1-q) = -\frac{k_B T}{V} \ln\left[1 - \exp\left(\frac{\mu - \epsilon_k}{k_B T}\right)\right] , \tag{2.153}$$

$$\bar{N}_k = \frac{\partial}{\partial\mu}(p_k V) = \frac{\exp\left(\dfrac{\mu-\epsilon_k}{k_B T}\right)}{1-\exp\left(\dfrac{\mu-\epsilon_k}{k_B T}\right)} = \frac{1}{\exp\left(\dfrac{\epsilon_k-\mu}{k_B T}\right)-1} \,. \tag{2.154}$$

This is the Bose–Einstein distribution for the mean occupation number of the one-particle state k. The condition $N = \sum_k \bar{N}_k$ provides the chemical potential μ and the internal energy follows by summing the energies over all states k.

In order to specify the formula for the energy we need to specify the concrete species with definite spin. We proceed as before, concentrating on the problem of phase transitions and Einstein condensation. However, we have to consider that, for bosons, the same state may be occupied by many particles. In particular, this is true for the lowest state, the ground state of the gas $p_x = p_y = p_z = 0$. We denote the mean occupation of the ground state by $\bar{N}_0$. Following Einstein, a macroscopic occupation of this state is possible. Therefore it is useful to separate the ground state occupation:

$$\bar{N}_0 = \frac{2s+1}{\exp(-\mu/k_B T)-1} = \frac{(2s+1)z}{z-1} \,. \tag{2.155}$$

Here $z = \exp(\beta\mu)$ is the fugacity. By integration in spherical coordinates and using the substitution $p = x\sqrt{2mk_B T}$, it follows that

$$N = \bar{N}_0 + \frac{2s+1}{\Lambda^3} V \frac{4}{\sqrt{\pi}} \int_{0^+}^{\infty} \mathrm{d}x\, x^2 \frac{z}{\mathrm{e}^{x^2}-z} \tag{2.156}$$

$$= \bar{N}_0 + \frac{2s+1}{\Lambda^3} V g_{3/2}(z) \,, \tag{2.157}$$

with the thermal wavelength

$$\Lambda = h/\sqrt{2\pi m k_B T} \,. \tag{2.158}$$

We have introduced here the Bose function

$$g_{3/2}(z) \equiv \frac{4}{\sqrt{\pi}} \int_0^{\infty} \mathrm{d}x\, x^2 \frac{z}{\mathrm{e}^{x^2}-z} = \sum_{l=1}^{\infty} \frac{z^l}{l^{3/2}} \,. \tag{2.159}$$

The series expansion is valid only within the radius of convergence. Since the signs of the contributions do not alternate as for Fermi gases, the convergence is restricted to $0 \le z \le 1$, whence the chemical potential of Bose gases is always negative. For the case $z = 1$, we may express the series in terms of Riemann's zeta-function. According to Riemann's definition we have

$$\zeta(z) = \sum_{\ell=1}^{\infty} \frac{1}{\ell^z} , \qquad g_{3/2}(1) = \sum_{\ell=1}^{\infty} \frac{1}{\ell^{3/2}} = \zeta(3/2) \approx 2.612 .$$

For the pressure, we get in a similar way,

$$pV = -Tk_\mathrm{B}(2s+1)\ln(1-z) + \frac{2s+1}{\Lambda^3} g_{5/2}(z) T k_\mathrm{B} V , \tag{2.160}$$

$$g_{5/2}(z) = \int_0^\infty \mathrm{d}x\, x^2 \ln\left(1 - z\mathrm{e}^{-x^2}\right) = \sum_{l=1}^{\infty} \frac{z^l}{l^{5/2}} . \tag{2.161}$$

We see once again that the series (2.161) converges only for $0 \le z \le 1$, which means that, due to $z = \exp(\mu/k_\mathrm{B}T)$, we always have $\mu < 0$. Bose–Einstein gases cannot have positive chemical potentials. Only photons, which are the limiting case, have zero chemical potential $\mu = 0$, and $z = 1$.

Referring to (2.157), we can draw some conclusions about occupation of the ground state. The function $g_{3/2}(z)$ has a vertical tangent at the point $z = 1$, but remains finite there. We can show that $g_{3/2}(1) \approx 2.612$ and find the relation

$$n = n_0 + \frac{2s+1}{\Lambda^3} g_{3/2}(z) , \tag{2.162}$$

$$n_0 = \frac{\bar{N}_0}{V} = \frac{2s+1}{V} \frac{z}{1-z} . \tag{2.163}$$

We see that, if $n\Lambda^3/(2s+1) > g_{3/2}(1)$ holds, the occupation $(\Lambda^3/V)z/(1-z)$ must have a finite positive value. This means that one value of the momentum $\mathbf{p} = \mathbf{0}$ is occupied by a finite macroscopic part of the bosons which are in the ground state.

This effect was first detected in 1925 by Einstein, after extending Bose's statistics to gases. We call this effect *Bose–Einstein condensation*. Very often the term 'Bose condensation' is used, but as a matter of fact, in Bose's work, we cannot find a single remark about the possibility of a condensation effect. For temperatures and densities outside the region of Bose–Einstein condensation, the ground state is occupied by only an infinitesimal fraction of the particles, whence

$$n = 0 + \frac{2s+1}{\Lambda^3} g_{3/2}(z) . \tag{2.164}$$

From these relations, we may deduce the phase diagram.

So how can one observe Bose–Einstein condensation? If we increase the density in a Bose–Einstein gas, we expect the phase transition at the point where

$$n = \frac{(2s+1)}{\Lambda^3} g_{3/2}(1)$$

is satisfied. Then occupation of the ground state begins with a finite number of gas particles, since from now on n_0 must have a positive value. At this point, the equation of state changes qualitatively. Below the phase transition we have

$$p = k_B T \frac{2s+1}{\Lambda^3} g_{5/2}(z) \ , \tag{2.165}$$

and after condensation has begun

$$pV = -k_B T (2s+1) \ln(1-z) \ . \tag{2.166}$$

Let us now calculate the relative occupation of the ground state $\bar{N}_0/N$ as a function of temperature. For the critical density as function of temperature, we obtain

$$n_{\text{cr}} = \frac{(2s+1)}{\Lambda^3} g_{3/2}(1) \ , \tag{2.167}$$

where $\Lambda = \Lambda(T)$. In this way, for the critical temperature and density, we find

$$T_{\text{cr}} = \frac{h^2 n^{2/3}}{2\pi m k_B \left[(2s+1) g_{3/2}(1)\right]^{2/3}} \ , \qquad \frac{\bar{N}_0}{N} = 1 - \left(\frac{T}{T_{\text{cr}}}\right)^{3/2} \ . \tag{2.168}$$

2.9.2 *Interacting Bose Gases and Phase Transitions*

As we have shown for non-interacting Bose particles, these systems may condense at $T = 0$ into a single quantum state with the minimal energy. For the homogeneous system, all particles then go into the quantum state with $\mathbf{p} = 0$. For interacting Bose gases which we discuss here, the situation is more complicated and we have to distinguish between the region far above the Bose condensation temperature and the region near to it and below it.

In previous sections, we have shown that, at not too low temperatures and not too high densities, boson gases including interactions are well described by virial expansions and the Beth–Uhlenbeck method. We have seen in detail how the thermodynamics of the He gas at $T < 100$ K is determined by the spin of the atoms forming the gas. For He-4 atoms, the spin is $s = 1$, i.e., the atoms are bosons, in contrast to He-3 atoms which have $s = 1/2$ and are therefore fermions. The role of the spins leads in particular to essential differences between the properties of the fermion gas He-3 and the boson gas He-4.

In an oft-cited paper by Kilpatrick et al. (1954), the second virial coefficients of He-3 and He-4 are calculated at closely spaced temperatures over the range

0.3–60 K using the Lennard-Jones 12-6 potential with parameters given by de Boer and Michels and based on a calculation of the phase shifts. A more recent calculation of the virial expansions for an He-4 gas in the region $5 < T < 100$ K was given by Costa et al. (2013). According to Costa et al. (2013), the ideal quantum term contributes approximately 11 percent at 3 K and less than 1 percent at 100 K. The He-4 diatomic molecule possesses a very weak bound state which contributes less than 0.1 percent at 5 K and much less at higher temperatures. To sum up, we may say that the properties of rare He-4 gases may be well described by a density expansion and the Beth–Uhlenbeck representation of the second virial coefficient. However, at still lower temperatures, below 3 K, and higher densities, the properties of He-4 gas can no longer be described by a density expansion and the Beth–Uhlenbeck representation of the second virial coefficient. This is characteristic for nearly all interacting Bose gases. Going down to even lower temperatures and up to higher densities, we reach regions where completely different methods have to be developed (Landau et al. 1980; Huang 2001).

Note that the problems connected with Bose–Einstein condensation and other low-temperature phenomena like superfluidity and superconductivity are of great interest to modern physics. In his pioneering work in 1925, Albert Einstein already pointed out that there could be links to phenomena like superfluidity and superconductivity.

Since cluster expansions were unable to describe the influence of interaction on phenomena near zero temperatures, several alternative methods applicable to that region were developed, e.g., by N.N. Bogoliubov, starting with his pioneering work in 1947 (see Bogoliubov 1947, 1952, 2005–2009). We may also mention beautiful work by Feynman (1954) and Brueckner and Sawada (1957) not forgetting the methods developed by Landau and his school (Landau et al. 1976, 1980).

Bogoliubov and other workers started from the model of a gas having a weakly repulsive potential with Fourier transform $\tilde{V}(p)$ and the idea that weak excitations may be modeled as a nearly ideal gas of bosonic quasi-particles. The ansatz for the Hamiltonian is

$$H = E_0 + \sum_p E(p) n_p \,, \tag{2.169}$$

where the n_p are the occupation numbers of the quasi-particles in momentum space and the $E(p)$ are the energies of elementary excitations, represented by

$$E(p) = \sqrt{\frac{N_0}{V} \frac{p^2}{2m} \tilde{V}(p) + \frac{p^4}{4m^2}} \,. \tag{2.170}$$

In a later paper by Belyaev (1958) the diagrammatic methods of field theory were applied to calculate several first order corrections to the particle spectrum and the ground state energy. The microscopic foundation of the Bogoliubov model is the momentum representation of the Hamiltonian, viz.,

$$\hat{H} = \frac{p^2}{2m} \hat{a}_p{}^\dagger \hat{a}_p + \frac{1}{2V} \sum \tilde{V}(|p_1 - p_1'|) \hat{a}^\dagger_{p_1} \hat{a}^\dagger_{p_2} \hat{a}_{p_2'} \hat{a}_{p_1'} \,, \tag{2.171}$$

where $p_1 + p_2 = p_1' + p_2'$ is stipulated. According to Bogoliubov's fundamental idea, due to the macroscopic occupation of the ground state, the operators a_0 are standard c-numbers and $(a_0)^2 = (a_0^\dagger)^2 = N_0$ is the macroscopic occupation of the ground state. After diagonalization of the Hamiltonian, Bogoliubov and also Brueckner and Sawada get the form

$$E(p) = \sqrt{\frac{N^2}{2V}\tilde{V}(0) + \frac{N_0}{2V}\sum_p \tilde{V}(p) + A_p} \,. \tag{2.172}$$

In the Bogoliubov theory, for small momenta, we get

$$E(p) = v_s p \,, \tag{2.173}$$

where v_s is the speed of sound given by

$$v_s^2 = \rho\tilde{V}(0)/2m \,. \tag{2.174}$$

The linear increase is determined by the speed of sound v_s, so that the corresponding excitations may be interpreted as phonons. Several theories, and in particular Landau's model, then provide a region of excitations corresponding to a minimum which Landau associates with 'rotons'.

Phonon–roton excitations, predicted by Landau and measured experimentally by neutron scattering, result in the critical velocity exceeding the observed values of superfluid flow breakdown by an order of magnitude. The Landau arguments associating the minimum in the energy of excitations with 'rotons' are notoriously difficult to understand. However, what seems to be evident from several theoretical approaches and experimental observations (Bobrov and Trigger 2013) is the existence of a minimum in the characteristic curve $E(p)$ which corresponds in Fig. 2.5 to a characteristic wave number $k_0 \sim 2\,\text{Å}^{-1}$. We notice in this connection that similar excitation curves are also observed in completely different systems consisting of active Brownian particles and dusty plasmas (Trigger and Zagorodny 2004; Trigger et al. 2015).

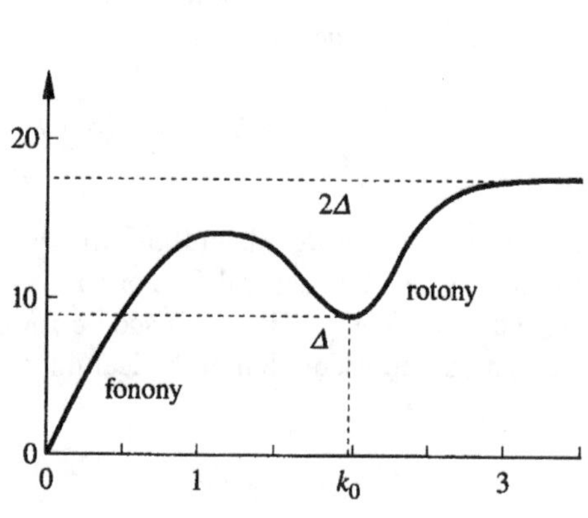

Fig. 2.5 Phonon and roton excitations in Landau's model for the dispersion curve of He-4 according to Huang (1987, 2001). The energy (in kelvin) is represented over the wave number k (in Å^{-1})

Phase Transitions in Low-Temperature Bose Gases

We consider again the noninteracting Bose–Einstein gas. For temperatures below the critical value $T < T_{\rm cr}$, the system consists of particles in the ground state and particles in states with finite momentum $|\mathbf{p}| > 0$. Therefore the momentum distribution is also a mixture of two parts, where the first is proportional to $\delta(\mathbf{p})$. The specific heat is shown in Fig. 2.6 as a function of temperature. The peak at $T = T_{\rm cr}$ is an edge, not a pole, and looks like the Greek letter λ. Therefore the transition is sometimes called the λ-transition.

Later we will discuss the fact that in experiments He-4 gases show a phase transition with a transition temperature $T_{\rm c} \approx 2.17$ K. Assuming a density $\rho \sim 0.15$ g/cm^3 for liquid He, we find a transition temperature around 3.2 K. This is sufficiently close to the experimental value. However, the experimentally observed transition in He-4 is not an 'Einstein condensation'. Analyzing the experiments (Fig. 2.7), we see a typically non-symmetric λ which is not similar to the 'edge' in Fig. 2.6. We may conclude that He-4 does not undergo a proper Einstein condensation, so we may ask which physical systems are the right candidates to manifest the condensation which Einstein predicted in 1925.

The Einstein condition

$$n\frac{h^3}{\sqrt{2\pi m k_{\rm B} T}^3} \approx 2.612(2s+1) \tag{2.175}$$

shows that we need low temperatures, high densities, small masses, and weak interactions. These conditions as formulated by Einstein are extremely difficult to fulfil. One needs very low temperatures in the range of milli- or even micro-kelvins and relatively high densities. Due to the enormous experimental difficulties, a convincing

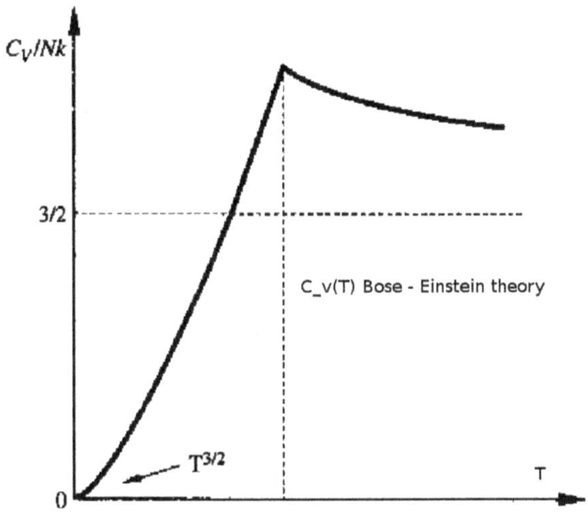

Fig. 2.6 Specific heat of a Bose–Einstein gas as function of temperature, showing a phase transition at $T \approx 3.2$ K

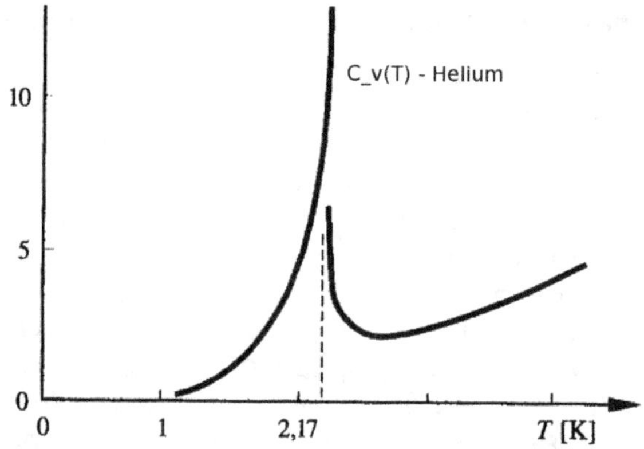

Fig. 2.7 Specific heat of helium-4 (in J/g K) against temperature (in degrees kelvin), with a so-called λ-point at $T \approx 2.17$ K

experimental verification of Einstein's prediction of a condensation of atomic gases at low temperature was confirmed only 70 years after the prediction.

Nowadays many groups around the world work in that field, and several Nobel Prizes have already been attributed. For gases of alkali atoms or atoms of other elements in magnetic traps, Eric Cornell reported in 1995 at the *International Conference of Laser Spectroscopy* on experiments at the University of Boulder which confirmed Einstein's prediction, as well as parallel experiments in the group run by Wolfgang Ketterle at MIT (Davis et al. 1995). For theoretical developments, see, e.g., Pethick and Smith (2001), and also Pitaevskii and Stringari (2003).

We note that Bose–Einstein condensation is related to two remarkable low-temperature phenomena (Greenberger et al. 2009):

- superfluidity, in which a liquid flows with zero friction, and
- superconductivity, in which electrons move through a material with zero electrical resistance.

A hypothetical system of special interest which we will discuss in Chap 10 is excitonic systems. In highly excited semiconductor systems, a Boson gas of electron–hole atoms may be created. These so-called excitons have a very small mass and are ideal candidates for Einstein condensates. A problem is, however, the existence of strong Coulomb interactions which may be responsible for other Coulomb-induced phase transitions (see Chaps. 6 and 8). Due to this difficulty the proof of a proper Bose–Einstein condensate in excitonic systems still seems to be lacking (see also Chap. 10).

Let us look now at helium at low temperatures. At very low temperatures, i.e., the region below 3 K for He-4, the He-4 gas shows a quite characteristic phase transition. This is evident, for example, by looking at measurements of the specific heat of He-4 gas in the region $1 < T < 5$ K (see Fig. 2.7). The data show a so-called lambda curve for the specific heat. The curve is reminiscent of the specific heat curve for an ideal Bose gas which we discussed above, but it differs too much in the details to allow us to conclude that the transition in He-4 is an Einstein transition.

The measurements show the specific heat of He-4 around the phase transition, which is located at

$$T_c = 2.172\ \mathrm{K}\ , \quad n_c \sim 2.16 \times 10^{22}\ \mathrm{cm}^{-3}\ , \quad v_c = 46.2\ \mathrm{Å}^3/\mathrm{atom}\ .$$

Around the phase transition, the He-4 fluid is a non-ideal Bose fluid in which interactions have an essential influence. As we have also seen, the simple Beth–Uhlenbeck theory is unable to describe the behavior near to the lambda phase transition. In the given region He-4 is a non-ideal Bose fluid. For this region, special methods have been developed, such as the Bogoliubov, Brueckner, Sawada, Feynman, and Landau methods (see Bogoliubov and Bogoliubov 1992; Bogoliubov 2005–2009; Pines and Nozieres 1967; Feynman 1954, 1972, Landau and Lifshits 1976, 1977, 1980; Bobrov and Trigger 2013).

References

A.A. Abrikosov, L. Gorkov, I.E. Dzyaloshinskyi, *Methods of quantum field theory in statistical physics, Moscow 1962* (Engl, Pergamon London, 1965)

J.A. Barker, D. Henderson, Perturbation theory and EOS for fluids. J. Chem. Phys. **47**, 2856 (1967)

N. Bessis, G. Bessis, G. Corbel, B. Dakhel, Bound state energies of the exponentially screened potentials. J. Chem. Phys. **63**, 3744–3749 (1975)

S.T. Belyaev, Energy spectrum of a non-ideal Bose gas. JETP **7**, 299 (1958)

E. Beth, G.E. Uhlenbeck, The quantum theory of the non-ideal gas II. Behaviour at low temperatures. Physica **4**, 915–924 (1937)

V.B. Bobrov, S.A. Trigger, Criterion of superfluidity, excitations and heat capacity singularity in superfluid helium. Progr. Theor. Exp. Phys. 04301 (2013)

V.B. Bobrov, A.G. Zagorodny, S.A. Trigger, Coulomb interaction and Bose-Einstein condensate (Russ.). Low Temp. Phys. **41**, 1154–1163 (2015)

N.N. Bogoliubov, On the theory of superfluidity. Akad. Nauk USSR, Phys. Ser. **11**, 1947 (1947)

N.N. Bogoliubov, N.N. Bogoliubov Jr., Introduction to Quantum Statistical Mechanics, New York Gordon and Breach 1992

N.N. Bogoliubov, Collected papers, vol. **1–12**, Fizmatlit Moscow (2005–2009)

K.A. Brueckner, K. Sawada, Bose-Einstein gas with repulsive interactions. Phys. Rev. **106**, 1117 (1957)

E.D.M. Costa, N.H.T. Lemes, M.O. Alves, R.C.O. Sebastiao, J.P. Braga, Quantum second virial coefficient calculation for the 4-He dimer on a recent potential. J. Braz. Chem. Soc. **24**, 363–368 (2013)

C.A. Croxston, *Liquid State Physics, A statistical mechanical introduction* (Cambridge University Press, Cambridge, 2009)

K.B. Davis, M.-O. Mewes, M.R. Andrews, N.J. van Druten, D.S. Durfee, D.M. Kurn, W. Ketterle, Bose-Einstein condensation in a gas of sodium atoms. Phys. Rev. Lett. **75**, 3969 (1995)

H.E. DeWitt, Thermodynamic functions of a partially degenerate, fully ionized gas. J. Math. Phys. **2**, 27 (1961)

H.E. De Witt, Statistical mechanics of high-temperature quantum plasmas beyond the ring approximation. J. Math. Phys. **7**, 616–626 (1966)

R. Dutt, U. Mukherjee, Improved approximation for the bound s-states of the static screened Coulomb potential. Z. Phys. A. Atoms Nucl. **302**, 199–201 (1981)

W. Ebeling, Statistical derivation of the mass action law or interacting gases and plasmas. Physica **73**, 573–584 (1974)

W. Ebeling, H.J. Hoffmann, G. Kelbg, Quantum statistics of high-temperature plasmas in thermodynamic equilibrium. Contr. Plasma Phys. **7**, 233 (1967)

W. Ebeling, W.D. Kraeft, D. Kremp, *Theory of Bound States and Ionisation Equilibrium in Plasmas and Solids* (Akademie, Berlin, 1976); extended Russ. translation Mir, Moscow, 1979

W. Ebeling, W.D. Kraeft, G. Röpke, On the quantum statistics of bound states in Rutherford's model. Ann. Physik (Berlin) **524**, 311–326 (2012)

W. Ebeling, W.D. Kraeft, G. Röpke, Bound states in Coulomb systems. Contr. Plasma Phys. **52**, 7–16 (2012)

R.P. Feynman, Atomic theory of the two-fluid model of liquid helium. Phys. Rev. **94**, 262 (1954)

R.P. Feynman, *Statistical Mechanics* (Benjamin, Reading Mass, 1972)

R.P. Feynman, A.R. Hibbs, *Quantum Mechanics and Path Integrals* (McGraw-Hill, New York, 1965)

M. Fisher, D.M. Zuckerman, Exact thermodynamic formulation of chemical association. J. Chem. Phys. **70**, 100 (1998)

V.E. Fortov, A.V. Ivlev, S.A. Khrapak, A.G. Khrapak, G.E. Morfill, Complex (dusty) plasmas: current status, open issues, perspectives. Phys. Rep. **421**, 1–103 (2005)

H.L. Friedman, *Ionic Solution Theory* (Interscience, New York, London, 1962)

H.L. Friedman, W. Ebeling, Theory of interacting and reacting particles. Rostocker Physikalische Manuskripte **4**, 33–48 (1979)

V.M. Galitskii, Energy spectrum of a Fermi gas. JETP **7**, 104 (1958)

V.M. Galitskii, A.B. Migdal, S.T. Belyaev, Applications of the quantum-field theory methods to the many-body problems. JETP **7**, 96 (1958)

M. Gellman, K.A. Brueckner, Phys. Rev. **106**, 364 (1957)

D. Greenberger, K. Hentschel, F. Weinert (eds.), *Compendium of Quantum Physics Concepts, Experiments, History and Philosophy* (Springer, Berlin, 2009)

D. Henderson, Analytic methods for the Percus-Yevick hard sphere correlation functions. Cond. Matter Phys. **12**, 127–135 (2009)

J.O. Hirschfelder, C.F. Curtiss, R.B. Bird, *Molecular Theory of Gases and Liquids* (Wiley, New York, 1954)

T.L. Hill, *Statistical Mechanics* (McGraw Hill, New York, 1956). Moskva 1960

K. Huang, *Introduction to Statistical Physics* (Taylor and Francis, Abingdon, 2001)

K. Huang, *Statistical Mechanics* (Wiley, New York, 1987)

G. Kelbg, H.J. Hoffmann, Quantenstatistik realer Gase und Plasmen. Ann. Physik **14**, 310–318 (1964)

K. Kilimann, W. Ebeling, Energy gap and line shifts for H-like ions in dense plasmas. Z. Naturforsch. **45a**, 613–617 (1990)

J.E. Kilpatrick, W.E. Keller, E.F. Hammel, N. Metropolis, Second virial coefficient of He 3 and He 4. Phys. Rev. **94**, 1103 (1954)

Y. Kano, N. Mishima, Ann. Phys. N.Y. **51**, 203 (1969)

W.D. Kraeft, W. Ebeling, D. Kremp, Phys. Lett. **29A**, 466 (1969)

W.D. Kraeft, D. Kremp, W. Ebeling, G. Röpke, *Quantum Statistics of Charged Particle Systems* (Berlin and Pergamon Press, New York, Akademie, 1986)

D. Kremp, W.D. Kraeft, Analyticity of the second virial coefficient as a function of the interaction parameter and compensation between bound and scattering states. Phys. Lett. A **38**, 167–168 (1972)

D. Kremp, M. Schlanges, W.D. Kraeft, Quantum Statistics of Nonideal Plasmas (Springer, 2005)

L.D. Landau, E.M. Lifshits, *Quantum mechanics: Non-relativistic theory, Nauka Moska 1976* (In Engl. Pergamon Press, Oxford, 1977)

L.D. Landau, E.M. Lifshits, Statistical Physics (Part I). Nauka Moska (1976), in German Berlin 1979, in Engl. Butterworth–Heinemann 1980

T.K. Langin, T. Strickler, N. Maksimovic, P. McQuillen, T. Pohl, D. Vrinceanu, T.C. Killian, Demonstrating universal scaling for dynamics of Yukawa OCP. Phys. Rev. E **93**, 023201 (2016)

H. Lehmann, W. Ebeling, Wave function properties and shifted energy levels of bound states in a plasma. Z. Naturforsch. **46a**, 583–589 (1991)

F.G. Lether, Analytical expansions and numerical approximations of Fermi-Dirac integrals. J. Sci. Computing **15**, 479 (2000)

E. Montroll, J. Ward, Quantum statistics of interacting particles. Phys. Fluids **1**, 55–72 (1958)

T. Morita, Equation of state of high temperature plasma. Prog. Theor. Phys. (Kyoto) **22**, 757–774 (1959)

N.F. Mott, Phil. Mag. **6**, 287 (1961)

T. Ott, M. Bonitz, L.G. Stanton, M.S. Murillo, Coupling strength in Coulomb and Yukawa one-component plasmas. Phys. Plasma **21**, 113704 (2014)

C.J. Pethick, H. Smith, *Bose-Einstein Condensation in Dilute Gases* (Cambridge University Press, Cambridge, 2001)

D. Pines, P. Nozieres, *The theory of quantum liquids, Benjamin, New York, Amsterdam 1966* (Russ. Transl. Mir, Moskva, 1967)

L.P. Pitaevskii, S. Stringari, *Bose-Einstein Condensation* (Clarendon Press, Oxford, 2003)

J.K. Percus, G.J. Yevick, Analysis of classical statistical mechanics by means of collective coordinates. Phys. Rev. **110**, 1103 (1958)

F. Perrot, M.W.C. Dharma-Wardana, Phys. Rev A **30**, 2619 (1984)

D. Kremp, W.D. Kraeft, W. Ebeling, Quantum statistics of the second virial coefficients and scattering theory. Physica **51**, 164–164 (1971)

D. Pines, P. Nozieres, *The Theory of Quantum Liquids* (Benjamin, New York, 1966); Russ. Translation Mir, Moskva (1967)

F.J. Rogers, H.C. Graboske, D.J. Harwood, Bound eigenstates of the static screened Coulomb potential. Phys. Rev. A **1**, 1577–1586 (1970)

M. Ross, Physics of dense fluids, Proc. Conf. Advances in High Pressure Studies, Corfu 1986, (Preprint UCRL-94997)

M. Ross, Phys. Rev. B **54**, 9589 (1996); **58**, 669 (1998)

J.C. Slater, *Introduction to Chemical Physics* (Mc Graw Hill, New York, 1939)

M. Steinberg, J. Ortner, W. Ebeling, Phys. Rev. E **58**, 3806 (1998)

S.A. Trigger, W. Ebeling, G.J.F. van Heijst, D. Litinski, Analysis of linear and nonlinear conductivity of plasma systems by Fokker–Planck equations, arXiv: 1412.4282 (2014), Plasma Physics (2015)

S.A. Trigger, A.G. Zagorodny, Brownian motion of grains and negative friction in dusty plasmas, Conden. Matter Phys. **7**(3(39)), 629–638 (2004)

G.E. Uhlenbeck, E. Beth, The quantum theory of the non-ideal gas I. Deviations from the classical theory. Physica **3**, 729–745 (1936)

A.A. Vedenov, A.I. Larkin, Equation of state of plasmas. Sov. Phys. JETP **9**, 806–821 (1959)

A. Wasserman, T.J. Buckholtz, H.E. De Witt, Evaluation of some Fermi-Dirac integrals. J. Math. Phys. **11**, 427 (1970)

R. Zimmermann, Many-particle theory of of highly excited semiconductors, Teubner Leipzig (1987)

Chapter 3
Coulomb Systems. Screening and Ionization Problems

The elementary quantum theory of screening effects is described using essentially classical semi-phenomenological methods, and tools developed by Debye–Hückel, Bogoliubov, Mayer, and other pioneers. Following Morita and Kelbg, quantum effects are modeled by nonsingular effective potentials.

3.1 Classical Systems with Coulomb Interactions

3.1.1 Long Range of Coulomb Interactions. Screening

The potential for the force between two charges of species a and b corresponding to pure Coulomb forces is, in Gaussian units, $V_{ab}(r) = e_a e_b / \varepsilon_r r$. In most cases we will assume without comment that the charges are in a vacuum so that $\varepsilon_r = 1$. The Coulomb model is a prototype system for describing real systems. The classical limit exists only for one-component systems of repulsive charges such as an electron gas. The case of opposite charges requires a quantum-statistical treatment. As discussed in Chap. 1, the Coulomb potential has a long-range character since it decays only with $1/r$, where r is the distance. For this reason Coulomb systems have a unique place in statistical mechanics and require special methods, as was seen first by Milner and Debye (see Chap. 1).

Joseph E. Mayer (1904–1983) was the first who saw in 1950 the significance of the Fourier transform of the Coulomb potential for solving problems of screening and cluster theory. The Fourier transform of the Coulomb potential is not well defined, so Mayer proposed to introduce a decay factor $\exp(-\alpha r)$ with some small α and found

W. Ebeling et al., *Quantum Statistics of Dense Gases and Nonideal Plasmas*, Springer Series in Plasma Science and Technology, DOI 10.1007/978-3-319-66637-2_3

$$\tilde{V}_{ab}(\mathbf{t}) = \int_0^\infty \frac{e_a e_b}{t} \sin(rt) \exp(-\alpha r) \mathrm{d}r = 4\pi \frac{e_a e_b}{t^2 + \alpha^2} \,. \tag{3.1}$$

Taking the limit $\alpha \to 0$, we get the Fourier transform of Coulomb potentials. There are some mathematical difficulties connected with extending the integrals over the whole volume, which for a proper definition of the thermodynamic limit has to be infinite $V \to \infty$. It is of some interest for this reason, that there exist other possibilities for the regularization of Coulomb integrals.

A standard way to regularize due to Bohm, Pines, and Nozieres, originally introduced for electron gases in solids, introduces Fourier transforms for finite volumes. In the pioneering work of Bohm and Pines, the Hamiltonian of a system of electrons on a background of uniform neutralizing positive charge is written in the form

$$H = \sum_{i=1}^{N} \frac{p_i^2}{2m} + 2\pi e^2 \sum_{\mathbf{k}}{}' \sum_{i,j} \frac{\exp\left[\mathrm{i}\mathbf{k} \cdot (\mathbf{r}_i - \mathbf{r}_j)\right]}{k^2} - 2\pi e^2 \sum_{k}{}' \frac{1}{k^2} \,, \tag{3.2}$$

where according to these authors (Bohm and Pines 1953):

> The first term corresponds to the kinetic energy of the electrons, the second to their Coulomb interaction and the third to a subtraction of their self energy. The prime in the summations over k denotes a sum in which $k = 0$ is excluded, and this takes into account the uniform background of positive charge of our system, and hence the overall charge neutrality of our system.

Bohm and Pines considered finite volumes and used Fourier series. The discrete Fourier transform is

$$V_{ab}(r) = \frac{1}{V} \sum_{\mathbf{q}} \exp(-\mathrm{i}\mathbf{q} \cdot \mathbf{r}) \tilde{V}_{ab}(\mathbf{q}) \,, \tag{3.3}$$

where the sum is over a discrete spectrum of Fourier wave modes $\mathbf{q}$ in the finite volume. This leads to (Pines and Nozieres 1966)

$$\tilde{V}_{ab}(\mathbf{q}) = \int_V \mathrm{d}(\mathbf{r}) V_{ab}(r) \exp(\mathrm{i}\mathbf{q} \cdot \mathbf{r}) = 4\pi \frac{e_a e_b}{q^2} \left[1 - \delta(\mathbf{q}, 0)\right] . \tag{3.4}$$

This formula shows that for Coulomb systems the term $\mathbf{q} = 0$ plays a special role, since it is completely missing in the series due to the Kronecker symbol. This property is of importance for Coulomb systems (Pines 1961; Pines and Nozieres 1966, 1967; Kremp et al. 2005).

Brown and Yaffé propose a dimensional regularization which starts from a representation of the Coulomb potential for an arbitrary dimension ν (Brown and Yaffé 2001):

$$\tilde{V}_{ab}^{\nu}(\mathbf{r}) = \frac{1}{(2\pi)^\nu} \int d^\nu \mathbf{t} \tilde{V}_{ab}(t) \exp(-\mathrm{i}\mathbf{t} \cdot \mathbf{r}) \,. \tag{3.5}$$

Dimensional continuation is a regularization method which uses the freedom in the dimension ν to make the Coulomb particle self-interactions for $\nu = 3$ vanish. We note that the dimensional regularization method is widely employed in relativistic quantum field calculations, and in particular also in the effective field theory of Coulomb systems (Brown and Yaffé 2001).

In a different approach based on relativistic electrodynamics Bobrov, Trigger, and Zagorodnyi developed in recent papers (Bobrov et al. 2015) a general model for systems with Coulomb interactions. These authors argue that parallel moving particles cannot transform momentum, and that therefore

$$\tilde{V}_{ab}(\mathbf{t} = \mathbf{0}) = 0 \tag{3.6}$$

should hold in any case.

According to our view, the models explained above, say for simplicity the Mayer model, Bohm–Pines model, Brown–Yaffé model, and Bobrov–Trigger model are not equal with respect to the regularization method for treating the 'zero modes', which correspond to particles moving parallel with the same velocity. We have also seen that this problem is connected with the old problem of the divergence of the self-energy of Coulomb particles. Furthermore, we should recall in this respect the following aspects of the whole problem:

- The description of the Coulomb interaction as a static potential force is an incomplete description. Actually, the Coulomb interaction is a dynamic electromagnetic interaction. According to classical electrodynamics, the force between two charges moving parallel with the same velocity v deviates from Coulomb's law, being given by

$$k = \frac{\pm e^2}{\varepsilon_{\mathrm{r}} r^2} \left(1 - \frac{v^2}{c^2}\right) . \tag{3.7}$$

  We see that the force between two equal charges is smaller than predicted by Coulomb's law and goes to zero when the velocity tends to the light velocity c.
- Increasing v up to relativistic velocities, the model of plasma point charges reaches its limits, since the elementary charges may then be decomposed into their constituent quarks.
- In a strict theory, a full description without the need for any regularization should be obtained in the framework of relativistic electrodynamics.

This is outside the scope of the present book, although quark plasmas will be considered in the last few chapters. On the other hand, a fully relativistic statistical thermodynamics exists only in a fragmentary way in the framework of quantum electrodynamics, and will surely lead to great calculational difficulties. In the density and temperature regions studied throughout most of this book, non-relativistic quantum statistics will suffice to describe nearly all essential properties. In order to remain within our framework of nonrelativistic quantum statistics, we will try to be as general as possible, and at least formally model-invariant.

In the first chapter we calculated the mean Coulomb energy by means of Debye's elementary screening concepts and arrived at the formula

$$\langle U_{\mathrm{c}}\rangle = -V(\kappa^3/8\pi)\,, \tag{3.8}$$

where $\kappa = (8\pi\beta ne^2/\varepsilon_{\mathrm{r}})^{1/2}$ is the reciprocal Debye length characterizing screening. Note that, strictly speaking, there is also a linear contribution in e^2. This contribution appears in the classical as well as the quantum-statistical theory. Carrying out the summation first, the linear term results in a zero contribution due to electroneutrality. However, carrying out the integration first, we find a sum of infinities which is indefinite. Here we will omit this term using electroneutrality and, for a discussion of the problems connected with the linear term, refer to Alastuey et al. (2015), Kraeft et al. (2015). The free energy is obtained from the internal energy by Debye's charging procedure, which leads to a prefactor 3/2.

There is an alternative approach which Bogoliubov developed in 1946. His main idea is based on the observation that, in units of the characteristic Debye length $r_{\mathrm{D}} = \kappa^{-1}$, the pair distribution and the correlation may be represented by a perturbation series, starting with

$$F_{ab}(r/r_{\mathrm{D}}) = 1 - \mu_{\mathrm{B}}\frac{\exp(-r/r_{\mathrm{D}})}{(r/r_{\mathrm{D}})}\,, \qquad \mu_{\mathrm{B}} = \frac{e^2}{k_{\mathrm{B}}Tr_{\mathrm{D}}}\,. \tag{3.9}$$

This suggests that μ_{B} is the proper expansion parameter. By expanding with respect to μ_{B}, Bogoliubov solved the called BBGKY hierarchy (Bogoliubov 1946). Details of this solution were given in Falkenhagen (1971). The screened potential satisfies the Bogoliubov integral equation

$$V_{ab}^{\mathrm{D}}(1,2) = V_{ab}(1,2) - \beta\sum_c n_c \int d\mathbf{r}_3\, V_{ac}(1,3)\, V_{bc}^{\mathrm{D}}(2,3)\,. \tag{3.10}$$

The physical meaning is that the screened potential V_{ab}^{D} is the sum of a direct contribution and a contribution to the potential stemming from the charges in the vicinity, which are distributed, as assumed by Debye, according to a linearized Boltzmann law. The Bogoliubov integral equation may be obtained from the Ornstein–Zernike equation by the substitutions $h(r) \to -\beta V^{\mathrm{D}}(r)$ and $c(r) \to \beta V(r)$.

The general solution of (3.10) is similar to the solution of the Ornstein–Zernike equation found by transformation to the Fourier space. Using Parseval's theorem, which transforms integrals of products to products of Fourier transforms, we find a closed algebraic equation:

$$\tilde{V}_{ab}^{\mathrm{D}}(\mathbf{k}) = \tilde{V}_{ab}(\mathbf{k}) + \sum_c n_c \tilde{V}_{ac}(\mathbf{k})\big[-\beta\tilde{V}_{bc}^{\mathrm{D}}(\mathbf{k})\big]\,. \tag{3.11}$$

The solution reads

$$\tilde{V}_{ab}^{\mathrm{D}} = \frac{e_a e_b}{k^2 + \kappa^2} , \qquad V_{ab}^{\mathrm{D}}(r) = \frac{e_a e_b \exp(-\kappa r)}{r} . \tag{3.12}$$

Finally, for the classical free energy, using the charging procedure again and Debye's result for the free energy and the classical chemical potential of particles of species a , we obtain

$$\mu_{\mathrm{a}} = \mu_{\mathrm{a}}^{\mathrm{id}} - \frac{e_a^2}{\varepsilon_{\mathrm{r}} k_{\mathrm{B}} T r_{\mathrm{D}}} , \qquad r_{\mathrm{D}} = 1/\kappa , \qquad \kappa^2 = 4\pi\beta \sum n_{\mathrm{a}} e_a^2 . \tag{3.13}$$

There is a different procedure for deriving the free energy due to Mayer and Haga, based on the summation of chain or ring diagrams of Coulomb lines. The Mayer method reproduces Debye's result in the classical low-density limit.

3.1.2 *Plasma Parameter Expansions and Prototype Models*

A prototype model for Coulomb systems is the one-component plasma (OCP). This is an artificial system consisting of negative charges, such as *electrons* with density n and charge $-e$, where e is the proton charge on a neutralizing smeared-out background of positive charges. Alternatively, we may consider positive charges on a background of negative charges. Since OCP systems are a prototype case, we will derive the basic results for the distribution functions and the thermodynamic functions now in a different and more systematic way, following the classic work of Bogoliubov performed in 1946 (Bogoliubov 2005–2009).

Bogoliubov proposed a procedure based on a systematic expansion with respect to a small parameter. He took into account the fact that the basic distance in a Coulomb system is the Debye length $r_{\mathrm{D}} = \kappa^{-1}$, and therefore introduced the dimensionless parameter $\mu_{\mathrm{B}} = e^2/r_{\mathrm{D}} k_{\mathrm{B}} T$, which is the Coulomb energy at the Debye distance in relation to the thermal energy. In order to check for the convergence of expansions with respect to μ_{B}, let us consider several higher order terms (Falkenhagen 1971). We consider a one-component plasma (OCP) on a neutralizing background.

In the first approximation with respect to the plasma parameter we obtained above, with a first order correlation function g_1 which is given by the Debye potential, the first order pair distribution is

$$F_2(r) = 1 + g_{\mathrm{B}}(r) , \qquad g_{\mathrm{B}}(r) = -\frac{e^2}{k_{\mathrm{B}} T r} \exp(-\kappa r) . \tag{3.14}$$

As we have shown, this approximation yields the limiting laws in thermodynamics. In order to proceed to the next order we have to study the 3-particle correlations in Bogoliubov's hierarchy (Schmitz 1966). His idea was to expand the correlation function g_B in a series with respect to the so-called plasma parameter $\mu_\mathrm{B} = e^2/k_\mathrm{B}Tr_\mathrm{D}$ in a systematic way

$$g_\mathrm{B} = g_1 + g_2 + g_3 + \cdots , \qquad g_n^k = O(\mu_\mathrm{B}^k) . \tag{3.15}$$

First, we write the classical limit of the equations for binary distributions with the first approximation

$$k_\mathrm{B}T g_1(1,2) + V(1,2) + n\int d\mathbf{r}_3 V_{ac}(1,3) g_1(2,3) = 0 . \tag{3.16}$$

From the hierarchy, the correlation function is

$$\begin{aligned} k_\mathrm{B}T\frac{\partial g_\mathrm{B}}{\partial \mathbf{r}_1} &+ 1 + g(1,2)\frac{\partial V(1,2)}{\partial \mathbf{r}_1} \\ &+ n\int d\mathbf{r}_3 \frac{\partial V(1,3)}{\partial \mathbf{r}_1}\left[g_\mathrm{B}(1,2) + g_\mathrm{B}(1,3) + g_\mathrm{B}(2,3) + g_3(1,2,3)\right] = 0 . \end{aligned} \tag{3.17}$$

Here $g_3(1,2,3)$ denotes a higher order correlation function. By starting an iteration with the first order $g_\mathrm{B} = g_1$, in the second order, we get the pair distribution (Schmitz 1966, Falkenhagen 1971)

$$\begin{aligned} F_2(1,2) = 1 &+ g_1(1,2) + \frac{1}{2}g_1^2(1,2) \\ &+ n\int d\mathbf{r}_3\left[g_1(1,3)^2 g_1(2,3) + g_1(1,3)g_1(2,3)^2\right] + \cdots . \end{aligned} \tag{3.18}$$

We remember that the first order solution was obtained by solving the integral equation in Fourier space, which reads in k-space

$$V_{ab}^\mathrm{D}(k) = \frac{V(k)}{1+\kappa^2 V(k)} , \qquad V(k) = \frac{4\pi e^2}{k^2} . \tag{3.19}$$

In this way, we get

$$V^\mathrm{D}(k) = \frac{4\pi e^2}{k^2+\kappa^2} , \qquad V_{ab}^\mathrm{D}(r) = -\frac{e_a e_b}{k_\mathrm{B}Tr}\exp(-\kappa r) . \tag{3.20}$$

There exist two basic prototype models which are not entirely realistic, but provide at least good approximations and are easier from the mathematical point of view.

(i) One-Component Plasmas (OCP) Model

In this model we study a one-component system of discrete charges on a smeared-out countercharge background. For convenience, we generally speak about an electron gas and consider the positive charges as a smeared-out background, although the situation is perfectly symmetric. The OCP-model may be treated in a completely classical way. In recent work, a very good parametrization of the correlation functions of the OCP has been given which is related to the Yukawa system (Desbiens et al. 2017). We will show later that, under many conditions, the quantum corrections are small and may be obtained by expansions with respect to $\hbar^2$ (Hoffmann 1968, Pollock and Hansen 1973).

(ii) Symmetric Two-Component Systems and RMA

Real plasmas have at least two components and consist of negative charges $e_- = -e$ with density n_- and positive charges $e_+ = e$ with density n_+. Due to electroneutrality, we have $n_+ = n_-$. As has been known since Bohr's theory, such plasmas of point charges need a quantum treatment, because of the singularity of the Coulomb force in $r = 0$. As we also know from Bohr's theory, which we explained in Chap. 1, the results depend crucially on the reduced mass, which is the mass corresponding to the relative electron–ion motion

$$\mu = m_{\rm ie} = \frac{m_+ m_-}{m_+ + m_-} \,. \tag{3.21}$$

In quantum statistical theory, the other relative masses m_{++} and m_{--} also come in. This makes the system very complicated and analytical calculations are often tedious or even impossible. On the other hand, only the reduced mass, the *Bohr mass* $\mu = m_{+-} = m_{\rm ie}$, is of any real relevance. The role of the other masses $m_{++} = m_{--}$ is small since collisions $++$ and $--$ occur rather seldom, except at very high temperatures, where interaction effects are generally very small anyway.

For this reason, the prototype symmetric two-component plasma model has been developed and is gaining importance in calculations (Lehmann and Ebeling 1996, Ebeling 2016). What makes the mass-symmetric TSP model so simple is the fact that, for reasons of symmetry and electroneutrality, all contributions of order e^2, e^6, e^{10}, etc., to thermodynamic functions cancel out, and only the orders e^4, e^8, e^{12}, and so on, contribute. In a certain approximation, one can neglect the mass differences and work with the reduced mass μ alone. This is the so-called reduced mass approximation (RMA), which leads to a drastic simplification of most mathematical problems in the quantum statistics of plasmas. The RMA is strictly true only for electron–positron plasmas, but provides an excellent approximation for nearly all two-component hydrogen-like plasmas, as long as the temperature is below 10^5 K.

3.1.3 OCPs and the Ion Sphere Model

As we have shown, the radial distribution for the OCP reads in lowest order of the Bogoliubov plasma parameter

$$F_2(r) = 1 - \mu_{\rm B} \frac{\exp(-\kappa r)}{\kappa r} \,. \tag{3.22}$$

For a second order calculation of the classical correlations of the electron gas, we refer to Schmitz (1966), Cohen and Murphy (1969), Krienke et al. (1975), Ortner (1999), and Kahlbaum (2000). To second order, the radial distribution for the OCP reads explicitly

$$F_2(r) = 1 - \mu_B \left\{ 1 + \mu_B \left[\frac{1}{6} + \frac{8}{3} \ln(3) \right] \frac{\exp(-\kappa r)}{\kappa r} - \frac{1}{8} \ln(3) \mu_B \exp(-\kappa r) - \frac{1}{2} \mu_B \frac{\exp(-2\kappa r)}{(\kappa r)^2} \right\} + \cdots . \tag{3.23}$$

Note that the additional terms completing the Debye-like term lead to logarithmic dependencies in the thermodynamic functions.

According to a theorem by Oskar Klein (1894–1977), who was one of the most gifted students of Svante Arrhenius and Niels Bohr, the excess parts of the thermodynamic functions depend only on one dimensionless parameter, which is either Γ or the Bogoliubov parameter $\mu_B = \sqrt{3}\Gamma^{3/2}$ (Klein 1919):

$$-\frac{F_{ex}}{N k_B T} = F_0(\Gamma) = F_1(\mu_B) , \qquad -\frac{U_{ex}}{N k_B T} = U_0(\Gamma) = U_1(\mu_B) . \tag{3.24}$$

As we have seen, for small plasma parameters the functions may be represented by Taylor series in the plasma parameters, which also contain logarithmic terms.

According to Cohen and Murphy (1969) and Ortner (1999), the internal energy per particle, which is classically equal to the Coulomb energy per particle, may be represented by the series

$$u_c = p_0 \Gamma^{3/2} + p_1 \Gamma^3 \ln \Gamma + p_2 \Gamma^3 + p_3 \Gamma^{9/2} \ln \Gamma + p_4 \Gamma^{9/2} + p_5 \Gamma^6 \ln^2 \Gamma + p_6 \ln \Gamma + p_7 \Gamma^6 + \cdots , \tag{3.25}$$

with the constants

$$p_0 = -.8660, \quad p_1 = -1.125, \quad p_2 = 1.3701, \quad p_3 = -.9243, \quad p_4 = .2350, \quad p_6 = -2.0959, \quad p_7 = +.0676. \tag{3.26}$$

The full analytical structure of this rather complex series is not known. However, it is very likely to be true that, as stated by Wigner, Onsager, Salpeter, DeWitt, and others, the asymptotic for large Γ is linear in Γ (Wigner 1934, Onsager 1937). Evidently, for larger Γ, the convergence of Ortner's series is rather slow and one should use Padé approximations (see, e.g., Kahlbaum in Kraeft and Schlanges 1996), or closed formulae, as proposed by Nordholm (1984) and Kaklyugin (1985).

Ion Sphere Theory and Onsager Bound for OCPs

In the last part, we discussed expansions for the energy of one-component plasmas with respect to plasma parameters. Due to the slow convergence of these expansions, the question of closed expressions also covering high densities seems urgent. As already mentioned, the OCP is one of the most important prototype systems for plasmas. We introduced the theorem due to Oskar Klein stating that the thermodynamic excess functions may depend only on one parameter: either we can take Γ or $\mu_{\mathrm{B}} = \sqrt{3}\Gamma^{3/2}$. The problem is to find just one analytic function, say of the Bogoliubov parameter μ_{B}, about which we already know a lot of properties (see the previous section).

In this respect, a very useful class of models is based on the idea of ion spheres, which goes back to Lars Onsager, Edwin Salpeter (1924–2008), Yasha Rosenfeld (1948–2002), and others. The basic idea of these models is, according to Onsager (1939), that the positive charge cloud around a negative charge $-e$ is concentrated in a sphere of Wigner radius d given by $4\pi d^3/3 = V/N = 1/n$. This sphere is sometimes called the Onsager atom. Onsager estimated the electric energy of this charged sphere and hence found a lower bound for the electrostatic energy of the OCP which has the form of a Madelung energy (Onsager 1939):

$$\frac{U_{\mathrm{c}}}{Nk_{\mathrm{B}}T} > -\frac{9}{10}\Gamma \,. \tag{3.27}$$

In order to extend the lower bound to formulae applicable for high densities, we follow an elegant derivation by Nordholm (1984) and a preprint (Ebeling 1998). The Nordholm theory is based on the study of the charge density $\sigma(r)$ in the vicinity of a central charge which is an electron with charge $-e$ in an OCP. Due to charge neutrality, the following condition holds for the charge density $\sigma(r)$:

$$-e = 4\pi \int_0^\infty \mathrm{d}r\, r^2 \sigma(r) \,. \tag{3.28}$$

In uniform plasmas, the charge density depends on the distance $r = |\mathbf{r}_1 - \mathbf{r}_2|$ and is related to the charge–charge correlator

$$Q(r) = S_{zz}(r) = e^2 \langle \delta N(\mathbf{r}_1) \delta N(\mathbf{r}_2) \rangle \,. \tag{3.29}$$

In contrast to the Onsager model, Nordholm admits a charge density outside the 'Onsager sphere', which in Nordholm's theory gets a free radius h. He uses the well-known fact that, at small densities, the Debye–Hückel theory predicts the charge density

$$\sigma(r) = -e\frac{\kappa^2}{4\pi r} \exp(-\kappa r) \,. \tag{3.30}$$

At high densities, the opposite charge should be concentrated in a small hole, the ion sphere, around the central charge, whose radius is of the order of the average distance d of the charges. Nordholm assumes that the charge density is constant and has the value $-ne\zeta$ inside a certain radius h, while decaying exponentially outside:

$$\sigma(r) = \begin{cases} -ne\zeta\,, & \text{if} \quad r < h\,, \\ -ne\zeta \dfrac{h}{r} \exp(\kappa h - \kappa r)\,, & \text{otherwise}\,. \end{cases} \tag{3.31}$$

In Nordholm's theory, the radius h and the charge parameter ζ are free constants (note that Nordholm orginally assumed $\zeta = 1$). The Coulomb energy per charge follows by integration:

$$\frac{-U_{\mathrm{c}}}{Nk_{\mathrm{B}}T} = \frac{\zeta}{2}\left[(\kappa h) + \frac{1}{2}(\kappa h)^2\right]. \tag{3.32}$$

The complete screening condition (3.28) enables us to exclude the radius h of the ion sphere and leads to a third order polynomial in h, viz.,

$$\zeta\left(\kappa^3 h^3 + 3\kappa^2 h^2\right) + 3\kappa h - \kappa^3 d^3 = 0\,. \tag{3.33}$$

The simple structure of the prefactors allows the explicit solution

$$h = \frac{1}{\kappa}\left\{\left[1 + \frac{1}{\zeta}(\kappa d)^3\right]^{1/3} - 1\right\}. \tag{3.34}$$

In order to calculate the Coulomb energy, we need the expression

$$(1 + \kappa h)^2 = \left(1 + \frac{d^3\kappa^3}{\zeta}\right)^{2/3} = \left(1 + \frac{3^{3/2}}{\zeta}\Gamma^{3/2}\right)^{2/3} = \left(1 + \frac{3}{\zeta}\mu\right)^{2/3}. \tag{3.35}$$

In this way, we get for the Coulomb energy per charge

$$\frac{U_{\mathrm{c}}}{Nk_{\mathrm{B}}T} = \frac{\zeta}{4}\left[\left(1 + \frac{3^{3/2}}{\zeta}\Gamma^{3/2}\right)^{2/3} - 1\right] = \frac{\zeta}{4}\left[\left(1 + \frac{3r}{\zeta}\mu_{\mathrm{B}}\right)^{2/3} - 1\right]. \tag{3.36}$$

In the case $\mu_{\mathrm{B}} \ll 1$, this leads to the Debye limit (independent of ζ):

$$\frac{U_{\mathrm{c}}}{Nk_{\mathrm{B}}T} = -\frac{\mu_{\mathrm{B}}}{2}\,, \tag{3.37}$$

and for $\Gamma \gg 1$, we get the asymptotic

$$\frac{U_{\mathrm{c}}}{Nk_{\mathrm{B}}T} = -\frac{3}{4}\zeta^{1/3}\Gamma\,. \tag{3.38}$$

We fit the constant $3\zeta^{1/3}/4$ to the Onsager bound of the Coulomb energy $-9\Gamma/10$ which has to be respected, and get

$$\frac{3}{4}\zeta^{1/3} = \frac{9}{10}\,, \qquad \zeta = \frac{216}{125} = 1.728\,. \tag{3.39}$$

With this value, an excellent fit of the numerical data was obtained by DeWitt (1976) in the whole range of available data. We note that the original Nordholm theory with $\zeta = 1$ leads to a slope which is too small by about 10 percent and is therefore unable to reproduce the numerical data.

The Nordholm theory provides a closed expression for the thermodynamic function of the OCP which is numerically very good in many respects. But the low density expansion is not correctly reproduced, and in particular the logarithmic contributions to the thermodynamics are not included, and this is a major disadvantage. However, this can be resolved by using the second order distribution function (3.24) instead of the Debye distribution, starting from the following expressions for the charge distribution (see Ebeling 1998):

$$\sigma(r) = -ne\left(\zeta - \alpha\frac{r^2}{h^2}\right)\,, \quad \text{if } r < h\,, \tag{3.40}$$

$$\begin{aligned}\sigma(r) = -ne\mu_{\mathrm{B}}\Bigg\{&1 + \mu_{\mathrm{B}}\left[\frac{1}{6} + \frac{8}{3}\ln(3)\right]\frac{\exp(-\kappa r)}{\kappa r} \\ &- \frac{1}{8}\ln(3)\mu_{\mathrm{B}}\exp(-\kappa r) - \frac{1}{2}\mu_{\mathrm{B}}\frac{\exp(-2\kappa r)}{(\kappa r)^2}\Bigg\}\,, \quad \text{if } r > h\,.\end{aligned} \tag{3.41}$$

This way the approximation can be made compatible with the full expansion by Cohen, Murphy, and Ortner (3.25). The price to pay is a rather complicated expression for the free parameters ζ and α. Later we will also give alternative closed expressions in the framework of collective variables.

3.2 Charged Hard Sphere Systems

3.2.1 Debye–Hückel Approximation

In the theory of ionic fluids, the model of charged hard spheres involving two kinds of charges with opposite sign is of similar importance to the hard sphere model for neutral fluids. Inside and outside the radius a, the potential is given by

$$V(r) = \begin{cases} \infty , & \text{if} \quad r < a , \\ \pm \dfrac{e^2}{\varepsilon_r r} , & \text{if} \quad r > a . \end{cases} \tag{3.42}$$

The symmetry of the short-range forces and the electroneutrality lead to many simplifications which allow analytical calculations (Falkenhagen et al. 1971). Debye and Hückel studied this model of hard charged spheres and used it to find a Poisson–Boltzmann equation for the screened potential:

$$\Delta\Phi(\mathbf{r}) = -4\pi en \Big\{ e \exp\big[- e\beta\Phi(\mathbf{r}) \big] - e \exp\big[+ e\beta\Phi(\mathbf{r}) \big] \Big\} . \tag{3.43}$$

Finally, by linearization we get the Debye equation

$$\Delta\Phi(\mathbf{r}) = \kappa^2 \Phi(\mathbf{r}) , \qquad \kappa^2 = 8\pi\beta n e^2/\varepsilon_r , \tag{3.44}$$

with the known solution

$$\Phi(r) = \frac{e \exp(\kappa a - \kappa r)}{\varepsilon_r (1 + \kappa a) r} , \tag{3.45}$$

$$\langle V_c \rangle = -\frac{V k_B T \kappa^3}{8\pi} \frac{1}{1 + \kappa a} , \qquad \mu_{ex} = -\frac{e^2 \kappa}{\varepsilon_r (1 + \kappa a)} . \tag{3.46}$$

Here κ is the reciprocal Debye length as defined earlier, but now taken for two-component systems. The rather simple Debye–Hückel approximation (DHA) is a quite powerful model. The pressure of the ionic fluid consists of ideal Coulomb and neutral system contributions, which in the first approximation is a hard core system:

$$p = p_{id} + p_{coul} + p_{hc} . \tag{3.47}$$

The ideal pressure is known. The Coulomb contribution is

$$p_{coul} = -\frac{k_B T}{24\pi} \kappa^3 \phi\big(\kappa a(T)\big) . \tag{3.48}$$

Here the Debye–Hückel pressure function $\phi(x)$ is defined by

$$\phi(x) = \frac{3}{x^3} \left[1 + x - \frac{1}{1 + x} - 2 \log(1 + x) \right] , \quad x = \kappa a(T) . \tag{3.49}$$

The corresponding expression for the free energy is

$$F = F_{id} + F_{hc} - k_B T V \frac{\kappa^3}{12\pi} \tau(\kappa a) . \tag{3.50}$$

Here the first two terms represent the free energy of the neutral component. The DHA expression for the Coulomb contribution of a mixture of charges all having the same diameter a reads

$$\tau(x) = \frac{3}{x^3}\left[\ln(1+x) - x + \frac{x^2}{2}\right] = 1 - \frac{3}{4}x + \frac{3}{5}x^2 - \cdots . \tag{3.51}$$

We see that, in this approximation, the Coulomb contribution is of order e^4, although at the end it is reduced by a factor of a half through the screening parameter κ. However, this is a rather technical aspect. From the physical point of view, the Debye contribution is basically to be considered as a second order contributionto the interaction, i.e., of order $O(e^4)$.

3.2.2 Mean Spherical and Hypernetted Chain Approximations

The mean spherical approximation (MSA) is in some sense a generalization of the Debye–Hückel theory, which is to be considered as a refinement of the second order perturbation in the interaction $O(e^4)$. The MSA has in fact a different physical origin, but leads to an equivalent mathematical structure. The idea of the mean sphere approximation can be described briefly as follows.

We start from the Ornstein–Zernike relation between the correlation function $h_{ab}(r)$ and the direct correlation $c_{ab}(r)$, which reads for a mixture

$$h_{ab}(r_{12}) = c_{ab}(r_{12}) + \sum_c n_c \int d\mathbf{r}_3 c(r_{13}) h(r_{23}) . \tag{3.52}$$

The direct correlation function has the property

$$c_{ab}(r_{12}) \longrightarrow \beta V_{ab}(r_{12}) , \quad \text{if} \quad |r_{12}| \longrightarrow \infty . \tag{3.53}$$

Waisman and Lebowitz (1972) proposed to extend this relation down to $r = a$, but at the same time to assume that the pair correlations in the region inside the sphere are zero:

$$c_{ab}(r) = \beta \frac{e_a e_b}{r} , \quad \text{if} \quad r > a , \qquad h_{ab}(r) = -1 , \quad \text{if} \quad r < a . \tag{3.54}$$

This provides us a closed system of equations which is solvable. In this mean sphere approximation, Waisman and Lebowitz (1972) derived explicit solutions, and in particular the following expression for the free energy function:

$$\tau(x) = \frac{3}{x^3}\left[2 + 6x + 3x^2 - 2(1+2x)^{3/2}\right] = 1 - \frac{3}{4}x + \cdots . \tag{3.55}$$

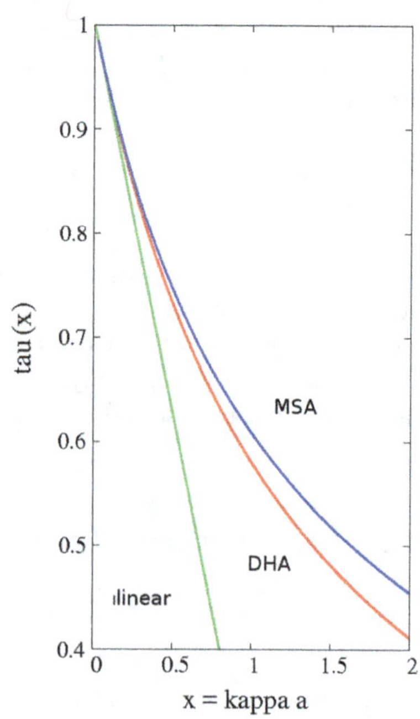

Fig. 3.1 Comparison of the characteristic function in the free energy of hard charged spheres in the MSA (*upper blue curve*), the DHA (*middle red curve*), and the linear approximation (*lower green curve*). We see that the agreement at low x is good, while at higher values the agreement is only qualitative, and the linear approximation is clearly insufficient

The Taylor expansion of the new τ function agrees in the first linear order with the Debye–Hückel function, but differences already appear in the second order (see Fig. 3.1). In any case, the two approximations belong to the same class of analytical functions. Because of the better agreement with numerical data for charged hard sphere systems, the MSA should be considered as a better founded theory (Blum 1975). A combination of the MSA with mass action laws, as well as an extension to non-equilibrium situations, has been considered in (Ebeling and Grigo 1982). The mixed case of classical MSA ions imbedded in a sea of quantum electrons has been considered in (Ebeling, Förster, Fortov et al. 1991).

In the hypernetted chain approximation (HNC), a different closure relation is proposed. In order to explain this scheme, for a mixture of charges $e_a, e_b, e_c, \ldots,$ with number densities $n_a, n_b, n_c, \ldots,$ we introduce a function

$$y_{ab} = \exp\left[\beta V_{ab}(r)\right] g_{ab}(r) \,. \tag{3.56}$$

Then, as a closure relation to solve the Ornstein–Zernike equation, the hypernetted chain equation is

$$\begin{aligned} \ln y_{ab}(r_{12}) &= \ln g_{ab}(r_{12}) + \beta V_{ab}(r_{12}) \\ &= \rho \int \left[h_{ac}(r_{13}) - \ln g_{ac}(r_{13}) - \beta U_{ac}(r_{13})\right] h_{bc}(r_{23}) \mathrm{d}\mathbf{r_3} \,. \end{aligned} \tag{3.57}$$

Rasaiah and Friedman generalized the HNC approximation and showed that it works better for Coulomb systems than the PY approximation. This is evidently connected with the fact that the long-range character of the Coulomb tails is better represented by HNC than by PY approximations (Rasaiah and Friedman 1969).

3.3 Quantum Debye–Hückel Theory of Screening

3.3.1 Quantum Debye–Hückel Approximation

The idea of effective potentials comes from the classical electrolyte theory of systems where short-range interactions supplement the Coulomb forces (Falkenhagen 1971). In the classical work of Debye and Hückel, we already find in the first papers an extension of the theory to the model of hard charged spheres with radius a. In the simplest approximation the effect of a hard charged sphere is just to exclude the region from $r = 0$ to $r = a$ from all integrations in normalization and calculation of the Coulomb energy.

Following Kelbg (1963, 1964), the main quantum effect on the correlations between charges is the 'softening' of the Coulomb charges at small distances due to Heisenberg effects. The strict theory due to Kelbg, and more recent extensions, will be given in Chap. 4, where we will start with model studies of the quantum effects at small distances between the particles, based on the Slater–Morita method. In order to introduce Kelbg's idea, we use here a simple semi-phenomenological approach following Ebeling et al. (1976) and starting from the basic relations of quantum mechanics, and in particular Heisenberg's uncertainty relation. Later we will optimize the model parameters in comparison with the strict theory.

The main quantum effect at temperatures which are of interest for ionization phenomena, i.e., $10^3 < T < 10^5$ K, comes from the Heisenberg uncertainty relation between coordinates and momenta:

$$\Delta r \Delta p \geq \hbar/2 \ .$$

Since in a thermal plasma the momentum of the charges is uncertain due to stochastic thermal motions, the location of point charges should also be distributed. In order to estimate the thermal uncertainty of the momentum, we start from the mean of the absolute value $|v_{\rm th}| = \sqrt{2k_{\rm B}T/\pi m}$. According to the Heisenberg condition, this gives $\delta p \sim 2|v_{\rm th}|$ for the uncertainty of the thermal momentum. In this way we estimate the momentum uncertainty of a particle participating in the thermal agitation to be

$$\delta p = 2\sqrt{\frac{2mkT}{\pi}} \ , \qquad \delta x = \frac{\hbar}{\delta p} = \frac{h}{8\sqrt{2\pi m_B kT}} \ . \tag{3.58}$$

Here δx is the size of the wave packet. The point particle electron thus appears from the point of view of Heisenberg's uncertainty principle like a kind of 'snow flake'. According to (3.58), the effective diameter of these snow flakes is estimated to be $\Lambda/8$, where Λ is the thermal wavelength of the electrons and m the electron mass:

$$\Lambda = \Lambda_{\mathrm{e}} = \frac{h}{\sqrt{2\pi m k_{\mathrm{B}} T}} . \tag{3.59}$$

The arguments leading to this estimate are based on Heisenberg's uncertainty principle and our knowledge of the thermal distribution of the momentum.

Identifying the length $\Lambda/8$ with an effective temperature-dependent diameter of the charges, we get a transition from a classical to a quantum Debye–Hückel radius

$$a \longrightarrow a(T) = \frac{\Lambda}{8} . \tag{3.60}$$

This leads to the so-called Lambda approximation of quantum effects, also known as the quantum Debye–Hückel approximation (QDHA). Here all quantum effects are expressed by just one characteristic length, the thermal de Broglie wavelength. Note that in the case where the charges themselves already have some intrinsic diameter $R > 0$, such as the radius of alkali ions, we may include the intrinsic length in the modified Debye–Hückel diameter

$$a(T) = \frac{\Lambda}{8} + R . \tag{3.61}$$

Note that an alternative concept for including the R parameter based on the MSA was also given in an earlier book (Ebeling et al. 1976, 1991).

In the framework of the simplest model of quantum effects like the QDHA, all quantum effects are modeled by just one temperature-dependent parameter $a(T)$. Further we note that even the effect of an external magnetic field B, which influences the Larmor frequency of the electrons $\omega_{\mathrm{c}}^{\mathrm{e}} = eB/m_{\mathrm{e}}$, may be included in the definition of an effective parameter $a(T)$. As shown in Steinberg et al. (1998), Ebeling et al. (2000, 2003), the influence of the magnetic field may be included through a field-dependent factor

$$a(T) = \frac{\Lambda}{8}\left(1 - \frac{x_{\mathrm{e}}^2}{48}\right) , \qquad x_{\mathrm{e}} = \frac{\hbar\omega_{\mathrm{c}}^{\mathrm{e}}}{2k_{\mathrm{B}}T} , \qquad \omega_{\mathrm{c}}^{\mathrm{e}} = \frac{eB}{m_{\mathrm{e}}} . \tag{3.62}$$

The magnetic field reduces the effective diameter.

Applying now the Debye–Hückel theory in this context leads to elementary expressions for the correlation function and the Coulomb energy:

$$g_{ab} = -\frac{\beta e_a e_b}{\varepsilon_r r} \frac{\exp\left[\kappa a(T) - \kappa r\right]}{1 + \kappa a(T)} , \tag{3.63}$$

$$\langle U_c \rangle = -V 2\pi \sum_{ab} n_a n_b \int_a^\infty dr\, r^2 \frac{e_a e_b}{r} g_{ab}(r) = -\frac{V k_B T \kappa^2}{8\pi\varepsilon_r k_B T\left[r_D + a(T)\right]} ,$$

where the Debye parameter κ and the Debye length r_D are defined by the relations $\kappa^2 = r_D^{-2} = 4\pi\beta(n_e e_e^2 + n_i e_i^2)$. As in the original Debye–Hückel theory, the modified denominator follows from the lower integration limit at $r = a(T)$. In this approximation, the chemical potentials of the electrons and the ions read

$$\mu_e = -\frac{e_e^2}{2\varepsilon_r\left[r_D + a(T)\right]} , \qquad \mu_i = -\frac{e_i^2}{2\varepsilon_r\left[r_D + a(T)\right]} . \tag{3.64}$$

The corresponding expression for the free energy follows from thermodynamic relations and is like the one for the classical case:

$$F = F_0 - k_B T V \frac{\kappa^3}{12\pi} \tau\big(\kappa a(T)\big) . \tag{3.65}$$

Here the first term represents the free energy of the system with zero charges.

The excess pressure functions are given in the QDHA by

$$\beta p = n_i + n_e + n_0 - \frac{1}{24\pi} \kappa^3 \phi\big(\kappa a(T)\big) , \tag{3.66}$$

where the familiar Debye–Hückel pressure functions $\tau(x)$ and $\phi(x)$ are defined as previously (see also Yeh et al. 1996).

3.3.2 Reduced Mass Approximation

So far we have left open which of the two masses is to be used in the definition of the effective radius of charges $a(T)$. In the more rigorous quantum statistical theory which will be developed in Chaps. 5 and 6, we will see that the relevant length is neither the thermal wavelength of electrons $\Lambda = \Lambda_e$, nor the thermal wavelength of the ions Λ_i, but rather the relative wavelength which is determined by the relative masses and describes the relative electron–ion motion. As we shall show, it was the reduced electron–ion mass μ which already played a basic role in Bohr's first theory of atoms and in all scattering theories. The reduced mass (or mean mass) is defined by $\mu = m_{ie} = m_e m_i/(m_e + m_i)$. The corresponding relative thermal wavelengths and the parameter $a(T)$ are defined by

$$\Lambda_{ei} = \Lambda = \frac{h}{\sqrt{2\pi\mu k_B T}} , \qquad a(T) = \frac{\Lambda_{ei}}{8} . \tag{3.67}$$

As we shall see in Chaps. 4 and 6, in the quantum statistical theory a different expression arises, although it is fully equivalent up to a constant. In the strict theory we will use $\Lambda_{\rm ei} = \hbar/\sqrt{2m_{\rm ei}k_{\rm B}T}$ as the definition of the relative thermal wavelength. This differs from the length $\Lambda_{\rm ie}$ by a constant factor.

We recall once again from Bohr's theory that the *reduced mass* μ is the central quantity in the theory of atomic bound states and scattering states, and we may expect therefore that this is also the relevant mass for most interaction and ionization problems. We now formulate one of the basic assumptions of this chapter:

Assumption about the Role of the Reduced Mass: In the region of intermediate temperatures $10^3 < T < 10^5$ K, which is the region of ionization phenomena, the behavior of the plasma is determined only by the interaction of opposite charges, by the corresponding reduced (relative) mass μ, and the relative wavelength Λ.

The mean mass assumption leads us to a very effective approximation, the *reduced mass approximation* (RMA). In practice, the RMA implies that, in all calculations of nonideality effects, all masses may be replaced by the relative electron–ion mass μ, which plays the role of a mean mass:

$$m_{\rm ee} \to m_{\rm ie} = \mu\ , \qquad m_{\rm ii} \to m_{\rm ei} = \mu\ , \qquad \frac{1}{\mu} = \frac{1}{m_{\rm e}} + \frac{1}{m_{\rm i}}\ . \tag{3.68}$$

This approximation is sometimes also called mean mass approximation. Here we will use the notation RMA having in mind that in fact that it is not the mean mass but the reduced mass which plays the basic role. Formally, the plasma behaves in the RMA like a mass-symmetric plasma with $m_{\rm ee} = m_{\rm ii} = m_{\rm ie} = \mu$. Further, we observe that one-component plasmas (OCP) and two-component plasmas (TCP) are given (up to trivial factors) by the same functions for the thermodynamic (excess) functions.

We will see later that the RMA is more than a plausible hypothesis, since in the end we will be able to show that the RMA is based on a strict quantum-statistical result, saying that, in the limit $k_{\rm B}T \ll I$ (where I is the ionization energy), the interactions are dominated by the reduced mass $\mu = m_{\rm ie}$ and the differences between the relative masses m_{ab} do not effect the ionization equilibrium and the thermodynamics (Ebeling et al. 1976, 2016). In this way, we will show that the RMA is not ad hoc, but based on strict mathematical properties of the so-called virial functions $Q(x)$ (Ebeling et al. 1976, Kraeft et al. 1986).

However, for the time being, working so far only with physical arguments, we shall take the MMA as a plausible physical assumption, expressing the idea that, in the region of ionization phenomena $10^3 < T < 10^5$ K, the plasma interactions and the thermodynamic excess quantities are fully decribed by the mean (reduced) mass μ. In other words, we assume that only the reduced mass μ and the reduced thermal wavelength $\lambda = \lambda_{\rm ie}$ come into the theory, so that the influence of the individual masses and individual relative lengths on excess quantities can be neglected.

Adopting the RMA, it makes sense to express Λ in terms of the relative (reduced) electron–ion wavelength and write

$$a \longrightarrow a(T) = \frac{\sqrt{\pi}}{4}\lambda\,, \qquad \lambda = \lambda_{\text{ie}} = \frac{\hbar}{2m_{\text{ei}}k_{\text{B}}T}\,. \tag{3.69}$$

From now on, we use the wavelength of relative motion of electron–ion pairs as the most relevant variable. The mean Coulomb energy and the excess chemical potential of TCP follow from (3.64) with the replacement (3.69):

$$\langle U_{\text{c}} \rangle = -\frac{Vk_{\text{B}}T}{8\pi}\frac{\kappa^3}{1+(\sqrt{\pi}/4)\kappa\lambda}\,, \qquad \mu_{\text{ex}} = -\frac{e^2}{1+(\sqrt{\pi}/4)\kappa\lambda}\,. \tag{3.70}$$

Here κ is as usual the reciprocal Debye length for two-component systems, with $\kappa^2 = 8\pi\beta ne^2/\varepsilon_{\text{r}}$.

The rather simple QDH approximation (QDHA or Lambda approximation) introduced above extends the Debye–Hückel theory, which is a very popular and effective tool from electrolyte theory (see, e.g., Falkenhagen et al. 1971) to quantum plasmas. This theory is not only simple and fully analytical, but it agrees rather well (at least to lowest order with respect to κ) with exact quantum-statistical results for small densities and low temperatures $T < 10^5$ K (Ebeling et al. 1976). The (nontrivial) proof of this statement will be delayed until Chap. 4.

Note that the validity of the QDH approximation is restricted to plasmas which are nondegenerate, i.e., $n^*\Lambda^3 \ll 1$, where n^* denotes the density of free particles. Another necessary condition for validity is that only a small fraction of the atoms should be bound in molecules. A great advantage of the QDH approximation is that it can easily be extended to Saha-type equations. But before going on to the treatment of ionization problems, we shall discuss an alternative effective potential approach which goes back to the work of Kelbg, Deutsch, and others. This approximation is also quite elementary, but it is effective and it also fits the known results of more rigorous quantum statistics if the free parameter is chosen in an appropriate way.

3.4 Slater Functions and Effective Potential Approach

3.4.1 Effective Potential Approach of Kelbg and Deutsch

The idea we used in the previous section was to define some volume around a charge where quantum effects, and in particular Heisenberg uncertainty, would dominate. The arguments in our elementary approach were based on the Heisenberg principle that the stochastic thermal momentum leads to a smearing out of the charge. Particles with distributed thermal momentum cannot interact like point charges. A similar idea was used in another effective potential approach (KEPA) introduced in the 1960s by Morita, Kelbg, Deutsch, and others, who based their work on a different simplified version of quantum-statistical theory.

The idea was to take into account complicated quantum effects using analytically simple and smooth pseudo-classical potentials derived from the Slater functions which modify the Coulomb potential in a semi-quantitative way at short distances. In contrast with the QDHA, the advantage of the KEPA is the analytic form of the new effective potential in comparison to the model of effective charged spheres. In Rostock in 1963/1964, Kelbg started from Morita's idea that effective potentials may be based on Slater functions, i.e., the diagonal elements of density matrices. In order to understand Kelbg's ideas, let us begin with a two-particle Slater function which we now know from the gas theory for point charges. We calculate the binary Slater sum of two charges belonging to species a and b:

$$S_{ab}(\mathbf{r}_1, \mathbf{r}_2) = (1 + \delta_{ab})! \frac{\Lambda_a^3 \Lambda_b^3}{(2s_a + 1)(2s_b + 1)} \times \sum_{\sigma_1, \sigma_2} \left\langle \mathbf{r}_1 \mathbf{r}_2 \sigma_1 \sigma_2 \right| \exp(-\beta \hat{H}_{ab}) \left| \mathbf{r}_1 \mathbf{r}_2 \sigma_1 \sigma_2 \right\rangle . \tag{3.71}$$

Neglecting exchange effects and using first order perturbation theory in the interaction parameter e^2, we arrive at expressions depending only on distance and, after linearization, define an effective potential following Kelbg (1963, 1964):

$$S_{ab}(r) = 1 - \beta \varphi_{ab}(r) . \tag{3.72}$$

More details of this nontrivial calculation will be given in Chap. 4. Here we need only two simple properties of the quantum function $\varphi_{ab}(r)$:

$$\varphi_{ab}(r) \longrightarrow \frac{e_a e_b}{r} , \quad \text{if } r \to \infty , \tag{3.73}$$

$$\varphi_{ab}(r) \longrightarrow \frac{\sqrt{\pi}}{\lambda_{ab}} , \quad \text{if } r \to 0 , \quad \lambda_{ab} = \frac{\hbar}{2m_{ab}k_\mathrm{B}T} . \tag{3.74}$$

Since the exact expression obtained by Kelbg, which contains hypergeometric functions, was too complicated for further calculations of thermodynamic functions, Kelbg proposed to use an approximate smooth potential with the correct asymptotic for both small and large distances (3.74) (Kelbg 1963, 1964):

$$\varphi_{ab}(r) = \frac{e_a e_b}{r} \big[1 - \exp(-\alpha_0 r)\big] , \qquad \alpha_0 = \frac{\sqrt{\pi m_{ab} k_\mathrm{B} T}}{\hbar} . \tag{3.75}$$

This form of soft Coulomb potential was first used by Kramers (1927) in electrolyte theory then a decade later by the pioneer of quantum chemistry Hans G.A. Hellmann (1903–1938) (Hellmann 1934, 1936), and by several authors in electrolyte theory (Falkenhagen 1971). Applications to quantum plasmas were first given by Kelbg (1963). Then the potential was worked out by Deutsch (1977) with the inclusion of exchange terms. Later, many applications were given by Baimbetov and his school in Almaty (Baimbetov et al. 1995). Therefore we will sometimes use the name

Kelbg–Deutsch potential, or in some contexts add Baimbetov's name to refer to the Kelbg–Deutsch–Baimbetov potential. The Fourier transform of the Kelbg–Deutsch potential is simple and reads

$$\varphi_{ab}(k) = \int \mathrm{d}\mathbf{k} \exp(-ikr)\varphi_{ab}(r) = \frac{4\pi e_a e_b \alpha_0^2}{k^2(k^2+\alpha_0^2)} \,. \tag{3.76}$$

Meanwhile, for the parameter α_0, several choices are in use. Let us first consider one-component plasmas (OCP) and symmetric two-component plasmas (TCP). As pointed out above, Kelbg fitted the height of the potential at zero distance to the first order quantum-statistical perturbation theory. For OCP, Deutsch used the value $\alpha_0 = \sqrt{mk_\mathrm{B}T}/\hbar$ where m is the electron mass (Deutsch 1977). The choice by Baimbetov, Ramazanov, and Arkhipov (Baimbetov 1995; Arkhipov et al. 2015) agreed with Kelbg's. On the other hand Bredow et al. (2015) used the Deutsch's choice (1977).

In recent work (Ebeling 2016), a different choice of the free parameter α_0 was proposed which uses a kind of average. For calibration, this new choice uses the available exact results for the thermodynamic functions. In other words the idea is to improve the agreement with those exact thermodynamic results by optimizing the parameter. According to this choice, for one-component plasmas, we take

$$\alpha_0^{\mathrm{OCP}} = \frac{4}{\sqrt{\pi}} \frac{\sqrt{mk_\mathrm{B}T}}{\hbar} \,. \tag{3.77}$$

The difference with the other choices is not so great. The value for α_0 is only around 30 percent higher than the value used by Kelbg and the Baimbetov group. However, the advantage with the new choice is that the first quantum corrections to classical thermodynamics are now exact, and the overall plasma thermodynamics for the higher orders is, for OCP and symmetric TCP, rather close to the known more rigorous results (DeWitt 1962, 1995, Alastuey and Perez 1992, 1996. Alastuey and Ballenegger 2010, Alastuey et al. 2008, 2015).

For simplicity, in the case of two-component plasmas, we consider only the charge-symmetric case $e_\mathrm{i} = -e$, $n_\mathrm{i} = n_\mathrm{e} = n$, e.g., H plasmas, and as recommended in the previous section, we use the reduced mass approximation (RMA). According to the results of earlier work (Ebeling et al. 1976, 1991, 2016), the RMA is valid for lower temperatures $10^4 < T < 10^5$ K and asymptotically exact for $I/k_\mathrm{B}T \gg 1$ up to order $O(\kappa\lambda)$. In the RMA, all relative thermal wavelengths are replaced by the electron–ion wavelength, so we need to introduce only one parameter α_0:

$$\alpha_0^{\mathrm{TCP}} = \frac{4\sqrt{2\mu k_\mathrm{B}T}}{\hbar\sqrt{\pi}} \,. \tag{3.78}$$

By using only one parameter α_0 for mixtures of charges, we avoid any matrix problem, and may work with a simple fully analytical theory. This is the Kelbg effective potential approximation (KEPA). It is identical with the QDHA up to the linear term, and comparable with the QDHA for simplicity, but wins the comparison by fitting the available exact results for higher densities in an even better way.

Screening and Thermodynamics

As before, we use Bogoliubov's pair correlation functions $g_{ab}(r)$ to take into account Debye screening. In contrast to Bogoliubov's original procedure described above, we have to replace the Fourier transform of the Coulomb potential by the Fourier transform of the Kelbg–Deutsch potential (3.76). We thus find the following algebraic equation in Fourier space:

$$k_B T \tilde{g}_{ab}(\mathbf{k}) + \tilde{\varphi}_{ab}(\mathbf{k}) + \sum_c n_c \tilde{\varphi}_{ac}(\mathbf{k}) \tilde{g}_{bc}(\mathbf{k}) = 0 . \tag{3.79}$$

With the ansatz $g_{ab}(k) = e_a e_b g(k)$ and stipulating as boundary condition that the screened potentials should converge for low densities to the effective potentials φ_{ab}, we get the Bogoliubov plasma correlation functions, similar to what is found in classical calculations for electrolytes (see Falkenhagen 1971):

$$g_{ab}(r) = -\frac{\beta e_a e_b}{\varepsilon_r r} \frac{\alpha^2}{s^2 - p^2} \left[\exp(-pr) - \exp(-sr) \right] , \quad \alpha = \alpha_0 . \tag{3.80}$$

Here we introduced the parameters

$$p = \frac{\alpha}{2} \left(\sqrt{1 + 2\frac{\kappa}{\alpha}} - \sqrt{1 - 2\frac{\kappa}{\alpha}} \right) , \quad s = \frac{\alpha}{2} \left(\sqrt{1 + 2\frac{\kappa}{\alpha}} + \sqrt{1 - 2\frac{\kappa}{\alpha}} \right) . \tag{3.81}$$

Note that for small densities, we have $p \approx \kappa$ and $s \approx \alpha$. The mean Coulomb energy is

$$\langle U_c \rangle = \frac{V}{2} \sum n_a n_b \int \mathrm{d}\mathbf{r} \frac{e_a e_b}{r} g_{ab}(r) . \tag{3.82}$$

Note that here we have to calculate the mean of the Coulomb potential and not the mean of the effective potential, as done in electrolyte theory (see Falkenhagen 1971). Averaging the effective potential would lead to an error here (as, e.g., in the appendix of Kraeft et al. 2015), since effective potentials are not physical energies. They are just auxiliary quantities and the mean does not have the meaning of a physical energy. The integrals in (3.82) are exactly solvable, and the result reads (Ebeling 2016)

$$U_c = -4 k_B T V (n \beta e^2)^2 \frac{\alpha^2}{s^2 - p^2} \left(\frac{1}{p} - \frac{1}{s} \right) . \tag{3.83}$$

The reader may easily reproduce the details of the calculation of the Bogoliubov plasma correlation functions for Kelbg–Deutsch effective interactions and the corresponding mean Coulomb energy by elementary integrations. The result for the Coulomb energy may be written in the simple analytical form

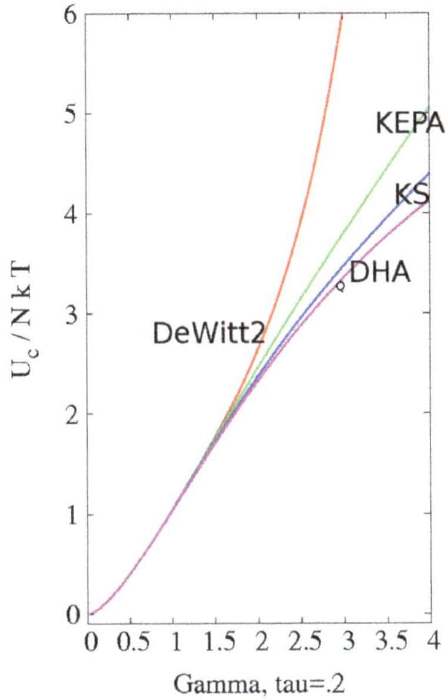

Fig. 3.2 Coulomb energy of a plasma for the temperature $\tau = 0.2$ ($T = 31\,500$ K) in QDHA, KEPA, and the second order DeWitt approximation. The lowest curve (*magenta*) shows the quantum Debye–Hückel aproximation (QDHA) and the second curve from the top (*green*) shows the effective potential approximation (KEPA) proposed here. For comparison, the upper curve (*red*) gives the second (quadratic) quantum correction to the limiting law according to DeWitt (1962, 1995). The third curve from the top (*blue*) shows the result of the (nonlinear) Kremp–Schmitz theory (KS), which will be discussed in Chaps. 4–6

$$U_c = -V k_B T \frac{\kappa^3}{8\pi} \frac{1}{\sqrt{1 + (\sqrt{\pi}/2)(\kappa\lambda)}} , \qquad \lambda = \lambda_{ie} , \tag{3.84}$$

where κ now has the value for two-component systems, viz., $\kappa^2 = 8\pi\beta n e^2$.

Figure 3.2 compares several approximations, including the QDHA and the KEPA, with two other approximations stemming from more rigorous quantum-statistical theories. To begin with we compare with a very early quantum-statistical theory developed by DeWitt (1962) and the theory developed by Kelbg (1963, 1964), Schmitz and Kremp (1967), Kremp and Schmitz (1967), and Kraeft et al. (2015). We follow here the discussion given in some recent work (Ebeling 2016) and use convenient units for temperature and nonideality, viz., $\tau = k_B T / Ry$ and $\Gamma = e^2/d = 2/\tau r_s$. Nowadays, the theory of DeWitt et al. (1962), Riemann et al. (1995) is considered in the literature to be exact up to quadratic order in the Taylor expansion with respect to $\kappa\lambda$ (Ebeling 2016). Therefore, at low densities (low Γ), we may consider the DeWitt theory as an appropriate standard for calibration. At higher values of Γ, DeWitt's Taylor expansion breaks down and the Kremp–Schmitz theory (or equivalent approximations) may possibly give an appropriate standard for calibration (see

Ebeling 2016). As we shall see, for $T \sim 30\,000$ K, the QDHA and the KEPA are in reasonably good agreement with more rigorous theories at least up to $\Gamma \sim 2$.

Comparing the different approximations for the mean Coulomb energy shows the following:

- The QDHA and the KEPA are in quite good agreement and agree also with the analytical theory of Kremp and Schmitz (1967), which uses the exact Kelbg potential.
- We see that the two theories agree with more rigorous theories at low densities and have a good shape at higher densities, close to the predictions of more advanced theories.

The free energy follows by Debye charging, and the pressure and chemical potential by differentiating the free energy with respect to the particle densities. The corresponding expression for the nonideal part of the free energy (excess free energy) obtained by the Debye charging procedure is

$$F_{\mathrm{ex}} = -V k_{\mathrm{B}} T \frac{\kappa^3}{12\pi} \tau\left(\frac{\kappa}{\alpha}\right) , \tag{3.85}$$

$$\tau(z) = \frac{6}{5\sqrt{1+2z}} - \frac{2(1+z-.5z^2-\sqrt{1+2z})}{5z^3\sqrt{1+2z}} , \tag{3.86}$$

where $\alpha = \alpha_0$ and

$$z = \kappa/\alpha_0 = \sqrt{\pi}\kappa\lambda/4 . \tag{3.87}$$

The result for the excess chemical potential of the free charges in electron–ion plasmas is quite similar to the mean Coulomb energy and reads

$$\mu_{\mathrm{ex}} = 2\mu_{\mathrm{i}} = 2\mu_{\mathrm{e}} = -\frac{e^2\kappa}{\sqrt{1+(\sqrt{\pi}/2)(\kappa\lambda)}} . \tag{3.88}$$

The result for the excess pressure is

$$\beta p_{\mathrm{ex}} = -\frac{\kappa^3}{24\pi}\phi(z) , \quad \phi(z) = \frac{3}{5\sqrt{1+2z}} - \frac{4(1+z-.5z^2-\sqrt{1+2z})}{5z^3\sqrt{1+2z}} . \tag{3.89}$$

In order to avoid any misunderstanding, let us note that the quantum-statistical Coulomb energy derived above is only in partial agreement with the expressions for the internal energy for this model potential known from electrolyte theory (see, e.g., Yukhnovsky and Holovko 1980). The reason is the different way of calculating the energy. For electrolytes, the term $e^2 \exp(-\alpha_0 r)/r$ is a physical potential corresponding to real forces which contribute to the energy. In the quantum case, as already stressed above, the complete interaction is given by e^2/r, while the second term depending on α_0 is not a physical interaction, but a correlation stemming from quantum effects, which does not contribute directly to the energy. A second reason for the differences is that the parameter α_0 depends on the temperature. This requires special care when using thermodynamic relations. Therefore a naive application of classical formulae for the calculation of internal energy and pressure may lead to incomplete or even incorrect results. The correct procedure, as demonstrated above, goes as follows:

- Find from correlation functions the Coulomb energy.
- Go by Debye charging of Coulomb energy to the free energy.
- Calculate other thermodynamic functions by standard relations.

The essential point is to get first the quantum-statistical Coulomb energy and then, using a Debye charging procedure, the free energy. Only then may one get the correct pressure, chemical potential, internal energy, and so on, from thermodynamic relations. It is not allowed to consider the Kelbg potential as a proper physical interaction. This would lead to incorrect relations for the thermodynamic functions (see Kraeft et al. 2015).

We would like to stress that this problem is not a deficiency of Kelbg's method, but a consequence of a possibly incorrect application of the thermodynamic relations, which can be applied to temperature-dependent potentials only with the greatest care. In this context, we also remind experts of the problems with the internal energy of electrolytes which are connected with the temperature-dependence of the potential due to the influence of temperature on the dielectric constant (Falkenhagen 1971).

3.4.2 Extensions by Wigner–Onsager Corrections

What we have improved so far in comparison to the simple QDHA is the effective potential. This was replaced by an improved model potential developed by Kelbg and Deutsch, which also leads to closed analytical expressions for the thermodynamics. We shall call this approximation the Kelbg effective potential approximation (KEPA) from now on. The open questions which remain are about the high orders and the asymptotic for large κ. In particular, we have left open the question of the accuracy of the thermodynamic functions at very high densities. The result of our study is:

- The convergence of Taylor expansions with respect to $\kappa\lambda$ which are now generally accepted up to quadratic order is not really so good.
- The asymptotic of the Coulomb energy per particle obtained so far is about $n^{1/4}$, which disagrees with the $n^{1/3}$ expected from Wigner's lattice effects.

The open problem left is the behavior at higher densities, which in the present version is not yet adequate.

According to the early work by Wigner (1934) and following several more recent analytical and numerical studies, the Coulomb energy per particle should converge at high densities to values of the order of the Wigner energy $Wi = e^2/d$, i.e., it should tend asymptotically to a $n^{1/3}$ law (DeWitt 1976). In other words, the Coulomb energy per unit $k_B T$ should be linear in Γ for large Γ. Laws of this type are typical for lattice-like structures (Wigner 1934, 1938, Berlin and Montroll 1952, DeWitt 1976).

An interesting idea developed by Kaklyugin and in more detail by Brilliantov shows how to apply Peter Debye's classical ideas about mode restriction of waves in solid states to plasma waves (Kaklyugin 1985, Brilliantov 1998). Kaklyugin suggested in 1985 that in plasmas there might also exist an upper limit of the k-modes in Fourier space $k_{\max}$. According to Debye, the number of excited waves, phonons in solids, should be finite. According to Kaklyugin and Brilliantov, the number of plasma wave modes (in the semi-classical region) is clearly limited by the smallest distance $d \sim n^{1/3}$. Kaklyugin's estimate for the largest wave number was based on an investigation of the dielectric function for large k values. He obtained

$$k_{\max} = (3\pi^2/4)n^{1/3} \approx 2n^{1/3} . \tag{3.90}$$

Brilliantov (1998) derived a different result for the constant, which we will use here. He followed Debye and argued that the number of Fourier waves in a finite system is limited by the total number of degrees of freedom in the system. He assumed that the total number of the physically different plasma wave modes with the wave vector **k** should be equal to $3N$. The number of modes is twice the number of the wave vectors, since the amplitude is complex:

$$3N = 2\frac{V}{8\pi^3}\int_0^{k_{\max}} \mathrm{d}k\, 4\pi k^2 , \qquad k_{\max} = (9\pi^2 n_e)^{1/3} . \tag{3.91}$$

The factor originates from replacing the summation by an integration in k space. For the classical Coulomb energy, this leads to an integral which was calculated in Chap. 2 and finally to the relation for the classical OCP energy:

$$\frac{-U_c}{Nk_B T} = \frac{9}{4}(.399\Gamma)^{1.5} \arctan\left[(0.399\Gamma)^{-0.5}\right] . \tag{3.92}$$

The asymptotic slope of -0.881Γ is fairly close to the numerical fit of simulation data which, according to DeWitt (1986), is given by -0.899Γ. The Brilliantov–Kaklyugin formula is also in good agreement with the numerical data, although we should note that the overall accuracy is not as good as for Nordholm's formula, given earlier.

In order to include quantum effects in the Brilliantov–Kalklyugin formula, we replace the classical Fourier spectrum used in Chap. 2 by the Fourier spectrum, including quantum corrections in the spirit of the pseudopotential after Kelbg and Deutsch. Including an upper limit in the integral for the mean Coulomb energy, we arrive at the formula

$$U_c = -4k_B T V(n\beta e^2)^2 \int_0^{k_{\max}} \mathrm{d}k \frac{\alpha^2}{s^2 - p^2}\left(\frac{1}{k^2 + p^2} - \frac{1}{k^2 + s^2}\right) . \tag{3.93}$$

Here we introduced the parameters $\alpha = 4/\sqrt{\pi}\lambda$ and $p \approx \kappa$, $s \approx \alpha$, where the precise definition of the parameters p and s is

$$p = \frac{\alpha}{2}\left(\sqrt{1+2\frac{\kappa}{\alpha}} - \sqrt{1-2\frac{\kappa}{\alpha}}\right), \quad s = \frac{\alpha}{2}\left(\sqrt{1+2\frac{\kappa}{\alpha}} + \sqrt{1-2\frac{\kappa}{\alpha}}\right). \tag{3.94}$$

In this form the integrals are exactly solvable and the result reads

$$U_{\mathrm{c}} = -4k_{\mathrm{B}}TV(n\beta e^2)^2 \frac{\alpha^2}{s^2-p^2}\left[\frac{1}{p}\arctan\left(\frac{k_{\max}}{p}\right) - \frac{1}{s}\arctan\left(\frac{k_{\max}}{s}\right)\right]. \tag{3.95}$$

The results obtained from (3.95) for the Coulomb energy of an OCP are represented in Fig. 3.3.

We see that, for small Γ, the new formula which includes quantum effects in KEPA follows the shape of the quantum statistical curve obtained above. On the other hand, for large Γ, the new formula (3.95) approaches a Wigner-type linear dependence on Γ. In this way, we develop a method to derive the expected Wigner

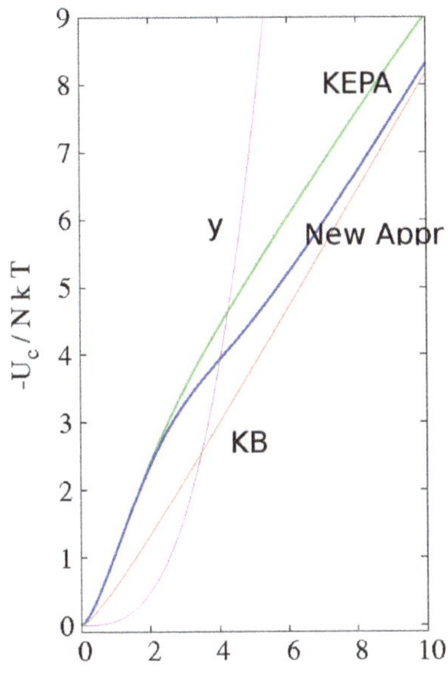

Fig. 3.3 Coulomb energy per particle for $\tau = 0.2$ in the QDHA (*upper green curve*) in comparison with the classical Kaklyugin–Brilliantov formula (KB *red*). The new aproximation (3.95) which combines both effects (New Appr *blue*) shows a transition between classical and quantum behavior in the region $\Gamma \sim 4$–8. The influence of degeneracy is shown by the lowest curve (y *magenta*) representing $y = n\Lambda_{\mathrm{e}}^3/2$. For $\tau = 0.2$, it makes no sense to use this method to calculate the energies beyond $\Gamma \sim 4$, since degeneracy effects then become relevant

behavior at large densities. In later chapters we will discuss the transition to the $n^{1/3}$-dependence which reflects the Wigner effects in more detail.

It should be borne in mind that so far our formulae, as represented in Fig. 3.3, include only weak degeneracy effects and lose validity for strong degeneracy, i.e., when $y = n_e \Lambda_e^3/2 \gg 1$. In the case of degenerate plasmas, the maximal k-value is in fact no longer determined by the smallest distance d, but instead by the Bohr distance a_B. This will be discussed later.

So Fig. 3.3 shows the quantum Coulomb energy per particle for $\tau = 0.2$ and $\tau = 0.02$ in comparison with the classical Kaklyugin–Brilliantov formula (Kaklyugin 1985, Brilliantov 1998). A smooth transition is possible from the energy coming from the quantum ring integral to the classical result of Kaklyugin–Brilliantov because of structural analogies between the formulae and curves (see Fig. 3.3). This suggests combining these methods by introducing a maximal wave number in the Kelbg effective potential method or more general ring integrals (see Chaps. 4–6).

Instead of the Kaklyugin–Brilliantov method, we may use a combination of effective potentials of Kelbg-type with the Nordholm method (see Sect. 3.1.3). With the assumption that the charge density is constant inside the radius h and given by the Kelbg theory outside, we arrive at the Coulomb energy

$$
\begin{aligned}
-\frac{U_c}{Nk_BT} &= \frac{\zeta}{2}\left[\frac{1}{2}\kappa^2 h^2 k_1(\kappa,\alpha,h)^2 + \kappa h k_1(\kappa,\alpha,h)\right] \\
&= \frac{\zeta}{4}\left\{\left[1 + \kappa h k_1(\kappa,\alpha,h)\right]^2 - 1\right\}, \qquad (3.96)
\end{aligned}
$$

with

$$
k_1(\kappa,\alpha,h) = \frac{1}{s} - \frac{1}{p} + \kappa^2 h \frac{\frac{1}{s}\exp(-ph) - \frac{1}{p}\exp(-sh)}{\exp(-ph) - \exp(-sh)}. \qquad (3.97)
$$

Since this expression is quite complicated, in a first approximation, it may be possible to omit the last terms which contain h in the form of an exponential function. Following the procedure given in Sect. 3.1.3 and using the continuity at $r = h$ and the condition of electroneutrality (complete screening), we then get a third order polynomial for the equation which determines h:

$$
h^3 + 3h^2 k_1(\kappa,\alpha,h) + 3hk_2(\kappa,\alpha,h) - \frac{1}{\zeta^3}d^3 = 0. \qquad (3.98)
$$

In the high Γ limit where Wigner effects matter, the approximate solution is

$$
h \approx \frac{d}{\zeta^3}. \qquad (3.99)
$$

Then from the condition that the Onsager bound

$$-\frac{U_{\rm c}}{Nk_{\rm B}T} \leq \frac{9}{10}\Gamma \qquad (3.100)$$

has to be satisfied, we find the best choice for the parameter ζ to be

$$\zeta = \frac{216}{125} = 1.728 \; . \qquad (3.101)$$

This leads us to the simple estimate

$$h_0 \approx 0.194d \; , \qquad (3.102)$$

which can be used for the zeroth order of the Coulomb energy in (3.96).

A next useful approximation which gives an improvement for smaller Γ is the estimate

$$h_0 \longrightarrow h_1 = \frac{1}{k_1(\kappa, \alpha, h_0)} \left\{ 1 + \frac{125}{216} \left[dk_1(\kappa, \alpha, h_0) \right]^{1/3} - 1 \right\} \; , \qquad (3.103)$$

and with (3.96), this leads to a second order estimate of the Coulomb energy, viz.,

$$-\frac{U_{\rm c}^{(1)}}{Nk_{\rm B}T} = \frac{216}{600} \left\{ \left[1 + \frac{125}{216} dk_1(\kappa, \alpha, h_0)^{1/3} \right]^{2/3} - 1 \right\} \; . \qquad (3.104)$$

The accuracy of these approximations in various regions of the (n, T) plane has still to be estimated.

To conclude this section, we may state that the different versions of the QDHA and the KEPA provide reasonably good representations of thermodynamics functions which are fully analytical, easy to handle, and nevertheless close to more rigorous theories in a wide parameter region. At present it is difficult to give a clear preference to either the QDHA or the KEPA. However, as stated in a recent work (Ebeling 2016), one advantage with KEPA is the better high-density asymptotic. In the next section we will give a systematic discussion of a combination of the QDHA and KEPA theories with the Saha approximation. The relationship between QDHA and KEPA and the more refined quantum theories of screening due to Macke, Gell-Man, Brueckner, Montroll, Ward, Vedenov, Larkin, DeWitt, Kelbg, and others (Berlin and Montroll 1952, Vedenov and Larkin 1959, DeWitt 1962, 1966, 1976, Kelbg 1963, 1964, Ebeling et al. 1976, Kraeft et al. 1986) will be discussed in Chaps. 5 and 6.

The theory we have developed so far can be applied to plasmas of point charges, i.e., hydrogen plasmas, or to some extent to electron–hole plasmas in solids. For applications to real plasmas like alkali-type plasmas and noble gas plasmas, we need some modifications which will be discussed in later sections.

3.5 Plasmons and Collective Mode Expansions

3.5.1 Plasma Wave Excitations

A characteristic feature of a plasma is the existence of a spectrum of collective oscillations or plasma waves (plasmons). The frequency and velocity of propagation of these waves are determined by the wave vector and the gross parameters of the plasma such as the density, the mean velocity spread, the magnetic field, etc., and this situation is a reflection of the fact that all the particles in the plasma are involved in the plasma oscillations.

The situation is different, however, when one examines the damping (or growth) of the oscillations. Damping (growth) is determined by the 'fine details' of the particle distribution in phase space, for example, by the derivative of the velocity distribution function. This situation reflects the specific role played by resonant particles moving with definite velocities. These particles are capable of exchanging energy with the waves and can thus amplify or damp it.

In this way Coulomb systems are not just gases with a special law of interaction, but are in some respects unique. The main reason is that the Coulomb energy is contained in the electric field between the particles. These fields are essentially classical and the local energy density is given by the square of the local field strength $(\mathbf{E}(\mathbf{r}, t)^2$. The electric field changes quickly in time and space, and in a uniform plasma, the average is zero at any coordinate and any instant of time. The correlations of the field $\mathbf{E}(\mathbf{r}, t)$ are thus the important factor determining the energy density.

The basic mode of existence of the electric field in a plasma are plasma waves, which have been known since the work of Langmuir and others. In some sense, the plasma waves or 'plasmons' are the characteristic quasi-particles of a plasma, in the same way as phonons are the basic quasi-particles of crystal lattices. The study of the fluctuating field in the plasma and its correlations is a very fruitful non-traditional view which was worked out in much detail by Bohm, Pines, and Nozieres (Bohm and Pines 1953, Pines 1961, Pines and Nozieres 1966), as well as by Klimontovich, Silin, and Rukhadse (Pines 1961, Klimontovich 1967, 1986, Silin and Rukhadse 1961). We will mainly use the methods developed by (Klimontovich 1967, 1986).

In many respects, this approach is also connected with the notation for collective excitations or plasma waves. The name *dielectric formalism* is sometimes used, since the dielectric function plays a primary role (Pines 1961, Pines and Nozieres 1966). An attractive feature of these methods is that the classical and quantum cases are treated in a very similar way. The physical reason for this similarity is of course that the electric field in a plasma is a classical object with a quantum dynamics determined by the quantum dynamics of the particles.

The energy density of the electric field is not integrable, i.e., the field energy of a Coulomb charge is infinite. This picture changes if we consider plasmas, e.g., the OCP, where the charges are imbedded in a uniformly distributed density of counter-charges, neutralizing all charges. Then instead of the Poisson equation determining the field, we get the Debye equation for the effective Debye field surrounding our central charge, as shown in Chap. 1. On the level of Fourier transforms the result for the potential and field is

$$V^{\mathrm{D}}(k) = 4\pi e k^2 \frac{1}{\varepsilon_{\mathrm{r}}(1+\kappa^2/k^2)} , \qquad \mathbf{E}(\mathbf{k}) = -\frac{4\pi e \mathbf{k}}{\varepsilon_{\mathrm{r}} k} \frac{1}{k^2+\kappa^2} . \tag{3.105}$$

We denote the denominator, which formally plays the same role as the relative dielectric constant, as a k-dependent dielectric function of the plasma:

$$\epsilon(k) = 1 + \kappa^2/k^2 . \tag{3.106}$$

Note that in the following we generally assume $\varepsilon_{\mathrm{r}} = 1$. The Coulomb fields try to avoid the separation of charges, and this may also lead to characteristic oscillations, as we have seen already in Chap. 1. If we move the electrons in a uniform electron plasma with density n_{e} by a small amount ξ in the x direction, then Newton's equation of motion requires

$$m_{\mathrm{e}} \frac{\mathrm{d}^2 \xi}{\mathrm{d}t^2} = -e\, \delta E . \tag{3.107}$$

Here δE is the extra field generated by the elongation of the charges. It satisfies Poisson's equation and this leads to the following equation for ξ:

$$\frac{\mathrm{d}^2 \xi}{\mathrm{d}t^2} + \omega_{\mathrm{pe}}^2 \xi = 0 , \qquad \omega_{\mathrm{pe}}^2 = \frac{4\pi e^2 n_{\mathrm{e}}}{m_{\mathrm{e}}} . \tag{3.108}$$

Here ω_{pe} is a characteristic frequency called the electron plasma frequency. Correspondingly, there exists a characteristic time τ_{pe} which characterizes the oscillatory electronic motions. The corresponding time for oscillations of the heavier plasma components, the ions is much shorter, due to the larger masses.

The simultaneous excitation of nonuniformities in space and time are called plasma waves, but also 'Langmuir waves'. These are the typical eigenmodes of the electron density in a plasma. As we will see later, the oscillations can be described as an instability in the dielectric function of a free electron gas. The frequency depends only weakly on the wavelength of the oscillation. Plasma waves were discovered by the American physicists Irving Langmuir and Levy Tonks in the 1920s. As we have shown, the oscillations are due to the restoring forces when the electrons are elongated relative to the background, the ions.

Let us consider an electrically neutral plasma in equilibrium, consisting of electrons and positively charged ions. If we displace all the electrons by a tiny amount with respect to the ions, the Coulomb force pulls the electrons back, acting as a restoring force. Here m_{e} is the electron mass. When the effects of the electron thermal speed $v_{\mathrm{e,th}} = \sqrt{k_{\mathrm{B}} T_{\mathrm{e}}/m_{\mathrm{e}}}$ are taken into account, the electron pressure acts as a restoring force, like the electric field, and the oscillations propagate with frequency and wave number related by the longitudinal Langmuir wave:

$$\omega^2 = \omega_{\text{pe}}^2 + \frac{3k_{\text{B}}T_{\text{e}}}{m_{\text{e}}}k^2 = \omega_{\text{pe}}^2 + 3k^2 v_{\text{e,th}}^2 \ . \tag{3.109}$$

This relation between the frequency of the waves and the wave number k is called the Bohm–Gross dispersion relation.

In a more general setting, the dispersion relation corresponds to the roots of the dielectric function. We define the dielectric function of the medium by the relation between the dielectric function and the electric field for wave-like excitations:

$$D(\omega, \mathbf{k}) = \epsilon(\omega, \mathbf{k}) E(\omega, \mathbf{k}) \ . \tag{3.110}$$

In Chap. 1 we already presented an expression for the dielectric function. This complex function determines most of the kinetic properties of plasmas, as known from many works, and as we will explain in detail in Chap. 7 (Pines 1961, Silin and Rukhadse 1961, Balescu 1963, 1965; Klimontovich 1967, 1982, 1986; Pines and Nozieres 1966). The dielectric function also plays a determining role in the collision term which appears in the kinetic equations of dense plasmas (Klimontovich 1964, 1967, 1982, 1986). The quasiparticle resulting from the quantization of these oscillations is called the plasmon. The theory of these quasi-particles in a plasma was worked out by Klimontovich and Silin (1952), and at nearly the same time by Bohm and Pines (1953). Essential contributions are due to Pines and Nozieres (1966) and (Ichimaru 1973, 1992).

3.5.2 Collective Mode Expansions

We concentrate first on non-degenerate or weakly degenerate plasmas, working in the canonical ensemble. For this region we find it quite convenient to use the methods developed by Kelbg half a century ago (Kelbg 1963, 1964). This method is based on others developed for classical electrolyte theory by Mayer, Haga, and Friedman, and applies the tools of collective coordinates due to Bohm, Pines, Zubarev, and Yukhnovsky. The analogy between classical and quantum-statistical calculations can be seen in the Morita representation of the quantum partition function:

$$F = F_{\text{id}}^B - k_{\text{B}}T \log Z(T, V, N) \ , \tag{3.111}$$

with

$$Z = \text{const.} \int \mathrm{d}r_1 \ldots \mathrm{d}r_N S(r_1 \ldots r_N) \ ,$$

where the Slater function, defined as the diagonal element of the density matrix, is given by

$$S(r_1 \ldots r_N) = \langle r_1, \ldots, r_N | \exp(-\beta H) | r_N, \ldots, r_1 \rangle \tag{3.112}$$

$$= \exp\left(-\beta \sum u_{ij} - \beta \sum u_{ijk}\right) . \tag{3.113}$$

These equations are nothing else than identical reformulations of the exact quantum-statistical partition function.

For Coulomb systems, the pair potential consists of a long range part V_{ab}, which contains a Coulomb tail and has a Fourier transformable part, and a short range part, viz.,

$$u_{ab}(r) = V_{ab}^{\mathrm{K}}(r) + V_{ab}'(r) . \tag{3.114}$$

The fact that the main part of the potential $V_{ab}^{\mathrm{K}}(r)$ has a Fourier transform

$$V_{ab}^{\mathrm{K}}(\mathbf{r}) = \frac{1}{V} \sum_k \exp(-\mathrm{i}\mathbf{k} \cdot \mathbf{r}) V_{ab}^{\mathrm{K}}(\mathbf{k}) \tag{3.115}$$

suggests a special mathematical technique for the evaluation of the classical or quantum partition function which is based on the method of collective variables. This method was developed for Coulomb systems by Bohm and Pines, Zubarev, Yukhnovsky, Kelbg (1964), and Pines and Nozieres (1966). It works in the classical as well as in the quantum case, and it leads typically to the following structure for the partition function:

$$\log Z(T, V, N) = \log Z_{\mathrm{id}}(T, V, N) + \log Z_0(T, V, N) \tag{3.116}$$

$$+ \log Z_{\mathrm{ring}}(T, V, N) + \sum_{n=2}^{\infty} \log Z_n(T, V, N) .$$

Here Z_0 is the lowest order term in the interaction, which is of order e^2, and is connected with the contribution of the zero wave vector $\mathbf{k} = 0$ to the Fourier transform (Kelbg 1964, Yukhnovsky and Holovko 1980, Brilliantov 1998, Ortner 1999):

$$\log Z_0(T, V, N) = -\frac{V}{2k_{\mathrm{B}}T} \sum n_a n_b V_{ab}^{\mathrm{K}}(\mathbf{k} = 0) . \tag{3.117}$$

This contribution is sometimes referred to as the self-energy of the long-range field (Brilliantov 1998). It contains the Hartree terms and possibly also some ideal quantum terms (depending on the definition of the term Z_{id}). The ring term comprises a set of contributions which describe screening effects typical of Coulomb systems, and it contains the Debye contribution and generalizations. The contributions $Z_0(T, V, N)$ and $Z_{\mathrm{ring}}(T, V, N)$ are mainly determined by the Fourier transform of the long-range part of the effective potential and are essentially determined by contributions of order e^4 in the interaction.

In order to give a general picture, we consider the one-component quantum plasma. In this case, Kelbg found the potential already given in an earlier section:

$$V_2^{\mathrm{K}}(r) = \frac{e^2}{r}\left\{1 - e^{r^2/\lambda^2} + \sqrt{\pi}\frac{r}{\lambda}\left[1 - \Phi\left(\frac{r}{\lambda}\right)\right]\right\} . \tag{3.118}$$

We have used here the standard definition of the error function $\Phi(x)$. For the Fourier transform, we get (Kelbg 1963, 1964, Ebeling et al. 1967, 1970, 1976)

$$\tilde{V}_2^{\mathrm{K}}(\mathbf{t}) = \frac{4\pi e_a e_b}{t^2}\exp\left(-\frac{\lambda^2 t^2}{4}\right) {}_1F_1\left(\frac{1}{2}, \frac{3}{2}, \frac{\lambda^2 t^2}{4}\right) , \tag{3.119}$$

where we use the well known hypergeometric functions. For the ring contributions in the partition function, we find (Kelbg 1962, Ortner 1999)

$$\log Z_{\mathrm{ring}}(T, V, N) = \frac{V}{4\pi^2}\int_0^\infty \mathrm{d}t\, t^2\Big\{w(t) - \ln\big[1 + w(t)\big]\Big\} , \tag{3.120}$$

$$w(t) = \frac{\kappa^2}{4\pi}\tilde{V}_2^{\mathrm{K}}(\mathbf{t}) . \tag{3.121}$$

The ring expression reduces in the zeroth approximation to the Debye result. Higher orders will be discussed in Chaps. 5 and 6.

The subsequent contributions to the partition function are a series with cluster structure (Ortner 1999), similar to those from Mayer expansions (Friedman 1962). The method of collective variables in combination with cluster-like expansions works in the classical and in the quantum case. Therefore it is sometimes called the 'semi-classical method'. Frankly speaking, we do not know what this could mean. An exact mathematical method for the evaluation of the partition function is in our view neither classical nor quantum-mechanical, it is just a mathematical method. The only relevant question is the region of good convergence of subsequent approximations using the method in question. For classical ionic solutions and the electron gas, the most extensive evaluation of the series (3.117) is due to Friedman (1962) and Ortner (1999) (see also Kahlbaum 1999, 2000).

3.6 Ionization Equilibrium Between Atoms, Electrons, and Ions

3.6.1 *Eggert–Saha Equation for Ideal Plasmas*

Plasmas are gases consisting of charged particles, mainly electrons and ions, created by the ionization of atoms at high density or high temperature. The development of the quantum statistics of ionization and plasmas around 1920 was intimately connected with the requirements of the scientific discipline of astrophysics. Although astronomy is as ancient as recorded history itself, it long remained separate from the study of physics. Only at the end of the 19th century was it discovered that, in the light from the Sun, a multitude of spectral lines were observed. Experiments with hot gases showed that the same lines could be observed in the spectra of gases, specific lines corresponding to unique chemical elements. In this way it was proved that the chemical elements found in the Sun were also found on Earth. Indeed, the element helium was first discovered in the spectrum of the Sun and only later on Earth, hence its name.

In order to obtain a theory of abundances, an ionization theory was developed by the work of the German physico-chemist John Eggert (1891–1973) and the Indian physicist Meghnad Saha (1893–1956). This work was built on the theoretical results obtained a few years earlier by Nernst, Sackur, and Tetrode and by Bohr. The Rutherford model provided a first understanding of why electron scattering can penetrate so deeply into the interior of atoms, a rather unexpected experimental result. Rutherford explained his scattering results as the passage of a high speed electron through an atom having a positive central charge $+Ne$, and surrounded by a compensating charge of N electrons.

According to the Bohr theory the radii of the orbits of electrons in an atom and the energies are

$$a_s = s^2 a_{\rm B} \,, \qquad a_{\rm B} = \frac{\hbar^2}{\mu e^2} \,, \qquad E_s = -\frac{\mu e^4}{2\hbar^2 s^2} \,, \tag{3.122}$$

where μ is the reduced mass. Here and in the following we denote the so-called main quantum number by s to avoid confusion with the density n. According to the Bohr model, there are infinitely many levels close to the series limit $s \to \infty$ (see Fig. 3.4).

In order to give a brief derivation of the Eggert–Saha equation, we consider a simple binary Coulomb system with n_+ positive ions (cations) and n_- negative charges (electrons) per cubic centimeter, where $n_+ = n_-$. The density of free atoms is $n_{\rm a}$. The total density is $n_0 = n_+ + n_{\rm a} = n_- + n_{\rm a}$. In the following we will use the plasma notations, i.e., we call the negative charges *electrons* and the positive charges simply *ions*. In the spirit of Nernst's and Eggert's ideas, we have a chemical equilibrium between electrons, ions, and atoms:

$$e^- + i^+ \longrightarrow a_0 \,, \qquad \mu_{\rm e} + \mu_{\rm i} = \mu_{\rm a} \,. \tag{3.123}$$

Here the μ_k are the chemical potentials for the corresponding species. For the atoms,

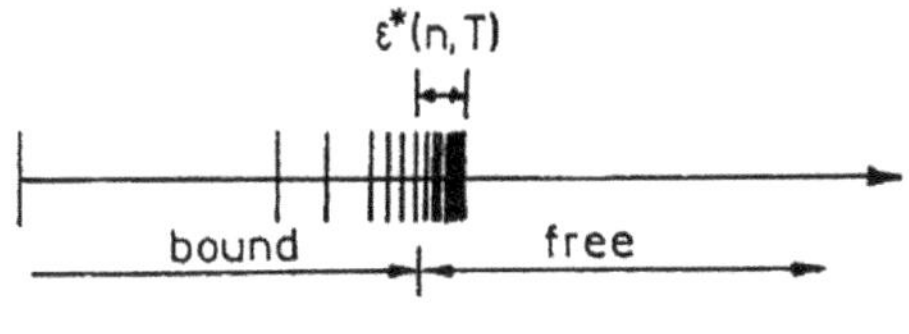

Fig. 3.4 Planck's diagram of the atomic levels of hydrogen. *On the left*, the ground state and the lower bound states, followed by the weakly bound discrete states, the series limit, and *on the right* the continuum

$$\mu_{\mathrm{a}} = I + k_{\mathrm{B}} T \ln n_{\mathrm{a}} + k_{\mathrm{B}} T \ln \Lambda_{\mathrm{a}}^3 , \tag{3.124}$$

where I is the ionization energy, and for the charges,

$$\mu_{\mathrm{e}} = k_{\mathrm{B}} T \ln n + k_{\mathrm{B}} T \ln \lambda_{\mathrm{e}}^3 , \qquad \mu_{\mathrm{i}} = k_{\mathrm{B}} T \ln n + k_{\mathrm{B}} T \ln \Lambda_{\mathrm{i}}^3 . \tag{3.125}$$

We mention that the last terms here are the constants connected with the work of Sackur and Tetrode, which contain Planck's constant $\Lambda_k = h/\sqrt{2\pi m_k k_{\mathrm{B}} T}$. The condition of chemical equilibrium implies the Eggert–Saha formula

$$\frac{n_{\mathrm{a}}}{n_{\mathrm{e}} n_{\mathrm{i}}} = \Lambda^3 \exp\left(\frac{I}{k_{\mathrm{B}} T}\right) , \qquad \Lambda = \frac{\Lambda_{\mathrm{e}} \Lambda_{\mathrm{i}}}{\Lambda_{\mathrm{a}}} . \tag{3.126}$$

In the work of Saha, this simple reasoning was extended to more complicated ionization phenomena, including multiple ionization. For quantitative astronomy, Saha's work was a real breakthrough (see Chap. 1).

A pioneer of applications to astrophysics and the physics of the Sun was Albrecht Unsöld (1905–1995). As a student of Sommerfeld in Munich, he learned all the details of the new theory of atoms. After a Rockefeller fellowship, he was an assistant at the Potsdam Observatory and the Mount Wilson Observatory in Pasadena, California. From 1930, he was an assistant at the Institute of Theoretical Physics at the University of Hamburg, and became finally Ordinarius Professor and Director of the Institute for Theoretical Physics and the Observatory at the University of Kiel. He is well known for his fundamental contributions to the theory of ionization phenomena in astrophysics (Unsöld 1948, 1955).

3.6.2 *Regularization of the Atomic Partition Function*

The problem of the divergence of the partition function of the Bohr atom was already seen by Bohr, but studied then in detail by Herzfeld and later by many other workers including Planck. For an atom with the energy levels E_{sl} depending on the main quantum numbers s, l, the atomic partition function is defined by

$$\sigma(T) = \sum_{s,l} \exp(-\beta E_{sl}) . \tag{3.127}$$

According to the definition (3.127), the terms in the atomic partition function for hydrogen diverge as s^2 (see Fig. 3.5). The problem of the divergence of the atomic partition function was already noticed by Bohr, and in particular by Herzfeld, but it took a long time before the mathematical background for a serious treatment of this problem was available. The first quantum-statistical solution of the Coulomb divergence problem of the partition function was given by Planck (1924).

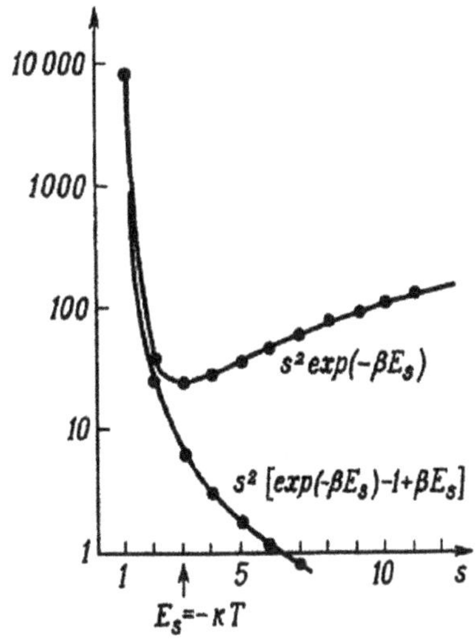

Fig. 3.5 Terms in the atomic partition function for $T \sim 17\,500$ K in comparison to the terms in the regularized formula (3.131). Up to terms of order $E_s^{\max} \approx k_B T$, the terms decrease monotonically and then start to increase as s^2. For comparison, we show the Brillouin–Planck–Larkin renormalized partition function

As briefly discussed in Chap. 1, Planck's idea was in fact to divide the partition function Σ into several contributions, as shown in Fig. 3.4:

$$\Sigma = \Sigma_1 + \Sigma_2 + \Sigma_3 , \tag{3.128}$$

where Σ_1 is the contribution of the scattering states, Σ_2 the contribution of the high bound states near the series limit, and Σ_3 the contribution of the lowest bound states. Next Planck proved that there was some compensation between the contributions of the high bound states and the scattering states, and arrived at the formula

$$\Sigma = \frac{V}{\Lambda^3} + \Sigma_3 . \tag{3.129}$$

Since the free contribution is not counted as an internal atomic contribution, we get

$$\sigma(T) = \Sigma_3 = \sum_{sl}^{E_{sl} < -\epsilon'} \mathrm{e}^{-\beta E_{sl}} . \tag{3.130}$$

In the more general case that the energy levels depend on density, which will be discussed later, we need an appropriate density- and temperature-dependent choice of the upper boundary $\epsilon'(n, T)$. At low density, $\epsilon'(n, T)$ should be near the terms in the partition function where the minimum of the terms in the sum is located.

A disadvantage of the Planck procedure or generalizations working with upper limits of the s-summations is the possibility of discontinuities, when levels cross the upper limit with changing density. A procedure which at least for hydrogen-like plasmas avoids this problem was worked out by 1930. Much later, in the 1960s, a more rigorous foundation was worked out in a letter by Larkin (1961) and in extended work by one of the present authors in collaboration with Kremp and Kraeft (Ebeling 1967, 1968, Ebeling et al. 1976). The regularization is based on subtracting the first divergent terms:

$$\sigma(T) = \sum_{s=1}^{\infty} s^2 \left(\mathrm{e}^{-\beta E_s} - 1 + \beta E_s \right) . \tag{3.131}$$

The formal justification for this subtraction is, that terms up to order e^4 in the charges are needed for the Debye screening procedure, so they cannot appear once more in the partition function. Otherwise this would be a double-counting of the terms of low order in e^2. In Chap. 4, this argument will be explained in more detail on the basis of rigorous quantum-statistical derivations.

The convergent (regularized) partition function given by (3.131) is called the Planck–Brillouin–Larkin (PBL) partition function. The deeper quantum-statistical justification for the procedure in (3.131) is based on the fact that the contributions up to $\mathcal{O}\left(e^4\right)$ are influenced by screening effects and have to be treated separately. An alternative way to express this is the observation that there are compensation effects between contributions from terms just below and just above the series limit (see Fig. 3.4). Due to these and other physical effects, the sum of contributions to any thermodynamic quantity is always finite (see Ebeling et al. 1976).

Planck's approach to the compensation effect between high level bound states and scattering states was essentially quasi-classical and difficult to reproduce. Planck's and Brillouin's calculations were reproduced and fully confirmed by one of these authors (Ebeling 1967). We will come back to the quantum-statistical foundation of PBL regularization in Chap. 6.

Lately, several papers have appeared which state that the PBL partition function is 'wrong' and should be replaced by other 'more correct' choices. For example Vorobjov and Khomkin (1971) and later Starostin et al. (2003, 2006) obtained a different regularized partition function. Starostin et al. even came to the conclusion that the PBL partition function is not correct (Starostin et al. 2003, 2006). Does this mean, that Planck, Brillouin, and Larkin are wrong in spite of the fact that nobody was able to find a concrete mathematical or physical error in their calculations. In our view, there is no error at all, there are just different interpretations of existing contributions either as bound state or free state terms.

Related questions arise in different fields of physics: If two different theories using other theoretical tools get different results for the same physical quantitiy, the question arises, whether one of them is necessarily wrong and the other is correct. In a famous historical controversy about the correctness of theoretical expressions for the mass action constant of ion association, Lars Onsager made clear that some problems are more complicated and that statements about correctness or incorrectness do not always have a precise meaning. We follow here Lars Onsager's views (see Hemmer et al. 1996) and come to the conclusion that statements about correctness of regularized partition functions do not have a direct physical meaning since the choice of Coulomb partition functions is to some extent open to choice (Ebeling et al. 1976, 2002, 2003).

Let us analyze once again the well known statement by Onsager for a related problem in electrolyte theory, which can be transferred to our problem. In a remark in a discussion at the electrolyte conference held in Montpellier in 1968, concerning the choice of the mass action constant for ion association Onsager said (see Falkenhagen et al. 1971, and Onsager in Hemmer et al. 1996, p. 249):

> Bjerrum's choice is good but we could vary it within reason. In a complete theory this would not matter; what we remove from one side of the ledger would be entered elsewhere with the same effect.

The problem of defining the mass action constant for electrolytic association is completely equivalent to our problem of the choice of a plasma partition function and the corresponding mass action constant. Applying Onsager's theorem to our problem, we come to the conclusion that there is indeed some freedom in defining the partition function, but in a complete theory this would not matter according to Onsager, because "what we remove from one side of the ledger would be entered elsewhere with the same effect". In correct theories, which consider "both sides of the ledger", the different choices of partition functions may be used (within certain limits) on equal rights. Following Onsager, we would prefer, to apply the terms *correct* or *wrong*, not to the bound state partition function, but only to the expression for complete thermodynamic functions, containing a sum of bound state and free state contributions. Following Onsager, the statement that a partition function of bound states (or the corresponding mass action constant) is correct or is wrong is meaningless in the context of a correct imbedding theory. The terms *correct* or *wrong* should be applied only to observable quantities, while partition functions correspond to certain contributions to observables but cannot be measured in isolation from other contributions. We thus conclude, modifying Onsager's statement that *the Planck–Brillouin–Larkin choice is good, but we could vary it within reason, and in a complete theory this would not matter.*

A different and quite meaningful question is, what is the most appropriate sum over states at higher densities. Possible concepts for the extension to finite densities will be discussed later. Note a related recent discussion of the relation to nearest-neighbor concepts by Khomkin et al. (2003, 2006, 2008). In order to make our view clear, we think that Onsager's opinion is correct and that the PBL - partition is for low densities the most appropriate choice. However it makes no sense to say that the PBL-partition is exact and also this does not mean that the PBL - partition is at any densities and for any investigation the best choice. This is a question which indeed has to be decided by a careful comparison with empirical data (Ebeling et al. 1973, 1976, 1989, Starostin et al. 2006, Khomkin et al. 2004, 2008, Lankin and Norman 2009, Däppen 2010).

3.7 Bound States and Ionization Equilibrium in Nonideal Plasmas

3.7.1 Weakly Nonideal EOS and Saha Equation

Let us repeat the basic ideas of the ionization theory but in a different form. As for any chemical reaction, the free energy as a function of the composition must have a minimum in thermodynamic equilibrium. For the ionization reaction

$$a \longrightarrow e^- + i^+ \,, \tag{3.132}$$

the free energy is a function of all particle numbers $F = F(T, V, N_\mathrm{e}^*, N_\mathrm{i}^*, N_\mathrm{a}^*)$. Here and in the following we associate the free particle numbers and densities with an asterisk. Note that sometimes, where the situation is unambiguous, we take the liberty to omit the stars. Since particles have the 'choice' of being in a free or a bound state, we have the conservation and neutrality conditions:

$$N_e = N_\mathrm{e}^* + N_\mathrm{a}^* \,, \quad N_i = N_\mathrm{i}^* + N_\mathrm{a}^* \,, \quad N_\mathrm{e}^* = N_\mathrm{i}^* \,, \quad N_e = N_i = N_0 \,.$$

For simplicity we consider here and in the following a binary Coulomb system with n_+ positive ions (cations) and n_- negative charges (electrons) per cubic centimeter, where $n_+ = n_-$. The density of neutrals is n_0. The total density of electrons is $n = n_+ + n_0 = n_- + n_0$. In the following we call the negative charges 'electrons' and the positive charges simply 'ions'.

As shown above, the densities of free and bound particles are connected by the mass action law (Saha equation). The condition of chemical equilibrium implies the Saha formula, which reads, with corrections with respect to nonideality,

$$\frac{n_\mathrm{a}}{n_\mathrm{e} n_\mathrm{i}} = \Lambda^3 \sigma(T) \exp\left(\frac{\mu_\mathrm{ex}}{k_\mathrm{B} T}\right) \,, \qquad \Lambda = \frac{\Lambda_\mathrm{e} \Lambda_\mathrm{i}}{\Lambda_\mathrm{a}} \,. \tag{3.133}$$

For completeness, we mention an alternative choice of the mass action law which is sometimes used in electrolyte theory (Justice et al. 1979):

$$\frac{n_\mathrm{a}}{n_\mathrm{i}} = n_o \Lambda^3 \sigma(T) \exp\left(\frac{\mu_\mathrm{ex}}{k_\mathrm{B} T}\right) \,, \qquad \Lambda = \frac{\Lambda_\mathrm{e} \Lambda_\mathrm{i}}{\Lambda_\mathrm{a}} \,. \tag{3.134}$$

The standard choice of the mass action constant for quantum plasmas is the Brillouin–Planck–Larkin expression, where $\sigma(T)$ is the atomic partition function

$$\sigma(T) = \sum_{sl} w_{sl}(2l+1) \exp\left(-\beta E_{sl}\right) \,. \tag{3.135}$$

The corresponding excess chemical potential is

$$\mu_{\mathrm{ex}} = -\frac{e^2\kappa}{\varepsilon_{\mathrm{r}}} G(\kappa\lambda) , \tag{3.136}$$

and the EOS assumes the form

$$\beta p = n_{\mathrm{e}} + n_{\mathrm{i}} + n_{\mathrm{a}} - \frac{\kappa^3}{24\pi} \Phi(\kappa\lambda) + O(n^{5/2}) . \tag{3.137}$$

Here $\kappa^2 = 4\pi(n_{\mathrm{e}} + n_{\mathrm{i}})e^2$ and n_{a} is the free atom density. Introducing the excess chemical potential we found before into the condition of chemical equilibrium, we get two different forms of the nonideal Saha equation:

- In QDHA. The nonideal Saha equation was already given in Ebeling et al. (1976) and reads

$$\frac{1-\alpha}{\alpha^2} = n_0 \Lambda_{\mathrm{e}}^3 \sigma(T) \exp\left(-\beta e^2 \frac{\kappa_0\sqrt{\alpha}}{1+\kappa_0\sqrt{\pi\alpha}\lambda/4}\right) . \tag{3.138}$$

  Here $a(T)$ is the effective quantum diameter of charges.
- In KEPA. The corresponding nonideal Saha equation for the degree of ionization α reads, in analogy to the quantum Debye–Hückel approximation (Ebeling 2016),

$$\frac{1-\alpha}{\alpha^2} = n_0 \Lambda_{\mathrm{e}}^3 \sigma(T) \exp\left(-\beta e^2 \frac{\kappa_0\sqrt{\alpha}}{\sqrt{1+\sqrt{\pi}\kappa_0\lambda_{\mathrm{ie}}\sqrt{\alpha}/2}}\right) , \tag{3.139}$$

  where once again $\kappa_0^2 = 8\pi n_0 e^2$ and n_0 is the total atom density.

The two Saha equations are quite similar in their general shape, but with some differences at higher densities.

With increasing density, the degree of ionization goes to zero and the relative pressure to fifty percent of the ideal pressure of an electron–ion gas. This is due to the fact that the number density of an atomic hydrogen gas is half the density of an electron–ion gas. This simple picture changes drastically if further effects are taken into account, in particular (Fig. 3.6):

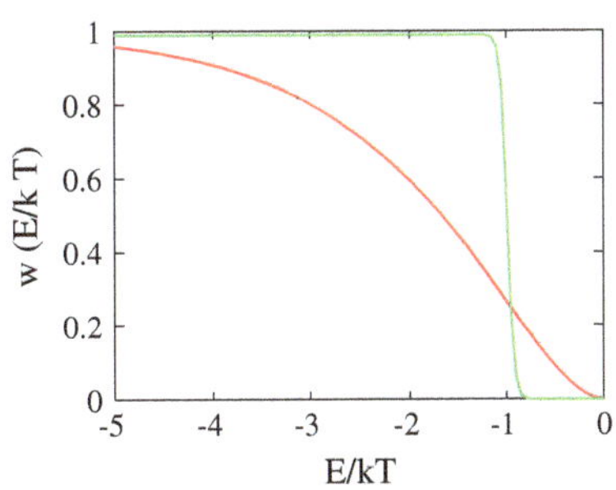

Fig. 3.6 Dependence of the weights on the bound state energies for the BPL partition function (*red curve*), showing a smooth decrease of the weights, and according to RRS (*green curve*), with a sharp cut in the weights at $-E \sim k_{\mathrm{B}}T$

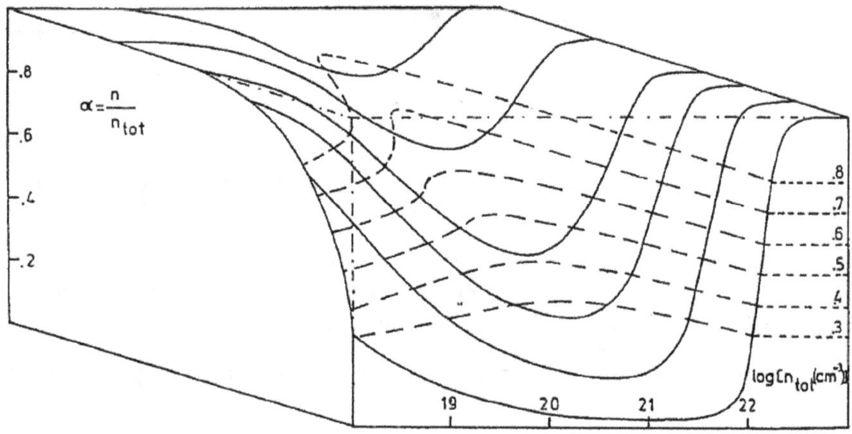

Fig. 3.7 The valley of bound states (atoms) of hydrogen in the temperature–density plane

- Fermi–Dirac contributions to the ideal free energy.
- Excluded volume effects.
- Hard-core contributions of the atoms to the free energy.
- Density-dependent shifts of the energy levels due to confinement.
- Including these effects at least partially, we observe that the degree of ionization goes up again with increasing density, and we get the overall picture of a 'valley of bound states'.

As an example, Fig. 3.7 shows the ionization of a hydrogen plasma as a function of density and temperature, restricting to regions where hydrogen molecules do not play a role.

3.7.2 *Nonideality in Atomic Partition Functions*

For hydrogen-like plasmas the contributions to the partition function diverge as s^2, where s is the main quantum number. As already discussed in Chap. 1, the most important possibilities for regularization are:

- Smooth regularization by omitting divergent terms in the exponential series:

$$w_{sl} = 1 - \exp(\beta E_{sl}) + \beta E_{sl} \exp(\beta E_{sl}) \ . \tag{3.140}$$

  Since the energies are negative, the factors disappear for all small energies near the series limit (see Fig. 3.6). For hydrogen, we get the Brillouin–Planck–Larkin partition function once again (Planck 1924, Brillouin 1930, 1931, Larkin 1960, Ebeling 1967, 1968).

- Regularization by omitting terms above a small negative bound ϵ^*. A good choice is a value around the minimum of the contributions to the partition function, or to use an equivalent cut criterion, as proposed, e.g., by Riewe, Rompe, and Steenbeck Riewe 1939, (Rompe and Steenbeck 1939):

$$\sigma(T) \approx \sum_{E_{sl}<-\epsilon^*} (2l+1)\exp(-\beta E_{sl}) \,. \tag{3.141}$$

  Here ϵ^* is an upper limit for the energy levels included in the partition function. A standard possibility is $\epsilon^* \sim k_B T$. Note that Riewe and Rompe used $s_{\max} \sim 5$ corresponding to about 20 000–30 000 K.

We have assumed so far that the levels are not influenced by the density and temperature of the plasma. This is certainly only a rough approximation, since the energy levels and their broadening are influenced by the conditions in the surrounding plasma, as is well known from observations of the spectra.

In strongly nonideal plasmas, the energy levels may be density- and temperature-dependent, and we may have to modify the atomic partition function. Since the ionization energy $I = |E_{10}|$ plays a special role here, it is sometimes useful to have a different way of writing this:

$$\sigma(T) = \exp(\beta I) \sum_{sl} w_{sl}(2l+1)\exp\left[\beta(I - E_{sl})\right] = \exp(\beta I)\sigma_1(T). \tag{3.142}$$

In general the energy levels are shifted with respect to the lines in vacuum and the shifts depend on the density and temperature:

$$E_{sl}(n, T) = E^0_{sl} + \delta\, E_{sl}(n, T) \,. \tag{3.143}$$

The question about the partition function in dense media is very difficult and has, in our view, not yet found a really satisfactory and consistent solution. In this area there are many approaches and many controversies. One of the first papers which studied the influence of electric micro-fields is due to Inglis and Teller (1939). Among the early papers, we should also mention those of Unsöld (1955), Ecker and Weizel (1956), and Mott (1974).

Khomkin and Shumikhin suggested choosing the weight factor on the basis of a nearest neighbor approximation (NNA), viz.,

$$w_s = \exp\left[-(n_e + n_i)v_s\right], \qquad v_s = \frac{4\pi}{3}(2a_B s^2)^3 \,, \tag{3.144}$$

where v_s corresponds to the volume of the excited state with main quantum number s. We should also mention a different approach to the question of how to separate bound and free states correctly, as put forward by Starostin and Roerich (2006).

Mott was one of the first to predict the disappearance of the hydrogen energy levels beyond a certain critical density, due to screening. In his pioneering work, Mott considered the Schrödinger equation for the ground state of the Debye potential:

$$\frac{\hbar^2}{2m_e}\Delta\psi_{100} + \frac{e^2\exp(-\kappa r)}{r}\psi_{100} + E_{100}\psi_{100} = 0\,. \tag{3.145}$$

Solving by means of a variational ansatz with a free parameter γ,

$$\psi_{100}(r,\gamma) = \frac{\gamma^3}{\sqrt{\pi}}\exp(-\gamma r)\,, \tag{3.146}$$

Mott came to the conclusion that the ground state shifts with density according to

$$E_{10}(\kappa) \approx E_{10}(0)(1-\kappa a_B)\,, \tag{3.147}$$

and disappears at a density corresponding to $\kappa a_B \sim 1$. Later numerical estimates found instead for the critical value $\kappa a_B \sim 1.2$ (Rogers et al. 1970) (Fig. 3.8).

We will see later that the linear shift with $\kappa \sim \sqrt{n_e}$ leads to some problems. We shall therefore discuss an alternative approach, leading only to a decrease of quadratic order (Ebeling 1974). The new argument is based on the fact that it is not

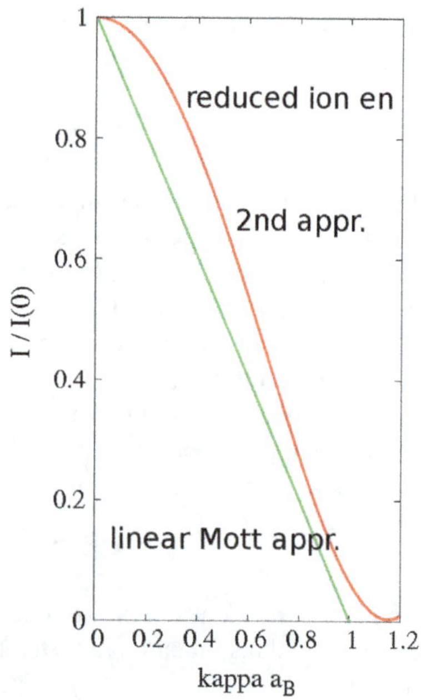

Fig. 3.8 Disappearance of ground state contributions with increasing Debye parameter. Comparison of the Mott approximation which is linear in κ and a second approximation (3.151) resulting from a charging procedure which starts with κ^2

the ground state energy itself which contributes to the free energy. In fact, we need a Debye charging procedure. Starting from (3.142), we see that the necessary free energy consists of contributions from the atoms and the free charges

$$F = F(T, V, N_e^*, N_i^*, N_a^*) = N_a^* I + N_a^* k_B T \ln \sigma_1 + F(T, V, N_e^*, N_i^*) \,. \quad (3.148)$$

Since the first term is a contribution to the free energy, it should be calculated by a Debye charging process:

$$\delta F_{100} = N_a^* \int_0^1 d\mathbf{r} \, \frac{e^2}{r} |\psi_{100}|^2 \,. \quad (3.149)$$

Using this ansatz we find

$$\delta F_{100} \approx |E_{10}| \left[1 - \frac{3}{2} (\kappa a_B)^2 + \cdots \right] \,. \quad (3.150)$$

A rather convenient fit formula for the effective ionization energy is

$$\delta I_{100} \approx |E_{10}| \left[1 - \frac{3}{2} (\kappa a_B)^2 + .565 (\kappa a_B)^4 \right] \,. \quad (3.151)$$

We now discuss another qualitative approximation which is relevant in the limit of high densities. We are referring to the so-called confined atom model (Graboske et al. 1969, Rogers et al. 1970, Bushman and Fortov 1983). The physical idea behind this is that the atomic levels are not formed in an infinite vacuum, but in a finite cage formed by other particles. The electron–ion pair forming a bound state has on average, at distance r_0, a neighbor which will deform or destroy the bound state orbit. The average radius is given by

$$4\pi N_0 r_0^3 / 3 = V \,. \quad (3.152)$$

This effect has two consequences:

- Bound states with larger radius $a_{sl} > r_0$ are on average not stable and have to be excluded from the partition function.
- The energy levels have to be shifted up, due to the different boundary conditions.

In order to give a simple estimate of the shift, we assume that the bound states are contained in a spherical box and that the wave function disappears at the boundary. Then a simple estimate follows from the Schrödinger equation for spherical boxes:

$$\delta E_{sl} = x_{sl} \frac{a_B^2}{r_0^2} Ry \,, \quad x_{10} = 3.14 \,, \quad x_{11} = 4.49 \,, \quad x_{12} = 5.76 \,, \quad x_{20} = 6.28 \,.$$

Assuming additivity of the box effect as a first approximation, we get

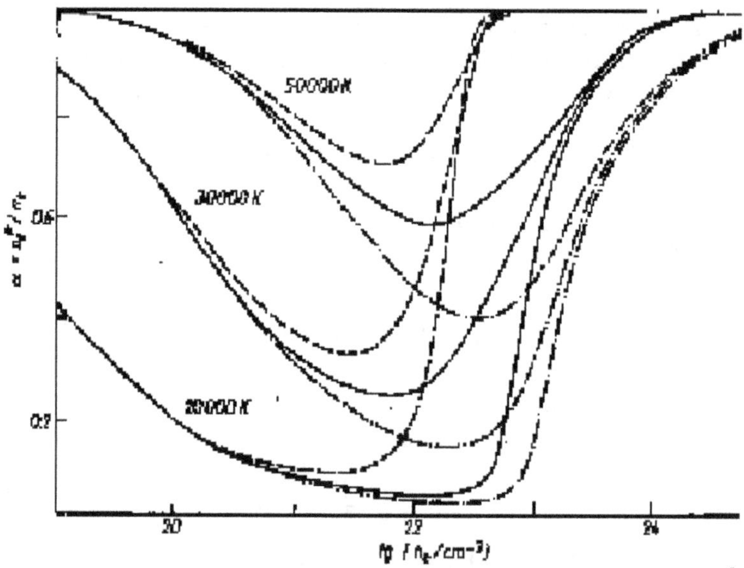

Fig. 3.9 Degree of ionization of H plasmas for the three temperatures $T = 50\,000$, $30\,000$, and $20\,000$ K. We show the influence of different physical effects influencing the valley of ionization. *Continuous curves* correspond to the nonideal Saha equation with excluded volume effects. *Dash-dotted curves* correspond to the same theory without the excluded volume effect. *Dashed curves* also include the shift in the energy levels due to confined atom effects

$$E_{sl} \sim E_{sl}(r_0 = \infty) + \delta\, E_{sl}(r_0)\,. \tag{3.153}$$

The shift is always positive and increases as $r_0^{-2} \sim n_0^{2/3}$. According to this simple estimate, the ground state level of H-plasmas is expected to disappear for $r_s = r_0/a_B < 2$. We thus expect the confinement effect to be essential only at densities above $n_0 > 10^{22}$–10^{23} cm^{-3}, i.e., if the density of the electrons approaches the electron density in metals. An application to hydrogen is shown in Fig. 3.9.

Note that the well-known astrophysicist Unsöld (1955) already recommended using the condition $a_{sl} > r_0$ as a criterion for the regularization of the partition function. However, for normal plasmas, the shifts due to confinement are evidently less relevant than the shift due to screening effects, already studied by Ecker and Weizel (1956) and Mott (1961), which are nowadays usually called Mott effects. According to Mott, the ground state of a hydrogen-like atom disappears for small values of the Debye radius $r_D < a_B$ (Mott 1974). A more precise calculation by Rogers et al. (1970) gives a smaller factor $r_D < 0.84 a_B$.

In more detail, the levels disappear one after the other, starting from the largest orbits. The hydrogen level s is expected to disappear if the orbit no longer has enough room inside the Debye radius, i.e., $r_D < s^2 a_B$. In early studies by Ecker and Weizel (1956) and by Jackson and Klein (1969), it was shown that the energy levels should not be calculated from solutions of the Schrödinger equation with the Debye potential, but instead from the shifted potential

$$V^{\mathrm{EW}}(r) = -e^2 \left[\frac{1}{r} \exp(-\kappa r) + \frac{1}{r_\mathrm{D}} \right] . \tag{3.154}$$

The idea behind this is that the natural reference level for the electron energy in a plasma is not $\epsilon = 0$ but $\epsilon = \mu_\mathrm{e}$, where μ_e is the chemical potential of an electron in the plasma. We remember that the Debye estimate for the chemical potential is $\mu_\mathrm{e} = -e^2/r_\mathrm{D}$. An electron in a plasma is always surrounded by a screening cloud of positive charge and therefore sits in a potential hole of energy $\mu_\mathrm{e} = -e^2/r_\mathrm{D}$. The chance to form a bound state with a positive charge is 'of interest for the electron' only if the level is deeper than the energy of the screening hole. In our view this is a strong argument in favor of the procedure proposed by Ecker, Weizel, Jackson, and Klein.

At the ICPIG conference in Berlin in 1977, it was shown that the Ecker–Weizel–Jackson–Klein approach may be derived on the basis of the Bethe–Salpeter equations (BSE) (Ebeling et al. 1977). The BSE are effective wave equations. In later papers, this approach was worked out by several authors (Röpke et al. 1978, Zimmermann et al. 1978, Kraeft et al. 1986, Kilimann and Ebeling 1990) and summarized in two books (Zimmermann et al. 1978, Kraeft et al. 1986, Kremp et al. 2005). Energy shifts for higher quantum states and for ions with $Z > 1$ have been discussed in Kilimann and Ebeling 1990, and the influence of Pauli blocking effects on the Mott transition in Ebeling et al. (2009), Redmer et al. (2010).

The main conclusion to be drawn is that the continuum decreases almost linearly in $a_\mathrm{B}/r_\mathrm{D} \sim n^{1/2}$, and that the low levels are nearly constant and change only at higher orders in the density. This will be explained in more detail in Chap. 6. Note that the continuum edge may be approximated by the excess part of the plasma chemical potential, which is given to zeroth order by the Debye potential (Kraeft et al. 1986, Kilimann et al. 1990, Ebeling et al. 2009, Redmer et al. 2012). The question of the density dependence of the discrete energy levels and the continuum is one of the more difficult issues of plasma physics, connected to the observations of level shifts and level broadening (Griem 1968).

We note that there are still several open problems, and in particular, the consistency of the density-dependent partition functions obtained in this way with higher order expressions for the thermodynamic functions. In this chapter, we will use mainly unperturbed energy levels and include a density-dependence of the partition functions only via the merging of levels into the continuum. We will show in Chaps. 5–7 that this approach is fully consistent with known rigorous results.

The density dependence of levels and the disappearance of energy levels have a strong influence on the degree of ionization $\alpha = n^*/n$, defined as the ratio of the number of free charges to the total number of charges, including those in bound states. Figure 3.9 shows the relative pressure, which reflects the behavior of α for three temperatures as a function of the total density of electrons (free and bound). The figure also shows the influence of some other effects, such as the excluded volume effect. We see that excluded volume effects and level shift effects raise the minimum and make the observed 'valley of ionization' flatter.

There exists a separate line of inquiry in the theory of ionization effects which goes back to the work of the astrophysicist Unsöld (1948, 1955) from Kiel. Of special interest in our context is the fact that he studied the lowering of the ionization energy and derived an $n^{1/3}$ law for this (Unsöld 1948, 1955). The estimate for the lowering derived by Unsöld (1948) is

$$\Delta I = 3(Z+1)^{2/3} e^2 \left(\frac{4\pi n_\mathrm{e}}{3}\right)^{1/3} . \tag{3.155}$$

For a more detailed discussion, see (Stolzmann and Ebeling 1998). Unsöld's result (1948, 1955) is clearly related to the high density asymptotic of the thermodynamic functions discussed below. We will discuss this topic in more detail in Chap. 6.

So far we have only taken Coulomb contributions into account. In the work of Richert et al. (1984) and Arkhipov et al. 2000, 2011, 2015 interactions with neutrals were also included using different methods. Let us note to end this section that the classical description of ion association may be treated in a quite analogous way (Falkenhagen et al. 1971, Ebeling and Grigo 1980, 1982). In this way we have seen that the classical and quantum cases might be treated in the present approximation by almost the same procedure.

3.8 Correlations in Noble Gas and Alkali Plasmas

3.8.1 Effective Potentials for Noble Gas and Alkali Plasmas

Noble gas plasmas are a standard topic of plasma research. This is in particular connected with the needs of traditional light source technologies, like helium and neon lamps. Noble gas plasmas are also a standard target for theoretical studies of plasma properties, such as the equation of state Ebeling et al. 1991, 2003, 2007, Fortov 2011). Note that noble gases like helium are a substantial component of our Universe and we need to know more about it in order to gain a deeper understanding of the properties of the dominant matter in the Universe (Fortov 2011, Fortov 2015).

During recent decades, a considerable amount of experimental and theoretical effort has been applied to describe the behavior of alkali metals in the liquid and plasma states produced by heating them (Richert et al. 1984, Hensel et al. 1985, Ebeling et al. 1987, Bonzel et al. 1989, Kuhlbrodt et al. 2005, Sizyuk et al. 2006). There have been important experimental and theoretical contributions to the understanding of the liquid–vapor phase transition of metals (Hensel and Pilgrim 2003, Kuhlbrodt et al. 2005, Hensel and Warren 2014).

It is well known that metal fluids are fundamentally different from normal insulating fluids in that the interparticle potentials are not quantities related to intrinsically atomic properties, but rather depend strongly on the nature of the electron gas which screens the Coulomb interaction. Furthermore, recent work has shown that, close to the metal–nonmetal transition, which occurs in the liquid when it is made to expand by heating to critical conditions, the nature of the interparticle interaction changes dramatically, from metallic to a van der Waals type interaction. Of particular interest for theoretical and experimental studies has been the liquid–vapor critical point (Hensel et al. 1985).

In recent years hot alkali plasmas have also been widely employed in many technical projects. For instance, lithium plasmas are planned for use in inertial confinement fusion devices, in solar power plants, for electrochemical energy storage, in magnetohydrodynamic power generators, and in many other applications. Recent advances in the field of extreme ultraviolet lithography have revealed that laser-produced Li plasmas are source candidates for next generation microelectronics (Sizyuk et al. 2006).

We begin by considering the class of noble gas plasmas. Here the outer electrons are rather loosely bound, while the core electrons are so tightly bound that the core is more ar less like a hard sphere. A similar picture may be applied for the description of noble gas atoms which are also like hard spheres. Note that in some approximation the ions may also be modeled as hard spheres with volume R_{i}.

We consider an elementary analytical model of hard sphere effects, if a plasma consisting of electrons with density n_{e} and (point-like) ions with density n_{i} is imbedded in a gas of atoms with radius R_{a} and density n_{a}. A rather simple model of the contribution of hard-core effects to the free energy is (Ebeling and Richert 1985)

$$F_{\mathrm{hc}} = V n_{\mathrm{a}} \frac{4\eta - 4\eta^2}{(1-\eta)^2} - V(n_{\mathrm{e}} + n_{\mathrm{i}}) \ln(1-\eta) \,. \tag{3.156}$$

Here the first term is the analytical expression for the free energy of hard cores proposed by Carnahan and Starling in 1969, while the second term represents the so-called excluded volume effect. The latter models the physical effect that, due to the Pauli principle, the plasma electrons cannot penetrate into the core of atoms which are already filled with electrons. In this elementary picture, both effects are described by just one parameter $\eta = 4\pi n_{\mathrm{a}} R_{\mathrm{a}}^3/3$, which gives the relative occupation of the total volume by atoms. The excluded volume effect takes into account the fact that the volume $4\pi V n_{\mathrm{a}} R_{\mathrm{a}}^3/3$ is not available for the electrons and ions and has to be excluded from the ideal free energy of electrons and ions.

To give an example, the atom radius for xenon is around $R_{\mathrm{a}} \sim 2$ Å and the ion radius is $R_{\mathrm{i}} \sim 1.8$ Å. A more precise choice would be the temperature-dependent radii proposed by Barker and Henderson (1967). We will see that the excluded volume effect is one of the most important reasons for the pressure ionization of matter. It is well known from observations that, at high density, e.g., in the interior of stars, atoms are always fully ionized. Many effects contribute to this, but in the end we will see that the thermodynamic effect due to the pressure of the free charges plays the dominant role.

Let us now discuss the screening effect. Within the quantum Debye–Hückel approach, the simplest way to take into account the fact that the ions may also have a radius is the replacement of the thermal wavelength, which played the role of an effective radius, by a sum

$$a \to a(T) = R_\mathrm{i} + \frac{\sqrt{\pi}}{4}\lambda\,, \qquad \lambda = \lambda_\mathrm{ie}\,. \tag{3.157}$$

From now on we shall use the wavelength of relative motion of electron–ion pairs as the most relevant variable. More precise approximations and a detailed analysis of the EOS of partially ionized xenon gases can be found in (Ebeling et al. 1988, 1991, 2003, 2007).

So far we have discussed the physics and the elementary theory of noble gas plasmas. However, alkali-type plasmas are more difficult, both theoretically and experimentally. In order to understand the difficulties of experiments with alkali plasmas, it may be enough to remember the reactive chemical nature of these elements. However, as explained above, for many reasons connected with applications, the investigation of static and dynamic properties of alkali plasmas is of great interest. For our purposes, it is important to note that there is a strong dependence of the 'effective' interparticle interactions on the thermodynamic state, and this considerably influences the properties of fluid metals, in particular the thermodynamic and kinetic features of the liquid–vapor transition (Hensel and Pilgrim 2003, Hensel and Warren 2014).

In some studies (Zimdahl and Ebeling 1977, Richert et al. 1984), we studied screened interactions and electric microfields in alkali plasmas. Other studies concentrated on the sudden increase in the conductivity connected with the transition to full ionization (Kuhlbrodt et al. 2005). Here we consider Li+, Na+, K+, Rb+, Cs+, and Be2+ plasmas and, for simplicity, neglect other ionization states for these plasmas. Thus, in alkali plasmas the concentrations of electrons and ions are presumed to be equal, i.e., $n_\mathrm{e} = n_\mathrm{i}$. It was shown that, in moderately coupled plasmas, the technique of expansion with respect to the plasma coupling parameter developed in the classical work of Bogoliubov provides sufficiently reliable expressions for thermodynamic and dynamic properties (Sadykova 2011).

The simplest effective potential for the description of the electron–ion interaction in alkali plasmas is the Hellmann potential (Richert et al. 1984):

$$V_\mathrm{ie}(r) = -\frac{e^2}{r}\big[1 - A\exp(-\alpha r)\big]\,. \tag{3.158}$$

In the case of cesium one finds, e.g., the parameters

$$\alpha = 1.082 \times 10^8\ \mathrm{cm}^{-1}\,, \qquad A = 1.5863\,. \tag{3.159}$$

Table 3.1 Ionization energies and characteristic parameters of the HGK potential

Element	Li	Na	K	Rb	Cs	Be
I [eV]	5.39	5.138	4.339	4.176	3.893	9.306
a [a.u.]	5.954	3.362	2.67	2.29	2.214	3.72
R [a.u.]	0.365	0.487	0.689	0.779	0.871	0.22

A more refined choice is the Hellmann–Gurski–Krasko potential:

$$\varphi_{ab}(r) = \frac{e_a e_b}{r}\left[1 - \exp(-\gamma r) + a\frac{r}{R}\exp\left(-\frac{r}{R}\right)\right], \tag{3.160}$$

$$\gamma = \alpha_0 + \frac{1}{R}, \qquad \alpha_0 = \frac{\sqrt{\pi m_{ab} k_B T}}{\hbar}. \tag{3.161}$$

Here R is a characteristic length of the HGK potential, which is determined for $\alpha_0 = 0$ from the experimental values of the energy levels and using the radial Schrödinger equation.

Further we assume that the other electrons are tightly bound. An essential piece of input information for our method is the pseudo-potential between the charges. In order to take into account the large deviations from the Coulomb law, we use here the Hellmann–Gurski–Krasko potential for the electron–ion interaction (see Table 3.1).

3.8.2 *Correlations and Thermodynamic Functions*

The HGK potential is the sum of several parts (Coulomb, Yukawa, and exponential potentials). These all have a Fourier transform and have all been investigate by ourselves in different connections. Summarizing the results, the Fourier transform of the HGK potential reads (Röpke et al. 1980, Richert et al. 1984)

$$\begin{aligned}\tilde{V}_{ab}(k) &= \frac{4\pi}{t}\int_0^\infty \mathrm{d}r\, r\sin(rt)\varphi_{ab}(r) \\ &= 4\pi e_a e_b\left[\frac{\gamma^2}{t^2(t^2+\gamma^2)} - \frac{a}{R}\frac{2R^3}{(1+R^2t^2)^2}\right].\end{aligned} \tag{3.162}$$

The mean Coulomb energy is given by

$$\langle U_c\rangle = \frac{V}{2}\sum n_a n_b \int \mathrm{d}\mathbf{r}\frac{e_a e_b}{r} g_{ab}(r). \tag{3.163}$$

When calculating the pair correlations, we may use the assumption that the third and final term in the HGK potential is short range and does not need to be screened, i.e., it can be represented by the bare potential. The first two terms have the shape attributed to them by Kelbg and Deutsch and hence require no extra treatment. Combining this leads to the correlation function

$$g_{ab}(r) = -\frac{\beta e_a e_b}{\varepsilon_r r}\frac{\gamma^2}{p^2 - s^2}\big[\exp(-pr) - \exp(-sr)\big] \\ + \beta e_a e_b \frac{a}{R}\exp\left(-\frac{r}{R}\right) . \tag{3.164}$$

Here we have introduced the same parameters s, p as in Sect. 3.4. Note that for small densities, $p \approx \kappa$ and $s \approx \gamma$. The result for the Coulomb energy may be written in the form

$$U_c = -V k_B T \frac{\kappa^3}{8\pi}\frac{\sqrt{1+\sqrt{\pi}\lambda/4R}}{\sqrt{1+\sqrt{\pi}\kappa/2\lambda+\sqrt{\pi}\lambda/4R}} + V k_B T \kappa^4 \frac{aR}{8\pi} , \tag{3.165}$$

where $\lambda = \lambda_{ie}$ and $\kappa^2 = 8\pi\beta n e^2$. The route to the free energy, pressure, and chemical potential is the same as described before.

In order to formulate the Saha equation, let us first discuss the question of the partition function. We will use the PBL schema in order to be consistent with the exact results for low-density plasmas, and approximate the discrete levels by the unperturbed levels. The lowering of the continuum edge is approximated by the excess chemical potentials (Kilimann and Ebeling 1990). In this way, we get the effective partition function

$$\sigma(T) \to \sigma_{eff}(n, T) = \sum_{E_{s,\ell} < \mu_{ex}} \Big\{\exp\big[-\beta E_{s\ell}(n, T)\big] - 1 + \beta E_{s\ell}(n, T)\Big\} , \tag{3.166}$$

where the $E_{s\ell}$ are the unperturbed Cs levels. This scheme is illustrated in Fig. 3.10.

EOS for Partially Ionized Nonideal Plasmas

In the previous section we gave two examples of the EOS, expressing the pressure of hydrogen and cesium plasmas. The corresponding expression for the pressure of the partially ionized plasma consists of ideal contributions together with Coulomb and van der Waals excess pressures:

$$p = p_{id} + p_{Coul} + p_{vdW} . \tag{3.167}$$

The ideal pressure in Boltzmann approximation is

$$p_{id} = (n_i + n_e + n_a)k_B T ,$$

and in the degenerate region is given by the formulae presented in this chapter. The van der Waals contribution can be approximated by

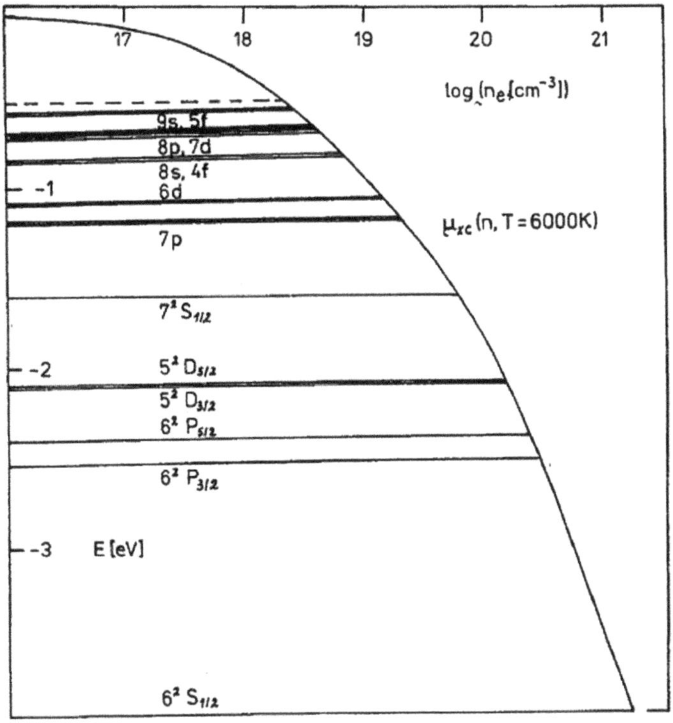

Fig. 3.10 Energy levels of cesium as a function of density and the merging into the continuum for the plasma temperature $T = 6000$ K

$$p = \frac{k_B T}{v - b} - \frac{a}{v^2}, \qquad v = \frac{1}{n}. \tag{3.168}$$

This contribution will be discussed in more detail in the next section. The ring term contribution reads, e.g., in the QDHA,

$$p_{\rm Coul} = -\frac{k_B T}{24\pi}\kappa^3 \phi\big(\kappa a(T)\big). \tag{3.169}$$

We denote here the number of free ions by n_i, the number of free electrons by n_e, and the number of free atoms by n_a. Further we denote the total atom density by n_0. We then have the following balance relations:

$$n_e + n_a = n_0, \qquad n_i + n_a = n_0, \tag{3.170}$$

and electroneutrality $n_e = n_i$ is also to be satisfied. For our partially ionized plasma, we introduce a degree of ionization by the relations

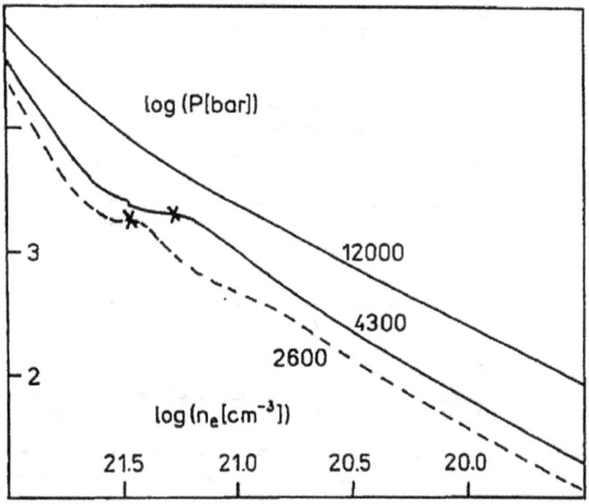

Fig. 3.11 Pressure versus reciprocal density for a Cs plasma. The *cross* for $T = 4300$ K denotes a possible critical point. The *dashed curve* for $T = 2600$ K includes polarization effects and shows a considerably lower critical point

$$\alpha = \frac{n_e}{n_0} = \frac{n_i}{n_0}\,, \qquad 1 - \alpha = \frac{n_a}{n_0}\,. \tag{3.171}$$

The excess pressure functions have already been given, but they must now be specified for partially ionized plasmas. In the QDH approximation, the EOS reads

$$\beta p = n_i + n_e + n_0 - \frac{1}{24\pi}\kappa^3 \phi(\kappa a)\,. \tag{3.172}$$

Here the Debye–Hückel pressure function $\phi(x)$ is defined by

$$\phi(x) = \frac{3}{x^3}\left[1 + x - \frac{1}{1+x} - 2\log(1+x)\right], \quad x = \kappa a(T)\,. \tag{3.173}$$

In the KEP approximation, the corresponding excess pressure is given by

$$\beta p_{id} = n_0(1+\alpha)\,, \qquad \beta p_{Coul} = -\frac{1}{24\pi}(\kappa_0\sqrt{\alpha})^3\phi(z)\,, \tag{3.174}$$

$$\phi(z) = \frac{3}{5\sqrt{1+2z}} - \frac{4(1 + z - .5z^2 - \sqrt{1+2z})}{5z^3\sqrt{1+2z}}\,, \quad z = \frac{\kappa_0\sqrt{\pi\alpha}\kappa_0\lambda}{4}\,. \tag{3.175}$$

Consider now the thermodynamic functions of Cs plasmas (Richert et al. 1984). For lower temperatures, the curve is observed to be nearly horizontal and we may guess a connection with the observed phase transition in cesium plasmas. Figure 3.11 shows the pressure and Fig. 3.12 shows the dependence of the ionization on the density and the influence of polarization effects (Richert et al. 1984). At lower pressures, we see the typical minimum of the degree of ionization, also observed for hydrogen plasmas. Then with increasing pressure, we see a sudden increase in α to the value $\alpha = 1$,

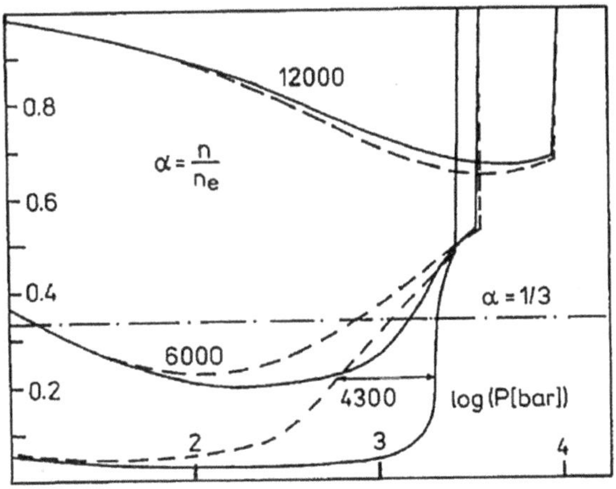

Fig. 3.12 Pressure dependence of the degree of ionization for Cs plasmas and the influence of polarization effects shown by *dashed curves*. The sudden increase of ionization at higher pressures is due to the disappearance of the ground state level. The corresponding sudden increase of conductivity has been studied experimentally and theoretically for noble gases, e.g., by Kuhlbrodt et al. (2005)

i.e., full ionization. Similar behavior is observed in experiments (Hensel et al. 1985, Bonzel et al. 1989, Hess 1992, Redmer et al. 2010, Hensel and Warren 2014). This sudden increase in ionization at higher pressures is determined by the disappearance of the ground state level we discussed before. The sharp edge of the curves at this pressure is surely an artefact. In improved theories, a smoother transition is to be expected.

3.9 Models of First Order Phase Transitions in Ionized Gases

3.9.1 *Van der Waals and Debye–Hückel–Bjerrum Models*

The theory of gases developed in the dissertation of van der Waals in 1873 may be considered as the starting point for the modern theory of phase transitions. At the same time, the van der Waals model still serves as the simplest model of a first order phase transition in fluids. As pointed out in Chap. 1, van der Waals' approach is based on a simple physical model of interactions between particles which takes into account short range repulsive as well as long range attractive forces. The model equation for the pressure reads

$$p = \frac{k_B T}{v - b} - \frac{a}{v^2} , \qquad v = \frac{1}{n} , \tag{3.176}$$

with $v = V/N$. The corresponding expressions for the specific internal energy and the entropy read

$$\frac{U}{N}=\frac{3}{2}k_{\mathrm{B}}T-na\,,\quad \frac{S}{N}=k_{\mathrm{B}}\ln\left[(v-b)\left(\frac{2\pi m k_{\mathrm{B}}T}{h^3}\right)^{3/2}\right]+\frac{5}{2}k_{\mathrm{B}}\,. \tag{3.177}$$

For isentropic processes, i.e., with $S=$ const., we get the relations

$$(V-Nb)T^{3/2}=\text{const.}\,,\qquad T(V-Nb)^{\gamma-1}=\text{const.}\,,$$

$$(p+n^2a)(N-Nb)^{\gamma-1}=\text{const.}$$

Below the critical temperature, the van der Waals model predicts the possibility of coexistence of two phases which differ from each other by the density of molecules. The critical point is the unique point where both

$$\frac{\partial p}{\partial V}=0\,,\qquad \frac{\partial^2 p}{\partial V^2}=0\,.$$

The critical parameters are

$$v_{\mathrm{c}}=3b\,,\qquad p_{\mathrm{c}}=\frac{a}{27b^2}\,,\qquad k_{\mathrm{B}}T_{\mathrm{c}}=\frac{8a}{27b}\,. \tag{3.178}$$

The universal constant

$$K_{\mathrm{c}}=\frac{RT_{\mathrm{c}}}{p_{\mathrm{c}}v_{\mathrm{c}}}=\frac{8}{3}$$

is often called the critical coefficient.

Using the dimensionless variables $\tilde{p}=p/p_{\mathrm{c}}$, $\tilde{v}=v/v_{\mathrm{c}}$, $\tilde{T}=T/T_{\mathrm{c}}$, we find the universal representation

$$\left(\tilde{p}+\frac{3}{\tilde{v}^2}\right)\left(\tilde{v}-\frac{1}{3}\right)=\frac{8k_{\mathrm{B}}\tilde{T}}{3}\,. \tag{3.179}$$

With the development of more rigorous theories, it became clear that the van der Waals approach is restricted to relatively weak attractive forces which either decay faster than $1/r^3$ or satisfy the Kac conditions. Therefore the conditions for applicability to Coulomb forces are not given. In 1943 Landau and Zeldovich discussed new possibilities for phase transitions, such as Coulomb effects in metal–insulator transitions in metal fluids in the vicinity of critical points. In this connection, we should also mention the related work by Mott on metal–insulator transitions in Coulomb systems.

In order to develop a statistical theory of Coulomb phase transitions, one needs good approximations for the thermodynamic functions in the regions where, according to first estimates, the phase transition is to be expected. As we have learned from the theory of phase transitions in neutral gases, simple expressions like the van der Waals equation may already give qualitatively correct results (Table 3.2).

Table 3.2 Critical temperature and critical volumes for several gases

Element	H_2	He4	N_2	CO_2	H_2O
T_c [K]	3.3	5.2	126	304	647
p_c [MPa]	1.3	0.23	3.4	7.4	22.1

We will now show that the Debye–Hückel nonideality approximations for the thermodynamic functions together with the mass action law provide qualitative approximations which might serve as a zeroth approximation for the description of Coulomb phase transitions (Ebeling et al. 1971, 1976). A first study of Coulomb transitions in quantum plasmas was given by Norman and Starostin in 1968 (Norman and Starostin 1968). The first order transition predicted by these authors was called the plasma phase transition (PPT). The PPT was predicted as a possible result of competition between effective Coulomb attraction and quantum repulsion in the partially ionized dense plasma (Norman and Starostin 1968, 1970).

The qualitative picture was similar to the van der Waals model, where the phase transition results from the balance between the long-range attraction between neutral molecules and their short-range repulsion. We remember that the two phases in the van der Waals model differ from each other by the density of molecules. The two phases in PPT have different number densities of charged particles and different degrees of ionization. Atoms, which are present in the two coexisting phases, were originally treated as an ideal gas.

In their first work, for a first estimate of the critical temperature, Norman and Starostin (1968) suggested using Debye–Hückel-type expressions $1 - c\Lambda/r_D$ for the chemical potential, where Λ is the electron thermal wavelength, r_D is the Debye radius, and c is a numerical coefficient estimated as $c \approx 0.1$ in the Norman–Starostin theory. The Debye–Hückel term describes an effective Coulomb attraction, while the term $c\Lambda/r_D$ represents an effective quantum repulsion due to the uncertainty principle in this approach. Norman and Starostin (1970) also used the original Debye–Hückel expression for the chemical potential, viz., $(1 - c\lambda/r_D)^{-1}$, which is closely related to our $\Lambda/8$ approximation discussed in the previous sections. In our QDHA, we estimated the coefficient c from the Heisenberg uncertainty relation and found $c = 1/8 = .125$. This is not so far from the estimate by Norman and Starostin, so the predictions are not so different.

Classical Charged Hard-Sphere Coulomb Systems

In parallel with the investigations on nonideal quantum plasmas, Coulomb transitions were studied in classical Coulomb systems. The first numerical results on phase transitions in ionic systems go back to the late 1960s. Voronzov et al. (1970) found a coexistence line and a critical point in the course of Monte Carlo studies of charged hard spheres imbedded in a dielectric medium. The first analytical estimate of the critical point and the coexistence line for electrolytes based on the Debye–Hückel theory was given by one of the authors (Ebeling 1971).

The result of the estimate for the critical Bjerrum parameter was

$$b_{\rm cr} = \frac{e^2}{\varepsilon_{\rm r} k_{\rm B} T} = 16 \, .$$

This corresponds to a (dimensionless) critical temperature

$$T^* = 1/16 = 0.0625 \, .$$

We used the classical expression for the excess part of the chemical potential

$$\mu_{\rm ex} = \mu_{+-} \approx -\frac{e^2}{\varepsilon_{\rm r} k_{\rm B} T (1 + \kappa a)} \tag{3.180}$$

and the stability condition (Ebeling 1971)

$$2 + n^* \frac{\partial(\beta\mu_{+-})}{\partial n^*} > 0 \, , \tag{3.181}$$

where n is the density of free charges.

In this way we found that instability is restricted to temperatures below the critical temperature, which is defined for Debye–Hückel systems by

$$T \le T_{\rm cr} = 16 \frac{e^2}{\varepsilon_{\rm r} k_{\rm B} a} \, . \tag{3.182}$$

The theory of this Coulomb instability was applied to nonideal semiconductor plasmas in other studies (Ebeling 1971, Ebeling et al. 1976). In a monograph the theory was extended and the spinodal curve was calculated for the parameter $\mu = e^2\kappa/k_{\rm B}T$ (Ebeling et al. 1976, 1979):

$$\mu_{1,2} = \left(\frac{b^2}{8} - b\right) \pm \left[\left(\frac{b^2}{8} - b\right)^2 - b^2\right]^{1/2} \, . \tag{3.183}$$

The theory of classical Coulomb phase transitions, including a comparison with numerical and experimental data, was further developed in pioneering papers by George Stell and Michael Fisher and coworkers (Stell 1995, Yeh et al. 1996, Levin and Fisher 1996, Fisher 1994). These authors also made an extensive investigation of the state-of-the-art in this field.

We mention also estimates in mean-sphere approximations and applications to the equilibrium and conductance of electrolytes (Ebeling and Grigo 1980, 1982), together more recent work in the field (Schröer 2011, Weingärtner and Schröer 2001). Furthermore, we note that extensive numerical work on the critical parameters of the restricted primitive model has led to more recent estimates e.g. by Panagitopoulos and Caillol:

$$T^* = 0.0489\,, \qquad \rho* = 0.976\,.$$

Summarizing these results for classical systems we may say that theory and experiment (Monte Carlo data as well as measurements on electrolytes) are in quite good agreement, and show that even the old estimate $T^* \approx 0.0625$ from 1971 was not very far from the later calculation results. There is no longer any doubt that a Coulomb phase transition exists in classical systems. We refer to the careful investigations of Fisher and his group (Levin and Fisher 1996) and Weingärtner and Schröer (2001).

The classical Coulomb phase transition, which is now more or less well understood, is due to a balance between a hard-core repulsion and a Coulomb attraction. In the quantum case, this transition, the PPT, is due to a balance between quantum repulsion between point charges and Coulomb attractions. In this respect, the PPT is a kind of quantum variant of the classical Coulomb transition. While the existence of a Coulomb transition in ionic systems is now well confirmed, the exact location of the PPT is still under discussion (Fortov 2013). However, there is now increasing evidence that a first order transition also exists for hydrogen and other high-pressure plasmas, e.g., from *ab initio* simulations (Morales et al. 2010), as well as from reverberating shocks with high explosives (Fortov et al. 2007, Fortov 2011, 2015, Knudson et al. 2001, 2015).

Including Both Coulomb and van der Waals Effects

In the Debye–Hückel type approximation derived in the previous sections, the interaction of charges with neutrals and the neutral–neutral interaction were not taken into account. This model might fail if the interactions with neutrals turn out to have an essential influence. In the following, we consider for concreteness a plasma system. We treated the neutrals as ideal particles in the simple approach given above, which is analytically solvable. The system of mass action laws might become rather difficult if we include the interactions with neutrals and/or chemical equilibria between neutrals, e.g., between atoms and molecules. A better starting point for these more complicated cases is as a rule the expression for the free energy, involving a minimization procedure.

Combining the Debye–Hückel model with a van der Waals expression and a contribution which takes into account the polarization of neutrals, we get for the free energy

$$\begin{aligned} F = {} & n_0 k_{\mathrm{B}} T \left[\ln \left(\frac{n_0 \Lambda_0^3}{\sigma(T)} \right) - 1 \right] + n_{\mathrm{i}} k_{\mathrm{B}} T \left[\ln \left(\frac{n_{\mathrm{i}} \Lambda_i^3}{g_i} \right) - 1 \right] \\ & + n_{\mathrm{e}} k_{\mathrm{B}} T \left[\ln \left(\frac{n_{\mathrm{e}} \Lambda_e^3}{g_e} \right) - 1 \right] - k_{\mathrm{B}} T V \frac{\kappa^3}{12\pi} \tau \left(\frac{\kappa \Lambda}{8} \right) \\ & - k_{\mathrm{B}} T n_0 \ln(1 - n_0 B) - A n_0^2 - W n n_0\,. \end{aligned} \tag{3.184}$$

Here the last three terms denote the contributions from the short range van der Waals interactions and the contributions from polarization terms. The polarization terms are of particular importance for the description of alkali and mercury plasmas (Hensel and Warren 2014, Redmer et al. 2010).

Three free constants now characterize our model:

- Repulsion of atoms B.
- Attraction of atoms A.
- Strength of atomic polarizability W.

In order to reduce the number of free parameters, we assume $W = 0$ in the following. We note that the free energy in this form defines a model plasma with properties intermediate between a van der Waals gas and a Debye–Hückel plasma. The equilibrium composition is given by the nonideal Saha equation

$$\frac{1-\alpha}{\alpha^2} = n\Lambda^3\sigma(T)\exp\left[-\frac{\Delta I(\alpha)}{k_B T}\right] . \tag{3.185}$$

Here

$$\begin{aligned}\Delta I(\alpha) = & \frac{e^2\kappa\sqrt{\alpha}}{k_B T\left[1+\kappa a(T)\sqrt{\alpha}\right]} - \frac{(1-\alpha)nB}{1-(1-\alpha)nB} \\ & +\log\left[1-(1-\alpha)nB\right] + (1-\alpha)nA - Wn(2\alpha-1)\end{aligned} \tag{3.186}$$

is the lowering of the ionization energy.

As is well known, real gases with attractive interactions show a first-order phase transition described in the (p, T) plane by a critical point C_1 and a coexistence line ending at C_1. Assuming that the Coulomb terms are omitted in our expressions for the free energy, a simple van der Waals-type expression is obtained with the critical point

$$T_{\mathrm{vdW}} = \frac{8a}{27k_B b} , \qquad n_{\mathrm{vdW}} = \frac{1}{3b} . \tag{3.187}$$

The critical temperature is about $10^1 < T_{\mathrm{cr}} < 10^3$ K for realistic values of the parameters. For example, $T_{\mathrm{cr}} \approx 33$ K for hydrogen. The degree of ionization is practically zero in this region, i.e., there is no interference with the PPT. On the other hand, at least for hydrogen, the PPT occurs in a region where the number density of neutrals is rather low. We may say therefore that the assumption $a = 0, b = 0$ is justified in this case. Then, as shown above, the PPT may be treated in an analytical way. In the Lambda approximation, we got an analytical expression for the critical point of the PPT in the previous section (Ebeling et al. 1976, 1985, Lehmann and Ebeling 1996). By introducing the Bohr radius a_B, we may transform the expression for the critical temperature and density to the form

$$T_{\mathrm{ppt}} = \frac{e^2}{8k_B a_B} , \qquad n_{\mathrm{ppt}} = a_B^{-3} . \tag{3.188}$$

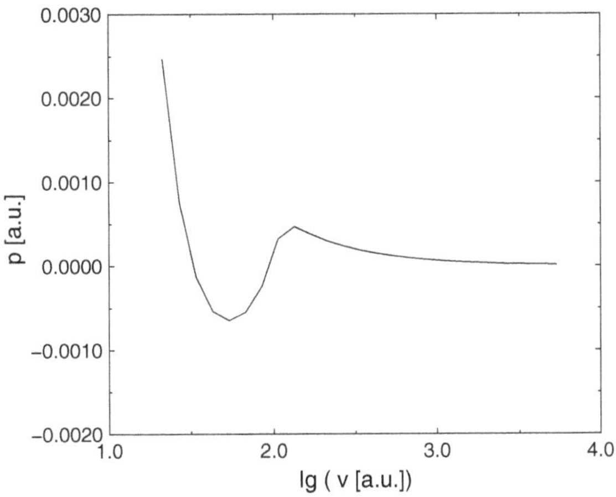

Fig. 3.13 Isotherm $T = 10\,000$ K for hydrogen plasmas, showing the PPT instability

What happens in a system where both types of interactions are present? Evidently, besides the classical first-order phase transition typical for neutral gases, a second first-order phase transition due to Coulomb forces may appear. The second one, which may occur only at rather high temperatures $T > 10^4$ K, is the PPT. Figure 3.13 shows the pressure isotherm of hydrogen plasmas at $T = 10\,000$ K. In the unstable region of the isotherm, the plasma is separated into two phases. The dense phase is highly ionized and the Coulomb interaction dominates over the van der Waals forces. The less dense phase is only weakly ionized and van der Waals interactions dominate.

Study of Model Plasmas

In order to investigate the qualitative influence of the van der Waals forces, we studied a model plasma with increasing values of the parameters a and b. Figure 3.14 shows the isotherm $T = 0.07$ and $a = b = 30$ a.u. In spite of the fact that these values are rather large, we cannot detect a separate van der Waals wiggle. Both attracting forces support each other and lead to one phase transition approximately in the same density region as the PPT. This is the situation we observe in nature for alkali plasmas. For a further increase in the values of the van der Waals parameters to $a = b = 100$ a.u., we observe two separate van der Waals loops, as shown in Fig. 3.15. This is the situation we observe for real noble gas plasmas. We should emphasize that Figs. 3.14 and 3.15 only have illustrative character. In reality we do not see two wiggles in one isotherm due to the drastic difference between the critical temperatures.

Summarizing these findings we may state that, depending on the values of the van der Waals parameters, we may obtain either a phase diagram with two first order transitions (hydrogen and noble gas plasmas) or a phase diagram where the two transitions merge into one (alkali plasmas). The existence of a phase diagram including a van der Waals type transition and a separate metal–insulator phase transition was discussed for the first time by Landau and Zeldovich in 1944. We repeat that the existence of two separate phase transitions requires the validity of the inequalities

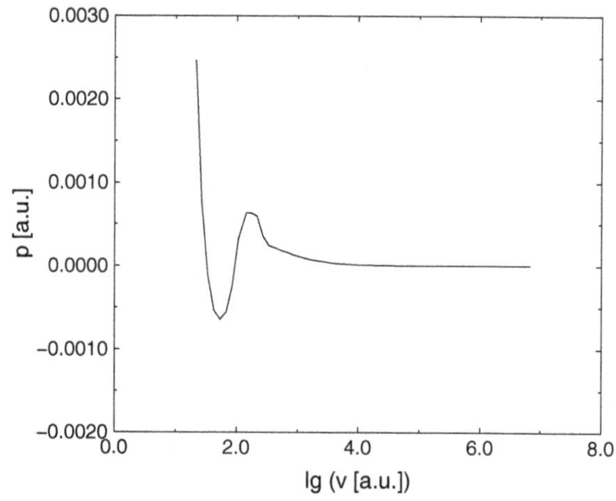

Fig. 3.14 The isotherm $T = 0.07$ a.u. for an alkali-like model plasma with one transition of mixed PPT–van der Waals type. The van der Waals parameters are $a = 30$ a.u., $b = 30$ a.u.

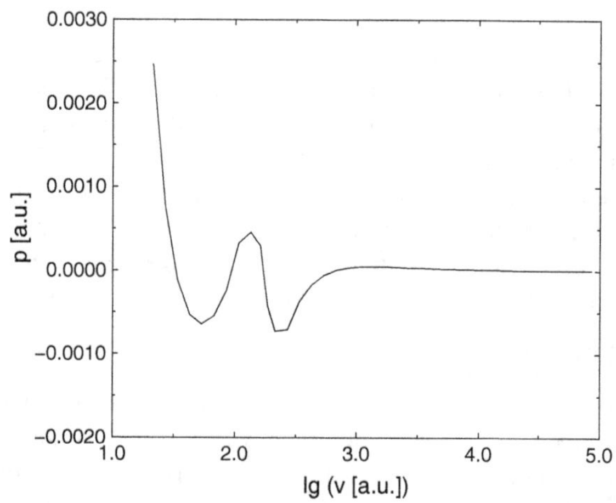

Fig. 3.15 The isotherm $T = 0.07$ a.u. for a noble gas-like model plasma with a PPT and a separate van der Waals transition. The van der Waals parameters are $a = 100$ a.u., $b = 100$ a.u.

$$T_{\mathrm{vdW}} \ll T_{\mathrm{ppt}} , \qquad \rho_{\mathrm{vdW}} \ll \rho_{\mathrm{ppt}} . \tag{3.189}$$

This is satisfied for hydrogen and for noble gas plasmas. In the case of alkali plasmas, the van der Waals and the PPT transition have about the same critical density and temperature. Therefore the two transitions merge into one with very specific properties (Redmer 1997, Hensel and Warren 2014).

3.9.2 Estimate of Critical Points in QDHA and KEPA

An interesting property which we mentioned already is that the expressions for thermodynamic functions, in particular the pressure, become unstable for higher densities. However, the accuracy of predictions depends on the quality of the approximations. Let us first consider the so-called quantum Debye–Hückel approximation (QDHA) or Lambda approximation (Ebeling et al. 1976). As already mentioned, quantum effects are expressed by one characteristic length, the thermal de Broglie wavelength Λ in combination with Debye–Hückel type approximations. The Lambda approximation agrees very well with the exact quantum-statistical results for small densities for $T < 10^5$ K (Ebeling et al. 1976, Kraeft et al. 1986). This approximation requires the plasma to be nondegenerate, i.e., $n^* \Lambda^3 \ll 1$ (where n^* denotes the density of free particles). A further condition is that only a small fraction of the atoms is bound in molecules, viz., $\beta_2 \ll 0.5$. Our choice of mass action constant for quantum plasmas is the Brillouin–Planck–Larkin expression, as introduced in Chap. 1, there are other variants under discussion (Khomkin et al. 2003, 2004, 2008, Starostin et al. 2003).

In this way, the classical and the quantum case may be treated in the present rough approximation by the same procedure. The easiest way to check for the stability of gas and plasma systems is to investigate the sign of the derivative of the pressure. The region where

$$\frac{\partial(\beta P)}{\partial n} < 0 \tag{3.190}$$

corresponds to thermodynamic instability. Equivalent to this is the condition (Norman and Starostin 1968, 1970)

$$\left(\frac{\partial n_{\mathrm{i}}}{\partial n}\right)_T < 0 \, , \tag{3.191}$$

which is valid only if the atoms are treated as an ideal gas. Instability is restricted to temperatures below the critical value, which is defined by

$$T \leq T_{\mathrm{cr}} = 16 \frac{e^2}{D_0 k_{\mathrm{B}} a(T)} \, . \tag{3.192}$$

We recall that, in the case of classical ionic solutions, the parameter a is to be identified with the diameter of ions, and in the case of dense quantum plasmas, we have to introduce here the effective quantum diameter $a(T) = \Lambda/8$. In this way, we have shown that the PPT is the analogue of the known classical Coulomb transition in ionic solutions. At least in the simplified theory given above, the classical transition in ionic solutions and the quantum PPT in dense plasmas are fully analogous. In both cases, the spinodal curve is given by

$$\left.\frac{e^2 \kappa}{k_{\mathrm{B}} T}\right|_{\mathrm{cr}} = \left(\frac{b^2}{8} - b\right) \pm \left[\left(\frac{b^2}{8} - b\right)^2 - b^2\right]^{1/2} \, , \tag{3.193}$$

where $b = e^2/k_{\mathrm{B}} T a(T)$. This formula yields the upper and lower densities of the two-phase region. In the case of hydrogen-like quantum plasmas, we find the estimate for the critical temperature to be

$$T_{\rm cr} \approx 12\,570\ \text{K}\,. \tag{3.194}$$

Note that in the case of electrolytic solutions or semiconductors, the critical temperature is much lower due to the appearance of the dielectric constant in the denominator. Another effect which decreases $T_{\rm cr}$ is the finite size of the ions. In order to describe the size of the ion cores in alkali or noble gas plasmas, we may introduce an effective ion radius R, e.g., we have $R \approx 1.69$ Å for Cs ions. Furthermore, we replace $\Lambda/8$ by

$$a(T) = \Lambda/8 + R\,. \tag{3.195}$$

Assuming that R is temperature-independent, we now find for the critical temperature the new estimate

$$T_{\rm cr} = \frac{2}{1 + (1 + 6R/\pi a_{\rm B})^{1/2}} \times 12\,570\ \text{K}\,, \tag{3.196}$$

where $a_{\rm B} = \hbar/e^2 m_{\rm e}$ is the Bohr radius. We see that a finite ion size may lead to an essential decrease in the critical temperature to values far below 10 000 K. For Cs plasmas, this estimate gives, e.g., $T_{\rm cr} \approx 6\,000$ K. Since this value is still far above the observed critical point $T_{\rm cr} \approx 2\,000$ K for cesium plasma, we have to search for additional effects, which bring the critical temperature down to the observed values.

Critical Point in the Effective Potential Approach

Let us now check the stability criterion $\partial p/\partial V < 0$ in Kelbg's effective potential approximation (KEPA). In this approximation, the stability criterion $\partial p/\partial V < 0$ leads to the inequality

$$\beta e^2 \kappa \Big[G(\kappa\lambda_{\rm ie}) + (\kappa\lambda_{\rm ie}) G'(\kappa\lambda_{\rm ie}) \Big] < 4\,. \tag{3.197}$$

This condition yields the inequality $\kappa\lambda_{\rm ie} < 0.35$. Correspondingly, within this simple model, for temperatures below 6 530 K, we have to expect a first order phase separation into two phases with different degrees of ionization, i.e., a so-called plasma phase transition. We may draw the following conclusions:

- The KEPA with an adapted α_0 parameter is in rather good agreement with known exact quantum-statistical results, and in particular with the DeWitt result, and provides a correct asymptotic.
- The KEPA for thermodynamic functions based on the Kramers–Kelbg–Deutsch pseudopotential is as simple as the quantum Debye–Hückel approximation $G(\kappa) = 1/\big[1+\kappa a(T)\big]$, but for a higher κ is in better agreement with known exact results, in particular for the asymptotic, and is therefore to be considered as an improvement.
- In comparison to the results for the quantum Debye–Hückel approximation which leads to instabilities below $T_{\rm c} \sim 12\,600$ K, the KEPA model leads to a thermodynamic critical point at $T_{\rm c} \approx 6\,350$ K. This is considerably lower than the QDHA prediction and comes closer to other recent results obtained with Pade approxi-

mations (Ebeling et al. 2009, Redmer et al. 2010) and other results obtained by molecular dynamics and density functional theories (Redmer et al. 2010, Bredow et al. 2013).

3.10 Discussion of Plasma Transitions in Theory and Experiment

3.10.1 Survey of Results on Plasmas and Ionic Fluids

In the preceding sections, we discussed several analytical approaches to the PPT resulting in critical temperatures for hydrogen in the regions of $T_{cr} \sim 6\,000$ K. Note that this result is considerably lower than in earlier estimates (Ebeling et al. 1976, 1991; Kraeft et al. 1986; Saumon and Chabrier 1989, 1991, 1992; Chabrier and Potekhin 1998). There exist many more advanced approaches.

In particular, we may mention the PACH method, developed by Beule et al. (1999). This is based on Padé approximations in combination with the chemical model, hence the name. This work gave a quite realistic calculation of the coexistence line for hydrogen plasmas. A coexistence pressure of $p \approx 115$ GPa was found in the cited work for $T = 10\,000$ K. The corresponding mass densities of the coexisting phases are $\rho_1 \approx 0.62$ and $\rho_2 \approx 0.82$ g/cm^3. The PACH-approach will be discussed in more detail in Chap. 7.

Experiments with shock-compressed hydrogen and deuterium plasmas have confirmed that a transition to a highly conducting state occurs around 140 GPa and 3 000 K (Weir et al. 1996; Nellis et al. 1999). This is approximately the pressure region where Beule et al. predict the coexistence line of the PPT. A corresponding behavior of conductivity data has been reported by Fortov et al. (2001). Pressure dissociation and ionization, which are almost independent of the temperature, become a dominant factor at 10^4 K and below in hydrogen and deuterium fluids studied experimentally at high pressures (Nellis et al. 1998, Cauble et al. 2001, Holst et al. 2013).

Since the PPT is located in regions of the density parameters where analytical methods and perturbation theories do not work properly, the best prospects for calculating the partition function are by numerical methods, and in particular by Monte Carlo methods (Zamalin et al. 1977, Ceperley 1995, Filinov et al. 2001, Morales et al. 2010). Several methods have been developed to perform quantum MC for dense plasmas (for surveys see Ceperley 1995, Filinov et al. 2004). Common to all approaches is the fermion sign problem, that is, the poor convergence of the simulations for Fermi systems at increasing values of the degeneracy parameter $n\lambda^3$, which characterizes the quantum 'extension' of the microparticle.

In the restricted PIMC method (RPIMC) (Ceperley 1995) special assumptions are made about the density operator ρ to reduce the sum over permutations to even (positive) contributions alone. This method crucially depends on knowledge of the 'nodes' of the density operator which can be seen, e.g., in the differences of the results computed with free or variational nodes. Furthermore, the presently used nodes seem to exclude the possibility of inhomogeneous plasma configurations, as they would appear in the case of a first order phase transition. One of the authors (VSF) has shown that this method does not reproduce the correct ideal Fermi gas limit (see Chaps. 9 and 10).

For these reasons, the alternative approach of direct fermion PIMC simulations (DPIMC) is of importance for the investigation of the PPT (Filinov et al. 2001, 2004). Obviously, the computational cost is higher, but the results continue to improve. This method now allows for direct fermion path integral Monte Carlo simulations of dense plasmas in a wide range of densities and temperatures. The pressure and energy of a degenerate strongly coupled hydrogen plasma have been computed using this concept (Filinov et al. 2001, 2004), as have the pair distribution functions in the region of partial ionization and dissociation Filinov et al. (2001). This scheme is rather efficient when the number of time slices (beads) in the path integral is less than or equal to 50 and has been found to work well for temperatures $k_B T > 0.1$ Ry. One of the most striking recent results is the observation of droplet formation below $T \sim 20\,000$ K in dense hydrogen, which confirmed previous chemical results about the plasma phase transition (Trigger et al. 2003).

For applications of different approaches to stellar systems and stars, we refer to the studies by Chabrier and Potekhin (1998) and Däppen (2010). We also mention the hypothetical phase transitions in multiple ionized plasmas which were treated, e.g., in (Förster et al. 1992, Ebeling et al. 1991, in particular for helium plasmas. A detailed study of He plasmas was given in an earlier book (Ebeling, Fortov et al. 1991, 2007). Here we considered mainly singly-charged ions in gaseous plasmas.

Plasma-like phase transitions in the optically excited electron–hole system in semiconductors were studied using similar methods (Ebeling 1971, Insepov and Norman 1972, Ebeling et al. 1976, Kilimann et al. 1977, Kraeft et al. 1986, Kremp et al. 2005). The experimental observations seem to suggest that these theoretical instabilities correspond to a real phase transition in semiconductors, but this topic will not be considered here in detail. We refer to Chaps. 9 and 10 for more details.

There is considerable debate about whether the existing model equations of state reproduce the recent experiments (Morales et al. 2010). We stress, however, that most theories do now predict a phase transition in all their variants.

Wave packet dynamics has also provided several hints regarding the existence of the plasma phase transition (Klakow et al. 1994, Ceperley 1996, Militzer et al. 2000). The validity of the thermodynamic expressions used been tested by comparing with molecular simulation and experimental data for nonideal plasmas and limiting expansions for weakly nonideal plasmas (Filinov et al. 2001, 2003). The problem continues to attract remarkable attention. Filinov (2003) claim that the present results of direct path integral Monte Carlo simulations provide first-principle support for such a phase transition in dense hydrogen. We mention these works only as part of the ongoing effort to simulate the PPT numerically.

Coulomb Phase Transition in Ionic Fluids

The plasma phase transition is closely analogous to a phase transition in classical ionic fluids and in electrolytes, as discussed briefly in a previous section. The existence of such a transition was first predicted on the basis of Monte Carlo simulations after 1970 by Voronzov-Veliaminov and coworkers (Voronzov et al. 1970) and on the basis of an analytical Debye–Hückel–Bjerrum theory by one of the present authors (Ebeling 1971). The first predictions of the critical point were still rather imprecise (Voronzov et al. 1970). The first analytical estimate of the critical b parameter gave $b_c \sim 16$ (Ebeling et al. 1970, 1976). Today most advanced simulations give $b_c \sim 20.45$.

In the meantime, there has been so many theoretical and experimental work on the phase transition in classical Coulomb systems that there are no doubts that this transition really exists. Many investigations on ionic solutions, in theory and experiment from Monte Carlo data to analytical estimates as well as measurements on electrolytes, show a first order Coulomb transition. So there is no longer any doubt that a Coulomb phase transition in classical systems does indeed exist, and all experts agree on this Fisher 1994, Weingärtner and Schröer 2001, Yeh et al. 1996.

According to our point of view, the well established existence of a first order phase transitions in classical ionic liquids is an important argument for the existence of a PPT. This clearly destroys the argument that the long-range character of Coulomb forces excludes first order phase transitions. What remains is only the statement that a phase transition in quantum plasmas is more complicated than the classical transition, since quantum effects play an important role in the transition.

A special discussion has focused on the problem of critical behavior and critical indices. In particular, for the case of classical Coulomb systems, much attention has been paid to this problem (Fisher 1994). For the PPT where quantum effects influence the critical point, the problem of critical indices seems to be still open. Michael Fisher (1994) clearly formulated the challenge to the critical phenomena theory:

> Critical behavior is thoroughly understood in Ising models alias lattice gases: But how far does that help our insight into the critical region of continuum models and real fluids. In particular, what causes some ionic solutions to exhibit van der Waals or 'classical' critical exponents, while other, so-called 'solvophobic' systems are of Ising-type?

It is evident that the difference between properties of various liquids under similar external conditions can be attributed principally only to the interaction potentials between the particles. However, the Kadanoff–Wilson theory of critical phenomena, which is based on the universality hypothesis, neglects interaction potential peculiarities and appeals only to large-scale properties, such as system dimension and Hamiltonian symmetry. So other approaches are needed, e.g., the integral equation method (Martynov 1999).

Experimental studies of classical electrolyte solutions revealed that the characteristics of Coulomb liquids differ from those of simple liquids. Even if crossover is observed from classical scaling laws to Ising scaling, it takes place much closer to the critical point. The Ginzburg parameter, which characterizes the size of the Ising region, turns out to be at least one or two orders of magnitude less than for simple liquids. So there is a remarkable difference in crossover range between simple and Coulomb liquids (Fisher 1994, Schröer et al. 1995, Stell 1995, Yeh et al. 1996, Weingärtner and Schröer 2001, Schröer 2011). Experimental study of crossover phenomena in real plasmas and in particular in cesium plasmas might help to determine whether the critical point is of the plasma or gas–liquid kind. Additional factors which do not exist in electrolytes, namely, quantum effects and charge–atom interaction, can influence the crossover phenomenon in PPT.

3.10.2 PPT in Hydrogen, Noble Gas, and Alkali Plasmas

Up to now the PPT in hydrogen plasmas has been observed several times, but not yet clearly enough identified and localized experimentally. Here we discuss some of the experimental aspects and for details refer the reader to articles and books by one of the present authors (Fortov et al. 2001, 2007, Fortov 2011, 2015)

The existence of a PPT in dense plasmas still seems to be less clear than the transition in classical Coulomb systems. This is mainly due to the fact that pressures predicted for the coexistence line are above 1 Mbar, which is still hardly reachable in experimental situations. However, in the last 20 years more and more experiments have been performed which cover the region of interest (Weir et al. 1996, Nellis et al. 1999, Ternovoi et al. 1999, Fortov et al. 2001, Cauble et al. 2001, Knudson and Hanson 2001, Knudson et al. 2015).

Experiments with shock-compressed hydrogen and deuterium plasmas have confirmed that a transition to a highly conducting state occurs around 140 GPa and 3 000 K (Weir et al. 1996, Nellis et al. 1999). This is approximately the pressure region where the coexistence line of the PPT is expected (Beule et al. 1999). A corresponding behavior of conductivity data has been reported (Fortov et al. 2001). Pressure dissociation and ionization, which are almost independent of the temperature, become a dominant factor at 10^4 K and below in hydrogen and deuterium fluids studied experimentally at high pressures (Nellis et al. 1999, Cauble et al. 2001). We will come back to this discussion in Chap. 6.

Most researchers now support the main statements about the PPT in spite of the fact that there are still enormous experimental and theoretical difficulties to be overcome. We also stress the great significance of this question for the solution of several astrophysical problems (Nettelmann et al. 2008). As far as can be seen now, the existing experiments for hydrogen, deuterium, argon, and other rare gas plasmas based on shock wave data are also in satisfactory agreement with the theory (Ebeling et al. 1991, 2002, 2007, Fortov 2011).

We now briefly discuss the state-of-the-art regarding the phase transitions observed in alkali plasmas. Here the experimental situation seems to be clear. The phase transitions and its properties are well investigated. However, as far as we can see, several theoretical questions remain open. As pointed out in previous sections, the theory predicts for this case that the PPT and the van der Waals transition will merge into a single transition. Near the critical point the degree of ionization is low, but different from zero. Coulomb interactions are present and influence the critical behavior, in spite of the fact that they do not play the dominant role. The area of parameters close to the critical point of cesium was carefully investigated (Hensel and Warren 2014). It was shown that near the critical point, the ionization landscape changes quickly and has a quite complicated structure. The theoretical predictions which take into account the charge–atom interaction (polarization effects) are in rather good agreement with the data (Redmer et al. 1995, 2010). We will come back to the problem of the PPT in Chaps. 6 and 10.

References

A. Alastuey, A. Perez, Virial expansion of the equation of state of a quantum plasma. Europhys. Lett. **20**, 19–24 (1992)

A. Alastuey, A. Perez, Virial expansions for quantum plasmas: Fermi-Bose statistics. Phys. Rev. E **53**, 5714–5728 (1996)

A. Alastuey, V. Ballenegger, F. Cornu et al., Exact results for thermodynamics of the hydrogen plasma: low-temperature expansions beyond Saha theory. J. Stat. Phys. **130**, 1119–1176 (2008)

A. Alastuey, V. Ballenegger, Pressure of a partially ionized hydrogen gas: numerical results from exact low temperature expansions. Contr. Plasma Phys. **50**, 46–53 (2010)

A. Alastuey, V. Ballenegger, W. Ebeling, Comment on direct linear term in the equation of state of plasmas. Phys. Rev. E. **92**, 047101 ff (2015), see also W.D. Kraeft, D. Kremp, G. Röpke, Phys. Rev. E. **92**, 047101 ff (2015)

F.B. YuV Arkhipov, A.E. Davletov, Baimbetov, Thermodynamics of dense high-temperature plasmas: semiclassical approach. Eur. Phys. J. D **8**, 299–304 (2000)

F.B. YuV Arkhipov, A.E. Davletov, Baimbetov, Self-consistent chemical model of partially ionized plasmas. Phys. Rev. E **83**, 016405 (2011)

A.B. YuV Arkhipov, A. Ashikbayeva, A.E. Askaruly, S. Davletov, V.V. Syzganbaeva, M. Tkachenko, Voronkov. Contrib. Plasma Phys. **55**, 381 (2015)

R. Balescu, *Statistical Mechanics of Charged Particles* (Wiley, London, 1963)

F.B. Baimbetov, M.A. Bekenov, T.S. Ramazanov, Effective potential of a semiclassical hydrogen plasma. Phys. Lett. A **197**, 157–158 (1995)

T.H. Berlin, E.W. Montroll, On the free energy of a mixture of ions: an extension of Kramers theory. J. Chem. Phys. **20**, 75 (1952)

D. Beule, W. Ebeling, A. Förster, H. Juranek, S. Nagel, R. Redmer, G. Röpke, Phys. Rev. B **59**, 14177 (1999)

J.S. Blakemore, *Semicionductor Statistics* (Pergamon Press, New York, 1962), *Solid State Physics* (Cambridge University Press, 1985)

L. Blum, Mol. Phys. **30**, 1529 (1975)

V.B. Bobrov, S.A. Trigger, A.G. Zagorodnyi, Special properties of the Fourier transform of Coulomb potential. Short Commun. Phys. FIAN **11**, 33 (2015)

V.B. Bobrov, A.G. Zagorodny, S.A. Trigger, Coulomb interaction and Bose-Einstein condensate (Russ.). Low Temp. Phys. **41**, 1154–1163 (2015)

N.N. Bogoliubov, *Problems of Dynamical Theory in Statistical Physics* (in Russ.) (Gostekhisdat, Moskva, 1946), Engl. in J. De Boer, G.E. Uhlenbeck, Studies in Stat. Mechanics, North Holland, Amsterdam 1962

N.N. Bogoliubov, *Selected Works: Part I Dynamical Theory, Part II Quantum and Classical Statistical Mechanics* (in Russ.) (Naukova Dumka, Kiev, 1969–1970), Engl. transl. Gordon and Breach, New York, 1991

H.P. Bonzel, A.M. Bradshaw, G. Ertl (eds.), *Physics and Chemistry of Alkali Metal Adsorption* (Elsevier, Amsterdam, 1989)

D. Bohm, D. Pines, A collective description of electron interactions. Phys. Rev. **92**, 609 (1953)

R. Bredow, W.-D. Th Bornath, R. Redmer, Kraeft. Contrib. Plasma Phys. **53**, 276 (2013)

L. Brillouin, *Les statistiques quantiques et leurs applications* (Paris, 1930)

N.V. Brilliantov, Accurate first-principle equation of state for the OCP. Contrib. Plasma Phys. **38**, 489 (1998)

L.S. Brown, L.G. Yaffé, Effective field theory of highly ionized plasmas. Phys. Rep. **340**, 1 (2001)

A.V. Bushman, I.V. Lomonossov, V.E. Fortov, *Equation of State of Metals at High Energy Densities* (Chernogolovka, 1992)

R. Cauble et al., Contrib. Plasma Phys. **41**, 279 (2001)

D.M. Ceperley, Rev. Mod. Phys. **65**, 279 (1995)

G. Chabrier, A. Potekhin, Phys. Rev. E **58**, 4941 (1998)

E.G.D. Cohen, T.J. Murphy, Phys. Fluids **12**, 1404 (1969)

W. Däppen, Accurate and versatile equations of state for the Sun and Sun-like stars. Astrophys. Space Sci. **328**, 139–146 (2010)

C. Deutsch, Nodal expansion in a real matter plasma. Phys. Lett. A **60**, 317–318 (1977)

N. Desbiens, P. Arnault, J. Clerouin, Parametrization of OCP, Physics of Plasmas 23, No. 9 (2017), https://doi.org/10.10163/1.4963388

H.E. DeWitt, Evaluation of the quantum-mechanical ring sum with Boltzmann statistics. J. Math. Phys. **3**, 1216–1228 (1962)

H.E. DeWitt, Statistical mechanics of high-temperature quantum plasmas beyond the ring approximation. J. Math. Phys. **7**, 616–626 (1966)

H.E. DeWitt, Asymptotic form of the classical one-component plasma. Phys. Rev. A **14**, 1290–1293 (1976)

H.E. DeWitt, M. Schlanges, A.Y. Sakakura, W.D. Kraeft, Low density expansion of the equation of state for a quantum electron gas. Phys. Lett. A **197**, 326 (1995)

W. Ebeling, Statistische Thermodynamik der gebundenen Zustände in Plasmen. Ann. Physik (Leipzig) **19**, 104–112 (1967)

W. Ebeling, Derivation of the free energy from exact scttering phase shifts (in German). Ann. Physik (Leipzig) **22**, 33–39 (1968), **22**, 383–391 (1969), **22** 392–401 (1969)

W. Ebeling, Equation of state and Saha equation of partially ionized plasmas. Physica **38** 378 (1968), **73**, 573–584 (1974)

W. Ebeling, On the possibility of diffusion instabilities in non-aqueous weak electrolytes. Z. Physik. Chem. **247**, 340 (1971), Quantum statistics of ionization and shielding effects. Phys. Stat. Sol. (b) **46**, 243 (1971)

W. Ebeling, *On the Thermodynamic Functions of One-Component Nonideal Plasmas* (Humboldt University, Berlin (Preprint Section of Physics, Statistical Physics, 1998)

W. Ebeling, The work of Baimbetov on nonideal plasmas and some recent developments. Contrib. Plasma Phys. **56**(3–4), 163–175 (2016)

W. Ebeling, R. Sändig, Theory of the ionization equilibrium in dense plasmas. Ann. Physik (Leipzig) **28**, 289–305 (1973)

W. Ebeling, M. Grigo, Mean spherical approximation-mass action law theory of ionic solutions. Annalen der Physik 37, 21 (1980). J. Solut. Chem. **11**, 151–167 (1982)

W. Ebeling, W. Richert, Nonideal hydrogen plasma. Ann. Physik (Leipzig) **39**, 362 (1982), Phys. Lett. A **108**, 80 (1985), Phys. Stat. Sol. (b) **128**, 167 (1985)

W. Ebeling, W. Richert, Pressure ionization of atoms in plasmas. Contrib. Plasma Phys. **25**, 431–436 (1985)

W. Ebeling, G. Norman, Coulomb phase transitions in dense plasmas. J. Stat. Phys. **110**, 861–877 (2003)

W. Ebeling, W.D. Kraeft, D. Kremp, Quantum statistical second virial coefficient for real gases and plasmas. Contrib. Plasma Phys. **10**, 237–263 (1970)

W. Ebeling, W.D. Kraeft, D. Kremp, Nonideal plasmas, *Proceedings of ICPIG. Invited Papers* (Berlin, 1977)

W. Ebeling, C.V. Meister, R. Sändig, Pressure ionization in nonideal alkali plasmas. Ann. Physik **36**, 321–400 (1978)

W. Ebeling, M. Steinberg, J. Ortner, Ionization equilibrium and EOS of a low temperature hydrogen plasma in a magnetic field. Eur. Phys. J. D **12**, 513–520 (2000)

W. Ebeling, S. Hilbert, H. Krienke, J. Mol. Liq. **96–97**, 409 (2002)

W. Ebeling, W.D. Kraeft, G. Röpke, Bound states in Coulomb systems, Contr. Plasma Phys. **52**, 7–16 (2012)

W. Ebeling, W.D. Kraeft, D. Kremp, K. Kilimann, Phys. Stat. Sol. b **78**, 241 (1976)

W. Ebeling, W.D. Kraeft, D. Kremp, G. Röpke, Energy levels in hydrogen plasmas and the Planck-Larkin partition function. Astrophys. J. **290**, 24–27 (1985)

W. Ebeling, A. Förster, W. Richert, H. Hess, Thermodynamic properties and plasma phase transitions of xenon at high pressure and high temperature. Physica A **150**, 159–171 (1988)

W. Ebeling, G.E. Norman, A.A. Valuev, I.A. Valuev, Contrib. Plasma Phys. **39**, 61 (1999)

W. Ebeling, W. Stolzmann, A. Förster, M. Kasch, Contrib. Plasma Phys. **39**, 287–306 (1999)

W. Ebeling, A. Förster, V. Fortov, V. Gryaznov, A. Polishchuk, *Thermophysical Properties of Hot Dense Plasmas (Teubner, Stuttgart, 1991)* (Russ. transl. Moskva, Ishevsk, 2007)

W. Ebeling, D. Blaschke, R. Redmer, H. Reinholz, G. Röpke, The influence of Pauli blocking effects on the Mott transition in dense hydrogen. J. Phys. A: Math. Theor. **42**, 214033 (2009)

G. Ecker, W. Weizel, Ann. Physik (Leipzig) **17**, 126 (1956)

H. Falkenhagen, *Theorie der Elektrolyte* (Hirzel, Leipzig, 1971)

H. Falkenhagen, W. Ebeling, Equilibrium properties of ionized dilute electrolytes, in *Ionic Interactions*, vol. I, ed. by S. Petrucci (Academic Press, New York, 1971)

V.S. Filinov, M. Bonitz, W. Ebeling, V.E. Fortov, Plasma Phys. Control. Fusion **43**, 743 (2001)

A.V. Filinov, M. Bonitz, W, Ebeling, Improved Kelbg potential for correlated Coulomb systems, J. Phys. A: Math. Gen. **36**, 5957–5962 (2003)

V.S. Filinov, M. Bonitz, V.E. Fortov, W. Ebeling, P. Levashov, M. Schlanges, Thermodynamic properties and plasma phase transition in dense hydrogen. Contrib. Plasma Phys. **44**, 388–394 (2004)

V.S. Filinov, P.R. Levashov, A.V. Botan, M. Bonitz, V.E. Fortov, Thermodynamic properties and electrical conductivity of strongly correlated plasma media. J. Phys. A: Math. Theor. **42**, 214002 (2009)

M.E. Fisher, J. Stat. Phys. **75**, 1 (1994)

M.E. Fisher, Y. Levin, Criticality in ionic fluids: Debye-Hückel. Bjerrum and beyond. Phys. Rev. Lett. **71**, 3826–3829 (1993)

M.E. Fisher, Y. Levin, Phys. Rev. Lett. **71**, 2138 (1993)

A. Förster, T. Kahlbaum, W. Ebeling, Equation of state and phase diagram of fluid helium in the region of partial ionization. Laser Part. Beams **10**, 253–262 (1992)

V.E. Fortov et al., Contrib. Plasma Phys. **41**, 215 (2001)

V. Fortov, M. Mochalov et al., Phase transition in a strongly nonideal deuterium plasma generated by quasi-isentropical compression at megabar pressures. Phys. Rev. Lett. **99**, 185001 (2007)

V.E. Fortov, *Extreme States of Matter* (in Russian) (FizMatGis, Moskva, 2009) (2015)

V.E. Fortov, Extreme States of Matter: On Earth and in the Cosmos, the Frontiers Collection (Springer, 2011)

V.E. Fortov, V.Y. Ternovoi, M.V. Zhernokletov, M.A. Mochalov et al., Phys. Rev. Lett. **99**, 185001 (2007)

H. Friedman, *Ionic Solution Theory* (Wiley Interscience, New York, 1962)

V.K. Gryaznov, I.I. Iosilevsky, Thermophysical properties of hydrogen plasmas to Megabars. Contrib. Plasma Phys. **56**, 200 (2016)

W. Ebeling, W.D. Kraeft, G. Röpke, Bound states in Coulomb systems. Contr. Plasma Phys. **52**, 7–16 (2012)

W. Ebeling, W.D. Kraeft, D. Kremp, *Theory of Bound States and Ionisation Equilibrium in Plasmas and Solids (Akademie-Verlag, Berlin, 1976)* (Extended Russ. transl. Mir, Moscow, 1979)

H. Hellmann, Über die Natur der chemischen Bindung, Acta Physicochem. USSR **1**, 333–353 (1934/1935)

H. Hellmann, Ein kombiniertes Näherungsverfgahren zur Energieberechnung im Vielelektronenproblem. Acta Physicochem. USSR **4**, 225–244 (1936)

V.E. Fortov, Equation of state from the ideal gas to the quark-gluon plasma (in Russ.) (Fizmatgis, Moskva, 2013), S.H. Glenzer, G. Gregori, R.W. Lee, F.J. Rogers, S.W. Pollaine, O.L. Landen. Phys. Rev. Lett. **90**, 175002 (2003)

A.V. Filinov, M. Bonitz, W, Ebeling, Improved Kelbg potential for correlated Coulomb systems. J. Phys. A: Math. Gen. **36**, 5957–5962 (2003)

H.C. Graboske, D.J. Harwood, F.J. Rogers, Phys. Rev. **186**, 210 (1969)

H. Griem, *Plasma Spectroscopy* (Mc Graw Hill, New York, 1968)

M. Grigo, W. Ebeling, An analytical calculation of EOS and critical point in a dense fluid of hard charged spheres. Ann. Physik (Leipzig) **37**, 21–30 (1980)

P.C. Hemmer, H. Holden, S. Kjelstrup Ratkje (eds.), *The Collected Works of Lars Onsager* (World Scientic, Singapore, 1996)

F. Hensel, S. Juengst, F. Noll, R. Winter, Localisation and metal-insulator transitions, ed. by D. Adler, H. Fritsche (Plenum Press, New York, 1985)

F. Hensel, W.W. Warren Jr., *Fluid Metals, The Liquid-Vapor Transition* (Princeton University Press, Princeton, 2014)

H. Hess, Physics of nonideal plasmas, ed. by W. Ebeling, A. Foerster, R. Radtke (B.G. Teubner, Leipzig, 1992)

B. Holst, R. Redmer, V.K. Gryaznov, V.E. Fortov, I.L. Iosilevskiy, Hydrogen and deuterium in shock wave experiments, ab initio simulations and chemical picture modeling. Eur. Phys. J. D **66**, 104 (2012)

H. Hoffmann, W. Ebeling, Quantum statistics of high-temperature plasmas in equilibrium. Contr. Plasma Phys. **8**, 43–56 (1968)

S. Ichimaru, *Basic Principles of Plasma Physics* (Addison-Wesley, Benjamin, London, 1973)

S. Ichimaru, *Statistical Plasma Physics* (Addison-Wesley, Redwood City, 1992)

D.R. Inglis, E. Teller, Astrophysics **90**, 430 (1939)

Z.A. Insepov, G.E. Norman, Sov. Phys. JETP **35**, 1198 (1972)

J.K. Jackson, L.S. Klein, Phys. Rev. **177**, 352 (1969)

J.C. Justice, W. Ebeling, Conductance theory of binary electrolytes for Hamiltonian models. J. Sol. Chem. **8**, 809–833 (1979)

A.S. Kaklyugin, Correlation energy of a non-debye plasma. Teplophys. Vyss. Temp. **23**, 217 (1985)

G. Kalman (ed.), *Strongly Coupled Coulomb Systems* (Pergamon Press, 1998)

T. Kahlbaum, Density expansions of the reduced distribution function and the excess free energy. Contrib. Plasma Phys. **39**, 181–184 (1999)

T. Kahlbaum, The quantum-diffraction term in the free energy for Coulomb plasma, effective-potential approach. J. Phys. France **10**, Pr5-455 (2000)

G. Kelbg, Quantenstatistik der Gase mit Coulomb-Wechselwirkung. Ann. Physik (Leipzig) **12**(219–224), 354–360 (1963)

G. Kelbg, Klassische statistische Mechanik der Teilchen-Mischungen mit sortenabhängigen weitreichenden Wechselwirkungen. Ann. Physik (Leipzig) **14**, 394–403 (1964)

G. Kelbg, *Methoden der statistischen Thermodynamik hochionisierter Plasmen, Plasmaphysik aund Gaselektronik III* (Akad. Verlag, Berlin, 1972)

A.L. Khomkin, I.A. Mulenko, High Temp. **41**, 275 (2003). arXiv.org/physics/0305118

A. Khomkin, A. Shumikhin, I. Mulenko, Basic chemical models of partially ionized plasma. Czech. J. Phys. **54**, Suppl. C 143 1b (2004)

A. Khomkin, A. Shumikhin, Features of chemical models of a nonideal atomic plasma at high temperatures (in Russ.). Phys. Plasmas **34**, 251–256 (2008)

K. Kilimann, D. Kremp, W.D. Kraeft, Phys. Lett A **61**, 393 (1977)

K. Kilimann, W. Ebeling, Energy gap and line shifts for H-like ions in dense plasmas. Z. Naturforschung **45a** 613–617 (1990)

O. Klein, Medd. Vedenskab. Nobelinst. **5**, 6 (1919)

Y.L. Klimontovich, *Statistical Theory of Nonequilibrium Processes in Plasmas* (Pergamon, Oxford, 1967)

Yu.L Klimontovich, *Statistical Physics* (in Russ) 1982 (Harwood, England, 1986)

M.D. Knudson, D.J. Hanson et al., Phys. Rev. Lett. **87**, 225501 (2001)

M.D. Knudson, M.P. Desjarlais, A. Becker, R.W. Lemke, K.R. Cochrane, M.E. Savage, D.E. Bliss, T.R. Mattsson, R. Redmer, Direct observation of an abrupt insulator-to-metal transition in dense liquid deuterium. Science **348**, 1455–1460 (2015)

W.D. Kraeft, D. Kremp, W. Ebeling, G. Röpke, *Quantum Statistics of Charged Particle Systems* (Akad. Verlag and Pergamon Press, Berlin, 1986)

W.D. Kraeft, M. Schlanges, *Physics of Strongly Coupled Plasmas* (World Scientific, 1996)

W.D. Kraeft, D. Kremp, G. Röpke, Direct linear term in the equation of state of plasmas. Phys. Rev. E **91**, 013108 (2015)

H.A. Kramers, Investigations on the free energy of a mixture of ions. Proc. Koninkl. Akademie v. Wetenschappen **30**, 145–158 (1927)

D. Kremp, G. Schmitz, Z. Naturforschung **22a**, 1366 (1967)

D. Kremp, M. Schlanges, W.D. Kraeft, *Quantum Statistics of Nonideal Plasmas* (Springer, Berlin, 2005)

H. Krienke, W. Ebeling, H.J. Czerwon, Wiss. Z. MNR Univ. Rostock **24**, 5 (1975)

S. Kuhlbrodt, R. Redmer, H. Reinholz, G. Röpke, B. Holst, V.B. Mintsev, V.K. Gryaznov, N.S. Shilkin, V.E. Fortov, Electrical conductivity of noble gases at high pressures. Contrib. Plasma Phys. **45**, 61–69 (2005)

A. Lankin, G. Norman, Density and nonideality effects in plasmas. Contrib. Plasma Phys. **49**, 723–731 (2009)

A.I. Larkin, Zh. eksp. teor. Fiz. (in Russ.) **38**, 1896 (1960)

B.P. Lee, M.E. Fisher, Phys. Rev. Lett. 76, 2906 (1996). Europhys. Lett. **39**, 3 (1997)

Y. Levin, M.E. Fisher, Criticality in the hard sphere ionic fluid. Physica A **225**, 164–220 (1996)

H. Lehmann, W. Ebeling, Coulomb phase transitions in symmetrical quantum systems. Phys. Rev. E **54**, 2451 (1996)

W. Lorenzen, B. Holst, R. Redmer, First-order liquid-liquid phase transition in dense hydrogen. Phys. Rev. B **82**, 195107 (2010)

W.R. Magro, D.M. Ceperley, C. Pierleoni, B. Bernu, Phys. Rev. Lett. **76**, 1240 (1996)

J.M. McMahon, M.A. Morales, C. Pierleoni, D.M. Ceperley, The properties of hydrogen and helium. Rev. Mod. Phys. **84**, 1612 (2012)

G. Martynov, Russ. J. Phys. Chem. **74/1** 149 (1999)

A.L. Mikhailov, A.S. Filimonov, A.A. Pyalling, V.B. Mintsev, V.K. Gryaznov, I.L. Iosilevskii, JETP **97**, 259 (2003)

B. Militzer, E.L. Pollock, Phys. Rev. E **61**, 3470 (2000)

B. Militzer, D.M. Ceperley, Phys. Rev. Lett. **85**, 1890 (2000)

M.A. Morales, C. Pierleoni, E. Schwegler, D.M. Ceperley, Evidence for a first-order liquid-liquid transition in high-pressure hydrogen from ab initio simulations. PNAS **110**, 12799–12803 (2010)

A.G. Moreira, M.M.T. da Gama, M.E. Fisher, J. Chem. Phys. **110**, 10058 (1999)

N. Mott, The transition to the metallic state. Phil. Mag. **6**, 287–309 (1961)

N.F. Mott, *Metal-Insulator Transitions* (Taylor and Francis, London, 1974)

W.J. Nellis, S.T. Weir, A.C. Mitchell, Phys. Rev. B **59**, 3434 (1999)

N. Nettelmann, B. Holst, A. Kietzmann, M. French, R. Redmer, D. Blaschke, Ab initio equation of state data for hydrogen, helium, and water and the structure of Jupiter. Astrophys. J. **683**, 1217 (2008)

S. Nordholm, Chem. Phys. Lett. **105**, 302 (1984)

G.E. Norman, A.N. Starostin, Teplopphys. Vys. Temp. (High Temp.) 6, 394 (1968). High Temp. **8**, 381 (1970)

L. Onsager, Electrostatic interaction of molecules. J. Phys. Chem. **43**, 189–196 (1939)
J. Ortner, Equation of states for classical; coulomb systems: use of the Hubbard-Schofield approach. Phys. Rev. E **59**, 6312–6320 (1999)
D. Pines, *The Many Body Problem* (Benjamin, New York, 1961)
D. Pines, P. Nozieres, *The Theory of Quantum Liquids*, Benjamin, New York, 1966 (Russ. Transl. Mir, Moskva, 1967)
M. Planck, Zur Quantenstatistik des Bohrschen Atommodells. Ann. Physik (Leipzig) **76**, 673–684 (1924)
L. Onsager, J. Chem. Phys. **43**, 189 (1939)
E.L. Pollock, J.P. Hansen, Phys. Rev. A **8**, 3110 (1973)
T.S. Ramazanov, K.N. Dzhumagulova, Phys. Plasmas **17**, 3758 (2002)
J.C. Rasaiah, H.L. Friedman, J. Phys. Chem. **48**, 2742 (1968), **50**, 395 (1969)
R. Redmer, Phys. Rep. **282**, 35 (1997)
R. Redmer, G. Röpke, Progress in the theory of dense strongly coupled plasmas. Contrib. Plasma Phys. **50**, 970–985 (2010)
R. Redmer, F. Hensel, B. Holst (eds.), *Metal-to-Nonmetal Transitions* (Springer, Berlin, 2010)
W. Richert, S.A. Insepov, W. Ebeling, Thermodynamic functions of nonideal alkali plasmas. Ann. Physik (Leipzig) **41**, 139–150 (1984)
J. Riemann, M. Schlanges, H.E. DeWitt, W.-D. Kraeft, EOS of the weakly degenerate one-component plasma. Physica A **219**, 423 (1995)
F.J. Rogers, H.C. Graboske, D.J. Harwood, Bound eigenstates of the static screened Coulomb potential. Phys. Rev. A **1**, 1577–1586 (1970)
R. Rompe, M. Steenbeck, Erg. exakt. Naturwiss. **18**, 275 (1939)
G. Röpke, W. Ebeling, W.D. Kraeft, Quantum-statistical conductance theory of nonideal plasmas, Physica **101 A**, 243–254 (1980)
G. Röpke, K. Kilimann, D. Kremp, W.D. Kraeft, Phys. Lett. A **68**, 329 (1978)
G. Röpke, K. Kilimann, D. Kremp, W.D. Kraeft, R. Zimmermann, Phys. Stat. Sol. (b) **88**, K59 (1978)
G. Röpke, T. Seifert, H. Stolz, R. Zimmermann, Phys. Stat. Sol. (b) **100**, 215 (1980)
S.P. Sadykova et al., Contrib. Plasma Phys. 47, 659, 1160 (2007), 49 76 (2009). Eur. Phys. J. D **61**, 117–130 (2011)
D. Saumon, G. Chabrier, Pressure ionization of hydrogen, Phys. Rev. Lett. **62**, 2397 (1989)
D. Saumon, G. Chabrier, Phys. Rev. A **44**, 5122 (1991), **46**, 2084 (1992)
G. Schmitz, Phys. Lett. **21**, 174 (1966)
G. Schmitz, D. Kremp, Z. Naturforschung **23a**, 1392 (1967)
W. Schröer, On the chemical and the physical approaches to ion association. J. Mol. Liq. **164**, 3–10 (2011)
V.P. Silin, A.A. Rukhadse, *Electromagnetic Properties of Plasmas and of Plasma-Like Matter (In Russ)* (Atomizdat, Moskva, 1961)
V. Sizyuk, A. Hassanein, T. Sizyuk, J. Appl. Phys. **100**, 103–106 (2006)
G. Stell, Criticality and phase-transitions in ionic fluids. J. Stat. Phys. **78**, 197–238 (1995)
A.N. Starostin, V.C. Roerich, R.M. More, How correct is the EOS of weakly nonideal plasmas. Contrib. Plasma Phys. **43**, 369–372 (2003)
A.N. Starostin, V. Roerich, Bound states in nonideal plasmas. Plasma Sour. Sci. Technol. **15**, 410–415 (2006)
M. Steinberg, J. Ortner, W. Ebeling, Phys. Rev. E **58**, 3806 (1998)
W. Stolzmann, W. Ebeling, New Pade approximations for free charges in plasmas. Phys. Lett. A **248**, 242–246 (1998)
V. Ya, Ternovoi, A.S. Filimonov, V.E. Fortov, S.V. Kvitov, D.D. Nikolaev, A.A. Pyalling. Physica B **265**, 6 (1999)
S.A. Trigger, W. Ebeling, V.S. Filinov, V.E. Fortov, M. Bonitz, JETP, **96**, 465–479 (2003). arXiv:physics/0110013
A. Unsöld, Z. Astrophys. **24**, 355 (1948)

A. Unsöld, *Physik der Sternatmosphären unter besonderer Berücksichtigung der Sonne (Springer, Heidelberg, 1938), 2* (Aufl, Heidelberg, 1955)

A.A. Vedenov, A.I. Larkin, Equation of state of plasmas (in Russ.). Zh. eksp. teor. Fiz. **36**, 1133 (1959)

A.A. Vedenov, A.I. Larkin, Equation of state of plasmas, Sov. Phys. JETP **9**, 806–821 (1959)

V.S. Vorobjov, A.L. Khomkin, Method of canonical transformation in the thermodynamics of partially ionized plasmas. Theor. Math. Phys. (Moscow) **8**, 109–118 (1971)

P.N. Voronzov-Veliaminov, A.M. Eliashevich et al., Teplophys. Vysokh. Temp. **8**, 277 (1970), **14**, 199 (1976)

H. Weingärtner, W. Schröer, Criticality of ionic fluids. Adv. Chem. Phys. **116**, 1–66 (2001)

H. Weingärtner, M. Kleemeier, S. Wiegand, W. Schröer, Coulomb and non-coulomb contributions to the criticality of ionic fluids. J. Stat. Phys. **78**, 169–196 (1995)

S.T. Weir, A.C. Mitchell, W.J. Nellis, Metallization of fluid molecular hydrogen at 140 GPa. Phys. Rev. Lett. **76**, 1860–1863 (1996)

E.P. Wigner, On the interaction of electrons in metals. Phys. Rev. **46**, 1002–1011 (1934), Trans. Far. Soc. **34**, 678 (1938)

S. Yeh, Y. Zhou, G. Stell, Phase separation of ionic fluids: an extended Ebeling-Grigo approach. J. Phys. Chem. **100**, 1415–1419 (1996)

I.R. Yukhnovsky, M.G. Holovko, Statistical Theory of Classical Equilibrium Systems (in Russ.) (Nauk., Dumka, 1980)

L.D. YaB Zeldovich, Landau. Zh Eksp. Teor. Fiz. **14**, 32 (1944)

W. Zimdahl, W. Ebeling, Theory of the ionization equilibrium in nonideal alkali plasmas. Ann. Physik (Leipzi) **34**, 9–22 (1977)

V.M. Zamalin, G.E. Norman, V.S. Filinov, The Monte Carlo Method in Statistical Thermodynamics (in Russ.) (Nauka, Moscow, 1977)

R. Zimmermann, K. Kilimann, D. Kremp, W.D. Kraeft, G. Röpke, Phys. Stat. Sol. (b) **90**, 175 (1978)

Chapter 4
Coulomb Correlations and EOS of Nondegenerate Nonideal Plasmas

As pointed out in Chaps. 1 and 3, the correlations in classical Coulomb systems such as electrolytes and quasi-classical plasmas were first studied by Milner, Debye, Hückel, Wigner, and others. Strictly speaking, there is no classical statistical theory of point charges, due to several divergencies which show when Debye's limiting laws are not applicable. The quantum-statistical theory of correlations in Coulomb systems is due to the work of Macke, Gell-Man, Brueckner, Bohm, Pines, Nozieres, Montroll, Ward, Klimontovich, Silin, Vedenov, Larkin, DeWitt, Kelbg, and others (Vedenov 1959; Vedenov and Larkin 1959; DeWitt 1962; Kelbg 1963, 1964; Ebeling et al. 1968; Ebeling et al. 1976, Klimontovich 1984, 1986). The most important results were obtained in different research centers, one of them founded in the 1960s by Kelbg in Rostock (Kelbg 1963, 1964; Ebeling et al. 1968; Ebeling et al. 1976; Kraeft et al. 1986). Field-theoretical approaches were developed to solve the problem of quantum screening by Montroll and Ward (1958) and by Vedenov and Larkin (1959). The Montroll–Ward method was further developed by DeWitt et al. (for the main results see DeWitt 1962; DeWitt et al. 1995; Riemann et al. 1995). An approach based on Feynman–Kac methods was developed by Alastuey and Perez et al. (see, e.g., Alastuey et al. 1992, 1996, 2008, 2015). For exact results see Lieb and Seiringer (2010), here we mainly follow the approach of the Kelbg school, but compare in detail with the results of other approaches.

4.1 Short-Range Quantum Effects in Low Density Plasmas

4.1.1 Pairs of Particles

The classical Debye–Hückel and Bogoliubov approximations which we have discussed so far treat pair correlations and screening effects of point charges. As we pointed out in the last chapter, systems of particles satisfying the Coulomb law like plasmas and electrolytic solutions differ essentially from standard gases with respect

W. Ebeling et al., *Quantum Statistics of Dense Gases and Nonideal Plasmas*, Springer Series in Plasma Science and Technology, DOI 10.1007/978-3-319-66637-2_4

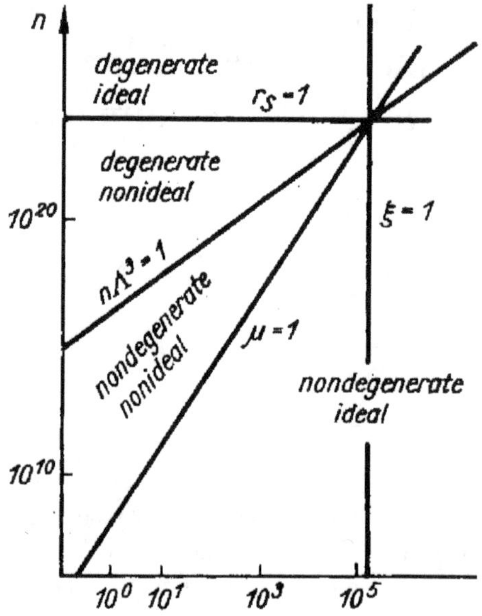

Fig. 4.1 Regions in the (n, T) plane (logarithmic scale) where, approaching from the right-hand side (ideality), electron degeneracy begins. For completeness we also show the border between weak and strong coupling $\mu_B = 1$. The theory which we discuss in the present chapter is to be applied only to the lowest part, viz., the triangle between the line $n_e \Lambda_e^3 = 1$ and the T axis

to their physical properties. In a gas, the two-particle correlations and the pair distribution function are essentially determined by the Boltzmann distribution law

$$F_{ab}(r) = \exp\big[-\beta V_{ab}(r)\big] . \tag{4.1}$$

At short range, the Boltzmann expression diverges for $(+-)$ interactions. However, at higher temperatures, where bound states are absent, the correlation is so weak that we may linearize the Boltzmann law. On the other hand, we have shown that the Coulomb law has to be replaced by a screened interaction law, with distribution

$$F_{ab}(r) = 1 - \beta V_{ab}^{\mathrm{D}}(r) = 1 - \frac{\beta e_a e_b}{\varepsilon_r r} \exp\big[-(r/r_{\mathrm{D}})\big] . \tag{4.2}$$

We define the Debye radius r_{D} and the Bogoliubov parameter μ_{B} by

$$r_{\mathrm{D}} = 1/\kappa , \qquad \kappa = \sqrt{4\pi\beta \sum_a n_a e_a^2/\varepsilon_r} , \qquad \mu_{\mathrm{B}} = \frac{\beta e_a e_b}{\varepsilon_r r_{\mathrm{D}}} . \tag{4.3}$$

Here ε_r is the dielectric constant of the medium, although we usually study the vacuum case, where $\varepsilon_r = 1$. The parameter μ_{B}, known as the nonideality parameter, was first introduced by Bogoliubov in 1946 as a characteristic parameter for series expansions of Coulomb systems, replacing the characteristic density parameter na^3 for gases.

The nonideality parameter μ_B and the degeneracy parameter $n_e \Lambda_e^3$ are represented in Fig. 4.1. As the name suggests, the first characterizes the deviations from plasma nonideality. Beyond the linear term in the plasma parameter, quantum effects play an essential role. In Chaps. 1 and 3, we already discussed the fact that, due to quantum effects, Coulomb particles cannot be considered as point charges, but should rather be viewed as a kind of snow flake with distributed charge. Here we approach the problem of short-range pair correlations in a more systematic way, based on the density matrix of pairs.

Recall that that the diagonal elements of the density matrix of a pair of particles of species a, b are given by the binary Slater sum $S_{ab}^{(2)}(r)$, which is defined as the diagonal representation of the density operator of pairs

$$S_{ab}^{(2)}(\mathbf{r}) = \left\langle \mathbf{r}_1\mathbf{r}_2 \left| \hat{\rho}_{ab} \right| \mathbf{r}_1\mathbf{r}_2 \right\rangle . \tag{4.4}$$

Here the species a and b are charges, where $ab \in \{++, --, -+, +-\}$. For small densities, the pair density operator is given by

$$\hat{\rho}_{ab} = \exp\left(- \beta \hat{H}_{ab} \right) , \qquad \hat{H}_{ab} = - \frac{\hbar}{2m_{ab}} \Delta + V_{ab} , \tag{4.5}$$

where $\hat{H}_{ab}$ is the Hamiltonian of relative motion of a pair, with relative mass m_{ab}.

Now we proceed to the representation by plane waves, as in Chap. 2 for Yukawa particles:

$$S_{ab}(\mathbf{r}) = C \int \mathrm{d}\mathbf{k} \exp\left(- \mathrm{i}\mathbf{k} \cdot \mathbf{r} \right) W_{ab}(\beta, \mathbf{k}) , \tag{4.6}$$

where $\mathbf{k} = \mathbf{p}/m$, while C is a constant and

$$W_{ab}(\beta, \mathbf{k}, \mathbf{r}) = \exp\left(- \beta \hat{H}_{ab} \right) \exp\left(+ \mathrm{i}\mathbf{k} \cdot \mathbf{r} \right) . \tag{4.7}$$

Differentiating we obtain the Bloch equation

$$\frac{\partial}{\partial \beta} W_{ab} = -\hat{H}_{ab} \exp\left(- \beta \hat{H}_{ab} \right) \exp\left(+ \mathrm{i}\mathbf{k} \cdot \mathbf{r} \right) = -\hat{H}_{ab} W_{ab} .$$

This was named after Felix Bloch (1905–1983, Nobel Prize 1952). For two interacting particles, it reads

$$\frac{\partial W_{ab}}{\partial \beta} = \frac{\hbar}{2m_{ab}} \Delta W_{ab} - V_{ab} W_{ab} . \tag{4.8}$$

The solution was found by Kelbg in 1963, using a similar perturbation theory to the one described for Yukawa gases in Chap. 2. The zero order approximation is

$$W_{ab}^0 = \exp\left(\mathrm{i}\mathbf{k} \cdot \mathbf{r} - \frac{\beta \hbar^2}{2m_{ab}} \mathbf{k}^2 \right) . \tag{4.9}$$

Including symmetry effects for Fermions, the corresponding zero order Slater function reads

$$S_{ab}(1,2)^{(0)}(r) = \left[1 - \frac{\delta_{ab}}{2s_a+1}\exp(-r^2/\lambda^2)\right] . \tag{4.10}$$

The first order problem leads to a diffusion problem with source (see Chap. 2). Solving this equation and including symmetry effects for fermions, we get to first order the Fourier transform (Ebeling, Hoffmann and Kelbg 1967)

$$\begin{aligned}\tilde{S}_{ab}(1,2)^{(1)}(t) = &-\beta\frac{4\pi e_a e_b}{t^2}\exp\left(-\frac{\lambda_{ab}^2 t^2}{4}\right){}_1F_1\left(\frac{1}{2},\frac{3}{2},\frac{\lambda^2 t^2}{4}\right) \\ &-\delta_{ab}\frac{2\pi e_a e_b \lambda_{ab}^2}{2s_a+1}\exp\left(-\frac{\lambda_{ab}^2 t^2}{4}\right){}_2F_2\left(\frac{1}{2},\frac{1}{2},\frac{3}{2},\frac{3}{2},\frac{\lambda_{ab}^2 t^2}{4}\right) ,\end{aligned} \tag{4.11}$$

where ${}_1F_1$ and ${}_2F_2$ are the well known hypergeometric functions. Higher approximations follow from iterations of the Bloch equation, viz.,

$$\frac{\partial W_{ab}^{n+1}}{\partial\beta} - \frac{\hbar^2}{2m_{ab}}\Delta W_{ab}^{n+1} = V_{ab}W_{ab}^{n} . \tag{4.12}$$

For a related discussion of higher orders in the perturbation series see Hoffmann and Kelbg (1966, 1967) and Chap. 2.

4.1.2 Kelbg Potential

Neglecting symmetry effects, after back-transformation in the first order approximation, we find the Kelbg potential

$$V_{ab}^{\mathrm{K}}(\mathbf{r}) = \frac{e_a e_b}{r}\int_0^1 \mathrm{d}\alpha\, \Phi\left[\frac{r}{2\lambda_{ab}\sqrt{\alpha(1-\alpha)}}\right] . \tag{4.13}$$

Carrying out the integral delivers the Kelbg potential in the explicit form

$$V_{ab}^{\mathrm{K}} = \frac{e_a e_b}{\epsilon_0 r}\left[1 - \exp\left(-\frac{r^2}{\lambda_{ab}^2}\right) + \sqrt{\pi}\frac{r}{\lambda_{ab}}(1 - \Phi(r)/\lambda_{ab})\right] . \tag{4.14}$$

The first order Slater sum is then given by

$$S_{ab}(1,2)^{(1)} = 1 - \beta V_{ab}^{\mathrm{K}}(r) , \qquad \Phi(x) = \frac{2}{\sqrt{\pi}}\int_0^x \exp(-t^2)\mathrm{d}t . \tag{4.15}$$

Here we shall mainly use the notation $\Phi(x)$ for the error function, but sometimes also erf(x), as often seen in the literature. In the nonlinear Kelbg approximation, we

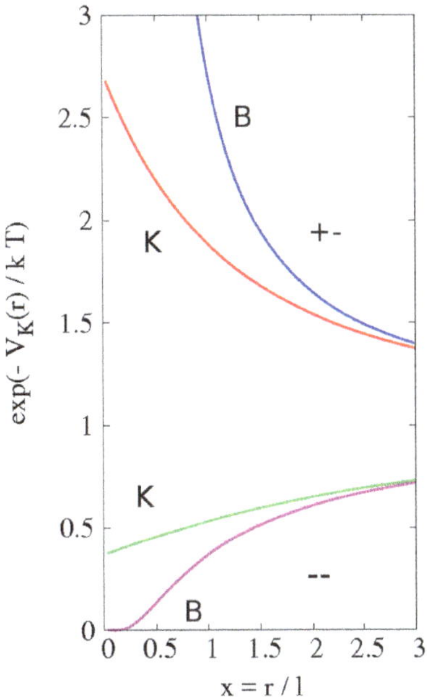

Fig. 4.2 Schematic view of the pair distribution function in the (nonlinear) Kelbg approximation as a function of the radius for (+−) pairs *red curve* and (−−) pairs *green curve*. The *upper* and the *lower curves* show the classical Boltzmann approximation for comparison

may write

$$S_{ab}(r) \approx \exp\left[-\beta V_{ab}^{\mathrm{K}}(r)\right] . \tag{4.16}$$

This function is shown in Fig. 4.2. We see that, in contrast to the Coulomb potential and the corresponding Boltzmann factor, the Kelbg potential and the corresponding Slater functions are finite at $r = 0$ due to quantum effects, in particular due to the Heisenberg uncertainty.

4.2 Screening in Weakly Coupled Plasmas

4.2.1 Screened Correlations in Nondegenerate Plasmas

In the last section we calculated the correlations between the particles for pairs imbedded in a plasma of zero density. Kelbg's idea was to interpret his pair function as an effective potential and extend the known classical procedures to the quantum case for the region of weak coupling using expansions with respect to the Bogoliubov parameter. He introduced his effective potential into the quasi-classical partition function of the Coulomb potential, and used first the methods of collective variables (Kelbg 1963, 1964). For an equivalent calculation based on the known Coulomb wave functions see Trubnikov and Elesin (1965).

On the basis of Kelbg's results, a similar procedure based on Bogoliubov's hierarchy for the density operators was proposed by Kremp and Schmitz in 1967 (Kremp et al. 1967, Ebeling 2016). Simplifying the derivation, in a first step we substitute the Kelbg potential for the Coulomb potential in Bogoliubov's equation for the lowest order correlation function. In the Fourier representation, according to (4.12) and neglecting symmetry, the Kelbg potential reads

$$V_{ab}^{\mathrm{K}}(k) = \frac{4\pi e_a e_b}{k^2} M(1; 3/2; -\lambda^2 k^2/4) \,, \tag{4.17}$$

where from now on we use the shorter notation $M(\alpha, \gamma; x)$ for our special hypergeometric functions, sometimes called the Kummer functions. Further, for simplicity, we use the reduced mass approximation and neglect the mass differences, working with the reduced mass μ instead of individual masses, and also with the reduced thermal wavelength $\lambda = \hbar/\sqrt{2\mu k_{\mathrm{B}} T}$.

Considering the Kelbg potential (without symmetry effects) as an effective potential, we may apply the Bogoliubov method described in more detail in Sect. 4.2.2. By replacing the Fourier transform of the Kelbg–Deutsch potential by the full Fourier transform (4.17), we get the following approximation for the screened correlation function in Fourier space

$$g_{ab}^{\mathrm{K}}(k) = -\beta \frac{4\pi e_a e_b M(1; 3/2; -\lambda^2 k^2/4)}{k^2 + \kappa^2 M(1; 3/2; -\lambda^2 k^2/4)} \,. \tag{4.18}$$

This result was already obtained by Kelbg using functional differentiation of the free energy (Kelbg 1984). It was then reproduced for the electron gas by Kremp and Schmitz (1967). Note that screening effects in fluid phases of charged particles give rise to a variety of exact sum rules for their correlation functions which will not be discussed here (Martin 1988). In their work, Kremp and Schmitz (1967) used methods due to Bogoliubov, Guernsey, Cohen, Hirota, and others. These authors applied a static approximation for the polarization function, taking into account only Matsubara frequencies $\nu = 0$ and also neglecting exchange effects. We note here that, as shown later by Cornu, Martin, Alastuey, and Ballenegger, the dynamical nonzero frequency components may contribute long-range algebraic tails (Cornu and Martin 1991; Cornu 1997; Alastuey 1999; Ballenegger et al. 2002). In the following we neglect these effects which lead to non-exponential screening, but we shall return to this topic later.

For a representation in coordinate space we have to perform the Fourier back-transformation. This leads to difficulties which may be overcome only by making further approximations. In the most drastic approximation, the Kummer function may be approximated in the nominator and the denominator by the useful Padé formula:

$$M\left(1, \frac{3}{2}; -\frac{1}{4}\lambda^2 k^2\right) = 1 - \frac{1}{6}\lambda^2 k^2 + \cdots \approx \frac{1}{1 + \lambda^2 k^2/6} \,. \tag{4.19}$$

By introducing a new (Kelbg-type) constant $\alpha = \sqrt{6}/\lambda$, we may carry out the integration and get the correlation function in coordinate space as a difference between two exponential terms (Falkenhagen 1971). This form of the correlation function is sometimes called Glauberman–Yukhnovsky–Kelbg approximation, named after those who first investigated it (Kelbg 1963; Falkenhagen 1971; Yukhnovsky and Holovko 1980):

$$g_{ab}^{\mathrm{K}}(r) = -\frac{e_a e_b \alpha^2}{r(p+s)(p-s)}\left[\exp(-sr) - \exp(-pr)\right], \tag{4.20}$$

$$s = \frac{\alpha}{2}\left(\sqrt{1+2\frac{\kappa}{\alpha}} + \sqrt{1-2\frac{\kappa}{\alpha}}\right), \quad p = \frac{\alpha}{2}\left(\sqrt{1+2\frac{\kappa}{\alpha}} - \sqrt{1-2\frac{\kappa}{\alpha}}\right). \tag{4.21}$$

We thus obtain simple expressions for the screened quantum correlation function in the first order of the Bogoliubov parameter, valid for weak coupling.

Thermodynamic Functions for Weakly Coupled Plasmas

The mean electrical energy is obtained by integrating the correlation function in the first Bogoliubov–Kelbg approximation:

$$\langle U_{\mathrm{c}} \rangle = \frac{1}{2}\sum n_a n_b \int \mathrm{d}\mathbf{r} \frac{e_a e_b}{r}\left[1 + g_{ab}(r)\right]. \tag{4.22}$$

Note that, as usual, the first term linear in e^2 gives zero due to electroneutrality. The integration is straightforward, and for binary plasmas, in the reduced mass approximation, we get

$$\langle U_{\mathrm{c}} \rangle = -N\frac{\kappa^2}{8\pi}\beta\frac{e^2\alpha^2}{p+s}. \tag{4.23}$$

These simple analytical expressions give the thermodynamic functions for weakly coupled OCPs and TCPs. Figure 4.3 shows the Coulomb energy of an electron gas found by evaluating the integral of the weak correlation function in the Bogoliubov–Kelbg approximation, comparing it with a classical estimate for the Coulomb energy from Monte Carlo calculations (DeWitt 1976). We see that, for the electron gas, the energy is surprisingly close to the classical result for the electron gas, and not far from the Wigner asymptotic going as $-a_{\mathrm{bcc}}e^2/d$.

4.2.2 Pair Correlations in Many-Component Systems

We now consider many-component equilibrium plasmas at relatively low densities and weak coupling. These plasmas still behave quasi-classically, and this allows us to treat the screening quasi-classically as above. In the many-component case, the Bogoliubov equation for the pair distribution functions reads

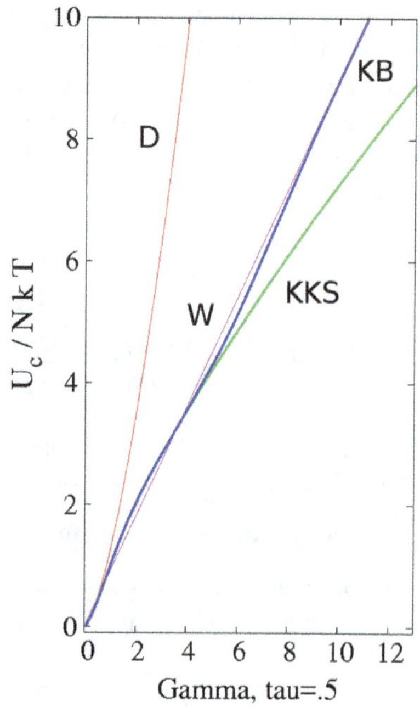

Fig. 4.3 Ring contribution to the Coulomb energy of an electron plasma for the temperature $\tau = 0.5$ (in Ry). The *red curve* some distance away from all the others is the Debye law, and the straight line (*magenta*) is the Wigner estimate. The curve close to it (*blue*) is the Kaklyugin–Brilliantov approximation and the lowest curve (*green*) shows this calculation when we use the Kelbg–Kremp–Schmitz evaluation of the ring terms. Note the remarkable result that the quantum corrections, which increase with the temperature, are in general relatively small, and are not far from Wigner's estimate for $\Gamma < 5, \tau < 0.5$

$$\frac{\partial}{\partial r_1} F_{ab} + F_{ab} \frac{\partial}{\partial r_1} V_{ab} + \sum_c n_c \int d\mathbf{r}_3 \frac{\partial}{\partial r_1} V_{ac}(13) F_{abc}(123) = 0 \,. \tag{4.24}$$

We introduce here the pair correlation functions and an approximation for the triple correlations by

$$F_{ab} = 1 + g_{ab} \,, \qquad F_{abc} = 1 + g_{ac} + g_{bc} + g_{ab} \,, \tag{4.25}$$

and find for the pair correlations

$$\begin{aligned} 0 = k_B T \frac{\partial g_{ab}}{\partial \mathbf{r}_1} + [1 + g_{ab}(1,2)] \frac{\partial V_{ab}(1,2)}{\partial \mathbf{r}_1} \\ + \sum_c n_c \int d\mathbf{r}_3 \frac{\partial V_{ac}(1,3)}{\partial \mathbf{r}_1} [g_{ab}(1,2) + g_{ac}(1,3) + g_{bc}(2,3)] \,. \end{aligned} \tag{4.26}$$

By integration, to lowest order with respect to the Bogoliubov parameter μ_B, we find

$$g_{ab}(1,2) = V_{ab}(1,2) + \sum_c \int d\mathbf{r}_3 V_{ac}(1,3) g_{bc}(2,3) = 0 \,. \tag{4.27}$$

The physical meaning is that the correlations are the sum of a direct contribution and a contribution stemming from the charges in the surroundings. Once again, we use a representation in Fourier space:

$$r g_{ab}(r) = \frac{1}{2\pi^2} \int_0^\infty \mathrm{d}k \sin(kr) k g_{ab}(k) \,. \tag{4.28}$$

Details of the solution in Fourier space are given in (Falkenhagen 1971) and the main steps have already been discussed in Sect. 3.1.

In the case of effective potentials like the Kelbg potential, except in the reduced mass approximation (RMA), the algebraic equations lead to matrix equations in Fourier space, even for simple charge-symmetric plasmas. For the case of electron–ion plasmas, Kelbg solved the problem by first calculating the free energy and the pair correlation function by functional differentiation (Kelbg 1964). The result of Kelbg's calculation is

$$g_{ab}^{\mathrm{K}}(k) = \frac{1}{Z(k)} \left\{ V_{ab}^{\mathrm{K}}(k) + \beta(n_a \delta_{\mathrm{i}a} + n_b \delta_{\mathrm{e}b}) \left[V_{\mathrm{ee}}^{\mathrm{K}}(k) V_{\mathrm{ii}}^{\mathrm{K}}(k) - V_{\mathrm{ei}}^{\mathrm{K}}(k)^2 \right] \right\} , \tag{4.29}$$

where, according to Kelbg (1964), the denominator is

$$Z(k) = 1 + \beta n V_{\mathrm{ee}}^{\mathrm{K}}(k) + \beta n V_{\mathrm{ii}}^{\mathrm{K}}(k) + \beta^2 n^2 \left[V_{\mathrm{ee}}^{\mathrm{K}}(k) V_{\mathrm{ii}}^{\mathrm{K}}(k) - V_{\mathrm{ei}}^{\mathrm{K}}(k)^2 \right] .$$

We note that several terms in the numerator disappear for fully symmetric plasmas. Using electroneutrality, the mean Coulomb energy in this approximation is given by

$$\begin{aligned} U_{\mathrm{c}} = C \int \mathrm{d}\mathbf{k} \frac{V_{\mathrm{ee}}(k)}{Z(k)} \Big\{ & V_{\mathrm{ee}}^{\mathrm{K}}(k) + \beta n \left[V_{\mathrm{ee}}^{\mathrm{K}}(k) V_{\mathrm{ii}}^{\mathrm{K}}(k) - V_{\mathrm{ei}}^{\mathrm{K}}(k)^2 \right] \\ & + V_{\mathrm{ii}}^{\mathrm{K}}(k) + \beta n \left[V_{\mathrm{ee}}^{\mathrm{K}}(k) V_{\mathrm{ii}}^{\mathrm{K}}(k) - V_{\mathrm{ei}}^{\mathrm{K}}(k)^2 \right] - 2 V_{\mathrm{ei}}^{\mathrm{K}}(k) \Big\} . \end{aligned} \tag{4.30}$$

Here the integral can be taken only numerically. Note that the matrix effects resulting from this complicated integral are in general rather small, except at $T > 10^5$ K, i.e., in a region where the correlations and their contribution to the thermodynamics are already very small. Hence, in nearly all cases, one may work in the reduced mass approximation (RMA), where matrix effects disappear, and we recover the rather simple expression (4.23). Note that, at sufficiently high densities and temperatures where bound states cannot survive in plasmas, the plasma system may be considered as a mixture of two Fermi fluids, exemplified by hydrogen under high pressure (Ebeling 1985).

The basic results we obtain in this chapter are always connected with exponential screening of the charges, which is just a generalization of Debye screening. It has been shown that, in higher order contributions, beyond the RPA, quantum fluctuations lead to a different type of correlations which lead to additional power law tails in the correlation functions (Cornu and Martin 1991; Cornu 1997). Just to provide a feeling for how quantum fluctuations may lead to correlations, let us just recall that, as is well known, the long tail in the van der Waals forces is also due to correlations between quantum fluctuations. These results about power law correlations are of great significance, but the practical relevance for strongly coupled plasmas has not yet been fully explored, except for a few cases, such as the study of charge correlations inside the core of the Sun (Cornu 1997).

4.3 Non-diagonal Pair Density Operators

4.3.1 Diagonal and Non-diagonal Pair Density Matrix

The Kelbg potential, which is the effective quantum version of the Coulomb potential, is closely related to the two-particle density matrix. For many applications, the Slater sum of pairs, which is the diagonal part of the pair density matrix, does not give sufficient information and we also need the non-diagonal representation, which depends on four coordinates. Later on we will study applications in the context of the path integral approach (see Filinov et al. 2000, 2001, 2003, 2004, and Chaps. 9–10). For this purpose, we need a modified representation for the pair density matrix:

$$\rho_{ab}^{(2)}(\mathbf{r}_1, \mathbf{r}_1', \mathbf{r}_2, \mathbf{r}_2') = C\Big\langle \mathbf{r}_1, \mathbf{r}_2 \Big| \exp(-\beta \hat{H}_{ab}) \Big| \mathbf{r}_1' \mathbf{r}_2' \Big\rangle . \tag{4.31}$$

In the linear approximation in e^2, we use the ansatz (Filinov et al. 2000, 2001, 2003, 2004, 2005)

$$\begin{aligned}\rho_{ab} = \frac{(m_a m_b)^{3/2}}{(2\pi\hbar\beta)^3} \exp\left[-\frac{m_a}{2\hbar^2\beta}(\mathbf{r}_1 - \mathbf{r}_1')^2\right] \exp\left[-\frac{m_b}{2\hbar^2\beta}(\mathbf{r}_2 - \mathbf{r}_2')^2\right] \\ \times\Big[1 - \beta U_{ab}(\mathbf{r}_1, \mathbf{r}_1', \mathbf{r}_2, \mathbf{r}_2' \mathbf{r})\Big] .\end{aligned} \tag{4.32}$$

Here $U_{ab}(\mathbf{r}_1, \mathbf{r}_1', \mathbf{r}_2, \mathbf{r}_2')$ is the off-diagonal two-particle effective potential, which is an appropriate generalization of the Kelbg potential. In a nonlinear generalization we propose the following representation for the two-particle density matrix:

$$\begin{aligned}\rho_{ab} = \frac{(m_a m_b)^{3/2}}{(2\pi\hbar\beta)^3} \exp\left[-\frac{m_a}{2\hbar^2\beta}(\mathbf{r}_1 - \mathbf{r}_1')^2\right] \exp\left[-\frac{m_b}{2\hbar^2\beta}(\mathbf{r}_1 - \mathbf{r}_1')^2\right] \\ \times \exp\big[-\beta U_{ab}(\mathbf{r}_1, \mathbf{r}_1', \mathbf{r}_2, \mathbf{r}_2')\big] .\end{aligned} \tag{4.33}$$

In the following, the off-diagonal matrix elements of the effective binary potentials are often approximated by the diagonal ones by taking the Kelbg potential at the center coordinate:

$$U_{ab}(r,r';\beta) \approx U_{ab}\left(\frac{r+r'}{2};\beta\right) .$$

Note that, in the relevant plasma parameter region, the protons can be treated classically, and U_{ii} may be approximated by the Coulomb potential.

We will now comment on these steps in more detail. First we calculate the effective potential by solving a Bloch equation by first order perturbation theory. The method has been described in detail in Chaps. 2–4, and we derived the function W^1_{ab} in first order. This function defines not only the first-order Slater sum, but also the first order off-diagonal quantum pair potential for Coulomb systems. This new potential depends on the interparticle distances $\mathbf{r}_{12}$ and $\mathbf{r}'_{12}$. As a result of our first-order perturbation theory, we get explicitly

$$U_{ab}(\mathbf{r}_{ab},\mathbf{r}'_{ab},\beta) \equiv e_a e_b \int_0^1 \frac{\mathrm{d}\alpha}{d_{ab}(\alpha)} \mathrm{erf}\left(\frac{d_{ab}(\alpha)}{2\lambda_{ab}\sqrt{\alpha(1-\alpha)}}\right) , \tag{4.34}$$

where

$$d_{ab}(\alpha) = \left|\alpha\mathbf{r}_{ab} + (1-\alpha)\mathbf{r}'_{ab}\right| .$$

We thus obtained a direct generalization of the representation of the usual Kelbg potential (4.13) which follows in the special case where the arguments coincide.

It is interesting to note that a simple approximation of the complicated integral over α by the length of the interval multiplied by the integrand in the center (Mittelwertsatz) leads to

$$U_{ab}(\mathbf{r}_{ab},\mathbf{r}'_{ab},\Delta\beta) \equiv \frac{e_a e_b}{d_{ab}(1/2)} \mathrm{erf}\left(\frac{d_{ab}(1/2)}{\lambda_{ab}}\right) . \tag{4.35}$$

The diagonal form leads to the so-called KTR potential due to Klakow, Toepffer, and Reinhard (1994) (see also Knaup et al. 2002, 2003):

$$U_{ab}(\mathbf{r}_{12},\beta) \equiv \frac{e_a e_b}{|\mathbf{r}_{12}|} \mathrm{erf}\left(\frac{|\mathbf{r}_{12}|}{\lambda_{ab}}\right) . \tag{4.36}$$

This diagonal KTP potential is often used in quasi-classical MD simulations (Klakow et al. 1994, Ebeling and Schautz 1997).

As we see again here, quantum effects remove the Coulomb singularities. In fact, two particles denoted by the indices i, j with masses m_i, m_j and thermal momentum corresponding to temperature T do not 'see each other' as point particles, but rather as charge clouds with an effective diameter

$$\lambda_{ij} = \frac{\hbar}{\sqrt{2m_{ij}k_B T}}, \qquad \frac{1}{m_{ij}} = \frac{1}{m_i} + \frac{1}{m_j}. \tag{4.37}$$

As already mentioned, the idea of effective potentials was first introduced in quantum chemistry by Hellman and Gombas, and in statistical physics by Morita. The idea was to model quantum effects using quasi-classical potentials. In the meantime, many forms of effective potentials have been. Several authors calculated the two-particle density matrix from the known wave functions. Here we may mention (Trubnikov and Elesin 1965, Rohde et al. 1968; Barker 1968, Storer 1968, Klimontovich et al. 1972, 1974). We may also mention several numerical approaches, and in particular, first applications of the Monte Carlo formalism (Zamalin et al. 1977; Zelener et al. 1981). Deutsch introduced exchange effects and developed useful approximations (Deutsch 1977) (see also Chap. 3). Hansen, McDonald, and Pollock (Hansen et al. 1976) developed first applications to molecular dynamics (see also Ortner et al. 1999, and Chap. 8), and Kleinert developed a variational approach (Kleinert 2009). We may also mention ad hoc modifications of the effective Kelbg potential for two particles of species i, j (Ortner et al. 1999). One possibility for correcting the Kelbg potential is the addition of a simple extra term of exponential shape which allows one to correct the height at the zero point (Ortner et al. 1999, Sadykova et al. 2007, 2009, 2011):

$$U_{ij}(r) = \frac{e_i e_j}{r}\left\{1 - \exp\left(-\frac{r^2}{\lambda_{ij}}\right) + \frac{\sqrt{\pi}r}{\lambda_{ij}}\left[1 - \operatorname{erf}\left(\frac{r}{\lambda_{ij}}\right)\right]\right\} \\ -k_B T \tilde{A}_{ij}(\xi_{ij}) \exp(-r^2/\lambda_{ij}^2). \tag{4.38}$$

Here the constant $\tilde{A}_{ij}$ is fitted to the exactly known height of the Slater sum at the zero point (Ortner et al. 1999, Sadykova et al. 2007, 2011).

We now discuss another, improved version, which is in some respects even more elegant. This method is based on the modification of just one parameter and has the form (Filinov et al. 2003)

$$U_{ij}(r) = \frac{e_i e_j}{r}\left\{1 - \exp\left(-\frac{r^2}{\lambda_{ij}}\right) + \frac{\sqrt{\pi}r}{\gamma_{ij}\lambda_{ij}}\left[1 - \operatorname{erf}\left(\gamma_{ij}\frac{r}{\lambda_{ij}}\right)\right]\right\}, \tag{4.39}$$

where λ_{ij} is the thermal wavelength given above, $\operatorname{erf}(x)$ is the standard error function, and γ_{ij} is a free parameter with $\gamma_{ij} = 1$ in Kelbg's original formula. We use the new parameter γ_{ij}. This does not change the properties of the Kelbg potential, but improves the properties of this potential. The original Kelbg potential has the property that at zero distance the potential is finite, i.e., $U_{ij}(0) = e_i e_j \sqrt{\pi}/\lambda_{ij}$. Of course, this is not correct. It is only a first order approximation in e^2. On the other hand the (nonlinear) Kelbg expression has the unique property that the first derivative is correct and in agreement with exact quantum-mechanical results (Kraeft et al. 1986).

In the modified Kelbg potential, we may use the freedom to correct the height at the zero point in agreement with the exact results and at the same time keep the derivative. Since our improved Kelbg potential has the property

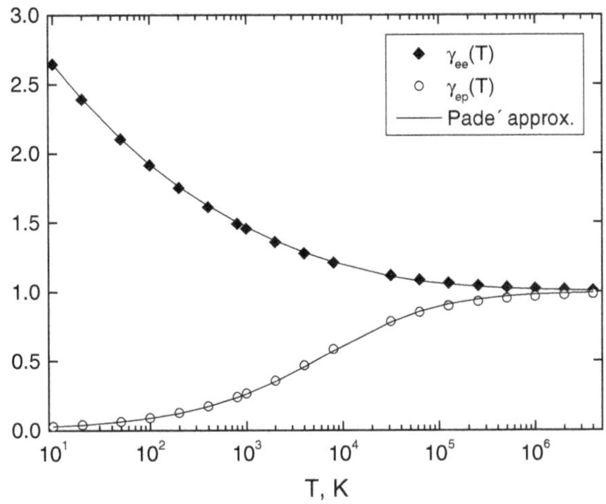

Fig. 4.4 Heuristic parameter $\gamma(T)$ which improves the Kelbg expression, providing exact values for the height and derivative at $r = 0$, obtained by exact solution of the two-particle problem (Filinov et al. 2004)

$$U_{ij}(r = 0) = \sqrt{\pi}\frac{e_i e_j}{\lambda_{ij}\gamma_{ij}} \,, \tag{4.40}$$

we can satisfy the relation

$$\exp\left(-\beta\frac{e_i e_j}{\lambda_{ij}\gamma_{ij}}\right) = S_{ij}(r = 0) \,, \tag{4.41}$$

with the exact Slater function from quantum mechanics. In other words, the new version agrees with the quantum mechanical relations at $r = 0$. This leads to the choice

$$\gamma_{ij} = -\sqrt{\pi}\beta\frac{e_i e_j}{\lambda_{ij}} \ln\left[S_{ij}(r = 0)\right] . \tag{4.42}$$

Using well known exact expressions for $S_{ij}(r = 0)$ (Barker 1969, Storer 1968, Rohde et al. 1968), we obtain excellent approximations for the Slater sum at all distances (Filinov et al. 2003).

Since Kelbg's expression and some generalizations are quite complicated due to the error functions, several useful approximations have been proposed (Zelener et al. 1981, Hansen and McDonald 1981, 2006. Figure 4.4 shows the fit parameter γ_{ij} as a function of temperature. We see that, for $T < 10\,000$ K, deviations from an unchanged Kelbg potential become essential. The parameters of effective potentials depend on the temperature, and in non-equilibrium this concept is not well defined. Klimontovich and coworkers have shown how effective potentials may be extended to non-equilibrium situations (Klimontovich et al. 1972, 1974).

4.3.2 Discussion of the Off-Diagonal Effective Potentials

On the basis of the previous calculations of the off-diagonal pair density matrix, we define the off-diagonal elements of effective potentials. In fact, we define the off-diagonal effective potential by means of the full binary density matrix

$$\rho_{ij}(\mathbf{r}_i, \mathbf{r}_j, \mathbf{r}_i', \mathbf{r}_j') = \rho_{ij}^{\text{ideal}} \exp\big[-\beta U_{ij}(\mathbf{r}_i, \mathbf{r}_j, \mathbf{r}_i', \mathbf{r}_j')\big]\,. \tag{4.43}$$

From the first order perturbation we have already calculated the off-diagonal Kelbg potential (Filinov et al. 2003, 2004). Let us discuss in more detail a simplification obtained by the 'Mittelwertsatz'

$$U_{ij}(\mathbf{r}_{ij}, \mathbf{r'}_{ij}) = \frac{e_i e_j}{|\xi\mathbf{r} + \xi'\mathbf{r}'|}\, \mathrm{erf}\left(\frac{|\xi\mathbf{r} + {}_{,}'\mathbf{r}'|}{2\lambda_{ij}\sqrt{\xi\xi'}}\right)\,, \tag{4.44}$$

with two numbers satisfying $\xi + \xi' = 1$. The diagonal elements of this expression are closely related to the effective potential used in wave packet dynamics (Klakow et al. 1994, see also Chap. 8). Figure 4.5 shows the off-diagonal effective potential of an electron–proton pair for three different angles ϕ between the vectors $\mathbf{r}$ and $\mathbf{r}'$, the magnitude of the second vector being fixed at $|\mathbf{r}'| = 0.25$, 1.0, and 2.0. In each panel, the upper curve corresponds to $\phi = 0$, the middle curve to $\phi = \pi/2$, and the lower curve to $\phi = \pi$. Solid lines show the 'exact' off-diagonal potential $U_{\text{pair}}(r)$, obtained using the matrix squaring technique (see Filinov et al. 2004). The dotted line represents the diagonal approximation

$$u_{ij}(\mathbf{r}_{ij}, \mathbf{r'}_{ij}) \approx \frac{1}{2}\big[u_{ij}(r_{ij}, r_{ij}) + u_{ij}(r_{ij}', r_{ij}')\big]\,.$$

One can see that, at distances $r' = 1.0$ and 2.0, the diagonal approximation becomes very crude and does not capture the dependence on ϕ.

The off-diagonal potential is exact only to first order in e^2. Spin and symmetry effects may be included by certain combinations of the expressions given above (Filinov et al. 2004). The results for the improved off-diagonal Kelbg potential are shown in Fig. 4.5 with open circles. The splitting of the integral into two parts and correction of the second term by the temperature-dependent parameter $\gamma_{ij}(T)$ extends applicability of the perturbation expression to rather low temperatures, e.g., $T =$ 31 250 K, corresponding to the ground state of the hydrogen atom. As one can see from Fig. 4.5, even at this temperature the deviations from the exact result are quite small. At temperatures higher than $T = 250\,000$ K, the agreement is excellent, and all angular dependencies of the off-diagonal potential are reproduced.

Applications of Effective Potentials

Up to now effective potentials have found their main application in the field of equilibrium statistics. The original idea due to Morita in 1959 was to express the diagonal density matrix in the form of Boltzmann factors with 2-particle, 3-particle,

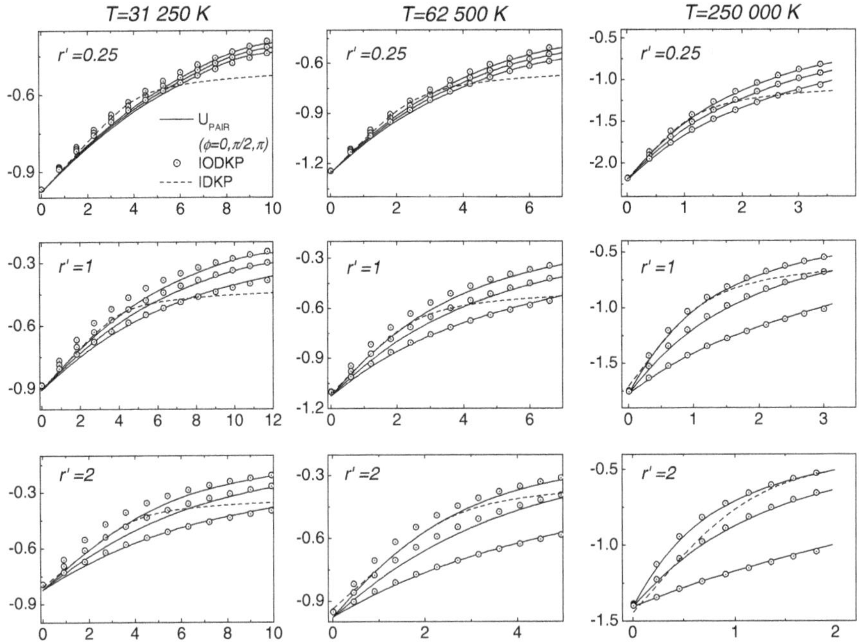

Fig. 4.5 Off-diagonal potential of an electron–proton pair as a function of the distance r and three angles between the vectors $\mathbf{r}$ and $\mathbf{r}'$. In each panel the *upper*, *middle*, and *lower curves* correspond to $\phi = 0, \pi/2$, and π, respectively. Distances are in units of the electron Bohr radius $a_B = \hbar^2/m_e e^2$ and the potential in the units of the electron Hartree $\text{Ha} = e^2/a_B$ (Filinov et al. 2004)

etc., effective potentials. The advantage with this procedure is that the calculation of the partition function may be reduced to a purely classical problem, although including many-particle interactions (Kraeft et al. 1986). On the other hand a well-known disadvantage with this method is that several many-particle effects, like exchange and bound state effects, are not easily described. Most successful applications have been given in the field of analytical thermodynamics. By means of cluster expansions, exact virial expansions for the thermodynamic functions of quantum plasmas could be found up to order $n^{5/2}$ in the density (Ebeling et al. 1976; Kraeft et al. 1986). Since the convergence of series is poor, the range of validity was extended by Padé methods. The available Padé formulas are widely used in applications in plasma and astrophysics. Good agreement of the Padé formulas has been shown with path integral Monte Carlo calculations (PIMC) (Filinov et al. 2001, Trigger et al. 2003).

PIMC simulations constitute an important field of applications of effective potentials (Zamalin et al. 1977, Ceperley 1995, Militzer and Pollock 2000, Filinov et al. 2001, Trigger et al. 2001). As we will explain in detail in Chaps. 9 and 10, PIMC calculations are based on high-temperature approximations for the density matrix and Trotter–Feynman expansions of the exponential operator. In this way, for applications, we need good expressions for high-temperature density matrices, and in particular their off-diagonal elements. This procedure greatly improves convergence and leads to effective calculations.

In the context of molecular dynamics, the first applications of effective potentials started with the pioneering work of Hansen and McDonald 1981. More recent MD simulations have shown that Kelbg-type effective potentials provide good thermodynamic functions, correlation functions, microfield distributions, and to some extent also non-equilibrium properties (Ortner et al. 1999, Morozov et al. 2009, 2012, Grabowski et al. 2013).

In conclusion, we may say that the method of effective potentials is nowadays a well established tool in the theory of dense quantum plasmas. We have extended the method here to non-diagonal contributions. In combination with quantum Monte Carlo methods, this approach gives quite a good description of thermodynamic properties and spatial correlations in strongly coupled Coulomb systems. In order to describe dynamical processes like transport phenomena, we require potentials with time-dependent parameters or potentials depending also on momentum (see Chap. 8).

4.4 Quantum Corrections in Thermodynamics

4.4.1 First Order Corrections to Classical OCP Results

In Sect. 3.1, we derived the correlation functions of the OCP by expansion with respect to the Bogoliubov parameter. The thermodynamic functions followed from the pair correlation functions by integration. As an expansion in the Bogoliubov parameter $\mu = \beta e^2 \kappa$, the free energy of the classical OCP has the form

$$F_{\rm cl} = F_{\rm B} - Nk_{\rm B}T\left[\frac{1}{3}\mu + \frac{1}{12}\mu^2\left(\ln\mu + 2C + \ln 3 - \frac{11}{6}\right) + O(\mu^3)\right]. \quad (4.45)$$

An equivalent writing of the free energy is

$$F_{\rm cl} = F_{\rm B} - Vk_{\rm B}T\left\{\frac{\kappa^3}{12\pi} + \frac{\pi}{3}n^2(\beta e^2)^3\left[\ln(3\beta e^2\kappa) + 2C - \frac{11}{6}\right] + O(n^{5/2})\right\}. \quad (4.46)$$

These results were obtained in many ways and by several authors, including Abe, Meeron, Cohen, Friedman, Guernsey, Yukhnovsky, and others. The complete calculation of the free energy up to and including order $n^{5/2}$ has been carried out by Cohen and Murphy 1969. A considerable effort is required to get the higher orders in μ. We already mentioned in Chap. 3 the exhaustive results derived by Ortner (1998) using the method of collective variables. We note also again that Wigner had already shown that, at low densities and low temperatures, the OCP forms a bcc lattice with correlation energy $u_c = -a_{bcc}\Gamma$. Later studies suggested more complicated formulae (DeWitt 1976, Stolzmann et al. 1998). The fit formula for large Γ due to DeWitt reads

$$\frac{U_c}{N_e k_B T} = -.89461\Gamma + 0.9165\Gamma^{1/4} - .5012 \,. \tag{4.47}$$

These analytical and numerical studies suggest that, for large Γ, the Coulomb energy per particle at finite temperatures should be asymptotically linear in Γ. There is an elegant way of deriving the law linear in Γ due to Kaklyugin (1995) and Brilliantov (1998), as already discussed in Chap. 3. To sum up, we may say that we have rather complete descriptions of the thermodynamic functions for the region were classical approximations work.

Quantum Corrections

In order to proceed to the quantum region, we may begin by using a 'trick'. Since for the electron gas there exist convergent expressions for the quantum and the classical free energy density, we may use the identity

$$F = F_{qu} = F_{cl} + (F_{qu} - F_{cl}) \,. \tag{4.48}$$

The second contribution does not have problems with screening and may be expressed up to second order in the density by using the real gas expansions as

$$F_{qu} - F_{cl} = -\frac{1}{2} k_B V n^2 \int \mathrm{d}\mathbf{r}_1 \,\mathrm{d}\mathbf{r}_2 \left[S_{qm}^{(2)} - S_{cl}^{(2)} \right] . \tag{4.49}$$

The value of this complicated integral will now be calculated. We first introduce the second virial coefficient by

$$F_{qu} - F_{cl} = -k_B V n^2 \delta B_2(T) \,, \tag{4.50}$$

$$\delta B_2(T) = 2\pi \int_0^\infty \mathrm{d}r\, r^2 \left[S_2(r) - \exp(-\beta e^2/r) \right] . \tag{4.51}$$

The easiest method to calculate the integrals uses expansions of the Slater sum with respect to Planck's constant $\hbar$. It is a textbook result that the density matrix for interactions $V(r)$ may be represented as a series in $\hbar^2$ (Landau and Lifshitz, 1979), viz.,

$$S_2(r) = \exp(-\beta V) \left\{ 1 + \frac{\beta}{6} \lambda^2 \left[-\Delta V + \frac{\beta}{2} (\nabla V)^2 \right] + \cdots \right\} ,$$

where we denote $\lambda^2 = \hbar^2 / m k_B T$ as usual. Note that the order $\hbar^4$ term decays as r^{-10} (Hoffmann and Ebeling 1968, Cornu 1997, Cornu and Martin 1991, Ebeling 1993). The integration can be carried out analytically and we obtain for the second virial contribution (Hoffmann and Ebeling 1968)

$$\delta B_2(T) = -\frac{\pi}{6}(\beta e^2)^3 \left[\frac{\lambda^2}{(\beta e^2)^2} + \frac{1}{5}\frac{\lambda^4}{(\beta e^2)^4} + \frac{4}{21}\frac{\lambda^6}{(\beta e^2)^6} + \cdots \right] . \tag{4.52}$$

Wigner–Kirkwood corrections to the classical free energy of the OCP can also be found in a paper by Pollock and Hansen (1973). For some higher order results and the so-called Hansen theorem, we refer to (Hansen and McDonald 1981, 2006). Expansions with respect to $\hbar^2$ work in the region of temperatures where the parameter $\eta = k_B T \lambda_{ee}/e^2$ is small, which means at low temperatures. The pressure of the electron gas including quantum corrections may be represented by

$$\frac{pV}{N_e k_B T} = 1 - \frac{1}{6}\mu_B - \frac{1}{12}\mu_B^2 \Big[\ln \mu_B + G_1(\eta) \Big] . \tag{4.53}$$

According to the findings above, which agree with an earlier result of DeWitt (1966), we obtain the representation

$$G_1(\eta) = 2C + \ln 3 - \frac{4}{3} - \frac{1}{2}\eta^2 + \frac{1}{16}\eta^4 + \frac{2}{21}\eta^6 + \cdots . \tag{4.54}$$

This is evidently a non-convergent asymptotic series. A successful guess of an analytic convergent Taylor series in the function η^{-1} with this asymptotic was found by Hoffmann and Ebeling (1968), viz.,

$$\begin{aligned} G_1(\eta) = & -\frac{3}{4}\sqrt{\pi}\eta^3 - \frac{3}{4}\sqrt{\pi}(1+\ln 2)\eta + \ln \eta + \frac{1}{2}C + \ln 3 + 0.411 \\ & +6\sqrt{\pi}\sum_{m=4}^{\infty}(-1)^m \frac{\zeta(m-2) - \zeta(m-1)(2^{-1} - 2^{1-m})}{2^m \eta^{m-3}\Gamma(1+m/2)} . \end{aligned} \tag{4.55}$$

The convergence of this expansion in η^{-1} is due to the Γ functions in the denominator. Note that $\zeta(n)$ is here the Riemann function, which we already used in Chaps. 2 and 3.

In order to give a strict derivation of such proper Taylor expansions in η^{-1}, we have to apply completely different methods (Ebeling 1968, 1969). These methods are based on expansions with respect to the charge strength e^2 and will be discussed in later sections. For physical and technical reasons, which will become clear later, instead of expansions in $\hbar$, we then have to use expansions in the parameter $\xi = -\beta e^2/\lambda$. As we will show later, using expansions in ξ, the integrals in the second virial coefficient (4.51) can be calculated exactly term by term, without making use of any approximations (Ebeling 1968). Incidentally, it is one of the specific properties of Coulomb systems that $\hbar^2$ expansions do not converge, since they are asymptotic series, while proper Taylor expansions exist only in the combination $e^2/\hbar$.

4.4.2 Higher Order Screening Contributions

In Sect. 4.2, we derived the mean Coulomb energy using first approximations for the correlation function. Following Kremp and Schmitz (1967), we neglected the contributions of nonzero Matsubara frequencies. As shown by Cornu, Martin, Ballenegger, and Alastuey, this leads to the omission of nonexponential contributions to the correlation effects in quantum systems (Martin 1988, Cornu and Martin 1991, Cornu 1997, Ballenegger et al. 2002). Moreover, this is related to the general breakdown of exponential screening by quantum fluctuations first derived in (Alastuey and Martin 1988, 1989). This is an effect beyond random phase approximations and leads to algebraic tails in the correlation functions (Alastuey and Martin 1989, Ebeling 1993, Cornu 1997, Alastuey 1999). As far as can be seen now, these terms influence only contributions beyond the order $n^{5/2}$ in the EOS.

Here we remain within the Kremp–Schmitz approximation and look again at the corresponding contributions to the ring integrals. Kremp and Schmitz (1967) and also Riemann et al. (1995) evaluated these integrals using different methods. All versions agree to order $\kappa\lambda$ and disagree to order $(\kappa\lambda)^2$. We recalculated the integral (Ebeling 2016) and, up to order $(\kappa\lambda)^2$, found expansions for the Coulomb energy, the free energy, and the electronic chemical potential. The Coulomb energy is

$$U_{\mathrm{R}} = -k_{\mathrm{B}}TV\frac{\kappa^3}{8\pi}\left(1 - \frac{\sqrt{\pi}}{4}\kappa\lambda + \frac{9}{48}\kappa^2\lambda^2 + \cdots\right) . \tag{4.56}$$

We refer to this expression as the ring approximation, hence the subscript R, and find the corresponding chemical potential and free energy to be

$$\mu_{\mathrm{e}}^{\mathrm{R}} = -\frac{1}{2}e^2\kappa\left(1 - \frac{\sqrt{\pi}}{4}\kappa\lambda + \frac{9}{48}\kappa^2\lambda^2 + \cdots\right) , \tag{4.57}$$

$$F_{\mathrm{R}} = -k_{\mathrm{B}}TV\frac{\kappa^3}{12\pi}\left(1 - \frac{\sqrt{3\pi}}{16}\kappa\lambda + \frac{9}{80}\kappa^2\lambda^2 + \cdots\right) . \tag{4.58}$$

We note the agreement of all authors up to order $O(\kappa\lambda)$, but some disagreement with respect to order $O(\kappa^2\lambda^2)$, although it is less than 10 percent.

In Chap. 3, we used a quantum Debye–Hückel approximation (the so-called QDHA or $\Lambda/8$-approximation). We may now prove that this approximation is also correct to first order in $O(\kappa\lambda)$ and is qualitatively correct for higher orders:

$$\mu_{\mathrm{e}}^{\mathrm{QDHA}} = -\frac{1}{2}e^2\kappa\frac{1}{1+\kappa a(T)} , \qquad a(T) = \frac{\sqrt{\pi}}{4}\lambda = \frac{1}{8}\Lambda_{\mathrm{e}} . \tag{4.59}$$

This and corresponding expressions for the free energy were already given in (Ebeling et al. 1976). In a more general representation, we introduce 'ring functions':

$$U^{\mathrm{R}} = -k_{\mathrm{B}} T V \frac{\kappa^3}{8\pi} G(\kappa\lambda) , \quad \mu_{\mathrm{e}}^{\mathrm{R}} = -\frac{1}{2} e^2 \kappa G(\kappa\lambda) ,$$

$$F^{\mathrm{R}} = -k_{\mathrm{B}} T V \frac{\kappa^3}{12\pi} R(\kappa\lambda) .$$

In the Kremp–Schmitz approximation, the integral for the function G may be rewritten as

$$\tilde{G}(\kappa\lambda) = \int_0^\infty \frac{dy}{1 + y^2 / M\big(1; 3/2; -y^2(\kappa\lambda)^2/4\big)} . \tag{4.60}$$

We calculated the integral with standard numerical mathematics for several points in the range $x = 0.1$–10, and estimated the asymptotic analytically. On the basis of the known properties of the hypergeometric function, we find the following asymptotic of the integral for large $\kappa\lambda$:

$$\tilde{G}(\kappa\lambda) \approx \frac{.8409}{\sqrt{\kappa\lambda}} . \tag{4.61}$$

Note that the numerical coefficient 0.8409 for the asymptotic is close to the result 0.639 obtained by Riemann et al. (1995). In this way, we constructed the following Padé formula connecting the expansion with the the numerical points and the asymptotic:

$$\tilde{G}(x) = \frac{1 + .1875x^2 + x^3}{1 + .4431x + .1963x^2 + 1.1892x^{7/2}} , \tag{4.62}$$

$$\tilde{R}(x) = \frac{1 + .1875x^2 + 3x^3}{1 + .4431x + .1963x^2 + 3.667x^{7/2}} . \tag{4.63}$$

Inspecting once again the earlier results for the ring function, we may conclude that our new ring function based on the work of Kremp and Schmitz is closely related to the DeWitt ring function as far as the expansions are concerned, except that the asymptotics of the two ring functions are not the same.

Figure 4.6 shows the resulting ring contribution of the Coulomb energy as a function of the convenient variables

$$\tau = k_{\mathrm{B}} T / Ry , \qquad \Gamma = e^2/d = 2/\tau r_{\mathrm{s}} .$$

For comparison, we show the simple Padé formula (QDHA) going to a constant for large Γ and the more refined Padé formula (4.63) resulting from the evaluation of the Kremp–Schmitz ring integral, which increases nearly linear in Γ as a straight line. The deviations of the Taylor expansion from the Padé formula for the ring function increase drastically with density and temperature, and even at $\tau \sim 0.5$, the Taylor expansion fails for $\Gamma > 1$. The thin line shows the increase in degeneracy effects, which grow linearly in the density, giving a cubic increase in Γ. This shows that the present theory fails for $\Gamma > 2$ in the considered range of temperatures $10^3 < T < 10^5$ K, since at higher densities, degeneracy effects are dominant.

In particular, we see that, for $\Gamma > 1$, expansions cut off after the linear or the quadratic term provide quite misleading results. This shows the poor convergence of the series. In our view the asymptotic of the ring terms is more important than the question of higher orders from $O(n^{5/2})$ and beyond. The formulae which we developed here work essentially in the region of non-degenerate and nonideal plasmas. Examining Fig. 4.6, we see that, with increasing Γ, we come quickly to the region of degeneracy where our formulae lose validity. The reason is that we increase Γ at fixed temperature, i.e., in Fig. 4.6, we move along vertical lines which rather quickly leave the non-degenerate nonideal region. With increasing temperature, the region of non-degenerate nonideal plasmas gets smaller and smaller.

In Chap. 6, we will come back to this problem and demonstrate that, for increasing $y = n_e \Lambda_e^3/2$, the chemical potential soon reaches a plateau. In this way, for low temperatures, a further increase in the values of the Coulomb energy, the free energy, and the chemical potential is blocked, and these functions reach saturation.

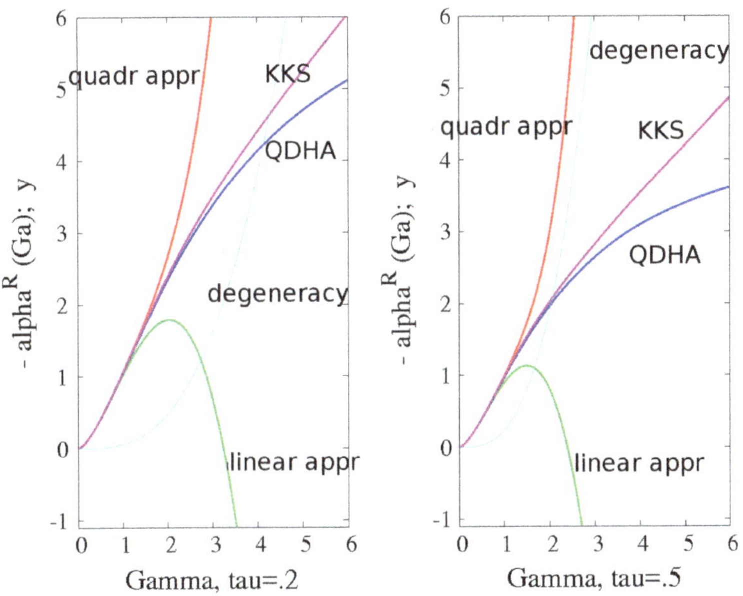

Fig. 4.6 Ring contribution to the chemical potential of an electron plasma for the temperatures $\tau = 0.2$ (*left*) and $\tau = 0.5$ (*right*), evaluated from the Kelbg–Kremp–Schmitz ring integral. The *curves below and above* show the result of the first linear and the second (quadratic) quantum correction to the limiting law, respectively. The *curves in the center* show the simple Padé formula corresponding to the QDHA going to a constant and the more refined Padé formula derived from the Kremp–Schmitz theory (4.63), which increases nearly as a straight line. The deviations of the Taylor expansion increase drastically with the temperature, and even at $\tau \sim 0.5$, the Taylor expansion fails for $\Gamma > 1$. The thin increasing line (*green*) shows that degeneracy effects grow according to a cubic law

4.4.3 Screening in Weakly Correlated Mixtures

Low density non-degenerate plasmas are characterized by a Debye-type screening. Then, in a linear approximation with respect to e^2, the binary density matrix and the correlation function can be expressed by the effective Kelbg potential $V_{ab}^{\mathrm{K}}(r)$. At zero density we have

$$\rho_{ab}(r) = 1 + g_{ab}(r)\ , \qquad g_{ab}(r) = -\beta V_{ab}^{\mathrm{K}}(r)\ . \tag{4.64}$$

To a first approximation, screening corresponds to replacing the Coulomb potential by the screened Debye potential, which is short-range, with range $r_{\mathrm{D}} = 1/\kappa$. In this Debye approximation, the correlation function reads

$$g_{ab}(r) = -\beta V_{ab}^{\mathrm{D}} \left\{ 1 - \mathrm{e}^{r^2/\lambda_{ab}^2} + \sqrt{\pi}\frac{r}{\lambda_{ab}}\big[1 - \Phi\left(r/\lambda_{ab}\right)\big] \right\} . \tag{4.65}$$

To calculate the free energy, we use the Debye charging procedure, which leads to the formula

$$F = F_{\mathrm{id}} - \frac{1}{2}k_{\mathrm{B}}T \sum_{ab} n_a n_b \int_0^1 \mathrm{d}\lambda\ V_{ab} g_{ab}(r;\lambda)\ . \tag{4.66}$$

Introducing the Debye approximation for the correlation function and carrying out the integration, we get for the free energy

$$F = F_{\mathrm{id}} - V\frac{4\pi}{3k_{\mathrm{B}}T\kappa}\sum_{ab} n_a n_b e_a e_b R(\kappa\lambda_{ab})\ , \quad \lambda_{ab}^2 = \frac{\hbar}{2m_{ab}k_{\mathrm{B}}T}\ , \tag{4.67}$$

where

$$R(x) = {}_1F_1\left(1, \frac{5}{2}, \frac{x^2}{4} + \frac{3\sqrt{(\pi)}}{x}\right)\left[1 - \exp\left(\frac{x^2}{4}\right)\right] . \tag{4.68}$$

The Taylor expansion of this function reads

$$R(x) = 1 - \frac{3\sqrt{\pi}}{16}x + \frac{1}{4}x^2 + \cdots\ . \tag{4.69}$$

The interaction part of the chemical potential of species a is given by

$$\mu_a^{int} = \frac{\partial F^{int}}{\partial N_a} = \frac{2\pi\beta^2 e_a^2}{\kappa}\sum_b e_b^2 G(\kappa\lambda_{ab})\ , \tag{4.70}$$

where

$$G(x) = \frac{\sqrt{\pi}}{x}\left\{1 - \exp\left(\frac{x^2}{4}\right)\left[1 - \Phi(x/2)\right]\right\} \approx \frac{1}{1 + \sqrt{\pi}x/4} . \tag{4.71}$$

For two-component systems, screening beyond the Debye approximation leads to difficult matrix problems. The required expression for the pair distribution function were obtained in 1964 by Kelbg using functional differentiation of the free energy (Kelbg 1964), and this has already been given in Sect. 4.2.2. Using the Kelbg pair distribution we get for the free energy

$$F^R = -Ck_BTV \int \frac{d(e^2)}{e^2} \int_0^\infty dk\, V_{ee}(k) \frac{N(k)}{Z(k)} , \tag{4.72}$$

where the numerator is given by

$$\begin{aligned} N(k) &= V_{ee}^K(k) + \beta n\Big[V_{ee}^K(k)V_{ii}^K(k) - V_{ei}^K(k)^2\Big] \\ &\quad + V_{ii}^K(k) + \beta n\Big[V_{ee}^K(k)V_{ii}^K(k) - V_{ei}^K(k)^2\Big] - 2V_{ei}^K(k) , \end{aligned} \tag{4.73}$$

and the denominator is the same as in Sect. 4.2.2. This integral is difficult to evaluate, but in most cases the approximation given above will be sufficient.

In the general case, only the first terms of the expansion with respect to κ are available. These are (Kraeft et al. 1986, Alastuey et al. 2015):

$$\beta F^R = -\frac{4\pi}{3\kappa}\sum_{ab} n_a n_b e_a^2 e_b^2 \left[1 - \frac{3\sqrt{\pi}}{16}\kappa\lambda_{ab} + O\big((\kappa\lambda_{ab})^2\big)\right] . \tag{4.74}$$

The problem with this expansion is, however, the poor convergence, even worse than for the case of the electron gas, due to the very different values of the thermal wavelengths $\lambda_{ee} \gg \lambda_{ii}$. For this reason, in the case of nonsymmetric systems or mixtures, it is recommended to use either the reduced mass approximation (RMA), or to use Padé approximations or numerical methods like nonlinear integral equations (Forster et al. 1992, Stolzmann and Ebeling 1998, Ebeling 2016, see also Chap. 6). For higher order contributions in the case of multi-component plasmas, see (Kahlbaum et al. 1998, 2000, Alastuey et al. 2015).

4.5 Virial Expansion in the Reduced Mass Approximation

4.5.1 *Free Energy and Pressure in the RMA*

The problem of bound states in plasmas is so complicated that it may be justified to begin by investigating a special case in which the problem is considerably simplified. We have in mind symmetrical plasmas with equal relative masses $m_{+-} = m_{++} =$

$m_{--} = \mu$. Strictly speaking, such plasmas do not exist in nature, except in extreme cases, such as electron–positron plasmas. For our model, we shall not require the masses to be equal, which would rule out all applications apart from the electron–positron plasma. We only require the relative masses to be close to the reduced mass, and this is satisfied to a good approximation in many plasmas in the interesting region $10^2 < T < 10^5$ K.

In this way, according to the arguments given in Chap. 3, we may consider the reduced mass approximation (RMA) to be a good model for two-component plasmas with temperatures in the range $10^2 < T < 10^5$ K. From the physical point of view, this means that collisions of equally charged particles play a minor role. In addition to the mass symmetry, we assume further that, in the considered temperature regime, exchange effects may be neglected. We refer here to arguments gathered in Chaps. 2 and 3 (see in particular Sect. 3.8).

Summarizing the previous results concerning the ring contributions, we may write for the free energy, up to the second virial term,

$$F = F_{\text{id}} - V k_{\text{B}} T \left[\frac{4\pi}{3\kappa} \sum_{ab} n_a n_b e_a^2 e_b^2 R(\kappa \lambda_{ab}) \right. \tag{4.75}$$

$$\left. + \frac{1}{2} \sum_{ab} n_a n_b \int d\mathbf{r} \left(S_{ab} - 1 - S_{ab}^1 - S_{ab}^2 \right) + \cdots \right] .$$

Here the additional terms S_{ab}^1, S_{ab}^2 under the integral, which are the two lowest orders, were subtracted, since they are already taken into account in the ring term and we have to avoid any double counting. Further we recall the ring functions determining the free energy, chemical potential, and pressure, which are defined as series beginning as

$$R(x) = 1 - \frac{3\sqrt{\pi}}{16} x + \cdots , \quad G(x) = 1 - \frac{\sqrt{\pi}}{4} x + \cdots ,$$

$$\Phi(x) = 1 - \frac{3\sqrt{\pi}}{8} x + \cdots .$$

We recall also that the Debye–Kelbg ring approximation already contains all contributions up to order e^4, so one has to make sure that these terms are not counted twice. So far our formula for the free energy is still entirely general and even covers multi-component plasmas with arbitrary masses, as long as no divergencies arise.

We now specialize to two-component plasmas with equal reduced masses $n_a = n$, $m_{ab} = \mu$, $\lambda_{ab} = \lambda$. Then (4.75) converts to the simpler form

$$F = F_{\text{id}} - V k_{\text{B}} T \frac{\kappa^3}{12\pi} R(\kappa \lambda) \tag{4.76}$$

$$- \frac{1}{2} n^2 G'' \int d\mathbf{r} \left(2S_{+-} + S_{++} + S_{--} - 4 - 2S_{++}^2 - S_{+-}^2 - S_{--}^2 \right) + \cdots .$$

Here we have taken into account the fact that terms of order e^2 and e^6 do not appear since they cancel out due to the symmetry in the masses and the antisymmetry in the charges. Symbolically, the operator G'' denotes the omission of all orders up to e^4, since they are already contained in the ring contribution.

The next step is inspired by the ideas of Max Planck 1924 and Brillouin in 1930. We separate the contributions of the lower lying bound states which are contained entirely in S_{+-} from the rest of the integral. Let us define the bound state contribution as

$$S^b_{+-}(r) = 2\pi^{1/2}\lambda^3_{ab}\sum_{l=0}^{\infty}(2l+1)\left\{\sum_{s=1}^{l-1}\left[\exp(-\beta E_s) - 1 + \beta E_s\right]|R_{sl}(r)|^2\right\},$$

where the energy levels are as usual $E_s = -me^4/2\hbar s^2$. The meaning of the pre-factor in front of the wave functions is that all levels with $|\beta E_s| < 1$ are suppressed in the sum. We recall that in the Planck paper we still find a discrete border at the level with $|E_s| \approx k_B T$. However, a few years later Brillouin introduced the smooth weight factor which we have used here, since it is confirmed by our calculations. Using the normalization of the radial wave functions, (4.77) implies finally,

$$\begin{aligned} F = F_{\text{id}} - Vk_BT\frac{\kappa^3}{12\pi}R(\kappa\lambda) \\ -4n^2\pi\sqrt{\pi}\lambda^3\sum_{s=1}^{\infty}s^2\left[\exp(-\beta E_s) - 1 + \beta E_s\right] - n^2B^{\text{f}}(T, e^2, \hbar) + \cdots . \end{aligned} \tag{4.77}$$

Here $B^{\text{f}}(T, e^2, \hbar)$ denotes the contribution of the free states. Max Planck found this contribution to be approximately zero in the classical limit $\hbar \to 0$. In some earlier work, we showed that the relation $B^{\text{f}}(T, e^2, \hbar = 0) = 0$ is exactly true in the classical limit (Ebeling 1967). We will give now a complete proof of this statement for mass-symmetric plasmas.

First we express the remaining contribution of the free states in the form

$$B^{\text{f}}(T) = 4\pi^{3/2}\sum_{\ell=0}^{\infty}(2\ell+1)G''\int_0^{\infty}\exp(-\lambda k^2)\frac{1}{\pi}\frac{\text{d}}{\text{d}k}\left(\delta_\ell^{++} + \delta_\ell^{--} + 2\delta_\ell^{+-}\right).$$

We now recall the known result for the phase shifts of Coulomb systems (Landau and Lifshits 1979), which we write in the form

$$\frac{\text{d}}{\text{d}x}\delta_\ell^{ab}(x) = \frac{\text{d}}{\text{d}x}\text{Im}\left[\ln\Gamma(\ell + 1 - \text{i}\xi_{ab}/2x)\right], \tag{4.78}$$

with $x = \lambda k$ and using once again the parameter $\xi_{ab} = -\beta e_a e_b/\lambda$. In the next step, we find (Ebeling 1968, 1969)

$$\frac{\mathrm{d}\delta_\ell^{ab}(x)}{\mathrm{d}x} = \frac{\xi_{ab}}{2x^2}\mathrm{Re}\big[\psi(\ell+1-\mathrm{i}|\xi_{ab}|/2x)\big] \, , \tag{4.79}$$

where $\psi(x) = \mathrm{d}\ln\Gamma(x)/\mathrm{d}x$ is a well investigated function. The relevant result in our context is that the derivative of the phase shift changes sign with the interaction parameter ξ_{ab}. Consequently, the sum under the integral for B^{f} cancels out and the integral itself disappears, i.e., $B^{\mathrm{f}} = 0$. Later we will find another analytical proof of this statement as a consequence of the formula for binary but not necessarily mass-symmetric plasmas.

Summarizing our result, we arrive at the following exact formula for the free energy density of low density plasmas with symmetric relative masses (or any binary plasma in the reduced mass approximation):

$$\begin{aligned} f = f_{\mathrm{id}} - k_{\mathrm{B}}T\frac{\kappa^3}{12\pi}R(\kappa\lambda) \\ -4n^2\pi\sqrt{\pi}\lambda^3\sum_{s=1}^{\infty}s^2\big[\exp(-\beta E_s) - 1 + \beta E_s\big] + O(n^{5/2}) \, . \end{aligned} \tag{4.80}$$

The equivalent result for the pressure reads

$$\begin{aligned} \beta p = \beta p_{\mathrm{id}} - \frac{\kappa^3}{24\pi}\big[R(\kappa\lambda) + xR'(\kappa\lambda)\big] \\ -4n^2\pi\sqrt{\pi}\lambda^3\sum_{s=1}^{\infty}s^2\big[\exp(-\beta E_s) - 1 + \beta E_s\big] + O(n^{5/2}) \, . \end{aligned} \tag{4.81}$$

This remarkable result says that the term supplementing the ring contributions is given by the PBL partition function, and that this is an exact result for plasmas with equal relative masses, all corrections to the PBL term cancelling out due to mass and charge symmetry. It is exactly this point which determines the unique role of the PBL partition function among all other possibilities. The deeper reasons can be summarized in two points:

- Only the PBL partition function gives the exact second order correction to the ring approximations for mass-symmetric plasmas without exchange effects.
- The rigorous theory for systems with different relative masses shows that the mass differences give only small corrections, which even vanish in the classical limit.

4.5.2 *Compatibility with the Mass Action Law Approach*

Our statement is that these results, including (4.80) and (4.81), are exact up to the given order $O(n^2)$ provided that (i) the relative masses are symmetric, and (ii) there are no exchange effects. A second, weaker form would be that, in the region where ionization phenomena occur, i.e., $10^2 < T < 10^5$ K, the relative masses are *approximately* equal and that exchange effects may be neglected. Therefore the statement is that results like (4.80) and (4.81) remain valid as an excellent approximation, even for plasmas in which the masses of the positive charges are much greater than the electron mass $m_+ > m_e$.

There have been several qualitative and quantitative arguments which we repeat briefly here. We already know from the quantum mechanics of Coulomb systems, in particular from the old Bohr theory, that most properties depend essentially only on the electron masses, while the heavy masses only give small corrections through the replacement of electron masses by reduced masses. Further, we argued that collisions between charges of the same sign are relatively seldom, so the differences between the relative masses existing in reality do not matter. In this way, we may use (4.80) and (4.81) as a very good approximation for any binary plasma, with $\lambda = \lambda_{+-}$, provided we remain at temperatures far below the ionization temperature $T \sim 10^5$ K. The reason for the restriction with respect to the temperatures is that, at very high temperatures, the difference between the masses plays a more important role, while on the other hand, the corrections due to nonideality are not relevant at all at high temperatures. Special systems where nonideality effects may be relevant at high temperatures are, for example, the atmospheres of stars (Riemann et al. 1995; Kalman et al. 1998).

At this point we should return to a discussion we began in Chap. 1, namely the old observation that the simple partition function of Coulomb atoms is divergent. Now we see that, in a quantum-statistical derivation of thermodynamic quantities, no divergence is seen at all. Instead of the naive Bohr–Herzfeld partition function, we find the regularized Planck–Brillouin–Larkin partition function

$$\sigma_{\mathrm{PB}}(T) = \sum_{s=1}^{\infty} s^2 \big[\exp(-\beta E_s) - 1 + \beta E_s \big] \,. \tag{4.82}$$

Our expression (4.81) contains precisely the PBL partition function. We thus arrive at the PBL procedure in a natural way. We may conclude that, in the limit of dilute plasmas, the PBL procedure is the most natural way to regularize the partition function.

This important result means that the pressure (4.81) in symmetric plasmas contains the PBL partition function and is exact up to order $O(n^2)$ in the density. Further, we find that plasmas in which the positive charges have larger masses do not give essentially different results.

We may show that our results are compatible with the mass action laws we used in Chap. 3. Investigating the result of the mass action formulae obtained there in the case of very low densities and expanding all functions for small density, we get for the EOS of binary plasmas

$$\beta p = n_{\mathrm{e}} + n_{\mathrm{i}} - \frac{\kappa^3}{24\pi}\left(1 - \frac{\sqrt{3\pi}}{8}\kappa\lambda + \cdots\right)$$
$$-4\pi^{3/2}\lambda^3 n_{\mathrm{e}} n_{\mathrm{i}} \sigma_{\mathrm{BPL}}(T) + O(n^{5/2})\,.$$

This result is fully compatible with (4.81). This statement is of great importance, so let us reformulate it again. The mass action approach is compatible with the exact results of quantum statistics if:

- The PBL distribution is used.
- The standard ring function expressions are used.

Two remarks are in order. First, this statement is valid for all plasmas with two-particle bound states, although the proof we have given so far only covers symmetric plasmas. A more general proof is still needed. Second, we do not exclude a priori other combinations of ring functions and partition functions. However, any other combination should satisfy the Onsager condition which says here that the sum should be compatible with the exact results for the total thermodynamic functions (Ebeling et al. 1968, 1974, 2002, 2003).

4.6 Low Density Expansions for Coulomb Systems

4.6.1 Virial Expansion for Arbitrary Mass Relations

Our aim is here to calculate the contributions of pair correlations in all orders of the interaction parameters of pairs $e_i e_j$. For the dimensionless interaction parameter we use

$$\xi_{ij} = -\frac{e_i e_j}{k_{\mathrm{B}} T \lambda_{ij}}\,. \tag{4.83}$$

Note that this parameter is proportional to the square root of the reduced masses and inversely proportional to the square root of the temperatures. Bound states are possible only if at least some of the interaction parameters are positive. According to Chap. 2, the free energy is given by

$$F(T, V, N, \xi_{ij}) = F_{\mathrm{id}}(T, V, N) - k_{\mathrm{B}} T \ln Q_{\mathrm{int}}(T, V, N, \xi_{ij})\,, \tag{4.84}$$

$$F_{\mathrm{id}}(T, V, N) = N k_{\mathrm{B}} T \left[\ln\left(\frac{n\Lambda^3}{2s+1}\right) - 1\right]. \tag{4.85}$$

This is an analytical function of the parameters $e_i e_j$, or more correctly of the dimensionless parameter ξ_{ij}. We see that these interaction parameters are positive for +− interactions and negative for ++ and −− interactions. Our aim is to represent the thermodynamic potentials by analytical functions of the interaction parameters.

In order to calculate the interaction contributions, we start from the cluster expansion for real quantum gases. We split the free energy into the contribution of a Boltzmann gas and an excess part which also contains all correlations, including those due to the symmetry properties of the particles:

$$\begin{aligned}
F(T,V,N) &= F_{\text{id}}^{\text{Bo}} + F_{\text{ex}} \\
&= Nk_{\text{B}}T\Big\{\ln\big[n\Lambda^3(2s+1)\big]-1\Big\} \\
&\quad -k_{\text{B}}TV\Bigg\{\frac{1}{2}\sum_{ab}n_an_b\int \mathrm{d}\mathbf{r}_1\mathrm{d}\mathbf{r}_2\Big[S_{ab}^{(2)}-1\Big] \\
&\quad +\frac{1}{6}\sum_{abc}n_an_bn_c\int \mathrm{d}\mathbf{r}_1\mathrm{d}\mathbf{r}_2\mathrm{d}\mathbf{r}_3\Big[S_{abc}^{(3)}-S_{ab}^{(2)}S_{ac}^{(2)}S_{bc}^{(2)}\Big]+\cdots\Bigg\}\,.
\end{aligned} \tag{4.86}$$

This expansion is complete and in principle already contains all terms, but all the integrals are divergent at large distances due to the Coulomb singularities. If we succeed in replacing Coulomb potentials by Debye potentials in a suitable way, no further divergencies will appear. In the standard way proposed by Mayer, this is achieved by summing an infinite number of Coulomb terms stemming from the higher integrals.

We will use here an equivalent procedure, sometimes called the Schmitz method, which is much simpler and goes the opposite way, expressing the Coulomb potential in terms of the Debye potential (Falkenhagen et al. 1971; Kraeft et al. 1986). We start from the Bogoliubov screening equation in the form

$$\beta V_{ab}(r) = -g_{ab}(r) - \sum_c \int \beta V_{bc}(r-r')g_{ac}(r')\,. \tag{4.87}$$

By iteration, formally, we get an infinite series of convergent integrals

$$\beta V_{ab} = -g_{ab}(r) + \sum_c \int \mathrm{d}3\,\beta g_{ac}g_{bc} - \sum_{cd}\int \mathrm{d}3\,\mathrm{d}4\,\beta g_{ac}g_{bc}g_{cd} + \cdots\,. \tag{4.88}$$

Introducing here the Debye approximation for the correlation function, we get

$$\begin{aligned}
F_{\text{ex}} &= -\frac{1}{2}k_{\text{B}}T\sum_{ab}n_an_b\int_0^1 \mathrm{d}\lambda V_{ab}g_{ab}(r;\lambda) \\
&\quad -k_{\text{B}}TV\Bigg\{\frac{1}{2}\sum_{ab}n_an_b\int \mathrm{d}\mathbf{r}_1\mathrm{d}\mathbf{r}_2\Big[S_{ab}^{(2)}-1+V_{ab}\int_0^1 \mathrm{d}\lambda g_{ab}(r;\lambda)\Big]\Bigg\} \\
&\quad +O(n_an_bn_c)\,.
\end{aligned}$$

We thus obtain the screened cluster expansion for the free energy density:

$$\beta f(T, n_a) = \beta f_{\mathrm{id}}^{\mathrm{Bo}} - \frac{\kappa^3}{12\pi} + \sum_{ab} n_a n_b B_{ab}(\kappa) + \sum_{abc} n_a n_b n_c B_{abc}(\kappa) + \cdots . \quad (4.89)$$

4.6.2 Screened Cluster Integrals

The screened cluster integrals are constructed in a completely analogous way to the screened cluster integrals for classical ionic solutions, as described by Meeron in 1950 and Friedman (1962). The representation is exact, but slowly convergent, and the orders in the density are

$$B_2(\kappa) = O(\kappa^0) , \qquad B_3(n) = O(\log \kappa) ,$$

$$B_4(n) = O(\kappa^{-2}) , \qquad B_5(n) = O(\kappa^{-3}) .$$

This estimate is only true when the third moments are zero. Otherwise, if $\mu_3 = \sum_a n_a e_a^3 \neq 0$, the estimate is even worse and gives (Friedman 1962)

$$B_2(n) = O(\log \kappa) , \qquad B_3(n) = O(\kappa^{-1}) ,$$

$$B_4(n) = O(\kappa^{-3/2}) , \qquad B_5(n) = O(k^{-4}) .$$

Friedman's estimates of the density order of the cluster integrals is strictly speaking invalid when there are long-range quantum correlations. However, in a study by Alastuey and Martin (1989), it was shown by means of Feynman–Kac expansions that there exist binary quantum correlations which decay at large distances according to a power law $\hbar^4/r^{10}$. The existence of such long-range power law correlations changes the Friedman density orders and leads to additional density orders (see Ebeling 1993), as demonstrated explicitly by Kahlbaum (1996, 1998, 2000).

In our case of diluted Coulomb plasmas, the second and third cluster integrals read explicitly (Ebeling 1967, 1968, 1969; Kraeft et al. 1986; Kahlbaum 1998, 2000)

$$B_{ab}(\kappa) = \frac{1}{2!} \int \mathrm{d}\mathbf{r}_2 \left(\Phi_{ab} - \frac{1}{2} g_{ab}^2 \right) , \quad (4.90)$$

$$B_{abc}(\kappa) = \frac{1}{3!} \int \mathrm{d}\mathbf{r}_2 \mathrm{d}\mathbf{r}_3 \Big[\Phi_{abc}(1,2,3) + 3 g_{ab}(1,2) \Phi_{ac} \Phi_{bc} \Big] , \quad (4.91)$$

with

$$\Phi_{ab} = S_{ab} \exp(g_{ab} + \beta V_{ab}) - 1 - g_{ab} , \quad (4.92)$$

$$\Phi_{abc} = \Phi_{ab}\Phi_{ac}\Phi_{bc} + \left(S_{abc} - S_{ab}S_{bc}S_{ac}\right)\exp\left(g_{ab} + \beta V_{ab} + g_{ac} + \beta V_{ac} + g_{bc} + \beta V_{bc}\right) . \tag{4.93}$$

For calculations of the third virial coefficient, see Czerwon (1972) and Kahlbaum (1996, 1998, 2000).

4.7 Exact Second Order Coulomb Virial Functions

4.7.1 Exchange Contribution to Coulomb Virial Functions

We start by calculating the exchange effects. The main contributions of exchange effects which we have considered so far were the Hartree–Fock effects (Chap. 2). The exchange effects somehow play a special role, since they are essentially quasi-interactions of short-range nature. Let us define the exchange contribution of thermodynamic functions as the difference between a system of Fermi–Dirac particles and a system of Boltzmann-like particles:

$$F^{\mathrm{exc}} = F^{\mathrm{F}} - F^{\mathrm{B}} . \tag{4.94}$$

At least at small densities, divergence problems due to screening effects do not play any role in the exchange contributions, and the free energy may be represented in the same way as for a quantum gas with short-range forces (see Chap. 2). We may thus write

$$F^{\mathrm{exc}} = -k_{\mathrm{B}}TV\left[\frac{1}{2}\sum_{ab} n_a n_b \int S_{ab}^{\mathrm{exc}}(1,2)\mathrm{d}1\,\mathrm{d}2 + O(n^3)\right] . \tag{4.95}$$

Here S_{ab} is what we previously called the Slater function, which is the diagonal part of the two-particle density operator, split for technical reasons into a direct and an exchange part:

$$S_{ab}(1,2) = S_{ab}^{\mathrm{dir}}(1,2) + S_{ab}^{\mathrm{exc}}(1,2) . \tag{4.96}$$

Here the direct part is defined by the Heisenberg quantum effects alone, which remain if exchange effects are not taken into account. Clearly, the exchange part contributes only for identical particles $a = b$. Using the known Coulomb wave functions with radial parts $R(r)$, for the fermionic exchange contribution, we may write

$$S_{ab}^{\mathrm{exc}} = -\sum_{l=0}^{\infty} \delta_{ab}\frac{(-1)^l(2l+1)}{2s_a+1}\int_0^{\infty}\frac{\mathrm{d}n_k}{\mathrm{d}k}\exp(-\lambda_{ab}k^2)|R_{kl}(r)^{\mathrm{exc}}|^2\mathrm{d}k . \tag{4.97}$$

Here $\mathrm{d}n_k/\mathrm{d}k$ is the density of states, which is exactly known for Coulomb systems. Using the normalization of the radial wave functions R_{kl} and the known expressions for the Coulomb phase shifts by ψ functions, we get for the exchange free energy of electrons

$$F^{\mathrm{exc}} = -k_{\mathrm{B}}TV\left[n_{\mathrm{e}}^2 2\sqrt{\pi}\mathrm{Re}\int_0^\infty \mathrm{d}x \frac{\exp(-x^2)}{x} f^{\mathrm{exc}}(\xi_{\mathrm{ee}}/x) + O(n_{\mathrm{e}}^3)\right], \quad (4.98)$$

with

$$f^{\mathrm{exc}}(y) = \frac{1}{2}y\sum_{l=0}^{\infty}(-1)^l(2l+1)\psi(l+1-\mathrm{i}y/2)\,. \quad (4.99)$$

By introducing the complex variable $z = x^2$, transforming to an integral in the complex plane with a contour C circumventing the positive axis including the zero point in the positive direction, we get the integral

$$F^{\mathrm{exc}} = -k_{\mathrm{B}}TV\left[n_{\mathrm{e}}^2\sqrt{\pi}\mathrm{Re}\int_C \mathrm{d}z \frac{\exp(-z)}{z^{1/2}} f^{\mathrm{exc}}(\xi_{\mathrm{ee}}/\sqrt{z}) + O(n_{\mathrm{e}}^3)\right]. \quad (4.100)$$

We now expand the ψ function in a Taylor series with respect to the interaction parameter ξ and integrate the convergent series term by term, making use of

$$\frac{1}{\Gamma(s)} = \frac{1}{2\pi\mathrm{i}}\int_C \mathrm{d}z(-z)^{-s}\exp(-z)\,. \quad (4.101)$$

This procedure yields the following representation of the exchange contributions of electrons to the free energy:

$$F^{\mathrm{exc}} = +k_{\mathrm{B}}TV\pi n_{\mathrm{e}}^2\lambda_{\mathrm{ee}}E(\xi_{\mathrm{ee}})\,, \quad (4.102)$$

where $E(x)$ can be represented either by a complex integral as introduced above or by a converging infinite series:

$$E(x) = \sum_0^\infty \frac{\sqrt{\pi}[1-2^{(2-n)}]\zeta(n-1)x^n}{2^n\Gamma(1+n/2)}\,, \quad (4.103)$$

as first derived from the exact Coulomb scattering phase shifts in (Ebeling 1968, 1969). The coefficients in these series are determined by the known exact Coulomb scattering phase shifts. Introducing the explicit values of the zeta and gamma functions, the first two terms in the exchange function series yield

$$E(x) = \frac{\sqrt{\pi}}{4} + \frac{\xi}{2} + \cdots\,. \quad (4.104)$$

These are the terms which are already contained in the (known to us) Fermi–Dirac and Hartree–Fock contributions. The rest are higher order contributions defined by transcendent functions first given in (Ebeling 1968) and later reproduced using different quantum-statistical methods (Ebeling et al. 1970, 1976, Kraeft et al. 1986, Kremp et al. 2005). The transcendent virial function $E(x)$ was defined here by a convergent infinite series in the coupling parameter, representing the contributions of ladder diagrams. Defining the electron interaction parameter by

$$\xi_{\mathrm{e}} = \frac{e^2}{k_{\mathrm{B}} T \Lambda_{\mathrm{e}}} = -\frac{\xi_{\mathrm{ee}}}{\sqrt{2\pi}} , \tag{4.105}$$

the free energy per electron may be rewritten by reordering with respect to the degeneracy parameter y to obtain

$$\varphi_{\mathrm{e}}^{\mathrm{exc}}(y) = y E_1(\xi_{\mathrm{e}}) - y^2 E_2(\xi_{\mathrm{e}}) + y^3 E_3(\xi_{\mathrm{e}}) - O(y^4) , \tag{4.106}$$

with the set of exchange functions

$$\begin{aligned} &E_1(x) = E\big(-\sqrt{2\pi}x\big)/\sqrt{2\pi} = 0.1768 - 0.5x + \cdots , \\ &E_2(x) = 0.065 - 0.3540x + \cdots , \quad E_3(x) = 0.000031 - 0.1825x + \cdots . \end{aligned} \tag{4.107}$$

For the series $E_2(x)$ and $E_3(x)$, we derived here only the first terms of expansions with respect to the interaction parameter by comparison with the Fermi–Dirac and the Hartree–Fock expressions for the free energy per electron given earlier. The full series expansion is known only for the first function $E_1(x)$, which, as we have said, is an exponentially fast decreasing function of the interaction parameter. In spite of our limited knowledge of the higher order exchange functions, there is some evidence that they are also exponentially decreasing with the interaction parameter, similarly to $E_1(x)$, i.e.,

$$E_2(x) \sim 0.065 \exp(-5.45x) , \quad E_3(x) \sim 0.000031 \exp(-5887x) . \tag{4.108}$$

Furthermore, there is some evidence that the relation between the interaction terms and the ideal terms gets smaller with increasing interaction strength. All this leads to the surprising (but not yet strictly proven) conclusion that, with increasing interaction strength, e.g., by decreasing the temperature, the exchange corrections decrease and the sum of ideal terms and exchange terms converges to the expression for a Boltzmann gas.

The hypothesis developed here says that, at least in certain regions, the interactions depress the exchange effects and make the thermodynamics more Boltzmann-like.

In other words, we found that, even approaching the degeneracy line of the electrons $n_e\Lambda_e^3 \to 1$, degeneracy effects do not increase. In fact, quite the opposite: they decay. The physical meaning is that the formation of degeneracy requires space for the overlap of wave functions or the corresponding thermal wavelength. Strong Coulomb repulsion depresses such effects. Looking again at the (n, T) plane shown in see Fig. 4.1, this means that, moving along a path from the right (very weak nonideality) to the left (stronger nonideality), the degeneracy effects decrease and electrons behave more like classical particles. However, this is is certainly not true for the quantum effects due to Heisenberg effects and bound state formation, but this will be studied in the following sections. The behaviour of two-component Coulomb systems consisting of light and heavy charges is extremely complicated and cannot be described just by a few lines in the (n, T) plane. Strictly speaking each of the regions shown in Fig. 4.1 needs its own theory. This will be demonstrated in the next part, where the bound state effects stemming from the direct contributions are incorporated.

4.7.2 Direct Contributions to Coulomb Virial Functions

The direct contribution to the second order virial expansion is much more difficult to calculate, since it depends on screening effects. The second order contributions are defined at the outset by

$$\beta F^{\text{dir}}(T, N_a) = -k_B T V \left[\frac{\kappa^3}{12\pi} + \sum_{ab} n_a n_b B_{ab}^{\text{dir}}(\kappa) + \cdots \right] , \tag{4.109}$$

with the direct contribution to the second order cluster being given by

$$B_{ab}^{\text{dir}}(\kappa) = \frac{1}{2!} \int d\mathbf{r}_2 \left[S_{ab} \exp(g_{ab} + \beta V_{ab}) - 1 - g_{ab} - \frac{1}{2} g_{ab}^2 - S_{ab}^{\text{exc}} \right] . \tag{4.110}$$

In order to calculate the terms beyond the first term stemming from ring diagrams, it is useful to expand with respect to the interaction e^2 and calculate term by term (see Ebeling 1968, 1969, Kraeft et al. 1986, Kahlbaum 1996, 1998, 2000). Equivalently, one can study the different contributions to the 'ladder diagrams' with more than two rungs (interaction lines between loops). We know that the Debye-type contributions stem from 'two-rung diagrams', i.e., diagrams having two interaction lines with an order of at least e^4. As can be seen from the definition of the Coulomb energy, one of the rungs stems from a screened potential and the other from a bare Coulomb potential. For these terms, screening is fundamental. The quantum effects which contribute to the non-classical corrections at order e^4 come from the scattering states.

In the third order beyond the ring terms, viz., e^6, we have to study three-rung contributions, containing three interaction lines. We study here these rather complicated contributions with the technique of effective potentials introduced by Morita and Kelbg (Kelbg 1963, 1964). As explained above, in this approximation, the Coulomb contribution in the binary Slater sum $S_{ab}(r)$ is replaced by a static screened Debye potential. In order to explain the procedure for the three-rung contributions, we have to replace the third order (unscreened) terms e^6/r^3 by Debye-type terms $e^6 \exp(-3\kappa r)/r^3$. Using the Friedman technique, several authors (see Hoffmann and Ebeling 1968, Ebeling et al. 1976, 1987: Kraeft et al. 1986, 1988; Kahlbaum 2000) thus derived an integral for the three-rung terms which corrects the long-range divergencies based on quasi-classical arguments:

$$F^{3R} = -k_B T V \frac{1}{2} \sum n_a n_b \int_0^\infty dr\, r^2 \left[S'''_{ab} + \frac{(\beta e_a e_b)^3 \exp(-\kappa r)}{6r^3} \right] . \tag{4.111}$$

This integral also follows from (4.110) by isolating the third order terms (Hoffmann and Ebeling 1968, Kraeft et al. 1986, Kahlbaum 1996). Carrying out the integration over the second contribution, using $Ei(x)$ functions, and assuming small κ, we arrive at the integral

$$F^{3R} = -k_B T V \frac{1}{2} \sum n_a n_b \lim_{R\to\infty} \int_0^R dr\, r^2 \Big\{ S'''_{ab}(r) + \frac{(\beta e_a e_b)^3}{6} \big[C + \ln(3\kappa R) \big] \Big\} ,$$

and obtain in the end an expression similar to the classical third order contribution, as first shown in (Hoffmann and Ebeling 1968). The correct third order contribution obtained by Hoffmann et al. includes a quantum term under the logarithm:

$$F^{L3} = -k_B T V \sum n_a n_b \frac{(\beta e_a e_b)^3}{6} \Big[\ln(\kappa \lambda_{ab}) + C/2 + \ln 3 - 1/2 \Big] . \tag{4.112}$$

For ladders with four or more rungs, the screening can be completely neglected to first order, and we may proceed as for the exchange contributions:

$$F^{LD} = -k_B T \left[\frac{1}{2} \sum_{ab} n_a n_b \int G'' S^{LD}_{ab}(1,2) d1 d2 + O(n^3) \right] . \tag{4.113}$$

Here S_{ab} is now the direct part of the two-particle Slater function and G'' denotes the operation of omitting the contributions with three or fewer interaction lines. Following the method of Beth, Uhlenbeck, and Kahn, the integral may be expressed in terms of the density of states dn_k/dk, which is exactly known for Coulomb systems. Introducing the known expressions for the phase shifts and transforming to a complex integral, we get the following structure:

$$F_{\rm LD} = +k_{\rm B}TV2\pi\sum_{ab} n_a n_b \lambda_{ab}\frac{1}{(2s+1)}Q_4(\xi_{ab})\,, \tag{4.114}$$

where $Q_4(x)$ is a complex integral which may be expanded into a converging infinite series (Ebeling 1968):

$$Q_4(x) = -\frac{\sqrt{\pi}}{2\pi \mathrm{i}}\frac{\xi^3}{8}\int_{0^-}^{\infty} \mathrm{d}t\frac{\exp(-t)}{(-t)^3/2}\left[C+\psi(1-\xi/2(-t)^{1/2})\right] \tag{4.115}$$

$$= \sum_{4}^{\infty}\frac{\sqrt{\pi}\zeta(n-1)x^n}{2^n\Gamma(1+n/2)}\,. \tag{4.116}$$

The coefficients in this series are derived from expansions of the ψ function, which itself comes from the exactly known Coulomb scattering phase shifts (Ebeling 1968, 1969). We note a parallel calculation based on the phase shifts by Kopyshev (1969). In a rather concise calculation which continues directly from the paper by Larkin (1960), Kopyshev derived similar results, but in a different form, but which at least agree in the asymptotic with the results derived here. The comparison with the independent calculations of virial functions by Alustuey and Perez (1992, 1996) is much easier, since the agreement is perfect (Ballenegger et al. 2015).

4.8 Discussion of Virial Functions and Thermodynamic Potentials

4.8.1 Analytical Properties of Virial Functions

By evaluating all integrals appearing in the virial expansions and ordering with respect to the order in the density, we get the following density expansion of the free energy density:

$$\beta f(n_a,T) = \sum_a n_a\left[\ln(n_a\Lambda_a^3)-1\right] \tag{4.117}$$

$$-\sum_{ab}2\pi n_a n_b\left[\frac{4\pi}{3\kappa}\beta^2 e_a^2 e_b^2 + \frac{1}{6}(\beta e_a e_b)^3\ln(\kappa\lambda_{ab}) + \lambda_{ab}^3 K_0(\xi_{ab};s_a)\right] + O(n^{5/2}\ln n)\,.$$

Here

$$K_0(\xi;s_a) = Q(\xi) + \frac{(-1)^{2s_a}}{2s_a+1}E(\xi) \tag{4.118}$$

is the so-called virial function, which was first defined in 1968 (Ebeling 1968), a result which differs only in minor points was obtained by Kopyshev (1968). Identical results were obtained using different methods by Kraeft et al. (1986), Alastuey and Perez (1992, 1996), and (Brown and Yaffé, 2001). The definition of the virial function was purposefully given here in analogy to the virial functions in the classical electrolyte theory, which depend on the so-called Bjerrum parameter. The classical virial function reads (Falkenhagen et al. 1971)

$$K_0(\xi) = -\frac{1}{2} - \frac{1}{2}\xi - \frac{1}{2}\xi^2 - \frac{1}{6}(C + \ln 3) + \sum_{m=4}^{\infty} \frac{\xi^m}{m!(m-3)} , \tag{4.119}$$

where

$$\xi = \xi_{ab} = -\frac{e_a e_b}{D_0 k_B T R_{ab}} ,$$

and R_{ab} is the minimal distance between the centers of two ions.

The introduction of classical virial functions goes back to the electrolyte theory and in particular to the work of Mayer (1950), Haga (1953), Ebeling and Kelbg (1966), and Kelbg et al. (1968). In this connection, let us return once again to the problem of 'direct linear terms in the equation of state of plasmas'. Recall that several authors have stated recently that "the introduction of direct linear terms leads to unjustified results for Coulomb systems" (see Kraeft, Kremp and Röpke 2015). This argument was rejected by pointing out that the same methods lead to identical results for the thermodynamics (Alastuey, Ballenegger and Ebeling 2015).

Let us reformulate the question under discussion. As direct linear contributions to the free energy density, we denote the contributions to the free energy linear in e^2 which do not stem from exchange effects. In the classical case, for charges with contact distance R_{ab}, these contributions are (Falkenhagen et al. 1971)

$$\delta f(n_a, T) = \sum_{ab} 2\pi n_a n_b \frac{1}{2} R_{ab}^2 (e_a e_b) . \tag{4.120}$$

The corresponding contribution in the quantum case is, according to our calculations,

$$\delta f(n_a, T) = \sum_{ab} 2\pi n_a n_b \frac{1}{6} \lambda_{ab}^2 (e_a e_b) . \tag{4.121}$$

In the classical case, the direct linear contribution disappears for symmetric ions $R_{ab} = R$ due to electroneutrality. For ions with different radii, the direct linear term does not disappear and gives measurable results. In the quantum case, the direct linear term gives for the strict Coulomb case

$$F^{\mathrm{H}} = -\frac{\pi}{3} V \sum_{ab} n_a n_b e_a e_b \lambda_{ab}^2 = -\frac{1}{6}\frac{\hbar^2}{k_B T} \sum_b e_b \sum_a \frac{e_a}{m_a} = 0 . \tag{4.122}$$

Because of the definition of relative masses, this term disappears again if electroneutrality holds. Note that this term appears in a natural way in the Kelbg procedure, coming from contributions to the Fourier integrals at $t = 0$. In the alternative approach given in (Kraeft and Kremp 1968, Kraeft et al. 2015), based on a calculation of the Coulomb energy and then applying the charging procedure, a direct linear term does not appear at all. In any case, however, these terms give zero contributions to the thermodynamic functions for electroneutral systems (Alastuey et al. 2015). The reader may ask why we need a discussion about zero terms! Is this just what Shakespeare might have referred to as much ado about nothing?

In our opinion, the discussion is not entirely meaningless since, by analogy with the classical case, there might be situations where linear terms give a measurable contribution to the thermodynamics. In order to give an explicit example, let us consider a plasma where the charges have different temperatures. As shown in (Klimontovich et al. 1972), the linear contributions to the correlations are then given once again by (4.12), the only change being that the relative wavelengths are now defined by

$$\lambda_{ab}^2 = \frac{\hbar^2 (m_a^{-1} + m_b^{-1})^2}{2(kT_a/m_a + kT_b/m_b)} \,. \tag{4.123}$$

For the Hartree contribution, this leads to the expression

$$F^{\mathrm{H}} = -\frac{\pi}{6} V \hbar^2 \sum_a \sum_b e_a e_b \frac{(1/m_a + 1/m_b)^2}{(kT_a/m_a + kT_b/m_b)} \,, \tag{4.124}$$

which disappears only for plasmas with identical temperatures for all components. A different example is provided by plasmas with short- and long-range forces, like alkali plasmas. The fact that there are special situations where the linear terms in question may contribute is the reason why we propose to keep the linear term, even for simple isothermal plasmas, where these terms give zero contributions.

We have to remember in this respect that the description of Coulomb interactions as a static potential force is an incomplete description. The Coulomb interaction is actually a dynamic electromagnetic interaction. Therefore, in order to decide the question on a higher level of the theory, we would need a full description in the framework of relativistic electrodynamics. This would go beyond the scope of the present book, except in the very last chapters. In recent papers (Bobrov et al. 2015), using a calculation based on relativistic electrodynamics, Bobrov, Trigger, and Zagorodnyi have developed a model based on the relativistic result that parallel particles cannot transform momentum. From these considerations, we give our final result for the Hartree–Fock term for isothermal plasmas in agreement with Ebeling et al. (1967); Kraeft et al. (1986); Alastuey et al. (2015), but with a slightly different contribution in comparison with Kraeft et al. (2015):

$$
\begin{aligned}
F^{\mathrm{HF}} &= 2\pi V \sum_{ab} n_a e_a n_b e_b \lambda_{ab}^2 \left[\frac{1}{6} + \frac{\delta_{ab}}{2(2s_a+1)}\right] \\
&= 2\pi k_{\mathrm{B}} T V \sum_{ab} n_a n_b \lambda_{ab}^3 \left[\frac{1}{6} + \frac{\delta_{ab}}{2(2s_a+1)}\right] \xi_{ab} \, .
\end{aligned}
\tag{4.125}
$$

The first of these (direct) linear terms eventually gives a zero contribution to the final results for the thermodynamics due to electroneutrality, except for (still hypothetical) special situations, such as non-isothermal plasmas, alkali plasmas, etc.

4.8.2 Virial Expansion of Thermodynamic Functions

By evaluating all integrals appearing in the virial expansions and ordering with respect to the orders in the density, we get the following density expansion of the free energy density up to second density order:

$$
\begin{aligned}
\beta f(n_a, T) = &\sum_a n_a \big[\ln(n_a \Lambda_a^3) - 1\big] - \frac{\kappa^3}{12\pi} \\
&- \sum_{ab} 2\pi n_a n_b \left[\frac{1}{6}(\beta e_a e_b)^3 \ln(\kappa \lambda_{ab}) + \lambda_{ab}^3 K_0(\xi_{ab}; s_a)\right] \\
&+ O(n^{5/2} \ln n) \, .
\end{aligned}
\tag{4.126}
$$

The corresponding expression for the pressure is

$$
\begin{aligned}
\beta p(n_a, T) = &\sum_a n_a - \frac{\kappa^3}{24\pi} \\
&- \sum_{ab} 2\pi n_a n_b \Bigg[\frac{1}{6}(\beta e_a e_b)^3 \ln(\kappa \lambda_{ab}) + \lambda_{ab}^3 K_0(\xi_{ab}; s_a) \\
&\qquad + \frac{1}{12}(\beta e_a e_b)^3\Bigg] + O(n^{5/2} \ln n) \, .
\end{aligned}
\tag{4.127}
$$

The function $K_0(\xi)$ represented graphically in Fig. 4.7 was introduced by analogy with similar functions in classical electrolyte theory (Falkenhagen et al. 1971). For quantum plasmas this so-called virial function is defined by

$$
K_0(\xi_{ab}, s_a) = Q(\xi_{ab}) + \frac{(-1)^{2s_a}}{2s_a + 1} \delta_{ab} E(\xi_{ab}) \, .
\tag{4.128}
$$

Here we have used the function $E(\xi)$ defined above and the new definition

$$Q(\xi) = -\frac{\xi}{6} + \frac{\sqrt{\pi}}{8}\xi^2 - \frac{1}{6}\left(\frac{C}{2} + \ln 3 - \frac{1}{2}\right)\xi^3 + \sum_{m\geq 4} \frac{\xi_{ab}^m}{2^m \Gamma(m/2+1)}\zeta(m-2) \, . \tag{4.129}$$

The shape of the function $Q(x)$ for negative and positive arguments is shown in Fig. 4.7. Note that the direct quantum contribution $Q(x)$ increases very quickly with increasing argument, in fact, exponentially. As shown first in Ebeling (1968), the asymptotic for large arguments is

$$Q(\xi) \approx -\frac{\xi^3}{6}\left(\ln|3\xi| + 2C - \frac{11}{6}\right) + \Theta(\xi) 2\sqrt{\pi}\sigma(T) \, . \tag{4.130}$$

Here $\sigma(T)$ is our already well studied PBL partition function and the function Θ is equal to unity for positive arguments and zero elsewhere. The fact that the PBL function appears in the asymptotic is another confirmation of our approach. Incidentally, the remaining term of order ξ^3 gives the correct classical asymptotic for the electron gas. In other words, our result is asymptotically consistent with all results known so far. Note that, if one wants to follow the arguments by Kraeft, Kremp, and Röpke, the linear term has to be omitted (Kraeft et al. 2015) and the function starts with a zero tangent. A disadvantage is, however, that the nice asymptotic of $Q(\xi)$ given by (4.130) is no longer valid, and one gets an additional linear term here. Finally, we observe that the function $K_0(\xi)$ found here (including the linear term) has also been confirmed by at least two independent groups (Alastuey and Perez 1992, 1996, Alastuey et al. 1996, 2008, 2010, 2015, Brown and Yaffé 2001).

Classical Electron Gas and Hydrogen as Special Cases

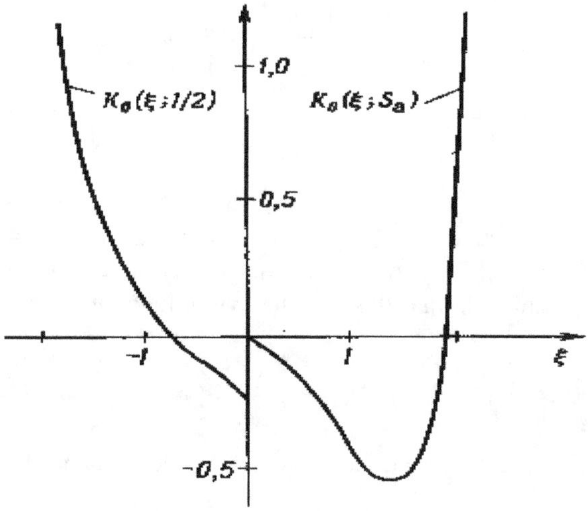

Fig. 4.7 The so-called virial function $K_0(\xi; s)$, which determines the deviations from Debye's limiting law for the free energy density of quantum plasmas

We consider now the special cases of a classical electron gas and fully ionized hydrogen. We can thus check the correctness of our calculations for known limits. First we study the electron gas, where there are no bound states at all. In this case all interaction parameters are zero, except for the electron–electron parameter, which is negative. The free energy density is

$$\beta f(n,T) = n\big[\ln(n\Lambda^3) - 1\big] - \frac{\kappa^3}{12\pi} \tag{4.131}$$
$$-2\pi n^2 \left[\frac{1}{6}\beta^3 e^6 \ln(\kappa\lambda) + \lambda^3 K_0(-\xi; s = 1/2)\right] + O(n^{5/2}\ln n)\,,$$

with $\lambda = \lambda_{ee}$ and $\xi = \xi_{ee}$. The near classical case of the electron gas corresponds to $\xi \to (-\infty)$ and the virial function has the asymptotic [see also (4.130)]

$$K_0(\xi; s) = -\frac{1}{6}\xi^3\left(2C - \ln 3 - \frac{11}{6} + \ln|\xi|\right)\,, \quad \xi < 0\,. \tag{4.132}$$

Introducing this into (4.132), we find for the free energy density

$$\beta f(n,T) = n\big[\ln(n\Lambda^3) - 1\big] \tag{4.133}$$
$$-nk_BT\left[\frac{1}{3}\mu_B + \frac{1}{12}\mu_B^2\left(\ln\mu_B + 2C + \ln 3 - \frac{11}{6}\right) + O(\mu_B^3)\right]\,.$$

We see that in the classical case the expansion parameter which is responsable for convergence is the plasma parameter introduced by Bogoliubov, viz., $\mu_B = e^2/k_BTr_D$. Our result for $\hbar \to 0$ is in full agreement with the known classical results for the free energy of the electron gas obtained by expansions with respect to μ_D. Further our result for the quantum electron gas is consistent with the known $\hbar$ expansions of the thermodynamic functions (Hoffmann and Ebeling 1968). In the present calculations, we considered only weakly non-ideal plasmas, where the expansions can be cut off after the second virial coefficient, i.e., after order n^2 in the EOS. For higher order contributions, including also order $n^{5/2}$, see, e.g., DeWitt et al. 1995, Riemann et al. 1995; Kahlbaum 2000, Alastuey et al. 2008, 2010, 2015. In order to check for convergence, we have to consider higher order terms (Kahlbaum 2000; Alastuey et al. 2015).

We consider a fully ionized hydrogen plasma in a temperature regime where bound states play a role. In order to test the influence of the second virial term, we calculate the equation of state in relation to the ideal EOS. Figure 4.8 represents the

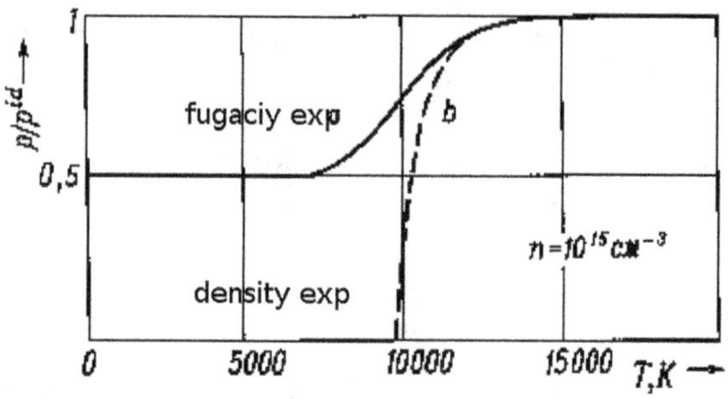

Fig. 4.8 Pressure of hydrogen plasmas compared with the ideal pressure distribution function for $n = 10^{15}$ cm^{-3} as a function of the temperature. The *dashed line* shows the result of a density expansion and the *full line* shows the corresponding fugacity expansion, each as functions of temperature

result of a calculation of the pressure in relation to the ideal pressure for one density $n = 10^{15}$ cm^{-3}, using (4.128) and comparing with the result of fugacity expansions which will be discussed in the next chapter. We observe a good agreement between density and fugacity expansions at higher temperatures, but strong divergence at temperatures below 10^4 K. Furthermore, for lower temperatures, we see that the pressure even assumes negative values in the density expansion. This is evidently an artefact due to an incorrect representation of the bound state contribution, which increases exponentially at lower temperatures. This leads to unphysical behaviour at low temperatures, and even to negative pressures. We will show later that this may be corrected by using either chemical descriptions (mass action laws) or fugacity expansions. In the presence of bound states, density expansions cut off after the second virial contribution have a limited range of validity with respect to temperature and density. For this reason we proceed to fugacity expansions in the next chapter.

Summary of Results

In this chapter, we derived the quantum-statistical corrections to Debye's law for the pressure and free energy to quadratic order in the plasma density $O(n^2)$. We claim that these results are exact within the non-relativistic theory and we think that this result is of some importance and needs careful checking, both experimentally and theoretically. We should note, however, that our statement is not without objections:

- Several authors, among them Starostin et al. (2003), have claimed that the Planck–Brillouin–Larkin partition function, which we have reproduced here, should be corrected.
- Other authors (Kraeft et al. 2015) claim that the linear contribution to the quantum virial function K_0 is wrong and should be omitted.

We note first that both these statements refer to intermediate steps of the theory, and not to the final result for thermodynamic quantities like pressure and free energy density. Our critics refer mainly to the definitions of functions (partition function and virial function). Our position is strongly influenced by Onsager's statement about the Bjerrum partition function. We believe, that it is useful here to recall once again this extremely important and famous remark (see p. 249 in Hemmer et al. 1996):

> Bjerrum's choice (for the partition function) is good but we could vary it within reason. In a complete theory, this would not matter; what we remove from one side of the ledger would be entered elsewhere with the same effect.

As a consequence, we may conclude by generalizing the Onsager statement: within a theory, definitions of functions could be varied within reason. What is missing in a definition would in a complete (and correct) theory enter elsewhere with the same effect. In other words, our strict position is that only a complete theory of certain physical quantities can be said to be 'right' or 'wrong'. Only measurable physical quantities can be checked experimentally. To use the predicates 'right' or 'wrong' for definitions within a theory does not make much sense. According to Onsager, we have here some freedom, and results which differ just by notation or a 'zero' are to be considered as equivalent, with the same value for describing reality.

Finally, we would like to reiterate that the behaviour of two-component Coulomb systems consisting of light and heavy charges is extremely complicated and very specific for each region in the (n, T) plane. Strictly speaking, each of the regions shown in Fig. 4.1 requires its own theory. The universal aspects studied in this chapter are the properties which appear at the lowest densities, where interactions and degeneracy effects start to grow and are still described by two universal 'virial functions' $Q(\xi)$ and $E(\xi)$ of the interaction parameters. However, as in the theory of real gases, even if their range of applicability is quite restricted, the second order virial contributions provide the seeds for any quantum-statistical theory with more extended range.

References

A. Alastuey, Breakdown of Debye screening in quantum Coulomb systems and van der Waals forces. Phys. A **263**, 271–283 (1999)

A. Alastuey, PhA Martin, Breakdown of Debye screening in quantum plasmas. Europhys. Lett. 6, 385 (1988). Phys. Rev. A **40**, 6485 (1989)

A. Alastuey, A. Perez, Virial expansion of the equation of state of a quantum plasma. Europhys. Lett. **20**, 19–24 (1992)

A. Alastuey, A. Perez, Virial expansions for quantum plasmas: Fermi-Bose statistics. Phys. Rev. E **53**, 5714 (1996)

A. Alastuey, V. Ballenegger, Pressure of a partially ionized hydrogen gas: numerical results from exact low temperature expansions. Contr. Plasma Phys. **50**, 46–53 (2010)

A. Alastuey, V. Ballenegger, W. Ebeling, Comment on direct linear term in the equation of state of plasmas. Phys. Rev. E **92**, 047101 (2015)

A. Alastuey, V. Ballenegger, F. Cornu, PhA Martin, Exact results for thermodynamics of the hydrogen plasma: low-temperature expansions beyond Saha theory. J. Stat. Phys. **130**, 1119–1176 (2008)

V. Ballenegger, A. PhA Martin, Alastuey. J. Stat. Phys. **108**, 169 (2002)

A.A. Barker, Radial distribution functions for a hydrogenous plasma in equilibrium. Phys. Rev. **171**, 168 (1968); **179**, 129–134 (1969)

N.V. Brilliantov, Contrib. Plasma Phys. **38**, 489 (1998) (ArXiv physics 1998)

L.S. Brown, L.G. Yaffé, Effective field theory of highly ionized plasmas, Phys. Reports **340**, 1 (2001), arXiv:physics/9911055

E.G.D. Cohen, T.J. Murphy, Phys. Fluids **12**, 1404 (1969)

F. Cornu, Exact algebraic tails of static correlations in quantum plasmas. Phys. Rev. Lett. **78**, 1464 (1997)

F. Cornu, PhA Martin, Electron gas beyond random-phase approximations: algebraic screening. Phys. Rev. A **40**, 4893 (1991)

H.-J. Czerwon, Diploma thesis, University Rostock 1972, see also H. Krienke et al., Wiss. Z. University Rostock MNR **24**, 5 (1975)

H.E. DeWitt, Evaluation of the quantum-mechanical ring sum. J. Math. Phys. **3**(6), 1216 (1962); **7**, 616 (1966)

H.E. DeWitt, Asymptotic form of the classical one-component plasma. Phys. Rev. A **14**, 1290–1293 (1976)

W. Ebeling, Equilibrium statistics of systems with bound states, in *Proceedings of the Van der Waals Centennial Conference on statistical mechanics*, ed. by C. Prins (North Holland Publication, Amsterdam, 1973), pp. 83–84

W. Ebeling, Quantum statistics of bound states in plasmas. Ann. Phys. (Leipzig) **19**, 104 (1967), **21**, 315 (1968); **22**, 383–391, 392–401 (1969)

W. Ebeling, Physica, EOS and Saha equation of plasmas, **38**, 378 (1968); **40**, 290 (1968); **43**, 293 (1969); **73**, 573 (1974)

W. Ebeling, Physica A **130**, 523 (1985)

W. Ebeling, Correlation functions and thermodynamic potentials for nonideal plasmas. Contr. Plasma Phys. **33**(5/6), 492–502 (1993)

W. Ebeling, S. Hilbert, On Saha's equation for partially ionized plasma and Onsager's bookkeeping rule. Eur. J. Phys. D **20**, 93–101 (2002)

W. Ebeling, F. Schautz, Simulations of the quantum electron gas using momentum-dependent potentials. Phys. Rev. E **56**, 3498 (1997)

W. Ebeling, H.D. Hoffmann, G. Kelbg, Quantenstatistik des Hochtemperaturplasma im thermodynamischen Gleichgewicht. Contr. Plasma Phys. **7**, 233–248 (1967)

W. Ebeling, W.D. Kraeft, D. Kremp, Quantum statistical second virial coefficient for real gases and plasmas. Contr. Plasma Phys. **10**, 237–263 (1970)

W. Ebeling, W.D. Kraeft, D. Kremp, *Theory of Bound States and Ionisation Equilibrium in Plasmas and Solids (Akademie-Verlag, Berlin, 1976)* (Extended Russ. translation Mir, Moscow, 1979)

W. Ebeling, S. Hilbert, H. Krienke, On Bjerrum's mass action law for electrolytes and Onsager's bookkeeping rule. J. Mol. Liq. **96**(97), 409–423 (2002)

W. Ebeling, H. Hache, M. Spahn, Thermodynamics of ionization and dissociation in hydrogen. Eur. Phys. D **23**, 265–272 (2003)

W. Ebeling, G.E. Norman, A.A. Valuev, I. Valuev, Contrib. Plasma Phys. **39**, 61 (1999)

W. Ebeling, A. Filinov, M. Bonitz, V. Filinov, T. Pohl, The method of effective potentials in the quantum-statistical theory of plasmas. J. Phys. A Math. Gen. **39**, 4309–4317 (2006)

W. Ebeling, The work of Baimbetov on nonideal plasmas and some recent developments. Contr. Plasma Phys. **56**(3–4), 163–175 (2016)

H. Falkenhagen, Theorie der Electrolyte, Teubner Leipzig (1971)

H. Falkenhagen et al., I. Equilibrium properties of ionized dilute electrolytes, II. Mass transport properties, in: *Ionic interactions*, ed. by S. Petrucci, (Academic Press, New York, 1971)

R.P. Feynman, A.R. Hibbs, *Quantum Mechanics and Path Integrals* (Graw-Hill, New York, 1965)

V.S. Filinov, V.E. Fortov, M. Bonitz, D. Kremp, Phys. Lett. A **274**, 228 (2000)

V.S. Filinov, M. Bonitz, W. Ebeling, V.E. Fortov, Plasma Phys. Control Fusion **43**, 743 (2001)

A.V. Filinov, M. Bonitz, W, Ebeling, Improved Kelbg potential for correlated Coulomb systems. J. Phys. A Math. Gen. **36** 5957–5962 PII: S0305-4470(03) 55108-3 (2003)

A.V. Filinov et al., Phys. Rev. E **70**, 046411 (2004)

V.S. Filinov, M. Bonitz, V.E. Fortov, W. Ebeling, H. Fehske, D. Kremp, W.D. Kraeft, V. Bezkrovniy, P. Levashov, Monte Carlo simulations of dense quantum plasmas. J. Phys. A Math. Gen. **38**, 1–9 (2005)

V.E. Fortov, *Extreme States of Matter* (FizMatGis, Moskva, 2009). (in Russian)

V.E. Fortov, *Extreme States of Matter: on Earth and in the Cosmos* (Springer, Berlin, 2011)

A. Förster, T. Kahlbaum, W. Ebeling, Equation of state and phase diagram of fluid helium in the region of partial ionization. Laser Part. Beams **10**, 253–262 (1992)

H.L. Friedman, *Ionic Solution Theory* (Interscience, New York, London, 1962)

P.E. Grabowski, A. Markmann, M.S. Murillo, C.A. Fichtl, D.F. Richards, V.S. Batista, F.R. Graziani, I.V. Morozov, I.A. Valuev, Wave packet spreading and localization in electron-nuclear scattering. Phys. Rev. E **87**, 063104 (2013)

R.I. Guernsey, Phys. Fluids **5**, 322 (1962)

J.P. Hansen, I.R. McDonald, Phys. Rev. A **23**, 2041 (1981)

J.P. Hansen, I.R. McDonald, *Theory of Simple Liquids*, 3rd edn. (Academic Press, New York, 2006)

J.-P. Hansen, I.R. McDonald, E.L. Pollock, Phys. Rev. A **11**, 1025 (1975)

E.J. Heller, J. Chem. Phys. **62**, 1544 (1975)

P.C. Hemmer, H. Holden, S. Kjelstrup Ratkje, *The Collected Works of Lars Onsager (with commentary)* (World Scientific, Singapore, 1996)

H.J. Hoffmann, G. Kelbg, Ann. Physik (Leipzig) **17**, 356 (1966); **18**, 186 (1967); **19**, 186 (1967)

H.J. Hoffmann, W. Ebeling, On the EOS of Fully Ionized Quantum Plasmas. Physica 39, 593 (1968); Contr. Plasma Phys. **8**, 43 (1968)

S. Ichimaru, *Statistical Plasma physics* (Addison-Wesley, Redwood City, 1992)

B. Jakob, P.-G. Reinhard, C. Toepffer, G. Zwicknagel. Phys. Rev. E **76**, 036406 (2007); L.P. Kadanoff, L.P. Baym, Quantum Statistical Mechanics. (New York, 1962)

T. Kahlbaum, The second virial coefficient for charged particle systems, in *Dynamik, Evolution, Strukturen*, ed. by J. Freund (Verlag Dr. Köster, Berlin, 1996)

T. Kahlbaum, Calculation of the virial expansion for quantum plasmas, in *Strongly Coupled Coulomb Systems*, eds. by G.J. Kalman, J.M. Rommel, K. Blagoev (Plenum, New York and London, 1998)

T. Kahlbaum, Advances in the calculation of virial expansions for quantum plasmas up to third order in the density, in *Strongly Coupled Coulomb Systems*, eds. by G.J. Kalman et al., (Plenum Press, New York, 1998); The quantum-diffraction term in the free energy for Coulomb plasma, effective-potential approach. J. Phys. France **10**, Pr5-455–459 (2000)

A.S. Kaklyugin, Teplophys. Vyss. Temp. **23**, 217 (1985)

G. Kalman, J. Rommel, K. Blagoev (eds.), *Strongly Coupled Coulomb Systems* (Plenum Press, New York, 1998)

G. Kelbg, Quantenstatistik der Gase mit Coulombwechselwirkung. Ann. Physik (Leipzig) **12**, 219–224, 354–360 (1963); **14**, 394–403 (1964)

G. Kelbg, *Einige Methoden der statistischen Thermodynamik hochionisierter Plasmen, Ergebnisse der Plasmaphysik aund Gaselektronik*, vol. III (Akademie Verlag, Berlin, 1972)

D. Klakow, C. Toepffer, P.-G. Reinhard, Phys. Lett. A 192, 55 (1994). J. Chem. Phys. **101**, 10766 (1994)

M. Knaup, P.-G. Reinhard, C. Toepffer, G. Zwicknagel, Wave packet molecular dynamics simulations of hydrogen at mbar pressures. Computer Phys. Comm. **147**, 202–204 (2002)

M. Knaup, P.-G. Reinhard, C. Toepffer, G. Zwicknagel, J. Phys. A Math. Gen. **36**, 6165 (2003)

H. Kleinert, *Path Integrals in Quantum Mechanics, Statistics, Polymer Physics, and Financial Markets*, 5th edn. (World Scientific, Singapore, 2009)

YuL Klimontovich, *Statistical Theory of Nonequilibrium Processes in Plasmas, Izdat MGU 1964. Engl. transl.* (Pergamon, Oxford, 1967)

Y.L. Klimontovich, *Statistical physics (in Russ)* (Nauka, 1982); Engl. transl (Harwood, New York, 1986)

W. YuL Klimontovich, Zh Ebeling, Eksp., Teor. Fiz. **42**, 146 (1972)

W.D. YuL Klimontovich, Kraeft. Teplofiz. Vyss. Temp. **12**, 239 (1974)

V.P. Kopyshev, Second virial coefficients of a plasma. Sov. Phys. JETP **28**, 684–686 (1969)

W.D. Kraeft, D. Kremp, W. Ebeling, G. Röpke, *Quantum Statistics of Charged Particle Systems* (Akademie-Verlag & Pergamon Press, Berlin & New York, 1986)

W.D. Kraeft, D. Kremp, G. Röpke, Direct linear term in the equation of state of plasmas. Phys. Rev. E **91**, 013108 (2015)

D. Kremp, W.D. Kraeft, W. Ebeling, Quantum statistics of the second virial coefficients and scattering theory. Physica **51**, 164 (1971)

D. Kremp, G. Schmitz, Z. Naturforschung **22a**, 1366, 1392 (1967)

D. Kremp, M. Schlanges, W.D. Kraeft, *Quantum Statistics of Nonideal Plasmas* (Springer, Berlin, 2005)

A.I. Larkin, Zh. eksp. teor. Fiz. (Russ.) **38**, 1896 (1960)

J.D. Landau, E.M. Lifshits, Statistical Physics (part I), Nauka Moska 1976, German transl Berlin 1979

E.H. Lieb, R. Seiringer, *The Stability of Matter in Quantum Mechanics* (Cambridge University Press, New York, 2010)

P.A. Martin, Sum rules in charged fluids. Rev. Mod. Phys. **60**, 1075 (1988)

J.E. Mayer, The theory of ionic solutions. Chem. Phys. **18**, 1426–1436 (1950)

B. Militzer, D.M. Ceperley, Phys. Rev. Lett. **85**, 1890 (2000)

E. Montroll, J. Ward, Quantum statistics of interacting particles. Phys. Fluids **1**, 55 (1958)

I.V. Morozov, I.A. Valuev, J. Phys. A Math. Theor. **42**, 214044 (2009); Contr. Plasma Phys. **52**, 140 (2012)

J. Ortner, Equation of states for classical; coulomb systems: use of the Hubbard-Schofield approach. Phys. Rev. E **59**, 6312–6320 (1999)

J. Ortner, I. Valuev, W. Ebeling, Electric microfield distribution in two-component plasmas. Theory Simulations Contrib. Plasma Phys. **39**, 311 (1999)

D. Pines, P. Nozieres, *The Theory of Quantum Liquids (Benjamin, New York, Amsterdam, 1966)* (Russ. Transl. Mir, Moskva, 1967)

M. Planck, Zur Quantenstatistik des Bohrschen Atommodells. Ann. Phys. **75**, 673–684 (1924)

E.L. Pollock, J.P. Hansen, Phys. Rev. A **8**, 3110 (1973)

R. Redmer, Phys. Rep. **282**, 35 (1997)

R. Redmer, G. Röpke, S. Kuhlbrodt, H. Reinholz, Phys. Rev. B **63**, 233104 (2001)

H. Reinholz, Ann. Phys. (Paris) **30**, 1 (2005)

J. Riemann, M. Schlanges, H.E. DeWitt, W.D. Kraeft, Equation of state of the weakly degenerate OCP. Phys. A **219**, 423–435 (1995)

G. Rohde, G. Kelbg, W. Ebeling, Binary slater sums and distribution functions. Ann. Physik (Leipzig) **22**, 1 (1968)

S.P. Sadykova at al., Contr. Plasma Phys. **47**, 659, 1160 (2007); **49**, 76 (2009)

S.P. Sadykova, W. Ebeling, I.M. Tkachenko, Static and dynamic structure factors with account of the ion structure for high-temperature alkali and alkaline earth plasmas. Eur. Phys. J. D **61**, 117–130 (2011)

G. Schmitz, D. Kremp, Quantenmechanische Verteilungsfunktion für ein Elektronengas. Z. Naturforschung A **23**, 1392–1395 (1967)

A.N. Starostin, V.C. Roerich, R.M. More, How correct is the EOS of weakly nonideal plasmas. Contr. Plasma Phys. **43**, 369–372 (2003)

A.N. Starostin, V.C. Roerich, Bound states in nonideal plasmas: Formulation of the partition function and application to the solar interior. Plasma Sources Sci. Technol. **15**, 410–415 (2006)

W. Stolzmann, W. Ebeling, New Padé approximations for the free charges in two-component strongly coupled plasmas based on the Unsöld-Berlin-Montroll asymptotics. Phys. Lett. A **248**, 242–246 (1998)

R.G. Storer, Path-integral calculation of the quantum-statistical density matrix for attractive Coulomb forces. J. Math. Phys. **9**, 964–970 (1968); Radial distribution function for a quantum plasma. Phys. Rev. **176**, 326–331 (1968)

S.A. Trigger, V.E. Filinov, V.E. Fortov, M. Bonitz, W. Ebeling, Internal energy of of high density hydrogen. Zh. eksp. tor. Fiz (USSR) **123**, 527–542 (2004)

B.A. Trubnikov, V.F. Elesin, Quantum correlation functions in a Maxwellian plasma. JETP **20**, 866–872 (1965)

A.A. Vedenov, A.I. Larkin, Equation of state of plasmas , Sov.Phys. JETP **9**, 806–821 (1959)

I.R. Yukhnovsky, M.G. Holovko, Statistical Theory of Classical Equilibrium Systems (in Russ.) (Nauk. Dumka, Kiev, 1980)

V.M. Zamalin, G.E. Norman, V.S. Filinov, *Method Monte Carlo in Statistical Physics* (Nauka, Moscow, 1977)

B.V. Zelener, G.E. Norman, V.S. Filinov, *Perturbation Theory and Pseudopotential Method in Statistical Physics* (Nauka, Moscow, 1981)

Chapter 5
Plasma Bound States in Grand Canonical and Mixed Representations

Plasmas with deep bound states are not well described by density expansions. Here fugacity expansions are a very useful alternative to density expansions. The quantum-statistical theory is formulated in the grand canonical ensemble. This provides representations of the pressure and the densities in terms of the fugacities, which serve as implicit variables. This has already been demonstrated for real gases in Chap. 2, where we showed the equivalence with chemical descriptions. Fugacity expansions are, in principle, completely equivalent to density expansions. They are just a useful alternative, but with a quite different range of convergence, and different applications. For systems with deep bound states, fugacity expansions are more quickly convergent and very useful for EOS calculations. Furthermore, we develop mixed representations which combine the advantages of density and fugacity expansions.

5.1 Fugacity Expansions of Thermodynamic Functions

5.1.1 Cluster Expansions in Fugacities

Derivations of screened fugacity expansions for plasmas were first given by Montroll and Ward (1958), including the Debye terms, then by Vedenov and Larkin (1959) with quantum corrections based on representations within the grand canonical ensemble. Expansions including higher order cluster contributions with bound state terms were derived by Kelbg et al. (1970), by Bartsch et al. (1971), and by Rogers (1974) using Mayer–Friedman-type cluster expansions, as well as by Kraeft et al. (1986) and Kremp et al. (2005) using Green function techniques. Independent derivations were given by Alastuey et al. (1992, 2008, 2010, 2015) using Feynman–Kac path integral representations, and by Brown and Yaffé (2001) using effective field theory, by Ichimaru (1992) using dielectric function methods and by Klimontovich (1967, 1986) by using his stochastic distribution functions.

Fugacity expansions express, at least in principle, the same physics as density expansions, but with completely different fields of applications. For example, we

W. Ebeling et al., *Quantum Statistics of Dense Gases and Nonideal Plasmas*, Springer Series in Plasma Science and Technology, DOI 10.1007/978-3-319-66637-2_5

showed in Chap. 2 that, for hard sphere systems and related systems with strong repulsive forces, the convergence of fugacity expansions is extremely poor. Using the equivalence of the two types of expansion, we write down a hypothetical fugacity series and check the correctness term by term by comparing with the well-founded density expansions from Chap. 4. According to the structure of the expansions with respect to density in Chaps. 3 and 4, we make the ansatz that the pressure can be expressed as a function of the fugacities in the form of an expansion, viz.,

$$\beta p(z_a, T) = \sum_a z_a + \frac{\kappa_g^3}{12\pi} + \sum_{ab} n_a n_b b_{ab}(\kappa_g) + \sum_{abc} n_a n_b n_c b_{abc}(\kappa_g) + \cdots \,, \tag{5.1}$$

where

$$z_a = \frac{2s_a + 1}{\Lambda_a^3} \exp(\beta\mu_a) \,, \quad \kappa_g^2 = 4\pi\beta \sum_a z_a e_a^2 \,. \tag{5.2}$$

and κ_g is the equivalent of the Debye κ in the grand canonical ensemble. Note that the fugacities are normalized here in such a way that

$$z_a \to n_a \,, \qquad \text{if} \quad n_a \to 0 \,.$$

Furthermore, we have the following relation between densities and fugacities:

$$\begin{aligned} n_a &= z_a \frac{\partial[\beta p(z_a, T)]}{\partial z_a} \\ &= z_a + \frac{1}{3}\beta e_a^2 \kappa_g + z_a \frac{\kappa_g^2}{12\pi} \frac{\partial \kappa_g}{\partial \kappa} + z_a \sum_b z_b b_{ab}(\kappa_g) + z_a \sum_{bc} z_b z_c b_{abc}(\kappa_g) \\ &\quad + z_a \sum_b b'_{ab}(\kappa_g) + z_a \sum_{bc} b'_{abc}(\kappa_g) + \cdots \,. \end{aligned} \tag{5.3}$$

The easiest way to get explicit expressions for the cluster coefficients $b_{ab}(\kappa_g)$, $b_{abc}(\kappa_g)$, etc., of the fugacity expansions is to expand them around the low density, low fugacity limits and to compare then term by term. According to Bartsch et al. (1971), starting from the known density expansions in Chap. 4, we get

$$\begin{aligned} \beta p(z_a, T) = \sum_a z_a + \frac{\kappa_g^3}{12\pi} + \sum_{ab} z_a z_b \Big[&\frac{\pi}{3} (\beta e_a e_b)^3 \ln(\kappa_g \lambda_{ab}) \\ &+ \frac{\pi}{2} \beta^3 e_a^2 e_b^4 + K_0(\xi_{ab}; s_a) \Big] + O(z^{5/2} \ln z) \,. \end{aligned} \tag{5.4}$$

The corresponding expansion of the density reads (Bartsch et al. 1971)

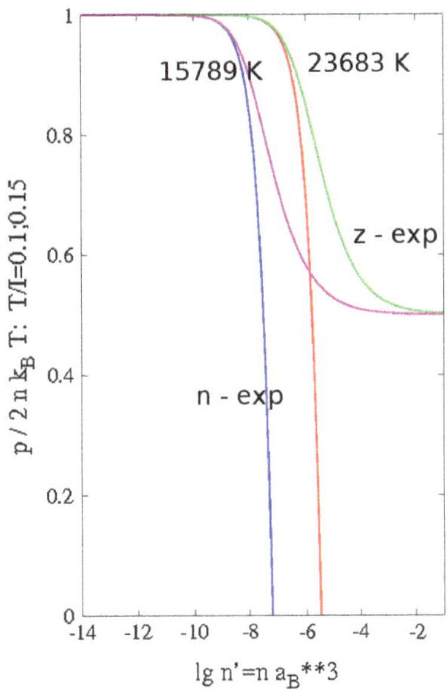

Fig. 5.1 Hydrogen pressure in units of total classical pressure $2nk_\mathrm{B}T$ at constant temperatures $T = 0.1I$ and $T = 0.15I$, where I is the ionization energy, as a function of the dimensionless density on a log scale $\log(na_\mathrm{B}^3)$. Density expansions up to second virial coefficients (rapidly decreasing *red* and *blue curves*) are compared with corresponding calculations based on fugacity expansions (*green* and *magenta curves* with saturation at $P/2nk_\mathrm{B}T = 0.5$)

$$n_a = z_a\Bigg[1 + \frac{\beta}{2}e_a^2\kappa_\mathrm{g} + \pi\frac{\beta^2}{2}\sum_b e_a^4 e_b^2 e_a^2 e_b^4 + \pi\frac{\beta^3}{3}\sum_b e_a^3 e_b^3 \ln(\kappa_\mathrm{g}\lambda_{ab}) \qquad (5.5)$$
$$+2\pi\sum_b z_b K_0(\xi_{ab}, sa) + \cdots\Bigg].$$

Recall that $z_a \to n_a$ in the small density limit. We shall henceforth write these fugacities without tildes.

Figure 5.1 compares a fugacity expansion for a binary plasma with a density expansion, both being cut off beyond quadratic terms. The comparison shows the equivalence of the two expansions at low densities, but a certain advantage of the fugacity expansion in the region of dimensionless densities $na_\mathrm{B}^3 \gg 1$ where bound states play a role. The main advantage is that the pressures do not go to unphysical negative values, but stay positive and approach a value which corresponds to the gas of hydrogen atoms. This is clearly a major step forward in the theoretical description of ionization phenomena, and in understanding the relation to the Saha equation (Ebeling 1974; Bartsch and Ebeling 1971; Förster et al. 1991, 1992; Alastuey and Ballenegger 2008, 2010).

We study now the higher order contributions in the fugacity expansions. By expanding around the limit of small densities/fugacities, for Coulomb plasmas with small fugacities (small densities), we get expressions for the cluster integrals $b_{ab}(\kappa_\mathrm{g})$ which read explicitly (Ebeling and Kelbg 1970, Bartsch and Ebeling 1971, Ebeling 1990, Vorobjov et al. 2000).

$$b_{ab}(\kappa_g) = \frac{1}{2!}\int \mathrm{d}\mathbf{r}_2\left[S_{ab}(r)\exp(G_{ab}+\beta V_{ab}) - 1 - g_{ab} - \frac{1}{2}g_{ab}^2\right] . \qquad (5.6)$$

In the grand canonical ensemble, the integral equation determining the screening function $G_{ab}(r)$ reads

$$G_{ab}(1,2) + \beta V_{ab}(1,2) + \sum_c z_c \int \mathrm{d}3\, \beta V_{bc}(1,3)G_{bc}(2,3) = 0 . \qquad (5.7)$$

An explicit numerical calculation of the cluster integrals in the grand canonical ensemble was performed by Rogers (1974). Däppen succeeded in showing that extended fugacity expansions allow an excellent description of measurements for the oscillation modes of the Sun (Däppen et al. 1988; Däppen 2010).

Here we have derived the fugacity expansion through the analogy with density expansions. A systematic quantum statistical approach to the pressure of plasmas in the grand canonical ensemble was given by Bartsch and Ebeling (1971), using Friedman's cluster expansions, while higher orders were later included by Rogers (1974) and by Friedman and Ebeling (1979), using systematic higher order cluster expansions. An alternative derivation based on the Green's function representations of the pressure was given by Kraeft et al. (1986) and by Kremp et al. (2005). The central relation for the pressure in the Green's function formalism reads (see Kraeft et al. 1986)

$$p(\beta,\mu_\mathrm{e},\mu_\mathrm{i}) = p_\mathrm{id} - \frac{1}{2V}\int_0^1 \frac{\mathrm{d}\lambda}{\lambda}\int \mathrm{d}1\,\mathrm{d}\bar{1}\; V(1\bar{1})G_2(1,\bar{1},1^{++},1^{+};\bar{t}_1 = t_1^{+}) ,$$

where $1 = \{\mathbf{p}_1, \sigma_1\}$ denotes momentum and spin variables. The Green's function representation works in the grand canonical ensemble, which immediately provides a series in the fugacities:

$$z_a = \frac{2s_a+1}{\Lambda_a^3}\exp(\beta\mu_a) = n_a \exp(\beta\mu_a^{\mathrm{ex}}) ,$$

rather than a series in the densities. The Green's function method can reproduce all the results obtained by Friedman–Kelbg cluster expansions and yields several additional results, e.g., concerning the elementary excitations and the bound state spectra in dense media (Kraeft et al. 1986; Kremp et al. 2005).

One disadvantage of the diagrammatic expansions used in the Green's function method is clearly the fact that no serious mathematical theorem about the density order of the Green's function diagrams for Coulomb systems exists so far. On the other hand, for cluster expansions, we may use the well-established theorems of Mayer, Haga, and Friedman about the density orders of higher order cluster diagrams (see Mayer 1950; Friedman 1962; Kraeft et al. 1986; Kahlbaum 2000), and we also have some knowledge of the density orders of Feynman–Kac diagrams (Alastuey et al. 1992, 2008, 2010, 2015).

5.1.2 Fugacity Representations and the Saha Equation

In order to understand the relation between the fugacity expansion for Coulomb systems and the Saha equation which we derived in Chap. 4, we have to remember our findings for two component Yukawa systems which we obtained in Sect. 3.9. Due to the screening phenomenon, the Coulomb system is more complicated than the Yukawa system, where screening is intrinsic (already contained in the definition of the potential).

As an illustration let us first consider the simple case of a system with positive and negative charges and denote the interactions by the indices ++, +−, −+, and ++. According to (5.4), the fugacity representation for the pressure is

$$\begin{aligned}\beta p(z_+, z_-, T) &= z_+ + z_- + \frac{\kappa_g^3}{12\pi} \\ &+ \sum_{ab} z_a z_b \Big[\frac{\pi}{3}(\beta e_a e_b)^3 \ln\left(\kappa_g \lambda_{ab} + \frac{\pi}{2}\right) \beta^3 e_a^2 e_b^4 \\ &+ K_0(\xi_{ab}; s_a) \Big] + O(z^{5/2} \ln z) \,. \end{aligned} \tag{5.8}$$

On the right-hand side we find a sum of rather difficult terms, but the dominating contribution comes, in the presence of (+−) bound states, from just one term based on

$$K_0(\xi_{+-}) \sim 2\sqrt{\pi} \sum_{s=1}^{\infty} s^2 \Big[\exp(-\beta E_s) - a + \beta E_s \Big] \,.$$

For all practical purposes, this contribution determines the second virial coefficient

$$B_{+-}^{(2)}(T) \approx 4\pi^{\frac{3}{2}} \lambda_{+-}^3 \sigma(T) \,, \tag{5.9}$$

where σ is the well-known Planck–Brillouin–Larkin (PBL) partition function. Since the terms diverge as s^2 in the original series in the atomic partition function, the divergent terms have been subtracted. In the canonical representation we have shown that the divergent terms are compensated by contributions from the scattering states, and this is also true in the grand canonical representation. Forgetting for a moment all other (much smaller) terms and keeping only the contribution arising from the PBL partition function, we get

$$\beta p = z_- + z_+ + 2B_{+-}^{(2)}(T) z_- z_+ \,. \tag{5.10}$$

Note that all the neglected contributions stem from scattering states. By differentiation, we get

$$n_- = z_- + 2B_{+-}^{(2)}(T) z_- z_+ \,, \qquad n_+ = z_+ + 2B_{+-}^{(2)}(T) z_- z_+ \,. \tag{5.11}$$

This set of equations can be interpreted in a 'chemical picture':

$$z_- \text{ density of free negative charges}\,,$$
$$z_+ \text{ density of free positive charges}\,.$$

In this chemical picture, the term $2B_{+-}^{(2)}(T) = K(T)$ is the 'mass action constant', and the term $z_- z_+ K(T) = n_0^*$ is the density of free atoms. As for an ideal gas of interacting electrons and protons, we may express the pressure by

$$\beta p = n_-^* + n_+^* + n_0^*\,, \quad n_- = n_-^* + n_0^*\,, \quad n_+ = n_+^* + n_0^*\,. \tag{5.12}$$

The corresponding mass action law (MAL) is

$$\frac{n_0^*}{n_+^* n_-^*} = K(T)\,. \tag{5.13}$$

This MAL describes a reaction

$$(+) \; + \; (-) \; \rightleftharpoons \; \text{hydrogen atom}\,. \tag{5.14}$$

In conclusion, we may state that, in the grand canonical ensemble, the contributions of the bound states are taken into account as in a chemical description (Hill 1956, 1960; Ebeling 1974; Friedman and Ebeling 1979).

In order to explain the chemical view in more detail, we study a hydrogen plasma of free protons and electrons with equal density $n_+^* = n_-^* = n^*$. In the ideal gas approximation, the pressure is

$$\beta p = 2n^* + n_0^* = 2n^* + K(T)n^{*2}, \quad n = n^* + n_0^* = n^* + K(T)n^{*2}. \tag{5.15}$$

The explicit solution of the corresponding quadratic equation (5.15) is

$$\frac{n^*}{n} = Z\big(nK(T)\big)\,, \qquad Z(x) = \frac{1}{2x}\left(\sqrt{1+4x} - 1\right)\,. \tag{5.16}$$

We introduced here a function $Z(x)$ which we call the Saha function, or later, the fugacity function. The ratio of the density of free charges to the total density of atoms is called the degree of ionization $\alpha = n^*/n$, and finally, we find the degree of ionization and the pressure to be

$$\alpha = Z\big[nK(T)\big]\,, \qquad \beta p = n(1+\gamma)\,, \qquad \gamma = Z\big[nK(T)\big]\,. \tag{5.17}$$

We see once again that the relative pressure (pressure divided by the classical ideal pressure) and the degree of ionization considered as functions of the total atom density (here denoted by n_0) behave in a quite similar way. On the basis of the findings in this section, we may state that the study of fugacity expansions has contributed much to the understanding of the Saha equation and hence also to the determination of a statistical theory of ionization phenomena (Ebeling 1974; Alastuey and Ballenegger 2008, 2010).

Finally, we note that the present interpretation of fugacity expansions may also be expanded to higher orders (Friedman et al. 1979). A systematic procedure has also been derived in (Alastuey et al. 2008), including corrections to the Saha theory, and used to study molecular dissociation of the hydrogen gas (Alastuey and Ballenegger 2012).

In a simpler framework, we now study the formation of hydrogen molecules, consisting of two protons and two electrons in a hydrogen plasma. Taking into account the basic contribution from the forth virial coefficient, we have

$$\beta p \approx z_- + z_+ + 2B_{+-}^{(2)}(T) z_- z_+ + 6B_{+-+-}^{(4)}(T) z_- z_+ z_- z_+ \; . \tag{5.18}$$

By differentiation we obtain an equation for the densities. Assuming full symmetry of the positive and negative particles, omitting tildes for simplicity, and using the abbreviations $z = z_- = z_+$, $B_2 = B_{+-}^{(2)} = B_{-+}^{(2)}$, $B_4 = B_{+-+-}^{(4)}$, and so on, the pressure and density become

$$\beta p = 2z + 2z^2 B_2(T) + 6z^4 B_4(T) \; , \quad n = z + 2z^2 B_2(T) + 12z^4 B_4(T) \; .$$

In the given approximation, we make the ansatz $z = nZ(n, T)$ and define the new variables $x = 2B_2(T)$, $y = 12B_4(T)$, and this leads to the function $Z(x, y)$ which gives the pressure

$$\beta p = n \left[1 + Z(x, y) - \frac{1}{2} y Z(x, y)^4 \right] . \tag{5.19}$$

We thus find a fourth order polynomial for the fugacity function $Z(n, T)$:

$$1 = Z + xZ^2 + yZ^4 \; . \tag{5.20}$$

The fugacity function $Z(x, y)$ depends on two dimensionless variables. Selecting the positive (physical) solution of the fourth order equation, we finally get the pressure in a closed form. The function $Z(x, y)$ may be defined by a Taylor series which starts with the terms

$$Z(x, y) = 1 - x - y + 2x^2 + 4y^2 + 2xy + \cdots \; . \tag{5.21}$$

This formula may be used for estimates in the region $x < 1$, $y < 1$. In the opposite case $x > 1$, $y > 1$, we may neglect the linear term and find from a biquadratic equation the analytical estimate

$$Z(x,y) \approx \left(\frac{x}{2y}\right)^{1/2} \left(\sqrt{1+\frac{y}{4x^2}}-1\right)^{1/2} . \tag{5.22}$$

When the asymptotic for one variable is finite and the other one goes to infinity, we have a simpler asymptotic. When x is large, which corresponds to the situation where many atoms are formed and just a few molecules, the asymptotic is

$$Z(x,y) \longrightarrow \frac{1}{x^{1/2}} , \quad \text{if} \quad x \to \infty , \tag{5.23}$$

and we find the pressure

$$\beta p = n \left[1 + \frac{1}{\sqrt{n B_2(T)}}\right] . \tag{5.24}$$

This means that in the case of atom formation, the number of atoms agrees with the original number of protons (or electrons) per unit volume. The second case corresponds to overwhelming formation of molecules, i.e.,

$$Z(x,y) \longrightarrow \frac{1}{y^{1/4}} , \quad \text{if} \quad y \to \infty , \tag{5.25}$$

and we get half of the value obtained for the case of atom formation:

$$\beta p = n \left\{1 - \frac{1}{\left[24 n^3 B_4(T)\right]^{1/4}}\right\} . \tag{5.26}$$

A simple Padé approximation for the fugacity function reads

$$Z(x,y) = \frac{1 + x^2 + 3y^3}{1 + x + y + 2x^{5/2} + 3y^{9/4}} . \tag{5.27}$$

A graphical representation is shown in Fig. 5.2. For higher accuracy, numerical calculations are needed. The figure also shows an example of the increase in pressure with increasing temperature (decreasing virial coefficients). Note that a systematic (exact) statistical theory for chemical association has been given by Fisher and Zuckerman (1998).

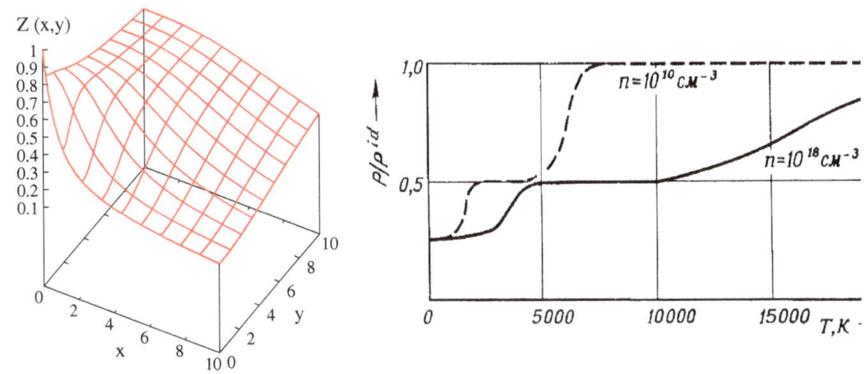

Fig. 5.2 Fugacity expansion for hydrogen plasmas. *Left* 3D plot of the fugacity function $Z(x, y)$. *Right* Relative pressure for the densities $n = 10^{10}$ cm^{-3} (*dotted line*), $n = 10^{18}$ cm^{-3} (*full line*) as a function of T. We note that $Z(n, T)$ increases with decreasing T. We see first an increase in the relative pressure with increasing T, due to the decay of molecules, and then a second increase due to the ionization of the atoms. Both processes lead to an increase in the total particle density

5.2 Combinations Between Canonical and Grand-Canonical Density Expansions

5.2.1 Structure of the Lower Order Terms in the Density Expansion

The aim of this section is to close the gap between density and fugacity expansions. Let us first recall our basic findings in Chap. 4, obtained on the basis of canonical ensemble representations. We used the method of canonical cluster expansions in order to derive low density EOS and exact virial functions determining the first corrections to the Debye law. These expansions differ in some respects from the virial expansions of real gases. We recall to begin with the result of the virial expansion:

$$F = F_{\text{id}}^{\text{B}} - k_{\text{B}} T V \left[\frac{\kappa^3}{12\pi} + \sum_{ab} n_a n_b \tilde{B}_{ab}(\kappa) + \sum_{abc} n_a n_b n_c \tilde{B}_{abc}(\kappa) + \cdots \right] . \tag{5.28}$$

Here the tilde indicates that screened potentials were included in the virial coefficients. In Chap. 5, we proceeded as follows. We applied a screening procedure only to the lowest orders of the interaction parameter e^2. Higher orders in e^2, beginning with e^8, already have a decay rate of e^8/r^4 and do not need any screening. Taking into account the fact that screening refers to the low orders, we split the second virial coefficients into two parts:

$$\tilde{B}_{ab} = G'\tilde{B}_{ab} + G''\tilde{B}_{ab} , \qquad G''\tilde{B}_{ab} = G''B_{ab} , \tag{5.29}$$
$$G'\tilde{B}_{ab} = O(1) + O(e^2) + O(e^4) + O(e^6) , \qquad G''B_{ab} = O(e^8) + \cdots .$$

Let us summarize first the result for the higher orders:

$$G''\tilde{B}_{ab} = 2\pi^{3/2}\lambda_{ab}^3(1+\beta e_a e_b \kappa) \tag{5.30}$$
$$\times \sum_{m\geq 4} \frac{\xi_{ab}^m}{2^m \Gamma(m/2+1)} \left[\zeta(m-2) \pm \delta_{ab} \frac{(1-2^{2-m})}{2s_a+1} \zeta(m-1) \right] .$$

For charge and mass symmetric systems (all masses are equal and the reduced masses are just half of the mass), the odd contributions cancel. In the case of hydrogen, taking into account the fact that $\beta I = \xi^2/4$, we get for the sum

$$\sum_{ab} G''\tilde{B}_{ab} = 2\pi\sqrt{\pi} \sum_{k=2}^{\infty} \frac{\zeta(2k-2)\xi^{2k}}{2^{2k}k!} = 4\pi\sqrt{\pi}\lambda_{\text{i.e.}}^3 \sigma_{\text{BPL}}(T) . \tag{5.31}$$

We see once again that the BPL partition function appears in a natural way as part of the full virial function. The lower order contributions have to be calculated term by term. Summarizing the result, we have

$$G'\tilde{B}_{ab} = 2\pi\lambda_{ab}^3(1+\beta e_a e_b \kappa)\left[-\frac{1}{6}\xi_{ab} - \frac{\sqrt{\pi}}{8}\xi_{ab}^2 - \frac{1}{6}\left(\frac{1}{2}C + \ln 3 - \frac{1}{2}\right)\xi_{ab}^3 \right]$$
$$+ \frac{\pi}{3}(\beta e_a e_b)^3 \Big\{ (1+\beta e_a e_b \kappa)\ln(3\kappa\lambda_{ab}) - \beta e_a e_b \kappa\big[1 - \ln(4/3)\big] \Big\}$$
$$\pm \delta_{ab}(1+\beta e_a e_b \kappa)\left[\frac{\sqrt{\pi}}{4} + \frac{\xi_{ab}}{2} + \frac{\sqrt{\pi}(\ln 2)}{4}\xi_{ab}^2 + \frac{\pi^2}{72}\xi_{ab}^3 \right] , \tag{5.32}$$

where $C = 0.577$ is Euler's constant. The final results for the equation of state of hydrogen may be expressed in the form of a density expansion:

$$\beta p(\beta, n) = 2n - 2n(8\pi\beta e^2)^{3/2} n^{1/2} - 2n^2 \Lambda^3 \sigma_{\text{BPL}}(T)\left(1 + \frac{3}{2}\beta e^2 \kappa\right)$$
$$-2n^2 K^*(T) + L(T) n^{5/2} + \mathcal{O}(n^3 \log n) . \tag{5.33}$$

Here we have introduced a function $K^*(T)$ which depends only on the temperature. This transcendent function, which is defined by the formulae above, may be calculated as an infinite series in ξ using the expressions for B_{ab} given above. Tables for hydrogen are also available (Ebeling et al. 1976). However, since the full expression is a quite long and not easy to handle, we will give here only a simple approximation which is good enough numerically:

$$K^*(T) = -\frac{1}{2}\Lambda^3 \beta I - \frac{1}{8\sqrt{2}}\Lambda^3 . \tag{5.34}$$

This expression is exact with respect to the largest asymptotic term $\mathcal{O}(I/k_B T)$ and has been checked with the available tables Ebeling et al. 1976. As a technical remark we note, that the terms not contained in the approximation (5.34) are given mainly by the contributions of the odd powers in ξ to the pressure. We shall not give here a detailed discussion since these terms, which disappear in the mean mass approximation, give only a small contribution to the pressure in the region $10^4 < T < 10^5$ K. Note that the complete term of order $n^{5/2}$ has also been computed in several studies (DeWitt 1962; Alastuey and Perez 1992; DeWitt et al. 1995; Brown and Yaffe 2001; Alastuey et al. 2015). We will not discuss this matter here in more detail, since we shall only use approximations for $L(T)$. Note that this function does not contain bound state contributions. Here we are interested mainly in bound state effects and restrict ourselves to the asymptotic results for the case $\xi_{\text{i.e.}}^2 \gg 1$, which is equivalent to $T \ll I/k_B$. For lower temperatures $T \ll I/k_B$, we get the following result for the pressure of hydrogen at low densities:

$$\begin{aligned}\beta p(\beta, n) = n\left(1 + \frac{1}{8\sqrt{2}} n \Lambda_e^3 + \cdots\right) + n - n(8\pi\beta e^2)^{3/2}\sqrt{n} - n^2 \Lambda^3 (\beta I) \\ + n^2 \Lambda^3 \sigma_{\text{BPL}}(T)\left(1 + \frac{3}{2}\beta e^2 \kappa\right) + \cdots + \mathcal{O}(n^{5/2}) + \mathcal{O}(n^3 \log n)\,.\end{aligned} \tag{5.35}$$

We thus show that the Brillouin–Planck–Larkin partition function appears in a natural way as a convergent infinite sum, starting with terms of order e^8. Here we follow Larkin's asymptotic approach for low temperatures, which was extended later in (Ebeling 1974; Alastuey et al. 2008; Ballenegger 2012). However, we include here more terms than in earlier chapters.

Let us make one further remark about the foundation of the Brillouin–Planck–Larkin partition function. As already mentioned in Chaps. 1 and 4, there seems to be a never-ending discussion about the BPL partition function. We repeat briefly our point of view which is based in particular on the results in Chap. 4 and rederived here:

- The quantum-statistical result for the contribution $\mathcal{O}(n^2)$ to the pressure is exact. The result of several groups working with independent methods confirm these results (see, e.g., Alastuey et al. 2015).
- The BPL partition function is asymptotically in full agreement with the second virial coefficient. There is a small freedom in the special choice of a partition function, as long as the asymptotic is not violated. However, so far there is no good reason, at least for the case of low and moderate densities, why we should leave the BPL partition functions and make a different choice.
- In the present context, the PBL contribution arises just as part of a sum and the splitting is not free of arbitrariness. The strongest argument in favour of the BPL function is the fact that (in lower orders) screening does not influence terms $O(e^6)$ in the BPL function.

- Onsager's bookkeeping theorem tells us that, in a correct theory, any small change in the partition function does not matter, since the final results for any physical observable will remain the same (Ebeling et al. 2002).

Some of the criticisms of the BPL partition function are based on intuitive arguments, others on physical reasoning (Starostin et al. 2003; Khomkin and Mulenko 2003, Khomkin and Shumikhin 2008). According to our position in spite of several quite interesting alternative results, there is no urgent need to revise the quantum-statistical foundations given in the work of Vedenov and Larkin and in subsequent papers (see Alastuey et al. 2015; Ballenegger 2012).

With respect to the intuitive arguments, we only want to say that the formal reasons for having no terms of order $O(e^0)$ and $\mathcal{O}(e^4)$ in the Planck–Brillouin–Larkin formula is that these terms were already built into the formulation of the ideal terms which are $\mathcal{O}(e^0)$ and the screening terms $\mathcal{O}(e^3)$ and $\mathcal{O}(e^4)$, which arise formally by summing up an infinite series in $\mathcal{O}(e^n)$. This procedure consumes some terms of order $\mathcal{O}(e^3)$, and all contributions to the partition function of order $\mathcal{O}(e^4)$ which can no longer appear in the exact expression for the partition function, if double counting of terms is strictly avoided.

Another, formally equivalent argument is based on the fact that the compensation of the terms below the series limit and above the series limit is absolutely essential in this context (Kraeft et al. 1969; Kremp et al. 2005, see also Chap. 2). We want to stress again that there is no unique choice for the partition function for hydrogen. In our view several other choices are also possible (see Chap. 3 and Ebeling and Hilbert 2002, Ebeling, Hilbert and Krienke 2002). However, it is essential that all possible choices should be compatible with the quantum statistical formulae given above. This means that, among the many admissible versions of partition functions, the essential criterion for correctness is Onsager's bookkeeping rule. In other words the question is (see Ebeling and Hilbert 2002): Is there a correct counting of all contributions or not? Is there a check for double-counting or not? As a matter of fact, all partition functions which work without a BPL or equivalent subtraction of first order terms may involve the problem of double-counting of certain diagrams of order $\mathcal{O}(e^n)$.

We note also a quite interesting alternative approach to our problem developed by Vorobjev, Mulenko and Khomkin (2000), which is also based on fugacity expansions but arrives at different conclusions with respect to the contributions of the excited states (Vorobjov et al. 2000; Khomkin et al. 2004). These authors show that a consistent use of the nearest neighbor approximation (NNA) for the atomic partition function makes it possible to understand the experimentally observed effect of significant overestimation of the contribution made by Coulomb interaction to the thermodynamics of a partially ionized plasma, as given by the Debye screening theory (Khomkin and Mulenko 2003, Khomkin et al. 2004, Khomkin and Shumikhin 2008). We note that an approach which considers the excited states as individual species was also presented in (Ebeling et al. 1987). In such an approach, the contributions of excited states are essentially reduced due to their large effective volume.

5.2.2 *Structure of Higher Order Contributions*

Using the asymptotic expression for $K^*(T)$, which is asymptotically exact for $k_B T \ll I$, the pressure may be written in the alternative form

$$\beta p(\beta, n) = n\left(1 + \frac{1}{8\sqrt{2}} n \Lambda_e^3 + \cdots\right) + \left\{n - \frac{\kappa^3}{24\pi}\left[1 - \frac{3\sqrt{\pi}}{8}(\kappa\lambda) + \frac{3}{10}(\kappa\lambda)^2\right] - n^2 \Lambda^3 \sigma_{\mathrm{BPL}}(T)(1 - \beta e^2 \kappa)\right\} + \mathcal{O}(n^3 \log n) . \tag{5.36}$$

Note that we combined here one of the two contributions from $K^*(T)$ with the ideal electron pressure and the other one with the Debye law. Furthermore, as usual, we introduced the mean thermal wavelength $\lambda = \lambda_{\mathrm{i.e.}}$, which is the thermal quantum length (see Chaps. 3 and 4). Note that the present results are not fully complete with regard to order $n^{5/2}$ and in higher order corrections. However, this does not seem to be relevant for the bound state problem. A formally equivalent form of the previous equation which better shows the general structure is the following:

$$\beta p(\beta, n) = n\left(1 + \frac{1}{8\sqrt{2}} n \Lambda^3 + \cdots\right) + n - n\gamma\left[(1 - \beta e^2 \kappa + \cdots) + \cdots\right] - \frac{\kappa^3}{24\pi}\left(1 - \frac{3}{2}\gamma + \cdots\right)\left[1 - \frac{3\sqrt{\pi}}{8}(\kappa\lambda) + \frac{3}{10}(\kappa\lambda)^2 + \cdots\right] + \mathcal{O}(n^{3/2}) , \tag{5.37}$$

with

$$\gamma = \Lambda^3 \sigma_{\mathrm{BPL}}(T) = 8\pi^{3/2} \lambda_{\mathrm{i.e.}}^3 \sigma_{\mathrm{BPL}}(T) . \tag{5.38}$$

Summarizing, we may state that there exists an asymptotically exact expression for the pressure of hydrogen in the limit of small densities, which is given above and represented in Fig. 5.1.

This approach is compatible with the extended Eggert–Saha theory, and this encourages us to proceed in this direction. Looking at the figures for the relative pressure, we see that the screening effects and the corresponding BPL effects tend to lower the pressure. However, we also see that this lowering becomes too great at increasing densities. The structure of (5.37) suggests that the expressions in brackets may just be the first terms of some infinite series with respect to $n\Lambda^3$, κ, and γ. The first series is evidently nothing but the ideal electron pressure, which may easily be extended to an infinite series using the known expression for the ideal Fermi pressure (see Chap. 2). There is no doubt that all these terms appear in the full cluster series. According to our general philosophy, the strong decay of the pressure observed in second order density expansions (Fig. 5.1) is due to the fact that important higher order terms corresponding to some infinite series were omitted in the derivation based on the canonical ensemble. It is no easy task to get round this difficulty. We mention that Alastuey et al. (2008) succeeded in deriving a Saha-type expression by summing up an infinite series of terms. Here we follow a similar idea, but choose an alternative route using fugacity expansions. As shown earlier (see Sect. 5.1.2 and Chap. 2), fugacity-type expansions are more or less equivalent to mass-action laws.

5.3 Combined Density–Fugacity Expansions

5.3.1 Partial Summation of Density Series

The fugacity expansions given in the previous section cover only the case of small fugacities, which in fact also means small densities. An important point is that small fugacities are not necessarily connected with small densities. For example, in the region of bound states, the region of small fugacities is larger than the region of small densities. Exploring this idea we are currently searching for expansions which combine the advantages of both types of expansion (Ebeling, Kraeft and Röpke 2011).

Recall once again that expressions for screened fugacity series were first written down by Montroll and Ward (1958), and nearly at the same time by Vedenov and Larkin (1959). Explicit calculations were then carried out by many workers (Kraeft et al. 1986). In the previous section we discussed the semi-classical fugacity series which were given by (Bartsch et al. 1971; Rogers 1974, 1988; Kraeft et al. 1986).

Here we are interested in the bound state effects. Therefore we may simplify our calculations by using approximations appropriate to the region $T \leq 157\,000$ K where the bound states appear. Furthermore, in order to simplify the quite complicated evaluation of these diagrams, we restrict the theory to the region of non-degenerate plasmas. As can be seen from the above formulae, the difference between the masses of electrons and ions (protons) does not play a role in the region $|\xi_{ab}| \ll 1$, and the plasma behaves asymptotically like a system with equal masses and equal relative De Broglie wavelengths $\lambda = \lambda_{\mathrm{i.e.}}$. In the lowest approximation, we find the following results for low temperature hydrogen plasmas, valid up to order $O(z^{5/2})$ (Ebeling et al. 1976, 2012):

$$\beta p = z_{\mathrm{e}} + z_{\mathrm{i}} + \frac{\kappa_{\mathrm{g}}^3}{12\pi} R(\kappa_{\mathrm{g}}\lambda) + 8\pi z_{\mathrm{e}} z_{\mathrm{i}} \lambda^3 \exp\left(1 + \beta e^2 \kappa_{\mathrm{g}}\right) \sigma_{\mathrm{BPL}}(T) + \cdots . \tag{5.39}$$

Here the grand-canonical screening quantity is defined by

$$\kappa_{\mathrm{g}}^2 = 4\pi\beta(z_{\mathrm{e}} + z_{\mathrm{i}}) e^2 ,$$

and the quantum-statistical ring function $R(x)$ is the same as introduced in Sect. 4.4 and defined by

$$R(x) = 1 - \frac{3\sqrt{\pi}}{16}x + \frac{1}{10}x^2 + \cdots . \tag{5.40}$$

The relations between densities and fugacities are

$$\begin{aligned} n_{\mathrm{i}} &= z_{\mathrm{i}} \frac{\partial(\beta p)}{\partial z_{\mathrm{i}}} \\ &= z_{\mathrm{i}} + \frac{\kappa_{\mathrm{g}}^2}{16} \frac{\partial}{\partial \lambda}\left[\kappa_{\mathrm{g}} \lambda f(\kappa_{\mathrm{g}} \lambda)\right] + 4\pi z_{\mathrm{e}} z_{\mathrm{i}} \left(1 + \beta e^2 \kappa_{\mathrm{g}} + \cdots\right) \lambda^3 \sigma_{\mathrm{BPL}}(T) , \end{aligned} \tag{5.41}$$

$$\begin{aligned} n_{\mathrm{e}} &= z_{\mathrm{e}} \frac{\partial(\beta p)}{\partial z_{\mathrm{e}}} \\ &= z_{\mathrm{e}} + \frac{\kappa_{\mathrm{g}}^2}{16} \frac{\partial}{\partial \lambda}(\kappa_{\mathrm{g}} \lambda f(\kappa_{\mathrm{g}} \lambda)) + 4\pi z_{\mathrm{e}} z_{\mathrm{i}} \left(1 + \beta e^2 \kappa_{\mathrm{g}} + \cdots\right) \lambda^3 \sigma_{\mathrm{BPL}}(T) . \end{aligned} \tag{5.42}$$

We will show that, going by iterations from the fugacity variable to the density, this result is consistent.

In order to proceed with this complicated system of equations, we consider the special case of non-degenerate hydrogen at low temperatures $T \leq I/k_{\mathrm{B}}$ and satisfying $n\Lambda^3 \leq 1$. As shown above, in this region, the difference between the electron and ion masses does not matter, and we may assume $z_{\mathrm{e}} = z_{\mathrm{i}} = z$. In this approximation, (5.39), (5.41), and (5.42) read

$$\beta p = 2z + \frac{\kappa_{\mathrm{g}}^3}{12\pi} R(\kappa\lambda) + 8\pi z^2 \lambda^3 \sigma_{\mathrm{BPL}}(T) + \cdots , \tag{5.43}$$

$$n = z + \frac{\kappa_{\mathrm{g}}^2}{16\pi} \frac{\partial}{\partial \lambda}\left[\kappa_{\mathrm{g}} \lambda R(\kappa_{\mathrm{g}} \lambda)\right] + 8\pi z^2 \lambda^3 \sigma_{\mathrm{BPL}}(T) \cdots . \tag{5.44}$$

In order to represent the pressure in terms of densities, we neglect the bound state contributions in a first step. Expressing first the fugacities in terms of the densities, we find

$$z = n \exp\left[-\frac{\beta e^2 \kappa}{2} G(\kappa\lambda)\right] , \tag{5.45}$$

where we have once again used the ring function

$$G(x) = 1 - \frac{\sqrt{\pi}}{4}x + \frac{1}{6}x^2 - \cdots . \tag{5.46}$$

By excluding the fugacities from the series step by step, we get the representation

$$\beta p = 2n - \frac{\kappa^3}{24\pi}\phi(\kappa\lambda) \,. \tag{5.47}$$

Here $\phi(x)$ is a transcendent function related to $R(x)$ and $R'(x)$, also already known from Chap. 4, namely the so-called pressure ring function. These functions were introduced as an analogue of the classical Debye–Hückel functions. Here the quantum ring functions are given in a kind of quantum Debye–Hückel approximation. The first terms of the series expansion read

$$\phi(x) = 1 - \frac{3\sqrt{\pi}}{8}x + \frac{3}{10}x^2 - \cdots \,. \tag{5.48}$$

More complete representations of this transcendent function may be obtained by methods similar to those described in Chap. 4 (see also Kraeft et al. 1986). For example, we may derive the infinite series

$$\phi(x) = 1 - \frac{\sqrt{\pi}}{3}\sum_{k=1}^{\infty}\frac{(k+1)(k+3)x^{2k-1}}{2^k k!} + \frac{1}{3}\sum_{k=1}^{\infty}\frac{(k+1)(k+3)x^{2k}}{2^k(2k-1)!!} \,. \tag{5.49}$$

The asymptotic behavior of this function is given by $\phi(x) = 3/x^2$. This suggests the convenient Padé representations, which may be used in numerical calculations:

$$\phi(x) \approx \frac{24}{24 + 9\sqrt{\pi}x + 8x^2} \,, \qquad G(x) \approx \frac{1}{1 + (\sqrt{\pi}/4)x} \,. \tag{5.50}$$

The saturating behavior of the nonlinear function $\Phi(x)$ is compared with its linear approximation in Fig. 5.3.

Let us summarize the results we have obtained so far. By means of grand-canonical methods we succeeded in obtaining some infinite series in the screening parameter κ, instead of just the first linear and quadratic terms, as we found for the density series [see (5.37)]. This series in κ leads to a saturating function (see Fig. 5.3) and very much improves behavior at larger densities, as we will show. In our representation (5.37), there is also a second series expansion in the density parameter γ. To sum up this series in γ, we proceed in a similar way. In order to understand the structure of this second series, we first study first the bound state contributions, neglecting those from the ring term and from the terms due to degeneracy. This yields a simple quadratic equation

$$n = z + 8\pi z^2 \lambda^3 \sigma_{\rm BPL}(T) \,. \tag{5.51}$$

This formula already shows that the fugacities of the electrons and the protons should be rather small in the bound state region where the partition function is large. The quadratic equation is easily solved and, after eliminating the fugacity z [or alterna-

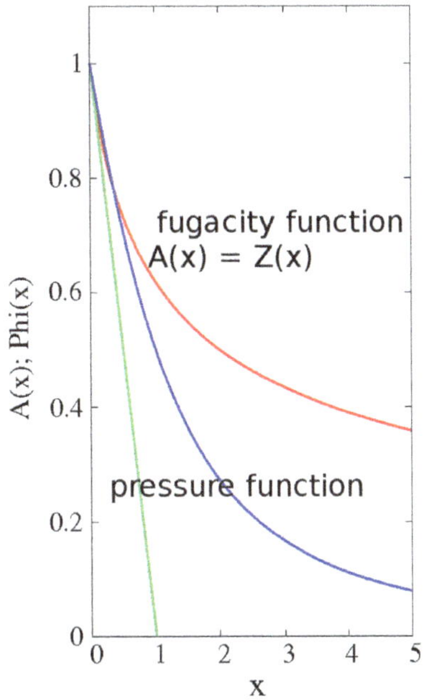

Fig. 5.3 The nonlinear ring function $\phi(x)$, which determines the contribution of screening to the pressure (*blue curve*). We also show the nonlinear fugacity $A(x) = Z(x)$ (*red curve*), which determines the contribution of the bound states to the pressure, and the linear approximation, which gives only the lowest order to the second virial coefficient (*thin curve*)

tively using $Z(x)$, sometimes denoted by $A(x)$], we find

$$z^0 = nZ(\gamma) , \quad Z(x) = \frac{1}{2x}(\sqrt{1+4x} - 1) , \quad \gamma = 8\pi n \lambda^3 \sigma_{\mathrm{BPL}}(T) . \tag{5.52}$$

We recall the series expansion and large x asymptotic of the nonlinear function $Z(x)$ given in last section. The saturating behavior of the nonlinear function $Z(x)$ is compared with its linear approximation in Fig. 5.3. This shows that, in the region of large partition functions, the fugacities disappear as

$$z^0 = \frac{1}{\sqrt{8\pi n \lambda^3 \sigma_{\mathrm{BPL}}(T)}} . \tag{5.53}$$

This is important for understanding why, in the region of bound states, the fugacity series has better convergence than the density series. The better convergence is due to the fact that the fugacities which include the bound state contributions disappear in the region where the bound state contribution is large. This is precisely the property we need to obtain good convergence in the region of bound states.

5.3.2 *Extended Representations of the EOS by Nonlinear Density Functions*

Here we consider the idea that any fugacity representation corresponds to an infinite series in the density and vice versa. Knowing which terms in the fugacity are needed to describe the formation of bound states, we now look for the corresponding density series and combine them with other terms in the density which are needed to describe short- and long-range forces.

If we understand the approximation given in the last subsection as the first step of an iteration, then the EOS is given in the zeroth order by the fugacity z^0 and the pressure p^0, and using a fugacity function $Z(x)$, we find

$$z^0 = nZ(\gamma) = nZ\big(8\pi n\lambda^3\sigma_{\mathrm{BPL}}(T)\big) , \tag{5.54}$$

$$\beta p^0 = n + z^0 = n\Big[1 + Z\big(8\pi z^2\lambda^3\sigma_{\mathrm{BPL}}(T)\big)\Big] . \tag{5.55}$$

This zeroth approximation just reproduces (in a new form) the known elementary EOS of an H-plasma already obtained in Sect. 5.1:

$$\beta p^0 = n\big[1 + Z(nK(T)\big] . \tag{5.56}$$

This elementary approximation already shows us that the Saha function $Z(\gamma)$ appears in the pressure in a quite natural way, arising from the fugacity expansion. We have tacitly introduced elements of the chemical picture without explicitly using the notation of chemical species. The new 'pseudo-chemical representation' includes the nonlinear Saha function $Z(x)$ which decays at large densities (see Fig. 5.3). It contains all terms in the density up to n^2, as well as several higher order terms in the density contained in the nonlinear $Z(\gamma)$, which corresponds to the Saha function defined by (5.52). We need this kind of function to reproduce an ideal Saha-type behavior. We will now discuss the problem of how to generalize this approach.

In a subsequent iteration, we include the contributions from Debye screening and, on the basis of the results derived in the previous section, we find the following expression for the fugacities:

$$z^1 = nZ\Big\{8\pi n\lambda^3 \exp\big[-\beta e^2\kappa G(\kappa\lambda)\big]\sigma_{\mathrm{BPL}}(T)\Big\} . \tag{5.57}$$

For the pressure, including the full electronic Fermi pressure consisting of Fermi–Dirac and Hartree–Fock contributions for electrons and Boltzmann approximations for the ions (see Chap. 3), we find a relatively simple formula which contains the three nonlinear functions discussed already in the foregoing section. In the first approximation, the pressure is expressed in the form

$$\beta p = p_{\mathrm{e}}^{\mathrm{FD}} + p_{\mathrm{e}}^{\mathrm{HF}} + p_{\mathrm{i}}^{\mathrm{B}} + nZ(\gamma^*) - \frac{\kappa^{*3/2}}{24\pi}\phi(\kappa^*\lambda) , \quad \lambda = \lambda_{\mathrm{i.e.}} . \tag{5.58}$$

Here we use the definitions of regularized values of the Saha parameter and the screening parameter

$$\gamma^* = 8\pi n\lambda^3 \exp\left[- \beta e^2 \kappa G(\kappa^* \lambda)\right]\sigma_{\mathrm{BPL}}(T) , \quad \kappa^* = \kappa Z^{1/2}(\gamma^*) . \qquad (5.59)$$

Equations (5.58) and (5.59) yield a closed and relatively simple formula for the pressure. One essential assumption is that all electrons (not only the free electrons) contribute to the ideal Fermi pressure.

Since the fugacity series contains more terms than the density series, we may expect some of the difficulties connected with the bad convergence of density expansions to be avoided, including the significant decrease in the pressure with increasing density or decreasing temperature shown in Fig. 5.1. Indeed, a representation of the pressure corresponding to a fugacity expansion shows a more reasonable behavior with increasing density, one which is much closer to the Saha-type behavior. As demonstrated by the curves shown in Fig. 5.4, the pressure calculated from fugacity expansions and improvements which at least in part include fugacity terms, goes to saturation at higher densities. The divergence which arises in the pure density expansions disappears. This is due to the fact that the fugacity is not fixed, but decreases with increasing values of the partition function, and hence limits growth. Evidently, the fugacity expansions provide a more correct description of the bound state contributions, as demonstrated in Figs. 5.4 and 5.5.

This result for the pressure according to the combined density–fugacity representation (5.58), which includes all contributions up to second order in the fugacity is shown for three temperatures in Fig. 5.5, which also contains a comparison with results obtained from a chemical picture of the plasma (see Beule et al. 1999, 2001, and Chap. 6).

The good convergence of the new expression, including the nonlinear functions $Z(x)$ and $\phi(x)$, based on the fugacity series is explained by the fact that the fugacities of the electrons and the protons are rather small in the bound state region. We notice once again that this result comes from an extension of the density expansion including all quadratic terms from the fugacity virial expansions and the full Fermi pressure of the electrons. The overall behavior of the new representation is much better than that of pure density or pure fugacity representations. In this way we may conclude that the most appropriate description of Coulomb systems is by density expansions including some contributions from the fugacity expansions. These additional contributions allow one to control the exponential increase in the second virial coefficient in the bound state region. The physics behind this is that the fugacity expansions give a good description of the saturable forces between bound states in Coulomb systems, while the additive long range Coulomb forces and screening effects are better represented by density series.

We note that the new theory based on (5.58) and (5.59) is consistent with the Saha theory, and in particular also with the Saha–Debye–Hückel theory developed in Chap. 3. Comparing the curves based on the extended theory, viz., Figs. 5.4 and 5.5, with Fig. 5.1, we see that the overall behavior of the new extended density representations is much better than that of the pure density or pure fugacity representations.

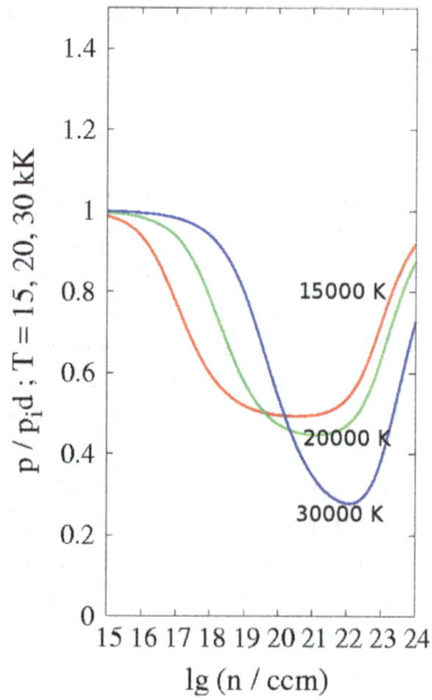

Fig. 5.4 Pressure relative to the ideal Fermi pressure as a function of the density in cm^{-3} on a log-scale, calculated from extended density representations, including the ring sum $\phi(x)$ and the Saha function $A(x)$, for the temperatures $T = 15\,000$ K (*lower left curve in red*), 20000 K (*curve in green*), 30000 K (*curve with largest amplitude in blue*). For low and high densities, the behavior corresponds to an ideal Fermi gas

Furthermore, we may state that the new extended analytical theory is, up to the density $n \approx 10^{21}$ cm^{-3}, in quite good agreement with earlier calculations based on the chemical picture and quite complex numerical codes (see Beule et al. 1999, 2001, and Chap. 7). Taking the more advanced PACH theory of Beule et al. (1999, 2001), which includes H_2 molecules, as a kind of standard for comparison, we may conclude that the simple analytical theory developed here fails at the highest densities, where molecules come into play and lead to deeper pressure minima (see Fig. 5.5 and Beule et al. 1999). At lower densities, however, we find good agreement with the Saha theory, and in particular with the extended Saha–Debye–Hückel theory developed in Chap. 4. Nevertheless, we note the following structural differences with the Saha theory:

- The exponent $\exp(\beta I)$ in the Saha equation is replaced by the PBL partition function in the quantum-statistical formula (5.58).
- The first Debye–Hückel screening function linear in κ is replaced here by a nonlinear ring function $\phi(x)$ describing the quantum statistical ring sums in the grand canonical expansion to a good approximation.
- The linear second virial coefficient is replaced by a nonlinear ladder function stemming from the representation in the grand ensemble.

We note that the grand canonical ensemble plays an essential role in our derivation, and also that the formula (5.58) corresponding to a second iteration is in some sense incomplete, e.g., in higher iterations, we would expect more κ terms to be replaced and more terms corresponding to a mass-action law to appear.

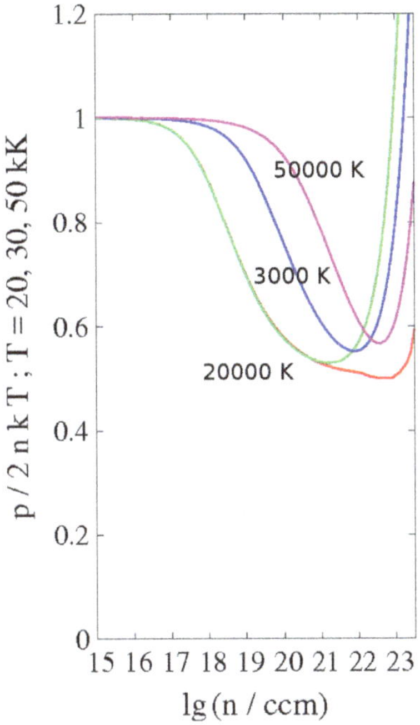

Fig. 5.5 Equation of state (EOS) of hydrogen represented by the ratio of the total pressure to the Boltzmann pressure. The pressure is calculated from the extended density representation, including the electronic Fermi pressure and the nonlinear ring and ladder functions $\phi(x)$, $A(x)$. We show the pressure $p/2nk_BT$ at constant temperatures $T = 20\,000, 30\,000, 50\,000$ K as a function of the density on a log scale. n is the proton density. The curve for the temperature $T = 20\,000$ K (*green*) is compared with the curve for an earlier calculation (*red*) based on the chemical picture and the Padé approximation (Beule et al. 1999)

We stress that several of the terms beyond n^2 and z^2 are based on extrapolations which still require further confirmation. However, in the low temperature region $T < I/k_B$ and non-degenerate plasmas, our rather simple formulae give a rather good behavior and a good description of the transition from low densities to the valley of bound states. On the other hand, several physical effects such as plasma phase transitions (see Chap. 4) are not yet described by the present approach. Evidently, this effect appears only in higher order terms or after transition to some chemical picture (Ebeling et al. 1976, 1985, 2002). However, this is not our aim here, and the reader is referred to Chap. 6.

In any case, the overall agreement of mixed representations with chemical descriptions is quite reasonable, the deviations increasing only at large densities beyond $n \approx 10^{21}$ cm^{-3}. The largest deviations appear near the minimum of the relative pressure for hydrogen at 20 000 K (the green line is the present theory and the red line, the lowest of the curves, comes from PACH) and may be interpreted as due to the formation of molecules which have been neglected so far in this section. In principle, we can include molecules by including contributions from the fourth order virial

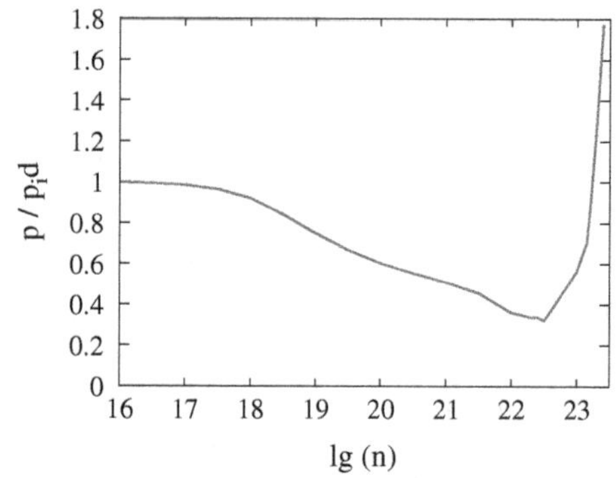

Fig. 5.6 Pressure of Hydrogen plasma including atoms and molecules on the ideal gas $2nk_BT$ ($n-$ total particle density). We include here the Fermi pressure, the ring sum $\phi(x)$ and the extended Saha function $Z(x, y)$ which takes into account basic contributions of the 4th virial coefficient for the temperature $T = 20000K$ in dependence on the density (cm^{-3}, log-scale))

coefficient, as described in an earlier section. This leads to closed equations of the form

$$\beta p = p_e^{FD} + p_e^{HF} + p_i^B - \frac{(\kappa^*)^{3/2}}{24\pi}\phi(\kappa^*\lambda)nZ(x^*, y^*) - \frac{1}{2}ny^*Z(x^*, y^*)^2 \,. \tag{5.60}$$

Here we have used the definitions

$$x^* = 8\pi n\lambda_{\text{i.e.}}^3 \exp\left[-\beta e^2\kappa^* G(\kappa^*\lambda_{\text{i.e.}})\right]\sigma_H \,, \tag{5.61}$$

$$y^* = 16\pi n^3\lambda_{HH} \exp\left[-2\beta e^2\kappa^* G(\kappa^*\lambda_{\text{i.e.}})\right]\sigma_{H2} \,. \tag{5.62}$$

Examining Fig. 5.6, we see a little shoulder of the relative pressure at $n \sim 10^{20}$ cm^{-3}, which reflects the formation of atoms, and a minimum at $n \sim 10^{22.5}$ cm^{-3} due to the formation of molecules. For still higher densities, all bound states break down due to the dominance of Fermi pressure, which increases steeply. However, in this region energy level shifts, which are not included, may start to play a role. These effects will be discussed in forthcoming sections.

Let us draw some conclusions. We have shown that density as well as fugacity expansions have both advantages and disadvantages:

- The density expansion describes screening effects well, but it fails to cope with the diverging contributions from the screening terms and from the BPL partition function.
- The fugacity expansions are smoother. They correspond to an infinite density series including the partition function σ. Short range interactions are not described properly. If σ is large, then the fugacity goes to zero, which guarantees finite

contributions to the pressure even at large densities. This is true for the screening contributions and for the bound state contributions.

- The combined density–fugacity expansion including nonlinear density functions given here provides useful smooth representations of the pressure. In this approach the most relevant effects are mapped to several nonlinear density functions such as $\Phi(x)$ and $Z(x, y)$. These functions have been derived here by extended density expansions which contain additional contributions corresponding to important terms in the fugacity expansions. This procedure yields Saha-like terms and chemical effects.

We may conclude that the mixed expansions from (Ebeling, Kraeft and Röpke 2012) combine the positive features of density and fugacity expansions, while avoiding most of the negative features. In order to demonstrate this we started here from the density expansions and used the fugacity expansions, to a large extent only to find the right continuation of the infinite series with respect to the γ parameter and the $\kappa\lambda$ parameter. In order to describe the transition to full ionization in the degenerate region as well, additional effects have to be taken into account, in particular bound state shifts and loss of symmetry between electrons. The disappearance of the bound states with increasing density will be discussed from a different point of view in one of the following sections (see also Kraeft et al. 1986; Ebeling et al. 2009; 2012).

Summarizing the results of this section, we have shown that transitions from full to partial ionization and back to full ionization at high densities occur in the temperature range 10^4–10^5 K, these being due to the formation of H atoms, while H_2 molecules can also be described in the physical picture. Ring and ladder diagrams in higher order approximations are taken into account, together with certain combinations of these basic diagrams. The main picture showing how the formation of bound states depends on the density at a fixed temperature is illustrated in Figs. 5.4, 5.5, and 5.6. At least the principal effects are correctly described, such as the formation of atoms and molecules and their destruction at very high densities.

Figure 5.5 compares the theory given here for $T = 20\,000$ K, $T = 50\,000$ K with the numerical results obtained in an advanced 'chemical picture' by minimization of the free energy (Beule et al. 1999, 2001). By chemical picture we mean here the Saha equation along with any description which starts from an expression for the free energy depending on the density of free charges, atoms, and molecules (Beule et al. 1999; Ebeling et al. 2005). Except for the range of very high densities $n \sim 10^{23}$, the agreement is reasonable.

On the basis of these observations, we recommend the use of the present alternative to the relative complicated chemical descriptions. The weakest point of the new and rather simple picture is at very high densities around the minimum of the relative pressure. We reiterate that the new approach is based on the use of fugacity functions stemming from summations of partial fugacity series and leading to density representations. This approach yields quite simple descriptions of bound state formation without using mass action law equations. From our long experience with chemical descriptions, our point of view is the following. In general, chemical descriptions lead to the highest accuracy in the thermodynamic functions and the EOS since they are based on minimization procedures. However, since chemical descriptions based on sets of coupled mass action laws are highly nonlinear systems, the solutions are never simple, and they are sometimes even numerically unstable or even chaotic.

In such cases it makes sense to investigate alternative procedures, such as our mixed density–fugacity representations. These take the form of density representations extended by certain nonlinear functions, which are sufficiently simple and always yield unique and stable results for the thermodynamic functions. They can explain several of the most important physical effects, like the lowering of the relative pressure by Coulomb interactions and bound state formation, and the final transition, beyond the valley of interactions and bound states, to ideal behaviour at very high densities. Due to its simplicity, the present simple approach by nonlinear density functions may be used to estimate the EOS on a pocket computer.

We shall now investigate a different physical effect, neglected so far: the fact that bound states in a plasma are influenced by interactions. In experiments, one observes shifts and broadening of the lines (Griem 1974; Kudrin 1974; Kilimann and Ebeling 1990).

5.4 Nonideality Effects in the Energy Spectrum

5.4.1 Energy Shifts in Effective Wave Equations

The study of level shifts and broadening due to interaction effects is of great significance for the investigation of plasmas (see, e.g., Ecker and Weizel 1956; Griem 1974; Kudrin 1974; Dharma-Wardana and Perrot 1992; Seidel et al. 1995; Arndt et al. 1996; Redmer et al. 2010). Our problem is to investigate the change in the two-particle energy levels due to interaction effects with a surrounding medium. Due to this effect the bound states may even disappear completely. The consequences of the Stark effect for line shift and broadening were first studied in early papers of Inglis and Teller, which we already discussed in Chap. 1.

A different physical effect which may also lead to the disappearance of levels is connected with the influence of screening at finite density. This effect was investigated in detail by Nevill Mott (1905–1996) who showed also that this effect plays an essential role in semiconductors (Mott 1961). The theory of the so-called Mott effect, the disappearance of energy levels due to screening effects, is in detail rather complicated (Kremp et al. 2005; Redmer et al. 2010), and is connected with the solution of the Bethe–Salpeter equation. The physical idea underlying this theory is, however, easy to understand (Ecker and Weizel 1956; Ebeling et al. 1976, 1977, 1985). The energy levels in a plasma are themselves not so much influenced by the density, so in first approximation, they are still the same as the unperturbed levels, i.e., $E_s' \approx -I/s^2$ for hydrogen. In the next approximation the bound state levels correspond to the so-called Ecker–Weizel potential, which has unshifted levels

$$V^{\mathrm{EW}}(r) = -\frac{e^2}{r}\Big[\exp(-r/r_{\mathrm{D}}) + 1/r_{\mathrm{D}}\Big] . \tag{5.63}$$

However, what comes down with the density is the series limit, given by the chemical potential of the charges, which is to a first approximation $E_\infty = 0 - e^2/r_{\mathrm{D}}$, where r_{D} is the Debye radius defined above. We see that, according to this estimate, the ground state level disappears at $r_{\mathrm{D}} < 2a_{\mathrm{B}}$. Later studies gave a more precise value of the Debye radius at which the levels disappear, viz., $r_{\mathrm{D}} < 0.84a_{\mathrm{B}}$. This idea goes back originally to Ecker and Weizel (1956) and was worked out by Theimer and Kepple (1970) and by Ebeling, Kraeft and Kremp (1977, 1985), and in more detail by Zimmermann (1987) and by Kremp et al. (2005). These authors developed the general idea that the continuum comes down with the density, so that the ionization energies are squeezed. Note that Mott-like effects may be of different origin (Redmer et al. 2010 and references therein). The disappearance of the bound states with increasing density has been discussed in many papers (see, e.g., Kraeft et al. 1986; Kremp et al. 2005; Ebeling et al. 2009; Zimmermann et al. 1978; Redmer et al. 2010).

Energy Shifts and Bethe–Salpeter Equations

The Bethe–Salpeter equation is named after Hans Bethe and Edwin Salpeter. This generalization of the Schrödinger equation describes the bound states of a two-body (or two-particle) system embedded in a medium. It is difficult to trace back to the origin of this fundamental equation. Evidently, it was first published in 1950 at the end of a paper by Yoichiro Nambu, but without derivation, and only then formulated by Bethe and Salpeter (1951). Due to its generality and its application in many branches of theoretical physics, the Bethe–Salpeter equation (it may be more correct to say the Nambu–Bethe–Salpeter-" equation) now appears in many different forms. The Bethe–Salpeter theory was used in particular for the development of a more advanced theory of the energy levels in plasmas (Zimmermann et al. 1978; Kremp et al. 2005; Ebeling et al. 1977, 1985, 2009, 2010, 2012; Redmer et al. 2010). This concept is currently found among other fruitful applications in the theory of nuclear bound states.

The bound states of pairs of particles play a major role in the properties of gases and plasmas. They are responsible for the formation of atoms, molecules, clusters, etc. Let us first consider the treatment of a pair bound state when the pair is isolated from its surroundings. In this way we have a many-body problem with $N = 2$ and a central force potential $U = U(r)$, with $r = |\mathbf{r}_1 - \mathbf{r}_2|$. The wave function is $\Psi(q, t) = \Psi(\mathbf{r}_1, \sigma_1, \mathbf{r}_2, \sigma_2, t)$, and the Schrödinger equation in 6-dimensional space reads

$$\mathrm{i}\hbar\frac{\partial\Psi}{\partial t} = \left(-\frac{\hbar^2}{2m_1}\Delta_1 - \frac{\hbar^2}{2m_2}\Delta_2\right)\Psi + U(r)\Psi . \tag{5.64}$$

Using the standard transformation to relative coordinates we find the unperturbed states. The energy eigenfunctions are given by Legendre polynomials. The ground state energy and the corresponding eigenfunction are

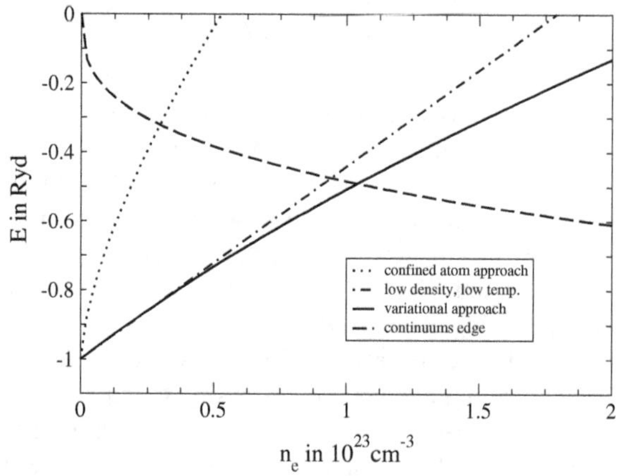

Fig. 5.7 Several first approximations to the density dependence of the effective ground state energy of hydrogen. We represent the low-temperature, low-density estimate according to (5.77) in comparison with the confined atom estimate (*dotted curve on the left*). The third curve (*continuous curve on the right*) corresponds to the variational approach. We have also shown the lowering of the continuum edge (*decreasing dashed curve*) according to (5.79). After Redmer et al. (2010)

$$E_1^0 = -\frac{\mu e^4}{2\hbar^2} , \qquad \psi_0(r) = C \exp\left(-\frac{r}{2a_B}\right) . \tag{5.65}$$

In the following, for convenience, we will use Rydberg units with $m_e/m_p \ll 1, m_e = 1/2, \hbar = 1$, and $e^2/4\pi\epsilon_0 = 2$, so that the binding energies of the isolated hydrogen atom are $E_{P,n}^0 = P^2/2m_p + E_n^0$ with $E_n^0 = -1/n^2$.

Effective Schrödinger Equation of Coulomb Pairs

After separating the center of mass motion, the Schrödinger equation of the electron–proton system in the momentum representation reads

$$p^2\phi_n(p) - \sum_q V(q)\phi_n(p+q) = E_n^0\phi_n(p) . \tag{5.66}$$

With the Coulomb interaction $V(q) = 8\pi/q^2$, the normalized wave function for the ground state $n = 1$ is

$$\phi_1(p) = \frac{8\sqrt{\pi}}{(1+p^2)^2} , \qquad \sum_p |\phi_1(p)|^2 = \int \frac{d^3p}{(2\pi)^3}|\phi_1(p)|^2 = 1 , \tag{5.67}$$

with the normalization conditions.

Imbedding the hydrogen atom in a plasma environment, the additional interactions with the medium can be treated in a quantum statistical approach by introducing concepts such as the self-energy, dynamical screening, and the spectral function. For these many-body quantities, special approximations can be performed which reflect different processes in the plasma. In this way, an effective wave equation is obtained (Kraeft et al. 1986):

$$p^2\psi_n(p) - \sum_q V(q)\psi_n(p+q) + \sum_q H^{\rm pl}(q)\psi_n(p+q) = E_n\psi_n(p) \;, \quad (5.68)$$

assuming the adiabatic limit $m_{\rm e}/m_{\rm p} \ll 1$, and neglecting the center of mass motion **P**. In general, the plasma Hamiltonian $H^{\rm pl}(q)$ will also depend on **P** and the energy, if dynamical and retardation effects are taken into account. The plasma Hamiltonian will shift the energy eigenvalues $E_n = E_n^0 + \Delta E_n$ and modify the wave functions $\psi_n(p)$. With increasing density, the influence of the plasma increases and the binding energies may merge into the continuum so that bound states disappear. This breakup of bound states is called the Mott effect and has important consequences for the macroscopic properties of the plasma.

In the framework of perturbation theory, various effects can be considered to contribute additively to the plasma Hamiltonian:

$$\begin{aligned} H^{\rm pl} = H^{\rm Hartree} + H^{\rm Fock} + H^{\rm Pauli} + H^{\rm MW} + H^{\rm Debye} \\ + H^{\rm polpot} + H^{\rm vdW} + \cdots \;. \end{aligned} \quad (5.69)$$

The first three contributions are of first order with respect to the interaction and determine the mean-field approximation, which is instantaneous in time and contains no dynamic contributions. These contributions are of similar structure and have to be considered simultaneously in order to find consistent approximations. Note that the electron–electron interaction is repulsive $V(q)$ and the electron–ion interaction is attractive $-V(q)$. The terms can be summarized as follows:

- The Hartree term contains the factor 2 due to spin summation (for abbreviation, only the momentum is given in the Fermi distribution function). However, this contribution vanishes for neutral plasmas where the densities $n_c = \sum_{p'} 2f_c(p')$ of the charge carriers $c = {\rm e, i}$ compensate one another.
- The Fock term as well as the Pauli blocking term describe exchange effects and refer only to the interaction between particles of identical species and spin.
- The origin of the Pauli blocking term is phase space occupation by free electrons according to the distribution function $f_{\rm e}(p)$. This phase space cannot be used to form a bound state. Thus the interaction of the electron with the ion is blocked if the final state is already occupied by a free electron of the same spin orientation.
- The following two terms of the plasma Hamiltonian are the Montroll–Ward term giving the dynamical screening of the interaction in the self-energy, and the (dynamical) screening of the interaction between the bound particles. These contributions are related to the polarization function and are of particular interest for plasmas due to the long-range character of the Coulomb interaction. In a consistent description, the terms should be treated simultaneously. A more detailed discus-

sion has been given in (Röpke et al. 1978, 1980; Kraeft et al. 1986; Kremp et al. 2005).

- The remaining contributions to the plasma Hamiltonian are of at least second order with respect to the interaction, e.g., the polarization potential, describing the interaction of a bound state with free charge carriers, as well as the van der Waals interaction, describing the interaction of a bound state with another bound state (Kraeft et al. 1986, Kremp et al. 2005; Redmer et al. 2010).

Evaluation of the Bound State Energy Shift

We focus on the influence of the mean-field contributions to the effective Schrödinger equation of pairs:

$$E_n\psi_n(p) = p^2\psi_n(p) - \sum_q V(q)\psi_n(p+q) \tag{5.70}$$
$$+\sum_q V(q)\Big[f_e(p)\psi_n(p+q) - f_e(p+q)\psi_n(p)\Big] .$$

For low densities, the perturbation due to the plasma Hamiltonian is small and the shift in the energy eigenvalues is obtained with the unperturbed wave functions as

$$\begin{aligned} E_n - E_n^0 &= \Delta E_n^{\text{Fock}} + \Delta E_n^{\text{Pauli}} \\ &= -\sum_{p,p'} \phi_n^*(p) V(\mathbf{p}' - \mathbf{p}) f_e(p')\phi_n(p) \\ &\quad + \sum_{p,p'} \phi_n^*(p) V(\mathbf{p}' - \mathbf{p}) f_e(p)\phi_n(p') . \end{aligned} \tag{5.71}$$

Here, inserting the Schrödinger equation, the Pauli blocking term can be rewritten as

$$\Delta E_n^{\text{Pauli}} = \sum_p \phi_n^*(p)(p^2 - E_n^0) f_e(p)\phi_n(p) . \tag{5.72}$$

A simple expression is found in the low-density limit, where the Fermi distribution with normalization $\sum_p f_e(p) = n_e/2$ is concentrated near $p = 0$. In the zero temperature limit, we have a Fermi sphere with Fermi momentum $p_F = (3\pi^2 n_e)^{1/3} \ll 1$. The energy shift in the ground state $\phi_1(p)$ given in (5.67) is

$$\Delta E_1^{\text{Pauli}} = \frac{1}{2} n_e(-E_1^0)|\phi_1(0)|^2 = 32\pi n_e . \tag{5.73}$$

Here we are using Rydberg units, where the dimensionless density of free electrons n_e is in units of a_B^3 and the ground state energy is $E_1^0 = -1$. The Pauli blocking shift is linear in the density. In the general case of arbitrary temperatures, we have to introduce the Fermi function and the integrals which arise cannot be carried out in a simple way:

$$\Delta E_1^{\text{Pauli}} = \frac{32}{\pi} \int_0^\infty \mathrm{d}p \frac{p^2}{(1+p^2)^3} f_{\mathrm{e}}(p) \,. \tag{5.74}$$

If we approximate the Fermi distribution by a Boltzmann distribution normalized to the same density, we obtain an analytical expression

$$\Delta E_1^{\text{Pauli}} \approx 32\pi n_{\mathrm{e}} G(T) \,, \qquad G(x) \approx \frac{1}{1+77/16x} \,, \tag{5.75}$$

where the function $G(T)$ expressing the temperature dependence is given in a simple approximation (Redmer et al., 2010). Here the Rydberg units of temperature are given by $\mathrm{Ry}/k_{\mathrm{B}} = 157\,886$ K, i.e., by the ionization temperature of hydrogen. In the asymptotic approximation $T \ll 1$, i.e., temperatures below 20 000 K where $G(x) \approx 1$, this leads back to the zero temperature expression for the shift given by (5.75).

Similar expressions can be given for the Fock term. In the low temperature, low density limit we get

$$\Delta E_1^{\text{Fock}} = -128 n_{\mathrm{e}} \int_0^\infty \mathrm{d}p \frac{1}{(1+p^2)^4} = -20\pi n_{\mathrm{e}} \,. \tag{5.76}$$

This partially compensates the Pauli shift, so that the total shift is

$$\Delta E_1^{\text{Fock}} + \Delta E_1^{\text{Pauli}} = 12\pi n_{\mathrm{e}} \,. \tag{5.77}$$

This is shown by the continuous curve in Fig. 5.7, indicating a rather steep shift of the bound state energy. Taking into account the temperature effects, the Fock shift is given by

$$\Delta E_1^{\text{Fock}} = -\frac{64}{\pi^2} \int_0^\infty \frac{p\,\mathrm{d}p}{(1+p^2)^4} \int_0^\infty \mathrm{d}k\, k \ln \frac{(p+k)}{|p-k|} f_{\mathrm{e}}(k) \,, \tag{5.78}$$

where $f_{\mathrm{e}}(k)$ is the Fermi distribution, which may be replaced by the Boltzmann distribution in a first approximation Due to phase space occupation, the bound state energy is shifted and may merge with the continuum of scattering states, indicating the breakup of bound states. Considering the continuum part of the spectrum describing scattering states in (5.71), only the Fock shift contributes to the energy shift. The lowest energy in the continuum occurs at $p = 0$ and is shifted by

$$\Delta E^{\text{Fock}}(p=0) = -\sum_q V(q) f_{\mathrm{e}}(q) = -4\frac{p_{\mathrm{F}}}{\pi} = -4\left(\frac{3n_{\mathrm{e}}}{\pi}\right)^{1/3} .$$

However, the two-particle continuum state can only be created at the Fermi momentum, since all states below that are occupied. Thus the continuum of scattering states, where the bound states become free, begins at p_{F}, where we have, in the zero temperature limit, the Fock shift

$$\Delta E^{\mathrm{Fock}}(p_{\mathrm{F}}) = -\sum_p V(\mathbf{p}-\mathbf{p}_{\mathrm{F}}) f_{\mathrm{e}}(p) = -2\frac{p_{\mathrm{F}}}{\pi} = -2\left(\frac{3n_{\mathrm{e}}}{\pi}\right)^{1/3} , \tag{5.79}$$

shown also in Fig. 5.7. Extrapolating these low-density results to higher densities, the ground state disappears at a density which corresponds in the first approximation to $n_{\mathrm{e}} \approx 0.015$. This leads to an average distance of $r_0 \approx 2a_{\mathrm{B}}$ and is, with respect to the density, lower than the Mott criterion. The Mott condition, viz.,

$$r_0 \approx a_{\mathrm{B}} , \qquad \frac{4\pi}{3} n_{\mathrm{e}} r_0^3 = 1 , \tag{5.80}$$

expresses the idea that atoms are destroyed if the mean distance of the electrons is less than or equal to the Bohr radius. We can give another discussion of the Mott criterion based on an alternative estimate of the binding energy shift.

In order to compare our expression for the shift, we may take, for example, the confined atom model (Graboske et al. 1969; Kraeft et al. 1986), which assumes that the atom is embedded in a hard sphere with radius r_0. In the first approximation, this theory gives the shift (in Rydberg units)

$$\Delta E_1^{\mathrm{ca}} = \pi^2 r_0^{-2} . \tag{5.81}$$

Correspondingly, the energy would already disappear at $r_0 \approx 3a_{\mathrm{B}}$, i.e., at a much smaller density (see also Fig. 5.7). Better estimates based on numerical solutions of the Schrödinger equation give a value of about $r_0 \approx 2a_{\mathrm{B}}$. Our first estimate is in the same region. The concept of the confined atom model is related to the concept of excluded volume, where a particular volume is attributed to each of the components which is not available for the other particles of the same species, thereby expressing the Pauli blocking mechanism. At the same time we also have a shift of the free electron energies due to Pauli blocking by electrons bound in atoms.

Limit of Strong Degeneracy

In the Fermi limit, we take the Fermi distribution function in the zero temperature limit, $f_{\mathrm{e}}(p) = \Theta(p_{\mathrm{F}} - p)$ and evaluate the Pauli blocking shift by integrating over the wave function $\Psi_1(p)$:

$$\begin{aligned}\Delta E_1^{\mathrm{Pauli}} &\approx \frac{4\pi}{(2\pi)^3}\int_0^{p_{\mathrm{F}}} \mathrm{d}p\, p^2(p^2+1)\psi_1^2(p) \\ &= \frac{4}{\pi}\left[\frac{p_{\mathrm{F}}(p_{\mathrm{F}}^2-1)}{(1+p_{\mathrm{F}}^2)^2} + \arctan(p_{\mathrm{F}})\right] ,\end{aligned}$$

with the Fermi momentum $p_{\mathrm{F}} = (3\pi^2 n)^{1/3}$ (all momenta are in Rydberg units). This formula also correctly reproduces the value $32\pi n_{\mathrm{e}}$ given in (5.75) in the low-density limit. In addition, we can calculate the temperature dependence of the Pauli blocking term. For this, we have to replace the zero-temperature Fermi function in

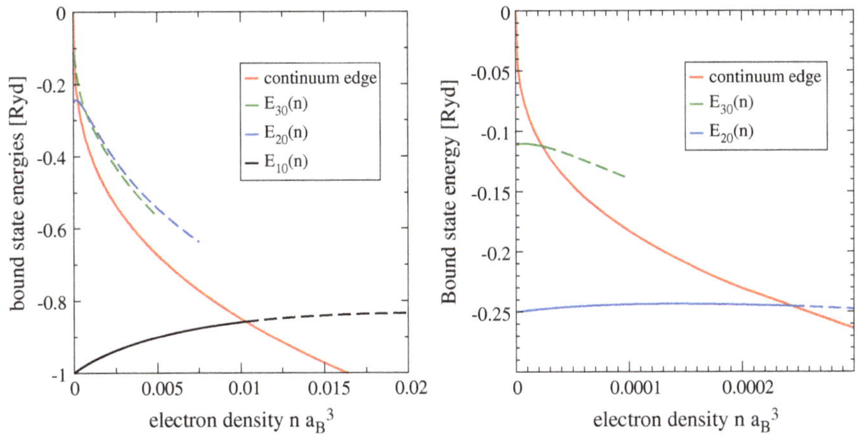

Fig. 5.8 Shifts of the ground state E_{10} (*black*) and two excited states E_{20} and E_{30} in the Pauli–Fock approximation (given in Ry units) and the lowering of the continuum level (*red*) as function of the density (given in units of Bohr cubes). From Redmer et al. (2010)

the interaction term by the finite temperature distribution. It can be seen that the temperature dependence of the Pauli blocking term becomes weak.

Even in the variational approach, the densities where the energy levels disappear and consequently full ionization occurs are evidently still too low to explain the observed effects. The Fock term occurs in the Bethe–Salpeter equation as a mean-field contribution and is of the same order as the Pauli blocking term. Even if the Fock term is not of primary importance for the disappearance of the bound state energy, it has to be included in the total shift of bound and scattering states to be consistent (so-called conserving approximations). A graphical representation of the Pauli shift and the negative Fock shift shows that the Pauli shift is above the Fock shift for $p_F < 1.35$, the difference being the remaining energy shift (Ebeling et al. 2009, 2010). Beyond the crossing point at $p_F \approx 1.35$, corresponding to the dimensionless density $n \approx 0.083$, the sum of the two shifts is negative. For a variational evaluation of energy shifts see (Ebeling et al. 2009, 2010).

To sum up, in this section the influence of the medium is studied in a mean-field approximation, i.e., by Pauli blocking and the Fock self-energy. The first of the contributions to the shift we refer to as Pauli blocking and the second as the Fock contribution. We consider the density region where Pauli–Fock shifts dominate, and remember that the Fock term and the Pauli blocking contributions have opposite signs and partially compensate each other (Fig. 5.8).

5.4.2 Hartree–Fock–Wigner Pressure at High Density

In the region of strongly degenerate electrons, the thermodynamics is to a good approximation determined by simple analytical approximations like the Hartree–Fock theory for the electrons and the Wigner lattice theory for the ions (protons). Looking at the (n, T) plane shown in Fig. 5.9, we now investigate the triangular region between $n_e \Lambda_e^3 \sim 1$ and $r_s \sim \Gamma_e \sim 1$ where the electrons are degenerate and strongly coupled $\Gamma_e > 1$, and the ions behave classically. Note that we have generalized the definition of the coupling parameter Γ_e here, including degeneracy effects by replacing classical by degeneracy temperatures:

$$k_B T \longrightarrow \Theta_e = \frac{2}{3} \langle \epsilon_{\text{kin}} \rangle \ . \tag{5.82}$$

This leads to the new definition

$$\Gamma_e = \frac{e^2}{\Theta_e d_e} \ . \tag{5.83}$$

The theory presented in the previous section provides isotherms of the pressure for the nondegenerate region. After crossing the line of degeneracy $n\Lambda_e^3 \approx 1$ to the degenerate regions above it, the previous theory is no longer valid, since degeneracy effects plays a role. Furthermore, the dissymmetry of the masses may be relevant. In this region the formulae given above are just an extrapolation which is exact only with respect to the (sometimes dominant) ideal Fermi contributions. For very dense, strongly degenerate plasmas, bound states do not exist, and the plasma behaves like a degenerate nonideal Coulomb gas.

Let us assume we start from a high density and decrease the density along an isotherm. The problem is then to establish at which density bound states will appear. In other words we have to ask at which densities the spectrum will have discrete levels corresponding to bound states of two particles, three particles, four particles, etc. Turning the question around we may ask for the density dependence of discrete states. In this way, the most important effect to be taken into account in the degenerate region $n\Lambda_e^3 > 1$ is the change brought about in the energy levels by effective density-dependent energy levels and the issue of merging of discrete levels in the continuum. This significantly affects the equation of state of hydrogen (Kraeft et al. 1986; Ebeling et al. 2009, 2010).

Figure 5.9 shows the region of interest here. It is the triangular region where the electrons and the protons are nonideal, i.e., $\Gamma_e > 1$ and $\Gamma_p > 1$, and where the electrons are degenerate, i.e., $n_e \Lambda_e^3 > 1$ and the protons are nondegenerate, i.e., $n_p \Lambda_p^3 < 1$. In particular, we are interested in the combination of degeneracy effects for the electrons and Wigner-like effects for the protons. There is some interest in this region for the investigation of solid state plasmas. In the region of strongly degenerate electrons, the thermodynamics is described to a good approximation by the Hartree–Fock approximation, and in that region the protons form a kind of a

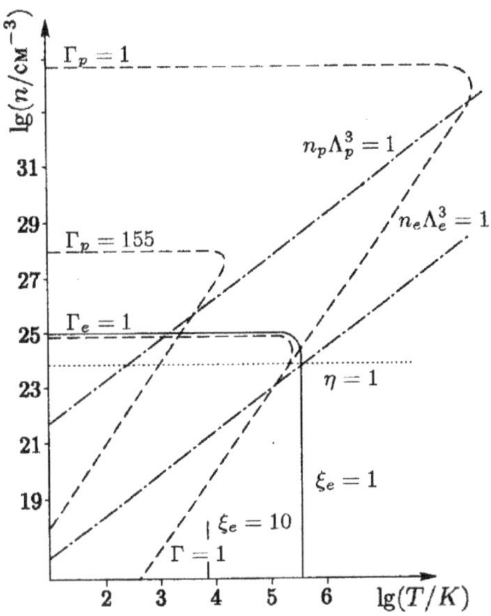

Fig. 5.9 The theory developed here refers to a triangular region of the parameters in an electron–proton gas where the electrons are degenerate but nonideal and the protons are non-degenerate. This is the triangle between $n_e \Lambda_e^3 \sim 1$ and $\Gamma_e \sim r_s \sim 1$ in the *bottom left*

Wigner lattice (Kraeft et al. 1986; Ebeling 2016). From (5.8), the grand-canonical pressure may be approximated by

$$p(\beta, z_e, z_i) = p^{id} + p_e^{HF}(z_e) + p_i^{Wi}(z_i) + k_B T z_e z_i \Lambda^3 \tilde{\sigma}(T) + \cdots , \quad (5.84)$$
$$\tilde{\sigma}(T) = \sum_{s\ell} \left[\exp(-\beta \tilde{E}_{s\ell}) - 1 + \beta \tilde{E}_{s\ell} \right] ,$$

with

$$n_e = z_e + z_e \frac{\partial \beta p_e^{HF}(z_e)}{\partial z_e} + z_e z_i \Lambda^3 \tilde{\sigma}(T) ,$$
$$n_i = z_i + z_i \frac{\partial \beta p_i^{Wi}(z_i)}{\partial z_i} + z_e z_i \Lambda^3 \tilde{\sigma}(T) .$$

We proceed as in the previous section. Neglecting first the bound state contribution and using the Hartree–Fock approximation, the fugacity of the electrons is given by

$$z_e^{HF} = n a_e^{HF} , \qquad a_e^{HF} = \exp\left[-\beta \frac{e^2}{\Lambda_e} I_{-1/2}(\alpha) \right] , \qquad \alpha = \beta \mu_e^{id} . \quad (5.85)$$

Here the $I_k(x)$ are the well-known Fermi integrals and a stands for activity coefficient. In the following, we use the high density limit including the first corrections. Useful approximations for the Hartree–Fock functions were given in Chap. 3. For convenience, in these calculations we used a more convenient Padé expression with a Sommerfeld term T^2 in the denominator ($t = k_B T/\text{Ry}$):

$$p_e^{HF}[\mathrm{Ry}] \approx -\frac{0.3059n}{r_s + 6.030t^2 r_s^5}, \qquad a_e^{HF} \approx \exp\left(-\frac{1.222}{tr_s + 0.1216t^3 r_s^5}\right).$$

The ionic contribution is more difficult at the high densities considered here. In the given density regime, where the ions are still nondegenerate, the pressure may be approximated in the framework of two-fluid models by Wigner-type expansions (Kraeft et al. 1986; Ebeling et al. 2012). In the density region which we study here, viz., $n \approx 10^{23}$–10^{24} cm^{-3}, we may use the simple analytical approximations

$$p_i^{Wi}[\mathrm{Ry}] \approx -\frac{0.5964n}{r_s}, \qquad a_i^{Wi} \approx \exp\left(-\frac{2.385}{tr_s + .1216t^{7/4} r_s^{7/4}}\right). \tag{5.86}$$

In contrast to the situation at low densities, the contribution of the bound states is not as big and may be treated as a perturbation. In the canonical ensemble, putting together the Hartree–Fock, Wigner, and ladder terms, we arrive at the following simple relation for the high density pressure of hydrogen:

$$\begin{aligned} p(\beta, n) \approx\ & p_e^{FD} + p_e^{HF} + nk_B T + p_i^{Wi} \\ & -k_B T n^2 a_e^{HF} a_i^{Wi} \Lambda^3 \sum \left[\exp(-\beta \tilde{E}_s) - 1 + \beta \tilde{E}_s\right], \end{aligned} \tag{5.87}$$

where p_e^{FD} is the Fermi–Dirac pressure of all electrons, including bound electrons. In this way, for the high density limit, we obtain a closed expression for the pressure of hydrogen expressed in the physical picture of densities. It remains to find the effective two-particle energy levels, which are now density dependent.

As an application, we calculated the pressure of hydrogen relative to the classical ideal pressure for the case of high densities beyond $n \sim 10^{23}$ cm^{-3} and at temperatures 10 000–20 000 K, using the extended Hartree–Fock and Wigner approximations given above (see Fig. 5.10). The energy shifts are taken into account. For comparison, we have given a curve for 10 000 K, obtained by Vorberger et al. (2011), and the result for $T = 20\,000$ K, obtained by a Padé approximation in combination with a mass action law (Ebeling 1990; Beule et al. 1999). The agreement of our new, rather simple approach with the previous ones (see, e.g., Saumon and Chabrier 1989; Beule et al. 1999; Kraeft et al. 2007; Vorberger et al. 2009, 2010, 2011) is satisfactory. The present theory may be considered as a simple extension of the approach given in Sect. 5.3.2 for densities beyond $n \sim 10^{23}$ cm^{-3}.

In certain regions of the density–temperature plane, the pressure at high densities is essentially determined by the Hartree–Fock approximation for the thermodynamic functions and the Hartree–Fock approximation for the energy shifts. We assumed here that in the high-density region these shifts provide the most important effects for the destruction of bound states. In particular, we contribute here to the theory of hydrogen at high pressures in the region where a Mott transition to full ionization has been predicted and where recent experiments have shown a transition from insulating behavior to metal-like conductivity. In order to understand this transition several effects have to be taken into account. We concentrated here on so-called Pauli

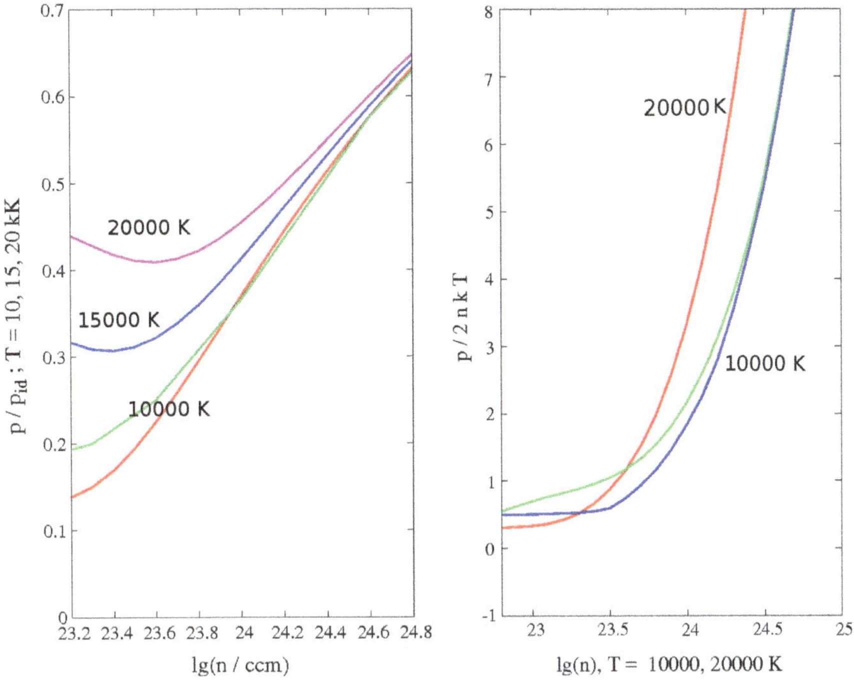

Fig. 5.10 Calculations of the high-density pressure based on the extended Hartree–Fock–Wigner approximation taking into account energy shifts. *Left* Pressure related to the ideal quantum-statistical pressure $p/p_{\rm id}$ for the temperatures $T = 10000$ (*bottom*), 15 000, and 20 000 K (*top*). Bound states developing below $n \sim 10^{23}$ cm^{-3} are not well described by this method. The second curve from the bottom (*green*) represents numerical and analytical results by Vorberger et al. 2011 for $T =$ 10 000 K. *Right* Pressure relative to the classical ideal pressure at temperatures $T = 10\,000$ (*blue*) and 20 000 K (*red*). We also present a comparison with a curve (*green*) from the PACH theory due to Beule et al. 1999 for $T = 10000$ K (PACH is the chemical picture with Padé approximations)

blocking effects, expressing the rule that states occupied by atomic electrons cannot be occupied by free electrons with the same spin state. This leads at high electron densities to the destruction of atomic states which need a relatively large amount of phase space. We calculated the energy shifts due to Pauli effects and discussed the Mott effects by solving effective Schrödinger equations for strongly correlated systems:

- For simplicity we neglected the formation of molecules which play a role at temperatures below 20 000 K (see Chap. 3 and Saumon and Chabrier 1989; Beule et al. 1999).
- In the temperature region which we studied here (10 000–50 000 K), no first order transitions were observed in this approximation.
- The approach based on the physical picture which we used in the present section, cannot be extended to lower temperatures, where H_2 molecules dominate and where the approach given in Sect. 5.3 may be more appropriate. However, in the high density region studied here, the Pauli–Fock contributions are most essential and the present picture is appropriate and quite simple.

References

A. Alastuey, V. Ballenegger, Pressure of a partially ionized hydrogen gas: numerical results from exact low temperature expansions. Contrib. Plasma Phys. **50**, 4653 (2010)

A. Alastuey, V. Ballenegger, Contrib. Plasma Phys. **52**, 95 (2012)

A. Alastuey, A. Perez, Virial expansion of the equation of state of a quantum plasma. Europhys. Lett. **20**, 19 (1992)

A. Alastuey, V. Ballenegger, F. Cornu et al., Exact results for thermodynamics of the hydrogen plasma: low-temperature expansions beyond Saha theory. J. Stat. Phys. **130**, 1119–1176 (2008)

A. Alastuey, V. Ballenegger, W. Ebeling, Comment on direct linear term in the equation of state of plasmas. Phys. Rev. E **62**, 047101–1 (2015)

S. Arndt, W.D. Kraeft, J. Seidel, Phys. Stat. Sol. B **194**, 601 (1996)

V. Ballenegger, Ann. Phys. (Berlin) **254**, 103 (2012)

G.P. Bartsch, Ebeling, Quantum statistical fugacity expansions for partially ionized plasmas in equilibrium, Contr. Plasma Phys. **11**, 393–403 (1971); see also Contrib. Plasma Phys. **15**, 25–32 (1975)

H.A. Bethe, E.E. Salpeter, A relativistic equation for bound state problems. Phys. Rev. **82**, 309 (1951)

D. Beule, W. Ebeling, A. Förster, H. Juranek, S. Nagel, R. Redmer, G. Röpke, Equation of state for hydrogen below 10 000 K: from the fluid to the plasma. Phys. Rev. B **59**, 14177 (1999)

D. Beule, W. Ebeling, R. Redmer, G. Röpke, Electrical conductivity in dense hydrogen fluid and metal plasmas. Control. Plasma Phys. **39**, 25–28 (1999)

D. Beule,W. Ebeling, A. Förster, H. Juranek, R. Redmer, G. Röpke, Isentropes and Hugoniots for dense hydrogen and deuterium. Phys. Rev. E **63**, 060202 R (2001)

L.S. Brown, L.G. Yaffe, Effective field theory of highly ionized plasmas. Phys. Rep. **340**, 1 (2001)

W. Däppen, Accurate and versatile equations of state for the sun and sunlike stars. Astrophys. Space Sci. **328**, 139146 (2010)

W. Däppen, D. Mihalas, D.G. Hummer, Astrophys. J. **332**, 261 (1988)

H.E. DeWitt, M. Schlanges, A.Y. Sakakura, W.D. Kraeft, Low density expansion of the equation of state for a quantum electron gas. Phys. Lett. A **197**, 326 (1995)

M.W.C. Dharma-Wardana, F. Perrot, Level shifts, continuum lowering, and the mobility edge in dense plasmas. Phys. Rev. A **45**, 5883 (1992)

W. Ebeling, Statistical derivation of the mass action law of interacting gases and plasmas. Physica **73**, 573–584 (1974)

W. Ebeling, Free energy and ionization of dense plasma of light elements, Contr. Plasma Phys. **29**, 238–242 (1989); **30**, 553–561 (1990)

W. Ebeling, H.D. Hoffmann, G. Kelbg, Quantenstatistik des Hochtemperaturplasmas. Contrib. Plasma Phys. **7**, 233–248 (1967)

W. Ebeling, W.-D. Kraeft, K. Kilimann, D. Kremp, Coexisting phases in an electron–hole plasma, Phys. Stat. Sol. **78**, 241–253 (1976)

W. Ebeling, W.-D. Kraeft, D. Kremp, *Nonideal Plasmas* (Proc. ICPIG XIII, Invited papers, Berlin, 1977)

W. Ebeling, W.D. Kraeft, D. Kremp, *Theory of Bound States and Ionization Equilibrium in Plasmas and Solids* (Akademie-Verlag, Berlin, 1976) (Extended Russ. translation Mir, Moscow, 1979)

W. Ebeling, W.D. Kraeft, D. Kremp, G. Röpke, Energy levels in hydrogen plasmas and the Planck–Larkin partition function–a comment. Astrophys. J. **290**, 24–27 (1985)

W. Ebeling, A. Förster, R. Redmer, T. Rother, M. Schlanges, ICPIG 18, Invited Lectures, Univ. of Wales (1987)

W. Ebeling, Free energy and ionization in dense plasmas of the light elements. Control. Plasma Phys. **30**, 53–561 (1990)

W. Ebeling, S. Hilbert, On Saha's equation for partially ionized plasmas and Onsager's bookkeeping rule, Eur. Phys. J. D **20**, 93–101 (2002)

W. Ebeling, S. Hilbert, H. Krienke, On Bjerrum's mass action law for electrolytes and Onsager's bookkeeping rule. J. Mol. Liq. **96/97**, 409–423 (2002)

W. Ebeling, H. Hache, H. Juranek, R. Redmer, G. Röpke, Pressure ionization and transitions in dense hydrogen. Contrib. Plasma Phys. **45**, 160–167 (2005)

W. Ebeling, D. Blaschke, R. Redmer, H. Reinholz, G. Röpke, The influence of Pauli blocking effects on the properties of dense hydrogen. J. Phys. A Math. Theor. **42**, 214033 (2009), arXiv:0810.3336v2

W. Ebeling, D. Blaschke, R. Redmer, H. Reinholz, G. Röpke, Pauli blocking effects and Mott transition in dense hydrogen. In: Redmer et al. (2010)

W. Ebeling, W.D. Kraeft, G. Röpke, Bound states in Coulomb systems—old problems and new solutions. Contrib. Plasma Phys. **52**, 7–16 (2012); Ann. Physik (Berlin) **524**, 311–326 (2012)

W. Ebeling, Work of Baimbetov on nonideal plasmas and developments, Contr. Plasma Phys. **56**, 163–175 (2016)

G. Ecker, W. Weizel, Zustandssumme und effektive Ionisierungsspannung eines Atoms im Inneren des Plasmas. Ann. Phys. (Leipzig) **452**, 126 (1956)

M. Fisher, D.M. Zuckerman, Exact thermodynamic formulation of chemical association. J. Chem. Phys. **109**, 7961–7981 (1998)

H.L. Friedman, *Ionic Solution Theory* (Interscience, New York, London, 1962)

H.L. Friedman, W. Ebeling, Theory of a fluid of interacting and reacting particles. Rostocker Physikal. Manuskripte **4**, 35–51 (1979)

A. Förster, T. Kahlbaum, W. Ebeling, Equation of state and pase diagram of fluid helium in the region of partial ionization. Laser Part. Beams **10**, 253–262 (1992)

H.C. Graboske, D.J. Harwood, F.J. Rogers, Phys. Rev. **186**, 210 (1969)

H.R. Griem, *Spectral Line Broadening in Plasmas* (Academic Press, New York, 1974)

T.L. Hill, *Statistical Mechanics* (Mc Graw Hill, New York, 1956). Russ. transl, Moskva, 1960

S. Ichimaru, *Statistical Plasma Physics* (Addison-Wesley, Redwood City, 1992)

G. Kelbg, W. Ebeling, Quantum statistics of thermal plasmas in equilibrium (in Russ.), 1–2, printed by Inst. Theor. Phys. Ukrain. Acad. Sci, Kiev (1970)

T. Kahlbaum, The quantum-diffraction term in the free energy for Coulomb plasma, effective potential approach. J. Phys. France **10**, 5-455 (2000)

A.L. Khomkin, I.A. Mulenko: High Temp. **41**, 275 (2003), arXiv.org/physics/0305118

A.L. Khomkin, A.S. Shumikhin, Features of the chemical models of nonideal atomic plasma at high temperatures. Plasma Phys. Rep. **34**, 251–256 (2008)

A. Khomkin, A. Shumikhin, I. Mulenko, Basic chemical models of partially ionized plasma, Czech. J. Phys. **54**, Suppl. C **143**, 1b (2004)

K. Kilimann, W. Ebeling, Energy gap and line shifts for H-like ions in dense plasmas. Z. Naturforsch. A **45**, 613–617 (1990)

Y.L. Klimontovich, *Statistical theory of nonequilibrium processes in plasmas, Izdat MGU 1964* (Engl. Pergamon, Oxford, 1967)

Y.L. Klimontovich, *Statistical Physics*, In Russ. (Nauka, 1982); Engl. (Harwood, New York, 1986)

W.D. Kraeft, D. Kremp, W. Ebeling, Complex representations of the quantum statistical second virial coefficient. Phys. Lett. A **29**, 466 (1969)

W.D. Kraeft, D. Kremp, W. Ebeling, G. Röpke, *Quantum Statistics of Charged Particle Systems* (Pergamon Press, New York, 1986)

W.-D. Kraeft, J. Vorberger, D.O. Gericke, M. Schlanges, Thermodynamic functions for plasmas beyond Montroll-Ward. Contrib. Plasma Phys. **47**, 253 (2007)

D. Kremp, M. Schlanges, W.D. Kraeft, Quantum Statistics of Nonideal Plasmas (Springer, Berlin, 2005)

L.P. Kudrin, Statistical Physics of Plasmas (in Russ.) (Nauka, Moskva, 1974)

J.E. Mayer, Theory of ionic solutions. J. Chem. Phys. **18**, 1426–1436 (1950)

E. Montroll, J. Ward, Quantum statistics of interacting particles. Phys. Fluids **1**, 55 (1958)

R. Redmer, B. Holst, F. Hensel (eds.), *Metal to nonmetal transitions* (Springer, Berlin, 2010)

F.J. Rogers, Statistical mechanics of Coulomb gases of arbitrary charge. Phys. Rev. A **10**, 2441–2456 (1974)

F.J. Rogers, Statistical mechanics of Coulomb gases of arbitrary charge. Phys. Rev. A **38**, 5007 (1988)

G. Röpke, K. Kilimann et al., Phys. Lett. A **69**, 329 (1978)

D. Saumon, G. Chabrier, Phys. Rev. Lett. **62**, 2397 (1989)

J. Seidel, S. Arndt, W.D. Kraeft, Energy spectrum of hydrogen atoms in dense plasmas. Phys. Rev. E **52**, 5387 (1995)

A.N. Starostin, V.C. Roerich, R.M. More, How correct is the EOS of weakly nonideal hydrogen plasmas? Contrib. Plasma Phys. **43**, 369–372 (2003)

O. Theimer , P. Kepple, Statistical Mechanics of a Partially Ionized Hydrogen Plasma, Phys. Rev. A **1**, 957 (1970)

A.A. Vedenov, A.I. Larkin, Equation of state of plasmas (in Russ.). Zh. eksp. teor. Fiz. **36**, 1133 (1959)

J. Vorberger, D.O. Gericke, Phys. Plasmas **16**, 082702 (2009)

J. Vorberger, D.O. Gericke, Th Bornath, M. Schlanges, Phys. Rev. E **81**, 046404 (2010)

J. Vorberger, D.O. Gericke, W.D. Kraeft, Equation of state for high density hydrogen, Poster SCCS 2011, (2011), arXiv:1108.4826

V.S. Vorobjov, I.A. Mulenko, A.L. Khomkin, The importance of excited states in the thermodynamics of partially ionized plasma, High Temp. **38**, 509–514 (2000), Teplofiz. Vyssokikh Temp. **38**, 533–538 (2000)

R. Zimmermann, K. Kilimann, W.D. Kraeft, D. Kremp, G. Röpke, Phys. Stat. Sol. (b) **90**, 175 (1978)

Chapter 6
Equations of State for Strongly Coupled Partially Ionized Plasmas

The aim of this chapter is to extend the equation of state (EOS) found in the last three chapters to higher densities. We separate the charges into free and bound ones, and the free charges are studied up to the fluid region with degenerate electrons. The atomic and molecular bound states are given a chemical description. We first investigate the free electron fluid on a neutralizing background, then study mass-symmetric non-symmetric plasmas and the influence of Wigner lattice effects. The main part is based on the chemical description. We minimize the free energy of multi-component plasmas using analytical approximations. We concentrate on the Coulomb contributions and show that, for small coupling, the thermodynamics of free charges satisfies the Debye law $n^{1/2}$ with quantum corrections, while for large coupling parameters, it satisfies the Wigner law $n^{1/3}$, with a slope modified by quantum corrections. We study the thermodynamics of mixtures of free and bound states and introduce advanced chemical models. These models are formulated using variational principles for the free energy density in combination with Padé approximations for the correlations of the free plasma constituents. A set of analytical formulae is developed for the thermodynamics of the free constituents of plasma systems, the composition being found by minimization of the free energy. The role of special effects such as excluded volume, magnetic fields, and bound energy shifts is studied. The present study is oriented toward analytical approaches. For studies of wide-range EOS by Monte Carlo methods, see Chaps. 9–10 of this book and recent papers like Dornheim et al. (2015a, 2016a). For numerical studies using molecular dynamics, see Chap. 8, and for numerical methods based on density functional theory (DFT), see, e.g., Holst et al. (2008, 2011), French et al. (2009), Dharmawardana (2015), and Knyazev and Levashov (2015).

W. Ebeling et al., *Quantum Statistics of Dense Gases and Nonideal Plasmas*, Springer Series in Plasma Science and Technology, DOI 10.1007/978-3-319-66637-2_6

6.1 Coulomb Fluid Models and Electrical Field Energy

6.1.1 Electrical Field Correlations and Coulomb Energy

The standard view on plasmas is that they consist of discrete particles interacting via Coulomb fields. However, as we know, the free charges in plasmas can also be considered as electromagnetic systems consisting of electric and magnetic fields distributed in space. This view is particularly useful in the case of strong coupling. The electromagnetic fields are fast-changing, essentially stochastic, and carry electrical energy. As long as the velocities are non-relativistic, the total Coulomb energy is contained in the Coulomb field between the particles. These fields are essential classical and the local Coulomb energy density is given by the square of the local field strength:

$$u_c(\mathbf{r}, t) = \frac{1}{8\pi}\left[\mathbf{E}(\mathbf{r}, t)\right]^2 . \tag{6.1}$$

The electric field in a plasma is a single system and it makes little sense to decompose it into electronic and ionic subsystems. This is the basis for several approaches, like the Gellman–Brueckner approach and the reduced mass approximation (RMA), which treat the field as one system without using any decomposition.

The electric field is fast-changing in time and space, and in a uniform plasma the average is zero at any space coordinate and at any instant of time. This means that the correlations of the field $\mathbf{E}(\mathbf{r}, t)$ are the important factor in determining the energy density. The study of the fluctuating field in the plasma and its correlations is a very fruitful non-traditional approach, which was worked out in much detail by Klimontovich, Pines, Silin, and Rukhadse (Pines, 1961; Klimontovich, 1964, 1967, 1982, 1986). In many respects, this approach is also connected with the notations of collective excitations or plasma waves. The technique is sometimes given the corresponding name of *dielectric formalism*, since the dielectric function plays a primary role (Pines, 1961; Pines and Nozieres, 1966). An attractive feature of these methods is that the classical and the quantum case are treated in a very similar way. The physical reason for this similarity is, of course, that the electric field in a plasma is essentially a classical object with a quantum dynamics determined by the quantum dynamics of the particles.

Note that, in a medium, the vacuum field or the corresponding potentials are divided by the dielectric constant of the medium ε_r. At lower densities, it is more convenient to calculate the Coulomb energy from the known binary correlation function $g_{ab}(r)$:

$$U_c = \frac{1}{2}\sum n_a n_b \int \mathrm{d}\mathbf{r}\, \frac{e_a e_b}{r} g_{ab}(r) . \tag{6.2}$$

For higher densities, it is more appropriate to integrate the energy density of the electrical fields, viz., $(1/8\pi)\left[\mathbf{E}(\mathbf{r}, t)\right]^2$ according to (6.1), directly in the coordinate space, or possibly in the Fourier space.

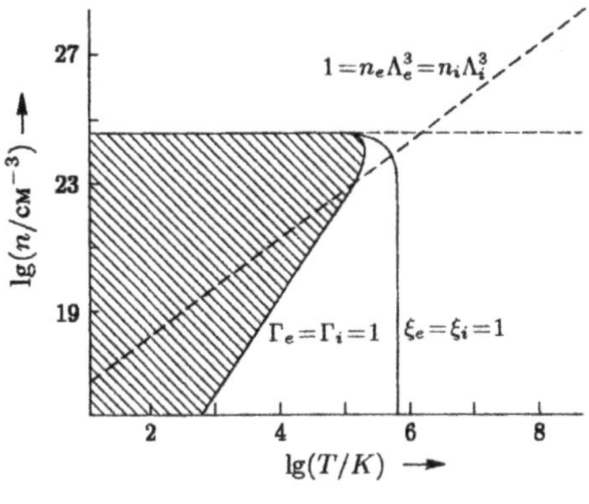

Fig. 6.1 Different regions of the parameters for an electron fluid and for symmetrical electron–positron fluids. The *dashed region* corresponds to stronger correlations, while the *rectangular region* marks the boundary of measurable interactions. The line $n_e \Lambda_e^3 = 1$ separates non-degenerate from degenerate plasmas. The region of stronger correlations is located inside $\Gamma_e = 1$

We begin by studying non-degenerate electron fluids on a compensating background, then symmetrical two-component fluids. First, we specify the parameter region in which we work. Figure 6.1 shows the corresponding density–temperature plane. We are interested in the density–temperature region where the electrons are strongly coupled, but still only weakly degenerate, i.e., we study the region below the line $n_e \Lambda_e^3 = 1$ in the diagram. As in Chap. 4, we start from the pair correlations in Fourier space given by Kelbg (1963) and Kremp et al. (1967) for the pair density matrix of plasmas (Schmitz and Kremp (1967)). Using Fourier representations, for the Coulomb energy of electron fluids, we get the integral (see Chaps. 3 and 4):

$$U_c = -4K_B T (n\beta e^2)^2 \int_0^\infty dk \frac{M(1; 3/2; -k^2\lambda^2/4)}{k^2 + \kappa^2 M(1; 3/2; -k^2\lambda^2/4)} . \tag{6.3}$$

The integral may be represented formally by a ring function depending on only one variable $\kappa\lambda$:

$$\langle U_c \rangle = -V K_B T \frac{\kappa^3}{8\pi} G(\kappa\lambda) , \qquad G(x) = 1 - \frac{\pi}{4} x + O(x^2) . \tag{6.4}$$

Several of the subsequent orders of the Taylor expansion are known (De Witt et al. 1995; Alastuey et al. 2015; Ebeling 2016).

In Sect. 4.4, we discussed several methods for evaluating the complicated integral for the Coulomb energy. Kremp et al. (1967) calculated the integral for U_c up to the linear order $O(\kappa\lambda)$, approximating the denominator by $k^2 + \kappa^2$. The integral was also treated by Riemann et al. (1995), but differently from Kremp and Schmitz, including a term $\kappa^2\lambda^2/8$ in the denominator. We showed that all versions agree to order $O(\kappa\lambda)$ and disagree slightly to order $O(\kappa^2\lambda^2)$. Furthermore, we evaluated the integral using standard numerical mathematics (Mathematica) for several points in

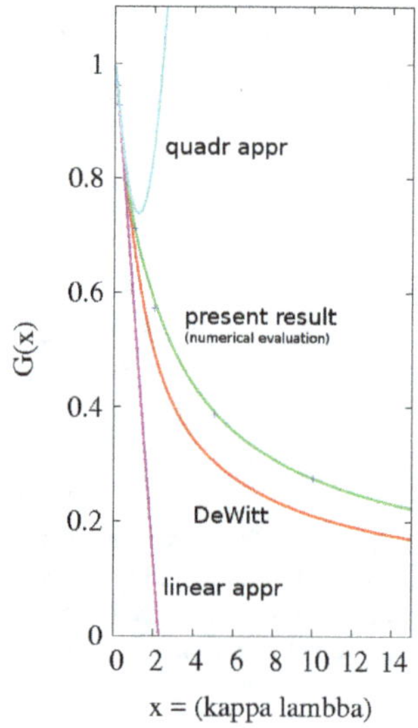

Fig. 6.2 Ring functions for the Coulomb energy and chemical potential. The *lower curve* is the first order expansion and *upper curve* the second order expansion in $(\kappa\lambda)^2$. The *green curve* shows a numerical evaluation of the integral and the *red curve* DeWitt's estimate for the function $G(x)$. The Taylor expansions, which diverge rapidly with the parameter $\kappa\lambda$, are useless for higher densities since they agree only up to quadratic order and have a quite different asymptotic

the range $x = 0.02$–1, checked the expansion in x, and estimated the coefficient of the second term in the Taylor expansion. We also checked the asymptotic by evaluating several points for $x = 1$–10, estimated the asymptotic analytically, and checked it numerically (Ebeling 2016). Figure 6.2 shows several results for the Coulomb ring function.

To represent the Coulomb energy, we have used the convenient variables $\tau = K_{\mathrm{B}}T/Ry$ and $\Gamma = e^2/d = 2/\tau r_s$. We see that, for $\Gamma > 1$, expansions cut off after the linear or the quadratic term yield quite misleading results which show the bad convergence of the Taylor series. In our view, finding the correct asymptotic of the ring terms is more important than the question of higher orders $O(x^3)$ and beyond. Our results are slightly different from those for the ring expressions obtained by Riemann et al. (1995). This question of the deeper reasons for the differences remains open. We guess that this issue is connected with the problem of whether the screening according to Kelbg, Kremp, and Schmitz is more or less complete than the screening based on the use of polarization functions according to Bohm and Pines, and in the thermodynamic Green function method (Kraeft et al. 1986).

We have shown that the concrete representation of the ring integrals plays an important role, and that several of the available ring expressions disagree slightly in higher orders and in the asymptotic.

6.1.2 Coulomb Energy of Dense Electron Fluids

The low temperature degenerate states of electron plasmas are of great interest for the physics of the cohesion forces in metals, and were thus already calculated in the

1950s by several researchers as Macke, Gellman and Brueckner, Pines and Nozieres, and others. In order to proceed, we will use discrete Fourier vectors, mainly following the derivation of Pines and Nozieres representing the Coulomb potential and the local density by (Pines and Nozieres 1966)

$$V(r) = \sum_q V_q \exp(-iqr) , \qquad \rho(r) = \sum_q \rho_q \exp(-iqr) . \tag{6.5}$$

In the ground state, the mean interaction energy is

$$U_c = \left\langle \Psi_0 \left| \sum_q v_q (\rho_q^+ \rho_q - N \right| \right\rangle , \tag{6.6}$$

and after transformation to the dielectric form this gives

$$U_c = -\sum_q \left[\int_0^\infty \frac{d\omega}{2\pi} \mathrm{Im} \frac{1}{\epsilon(\omega, \mathbf{q})} - N \frac{2\pi e^2}{q^2} \right] . \tag{6.7}$$

The final result of the calculations of Gellman and Brueckner for the energy of an electron in dense electron gases is (in *Ry*)

$$\varepsilon = \left[\frac{2.21}{r_s^2} - \frac{0.916}{r_s} + 0.622 \ln(r_s) - 0.412 \right] . \tag{6.8}$$

We consider now the weakly degenerate quantum case. Recall first that, in the quantum case, a static plasma does not exist, because at least the zero-point quantum oscillations are always present. It makes sense therefore to include the time variable in the classical picture.

The expressions given so far are valid for $r_s < 1$, and in some approximations, they are still valid for $r_s < 5$. For higher values, a better approximation is given by the Wigner energy formula, which is valid for $r_s \gg 1$:

$$\varepsilon_c = -\frac{0.8755}{r_s} Ry . \tag{6.9}$$

By adding the exchange energy, we find the total correlation energy per particle for $r_s \gg 1$ to be

$$\varepsilon_c = -\frac{1.792}{r_s} Ry . \tag{6.10}$$

Wigner tried to find an estimate for $r_s \sim 3$–5, which connects the lattice limit with an estimate for the energy in the metallic region $r_s \sim 4$, and proposed the simple formula

$$\varepsilon_c = -\frac{0.8755}{r_s + 7.8} Ry . \tag{6.11}$$

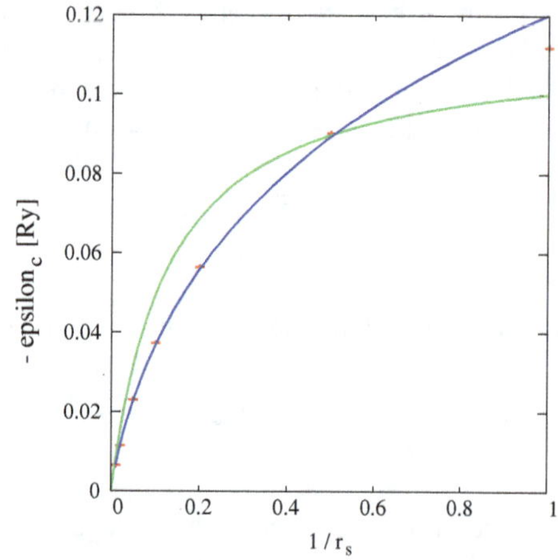

Fig. 6.3 The interpolation formula for the correlation energy proposed by Richert and Ebeling (1984) is in rather good agreement with the early QMD results for $T = 0$ due to Ceperley and Alder (1980), denoted by *crosses*. On the other hand, we see that the original Wigner formula (*green curve*) is just a rough estimate

There exist many approaches to construct interpolation formulae between the Wigner estimate for $r_s \gg 1$ and the Gell-Man–Brueckner results for $r_s \ll 1$ (Kraeft et al. 1986). A quite useful and simple approximation for the exchange and correlation energy measured in Ry is Richert and Ebeling (1984)

$$\varepsilon_{xc}(r_s, T = 0) = -\frac{0.916}{r_s} - 0.1244 \ln\left(\frac{1 + 0.3008 r_s^{1/2} + 2.117 r_s^{-1/2}}{1 + 0.3008 r_s^{1/2}}\right).$$

Note that in order to obtain the full energy, one has to add the Fermi energy $\varepsilon_F = 2.21/r_s^2$. The good agreement between this approximation and available QMC results for the ground state energy of the electron gas and the Richert formula is presented in Fig. 6.3. Note that, in the same approximation, the exchange correlation part of the chemical potential reads

$$\mu_{xc}(r_s, T = 0) = -\frac{1.222}{r_s} - 0.1244 \ln\left(\frac{1 + 0.2664 r_s^{1/2} + 2.501 r_s^{-1/2}}{1 + 0.3664 r_s^{1/2}}\right) Ry.$$

Good approximations for the ground state energy of electron systems are of great value for the theory of metals (Callaway 1964, Perrot and Dharmawardana 1984, 2000). They are needed to calculate the cohesion energy, which is a fine balance between positive and negative contributions, whence accurate values for the Coulomb energy are required. In conclusion, looking in particular at Fig. 6.4, we may say that the lattice-like dependence as $n^{1/3}$ seems to be quite typical for higher densities. We will come back to this problem in a later section.

We now discuss the influence of temperature. For the ideal part and the Hartree–Fock contribution from the electron gas, this problem has already been solved in Sects. 2.6 and 2.7. We derived there a formula for the exchange energy of the electron

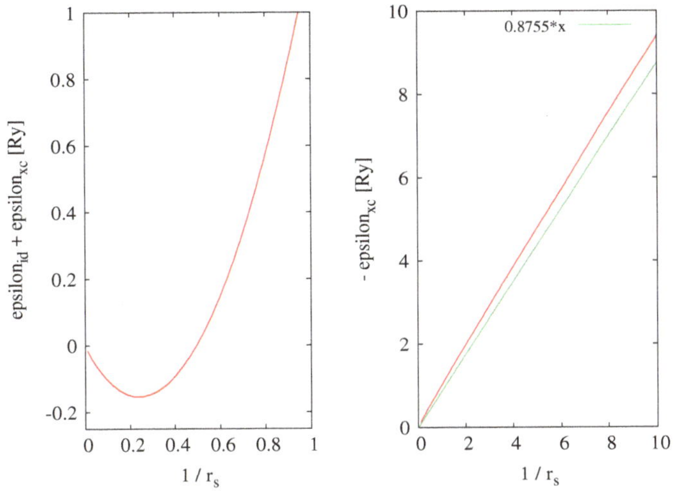

Fig. 6.4 The total energy per particle including the Fermi energy and the correlation energy shows a minimum in the region $r_s \approx 0.2$–0.3, which is important for understanding the cohesion of metals (*left*). The correlation energy per particle is in a large region of higher densities, close on average to the Wigner estimate (6.9) (*green curve*) $\varepsilon_c \sim n^{1/3}$ for the Coulomb lattice energy (*right*)

gas and showed also that these expressions are compatible with the approximations for low density and high degeneracy [see also the important work of Perrot and Dharmawardana (1984, 2000), as well as Dharmawardana (2015)]. Let us consider now the next contribution which is the Gellman–Brueckner result in the degenerate low temperature case and the Debye–Hückel result with quantum corrections in the non-degenerate high-temperature case.

In the following we use the dimensionless parameters temperature τ, the Bogoliubov parameter μ_B, and the density $\bar{n}$, written in the convenient forms

$$\tau = \frac{K_B T}{Ry} , \qquad \mu_B = \frac{e^2}{r_D K_B T} = \frac{2\sqrt{6}}{(r_s \tau)^{3/2}} , \qquad \bar{n} = \frac{6\sqrt{\pi}}{r_s^3 \tau^{3/2}} .$$

Then an appropriate Padé formula, interpolating between the results for the correlation ground state energy and the QDHA, gives for the chemical potential Richert and Ebeling (1984)

$$\mu_c = \frac{\bar{n}^2 \mu_c(r_s, \tau = 0) - 0.5\mu_B \tau - .125(\mu_B \tau)^2}{1 + .3133\mu_B \tau^{1/2} + \bar{n}^2} . \tag{6.12}$$

Note that, in the work by Richert and Ebeling (1984), one also finds formulae including higher order contributions stemming from the virial functions $K_0(\xi)$. The result (6.12) may be combined with the result for the exchange chemical potential to obtain the exchange correlation chemical potential $\mu_{xc} = \mu_x + \mu_c$. An argument due to Zim-

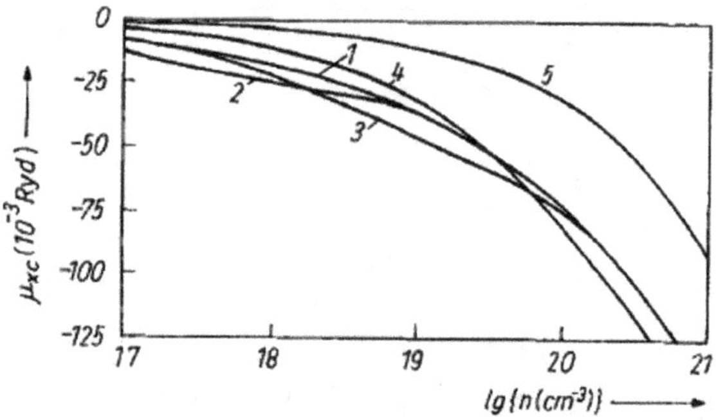

Fig. 6.5 Interpolation formula for the correlation energy proposed by Richert and Ebeling (1984) for finite temperatures (in kelvin): [1] $T = 0$, [2] $T = 100$, [3] $T = 500$, [4] $T = 3430$, [5] $T = 50,000$

mermann et al. (1984, 1988) suggests the existence of certain compensation effects between the exchange and the correlation contributions. This suggests keeping the two contributions together in one formula. In this way we arrive at the following expression for the exchange correlation chemical potential

$$\mu_{\text{xc}} = \frac{\bar{n}^2 \mu_{\text{xc}}(r_s, \tau = 0) - 0.5\mu_{\text{B}}\tau - .125(\mu_{\text{B}}\tau)^2}{1 + .3133\mu_{\text{B}}\tau^{1/2} + \bar{n}^2} \,. \tag{6.13}$$

A set of curves is shown in Fig. 6.5 for a range from low to high temperatures.

In a recent study by Groth et al. (2016), it has been shown that the present formulae are not very accurate in the metallic region. Somewhat better in this region are the fit formulae due to Tanaka and Ichimaru (1985, 1986). The fits by Tanaka et al. were obtained on the basis of a generalization of the scheme developed earlier by Singwi, Tosi, Land, and Sjölander (1968), in combination with the solution of the HNC integral equations. By a careful fit of the available data, Tanaka and Ichimaru (1985, 1986) derived the following representation of the internal energy per particle:

$$\varepsilon_{\text{int}} = -K_{\text{B}}T \frac{a(\Theta)\Gamma + b(\Theta)\Gamma^{3/2} + d(\Theta)\Gamma^{5/2}}{1 + e(\Theta)\Gamma^{1/2} + \Gamma} \,, \tag{6.14}$$

where a, b, c, d, e are simple rational functions of the dimensionless temperature, e.g.,

$$a(x) = \frac{9}{4\pi^2} \frac{.75 + 3.04363x^2 - .09227x^3 + 1.7035x^4}{1 + 8.31051x^2 + 5.1105x^4} \tanh\left(1/x\right) \,, \tag{6.15}$$

with

$$\Theta = \frac{2m_e K_{\text{B}} T}{\hbar^2 (3\pi^2 n_e)^{2/3}} \,: \tag{6.16}$$

In the lowest order in Γ, the following accurate fitting formula obtained by Perrot and Dharmawardana (1984) may be used:

$$\varepsilon_{\text{int}} = -K_{\text{B}} T a(\Theta)\Gamma \,. \tag{6.17}$$

In conclusion, we may say that the temperature corrections to the zero-temperature correlation energy are quite relevant, since they are not small. In particular, for smaller densities, they may go to both sides of the ground state energy. The problem with Padé-type interpolations is that the intermediate region remains uncertain to some extent. However, there are many possibilities to interpolate between the known results for the degenerate and the non-degenerate electron gas, corresponding to quite different formulae which are all in qualitative agreement, but with some quantitative differences (Zimmermann et al. 1984, 1988; Ichimaru 1992, 1993; Tanaka and Ichimaru 1985; Stolzmann and Blöcker 1996, 1998; Dornheim et al. 2016c). In papers by Lehmann and Ebeling 1996, an improvement on (6.13) was given which reads (same notation as above)

$$\mu_{\mathrm{xc}} = \frac{U(r_{\mathrm{s}}, \tau = 0) - 3.4634 r_{\mathrm{s}}^{-3/2} \tau^{-1/2} - 3 r_{\mathrm{s}}^{-3} \tau^{-1}}{1 + R(r_{\mathrm{s}}, \tau) + U(r_{\mathrm{s}}, \tau)} . \tag{6.18}$$

$$R(r_{\mathrm{s}}, \tau) = 6.8 \ln(1 + 0.3685 r_{\mathrm{s}}^{-3/2} \tau^{-1}) , \tag{6.19}$$

$$U(r_{\mathrm{s}}, \tau) = 3.1897 (r_{\mathrm{s}} \tau)^{-3} + 169.64 r_{\mathrm{s}}^{-6} \tau^{-3} . \tag{6.20}$$

It was shown that this formula is also in good agreement with molecular dynamics data for finite temperatures (see Ebeling and Schautz 1977 and Chap. 8). We note that recent *ab initio* QMC calculations (Dornheim et al. 2015b, 2016b; Groth et al. 2016) provide much greater accuracy than the 'old data' available in the 1980s and may be used to derive more precise analytical formulae. We note that there already exist rather precise analytical formulae obtained by the parametric fit of available numerical data, which also use available theoretical formulae for limiting cases, e.g., the formulae given by Ichimaru (1993) and by Perrot and Dharmawardana (2000). A critical discussion of accuracy may be found in Dharmawardana (2015), Groth et al. (2016), and Desbiens et al. (2017).

There can be no doubt that adapted formulae that agree very well with the numerical data are quite useful for applications in many fields. However, we would express the opinion that better agreement does not necessarily mean that the corresponding theoretical basis is better. The very good agreement with data was in part obtained by fitting 20 and more free parameters. The relatively simple Padé formulae given above do not contain more than one or two free parameters and have the greatest uncertainty around $n\Lambda^3 \sim 1$. In this context, we also mention the old consensus: "the smaller the number of 'free parameters' in the model, the better". Many physicists still remember an anecdote concerning Enrico Fermi who said in a discussion with a student: "I remember my friend Johnny von Neumann used to say, with four parameters I can fit an elephant, and with five I can make him wiggle his trunk." More seriously, according to Fermi and von Neumann, a formula with more than 2 or 3 free parameters should be considered just as a numerical fit, not a theory.

6.2 Chemical Potential and Internal Energy of Dense Electron–Ion Fluids

6.2.1 Reduced Mass Approximation and Beyond

The subsystem of free charges in a plasma consists of the electrons which are not tightly bound to nuclei forming atoms or molecules. We represent the chemical potential of the plasma of free charges as the sum of ideal, Hartree–Fock and correlation contributions:

$$\mu_{\text{plasma}} = \mu_{\text{e}}^{\text{id}} + \mu_{\text{i}}^{\text{id}} + \mu_{\text{e}}^{\text{HF}} + \mu_{\text{i}}^{\text{HF}} + \mu_{\text{c}} \,. \tag{6.21}$$

The first four terms, i.e., the ideal Fermi–Dirac and the Hartree–Fock terms are exactly known, as described in detail in Chap. 3.

To find approximations beyond Hartree–Fock, we have to consider the rather complex landscape of the parameters in a two-component plasma. To be more specific, we consider an electron–proton system, while other two-component plasmas are rather similar. To specify the parameter region where we are working in, we look again at the density–temperature plane. For the case of a symmetrical plasma, the (n, T) plane looks like the one for the OCP which is shown in Fig. 6.1. The corresponding regions for an electron–proton plasma, which are much more complex, are shown in Fig. 6.6. In a specific region in the (n, T) plane, the plasma does not 'feel' the large differences between the masses, but only the reduced mass of the $+-$ interactions. This is the lower right region in Fig. 6.6 with moderate densities and moderate temperatures. It is the region where the reduced mass approximation (RMA) works.

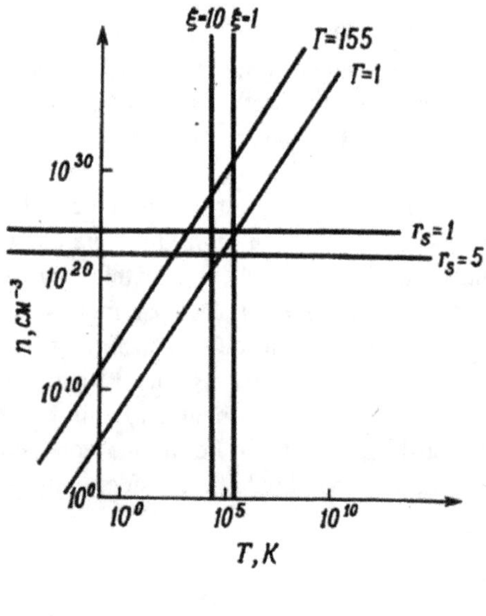

Fig. 6.6 Different regions of the parameters in an electron–proton gas. The *curve* shows a transition between Debye-like and Wigner-like behavior in the region $\Gamma > 1$

In the RMA, the correlation potentials depend only on the reduced electron–ion mass $\mu = m_e m_i/(m_e + m_i)$ and the corresponding relative thermal wavelength λ. All expressions have the same analytical shape as for the OCP, and only two parameters change:

$$\lambda_{ee} \longrightarrow \lambda = \lambda_{i.e.} = \frac{1}{\sqrt{2}}\lambda_{ee} \, , \qquad \kappa_e \longrightarrow \kappa_{i.e.} = \sqrt{2}\kappa_e \, . \tag{6.22}$$

Denoting the density of the free charges by $n^* = n_e^* = n_i^*$, in the RMA, the (total) chemical potential (Kraeft et al. 1986; Lehmann and Ebeling 1996) is given by

$$\mu_c^{TCP} = \mu_c^e + \mu_c^i = 2\mu_c^{OCP}(n^*, \tau/2) \, . \tag{6.23}$$

Note that the expressions derived by Lehmann and Ebeling 1996 were originally valid only for the case of mass-symmetric plasmas, but as we have already stressed several times, the RMA makes it possible to reduce any plasma in some region to a mass-symmetric one. Using the scaling laws, in the RMA, we get the same expressions for the free charges in an electron–ion plasma as for the symmetrical TCP, namely (6.23). This way the RMA reduces the chemical potential of mass-unsymmetrical plasmas to that of the OCP.

Summarizing, we may say that, in the ring contributions, the effect of the asymmetry of the masses on the correlation energy is quite weak and can be neglected. The main effects of mass asymmetry are contained in the ideal and the Hartree–Fock terms (see Chap. 2). Strict extensions to asymmetric plasmas based on Padé approximations were given in (Ebeling 1989, 1990; Ebeling 1989, 2007; Förster et al. 1992).

6.2.2 *Wigner Lattice Effects for Strong Coupling by Mode Restriction*

The quantum effects described so far are relatively small in the region of nondegenerate plasmas. The changes due to Wigner effects in the strongly coupled region where $\Gamma > 1$ are more dramatic. According to Wigner (1934) and others Unsöld (1955), for nondegenerate plasmas at low densities, a lattice-like configuration is most favorable with respect to the free energy density, and this results in a typical $n^{1/3}$ density dependence. A first systematic statistical derivation of such laws was made by Berlin and Montroll (1952). In order to overcome the mathematical difficulties, these authors used a rather artificial regularization of the $+-$ divergencies at small distances, based on the scaling already proposed by Kramers. A simpler method to obtain laws of type $n^{1/3}$ was developed by Kaklyugin (1985) and Brilliantov (1998), based on the philosophy of plasma waves with mode restrictions (see Chap. 3).

In the last section, we started from available (reliable) results for the ground state energy of the electron gas and for the region of strongly degenerate electron fluids. We then interpolated to the region where the quantum Debye–Hückel approximation holds and to regions where the Wigner approximation for lattice-like configurations holds. The region in-between the limiting cases remains to some extent uncertain in Padé-type interpolations. In order to approach the uncertain region, we start now from a different and more or less independent side. We include lattice effects directly into the ring contributions obtained in (Kelbg 1963; Schmitz and Kremp 1967; De Witt et al. 1995), using the method of mode restrictions (Kaklyugin 1985; Briliantov 1998; Ebeling 2016). This method, which we already discussed in Chap. 4, involves restricting the k-modes by excluding the shortest plasma waves.

We begin once again from the expressions for $g^{\mathrm{K}}(k)$ and U_{c} for the pair density matrix of plasmas which we derived in Sect. 4.4, and use for electron–ion plasmas the reduced mass approximation. This means that there appears only one $\lambda = \lambda_{\mathrm{i.e.}}$. The only difference with respect to the OCP is that, in the new expression, there appears a screening parameter $\kappa^2 = 4\pi\beta(n_{\mathrm{e}} + n_{\mathrm{i}})e^2$. In Sect. 4.4, we discussed several methods for evaluating the complicated integral for the Coulomb energy. Here we start from the Taylor series of the Kummer functions $M = 1 - \lambda^2 k^2/6 + \cdots$ and their known decay according to k^{-2} at large arguments, then use the Padé approximation

$$M(1; 3/2; -\lambda^2 k^2/4) \approx \frac{1}{1 + \lambda^2 k^2/6} . \tag{6.24}$$

Introducing now an upper limit for the k integration, and thereby restricting the contributions of the short-range modes, we get (Ebeling 2016)

$$U_{\mathrm{c}} = -4K_{\mathrm{B}}TV(n\beta e^2)^2 \int_0^{k_{\max}} \mathrm{d}k \frac{\alpha^2}{p^2 - s^2} \left(\frac{1}{k^2 + p^2} - \frac{1}{k^2 + s^2} \right) . \tag{6.25}$$

Here we have introduced the parameters $\alpha = 4/\sqrt{\pi}\lambda$ and $p \approx \kappa$, $s \approx \alpha$, the precise definition being given in Sect. 4.4 and the cited work. The integrals can be evaluated and the result is

$$U_{\mathrm{c}} = -4K_{\mathrm{B}}TV(n\beta e^2)^2 \frac{\alpha^2}{p^2 - s^2} \left[\frac{1}{p} \arctan\left(\frac{k_{\max}}{p} \right) - \frac{1}{p} \arctan\left(\frac{k_{\max}}{p} \right) \right] . \tag{6.26}$$

The corresponding Coulomb energy of an OCP, valid also for a related symmetric TCP, are represented in Fig. 6.7. We see that, for small Γ, the new formula, which includes quantum effects in the effective potential approximation, implies the shape of the quantum statistical curve obtained in Sect. 4.2. On the other hand, for large Γ, the new formula (6.26) approaches a Wigner-type linear dependence on Γ. In this way we develop a method to derive the expected Wigner behavior at high densities (Ebeling 2016).

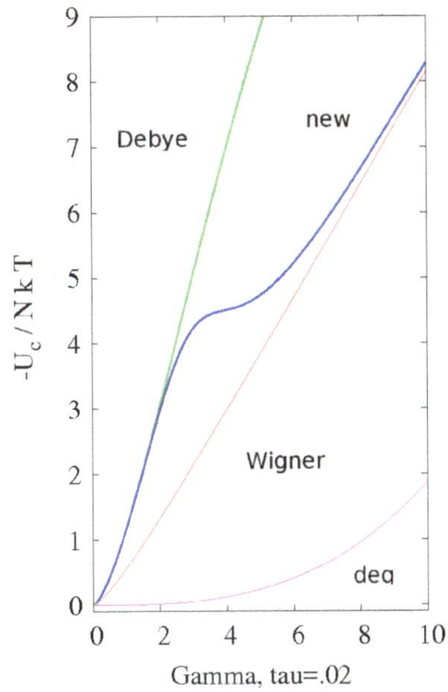

Fig. 6.7 Comparison between the Coulomb energy per particle for a rather low temperature $\tau = 0.02$ (*green curve*), as a function of the parameter $\Gamma \sim n^{1/3}$, the classical Kaklyugin–Brilliantov formula (*red curve*), and a combined theory including both effects as described in (6.26). The curve representing the new theory shows a transition between Debye-like and Wigner-like behavior in the region $\Gamma \sim 2$–8. The influence of degeneracy is shown by the lowest curves (*magenta*), with degeneracy $y = n\Lambda_e^3/2$. We see that, for $\tau = 0.02$, degeneracy is still practically absent in the given region

We note that the results represented in Fig. 6.7 so far include only weak degeneracy effects and lose validity for strong degeneracy, i.e., at $y = n_e\Lambda_e^3/2 \gg 1$. In the case of degenerate plasmas, the maximal k value is in fact no longer determined by the smallest distance d, but rather by the Bohr distance a_B. The figure compares the quantum Coulomb energy per particle for $\tau = 0.02$ (about 3140 K) with the original, entirely classical Kaklyugin–Brilliantov formula (see Brilliantov (1998). A smooth transition from the energy obtained from the quantum ring integral to the classical result of Kaklyugin–Brilliantov follows from the structural analogies of the formulae (see Fig. 6.7).

6.2.3 *Internal Energy of Free Charges Using HNC Calculations*

In the last subsection we presented a simple approach to electron–ion plasmas based on the mean mass approximation (MMA), which avoids matrix calculations. A problem with this simple approximation is the sometimes incorrect behavior at high Γ, where for physical reasons linearity in Γ is to be expected. We note first that a density dependence like $n^{1/3}$ was predicted, not only by Wigner (1934) and Berlin and Montroll (1952), but also by several astrophysicists such as Unsöld (1948, 1955), by estimating the Coulomb energy between the free ions in a plasma. Unsöld's estimate was based on a calculation of the potential threshold between two ions in a plasma, and led to a Coulomb energy (Unsöld 1948, 1955; Stolzmann and Ebeling 1998)

$$U_{\text{Uns}} \approx N K_{\text{B}} T a \Gamma \ , \qquad a \approx 1.79 \ . \tag{6.27}$$

In the last section, we have shown how a similar dependence linear in Γ may be obtained by introducing short wavelength corrections, as proposed by Kaklyugin and Briliantov. Here we use a different and perhaps more realistic route for calculating the internal energy, first separating the free charges in the plasma and then using HNC approximations.

Berlin and Montroll calculated the partition function for a nonsingular potential with special scaling. In 1927, for the treatment of point ions in electrolyte theory, Kramers proposed the potential

$$V_{kl} = \frac{e_k e_l}{r_{kl}} \left[1 - \exp(\alpha_{kl} r_{kl}) \right] . \tag{6.28}$$

We remember that this potential was used by Kelbg, Deutsch, and others to model short-range quantum effects. Kramers' aim was to give a more exact derivation of the Debye formula, and he made the choice

$$\alpha_{kl} = \frac{\lambda}{R_{kl}} \ , \tag{6.29}$$

where λ is a parameter and R_{kl} the distance from a point outside the system. Kramers argued that the additional forces have a negligible influence, but avoid divergence, while they guarantee that the thermodynamic functions satisfy Klein's theorem due to their scaling properties, and depend only on the combination $n^{1/3}/T$. Kramers succeeded in deriving Debye's limiting law, but the theory breaks down beyond a critical concentration Kramers (1927). In a pioneering work, Berlin and Montroll derived the thermodynamic functions for a system interacting through a Kramers potential for all densities, and found expressions depending only on Γ and satisfying Debye's law for small densities and Wigner's law beyond a critical density. In a tour de force, Berlin and Montroll (1952) derived the following exact analytical result for the interaction part of the internal energy:

$$\frac{U_{\text{in}}(\Gamma)}{N K_{\text{B}} T} = \begin{cases} -\left(\dfrac{3}{2}\right)^{1/2} \Gamma^{3/2} \ , & \text{if} \quad 0 < \Gamma < \Gamma_{\text{c}} \ , \\ -a\Gamma + c \ , & \text{if} \quad \Gamma > \Gamma_{\text{c}} \ , \end{cases} \tag{6.30}$$

with $a = (3/2)^{4/3} \approx 1.717$, $c = 1/2$, and $\Gamma_{\text{c}} = (2/3)^{1/3} \approx 0.8736$.

We note that Berlin and Montroll's result is surprisingly close to Unsöld's estimate 1948. Subsequently, several efforts have been made to implement this so-called Berlin–Montroll–Unsöld asymptotic in the thermodynamic theory (Stolzmann and Ebeling 1998). In Chaps. 2 and 4 and in the previous section, we presented an ad hoc solution of this problem based on the idea of mode restrictions due to Debye, Kaklyugin, and Brilliantov. However, in order to find a more fundamental solution to this problem, we have to improve the pseudo-potentials at small distances and the methods for calculating the distribution functions at high densities. In fact, for high densities, only numerical methods provide a way to calculate the distribution functions with sufficient accuracy, e.g., molecular dynamics or the solution of integral equations of HNC-type (Sandig 1975; Ebeling et al. 1999a, Sadykova and Ebeling 2007; Bredow et al. 2013).

Here we will use an HNC approach in combination with pseudo-potential methods, a procedure which belongs nowadays to standard methods (Ebeling et al. 1999a, Bredow et al. 2013; Erimbetova et al. 2013). In order to derive improved pseudo-potentials, we correct the Kelbg potential for small densities by including higher orders in the perturbation theory for the free states. The contributions of the bound states are excluded. Since the nonsingular model potential due to Kramers, Berlin, and Montroll was based only on scaling arguments, without a clear physical derivation, we start again from the general theory of effective potentials, as explained in Chaps 3–5. In this way, we obtain effective potentials for the interactions between the free charges in an electron–ion plasma which are nonsingular.

In a first approach we start from the classical limit of the Slater sum for the free states, which was obtained by following Planck's procedure in (Ebeling 1968; Sandig 1975):

$$S^{\mathrm{f}}_{\mathrm{ei}}(r) = \exp(y)\big[1 - \Phi(y^{1/2})\big] + \frac{2}{\sqrt{\pi}} y^{1/2} + \frac{4}{3\sqrt{\pi}} y^{3/2} + \frac{8}{15\sqrt{\pi}} y^{5/2} \,, \tag{6.31}$$

$$y = Ze^2/K_{\mathrm{B}}Tr \,. \tag{6.32}$$

For the effective potential of the interaction of free charges, it follows classically that

$$\beta\varphi^{\mathrm{f}}_{\mathrm{ei}} = -\ln S^{\mathrm{f}}_{\mathrm{ei}}(r) \,, \tag{6.33}$$

$$\varphi^{\mathrm{f}}_{\mathrm{ei}} = -K_{\mathrm{B}}T \ln\Big\{ \exp(y)\big[1 - \Phi(y^{1/2})\big] + \frac{2}{\sqrt{\pi}} y^{1/2} + \frac{4}{3\sqrt{\pi}} y^{3/2} + \frac{8}{15\sqrt{\pi}} y^{5/2} \Big\} \,.$$

Quite useful for calculation is the asymptotic formula for large y, i.e., small r values:

$$\exp(y)\big[1 - \Phi(y^{1/2})\big] = \frac{1}{\pi} \sum_{k \geq 0} (-1)^k \frac{\Gamma(k/2 + 1)}{y^{k+1/2}} \,.$$

The classical effective potential is temperature dependent and has only a weak logarithmic singularity at $r \to 0$, $y \to \infty$. In a rough approximation, we may use

$$\varphi_{\mathrm{ei}}^{\mathrm{f}} = -\frac{5}{4} K_{\mathrm{B}} T \ln \left(1 + \frac{4}{5} \frac{e^2}{K_{\mathrm{B}} T r} \right) . \tag{6.34}$$

For the electron–electron and the proton–proton interaction the standard Coulomb potentials are applied. Sandig (1975) first carried out simulations based on the HNC integral equations (see Chap. 2) in the form developed and used by Allnatt, Rasaiah, and Friedman for Coulomb systems. More extended calculations based on HNC equations (see Chap. 2) were performed by Kasch et al. (see Kraeft and Schlanges 1996, Ebeling et al. 1999b). The Allnatt–Rasaiah–Friedman equation for the pair distribution function F_{ab} may be written in the short form (Rasaiah and Friedman, Sandig 1975)

$$F_{ab} = 1 + h_{ab} = \exp(g_{ab} + \tau_{ab}) , \tag{6.35}$$

$$\tau = X * h + g * X + g * X * h , \qquad X = h - g - \tau , \tag{6.36}$$

where g_{ab} is the usual statically screened Coulomb potential. The asterisk indicates summation over the species at the cutting point and the performance of the convolution integral. The essence of the approximation is that bridge contributions are neglected. Meanwhile standard algorithms exist for the numerical solutions.

The results for the energy depend only on the nonideality parameter Γ and may be represented in the form (6.30) used by Berlin and Montroll. Note that more general results for the scaling properties of Coulomb systems were obtained in Rosenfeld's density functional theory (Rosenfeld 1989, 1994). The fact that once again we get here a dependence solely on Γ is connected with the scaling properties of our classical potentials, which depend only on $y = Ze^2/K_{\mathrm{B}}Tr$. Following Rosenfeld and Tarazona 1998, we proposed for our case the asymptotic of the internal energy and the excess free energy (Ebeling, Stolzmann, Förster and Kasch 1999b):

$$\frac{U_{\mathrm{in}}}{N K_{\mathrm{B}} T} = -a\Gamma + b\Gamma^{3/2} + c , \quad \frac{F_{\mathrm{ex}}}{N K_{\mathrm{B}} T} = -a\Gamma + \frac{5}{2} b\Gamma^{5/2} + c \ln \Gamma + d . \tag{6.37}$$

When fitting the results of the HNC calculations, a problem appears due to the temperature dependence of the effective potential. By taking the average of the potential $\varphi_{\mathrm{ei}}^{\mathrm{f}}$ and after fitting the data the values, we obtained

$$a = 2.25 , \qquad b = 0 , \qquad c = 0.30 .$$

However, these results need revision. Debye showed for ionic solutions that only the average of the effective potential gives the proper internal energy. According to Debye, the effective potential is in general

$$u_{ab}^{\text{eff}} = \frac{\partial(\beta\varphi_{ab})}{\partial\beta} , \tag{6.38}$$

and in our case we have

$$u_{ab}^{\text{eff}} \approx -\frac{e^2}{(r + 4\beta e^2/5)} . \tag{6.39}$$

By averaging the effective potential, we find the new constants (Kasch in Kraeft and Schlanges 1996; Ebeling et al. 1999a)

$$a = 1.72 , \qquad b = 0 , \qquad c = 0.35 . \tag{6.40}$$

After this revision, we see that the energies contained in all estimates due to Unsöld (1948), Berlin and Montroll (1952), and Kasch (in Kraeft and Schlanges 1996) are in reasonable agreement. Let us summarize the result of Berlin and Montroll, which agrees with Unsöld's estimate and the HNC results of Sändig, Kasch et al.:

> The thermodynamics of classical free charges with regularized interactions satisfies the Debye law for small coupling and the Wigner law, or more precisely, a Berlin–Montroll–Rosenfeld asymptotic, beyond a critical coupling. We will show in the following that, in the case of quantum plasmas, the structure remains essentially the same, although the Debye law and the Wigner slope are significantly modified by quantum corrections.

For explicit numerical treatment of the free charges of quantum plasmas we proceed as follows. First, we split the pair density matrix or correlation function of pairs into a bound state and a free state part. The new pseudo-potential for electron–ion plasmas is derived from the free state contribution, which is found from the exact Coulomb wave functions (Sadykova and Ebeling 2007; Ebeling and Sändig 1973). The idea is to improve the Kelbg potential by a correction for small distances, using an additional term which is formally similar to the second part of the Deutsch potential. In this way, the new effective interaction potential for free charges, which includes quantum effects, consists of two parts:

$$\varphi_{ab}^{\text{f}}(r) = \varphi_{ab}^{\text{K}}(r) - K_{\text{B}}T\tilde{A}_{ab}(\xi_{ab})\exp\left(-\frac{r^2}{\lambda_{ab}^2}\right) , \qquad \xi_{ab} = -\frac{e_a e_b}{K_{\text{B}}T\lambda_{ab}} . \tag{6.41}$$

Here $\tilde{A}_{ab}(\xi_{ab})$ is a quite complicated expression, given by the exact wave functions of the free states. The amplitude $\tilde{A}_{ab}(\xi_{ab})$ is defined in such a way that, in the limit of small distances and zero densities, viz., $r \to 0$ and $n_{\text{e}} \to 0$, respectively, the function $\varphi_{ab}^{\text{f}}(r = 0)$ and the first derivative with respect to r have the exact quantum statistical values. Thes exact values are given by the density matrix of free particles, defined as the difference between the total density matrix and the bound state part at zero density. This quantum statistically well defined expression is used to define our new pseudo-potential $\varphi_{ab}^{\text{f}}(r)$ by

$$F_{ab}^{\text{f}} = F_{ab} - F_{ab}^{\text{b}} = \exp(-\beta\varphi_{ab}^{\text{f}}) , \quad \text{for} \quad n_a \to 0 . \tag{6.42}$$

The bound state contribution for electron–ion pairs is given by the Brillouin–Planck–Larkin expression for the two-particle bound states of electrons and ions:

$$F_{\mathrm{ei}}^{\mathrm{b}}(1,2) = C\sum\left[\exp(-\beta E_s) - 1 + \beta E_s\right]|\psi_{\mathrm{ei}}^{s}(1,2)|^2 . \tag{6.43}$$

Using the pseudo-potentials $\varphi_{ab}^{\mathrm{f}}(r)$ defined above, the distribution functions are calculated by the HNC method for Coulomb particles described in Chaps. 2–3. Note that these pseudo-potentials for the interaction of free charges and HNC solutions were discussed in (Ebeling and Sändig 1973, Sadykova and Ebeling 2007). The pseudo-potentials $\varphi_{ab}^{\mathrm{f}}(r)$ are not singular, so divergencies no longer arise at $r = 0$. They thus belong to the class of potentials studied by Berlin and Montroll. They are constructed in such a way that the values at zero distance and the first derivatives are quantum-statistically exact, since they are defined by the exact Coulomb wave functions. By calculating the pair distribution and the energy using molecular dynamics methods or using HNC integral equations (Ebeling and Sändig 1973), respectively, we reproduced the functional dependence found by Berlin and Montroll for larger (overcritical) Γ, although with different and slightly temperature-dependent coefficients:

$$\frac{U_{\mathrm{in}}(\Gamma)}{N K_{\mathrm{B}} T} = -a(\tau)\Gamma + c(\tau) . \tag{6.44}$$

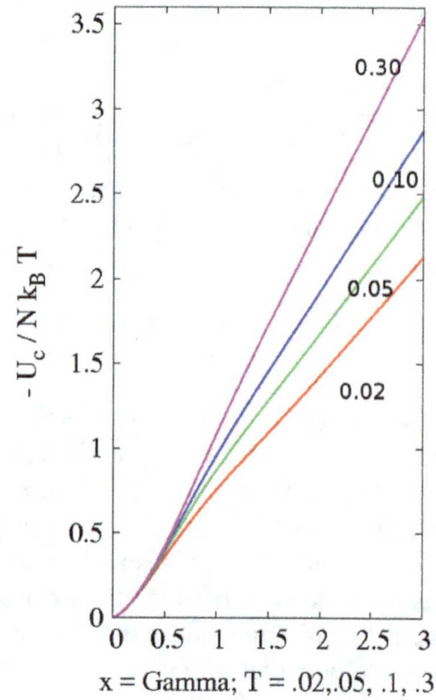

Fig. 6.8 Coulomb energy per $N K_{\mathrm{B}} T$ for the free component of an electron–ion plasma. We show our extended Berlin–Montroll formula (6.46) for the temperatures $\tau = .02$ (*red*), $\tau = .05$, $\tau = .1$, and $\tau = .3$ (*magenta*). The curves correspond to the quantum statistical estimates from ring integrals for low Γ and to the asymptotic from solution of the HNC equations for the pseudo-potential of free charges in the high Γ region

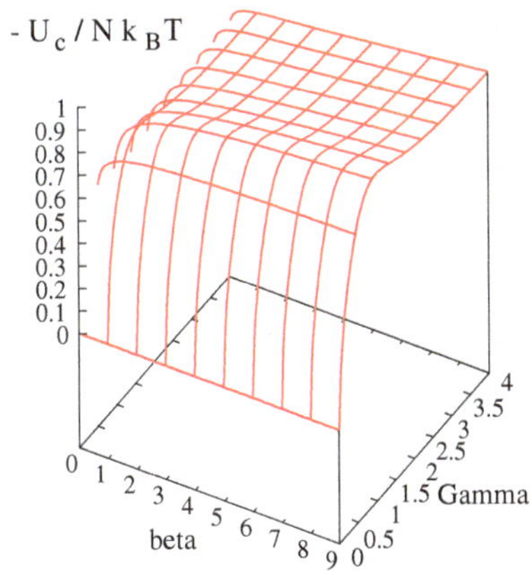

Fig. 6.9 Two-dimensional plot of the Coulomb energy per NK_BT for the free component of electron–ion plasmas. We show our generalized Berlin–Montroll formula (6.46) for quantum plasmas as a function of Γ and $\beta = 1/\tau$. The low density (small Γ) region is the region of quantum effects in the ring approximation. For high Γ, the energy corresponds to the $n^{1/3}$ slope given by the asymptotic from solutions of the HNC equations for pseudo-potentials

The differences with the original Berlin–Montroll formula are due to the quantum corrections we included here. A new fit of the empirical HNC data leads us to the following choice of coefficients:

$$a(\tau) = 0.72 + 1.0 \exp(-8\sqrt{\tau}) \,, \qquad c(\tau) = 0.35 \exp(-8\sqrt{\tau}) \,. \tag{6.45}$$

Collecting the available results of the simulations and imposing the limiting law leads to a simple Padé approximation covering a wide range of temperatures and densities in the nondegenerate or weakly degenerate region (Ebeling 2016):

$$-\frac{U_{\text{in}}}{NK_BT} = \frac{1.225\Gamma^{1.5} + \Gamma^3\big[a(\tau)\Gamma - c(\tau)\big]}{1 + 1.0854\tau^{.5}\Gamma^{1.5} + \Gamma^3} \,. \tag{6.46}$$

The shape of this function of Γ and τ is given in Figs. 6.8 and 6.9.

6.3 Free Energy of Dense Plasmas

6.3.1 Main Contributions and Limits of the Free Energy

The contribution of the Coulomb forces to the free energy of electron–ion plasmas is the most important part of the total free energy needed for the transition to a chemical description based on minimization of the free energy. We now specify for this investigation the various contributions to the free energy density $f = F/V$ in a chemical description of the plasma. We assume the following structure (Beule et al. 1999b, 2001):

$$f = f_{\mathrm{e}}^{\mathrm{id}} + f_{\mathrm{i}}^{\mathrm{id}} + f_{\mathrm{a}}^{\mathrm{id}} + f_{\mathrm{m}}^{\mathrm{id}} + f_{\mathrm{hc}} + f_{\mathrm{coul}} \,. \tag{6.47}$$

The first 3 terms describe the ideal contributions of the electrons, ions, atoms, and molecules, while f_{hc} includes the hard core interactions of the neutral components. The last term represents the electric contribution of the free plasma. The free plasma contributions to the chemical potential and energy were already treated in the last few sections.

We split the free energy density for technical reasons into three contributions,

$$f_{\mathrm{coul}} = f_{\mathrm{ee}} + f_{\mathrm{ii}} + f_{\mathrm{i.e.}} \,, \tag{6.48}$$

representing the Coulomb interactions and the electron–electron, ion–ion, and ion–electron contributions. For these terms, different approximations exist, such as the quantum Debye–Hückel approximation (QDHA) and the Padé approximations, described in part in the last few sections. For the first qualitative estimates, we may use the simple closed expressions given by the quantum Debye–Hückel approximation (QDHA), which was discussed in detail in Chap. 3:

$$f_{\mathrm{coul}} \approx f_{\mathrm{DH}} \,. \tag{6.49}$$

In the QDHA, screening and quantum effects are both taken into account in a first approximation. In the QDHA or $\Lambda/8$ approximation introduced in Chap. 3, we have

$$f_{\mathrm{coul}} = -k_{\mathrm{B}}T \frac{\kappa^3}{12\pi} \tau\big(\kappa a(T)\big) \,. \tag{6.50}$$

Here the reciprocal Debye radius is defined by

$$\kappa = \sqrt{\frac{2n_{\mathrm{i}}e^2}{\varepsilon_0 k_{\mathrm{B}}T}} \,, \tag{6.51}$$

and the effective thermal electron radius is given by

$$a(T) = \frac{\Lambda_{\mathrm{e}}}{8} \,, \tag{6.52}$$

where Λ_{e} is the electron thermal wavelength defined above. Furthermore, the function τ has the standard Debye–Hückel form derived in Chaps. 3 and 4, viz.,

$$\tau(x) = \frac{3}{x^3}\left[\ln(1+x) - x + \frac{x^2}{2}\right] \,. \tag{6.53}$$

In the case of strong interactions, say $\Gamma > 1$, we expect a dependence that is linear in Γ, viz.,

$$f_{\mathrm{coul}} = -n K_{\mathrm{B}} T a \Gamma \ . \tag{6.54}$$

In an early classical estimate, Unsöld found the value for the constant to be $a \approx 1.78$. In the statistical mechanical calculation by Berlin and Montroll, the value $a \approx 1.72$ was obtained, and in our semi-classical HNC calculation, we reproduced essentially the Berlin–Montroll result with quantum corrections. This way we found a quantum-modified Debye law for $\Gamma < 1$, as decribed in Sect. 6.2.1 and for larger Γ, say for $\Gamma \leq 1$, a law like

$$\frac{F_{\mathrm{ex}}(\Gamma)}{N K_{\mathrm{B}} T} = -\big[0.72 + 1.0 \exp(-8\sqrt{\tau})\big]\Gamma + 0.35 \exp(-8\sqrt{\tau}) \ln(\Gamma) + 0.8 \ . \tag{6.55}$$

The inclusion of further effects, such as the Hartree–Fock effect for the electronic component and others, will be discussed in the next part using Padé approximations.

6.3.2 *Padé Approximations for the Plasma Free Energy*

According to Wikipedia, we define a Padé approximant as the 'best' approximation of a function by a rational function of given order under this technique. In general, the source function is given as a power series which is known up to a certain order, and in some cases we have information about the asymptotic at large arguments. The task is to search a rational approximation of this power series which also reflects any additional information, e.g., about the asymptotic. The power series for the approximant should agree with the power series of the function it is approximating. The technique was developed around 1890 by Henri Padé, but goes back to Georg Frobenius, who introduced the idea and investigated the features of rational approximations of power series. The Padé approximant often gives a better approximation of functions than truncated Taylor series, and it may still work when the Taylor series does not converge. Since the statistical thermodynamics of plasmas often starts with power series in the density, Padé approximants are used extensively in representations of thermodynamic functions of plasmas. In particular, the higher approximations are most usefully represented in the form of Padé formulae (Kraeft et al. 1986; Ebeling 1990, Ebeling et al. 2007).

But before we go into the details, let us analyze the main results of the last section. Summarizing, we may say that, at the lowest densities, the Debye law gives the main contribution, although it is strongly modified by quantum effects. However, for low Γ, the quantum effects give rather small corrections to the classical limiting law. The main quantum effects show up in the slope of the Wigner-like behavior reached for large Γ. The Coulomb energy obtained from the new pseudo-potential (6.41) which is described by the Padé formula (6.46) is in good agreement with data points obtained earlier from HNC calculations (Ebeling et al. 1999a). The agreement with the data from molecular dynamics calculations (Ortner et al. 1997, 1999) is only qualitative. The data from molecular dynamics have a similar shape to the HNC data, but are, e.g., at $\Gamma = 3$, 20 to 30 percent higher.

Note that the results of extensive HNC calculations for the free energy where presented by Tanaka et al. (Tanaka and Ichimaru 1985, 1986) and (1992, 1993). The overall landscape of the Coulomb energy of the free charges with respect to Γ and the reciprocal temperature $\beta = 1/\tau$ (Fig. 6.9) shows first a fast increase with Γ according to the limiting law, then a change in the slope due to quantum effects, and finally reaches a Wigner-like slope, weakly increasing with $n^{1/3}$. This looks quite similar to the landscape for the degenerate region obtained in (Kraeft et al. 1986). In contrast to the landscapes obtained from simpler methods, based on the introduction of a maximal wave number as explained in Sect. 6.2.2, we find now that the coefficient of the Wigner slope $n^{1/3}$ is determined by quantum effects and decreases strongly with temperature. Summarizing the most important results from the last few sections, we may state that, evidently at higher densities $\Gamma > 1$, the free charges in a plasma are no longer determined by Debye-like laws, but by Wigner-type laws with a Berlin–Montroll–Rosenfeld asymptotic. In other words, at high (but still nondegenerate) densities, the thermodynamic functions follow cubic laws with density dependence $n^{1/3}$ and prefactors determined by quantum effects.

We now study Padé approximations for the constituents of the free energy density. For the contribution of the electron–electron interactions, there exist many versions (Perrot and Dharmawardana 1984; Richert and Ebeling 1984; Kraeft et al. 1986; Stolzmann and Blöcker 1996, 1998; Potekhin and Chabrier 2013). In order to give an example, we discuss here the Padé approximations proposed by Förster et al. (1992), Ebeling et al. (1991, 2007), and Stolzmann and Blöcker (1996), Stolzmann and Ebeling (1998) For the electron–electron contribution, the authors used

$$f_{\mathrm{ee}} = -n_{\mathrm{e}} \frac{\tau a_0 \Gamma^{3/2} - a_2 \Gamma^6 \varepsilon_{\mathrm{c}}(r_s, \tau)}{1 + a_1 a_3 \Gamma^{3/2} + a_2 \Gamma^6} . \tag{6.56}$$

The contribution of the electron–electron interaction may be improved by using an integration of the many parameter fit formula (6.14) derived by Tanaka and Ichimaru (1985, 1986):

$$f_{\mathrm{ee}} = -n_{\mathrm{e}} K_{\mathrm{B}} T \int_0^{\Gamma} \frac{\mathrm{d}\Gamma}{\Gamma} (\beta \varepsilon_{\mathrm{int}}) . \tag{6.57}$$

For the ion–ion contribution, the authors propose

$$f_{\mathrm{ii}} = -n_{\mathrm{i}} \tau \frac{b_0 \Gamma^{3/2} - b_2 \Gamma^{9/2} \epsilon_{\mathrm{ii}}}{1 + b_1 \Gamma^{3/2} + b_2 \Gamma^{9/2}} , \tag{6.58}$$

where ϵ_{ii} is the classical expression as found from MC data. For the contribution of the ion–electron interactions, the following expression has been used:

$$f_{\mathrm{i.e.}} = -n_{\mathrm{i}} \tau \frac{c_0 \Gamma^{3/2} + c_2 \Gamma^6 \varepsilon_{\mathrm{i.e.}} + c_3 c_4 \Gamma^{7/2}}{1 + c_1 \Gamma^{3/2} + c_3 \Gamma^{9/2}} . \tag{6.59}$$

These forms can be applied to electron–ion plasmas with arbitrary charge number Z, such as for hydrogen $Z = 1$ and helium $Z = 2$. For the corresponding many-parameter fits of the ion–electron and ion–ion contributions for wide ranges of densities and temperatures, see, e.g., the papers (Ichimaru et al. 1987, Chabrier and Potekhin 1998), and for more information on Padé formulae, see Stolzmann and Blöcker (1996); Stolzmann and Ebeling (1998). For further development of HNC calculations, see, e.g., Bredow et al. (2013, 2015).

6.4 Advanced Chemical Models Including Bound States

6.4.1 *Free Energy in the Chemical Picture*

From now on we study many-component systems with bound states described in the framework of chemical models. We will show that this description provides a relatively simple formalism and in general gives a good representation of the properties of real plasmas. This is basically connected with the fact that chemical models are based on variational principles for the free energy of the plasma. However, we will also point out the specific limitations of chemical models, in particular in cases where the variational calculation does not lead to well defined minima of the free energy.

Let us first present our chemical approach to the free energy of dense hydrogen plasmas which was recently applied to temperatures between 2000 and 10,000 K (Beule et al. 1999b, 2001). The effects of pressure dissociation, $H_2 \rightleftharpoons 2H$, and ionization, $H \rightleftharpoons e + p$, are taken into account to explain the transition from a molecular fluid at low temperatures and pressures, through a partly dissociated, warm fluid at medium temperatures of a few thousand kelvin, to a fully ionized, hot plasma above 10,000 K. Basically, the hydrogen system consists of two main components, the plasma (electrons and protons) and the neutral fluid (atoms and molecules). Correspondingly, the free energy for a two-component system of neutral (F_{nl}) and charged particles (F_{pl}) has the form

$$F(V, T, N) = F_{\mathrm{nl}}(V, T, N) + F_{\mathrm{pl}}(V, T, N) . \tag{6.60}$$

In this section, following (Beule et al. 1999b, 2001), we combine the available results for the fully ionized plasma domain (see previous sections) with improved data for the dense, neutral fluid calculated in a dissociation model (see Bunker et al. 1997). Both contributions to the free energy are split into ideal and interaction parts. We take into account the interactions in the neutral and the charged subsystem, whereas the interaction between charges and neutrals is accounted for by the reduced-volume concept. For the purpose of this comparison, it it useful to consider the free energy per proton, which is denoted here by $\phi = F/K_{\mathrm{B}}TN_{\mathrm{p}}$, or the free energy density $f = F/V$.

Instead of going into details, we discuss here only the principal structure of the free energy in the chemical picture as a function of the variational parameters. We consider a hydrogen plasma at fixed temperature T and proton density n. We take into account ionization and dissociation processes

$$\mathrm{H} \Longleftrightarrow p + e^- , \qquad \mathrm{H}_2 \Longleftrightarrow \mathrm{H} + \mathrm{H} . \tag{6.61}$$

For simplicity, the formation of other possible species as H_2^+ and H^- species will be neglected. The degrees of ionization and dissociation will be defined by (Beule et al. 1999b)

$$\alpha = \frac{n_\mathrm{i}}{n_\mathrm{i} + n_\mathrm{a} + 2n_\mathrm{m}} , \qquad \beta_\mathrm{d} = \frac{n_\mathrm{a}}{n_\mathrm{a} + 2n_\mathrm{m}} , \tag{6.62}$$

$$\beta_\mathrm{a} = \frac{n_\mathrm{a}}{n_\mathrm{i} + n_\mathrm{a} + 2n_\mathrm{m}} , \qquad \beta_\mathrm{m} = \frac{2n_\mathrm{m}}{n_\mathrm{i} + n_\mathrm{a} + 2n_\mathrm{m}} . \tag{6.63}$$

These are the variational parameters of our problem. The free energy has to be minimized with respect to them. We note that β_d is the degree of dissociation of molecules into atoms, β_a is the relative amount of protons bound in atoms and β_m the relative amount bound in molecules. Due to the balance relation for the total proton density we get

$$n = n_\mathrm{i} + n_\mathrm{a} + 2n_\mathrm{m} , \quad \beta_\mathrm{a} = \alpha(1 - \beta_\mathrm{d}) , \quad \beta_\mathrm{m} = (1 - \alpha)(1 - \beta_\mathrm{d}) . \tag{6.64}$$

In other words, only one of the β parameters is independent, and we have some freedom. It can be shown that atoms appear only in a rather narrow region of the temperature–density plane. Therefore, we may in some approximation assume $\beta_\mathrm{a} = 0$, whereupon α remains as the only free parameter. The condition of neutrality requires that electron and ion densities are always equal $n_\mathrm{e} = n_\mathrm{i}$. The equilibrium composition is found by minimizing the free energy. Examples are shown in Fig. 6.10.

We performed classical Monte Carlo simulations for partially dissociated fluid hydrogen and for a grid of temperature and density points in the region $T = (2\text{–}10) \times 10^3$ K and $\varrho = (0.2\text{–}1.1)$ g/cm^3. Effective pair potentials of the form exponential-6 were used to model the interactions between molecules and atoms in the dense fluid. The dissociation equilibrium $H_2 \rightleftharpoons 2H$ was solved taking into account the correlation parts of the chemical potentials using variational fluid theory. The Monte Carlo data for the interaction contribution can be interpolated accurately within an eight-parameter fit with respect to density and temperature, leading to an analytical expression for the free energy density (see Beule et al. 1999b). The contribution of Coulomb interactions is described here by Padé approximations, as discussed in the previous sections (see also Kraeft et al. 1986; Stolzmann and Ebeling 1998; Ebeling 1989, 2007). The interaction contributions of the charged component account for correlations and exchange in the electronic and ionic subsystems, and electron–ion screening. They are treated in terms of Padé approximations based on analytical results for the quantum virial expansion. The comparison shows that the

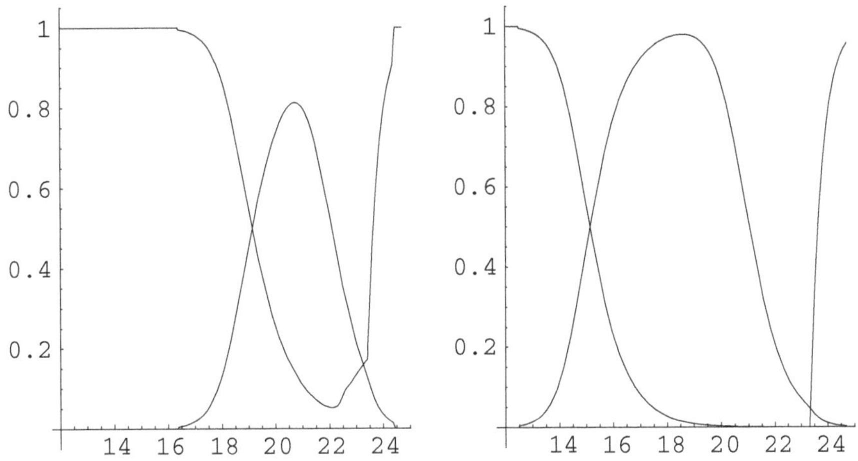

Fig. 6.10 Degree of ionization and degree of dissociation at $T = 10,000$ K (*left*) and $T = 20,000$ K (*right*)

two quantities behave in a similar way, showing a minimum at intermediate densities. However, we also see that the composition depends in a much smoother way on the density than on the degree of ionization.

Here and in the last two chapters, we have studied the problem of bound states in different approximations and may compare several methods for studying the thermodynamics of dense hydrogen plasmas:

- density expansions,
- fugacity expansions,
- summed density (or mixed) representations,
- chemical descriptions based on minimization of the free energy or Saha equations.

In short, we may conclude that the results are the same in qualitative respects, but differ in the details. So far our work has concentrated on hydrogen. In order to use the hydrogen EOS for deuterium, mass scaling can be applied for the interpolation formula of the interaction contributions, i.e., it is assumed that the same particle numbers for hydrogen and deuterium lead to the same degree of dissociation and to identical interaction contributions to the thermodynamic function of the neutral fluid for a given temperature.

At very high densities the description of ionization and dissociation by mass action laws and the Saha equation fails in special cases and might possibly give incorrect results. The reason is that, at high densities, the bound states are not well defined. However, we may always find the equilibrium values of the degrees of ionization and dissociation by minimization of the free energy at fixed total mass (Beule et al. 1999b, Ebeling et al. 2003, 2005, 2007). The minimization procedure for functions involves some intrinsic difficulties which may be connected with the shape of the function under study.

Geometry of the Free Energy Landscape

The idea of applying geometrical methods in thermodynamics goes back to the pioneering papers of Gibbs published in 1892. We apply Gibbs' method to plasmas here. We will study different approximations to the thermodynamic functions and the influence of several effects. Our approach is limited to the region $T > 10,000$ K. A more advanced approach for still higher densities will be given in the next section. Instead of the Saha equations we will use free energy minimization (Ebeling et al. 2000, 2003).

We also include here in part the effects of magnetic fields, which are of increasing interest, not only for astrophysical applications, but also for plasmas confined in magnetic traps. We refer to the results on magnetic field effects given in (Steinberg et al. 2000a). Note that the magnetic field introduces an anisotropy into the system. This changes the ideal gas contribution, the bound state energies, and the contributions due to interactions of the free particles. Here we will take into account the ideal gas corrections in an exact way, and the bound state and interaction corrections in approximations which include the quadratic orders in the field $O(B^2)$. From a practical standpoint, this restricts our approach to fields $B < 10^5$ T.

First we explore the geometry of the free energy landscape as a function of the degrees of ionization and dissociation. On this basis we are able not only to describe the chemical equilibrium, but also to get an idea about the fluctuations around the equilibrium values. We begin by defining the basic variational parameters of the system. In the following, we consider hydrogen plasmas at fixed temperature T and proton density n. Our aim is to study ionization and dissociation processes:

$$\mathrm{H} \Longleftrightarrow p + e^- \,, \qquad \mathrm{H}_2 \Longleftrightarrow \mathrm{H} + \mathrm{H} \,. \tag{6.65}$$

For simplicity, the formation of H_2^+ and H^- species will be neglected. In order to define the thermodynamics of the system we introduce the free energy and its density, viz.,

$$F(T, V, \{N\}) \,, \qquad f(T, n) = \frac{F}{V} = f(T, n_\mathrm{e}, n_\mathrm{i}, n_\mathrm{a}, n_\mathrm{m}) \,, \tag{6.66}$$

besides the degrees of ionization and dissociation as in (Beule et al. 1999b). The free energy density depends on only four independent parameters

$$f \longrightarrow f(T, n, \alpha, \beta) \,. \tag{6.67}$$

If we include magnetic field effects, we have five independent parameters

$$f \longrightarrow f(T, n, B, \alpha, \beta) \,. \tag{6.68}$$

The equilibrium composition is found by the minimization procedure

$$\frac{\delta f}{\delta \alpha} = 0 \,, \qquad \frac{\delta f}{\delta \beta} = 0 \,. \tag{6.69}$$

Instead of β, we may use $\beta_{\rm a}$ or $\beta_{\rm m}$ as variational parameters. The relation

$$\alpha + \beta_{\rm a} + \beta_{\rm m} = 1 \tag{6.70}$$

provides a way to visualize the free energy surface on a simplex.

We will show that in the region of partial ionization and dissociation, the QDHA yields a reasonable description of ionization and dissociation phenomena. The interactions of the neutral components are taken into account in the first approximation by hard core approximations including the effects of reduced volume (Beule et al. 1999b). We will now discuss these contributions in more detail:

- Ideal free energy of electrons $f_{\rm e}^{\rm id}$. We use Fermi–Dirac statistics including excluded volume corrections representing the free energy in the following way:

$$f_{\rm e}^{\rm id\,ex} = n_{\rm e} k_{\rm B} T z \left(\frac{1}{2} \frac{n_{\rm e} \Lambda_{\rm e}^3}{1 - \eta_0} \right) \,. \tag{6.71}$$

  The function $z(x)$ which is due to Zimmermann and was described in Chap. 2 (see also Ebeling et al. 1991, 2007) interpolates between the low density limit $z(x) \approx \ln x - 1$ and the high density limit $z\,(x) \propto x^{2/3}$. Furthermore, $\Lambda_{\rm e} = h/\sqrt{2\pi m_{\rm e} K_{\rm B} T}$ is the thermal wavelength of the electrons, and we will now use the thermal wavelengths of ions, atoms, and molecules, denoted $\Lambda_{\rm i}$, $\Lambda_{\rm a}$, and $\Lambda_{\rm m}$, respectively. The excluded volume factor $\eta_0 < 1$ expresses the fact that some part of the total volume V which is occupied by atoms and molecules is not accessible to the electrons. The accessible volume $(1 - \eta_0)V$ is smaller than the total volume and the effective densities are higher. A strict theory for η_0 is not yet available. We will use here

$$\eta_0 = \frac{4}{3} \pi R_0^3 (n_{\rm a} + 2 n_{\rm m}) \tag{6.72}$$

  as the definition of the packing parameter with an estimate for the atomic radius $R_0 = 0.78$.
- Ideal free energy of ions $f_{\rm i}^{\rm id}$. This is given by the classical term

$$f_{\rm i}^{\rm id} = n_{\rm i} k_{\rm B} T \left[\ln(n_{\rm i} \Lambda_{\rm i}^3) - 1 \right] - n_{\rm i} K_{\rm B} T \ln(1 - \eta_0) \,. \tag{6.73}$$

  The second term describes the excluded volume effect for classical bare protons. The parameter $x_{\rm i} = \hbar \omega_{\rm c}^{\rm i} / 2 K_{\rm B} T$ $(\omega_{\rm c}^{\rm i} = eB/m_{\rm i})$ is very small $x_{\rm i} \ll 1$ for small B. Therefore, magnetic field effects can be neglected for protons.

- The ideal atom contribution $f_{\mathrm{a}}^{\mathrm{id}}$. We assume Boltzmann statistics, i.e.,

$$f_{\mathrm{a}}^{\mathrm{id}} = n_{\mathrm{a}} k_{\mathrm{B}} T \left[\ln \left(\frac{n_{\mathrm{a}} \Lambda_{\mathrm{a}}^3}{\sigma_{\mathrm{a}}} \right) - 1 \right] , \tag{6.74}$$

including the internal states by means of a BPL partition function

$$\sigma_{\mathrm{a}}(\xi) = \sum_{n=1}^{\infty} n^2 \left[\exp \left(\frac{\xi}{2n} \right)^2 - 1 - \left(\frac{\xi}{2n} \right)^2 \right] , \tag{6.75}$$

where $\xi = 2\sqrt{\mathrm{Ry}/K_{\mathrm{B}}T}$, with Ry $\approx$ 13.598 eV the ionization energy of hydrogen. Including the magnetic field, we have to replace the atomic partition function σ_{a} by an effective partition function, depending on the magnetic field, which will be discussed later.

- The molecular contribution $f_{\mathrm{m}}^{\mathrm{id}}$. We will use the representation (Beule et al. 1999b)

$$f_{\mathrm{m}}^{\mathrm{id}} = n_{\mathrm{m}} k_{\mathrm{B}} T \left[\ln \left(\frac{n_{\mathrm{m}} \Lambda_{\mathrm{m}}^3}{\sigma_{\mathrm{m}}} \right) - 1 \right] + f_{\mathrm{m}}^{\mathrm{vib}} + f_{\mathrm{m}}^{\mathrm{rot}} , \tag{6.76}$$

including the internal molecular and atomic states,

$$\sigma_{\mathrm{m}} = \sigma_{\mathrm{a}}^2 \exp \left(\frac{4.746 \mathrm{eV}}{k_{\mathrm{B}} T} \right) , \tag{6.77}$$

and the contributions of vibrational and rotational states (Beule et al. 1999b)

$$f_{\mathrm{m}}^{\mathrm{vib}} = n_{\mathrm{m}} k_{\mathrm{B}} T \left[\ln \left(1 - \mathrm{E}^{-T_{\mathrm{v}}/T} \right) + \frac{T_{\mathrm{v}}}{2T} \right] , \quad T_{\mathrm{v}} = 6210\,\mathrm{K} , \tag{6.78}$$

$$f_{\mathrm{m}}^{\mathrm{rot}} = n_{\mathrm{m}} k_{\mathrm{B}} T \left[\ln \frac{T_{\mathrm{r}}}{T} - \frac{T_{\mathrm{r}}}{3T} - \frac{1}{90} \left(\frac{T_{\mathrm{r}}}{T} \right)^2 \right] , \quad T_{\mathrm{r}} = 85\,\mathrm{K} .$$

- The Debye–Hückel contribution which may be used as a first approximation f_{DH}. Using here the so-called QDHA or $\Lambda/8$ approximation described in Chap. 3, we have

$$f_{\mathrm{DH}} = -k_{\mathrm{B}} T \frac{\kappa^3}{12\pi} \tau \left(\kappa a(T) \right) , \tag{6.79}$$

where the quantities and functions were defined above.

- The hard-core contribution f_{hc}. For this term we may use the Carnahan–Starling approximation (see Chap. 2):

$$f_{\mathrm{hc}} = k_{\mathrm{B}} T \frac{(n_{\mathrm{a}} + 2n_{\mathrm{m}}) \left(4\eta - 3\eta^2 \right)}{(1 - \eta)^2} . \tag{6.80}$$

The volume fraction of the neutrals is here defined by the effective packing parameter

$$\eta = \frac{4\pi}{3} R^3 \left(n_\mathrm{a} + 2n_\mathrm{m}\right). \tag{6.81}$$

Following the work of Beule et al. (1999b, 2001) we may fix the effective packing radius of atoms at a mean value $R = 0.37$. This is just a rough approximation which might be refined by taking into account different radii for the atoms and molecules, and by introducing temperature-dependent radii based on fluid variational theory (Juranek et al. 1999).

A more advanced description of the pressure of the neutral subsystem p_0 was given by (Beule et al. 1999b). The pressure was split into the ideal contributions of the atoms and molecules and an interaction contribution, viz.,

$$p_0 = \frac{\beta \rho_0 K_\mathrm{B} T}{m_\mathrm{H}} + \frac{(1-\beta)\rho_0 K_\mathrm{B} T}{2m_\mathrm{H}} + p_0^\mathrm{int}(\rho_0, T). \tag{6.82}$$

The Monte Carlo data can be interpolated accurately using an expansion of the interaction contribution with respect to the density and temperature (Beule et al. 1999b):

$$p_0 = \sum_{i,j} \left(c_{ij} T^{j/2} \rho_0^i + C_{ij} T^{j/2} \rho_0^i \ln \rho_0 \right). \tag{6.83}$$

The advantage of this expansion is that all other thermodynamic quantities including their derivatives are now available in a similar analytical form. An eight-parameter fit is sufficient to reproduce the Monte Carlo data within typical errors of less than 4 percent for the pressure and less than 0.4 eV/atom for the internal energy within the region 0.2 g/cm^3 $< \rho_0 <$ 1.1 g/cm^3 and $T = (2\text{–}10) \times 10^3$ K. The respective coefficients, derived from a simultaneous least squares fit of the pressures and internal energy densities, were given by Beule et al. (1999b).

We note also that several works include the magnetic field B. Magnetic field effects are relevant for the physics of magnetized plasmas, e.g., in many cosmic objects (Steinberg et al. 2000b, Potekhin and Chabrier 2013). The first and main effect of an external magnetic field is through the ideal contributions (Ebeling et al. 2000):

$$n_\mathrm{e} \Lambda_\mathrm{e}^3 \rightarrow n_\mathrm{e} \Lambda_\mathrm{e}^3 \frac{\tanh(x_\mathrm{e})}{x_\mathrm{e}}, \tag{6.84}$$

where $x_\mathrm{e} = \hbar\omega_\mathrm{c}^\mathrm{e}/2K_\mathrm{B}T$ and $\omega_\mathrm{c}^\mathrm{e} = eB/m_\mathrm{e}$. Another correction comes from the screening contributions due to the fact that the field changes the screening cloud. An advantage of the QDHA is that magnetic field effects may easily be taken into account by a factor Ebeling and Hilbert 2002a, Ebeling et al. 2003:

$$a(T) \longrightarrow a(T, x_\mathrm{e}) = a(T)\left(1 - \frac{1}{48} x_\mathrm{e}^2\right). \tag{6.85}$$

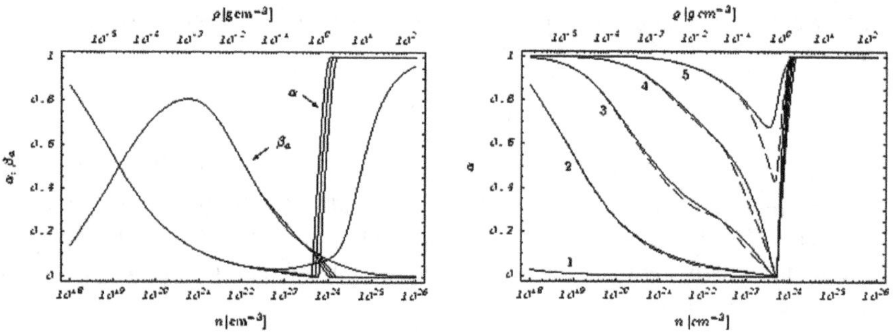

Fig. 6.11 Degree of ionization and dissociation. *Left*: α, β for different radii R_0 at a temperature of $T = 2 \times 10^4$K. *Right to left*: $R_0 = 0$, $R_0 = 0.73$, $R_0 = 0.78$, and $R_0 = 0.83$. *Right*: Degree of ionization as a function of the total proton density for different temperatures: (1) $T = 10^4$ K, (2) $T = 2 \times 10^4$ K, (3) $T = 3 \times 10^4$ K, (4) $T = 5 \times 10^4$ K, (5) $T = 10^5$ K. The *solid line* represents the calculation with the QDHA and the *dashed line* the calculation with Padé approximation

The magnetic field reduces the effective diameter. This reflects the fact that the magnetic field localizes the particles perpendicularly to the field. Studying the influence of magnetic fields, we have shown that a remarkable influence on the degree of ionization begins at $B \sim 10^4$ T, and the effect increases with the magnetic field strength. At fields $B \gtrsim 5 \times 10^4$ T, the degree of ionization and also the conductivity and other related quantities are substantially higher than for the field free case, but then the present approach is no longer valid (see also Potekhin and Chabrier 2013). Under normal conditions, the magnetic field effects are very small and may be neglected here.

We are interested in partial ionization and dissociation, i.e., in the region of the density–temperature plane where electrons, ions, atoms, and molecules exist. In this region, electrons may exhibit quantum effects, but ions, atoms, and molecules are still classical. The excluded volume effect, i.e., the effect that the electrons and ions cannot penetrate neutral particles, is taken into account in the ideal contributions. The excluded volume factor η_0 is defined above with a fixed radius R_0. If this effect is not considered, we find a lower ionization rate α at higher densities, as shown in Fig. 6.11.

Fig. 6.11 (left) shows that a small difference in the radius R_0 has an influence on the degree of ionization and dissociation. For larger radii, α begins to increase at smaller densities, since the effective volume $V_{\text{eff}} = (1 - \eta_0)V$ available for the electrons decreases with increasing η_0, whereupon the effective electron density $n_e^{\text{eff}} = n_e / V_{\text{eff}}$ increases. Therefore, the system tends to higher degrees of ionization. The transition between low and complete ionization takes place in a very narrow density range. For high densities, the excluded volume effect has to be considered and may have to be further refined by taking into account temperature dependencies of the radius and introducing occupation probabilities.

The non-ideal effects of the free charged particles can be described more correctly by means of the Padé approximation (see, e.g., Ebeling and Richert 1985,

Stolzmann and Ebeling 1998, 2000). Figure 6.11 (right) shows the ionization and dissociation rate as a function of the proton density, calculated using the QDHA and the Padé formulae given in (Stolzmann and Ebeling 1998). For the density range 10^{22}–10^{25} cm^3 and the temperature range 10^4–5×10^4 K, we find only small deviations of the QDHA from the Padé approach. For higher temperatures, above 10^5 K, the deviations are much greater and the QDHA is less accurate. In the following, we will restrict our study to temperatures well below 10^5 K, so we will be able to justify using the QDHA to simplify the formulae and reduce calculational effort.

6.4.2 *Geometry of the Free Energy Landscape*

On the basis of the formulae given previously, we represented the geometry with respect to $f(\alpha, \beta)$ at fixed T, n and starting with $B = 0$ (Figs. 6.12 and 6.13). We

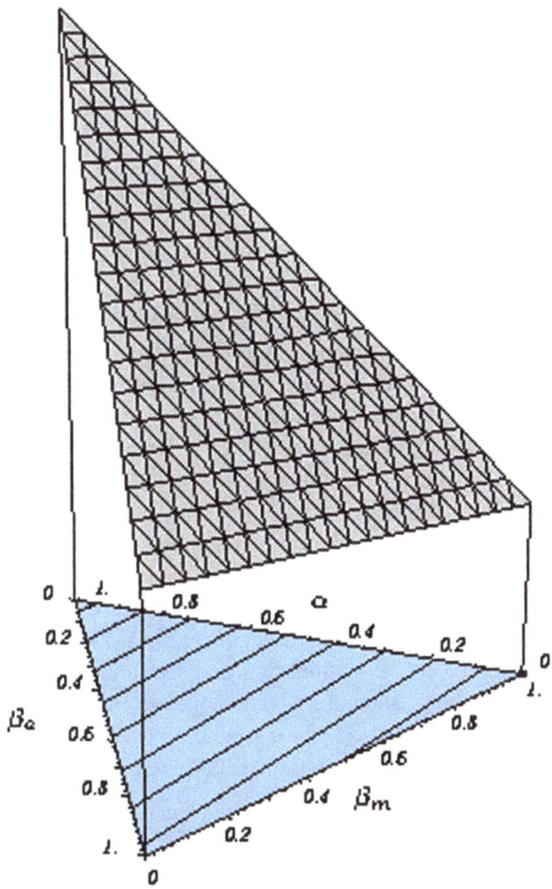

Fig. 6.12 Free energy density at $T = 300$ K, $n = 2 \times 10^{19}$ cm^{-3}, represented on the simplex α, β_a, β_m. The minimum of the free energy is at the corner $\alpha = 0$, $\beta_a = 0$, $\beta_m = 1$, i.e., in equilibrium, we find practically only molecules

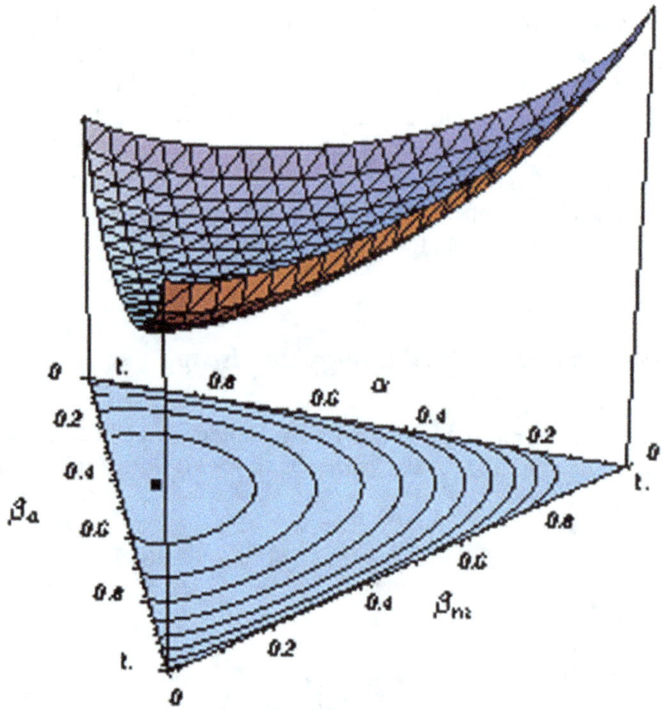

Fig. 6.13 Free energy density at $T = 4 \times 10^4$ K, $n = 10^{22}$ cm^{-3}. The minimum is found in a region where electrons, ions, atoms, and molecules are all present

used the variables α, β_a, and β_m which allow a representation on a simplex according to (6.70). The minimum of the free energy corresponds to the stationary degrees of ionization and dissociation. The form of the landscape around the minimum determines the thermal fluctuations.

Depending on the values of the temperature and the proton density, we find different locations of the minimum. At $T = 300$ K, $n = 2 \times 10^{19}$ cm^{-3}, we observe a minimum in the corner $\alpha = 0$, $\beta_\mathrm{a} = 0$, $\beta_\mathrm{m} = 1$, meaning that only molecules exist (see Fig. 6.12). At different temperatures and proton densities, e.g., $T = 4 \times 10^4$ K, $n = 10^{22}$ cm^{-3}, we observe a minimum in the center of the simplex. This corresponds to a situation where free electrons and ions exist with bound states (atoms and molecules) under the same conditions (see Fig. 6.13). With the procedure described above, we calculated the stationary degrees of ionization and dissociation for a lattice of points in the temperature–density plane. Figure 6.14 shows a cross-section for the degrees of ionization and dissociation at three temperatures. The minimum of the degree of ionization at intermediate densities accompanied by a maximum of the degree of dissociation is typical.

In this way, we also obtained several landscapes, for example, of the fraction of protons bound in molecules as a function of the density and temperature (Fig. 6.15).

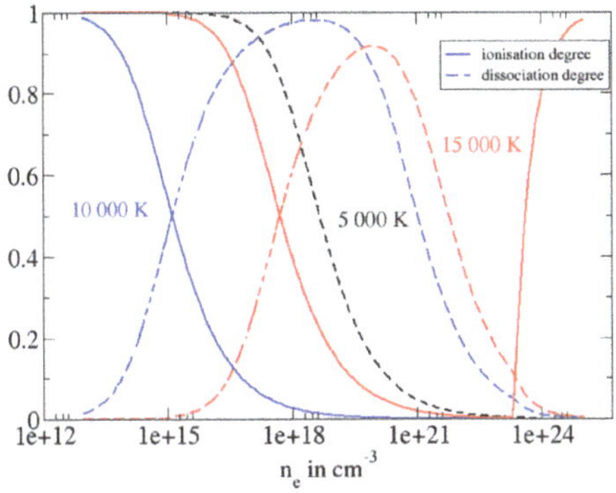

Fig. 6.14 Ionization isotherms at $T = 5000$, 10,000, 15,000 K as a function of the proton density

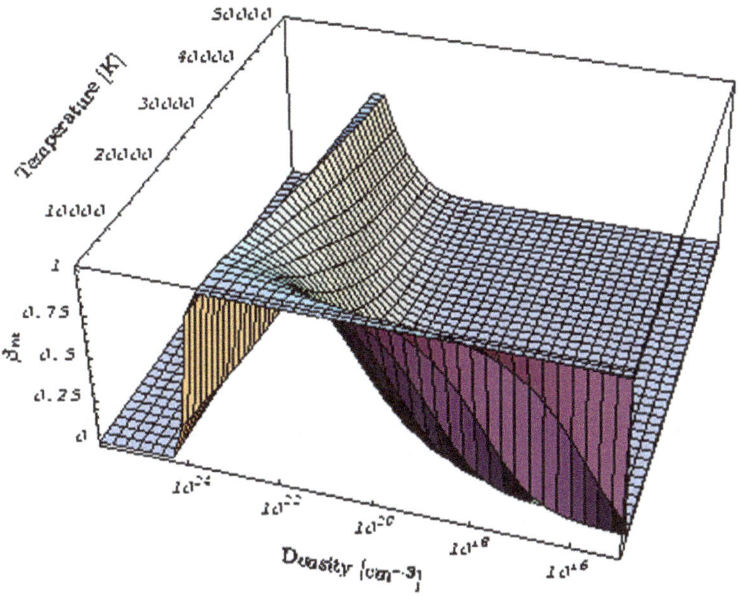

Fig. 6.15 Fraction of protons bound in molecules $\beta_{\rm m} = (1 - \alpha)\,(1 - \beta)$

We observe full ionization ($\alpha = 1$ on a plateau in the upper right corner and at the left edge of the diagram) at very high densities and in the region of lower densities, but high temperatures. At sufficiently low densities, we find only bound states when the temperature is low ($\alpha = 0$ in a region beginning at the lower edge of the diagram). In the intermediate region of proton densities we find a valley of bound states stretching over the whole temperature range, with low ionization occurring only at very high temperatures.

We now consider the formation of molecules (see Fig. 6.15). We observe similar features to those above in the corresponding parameter regions. At very high densities (left edge), no molecules exist, and likewise in the region of lower densities and high temperatures (upper right corner). When the density is sufficiently low and the temperature is low as well, we find that the bound states observed in our calculations are practically only made up of molecules ($\beta_{\mathrm{m}} \approx 1$ at the lower edge). In the intermediate density region, we find a mountain of molecules ($\beta_{\mathrm{m}} = 1$) and a decline in the degree of dissociation with rising temperature.

To summarize our findings, from a detailed study of the free energy landscape, we developed a new description of ionization and dissociation phenomena. The relatively simple approximation QDHA used for the interaction term in the free energy allows one to study the dispersion of the degrees of ionization and dissociation and also magnetic field effects. The QDHA may be replaced by more refined approximations.

6.5 Thermodynamics of High-Pressure Plasmas

6.5.1 Advanced Chemical Models with Energy Shifts

The thermophysical properties of dense hydrogen and other gases at high pressures were studied in the past by many authors, starting with Wigner, Huntington, Abrikosov, and others. Of special interest was the transition of molecular non-conducting hydrogen to a highly conducting phase at high pressures, this being considered as a Mott-type transition. This section is largely based on the papers by Beule et al. (1999b), also taking into account more recent approaches to include the effects of energy shifts due to the high pressure (Ebeling et al. 2009a; Redmer and Röpke 2010).

Basic studies of the effect of energy shifts on the properties of dense plasmas go back to the 1970s and 80s (see Kraeft et al. 1986). However, a detailed study of the influence of energy shifts on the thermodynamic properties of dense plasmas has rarely been considered. In particular, we will consider the effects of Pauli's exclusion principle on the thermodynamic properties of hydrogen plasma. Previous studies of dense hydrogen included several assumptions about the nature of the high-density phase (Ebeling et al. 1976, Kraeft et al. 1986, Saumon and Chabrier 1989, Ebeling 1989; Ebeling et al. 2007, Kremp et al. 2005). In fact, many questions remained open, in particular that about the nature of the highly conducting, possibly even superconducting state.

While metallization of solid hydrogen near $T = 0$ K has not been clearly demonstrated so far for pressures up to 300 GPa (Da Silva et al. 1997), a metal-like behavior has been observed in fluid hydrogen and deuterium by shock-compression experiments driven by a two-stage light gas gun and powerful lasers at Mbar pressures (1 Mbar=100 GPa) (Weir et al. 1996; DaSilva and Celliers 1997; Collins et al. 1998). For instance, metal-like conductivities have been observed in these pioneering experiments around 140 GPa and 3000 K (Weir et al. 1996) in a warm dense fluid. First signatures of a possible phase transition which is due to the electronic nonmetal-to-metal transition in this region were found in another recent shock-compression experiment (Fortov et al. 2007; Fortov 2013). These findings have an enormous impact on our current understanding of the behavior of hydrogen at ultra-high pressures, and this is relevant for models of planetary and stellar interiors, and also inertial confinement fusion studies.

In the present section, we will show following Beule et al. 1999b, 2001, Ebeling et al. 2009, Redmer and Röpke 2010, that one of the important effects leading to the destruction of bound states like atoms and molecules is Pauli blocking. Due to the Pauli exclusion effect, free electrons in the plasma cannot penetrate the interior of atoms and molecules since these bound states are already occupied by electrons. At high densities this leads to an enormous pressure acting on the neutrals which will finally lead to ionization. Several approaches to incorporate this effect will be discussed here, including perturbation theory as well as variational methods. However, polarization effects are not considered. The effective energy levels of hydrogen, which depend on density and weakly also on temperature, are introduced into the thermodynamic functions in the chemical picture.

In order to calculate the ionization/dissociation equilibrium, we use the thermodynamic variational principle and perform a minimization of the free energy with respect to the composition. This approach, which is in principle equivalent to the standard method based on Saha equations, has some advantages and leads to additional information in special cases. We constructed the thermodynamic functions of hydrogen by using a chemical approach to the free energy which recently has been applied to temperatures between 2000 and 10,000 K (Beule et al. 1999b, 2001). The effects of pressure dissociation $H_2 \rightleftharpoons 2H$, and ionization $H \rightleftharpoons e + p$, have been taken into account in order to explain the transition from a molecular fluid at low temperatures and pressures, through a partly dissociated warm fluid at medium temperatures of a few thousand kelvin, to a fully ionized hot plasma above 10,000 K.

Hydrogen basically consists of two main components, the plasma (electrons and protons) and the neutral liquid consisting of atoms and molecules. Correspondingly, the free energy for a two-component system of neutral ($F_{\rm nl}$) and charged particles ($F_{\rm pl}$) reads

$$F(V, T, [N]) = F_{\rm bs} + F_{\rm nl}(V, T, [N]) + F_{\rm pl}(V, T, [N]) + F_{\rm int}(V, T, [N]) \,. \tag{6.86}$$

The first term is the bound state contribution. The 'normal' free energy contributions of the neutral liquid $F_{\rm nl}$ and the plasma $F_{\rm pl}$ are completed by a contribution due to interactions between the neutrals and the charges $F_{\rm int}$.

Before we continue with the discussion of the various contributions to the free energy $F(T, V, N)$, let us state that any such partitioning of F within the chemical picture is subject to a condition. Depending on the particular choice we use, the contributions may change, and the condition is that the total free energy F should be invariant with respect to all possible model pictures. This statement is based on the constraint that no physical effect, including the Pauli blocking effects which we study here, should be accounted for twice. In any case, a double counting of contributions has to be strictly avoided (see Chaps. 4 and 5).

The bound state contribution to the free energy (6.86) is given in the form

$$F_{\mathrm{bs}} = N_{\mathrm{a}}^{*} E_{1}^{\mathrm{a}} + N_{\mathrm{m}}^{*} E_{1}^{\mathrm{m}} + N_{\mathrm{a}}^{*} K_{\mathrm{B}} T \ln \sigma'(T) + N_{\mathrm{m}}^{*} K_{\mathrm{B}} T \ln \sigma'_{\mathrm{m}}(T) , \tag{6.87}$$

where the E_1 are the ground state energies and the N^* denote the particle densities of atoms or molecules. Here we have separated out the shifted energy of the ground state to study the effects of a ground state shift in detail. The reduced atomic Brillouin–Planck–Larkin partition function (excluding the ground state) is now given by

$$\sigma'(T) = 4 \exp\left(\frac{E_{1}^{\mathrm{a}}}{K_{\mathrm{B}} T}\right) \sum_{n} \left[\exp\left(-\frac{E_n}{K_{\mathrm{B}} T}\right) - 1 + \frac{E_n}{K_{\mathrm{B}} T}\right] . \tag{6.88}$$

The molecular partition function $\sigma'_{\mathrm{m}}(T)$ has been discussed in a previous section (Beule et al. 1999b).

We now discuss the contributions of the energy shifts, assuming that the shift is the same for bound states in atoms and bound states in molecules, so that we can write for the bound state contribution

$$F_{\mathrm{bs}} = (N_{\mathrm{a}}^{*} + 2N_{\mathrm{m}}^{*})\Delta E_1 + F_{\mathrm{bs}}^{0} , \tag{6.89}$$

where ΔE_1 is the shift we calculated in Chaps. 3 and 5 and F_{bs}^{0} is the bound state contribution we calculated so far without taking into account the contribution of shifts. The shift of the atomic levels is approximated by the sum of Pauli and Fock terms, viz.,

$$\Delta E_1 = \Delta E_{1}^{\mathrm{Pauli}} + \Delta E_{1}^{\mathrm{Fock}} , \tag{6.90}$$

for which the explicit expressions were derived in Chap. 5. In the region of the transition from a neutral liquid to a plasma state, the molecules, which first dissociate into atoms, are less relevant than the atoms. To be quite clear, we use the approximation that molecules are simply composites of two atoms, i.e., the shifts are additive, and for molecules we have a double atomic shifts. The treatment of the neutral liquid contribution F_{nl} to the free energy (6.86) is as proposed in (Beule et al. 1999b) based on data for the dense fluid calculated within a dissociation model.

In earlier work we performed classical Monte Carlo simulations for partially dissociated fluid hydrogen for a grid of temperature and density points in the region of $T = (2\text{–}10) \times 10^3$ K and $\varrho = 0.2\text{–}1.1$ g/cm^3. Effective pair potentials of the exponential-6 form have been used to model the interactions between the molecules and the atoms in a dense fluid. The dissociation equilibrium $H_2 \rightleftharpoons 2H$ has been solved taking into account the correlation parts of the chemical potentials using fluid variational theory. The Monte Carlo data for the interaction contribution can be interpolated accurately within an eight-parameter fit with respect to density and temperature, leading to an analytical expression for the free energy density (see Beule et al. 1999b, 2001). The resulting pressure of the neutral liquid term can be written in the form

$$p_{\rm nl} = \frac{1}{2}(1+\beta)nK_{\rm B}T + p_{\rm int} \,. \tag{6.91}$$

Here n is the total proton density and β the degree of dissociation of molecules into atoms. Since the fit formula is rather soft, we have used the Carnahan–Starling expression for hard spheres at temperatures beyond 15,000 K, with an interpolation between the the two cases (Ebeling, Hache and Spahn 2003).

The plasma contribution $F_{\rm pl}$ to the free energy (6.86) has been built from several results for the fully ionized plasma domain in the form of Padé approximations (Ebeling and Richert 1982, 1985, Stolzmann and Ebeling 1998). For plasmas of the light elements, including mixtures, Padé formulae have been given in Ebeling (1990). For the ideal free electron term, taking Pauli blocking effects into account, we have to write

$$F_{\rm pl,id}(V,T,[N]) = N_{\rm e}K_{\rm B}Tf\left(\frac{N_{\rm e}\Lambda_{\rm e}^3}{2V*}\right) \,, \tag{6.92}$$

where V^* is the volume available to free charged particles. Further, $f(y)$ is the Fermi function, which is approximated by the Zimmermann formula (see Chap. 2 and Röpke 1978, 1980). The interaction part of the plasma component is represented by the standard Padé approximations developed in Sect. 6.4 (see also Stolzmann and Ebeling 1998 and Ebeling 1989; Ebeling et al. 2007). We have calculated the chemical equilibrium between the charges and neutrals (atoms and molecules) by means of a numerical variational procedure based on direct minimization of the free energy (for details, see, e.g., Ebeling, Hache and Spahn 2003), including the description of a numerical code based on Mathematica. Due to several numerical advantages, we prefer to minimize the free energy rather than use the Saha equation approach. The interactions of electrons with neutrals (bound states) is described by the idea of energy shifts which are due to the interaction of free electrons with bound electrons. This will be described in the following section.

6.5.2 *Methods for Minimizing the Free Energy*

The method of Saha equations used so far in his book has some serious limitations. The Saha equation is a nonlinear equation which may have no solutions, in particular for strongly coupled systems. Furthermore, iteration procedures may lead to numerical (chaotic) instabilities. Another point is that solutions at the boundary, e.g., $\alpha = 0$ or $\beta = 0$, may not be included in the resulting set of solutions. Therefore, the method which uses minimization of the free energy, which is in principle mathematically equivalent, may provide a more complete and more stable set of solutions for the chemical decomposition.

We consider a hydrogen plasma at fixed temperature T and proton density n, including as before the processes of ionization and dissociation. The variables α and $\beta_{\rm d}$ or $\beta_{\rm a}$ are the variational parameters of our problem. The free energy has to be minimized with respect to these. We note that β_d is the degree of dissociation of molecules into atoms, β_a is the relative amount of protons bound in atoms, and β_m the relative amount bound in molecules. Due to the balance relation for the total proton density, we have the relation

$$\beta_d = \frac{\beta_a}{\beta_a + \beta_m} . \tag{6.93}$$

In other words, only one of the β parameters is independent, so we have the freedom to choose α and β_m as the independent parameters and make use of the simplex relation $\alpha + \beta_a + \beta_m = 1$. Incidentally, we already know that atoms appear only in a rather narrow region of the temperature–density plane. In broad regions of the density–temperature plane, we may use the approximation $\beta_a = 0$, whereupon α is the only remaining free parameter. The neutrality condition requires the electron and ion densities always to be equal, i.e., $n_e = n_i$.

The free energy per proton is our basic quantity, depending on only four independent parameters. The variational parameters are the degrees of ionization/dissociation. From now on, we denote the variational parameter for the degree of ionization by x and the variational parameter for the degree of dissociation by y. The values corresponding to the minima of F are the real degrees of ionization/dissociation

$$\alpha = x_{min} , \qquad \beta = y_{min} . \tag{6.94}$$

Here β stands for one of the three dissociation parameters introduced above. We are free in this choice. We calculated the chemical equilibrium using a numerical variational procedure based on direct minimization of the free energy. The calculus we used, based on *Mathematica*, was developed by Hache and Spahn and first presented in (Ebeling et al. 2003).

In the numerical procedure, we prefer to minimize the free energy rather than follow the Saha approach, because there are several numerical advantages. The results of the two approaches (minimization and Saha approach) are qualitatively similar, but differ in the details. The characteristic picture that emerges is high ionization at very low and very high densities. The region of bound states forms a valley at intermediate densities. The transition to full ionization at high densities (pressure ionization) occurs in a small region of densities near 5–6 $\times$ 10^{24}cm^{-3}.

Saha equations provide a more explicit formulation of the minimum condition. The Saha equation finds one concrete minimum located inside the corner of the degrees of ionization and dissipation. There are no minima at the boundaries, e.g., when $\alpha = 1$ or $\beta = 0$. Furthermore, density-dependent energies cannot be used in Saha equations without involving several complicated generalizations. The direct minimization of the free energy used here finds all existing minima, including multiple minima and those at the boundaries, and has no difficulties with energy shifts. Furthermore, we mention that the iterative solution of Saha equations with shifts, which are highly nonlinear, may lead to serious numerical instabilities and convergence to spurious solutions. In particular, these difficulties may appear, if the Saha equation contains mixed nonlinearities with respect to electronic and atomic densities in the exponent.

This is precisely what happens for our Pauli blocking effects. In our case, we thus consider direct minimization of the free energy as a more reliable method, since it provides more information and contains the solutions of the Saha equation as special cases. The transition density is between 2.5×10^{-23} and 4.5×10^{-24} protons cm^{-3}. This means we are in the region around 0.5 $\mathrm{g\,cm}^{-3}$ for hydrogen and 1 $\mathrm{g\,cm}^{-3}$ for deuterium. At 5000 K, the transition occurs in the region of densities around $n = 3 \times 10^{-23}\ \mathrm{cm}^{-3}$. In order to show what happens during this transition, we may study once again the behavior of α and β_a (percentage of protons bound in atoms) as shown in Fig. 6.10. Examples of our results for the pressure are shown in Figs. 6.16, 6.17, and 6.18. We see a phase transition of first order below 6450 K (red curve) and also a rather weak diffusive transition in the relative pressure up to temperatures of 20,000 K.

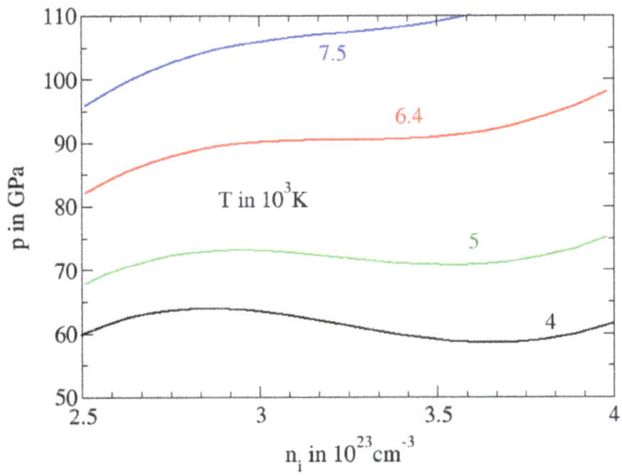

Fig. 6.16 Pressure isotherms at $T = 4000$ K (*black curve*), 5000 K, 6400 K, and 7500 K (*blue curve*). According to this calculation, the critical point of the first order transition is around $T = 6400$ K

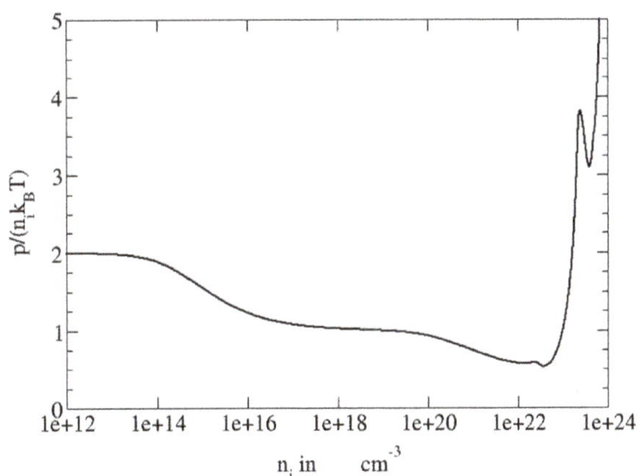

Fig. 6.17 Isotherms of the relative pressure $P/n_\mathrm{i} K_\mathrm{B} T$, showing several transitions. The shoulder we observe at $n \sim 10^{18}\ \mathrm{cm}^{-3}$ corresponds to soft transitions between atoms and molecules. Beyond the minimum, the molecules break down. Temperature 10,000 K

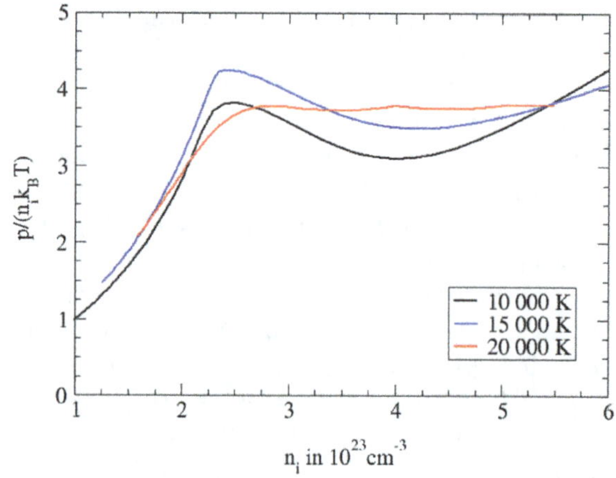

Fig. 6.18 Isotherms of the relative pressure at $T = 10,000$, 15,000, 20,000 K. We see a critical point of the soft transition around 20,000 K

Thermodynamic Functions and Phase Transitions

As stated above, the pressure exhibits a first order transition below 6450 K. The transition density lies in all cases between 2.5×10^{23} and 4.5×10^{23} protons cm^{-3}. The transition to full ionization is softer than in our previous approach (Ebeling et al. 1976, Beule et al. 1999b, 2000). We note that, in earlier work, we found a first order transition at even higher temperatures (Beule et al. 1999b). In the meantime, the existence of a hard transition has been demonstrated in isentropic shock experiments (Fortov et al. 2007, Fortov 2011). Later we will compare our results with part of the isentropic EOS data provided by Fortov et al. (2007) and with an isotherm obtained in our earlier work (Beule et al. 1999b, 2001). We see that the horizontal part of the EOS occurs at the same densities, but at lower pressures than in the experiment and in the theoretical approach of Beule et al. (1999b, 2001). This point needs further discussion.

We now consider the relative pressure. Examples of our results for the relative pressure (in relation to the reference pressure of a fully ionized plasma) are shown in Fig. 6.17. The isochore of the relative pressure at total density $n = 10^{10}\,\mathrm{cm}^{-3}$ exhibits two soft transitions connected at low density with the formation of atoms and molecules. The corresponding isotherms of the relative pressure at $T = 10,000$ K exhibit three soft transitions at $n_\mathrm{i} \approx 10^{15}\,\mathrm{cm}^{-3}$ (formation of atoms), at $n_\mathrm{i} \approx 10^{21}\,\mathrm{cm}^{-3}$ (formation of molecules), and at $n_\mathrm{i} \approx 10^{23}\,\mathrm{cm}^{-3}$ (connected with the pressure ionization).

Most interesting is the weak diffusive transition in the relative pressure beyond $10^{23}\,\mathrm{cm}^{-3}$, which appears up to $20,000$ K. In this region the pressure isotherm is nearly horizontal. The derivative is rather small, but never reaches zero, i.e., the system is not unstable. The wiggle in the relative pressure is a sign that the derivative is very small, but may still be positive. We believe that under experimental conditions, it might be rather difficult to see a difference between a proper first order phase transition, where the derivative is really zero, and a fast diffusive transition of the type shown in Fig. 6.18. The critical point of the soft transition is around 20,000 K. The isochore of the relative pressure exhibits several soft transitions. At total density $n = 10^{10}$ cm^{-3}, we have transitions corresponding to the formation of atoms and molecules.

In order to understand these transitions several effects have to be taken into account. We concentrate here on so-called Pauli shifts and Fock shifts of the ground state energy level. Both blocking effects express the rule that states occupied by atomic electrons cannot be occupied by free electrons with the same spin direction. At high electron densities, this leads to the destruction of atomic states which require a relatively large amount of phase space. We calculated the lowest order energy shifts, known as the Pauli and Fock shifts. These terms are calculated by solving effective Schrödinger equations for strongly correlated systems.

We conclude with two methodological remarks. The first is connected with the chemical description. The results in this section are based on the chemical picture. As is well known, there are two variants of the chemical picture:

- The calculations of the chemical equilibrium are based on the concept of Saha-type equations, as shown in Chaps. 2–4. Saha equations are just reformulations of the equilibrium conditions for the chemical potentials of the species.
- Alternatively, as shown in this chapter, one can use a direct minimization of the free energy by varying the degrees of ionization and dissociation.

The two methods are based on the same variational principle for the free energy of the plasma. The results are qualitatively similar, but differ in the details. Saha equations provide a more explicit formulation of the minimum condition. The solution of a Saha equation gives a concrete minimum located inside the corner of the degrees of ionization and dissipation where the derivative of the free energy is zero. There are no minima at the boundaries, e.g., $\alpha, \beta = 0$ or $\alpha, \beta = 1$, since as a rule the derivative is not zero there.

Furthermore, the standard Saha equation leads to difficulties in the case of density-dependent energy levels, and one must then use generalized Saha equations with additional terms. The direct minimization of the free energy used in the present chapter finds all possible minima, including all relative minima and those at the boundaries. Therefore we may consider the direct minimization of the free energy used here to be a more general and more advanced description. It provides more information and always contains the solutions of the Saha equation as special cases.

The second remark concerns the role of Pauli blocking effects. We have shown that at high densities the changes in the energy levels are essential for the ionization equilibria and determine the details of the high pressure behaviour. The transitions to highly conducting states occur at densities around $2.5–4.5 \times 10^{23}$ protons cm^{-3}. This means we are in the region around 0.5 g cm^{-3} for hydrogen and 1 g cm^{-3} for deuterium. The corresponding pressures lie in the region of $0.5–1.2 \times 10^{11}$ Pa, i.e., above a Mbar. In the present model, the transition to a highly conducting phase is a thermodynamic phase transition of first order at $T < 6450$ K, and in the region 6,450–20,000 K, a diffusive phase transition.

6.6 Hydrogen-Like and Helium-Like Plasmas at Ultrahigh Pressures

6.6.1 Hydrogen and Deuterium Hugoniots and Isentropes

In previous sections, we calculated the free energy and the pressure of hydrogen plasmas under high pressures on the basis of the chemical picture. Other thermodynamic functions may be calculated from the free energy (6.86) by differentiation. For instance, the entropy follows by differentiation with respect to the temperature:

$$S(T, V, N) = -\frac{\partial F(V, T, N)}{\partial T} .$$

Combining this expression with the isothermal EOS $p = p(\rho, T)$, we may get the isentropic (adiabatic) EOS $p = p(s, T)$, where ρ is the total mass density and $s = S/N_p K_B$ the specific entropy per proton, N_p which is the total number of protons in the plasma including the protons bound in H and in H_2. We may thus calculate the isentropic EOS (Beule et al. 1997, 2001).

Here we consider hydrogen-like plasmas, such as deuterium, which are of interest for fusion problems. In the last section, the standard chemical model was improved in several respects to allow applications to very dense systems. We replaced the standard excluded volume approximation used in our earlier work by a more rigorous approach based on effective wave equations, including symmetry effects. Energy shifts were calculated in several approximations including perturbation theory and variational approaches. This allowed us to include the interactions between electrons and neutrals in a more systematic way. We calculated the shift in the ground state energy due to Pauli exclusion, which prevents the electron from penetrating atoms. In particular, we investigated the influence of energy shifts which have so far been neglected in most thermodynamical calculations. Furthermore, we studied in detail the region where a Mott transition was expected and where recent experiments and simulations have shown a transition from insulating behavior to metal-like conductivity (Fortov et al. 2007; Morales et al. 2010; Redmer and Röpke 2010; Savage et al. 2015). Ionization and dissociation equilibria are treated within an advanced chemical approach based on the assumption that the system is a gas mixture of chemical species.

We have shown that the effects resulting from the Pauli exclusion of electrons from the interior of atoms have a major influence in the high pressure region. We presented explicit calculations of the ionization and dissociation equilibria in the region $5000 < T < 20,000$ K, 2–6 $\times$ 10^{23} cm^{-3}. What we see is a first order transition from a neutral to a conducting phase at $T < 6450$ K and a diffusive transition at 6,450–20,000 K. At still higher temperatures, the transitions to a highly ionized plasma are very smooth and no wiggles are seen (see Fig. 6.19).

We now describe a detailed study of the situation near the coexistence region, following Beule et al. (1999b). The free energy expression for a two-component system of neutral (F_0) and charged ($F_\pm$) particles, viz.,

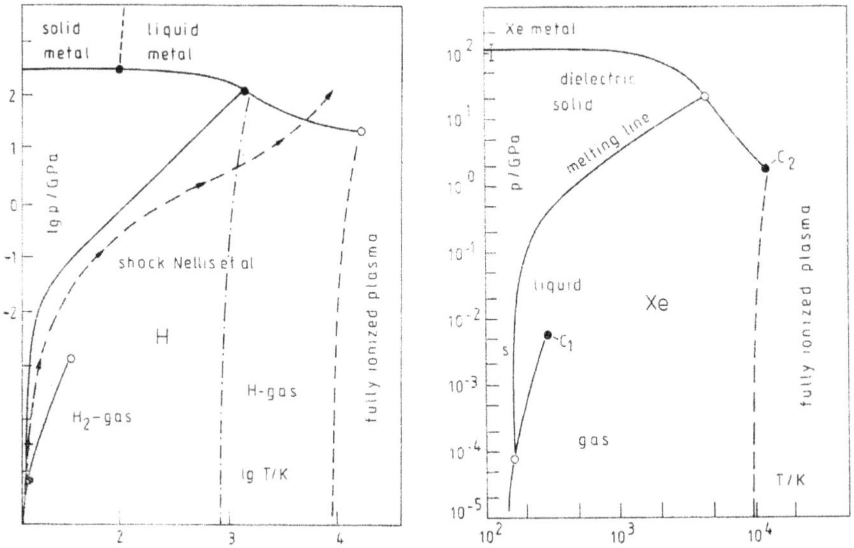

Fig. 6.19 Phase transitions in hydrogen gases (*left*) and in xenon plasmas (*right*), including the still hypothetical PPT transitions at high temperatures and pressures (Haronska et al. 1987; Kraeft et al. 1988)

$$F(V, T, N) = F_0(V, T, N) + F_{\pm}(V, T, N) , \tag{6.95}$$

combines results for the fully ionized plasma domain (Ebeling and Richert 1982, 1985), with improved data for the dense neutral fluid calculated within a dissociation model (Bunker et al. 1997). Both contributions to the free energy are split into ideal and interaction parts. We take into account the interactions in the neutral and in the charged subsystem, respectively, whereas the interaction between charges and neutrals is accounted for by the reduced volume concept.

The essential point in the calculations by Beule et al. (1999b) is a rather precise calculation of the thermodynamics of the neutral part based on pair potentials of the exponential-6 form to model the interactions between the molecules and atoms in the dense fluid. The dissociation equilibrium $H_2 \to 2H$ has been solved, taking into account the correlation parts of the chemical potentials using fluid variational theory for details (see Bunker et al. 1997). Dissociation is already important in the dense fluid above 20 GPa due to the lowering of the dissociation energy with increasing density. The degree of dissociation is defined as in Sect. 6.3. We have shown that the results for the pressure of the neutral fluid as a function of density are in reasonable agreement with the available shock-wave experiments, which have reached temperatures around 3000 K (Bunker et al. 1997), and with the similar linear-mixing model of Ross (1996). For densities and temperatures leading to a low degree of dissociation, the pressure agrees well with the data given by Saumon and Chabrier (1989, 1991, 1992). Systematic deviations occur with an increasing fraction of dissociated molecules for higher temperatures and/or densities. The pressure of the neutral subsystem was modeled as in Beule et al. (1999b). On the basis of the above-mentioned formulae, a detailed study of the situation near the phase transition line in the region $2000 < T < 10,000$ K was also given by Beule et al. (1999b).

We now study the plasma contribution. The entropy and other thermodynamic functions can be calculated from the free energy by differentiating. For instance, the isothermal EOS follows by differentiating with respect to the volume, and the entropy by differentiating with respect to the temperature:

$$p = -\frac{\partial F}{\partial V}, \qquad S = -\frac{\partial F}{\partial T}. \tag{6.96}$$

From the expression for the pressure, we calculated the coexistence line in the standard way. Figure 6.20 shows the coexistence pressure and Fig. 6.21 the coexistence region in the temperature–density plane.

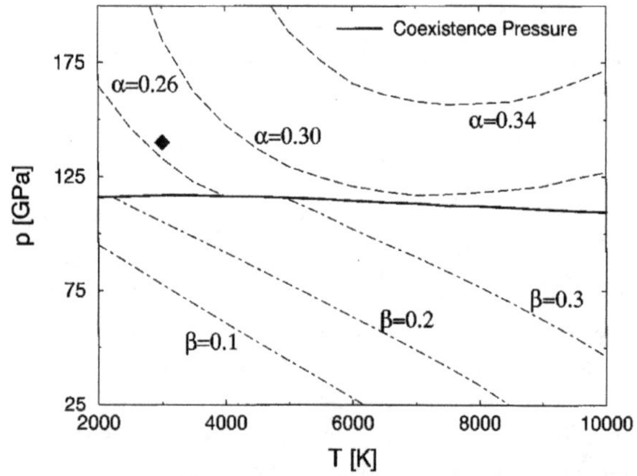

Fig. 6.20 Coexistence pressure as a function of the density for the PPT in hydrogen plasmas

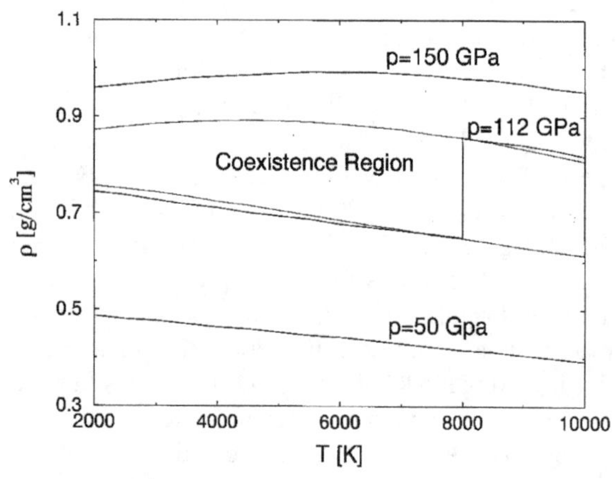

Fig. 6.21 Coexistence region for the plasma phase transition in hydrogen plasmas at high pressures. The transition observed in multishock experiments is marked by a *diamond* (Weir et al. 1996)

Combining the expressions for F and S, we get the pressure along an isentrope $p = p\,(s = \text{const.}, T)$, where $s = S/N_p K_B$ is the specific entropy per proton, with N_p the total number of protons in the plasma, including protons bound in H and H_2. Simple expressions for the isentropes were calculated by Beule et al. (1997). Here we used the expressions for the free energy given in (Beule et al. 1999b). Figure 6.24 shows a set of isentropes at pressures below and near to the coexistence line.

To compare the results of the present EOS directly with the available experimental data, we also calculated the pressure p, the degree of dissociation β, and the temperature T along a Hugoniot curve of D_2, using an initial pressure and density typical for single shock experiments (Beule et al. 2001). Temperatures up to 10,000 K correspond to a maximum density of about 0.8 g/cm^3. In this range, our model is in good agreement with the experimental data, and the linear mixing model of Ross (see Chap. 2). Large deviations from the behavior predicted by the often used SESAME tables occur due to dissociation. Tight-binding molecular dynamics (TBMD) simulations do not reproduce the high compression between 30 GPa and about 100 GPa that is observed experimentally. Applying the hydrogen EOS given by Saumon and Chabrier (1989, 1992) for the Hugoniot curve, a rather abrupt compression occurs at about 70 GPa which does not appear in the experimental data. Path integral Monte Carlo (PIMC) simulations become increasingly reliable as the temperature increases, but fail to reproduce the neutral fluid behavior at low temperatures.

We proposed here an EOS for dense hydrogen plasma for $T < 10^4$ K which gives realistic behavior for low temperature neutral molecular fluids and high temperature fully ionized plasmas. A thermodynamic instability occurs in the intermediate region which is connected with a first-order phase transition. The plasma phase transition describes a transition from a partially dissociated fluid to a partially ionized plasma. We derived the corresponding coexistence line and the coexistence region. The coexistence line may end in a critical point as predicted by other theories. For a detailed study of the region near the critical point, the present EOS has to be extended to higher temperatures. Furthermore, the charged–neutral interaction must also be considered.

Compared with other approaches, the correlations in the dense neutral component are now considered in a systematic way, going beyond rather approximate methods such as the hard-sphere reference system or perturbation theory. The applicability of the dissociation model for not too high temperatures and densities has been demonstrated by comparing the proton–proton distribution functions with results from computer simulations. As discussed above, the present EOS yields a first-order phase transition with a corresponding separation between two phases of different mass density and degree of ionization. This so-far hypothetical phase transition may be closely related to the experimentally observed electronic transition to metallic-like conductivities in fluid hydrogen. The present EOS yields a transition pressure of about 110–120 GPa in that region. Note that numerical simulations with PIMC also support the occurrence of a first-order phase transition in hydrogen (see Chaps. 9–10).

In order to check the accuracy of the present EOS, we calculated the corresponding Hugoniot curves and compared with the data available from recent shock-compression experiments, as well as other theories (Beule et al. 2001). We showed

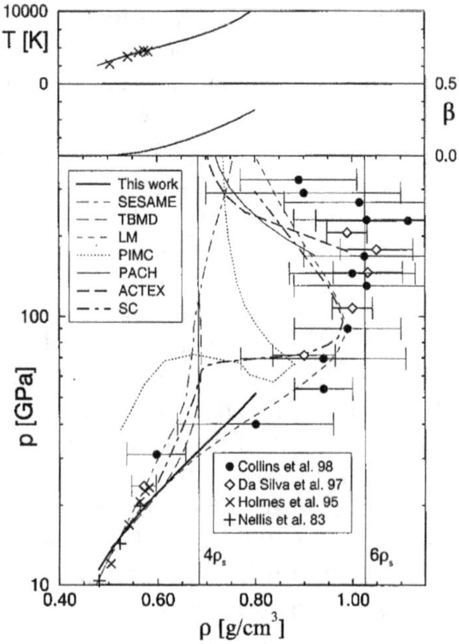

Fig. 6.22 Hugoniot of deuterium plasmas at high temperatures and pressures, calculated in the PACH approximation (Beule et al. 1999b) and compared with other theoretical and experimental results

that, in the density region studied here ($\rho \sim 0.8$ g/cm^3), our model provides a good representation of the experimental data points and compares well with the results of the linear-mixing model of Ross (1996, 1998). Figure 6.22 shows the pressure p, degree of dissociation β, and temperature T along a single-shock Hugoniot curve of D_2 versus density, where L and d denote data from laser-driven shock-wave experiments and from gas gun experiments (Beule et al. 2001). The present results (thick solid line) are compared with Hugoniot curves derived from a semi-empirical linear-mixing model (Ross 1996, 1998), the SESAME tables, tight-binding molecular dynamics (Lenosky et al. 1997), and path integral Monte Carlo simulations PIMC (Militzer and Ceperley 2000), from Padé approximations in a chemical picture called PACH (Ebeling 1990; Beule et al. 1999b), the activity expansion method ACTEX (Rogers and Young 1997), and the hydrogen equation of state by Saumon and Chabrier (1989, 1992).

Note that studies of the ionization equilibrium in dense hydrogen plasma are also uses as a basis for understanding the electrical conductivity transported by the free charges in the plasma (Beule et al. 1999b).

So far our work has concentrated primarily on hydrogen plasma. In order to use the hydrogen EOS for deuterium, mass scaling can be applied for the interpolation formula of the interaction contributions, i.e., it is assumed that the same particle numbers for hydrogen and deuterium lead to the same degree of dissociation and to identical interaction contributions to the thermodynamic function of the neutral fluid for a given temperature.

Deuterium and deuterium–tritium plasmas are quite similar to hydrogen plasmas, since the only essential difference is in the masses. Deuterium atoms have two times the mass of hydrogen atoms and tritium atoms three times the mass. Deuterium plasmas and deuterium–tritium plasmas from the standard TOKAMAK machines to exotic versions like laser-induced ultra-dense deuterium are of great interest for future thermonuclear devices (Lindl 1995; Holmlid 2015a; Hu et al. 2015). Recently, expressions for the free energy have been derived and the isothermal equation of state, the isentropes, and the Hugoniot curve have been calculated. We assume that isentropes and Hugoniot curves are theoretical idealizations of the actual compression process. While single-shock compression and the initial shock of a multishock compression are described by Hugoniot curves, the subsequent multishock compression process is close to an isentrope. As is well known, the pressure increases much faster with the temperature along isentropes than along Hugoniot curves. Therefore, the calculation of a set of Hugoniot curves and isentropes gives some idea of how the actual multishock compression process proceeds in the pressure–temperature plane. Nearly isentropic parts of the compression curve may also cross the coexistence line of the plasma phase transition PPT, which has been predicted to follow almost an isobar. Other thermodynamic functions can be calculated from the free energy by differentiation as shown above.

Isentropic Compressions

The highest pressures my be reached in experiments with isentropic compressions, i.e., compressions at constant entropy density (Beule et al. 1997, 2001). The entropy of a hydrogen plasma per proton is given by

$$s(T, \rho) = -\frac{\partial F(V, T, N)}{N_{\mathrm{p}} \partial T} .$$

Combining this expression with the isothermal EOS as a function of the mass density $p = p(\rho, T)$, we get the isentropic (adiabatic) EOS $p = p(s, T)$:

$$p = p(\rho, T) , \qquad p = p(s, T) . \tag{6.97}$$

Here ρ is the total mass density and $s = S/N K_{\mathrm{B}}$ the specific entropy per proton, where N_{p} is the total number of protons in the plasma including the protons bound in H and in H_2. In earlier work we calculated the isentropic EOS using the chemical picture in the ideal approximation and in the simplest nonideal approximation (QDHA) (Beule et al. 1997) and in the more advanced chemical description (Beule et al. 2001). The most relevant part of the entropy is the ideal contribution which is (Beule et al. 1997)

$$s_{\mathrm{id}} = \frac{5}{2}(1 + \alpha) - \alpha \ln \frac{\alpha n \Lambda_{\mathrm{e}}^3}{2} - \alpha \ln \frac{\alpha n \Lambda_{\mathrm{i}}^3}{2} - (1 - \alpha) \ln \frac{(1 - \alpha) n \Lambda_0^3}{4} . \tag{6.98}$$

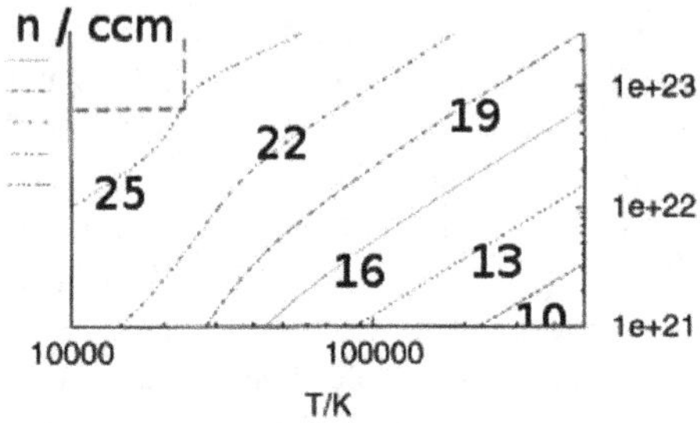

Fig. 6.23 Hydrogen plasmas in the (n, T) plane. Lines of constant entropy per particle for some lower entropy values $s = 10, 13, 16, 19, 22, 25$, calculated by Beule et al. using the PACH approximation (Kraeft and Schlanges 1996)

Our calculations show that the corrections due to nonideality effects are rather small for the entropy. For this reason, in a first approximation, we can use the ideal approximation $s_{\text{id}} = \text{const.}$ to calculate the isentropes. To estimate the nonideality corrections and calculate the pressure along the isentropes, we have the expressions for the free energy and the pressure given in Beule et al. (1999b, Beule et al. 2001). Figure 6.23 shows a set of curves of constant specific entropy $s = S/NK_BT$, calculated with nonideality corrections in the PACH approximation.

We are convinced that the entropy of plasmas urgently needs further study since, e.g., the lines of constant entropy in the (n, T) and (p, T) planes are absolutely fundamental for a deeper understanding of the physical properties of dense plasmas. We need this knowledge, not only to plan compression experiments, but also to understand the structure of stars and planets. Figure 6.23 shows a set of isentropes in the (p, T) plane, calculated for some lower values of the specific entropy $s \leq 10$ in PACH, while Fig. 6.24 shows data for high specific entropy densities, corresponding to high pressure plasmas, together with experimental data for isentropic compression.

To compare with experimental results, we calculated the single-shock Hugoniot curve of deuterium D_2 with initial conditions taken from the experiments, for liquid deuterium. Figure 6.25 shows the resulting Hugoniot curve together with experimental data and other theoretical calculations. We also mention extensive calculations of the EOS, and shock Hugoniots for hot dense fluid deuterium due to Zaghloul (2015). At high pressures and temperatures, the Hugoniot curve enters the region of highly ionized plasma. The EOS used there reduces to the previous PACH model (thick curve), which describes thermodynamic properties by Padé approximations in the chemical picture (Beule et al. 1999b, 2001). The activity expansion method ACTEX of Rogers and Young (1997) is also indicated in that region. Multishock experiments allow one to generate higher pressures at lower temperatures than single-shock experiments.

Analysis of the experimental data has shown that the first of the reverberating shocks is a strong shock, while the successive reverberations are much weaker and can therefore be approximated by an isentrope. For this case, entropy production during the shock is small and the increase in pressure along the Hugoniot curve is only slightly higher than along the corresponding isentrope. We refer again to Fig. 6.24,

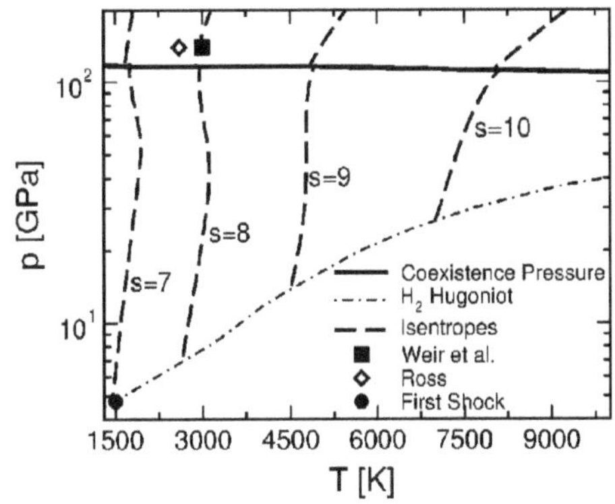

Fig. 6.24 Isentropes of deuterium plasmas up to high pressures and coexistence line of the PPT for hydrogen, from calculations by Beule et al. (1999b, 2001). The *square* marks the location of the insulator-to-metal transition, as found in the multishock experiments (Weir et al. 1996). The idea is that the plasma follows the Hugoniot during the first shock. The final state of the first shock is marked by a *closed circle*. In the second shock, the plasma should follow approximation an isentrope in some approximation. The transition point (insulator-to-metal) calculated in the linear mixing model is marked by a *diamond* (Ross 1996)

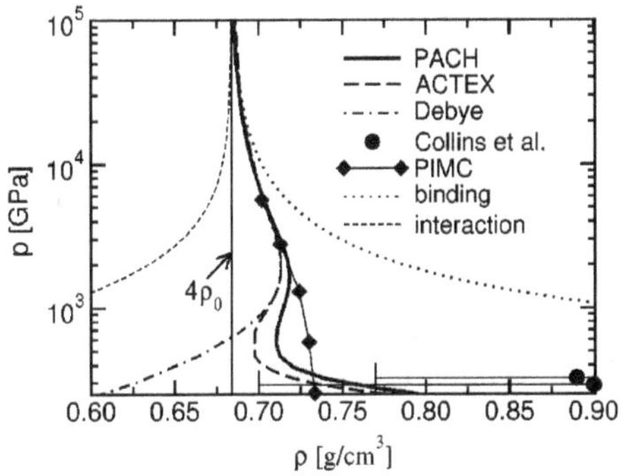

Fig. 6.25 Shock adiabats of deuterium plasmas for ultrahigh pressures, as obtained from the PACH approximation (*solid line*) (Beule et al. 2001) and ACTEX methods (*dotted line*) (Rogers and Young 1997), path integral Monte Carlo PIMC simulations (Militzer and Ceperley 2000), and the Debye approximation (*dash-dotted line*). The highest experimental pressures (Collins and DaSilva 1998) are marked by a *full circle*. The *dotted curve* corresponds to an ideal plasma model

which shows isentropes for different values of specific entropy together with the single-shock Hugoniot curve and the coexistence line of the PPT in hydrogen. After the first shock, pressure and temperature reach values of about 4.7 GPa and 1450 K, respectively. The isentrope starting from these conditions ($s = 7$) reaches a pressure of 140 GPa at a temperature below 2000 K and a density of about 0.9 g/cm^3. These values are based on the extrapolation of the present model to temperatures below 2000 K. Note that the small backward movement in the upper part of the isentrope is evidently an artefact of the calculations.

The linear-mixing model determines a temperature of 2600 K at a density of 0.7 g/cm^3. The lower temperature in our calculation is consistent with the higher degree of dissociation and the additional ionization found in the PACH approach. The coexistence line located at about 140 GPa is weakly temperature-dependent and falls only slowly. The Hugoniot curve increases with a relatively small derivative which is a decreasing function of the temperature. The isentropes are steeply increasing and, at small temperatures, nearly vertical in this presentation.

We conclude from Fig. 6.24 that the Hugoniot curve is unlikely to be able to cross the coexistence line of the PPT, but that isentropes may cross it due to their steep slope. Below temperatures of 10,000 K, the coexistence line lies substantially above the single-shock Hugoniot curve for hydrogen. For applications of EOS calculations to planets and other astrophysical problems, see (Nettelmann et al. 2012, 2013, Becker et al. 2013).

6.6.2 Helium and Other Plasmas of Light Elements

The study of helium plasmas, lithium plasmas, carbon plasmas and other plasmas of light elements with highly charged ions is of great interest for the physics of stars and planets (Nettelmann et al. 2008, French et al. 2009). To give an example, let us look at the biggest planet in our planetary system. Jupiter comprises more than 85 percent hydrogen and helium by mass. Among the smaller constituents, oxygen contributes the largest fraction. About 90 percent of the planetary material in Jupiter is in a high-pressure state beyond the 1 Mbar level and up to about 70 Mbar in the center, and the temper- ature varies by 2 orders of magnitude from the cold and dilute outer envelope up to about 20,000 K in the deep interior. Therefore the equation of state of hydrogen, helium, water, and other light elements significantly affects internal structural models of giant planets. The interior covers the transition from weak to strong ion coupling, and from weak to moderate electron degeneracy. Giant planets like Jupiter are an excellent example for the applications of the physics of non-ideal plasmas (Nettelmann et al. 2008) and are therefore particularly suitable for study.

For studies of the EOS at low densities see, e.g., (Ebeling 1989, 1990), and for studies of dense systems including mixtures (Ebeling 1989; Ebeling et al. 2007; Saumon and Chabrier 1992; Nettelmann et al. 2008; French et al. 2009). With respect to phase transitions the situation in helium is much more complicated than in hydrogen and not yet fully explored. The possible situation in the pressure–temperature plane is shown in Fig. 6.26 and typical isotherms are shown in Fig. 6.27 due to Förster et al. (1992, 1998) and Ebeling et al. (2007). For more calculations on helium plasmas, see (Lorenzen et al. 2011) and Chap. 10.

Studies of elements with higher nuclear charge and mixtures of them are even more complicated and require special studies in which essentially all elements of the theory developed here find applications. Applications of the theory developed here to highly ionized carbon created in capillary discharge plasmas were made by (Förster et al. 1998). These authors calculated the abundances of carbon ions in

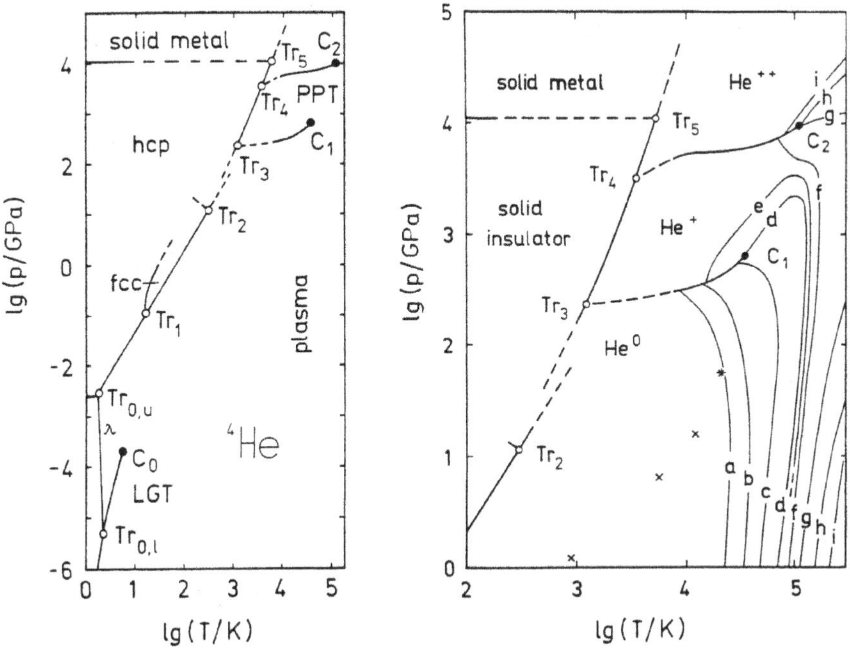

Fig. 6.26 Phase transitions in helium. *Left*: Phase coexistence lines at lower temperatures. *Right*: Still hypothetical transitions at high temperatures and pressures. From Förster et al. 1992

Fig. 6.27 Pressure as a function of density for helium plasmas, showing two phase transitions. From Förster et al. (1992)

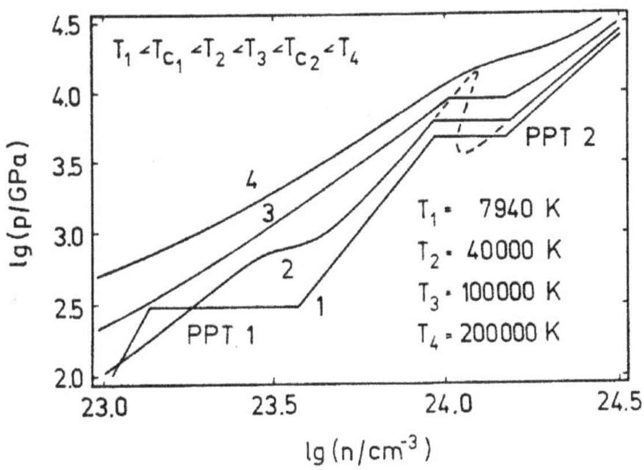

capillary discharges of $(CH_2)_n$ plastics, something of great relevance for technological applications, considering their dependence on temperature and time and comparing with data from experiments. We mention also recent ab initio calculations of thermodynamics, transport, and optical properties of $(CH_2)_n$ plastics (Knyazev and Levashov 2015). Formally, these plasmas are mixtures of protons, carbon ions, and electrons. Note that there may also be quite relevant applications of related plasmas in experiments targeting inertial confinement fusion (Hu et al. 2016).

Dense Nuclear Fusion Plasmas

We conclude with some remarks about the state of the art in the realization of nuclear fusion in strongly coupled plasmas (Ichimaru 1993; Nakai et al. 1997; Freiberg 2007; Potekhin and Chabrier 2013). Fusion experiments have been carried out in gas plasmas and in solid state plasmas. Gaseous fusion reactions are divided into two main classes: the thermonuclear and the pycno-nuclear reactions. In gases of low density and high temperature the Coulomb-barrier penetration probability is only slightly affected by electron screening, and the main contribution to the reaction probability comes from the relatively few fast-moving nuclei at the energy of the Gamow peak. These are the temperature-sensitive thermonuclear reactions now being studied in Tokamak and stellarator devices.

As the density increases and the temperature decreases, the potential barriers are depressed by electron screening (Ichimaru 1992; 1993, 1994; Potekhin and Chabrier 2013) and the Gamow peak is increased in height and displaced towards a lower energy. A new hypothesis assumes that long tails of the velocity distribution of Levy-type essentially increase the reaction rates (Ebeling, Romanovsky and Sokolov 2009). At high densities and low temperatures, when the potential barriers are depressed by a larger amount, the Gamow peak is shifted across the origin and only its tail-end remains. The reaction probability is density sensitive, and the main contribution comes from the relatively abundant slow-moving nuclei. These reactions are no longer thermonuclear and are more aptly described as 'pycno-nuclear' (after Cameron). At still higher densities and lower temperatures the nuclear gas becomes degenerate and the reaction probability is independent of the temperature of the gas; these reactions are referred to as 'cryo-nuclear'.

The present results for pycno-nuclear reactions differ in detail from those obtained by previous authors. Although the results can be readily adapted and extended in the case of astrophysical applications, the main emphasis is on hydrogen-isotope reactions with the possibility in mind of achieving the controlled release of nuclear energy from dense media in the laboratory. For example, in the typical thermonuclear working region of $10^{14} < n < 10^{16}$ cm^{-3}, $10^8 < T < 10^9$ K, where n is the ion number density and T the temperature, the burn-up time of the nuclear fuel is the same as in the cryo-nuclear working region of $3 \times 10^4 < \rho < 10^5$ g/cm^3, $T < 10^5$ K (where ρ is the mass density of the gas).

Inertial confinement fusion (ICF) is an approach to fusion that relies on the inertia of the fuel mass to provide confinement. To achieve conditions under which inertial confinement is sufficient for an efficient thermonuclear burn, a capsule (generally a spherical shell) containing thermonuclear fuel is compressed in an implosion process to conditions of high density and temperature. ICF capsules rely on either electron conduction (direct drive) or X rays (indirect drive) for energy transport to drive an implosion. In direct drive, laser beams (or charged particle beams) are directed at a target. The laser energy is transferred to electrons by means of inverse bremsstrahlung or a variety of plasma collective processes. In indirect drive, the driver energy (from laser beams or ion beams) is first absorbed in a high-Z enclosure, a so–called 'hohlraum', which surrounds the capsule. The material heated by the driver emits X rays, which drive the capsule implosion. For optimally designed targets, 70–80 percent of the driver energy can be converted to X rays. The optimal hohlraum geometry depends on the driver.

The ignition target requirements are difficult to fulfil. There are the conditions for hohlraum energetics, radiation symmetry, hydrodynamic instabilities, laser plasma interaction, and pulse shaping. Other developments such as the conceptual design study of a laser fusion power plant KOYO in Japan, take into account the recent progress in implosion physics research and reactor-related technologies. Another interesting direction is the search for fusion in ultra-dense deuterium (Holmlid 2015a). However, all these interesting fields lie outside the scope of this volume and the reader is referred to the original papers. For more about the physics, see Chap. 1 and for more information on theoretical problems connected with fusion studies, see Ichimaru (1993), Lindl (1995) and Hu et al. (2015).

References

A. Alastuey, Perez, Virial expansions for quantum plasmas: Fermi-Bose statistics. Phys. Rev. E **53**, 5714–5728 (1996)

A. Alastuey, V. Ballenegger, W. Ebeling, Comment on direct linear term in the equation of state of plasmas. Phys. Rev. E **92**, 047101 (2015)

A. Ashcroft, N.D. Mermin, *Solid State Physics* (Holt, Rinehardt Winston, Philadelphia, 1976)

J.A. Barker, D. Henderson, J. Chem. Phys. **47**, 4714 (1967)

A. Becker, N. Nettelmann, B. Holst, R. Redmer, Isentropic compression of hydrogen: Probing conditions deep in planetary interiors. Phys. Rev. B **88**, 045122 (2013)

T.H. Berlin, E.W. Montroll, On the free energy of a mixture of ions: an extension of Kramer's theory. J. Chem. Phys. **20**, 75 (1952)

D. Beule, W. Ebeling, A. Förster, Adiabatic equation of state and ionization equilibrium. Physica A **226**, 719–728 (1997)

D. Beule, W. Ebeling, A. Förster, H. Juranek, S. Nagel, R. Redmer, G. Röpke, Equation of state for hydrogen below 10 000 K: From the fluid to the plasma. Phys. Rev. B **59**, 14177 (1999a)

D. Beule, W. Ebeling, R. Redmer, G. Röpke, Electrical conductivity in dense hydrogen fluid and metal plasmas. Contrib. Plasma Phys. **39**, 25–28 (1999b)

D. Beule, W. Ebeling, A. Förster, H. Juranek, R. Redmer, G. Röpke, Isentropes and Hugoniot for dense hydrogen and deuterium. Phys. Rev. E **63**, 060202 R (2001)

D. Bohm, D. Pines, A collective description of electron interactions. Phys. Rev. **92**, 609 (1953)

H.P. Bonzel, A.M. Bradshaw, G. Ertl (eds.), *Physics and Chemistry of Alkali Metal Adsorption, Materials Science Monographs* (Elsevier, Amsterdam, 1999)

R. Bredow, W.-D. Th Bornath, R. Redmer, Kraeft, Hypernetted chain calculations for multi-component and non-equilibrium plasmas. Control. Plasma Phys. **53**, 276 (2013)

R. Bredow, W.D. Th Bornath, M.W.C. Kraeft, R. Redmer, Dharmawardana, Classical-map hypernetted chain calculations for dense plasmas. Contrib. Plasma Phys. **55**, 222–229 (2015)

N.V. Brilliantov, Accurate first-principle equation of state for the OCP. Contrib. Plasma Phys. **38**, 489 (1998)

L. Brillouin, *Les Statistiques Quantiques et leurs Applications, (Paris 1930)* (German transl, Berlin, 1931)

A. Bunker, S. Nagel, R. Redmer, G. Röpke, Phys. Rev B **56**, 3094 (1997); Control. Plasma Phys. **37**, 115 (1997)

A.V. Bushman, I.V. Lomonossov, V.E. Fortov, *Equation of state of metals at high energy densities* (Chernogolovka, 1992)

J. Callaway, *Energy Band Theory* (Acad. Press New York 1964) (Russ. transl. Mir Moska 1968)

D.M. Ceperley, B. Alder, Phys. Rev. Lett. **45**, 566 (1980)

G. Chabrier, A. Potekhin, Phys. Rev E **58**, 4941 (1998)

G.W. Collins, L.B. DaSilva et al., Science **281**, 1178 (1998)

P. Debye, E. Hückel, Theorie der Elektrolyte I. Physik. Z. **24**, 185 (1923)

L.B. DaSilva, P. Celliers et al., Phys. Rev. Lett. **78**, 483 (1997)

H.E. De Witt, Phys. Rev. A **14**(816), 1290 (1976)

H.E. De Witt, M. Schlanges, A.Y. Sakakura, W.D. Kraeft, Phys. Lett. A **197**, 326 (1995)

N. Desbiens, P. Arnault, J. Clerouin, Parametrization of OCP. Phys. Plasmas **23**(9) (2017), https://doi.org/10.10163/1.4963388

M.W.C. Dharmawardana, A review of studies on strongly-coupled Coulomb systems since the rise of DFT and SCCS 1977. Contrib. Plasma Phys. **55**, 85–101 (2015)

T. Dornheim, S. Groth, A. Filinov, M. Bonitz, Permutation blocking path integral Monte Carlo: a highly efficient approach to the simulation of strongly degenerate non-ideal fermions. New J. Phys. **17**, 073017 (2015a)

T. Dornheim, T. Schoof, S. Groth, A. Filinov, M. Bonitz, Permutation blocking path integral monte carlo approach to the uniform electron gas at finite temperature. J. Chem. Phys. **143**, 204101 (2015b)

T. Dornheim, S. Groth, T. Schoof, C. Hann, M. Bonitz, Ab initio quantum Monte Carlo simulations of the uniform electron gas without fixed nodes: The unpolarized case. Phys. Rev. B **93**, 205134 (2016a)

T. Dornheim, H. Thomsen, P. Ludwig, A. Filinov, M. Bonitz, Analyzing quantum correlations made simple. Contrib. Plasma Phys. **56**, 371 (2016b)

T. Dornheim, S. Groth, T. Sjostrom, F.D. Malone, W.M.C. Foulkes, M. Bonitz, Ab Initio quantum monte carlo simulation of the warm dense electron gas in the thermodynamic limit. Phys. Rev. Lett. **117**, 156403 (2016c)

W. Ebeling, Equation of state and Saha equation of partially ionized plasmas. Physica **38**, 378 (1968)

W. Ebeling, Free energy and ionization in dense plasmas of the light elements. Control. Plasma Phys. **30**, 53–561 (1990)

W. Ebeling, The work of Baimbetov on nonideal plasmas and some recent developments. Control. Plasma Phys. **56**(3–4), 163–175 (2016)

W. Ebeling, R. Sändig, Ann. Physik (Leipzig) **28**, 289 (1973); Proc. ICPIG Contr. Papers **11**, 421 (1973)

W. Ebeling, W.D. Kraeft, D. Kremp, *Theory of Bound States and Ionisation Equilibrium in Plasmas and Solids, (Akademie-Verlag, Berlin, 1976)* (Extended Russ. transl, Mir Moscow, 1979)

W. Ebeling, C.V. Meister, R. Sändig, Pressure ionization in nonideal alkali plasmas. Ann. Physik **36**, 321–400 (1978)

W. Ebeling, V.F. Fortov, YuL Klimontovich, et al. (eds.), *Transport Properties of Dense Plasmas* (Birkhäuser, Boston, 1984)

W. Ebeling, Free energy and ionization of dense plasma of light elements, Contr. Plasma Physics **29**, 238–242 (1989); **30**, 553–561 (1990), see also Inside the Sun, Proc. 121st Coll. Int. Astron. Union, Versailles 1989

W. Ebeling, W. Richert, Pressure ionization of atoms in plasmas, Ann. Physik 39, 362 (1982); Contr. Plasma Phys. **25**, 431–436 (1985)

W. Ebeling, W. Richert, Phys. Lett. A **108**, 80 (1985); Phys. Stat Sol. (b) **128**, 167 (1985)

W. Ebeling, F. Schautz, Simulations of the quantum electron gas using momentum-dependent potentials. Phys. Rev. E **56**, 3498 (1997)

W. Ebeling, W.D. Kraeft, D. Kremp, G. Röpke, Physica A **140**, 160–168 (1986)

W. Ebeling, A. Förster, W. Richert, H. Hess, Thermodynamic properties and plasma phase transitions of xenon at high pressure and high temperature. Physica A **150**, 159–171 (1988)

W. Ebeling, G.E. Norman, A.A. Valuev, I.A. Valuev, Contr. Plasma Phys. **39**, 61 (1999a)

W. Ebeling, W. Stolzmann, A. Förster, M. Kasch, Contr. Plasma Phys. **39**, 287–306 (1999b)

W. Ebeling, G. Norman, Coulomb Phase Transitions in Dense Plasmas. J. Stat. Phys. **110**, 861–877 (2003)
W. Ebeling, S. Hilbert, On Saha's equation for partially ionised plasmas and Onsager's bookkeeping rule. Eur. Phys. J. D **20**, 93–101 (2002a)
W. Ebeling, H. Hache, M. Spahn, Thermodynamics of ionization and dissociation in hydrogen plasmas including fluctuations and magnetic fields. Eur. Phys. J. D **23**, 265–272 (2003)
W. Ebeling, H. Hache, H. Juranek, R. Redmer, G. Röpke, Control. Plasma Phys. **45**, 160 (2005)
W. Ebeling, D. Blaschke, R. Redmer, H. Reinholz, G. Röpke, The influence of Pauli blocking effects on the Mott transition in dense hydrogen. J. Phys. A: Math. Theor. **42**, 214033 (2009a)
W. Ebeling, A. Förster, V.E. Fortov, V.K. Gryaznov, A.Ya. Polishchuk, Thermophysical Properties of Hot Dense Plasmas, Teubner-Verlag, Stuttgart-Leipzig, 1991; Russ. transl. R &C Dynamics 2007
W. Ebeling, A. Förster, R. Redmer, T. Rother, M. Schlanges, *ICPIG 18* (University of Wales, Invited Lectures, 1987)
W. Ebeling, W.D. Kraeft, D. Kremp, *Theory of Bound States and Ionisation Equilibrium in Plasmas and Solids, (Akademie-Verlag, Berlin, 1976)* (Extended Russ. transl. Mir, Moscow, 1979)
W. Ebeling, M. Steinberg, J. Ortner, Ionization equilibrium and EOS of low-temperature hydrogen plasmas. Phys. Rev. E **61**, 2290 (2000)
J. Ebeling, S. Hilbert, H. Krienke, On Bjerrum's mass action law for electrolytes and Onsager's bookkeeping rule. J. Mol. Liquids **96/97**, 409 (2002)
W. Ebeling, MYu. Romanovsky, I.M. Sokolov, Velocity distributions and kinetic equations for plasmas including Levy type power law tails. Contr. Plasma Phys. **49**, 704–712 (2009)
L.T. Erimbetova, A.E. Davletov, Y.S. ZhA Kudyshev, Mukhametkarimov. Contrib. Plasma Phys. **53**, 414 (2013)
H. Falkenhagen, *Theorie der Elektrolyte Hirzel* (Leipzig 1971)
M.E. Fisher, J. Stat. Phys. **75**, 1 (1994)
M.E. Fisher, Y. Levin, Phys. Rev. Lett. **71**, 2138 (1993)
V. Fortov, M. Mochalov et al., Phys. Rev. Lett. **99**, 185001 (2007)
V.E. Fortov, *Extreme States of Matter (in Russian)* (FizMatGis, Moskva, 2009,2015)
V.E. Fortov, *Extreme States of Matter: On earth and in the cosmos, the Frontiers Collection* (Springer, Berlin, 2011)
V.E. Fortov, *Equation of state from the ideal gas to the quark–gluon plasma (in Russ.)* (Fizmatgis, Moskva 2013)
A. Förster, T. Kahlbaum, W. Ebeling, High Press. Res. **7**, 375 (1991)
A. Förster, T. Kahlbaum, A. Rickert, Z. Physik D Suppl. **21**, 171 (1991)
A. Förster, T. Kahlbaum, W. Ebeling, Equation of state and pase diagram of fluid helium in the region of partial ionization. Laser Part. Beams **10**, 253–262 (1992)
A. Förster, D. Beule, H. Conrads, W. Ebeling, Highly ionized carbon in capillary discharge plasma. Control. Plasma Phys. **38**, 655–660 (1998)
J.P. Freiberg, *Plasma Physics and Fusion Energy* (Cambridge University Press, Cambridge, 2007)
M. French, T.R. Mattsson, N. Nettelmann, R. Redmer, Equation of state and phase diagram of water at ultrahigh pressures as in planetary interiors. Phys. Rev. B **79**, 054107 (2009)
S.H. Glenzer, R. Redmer, Rev. Mod. Phys. **81**, 1625 (2009)
S. Groth, T. Dornheim, M. Bonitz, Free energy of the uniform electron gas: Testing analytical models against first principle results, Control. Plasma Phys. submitted (2016)
P. Haronska, D. Kremp, M. Schlanges, Equation of state, composition and phase transitions in hydrogen plasma. Wiss. Z. Uni Rostock **36**, 98–102 (1987)
F. Hensel, S. Juengst, F. Noll, R. Winter, in *In Localisation and Metal Insulator Transitions*, ed. by D. Adler, H. Fritsche (Plenum Press, New York, 1985)
F. Hensel, W.W. Warren Jr., *Fluid Metals, The Liquid-Vapor Transition* (Princeton University Press, Princeton, 2014)
N.C. Holmes, M. Ross, W.-J. Nellis, Phys. Rev. B **52**, 15833 (1995)

L. Holmlid, Heat generation above break-even from laser-induced fusion in ultra-dense deuterium, AIP Advances **5**, no. 087129 (2015a)
L. Holmlid, MeV particles in a decay chain process from laser-induced processes in ultra-dense deuterium D(0). Int. J. Mod. Phys. E. Nucl. Phys. **24**, 1550026 (2015b)
B. Holst, R. Redmer, M.P. Desjarlais, Phys. Rev. B **77**, 184201 (2008)
B. Holst, M. French, R. Redmer, Electronic transport coefficients from ab initio simulations and application to dense liquid hydrogen. Phys. Rev. B **83**, 235120 (2011)
S.X. Hu, V.N. Goncharov, T.R. Boehly, R.L. McCrory, S. Skupsky, L.A. Collins, Impact of first-principles properties of deuterium-tritium on inertial confinement fusion target designs. Phys. Plasmas **22**(5), 056304 (2015)
S.X. Hu et al., First-principles investigations on ionization and thermal conductivity of polystyrene for inertial confinement fusion applications. Phys. Plasmas **23**, 042704 (2016)
S. Ichimaru, *Basic principles of plasma physics* (Addison-Wesley, Benjamin London, 1973)
S. Ichimaru, *Statistical Plasma Physics: II* (Condensed Plasmas, Addison-Wesley, Reading, MA, 1994)
S. Ichimaru, S. Tanaka, Phys. Rev. A **32**, 1790 (1985)
S. Ichimaru, H. Iyetomi, S. Tanaka, Phys. Reports **149**, 91 (1987)
S. Ichimaru, *Statistical Plasma physics* (Addison-Wesley, Redwood City, 1992)
S. Ichimaru, Nuclear fusion in dense plasmas. Rev. Mod. Phys. **65**, 255 (1993)
H. Juranek, R. Redmer, G. Röpke, V.E. Fortov, A.A. Pyalling, A comparative study for the equation of state of dense fluid hydrogen. Contrib. Plasma Phys. **39**(3), 251–261 (1999)
T. Kahlbaum, The quantum-diffraction term in the free energy for Coulomb plasma, effective-potential approach. J. Phys. France **10**, Pr5-455 (2000)
A.S. Kaklyugin, Teplophys. Vyss. Temp. **23**, 217 (1985)
G. Kalman (ed.), *Strongly Coupled Coulomb Systems* (Pergamon Press, 1988)
G. Kelbg, Quantenstatistik der gase mit coulomb-wechselwirkung. Ann. Physik (Leipzig) **12**(219–224), 354–360 (1963)
K. Kilimann, W. Ebeling, Energy gap and line shifts for H-like ions in dense plasmas, Z. Naturforschung **45a**, 613–617 (1990)
YuL Klimontovich, *Statistical Theory of None Quilibrium Processes in Plasmas, Izdat MGU 1964* (Engl. transl. Pergamon, Oxford, 1967)
Yu.L. Klimontovich, Statistical physics, in Russ. Nauka 1982; (Engl. Harwood New York 1986)
M.D. Knudson, M.P. Desjarlais, A. Becker, R.W. Lemke, K.R. Cochrane, M.E. Savage, D.E. Bliss, T.R. Mattsson, R. Redmer, Science **348**(6242), 1455–1460 (2015)
D.V. Knyazev, P.R. Levashov, Ab initio calculation of thermodynamic, transport, and optical properties of CH_2 plastics. Phys. Plasmas **22**, 035303 (2015)
W.D. Kraeft, D. Kremp, G. Röpke, Direct linear term in the equation of state of plasmas. Phys. Rev. E **91**, 013108 (2015)
W.D. Kraeft, D. Kremp, W. Ebeling, G. Röpke, *Quantum Statistics of Charged Particle Systems* (Akademie-Pergamon Press, Berlin, 1986)
W.D. Kraeft, M. Schlanges, *Physics of strogly coupled plasmas* (World Scientific, Singapore, 1996), pp. 49–58
W.D. Kraeft, W. Ebeling, D. Kremp, G. Röpke, Thermodynamic functions, ionization equilibrium and phase transitions in Coulomb systems. Ann. Physik (Leipzig) **45**, 429–442 (1988)
H.A. Kramers, Investigations on the free energy of a mixture of ions. Proc. Koninkl. Akademie v. Wetenschappen **30**, 145–158 (1927)
D. Kremp, G. Schmitz, Z. Naturforschung **22a**, 1366 (1967)
D. Kremp, M. Schlanges, W.D. Kraeft, *Quantum Statistics of Nonideal Plasmas* (Springer, Berlin, 2005)
J.D. Kress, B. Militzer, Impact of first principles properties of deuterium-tritium on inertial confinement fusion target designs. Phys. Plasmas **22**, 056304 (2015)
A. Lankin, G. Norman, Density and nonideality effects in plasmas. Control. Plasma Phys. **49**, 723–731 (2009)

A.I. Larkin, Zh. eksp. teor. Fiz. (in Russ.) **38**, 1896 (1960)

B.P. Lee, M.E. Fisher, Phys. Rev. Lett. 76, 2906 (1996); Y. Levin, M.E. Fisher, Phys. A 225, 23 (1996). Europhys. Lett. **39**, 3 (1997)

H. Lehmann, W. Ebeling, Coulomb phase transitions in symmetrical quantum systems, Phys. Rev. E **54**, 2451–2457 (1996); see also Ann. Physik (Leipzig) **45**, 529 (1988)

T.J. Lenosky, J.D. Kress, L.A. Collins, I. Kwon, Phys. Rev. B **55**, R11 907 (1997)

J. Lindl, Development of the indirect-drive approach to inertial confinement fusion and the target physics basis for ignition and gain. Phys. Plasmas **2**, 3933 (1995), https://doi.org/10.1063/1.871025

W. Lorenzen, B. Holst, R. Redmer, Metallization in hydrogen-helium mixtures. Phys. Rev. B **84**, 235109 (2011)

W.R. Magro, D.M. Ceperley, C. Pierleoni, B. Bernu, Phys. Rev. Lett. **76**, 1240 (1996)

B. Militzer, D. Ceperley, Phys. Rev. Lett. **85**, 1890 (2000)

A.G. Moreira, M.M. da Gama, M.E. Fisher, J. Chem. Phys. **110**, 10058 (1999)

S. Nakai, K. Mima, Y. Kitagawa, S. Sakabe, Y. Izawa, M. Nakatsuka, M. Yamanaka, H. Fujita, T. Jitsuno, T. Kanabe, N. Miyanaga, K. Tsubakimoto, S. Motokoshi, H. Kiriyama, H. Matsui, H. Takabe, T. Norimatsu, M. Takagi, Y. Kozaki, T. Yamanaka, C. Yamanaka, H. Kan, T. Hiruma, Development of inertial fusion energy. AIP Conf. Proc. **406**, 84 (1997), https://doi.org/10.1063/1.53510

M.A. Morales, C. Pierleoni, E. Schwegler, D.M. Ceperley, Evidence for a first-order liquid-liquid transition in high-pressure hydrogen from ab initio simulations. PNAS **110**, 2799–2803 (2010)

W.J. Nellis, S.T. Weir, A.C. Mitchell, Phys. Rev. B **50**, 3434 (1999)

N. Nettelmann, B. Holst, A. Kietzmann, M. French, R. Redmer, D., Ab initio equation of state data for hydrogen, helium, and water and the internal structure of Jupiter. Astrophys. J. **683**, 1217 (2008)

N. Nettelmann, A. Becker, B. Holst, R. Redmer, Jupiter models with improved ab initio hydrogen equation of state (H-REOS.2). Astrophys. J. **750**, 52 (2012)

N. Nettelmann, R. Püstow, R. Redmer, Saturn layered structure and homogeneous evolution models with different EOS. Icarus **225**, 548–557 (2013)

G.E. Norman, A.N. Starostin, High Temp. 6, 394 (1968). High Temp. **8**, 381 (1970)

J. Ortner, F. Schautz, W. Ebeling, Quasiclassical molecular-dynamics simulations of the electron gas. Phys. Rev. E **56**, 4665–4670 (1997)

J. Ortner, I. Valuev, W. Ebeling, Semiclassical dynamics and time correlations in two-component plasmas. Control. Plasma Phys. **39**, 311–321 (1999)

F. Perrot, M.W.C. Dharmawardana, Exchange and correlation potentials at finite temperatures. Phys. Rev. A **30**, 2619 (1984)

F. Perrot, M.W.C. Dharmawardana, Spin-polarized electron liquid at arbitrary temperatures. Phys. Rev. B **62**, 16536 (2000)

D. Pines, *The many-body problem. A lecture note*, (Benjamin, New York, 1961) (Russ. transl. Moskva 1963)

D. Pines, P. Nozieres, *The theory of quantum liquids, (Benjamin, New York, Amsterdam 1966)* (Russ. Transl. Mir, Moskva, 1967)

M. Planck, Zur Quantenstatistik des Bohrschen Atommodells. Ann. Physik (Leipzig) **75**, 673–684 (1924)

A.Y. Potekhin, G. Chabrier, Electron screening effect on stellar thermonuclear fusion, Contr. Plasma Physics **53**, 397–405 (2013); Equation of state for magnetized Coulomb plasmas, Astronomy and Astrophysics **550**, A43 (2013)

J.C. Rasaiah, H. Friedman, Integral equation methods in the computation of equilibrium properties of ionic solutions. J. Chem. Phys. **48**, 2742 (1968)

R. Redmer, Physical properties of dense, low-temperature plasmas. Phys. Rep. **282**, 35 (1997)

R. Redmer, G. Röpke, Progress in the theory of dense strongly coupled plasmas. Control. Plasma Phys. **50**, 970–985 (2010)

W. Richert, W. Ebeling, Thermodynamic functions of thc electron fluid for a wide density-temperature range. Phys. Stat. Sol. (b) **121**, 623–639 (1984)

J. Riemann, M. Schlanges, H.E. DeWitt, W.D. Kraeft, EOS of weakly degenerate OCP. Physica A **219**, 423 (1995)

F.J. Rogers et al., Astrophys. J. **310**, 723 (1986); ibid. **456**, 902 (1996)

F.J. Rogers, D.A. Young, Phys. Rev. E **56**, 5876 (1997)

G. Röpke, W. Ebeling, W.D. Kraeft, Quantum-statistical conductance theory of nonideal plasmas. Physica A **101**, 243–254 (1980)

G. Röpke, K. Kilimann, D. Kremp, W.D. Kraeft, Phys. Lett. A **68**, 329 (1978)

G. Röpke, K. Kilimann, D. Kremp, W.D. Kraeft, R. Zimmermann, Phys. Stat. Sol. (b) **88**, K59 (1978)

G. Röpke, T. Seifert, H. Stolz, R. Zimmermann, Phys. Stat. Sol. (b) **100**, 215 (1980)

Y. Rosenfeld, Phys. Rev. Lett. **63**, 980 (1989); **72**, 3831 (1994)

Y. Rosenfeld, P. Tarazona, Molec. Phys. **95**, 141 (1998)

M. Ross, Phys. Rev. B **54**, 9589 (1996); **58**, 669 (1998)

S. Sadykova, W. Ebeling, Control. Plasma Phys. **47**(659), 1160 (2007)

S. Sadykova, W. Ebeling, I. Valuev, I. Sokolov, Contrib. Plasma Phys. **49**, 76 (2009)

S.P. Sadykova, W. Ebeling, I.M. Tkachenko, Eur. Phys. J. D **61**, 117–130 (2011)

M.N. Saha, Phil. Mag. 40, 472 (1920). Z. Phys. **6**, 40 (1921)

R. Sändig, Thermodynamic functions of dense plasmas using Allnatt integral equation, Phys. Lett. **51** A, 181–183 (1975)

D. Saumon, G. Chabrier, Phys. Rev. Lett. **62**, 2397 (1989); Phys. Rev. A **44**, 5122 (1991); **46**, 2084 (1992)

M.E. Savage, D.E. Bliss, T.R. Mattsson, R. Redmer, Direct observation of an abrupt insulator-to-metal transition in dense liquid deuterium. Science **348**, 1455–1460 (2015)

G. Schmitz, D. Kremp, Z. Naturforschung **23a**, 1392 (1967)

K.S. Singwi, M.P. Tosi, R. Land, A. Sjölander, Phys. Rev. **176**, 589 (1968)

V. Sizyuk, A. Hassanein, T. Sizyuk, J. Appl. Phys. **100**, 103106 (2006)

M. Steinberg, J. Ortner , W. Ebeling, The equation of state of magnetized plasma, J. de Physique **4**, 377 ff (2000a)

M. Steinberg, J. Ortner, W. Ebeling, Ionization equilibrium and equation of state of low temperature hydrogen plasma in weak magnetic fields. Eur. J. Phys. D **12**, 513–520 (2000b)

W. Stolzmann, T. Blöcker, EOS of multi-component plasmas. Phys. Lett. A **221**, 99 (1996)

W. Stolzmann, W. Ebeling, New Padé approximations for the free charges in two-component strongly coupled plasmas based on the Unsöld-Berlin-Montroll asymptotics. Phys. Lett. A **248**, 242–246 (1998)

S. Tanaka, S. Ichimaru, Parametrized equation of state for dense hydrogenic plasmas. Phys. Rev. A **32**, 3756–3757 (1985)

S. Tanaka, S. Ichimaru, Thermodynamics and correlational properties of finite-temperature electron liquids in the STLS approximation. J. Phys. Soc. Jpn. **55**, 2278–2298 (1986)

A. Unsöld, Z. Astrophys. **24**, 355 (1948)

A. Unsöld, *Physik der Sternatmosphären unter besonderer Berücksichtigung der Sonne, Springer, Heidelberg 1938, 2* (Aufl, Heidelberg, 1955)

S.T. Weir, A.C. Mitchell, W.J. Nellis, Phys. Rev. Lett. **76**, 1860 (1996)

R. Hou, Yong, Bredow, Y. Yianmin, R. Redmer, Average-atom model combined with the hypernetted chain approximation applied to warm dense matter. Phys. Rev. E **91**, 033114 (2015)

E.P. Wigner, Interaction of electrons in metals. Phys. Rev. **46**, 1002–1011 (1934)

L.D. YaB Zeldovich, Landau. Zh Eksp. Teor. Fiz. **14**, 32 (1944)

M.R. Zaghloul, Tables of equation-of-state, thermodynamic properties, and shock Hugoniot for hot dense fluid deuterium, Phys. Plasmas **17**, 062701 (2010); **22**, 112709 (2015)

M.R. Zaghloul, Dissociation and ionization equilibria of deuterium fluid over a wide range of temperatures and densities. Phys. Plasmas **22**, 062701 (2015)

W. Zimdahl, W. Ebeling, Theory of the ionization equilibrium in nonideal alkali plasmas. Ann. Physik (Leipzig) **34**, 9–22 (1977)

Chapter 7
Kinetic Equations and Fluctuations in Nonideal Gases and Plasmas

This brief survey of kinetic and transport theory starts with the stochastic approach, which is mathematically simpler than kinetic equations of Boltzmann and Lenard–Balescu type. We continue with Lorentz approximations and then discuss the derivation of quantum Boltzmann equations using the Bogoliubov method. The next topic is fluctuation–dissipation theorems, including a discussion of recent controversies in this field, and several applications to plasmas. For more detailed introductions to stochastic and transport problems, the reader is referred to other books (Balescu 1975; Klimontovich 1964, 1967, 1986; Ebeling et al. 1984; Ebeling and Sokolov 2005; Ebeling at al. 2008; Feistel and Ebeling 2011; Dittrich et al. 1998; Bonitz 2000, 2016; Kremp et al. 2005).

7.1 Stochastic Kinetics

7.1.1 Smoluchowski–Fokker–Planck and Master Equations

The starting point of quantum statistics is intimately connected with the work of Boltzmann on the one hand, but also the mathematical and physical theory of stochastic processes. The stochastic description put forward by Einstein in 1905/1906, Smoluchowski in 1906, and Langevin in 1908 is so different from those of Boltzmann and Gibbs, since it dispenses from the description of the system evolution in phase space and relies from the very beginning on probabilistic concepts (Ebeling and Sokolov 2005; Ebeling et al. 2008).

Parallel to Einstein's work on diffusion and mean square displacement, Marian Smoluchowski developed in Lemberg and Krakow the basic tools of the stochastic theory in physics. Mark Kac, in an essay about Smoluchowski, put it as follows:

> While directed towards the same goal, how different the Smoluchowski approach is from Boltzmann's. There is no dynamics, no phase space, no Liouville theorem—in short none of the usual underpinnings of statistical mechanics. Smoluchowski may not have been aware

W. Ebeling et al., *Quantum Statistics of Dense Gases and Nonideal Plasmas*, Springer Series in Plasma Science and Technology, DOI 10.1007/978-3-319-66637-2_7

of it, but he began writing a new chapter of statistical physics, which in our time goes by the name of stochastic processes.

A synthesis of the various approaches leading to our present understanding of how the properties of stochastic motions are connected to the deterministic dynamics of the system and its heat bath were understood much later in work by Mark Kac, Robert Zwanzig, and others.

We take the general view developed by Neimark (see Ebeling and Sokolov 2005; Feistel and Ebeling 2011) and start from a general dynamical system with a state at time t given by a set (a vector) of time-dependent parameters $x_i(t)$ (coordinates), viz., $x(t) = [x_1(t); x_2(t); \ldots; x_n(t)]$. The state vectors $\mathbf{x}(t)$ span a vector space X which will be called the state space or phase space of the system. Let us assume now that the state at time t, which is $\mathbf{x}(t)$, and the state at a later time $t+\delta t$, which is $\mathbf{x}(t+\delta t)$, are connected in a causal way. In particular, we assume that the state $\mathbf{x}(t+\delta t)$ is given as a function or a more general mapping of the state $\mathbf{x}(t)$ and certain parameters which we denote by the vector $\mathbf{u}(t)$. The set of all possible parameter values $\mathbf{u}$ forms the control space of the system. The parameters $\mathbf{u}(t)$ take into account the influence of the environment, and possibly also external control actions. We assume that the connection between $\mathbf{x}(t)$ and $\mathbf{x}(t+\delta t)$ is given by a dynamical map $\mathbf{T}$ such that

$$\mathbf{x}(t+\delta t) = \mathbf{T}\big(\mathbf{x}(t); \mathbf{u}; \delta t\big) . \tag{7.1}$$

According to Neimark, the set ($\mathbf{X}$; $\mathbf{T}$) is called the dynamical model of the system (Feistel et al. 2011).

The choice of phase space is not unique and is in general a rather difficult problem. Even if we have a set of observations and measurements, there are still many possibilities for the choice The description of processes on the deterministic level is in some sense incomplete, since stochastic influences are always present under real conditions. We will introduce here only a couple of the main aspects of stochastic descriptions. Details and examples of stochastic descriptions can be found elsewhere (Feistel and Ebeling 2011).

Stochastic models are important for several reasons:

- Mesoscopic or macroscopic variables are always representative of a large number of microscopic degrees of freedom which are subject to thermal fluctuations.
- All variables depending on the number of particles are necessarily discrete. This gives rise to a special kind of noise, so-called shot noise.
- The intrinsic quantum character of the microscopic dynamics leads to stochastic effects.
- A mesoscopic or macroscopic system is as a rule imbedded in very complex surroundings, leading to stochastic external forces.

Due to these stochastic influences, the future state of a dynamical system is in general not uniquely specified by its previous state. In other words the dynamic map $\mathbf{T}$ is non-unique. A given initial point $\mathbf{x}(0)$ may be the source of several different trajectories. The choice between the different possible trajectories is a random event. In this way the term trajectory loses its precise meaning and should be supplemented in terms of probability theory. Stochastic models will be described in detail later.

In the framework of stochastic descriptions, we describe the state of the system at time t by a probability density $P(\mathbf{x}; t; u)$. By definition, $P(\mathbf{x}; t; u)\mathrm{d}x$ the probability of finding the trajectory at time t in the interval $(\mathbf{x}; \mathbf{x} + \mathbf{dx})$. Instead of the deterministic equation for the state, we now get a differential equation for the probability density $P(\mathbf{x}; t; u)$. A standard model is the Smoluchowski–Fokker–Planck equation

$$\frac{\partial P(\mathbf{x}, t)}{\partial t} = \frac{\partial}{\partial \mathbf{x}}\left[F(x, t; u)P(x, t) + D\frac{\partial P(x, t)}{\partial \mathbf{x}}\right]. \tag{7.2}$$

Here $F(x, t; u)$ is the deterministic driving force and D the diffusion coefficient given by the correlations of the stochastic forces. In a more general case where the transitions have nonlocal character, by balancing the input and output, we obtain an integral equation for the probability density:

$$\frac{\partial P(\mathbf{x}; t)}{\partial t} = \int \mathrm{d}x_0\left[W(x|x_0)P(\mathbf{x}_0; t) - W(x_0|x)P(\mathbf{x}; t)\right]. \tag{7.3}$$

In mathematics, this is known as the Chapman–Kolmogorov equation, and in physics as the Pauli equation or master equation. It plays a fundamental role in the theory of stochastic processes. The integration is carried out over all possible states x_0 which are attainable from the state x by a single jump. It is a linear equation with respect to $P(x, t)$, and uniquely determines the evolution of the probability density. The right-hand side consists of two parts: the first stands for the gain in probability and the second describes the loss due to reverse events. Using a Taylor expansion with respect to the moments of the transition probability, one can transform (7.3) into an infinite Taylor series. This is the so-called Kramers–Moyal expansion. If this expansion ends after the second contribution, we arrive once again at the Smoluchowski–Fokker–Planck equation.

In short, we may say that many physical phenomena can be explained only by taking into account fluctuations. The stochastic approach on a mesoscopic level often provides more elegant solutions than the microscopic statistical approach. Inclusion of fluctuations of the macroscopic variables does not necessarily increase the number of relevant variables, but only changes their character by transforming them into stochastic variables. The main difference compared with the deterministic models is the permeability of separatrices. This statement concerns in particular non-chaotic dynamics, for instance, if dealing with one or two order parameters. Stochastic realizations reach (or cross) unstable points, saddle points, and separatrices with a certain probablity, which is impossible in the deterministic description. Stochastic effects make it possible to escape regions of attraction around stable manifolds. Physical situations which make use of this circumstance are, e.g., nucleation or chemical reactions, where energetically unfavorable states have to be overwhelmed. For the quantum theory of stochastic processes, see, e.g., (Dittrich et al. 1998). Here we will discuss only one special topic.

Landau Theory of Quasi-particles and Ginzburg–Landau Kinetics

The Landau theory is a semi-phenomenological quantum approach, similar to the Smoluchowski–Fokker–Planck theory. The essence of the Landau theory is the idea that one can have a one-to-one correspondence between the elementary excitations of an interacting system and those of an appropriate noninteracting system (Landau and Lifshits 1977, 1980; Inkson 1984). The Ginzburg–Landau theory, named after Vitaly Ginzburg (1916–2009, Nobel Prize 2003) and Lev Landau (1908–1968, Nobel Prize 1962), is a theory used to describe superconductivity. In its initial form, it was postulated as a phenomenological model for phase transitions of second order, which developed into a theory of type I superconductors, without examining their microscopic properties.

Later, a version of the Ginzburg–Landau theory was derived from the Bardeen–Cooper–Schrieffer microscopic theory by Lev Gorkov (1959), thus showing that it also arises in some limit of the microscopic theory, and thereby giving a microscopic interpretation of all its parameters. We mention the Ginzburg–Landau kinetics here, since the simplest equation for relaxation around the phase transition at a temperature T_c has the form of a Smoluchowski–Fokker–Planck equation (see, e.g., Klimontovich 1995, 1999, 2001)

$$\frac{\partial X(R,t)}{\partial t} = -\gamma \left[\frac{(T - T_c)}{T_c} + bX^2(R,t) \right] + D \frac{\partial^2 X(R,t)}{\partial R^2} . \tag{7.4}$$

Here $X(R,t)$ is an order parameter of a second order phase transition. In the quantum case, the complex wave function satisfies

$$\frac{\partial \psi(R,t)}{\partial t} = -\gamma \left[\frac{(T - T_c)}{T_c} + b|\psi(R,t)|^2 \right] \psi + D \frac{\partial^2 \psi(R,t)}{\partial R^2} . \tag{7.5}$$

Note that these equations may be derived from a functional, viz., an effective Hamiltonian or an effective free energy named after Ginzburg and Landau, by functional derivation: of

$$\frac{\partial \psi(R,t)}{\partial t} = -C \frac{\delta H_{\text{eff}}}{\delta \psi(R,t)} . \tag{7.6}$$

This formalism is based on Landau's previously established theory of second-order phase transitions. Ginzburg and Landau argued that the free energy functional F_{eff}, of a superconductor near the superconducting transition can be expressed in terms of a complex order parameter field ψ, which is nonzero below a phase transition into a superconducting state and is related to the density of the superconducting component.

Stochastic Master Equations for Discrete Variables

Another large class of Markovian processes contains systems with a discrete state space. This concerns the atomic processes or extensive thermodynamic variables, like particle numbers in chemically reacting systems or in grand canonical ensembles where the particle number N is a variable. Let us consider a system with a finite spectrum of s states numbered by $i = 1, \ldots, s$ and possible transitions between them. Correspondingly, we define the probabilities $p_i(t)$ of the (mixed) states of the system. Probabilities have to satisfy some conservation rules such as imposed by normalization:

$$\sum_{i=1}^{s} p_i(t) = 1 . \tag{7.7}$$

The evolution of probabilities in time should conserve this normalization. A quite general kinetic equation which meets this condition is the *Pauli equation*

$$\frac{\mathrm{d}}{\mathrm{d}t} p_m(t) = \sum_{m'}^{s} \left[W_{mm'} p_{m'}(t) - W_{m'm} p_m(t) \right] . \tag{7.8}$$

where we introduce the transition rates from state m' to state m, viz.,

$$W_{m \leftarrow m'} = W_{mm'} = W_{m'm} .$$

According to the Pauli postulate, the transition rates are symmetric. This expresses the reversibility of the microscopic quantum mechanics and is closely connected to Onsager's symmetry postulate in nonequilibrium thermodynamics (Ebeling and Sokolov 2005). The time evolution is irreversible and in the limit of large times $t \gg 1/R_m$ leads to an equal (microcanonical) distribution:

$$p_m(t) \to \frac{1}{s} \quad \text{(for all m)}. \tag{7.9}$$

Inserting this into the Pauli equation, we see that we have indeed found a stationary solution. The equal distribution corresponds to a microcanonical ensemble. The stationary solution corresponding to a canonical ensemble reads

$$p_m^{(0)} = \frac{1}{Q} \exp\left(-\beta E_m\right) . \tag{7.10}$$

The requirement that this be a stationary solution leads us to different temperature dependent transition probabilities $W_{mm'}(\beta)$ which satisfy a condition of *detailed balance*:

$$W_{mm'}(\beta)\mathrm{e}^{-\beta E_{m'}} = W_{m'm}(\beta)\mathrm{e}^{-\beta E_m} . \tag{7.11}$$

There exists a whole class of transition matrices which meet these conditions. The simplest possibility which models the embedding of the system into a heat bath of temperature T is the Monte Carlo transition probability

$$W_{mm'}(\beta) = W_{mm'} \begin{cases} 1\,, & \text{if} \quad E_m < E_{m'} \\ \exp\left[-\beta(E_m - E_{m'})\right], & \text{if} \quad E_m > E_{m'}\,, \end{cases} \tag{7.12}$$

where $\beta = 1/k_B T$ is the reciprocal temperature and k_B is the Boltzmann constant. The essence of this algorithm is that transitions are always carried out if they are connected with a lowering of the energy. In the opposite case, the transition is carried out with some smaller probability which contains a Boltzmann factor for the energy step.

7.1.2 Stochastic Kinetics of Pauli and Tolman

We calculate the transition probabilities starting from the *von Neumann equation* for the density operators:

$$\mathrm{i}\hbar \frac{\partial}{\partial t} \hat{\rho}(t) = \left[\hat{H}, \hat{\rho}\right], \tag{7.13}$$

where $\hat{\rho}(t)$ is the time-dependent density operator. The matrix elements are calculated with a complete set of eigenfunctions $\rho_{mn} = \langle m|\hat{\rho}(t)n\rangle$. In the following, the diagonal elements will be denoted by $P_m(t) = \langle m|\hat{\rho}(t)m\rangle$ and we will prove that under certain conditions these probabilities satisfy the Pauli equations. We mention that this is connected with some deep problems due to the fact that the von Neumann equation is reversible, in contrast to the *irreversible* Pauli equation. This transition is not trivial (Tolman 1938; Zubarev et al. 1996, 1997; Kremp et al. 2005). We write the Hamilton operator of our system in the form

$$\hat{H} = \hat{H}_0 + \lambda \hat{V}\,, \tag{7.14}$$

assuming a basic term which defines the energy eigenfunctions

$$\hat{H}_0|m\rangle = E_m^0|m\rangle\,, \tag{7.15}$$

together with some perturbation proportional to a small parameter λ. The matrix elements

$$H_{mm'} = E_m^0 \delta_{mm'} + \lambda V_{mm'} \tag{7.16}$$

consist of the basic contributions and the perturbations $\hat{V}$. We assume $\mathrm{Tr}(\hat{V}) = 0$. We multiply the von Neumann equation on both sides with the energy eigenfunctions and get for the diagonal elements

$$\frac{\partial}{\partial t} \rho_{mm} = -\frac{\mathrm{i}}{\hbar} \lambda \sum_{m''} (V_{mm''} \rho_{m''m} - \rho_{mm''} V_{m''m})\,. \tag{7.17}$$

The transitions between different states are due to the interactions. The parameter λ measures the strength of these interactions. For $\lambda \to 0$, no transitions are possible. The derivative $\partial P_m(t)/\partial t$ depends on diagonal as well as non-diagonal elements of the density matrix.

Let us assume in some approximation that the diagonal terms $\rho_{mm}(t)$ contain the essential information about the system and the non-diagonal terms $\rho_{mm'}$ are just a small, less relevant perturbation. The non-diagonal terms $\rho_{mm'}(t)$ are assumed to relax quickly to a stationary value on a time scale which is much smaller than the time scale of the diagonal elements. In this 'adiabatic approximation', $\partial \rho_{mm'}(t)/\partial t = 0$ and we get finally

$$\rho_{mm'} = -\mathrm{i}\frac{\lambda/\hbar}{\gamma_{mm'} + \mathrm{i}(E_m^0 - E_{m'}^0)/\hbar} V_{m'm}\left(\rho_{m'm'} - \rho_{mm}\right) , \tag{7.18}$$

where we have taken into account the symmetry of the perturbation matrix. Now we have expressed all non-diagonal elements in terms of diagonal elements and may derive a closed equation for the diagonal matrix elements. Identifying $\rho_{mm}(t) = P_m(t)$ and $\rho_{m'm'} = P_{m'}(t)$ and comparing with the Pauli equation, we get

$$\frac{\partial}{\partial t} P_m(t) = \sum_{m' \neq m} \left[W_{mm'} P_{m'}(t) - W_{m'm} P_m(t) \right] . \tag{7.19}$$

The transition rates are proportional to $(\lambda/\hbar)^2$. For a shorter expression, we may use the identity due to Sochotzki and Plemelj (also called the Dirac identity):

$$\frac{1}{z + \mathrm{i}0} = \mathrm{VP}\left(\frac{1}{z}\right) - \mathrm{i}\pi\delta(z) . \tag{7.20}$$

Introducing this above, we get (in the limit $\gamma \to 0$) the Pauli transition rates

$$W_{m'm} = \lambda^2 \frac{2\pi}{\hbar} |V_{m'm}|^2 \delta\left(E_m^0 - E_{m'}^0\right) . \tag{7.21}$$

In this way we find that the rates for the transition from the state m to the state m' are proportional to $|V_{m'm}|^2/\hbar$. Here the delta function expresses the law of energy conservation during the transition. Our result (7.21) is called *Fermi's golden rule.*

In conclusion, we get the Pauli equation in the form which follows by perturbation theory:

$$\frac{\partial}{\partial t} P_m(t) = \frac{2\pi}{\hbar} \sum_{m' \neq m} \left\{ |\lambda V_{m'm}|^2 \left[P_{m'}(t) - P_m(t) \right] \delta\left(E_m^0 - E_{m'}^0\right) \right\} . \tag{7.22}$$

Due to the existence of factors keeping the energy constant, the Pauli equation in the given form is restricted to processes on the energy shell, i.e., in the microcanonical ensemble. This is connected with our limit $\gamma \to 0$. We now introduce the heat bath in an explicit way.

Pauli–Tolman Kinetics Including a Heat Bath

In the presence of a heat bath, the delta function has to be replaced by a temperature-dependent function in order to provide the correct stationary distribution. This procedure is not unique, since there exist many different heat reservoirs. On the other hand, there are several sufficiently general standard methods (Tolman 1938; Böttger and Bryksin 1985). Let us write the transition probabilities in the form

$$W_{m'm} = \frac{\lambda^2}{\hbar} |V_{m'm}|^2 E(m, m') \,. \tag{7.23}$$

The factor E expresses energy conservation in microscopic transitions and initially contains a delta function $\delta(E_n - E_{n'})$. Under the influence of a heat bath, the delta function has to be replaced by a distributed temperature function (Tolman 1938; Böttger and Bryksin 1985; Chetverikov et al. 2011, 2013, 2014, 2015):

$$E(n, n') = \exp\left[-\frac{\beta}{2}(E_n - E_n')\right] F(E_n - E_n') \,, \tag{7.24}$$

where $F(E_n - E_n')$ is an even function of the difference between the energy levels. There are several variants of this even function $F(x)$ in the literature. The simplest is the Monte Carlo procedure which corresponds to the function

$$F(E_n - E_{n'}) = \exp\left(-\frac{\beta}{2}|E_n - E_n'|\right) \,. \tag{7.25}$$

Statistical derivations of the thermal factors may be based on certain microscopic models of the heat bath. In the simplest case, one gets the Lorentz profile

$$F(E_n - E_{n'}) = C \frac{\gamma}{\gamma^2 + \left(|E_n - E_{n'}|/\hbar\right)^2} \,. \tag{7.26}$$

In the limit of small damping, we come back to the delta function in the Pauli expression for the transition probabilities.

Tolman's Transition Kinetics for Electrons Including Fermion Effects

As shown by Tolman (1938), the Pauli approach can be generalized to include the influence of spin and symmetry effects (Chetverikov et al. 2013). We follow once again Pauli's procedure in the version due to Tolman (1938). Considering electrons, we take into account the fact that they are fermions, hence not allowed to occupy a quantum state already occupied by an identical particle. In principle, there exists the possibility of two electrons pairing to form a boson, but this is a special effect (Brizhik et al. 2012). In a first approach, we study electrons without spin, or equivalently, we consider the case of very low density, where double occupation cannot occur due to the low probability that two electrons meet in the same place. Taking into account symmetry effects for (non-interacting) fermions without spin, the Boltzmann equilibrium distributions are replaced by Fermi distributions

$$p_n^0 = \frac{1}{\exp\left[\beta(E_n - \mu)\right] + 1} \,. \tag{7.27}$$

Here the plus one in the denominator expresses the fermionic character.

The chemical potential μ marks the boundary between the occupied and the non-occupied states. Following a procedure described by Tolman and later in more detail by Van Hove, the master equation may be generalized to include boson or fermion symmetry effects. The idea is to change the transition probabilities in a way that depends on the occupation of the target state. We explain this procedure for fermions with zero spin. In order to include the Fermi principle, we introduce the modified transition probabilities

$$\tilde{W}_{nn'} = (1 - p_n) W_{nn'} \,. \tag{7.28}$$

The prefactor reduces the probability of the transition according to the occupation of the target state. We thus arrive at a nonlinear master equation

$$\frac{\mathrm{d}p_n}{\mathrm{d}t} = \sum \left(\tilde{W}_{nn'} p_{n'} - \tilde{W}_{n'n} p_n \right) \,. \tag{7.29}$$

This modification of the transition probabilities takes into account the fermionic character. If the spin differs from zero, the prefactor appears only for transitions to states with the same spin direction. We may convince ourselves that the solution is now given by the Fermi function. The appearance of products like $(1 - p_n) p_{n'}$ leads to the fact that effective hopping is restricted to transitions between states close to the Fermi surface. The meaning is as follows. Consider the transitions $n' \to n$. These transitions occur with weight $p_{n'}$ if and only if the state n is free, and this is expressed by the weight factor $1 - p_n$. Figure 7.1 shows that, according to this weight factor, the states near the Fermi surface contribute most to transport.

The new probabilities still lie between zero and one, i.e., $0 < p_n < 1$, but they are normalized in a different way, viz.,

$$\sum_{n=1}^{N} p_n = N_\mathrm{e}, \tag{7.30}$$

where $N_\mathrm{e} < N$ is the total number of free electrons in the system, and p_n expresses the probability of finding one electron in the state $1 \leq n \leq N$, bearing in mind that we assume one state per atom in our model. This one electronic state per atom may be occupied or not. The prefactors on the right-hand side of (7.29) ensure that the

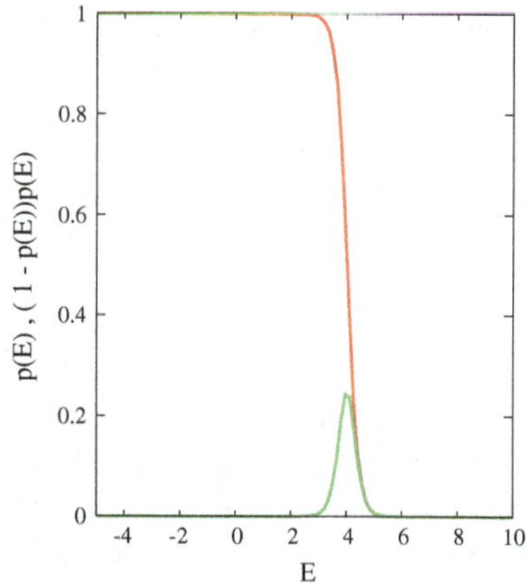

Fig. 7.1 The weight for transitions n' to n is represented by the function $p(E)\left[1 - p(E)\right]$. The maximum corresponds to the Fermi surface on which transitions have the highest weight

probabilities p_n cannot grow larger than unity. So far, the transition probabilities are the same as for the Boltzmann case discussed above. The only general condition they have to satisfy is

$$\frac{W(n', n)}{W(n, n')} = \frac{\exp(-\beta E_n')}{\exp(-\beta E_n)} \ . \tag{7.31}$$

In equilibrium, the normalization defines the chemical potential by

$$\sum_n \frac{1}{\exp\left[\beta(E_n - \mu)\right] + 1} = N_\mathrm{e} \ . \tag{7.32}$$

We note that the states $E_n = \mu$ play a very special role, not only in equilibrium as shown in Chap. 1, but also for nonequilibrium and transport.

By taking into account the spin, we can easily modify the distribution in such a way that two electrons occupy a given site. This corresponds to the formation of electron pairs (Brizhik et al. 2012). We note that the states $E_n = \mu$ play a very special role, not only in equilibrium, but also for transport.

With increasing density, the Fermi level increases, and increasing Fermi levels may lead to percolation of the electronic density corresponding to a sudden increase of diffusion, conductivity, and other transport properties, as shown in (Chetverikov et al. 2013). Note that the master equations are not closed. They may still depend on the coordinates of particles to which the electrons are bound, something we will investigate later. A typical result of the distribution obtained from simulations is shown in Fig. 7.2 (Chetverikov et al. 2013).

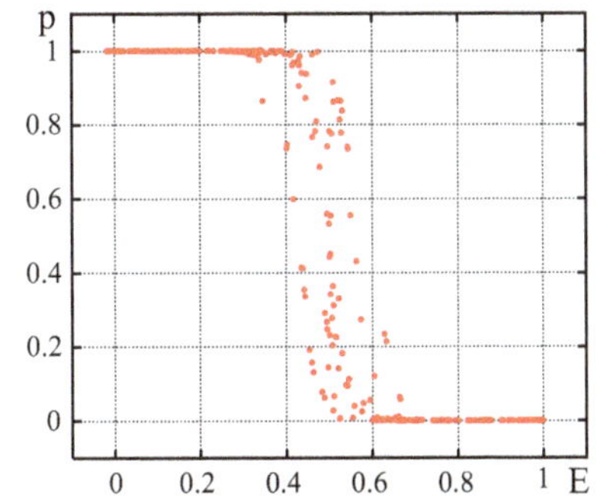

Fig. 7.2 Example of a Fermi probability distribution obtained from the simulations of the energy distribution of Fermi electrons at a low temperature $T = 0.001$ for the intermediate doping density $\nu_e = 200/400 = 0.5$ (after Chetverikov et al. 2013). We observe the formation of a Fermi edge and can identify the Fermi energy

7.2 Quantum Kinetics and Transport Theory

7.2.1 *Lorentz Kinetics and Relaxation Approximation*

The Lorentz gas is a model for a system of light particles like electrons moving on a background of heavy scatterers. Although rather simple, this model provides important insights. So far we have considered systems with discrete states, such as atomic or molecular systems with discrete energy levels. Now we study the transition to another sufficiently simple kinetic equation, the so-called Lorentz approximation. We study an ideal gas which consists of very light particles. We may think of electrons moving in a force field of heavy scatterers with fixed positions. In some sense, this is just the opposite case to the tight-binding electrons discussed above. We assume that our electron gas which interacts with unmovable scatterers is enclosed in a rectangular box with sides a_i and volume $V = \prod_i a_i$. The momentum eigenfunctions are then given by

$$\langle \mathbf{x}|\mathbf{p}\rangle = \frac{1}{\sqrt{V}} \exp\left(\frac{\mathrm{i}}{\hbar}\mathbf{p}\cdot\mathbf{x}\right) , \tag{7.33}$$

with quantized momentum eigenvalues. The Hamilton operator has the form

$$\hat{H} = \hat{H}_0 + \lambda\hat{V} , \qquad \hat{H}_0 = \sum_i \frac{\hat{p}_i^2}{2m} . \tag{7.34}$$

In addition to the ideal gas term, there is a small interaction $\hat{V} = V(\mathbf{r}-\mathbf{R})$ describing the scattering on heavy scatterers, where the position of the scatterer is denoted by the vector $\mathbf{R}$. We define the Lorentz gas as a system of electrons or other light particles on a background of heavy scatterers, with the momentum distribution

$$P_m(t) = \rho_{mm}(t) = \langle m|\hat{\rho}(t)m\rangle = \langle \mathbf{p}|\hat{\rho}(t)\mathbf{p}\rangle \sim f(\mathbf{p}, t) \ . \tag{7.35}$$

We denote the kinetic distribution function for the electrons by $f(\mathbf{p}, t)$. Then the Pauli equation reads

$$\frac{\partial}{\partial t} f(\mathbf{p}, t) = \frac{2\pi}{\hbar} n_s \int \frac{d\mathbf{p}'}{h^3} |\lambda V(\mathbf{p}, \mathbf{p}')|^2 \left[f(\mathbf{p}', t) - f(\mathbf{p}, t) \right] \delta\left(\frac{\mathbf{p}'^2}{2m} - \frac{\mathbf{p}^2}{2m} \right) , \tag{7.36}$$

where n_s is the density of scatterers and $V(\mathbf{p}', \mathbf{p})$ is the Fourier transform of the potential. By approximating the delta function, using spherical coordinates in the momentum space and the differential $d\mathbf{p} = d\Omega\, dp\, p^2$, we get

$$\frac{\partial}{\partial t} f(\mathbf{p}, t) = \frac{2\pi}{\hbar} m n_s \int d\Omega \int_{p=p'} dp' p' |\lambda V(\mathbf{p}, \mathbf{p}')|^2 \left[f(\mathbf{p}', t) - f(\mathbf{p}, t) \right] \delta(p - p') \, .$$

Finally, with the scattering cross-section $\sigma(\Omega) = \text{const.} |\lambda V(\mathbf{p}, \mathbf{p}')|^2$, we get Lorentz' quantum kinetic equation

$$\frac{\partial}{\partial t} f(\mathbf{p}, t) = n_s \int_{p=p'} d\Omega \frac{p}{m} \sigma(\Omega, p) \left[f(\mathbf{p}', t) - f(\mathbf{p}, t) \right] . \tag{7.37}$$

Like the Pauli equation, this kinetic equation contains only second order terms in the potential.

Isotropic Scatterers and the Relaxation Approximation

As a simpler special case, we now study *isotropic scatterers* for which $\sigma(\Omega, p) \to \sigma(\vartheta, p)$ is independent of the angle φ. For weak deviations from equilibrium, we may expand $f(\mathbf{p}, t)$ up to first order deviations and write

$$\frac{\partial}{\partial t} f(\mathbf{p}, t) = f^{(0)}(p, t) + f^{(1)}(p, t) \cos(\vartheta) \ . \tag{7.38}$$

Transforming the coordinate system, we get $\vartheta = \pi \to f(\mathbf{p}, t) = f^{(0)}(p, t)$ and

$$\frac{\partial}{\partial t} f(\mathbf{p}, t) = n_s \int d\Omega \frac{p}{m} \sigma(\Omega, p) f^{(1)} \left[\cos(\vartheta) - 1 \right] = -f^{(1)}(p, t) \frac{1}{\tau(p)} \, .$$

Here we have introduced the relaxation time $\tau(p)$, which is related to the transport cross-section q by

$$\tau(p)^{-1} = n_s q \ , \qquad q = 2\pi \int_0^{\vartheta} d\vartheta' \left[1 - \cos(\vartheta') \right] \sin(\vartheta')\, \sigma(\vartheta', p) \, . \tag{7.39}$$

The relaxation time $\tau(p)$ is the time the system requires for the decay of an initial deviation from equilibrium $f^{(1)}(p, 0)$. In the standard case where the equilibrium distribution (perhaps a Boltzmann or Fermi distribution) is not explicitly time-dependent, viz., $f^{(0)}(p, t) = f^{(0)}(p)$, we have a solution

$$f^{(1)}(p, t) = f^{(1)}(p, t_0) \exp\left[-\frac{(t - t_0)}{\tau(p)}\right] . \tag{7.40}$$

The simple kinetic equation (7.40) is often called the *relaxation approximation*. It describes situations where initial deviations from equilibrium are exponentially damped out. For the treatment of transport processes in the relaxation approximation, see (Ebeling, Fortov et al. 1984).

7.2.2 Bogoliubov Quantum Kinetic Theory

We present here in brief a more general approach to kinetic equations which is based on Bogoliubov's hierarchy for the reduced density operators. In the previous section, we derived and discussed several special kinetic equations for electrons and other particles which share the fact of being linear with respect to the probabilities or the distribution functions $f(\mathbf{p}, t)$. Prototypes of this type of linear kinetic equation are the Pauli master equation, the Lorentz equation, and the diffusion equations in coordinate and in velocity space. On the other hand it is well known that the Boltzmann equation, derived at the end of the nineteenth century for classical gases, is in fact nonlinear. It contains collision integrals which are quadratic functionals of the distribution functions $f(\mathbf{p}, t)$. This means that we should also expect more general quantum kinetic equations to be nonlinear. In order to derive such equations, we follow the method due to Bogoliubov, which we have already discussed briefly in Chap. 2 (Bogoliubov 2005–2009)

As we have shown, complete information about the quantum statistical state is contained in the density operator $\hat{\rho}_N(1, \ldots, N, t)$. The time dependence of this operator is given by the *von Neumann equation*

$$i\hbar \frac{\partial}{\partial_t} \hat{\rho}(t) = \left[\hat{H}, \hat{\rho}(t)\right] . \tag{7.41}$$

When calculating physical quantities like the kinetic energy, the potential energy, correlation functions, and many others, we do not need to have complete information. For this reason several researchers as Yvon, Bogoliubov, Bern, Green, Kirkwood, and others proposed to introduce the so-called reduced distribution functions. We introduce these in the same way as Bogoliubov (2005–2009), via the reduced density operators

$$\hat{F}_s(1,\ldots,s) = V^s \hat{\rho}_s(1,\ldots,s) \\ = \mathrm{Tr}_{(s+1,\ldots,N)}\left[V^s \hat{\rho}(1,\ldots,N)\right] . \tag{7.42}$$

Let us consider a gas with only one component and Hamiltonian $\hat{H} = \hat{T} + \hat{U}$. By partial trace operations, we get the hierarchy of operator equations

$$\mathrm{i}\hbar \frac{\partial}{\partial t} \hat{F}_s(1,\ldots,s,t) = \left[\hat{H}_s, \hat{F}_s(1,\ldots,s,t)\right] \\ + n\mathrm{Tr}_{(s+1,\ldots,N)}\left(\left[U(1,2,\ldots,N), \hat{F}_{s+1}(1,\ldots,s+1,t)\right]\right) .$$

We require specific methods to decouple the hierarchy in order to find a closed equation for $\hat{F}_1(1,t)$, and in particular for the momentum representation $f(\mathbf{p},t) = \text{const.}\langle \mathbf{p}_1 | \hat{F}_1(1,t) \mathbf{p}_1 \rangle$. In order to proceed, we use a *binary collision approximation*. This leads to

$$\mathrm{i}\hbar \frac{\partial}{\partial t} \hat{F}_1(1,t) = \left[\hat{H}_1, \hat{F}_1(1,t)\right] + n\mathrm{Tr}_{(2,\ldots,N)}\left(\left[V(1,2), \hat{F}_2(1,2,t)\right]\right) ,$$

$$\mathrm{i}\hbar \frac{\partial}{\partial t} \hat{F}_2(1,2,t) = \left[\hat{H}_2, \hat{F}_2(1,2,t)\right] + \{\text{higher orders}\} .$$

In the case of rare gases or plasmas, we may neglect higher orders $O(n)$. However, we still have to supplement the second equation with initial conditions. In order to find the initial conditions which lead to binary collisions, we consider first the characteristic times. The interaction time during a binary collision is of order $\tau_{\text{corr}} \approx 10^{-12}$–$10^{-13}$ s. The time between two collisions is, however, $\tau_{\text{coll}} \approx 10^{-10}$–$10^{-12}$ s. Using the condition $\tau_{\text{coll}} \gg \tau_{\text{corr}}$, we may assume that the incoming particles are independent, i.e.,

$$\lim_{t \to -\infty} \hat{F}_2(1,2,t) = \hat{F}_1(1,t)\hat{F}_1(2,t) . \tag{7.43}$$

The so-called Bogoliubov condition corresponds to Boltzmann's assumptions about binary collisions. From the physical point of view, this is the source of the irreversibility in the final kinetic equations. In other words, the condition (7.43) breaks the time symmetry.

As before, we split the two-particle density operator for classical situations into two parts

$$\hat{F}_2(1,2,t) = \hat{F}_1(1,t)\hat{F}_1(2,t) + \hat{G}(1,2,t) . \tag{7.44}$$

Here we introduce the quantum-statistical correlation operator $\hat{G}(1,2,t)$. The Bogoliubov condition for the correlation operator reads

$$\lim_{t_0 \to -\infty} \hat{G}(1,2,t) = 0 . \tag{7.45}$$

For the correlation operator we find

$$i\hbar\frac{\partial}{\partial t}\hat{G}(1,2,t) = \left[\hat{H}_1(1)+\hat{H}_1(2), \hat{G}(1,2,t)\right] + \left[V(1,2), \hat{F}_1(1,t)\hat{F}_1(2,t)\right] + \cdots . \tag{7.46}$$

The homogeneous equation corresponding to (7.46) has the solution

$$\hat{G}_{\text{hom}} = \exp\left(-\frac{i}{\hbar}\hat{H}_2^0 t\right)\hat{G}(1,2,t)\exp\left(\frac{i}{\hbar}\hat{H}_2^0 t\right), \tag{7.47}$$

and the general solution taking into account the initial condition is

$$\begin{aligned}\hat{G}(1,2,t) &= \hat{G}_{\text{hom}}(1,2,t_0) \\ &\quad -\frac{i}{\hbar}\int_0^{t-t_0} dt' e^{-i\hat{H}_2^0 t'/\hbar}\left[V(1,2), \hat{F}_1(1,t')\hat{F}_1(2,t')\right]e^{i\hat{H}_2^0 t'/\hbar} \\ &= -\frac{i}{\hbar}\int_0^{\infty} dt' e^{-i\hat{H}_2^0 t'/\hbar}\left[V(1,2), \hat{F}_1(1,t')\hat{F}_1(2,t')\right]e^{i\hat{H}_2^0 t'/\hbar} .\end{aligned} \tag{7.48}$$

We thus get a closed equation for $\hat{F}_1$:

$$\begin{aligned}i\hbar\frac{\partial}{\partial t}\hat{F}_1(1,t) &= \left[\hat{\hat{H}}_1, \hat{F}_1(1,t)\right] \\ &\quad -\frac{i}{\hbar}n\int_0^{\infty} dt' \mathrm{Tr}_{(2)}\left(\left[V(1,2), \left\{e^{-i\hat{H}_2^0 t'/\hbar}\right.\right.\right. \\ &\quad \left.\left.\left.\times\left[V(1,2), \hat{F}_1(1,t-t')\hat{F}_1(2,t-t')\right]e^{i\hat{H}_2^0 t'/\hbar}\right\}\right]\right) .\end{aligned} \tag{7.49}$$

Here we have introduced the Hamilton operator of the self-consistent field $\hat{\hat{H}}_1$. In order to find a standard form for the kinetic equation, we assume uniformity. Using completeness relations, we may transform this to

$$\begin{aligned}&i\hbar\frac{\partial}{\partial t}\langle \mathbf{p}_1|\hat{F}_1(1,t)\mathbf{p}_1\rangle \\ &= n\int d\mathbf{p}_1' d\mathbf{p}_2' d\mathbf{p}_2 \left\{\langle \mathbf{p}_1\mathbf{p}_2|V(1,2)\mathbf{p}_1'\mathbf{p}_2'\rangle\langle \mathbf{p}_1'\mathbf{p}_2'|\hat{G}(1,2,t)\mathbf{p}_1\mathbf{p}_2\rangle\right. \\ &\quad \left. -\langle \mathbf{p}_1\mathbf{p}_2|\hat{G}(1,2,t)\mathbf{p}_1'\mathbf{p}_2'\rangle\langle \mathbf{p}_1'\mathbf{p}_2'|V(1,2)\mathbf{p}_1\mathbf{p}_2\rangle\right\} .\end{aligned} \tag{7.50}$$

Finally, after some further transformations and using the completeness relations and $\langle \mathbf{p}|\hat{F}_1(1,t)\mathbf{p}\rangle = \hat{F}_1(\mathbf{p},t)\delta(\mathbf{p}-\mathbf{p}')$ again, we get

$$
\begin{aligned}
&\langle \mathbf{p}_1\mathbf{p}_2|\hat{G}(1,2,t)\mathbf{p}_1'\mathbf{p}_2'\rangle \qquad (7.51)\\
&= -\frac{\mathrm{i}}{\hbar}\int_0^\infty \mathrm{d}t' \left\{ \mathrm{e}^{-\mathrm{i}[E(p_1)+E(p_2)-E(p_1')-E(p_2')]t'/\hbar} \langle \mathbf{p}_1\mathbf{p}_2|V(1,2)\mathbf{p}_1'\mathbf{p}_2'\rangle \right.\\
&\qquad \left. \times\left[\hat{F}_1(\mathbf{p}_1',t')\hat{F}_1(\mathbf{p}_2',t') - \hat{F}_1(\mathbf{p}_1,t')\hat{F}_1(\mathbf{p}_2,t')\right]\right\} .
\end{aligned}
$$

Introducing (7.51) into (7.50), carrying out the time integration, and using symmetry, we get

$$
\begin{aligned}
\frac{\partial}{\partial t} f(\mathbf{p},t) = \frac{1}{\hbar}\int \mathrm{d}\mathbf{p}_1'\mathrm{d}\mathbf{p}_2'\mathrm{d}\mathbf{p}_2 &\Big\{ |V(p_1-p_2)|^2 \delta\left(\mathbf{p}_1'+\mathbf{p}_2'-\mathbf{p}_1-\mathbf{p}_2\right)\\
&\times\delta\big(E(p_1')+E(p_2')-E(p_1)-E(p_2)\big)\\
&\times\left[f\left(\mathbf{p}_1',t\right) f\left(\mathbf{p}_2',t\right) - f\left(\mathbf{p}_1,t\right) f\left(\mathbf{p}_2,t\right)\right]\Big\} . \qquad (7.52)
\end{aligned}
$$

Once again, we have introduced $\hat{F}_1(\mathbf{p},t) = \text{const.} f(\mathbf{p},t)$.

Note that our final kinetic equation (7.52) contains quadratic terms in $f(\mathbf{p},t)$, like the original Boltzmann equation The equation is thus nonlinear, in contrast to the Lorentz equation. The reason is that we no longer fix the scatterers at given places. In the new model the scatterers are moving, have their own distribution function, and interact with one another. The role of the delta functions in the kinetic equation (7.52) is to guarantee the conservation of energy and momentum in scattering processes. The quantum mechanical Boltzmann equation derived above may be used to formulate a quantum transport theory. In spite of the fact that our derivation is formally on the same level of correlation as the equilibrium theory, several effects of nonideality with respect to transport quantities are still missing, as can be seen in the missing nonideality terms in the transport equations for energy, pressure, etc. (Klimontovich 1984). Quantum kinetic equations including also these nonideality terms were derived in Klimontovich and Ebeling 1972, while the contributions of binary bound states were included in Ebeling 1976, Kremp et al. 2005.

7.3 Irreversibility, Boltzmann, and Kullback Entropy. H-Theorems

7.3.1 *Entropies and Pauli Dynamics*

We begin by introducing the concepts of *Shannon entropy*, *Kullback information*, and *H-theorem*. The kinetic equations derived above describe the irreversible evolution of probability distributions toward a stationary one. This irreversibility is connected with the second law and was first described by Boltzmann through a so-called H-theorem. In order to give an illustration of the H-theorem for the Pauli equation,

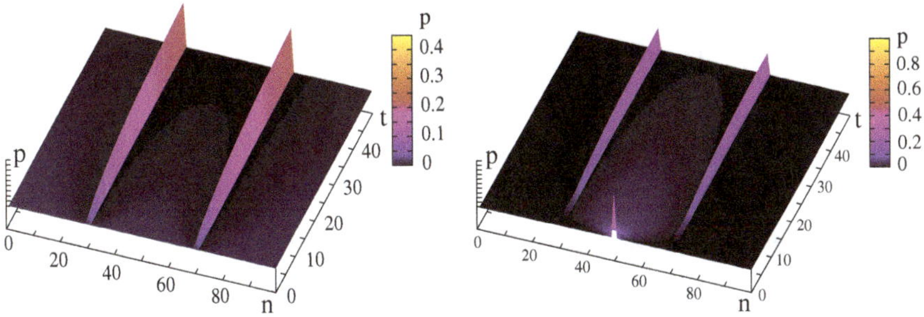

Fig. 7.3 Pauli dynamics of the electron density for a one-dimensional state space with a lattice of 100 discrete states $n = 1, 2, \ldots, 10$ and two energy wells located at $n = 30$ and $n = 70$. The temperature is $T = 0.01$ for the two initial electron distributions. The irreversible dynamics illustrates the H-theorem. We study the dynamics of the electron density for the case where the fixed wells have depth $E_n = -2$. Left: The initial electron density is uniform. Right: We study a 'localized' (delta-like) initial condition, which splits into two peaks around the wells. Parameter values $T = 0.01$, $N = 100$, $\tau = 10$

we study an example with Monte Carlo transitions (we consider only next nearest transitions):

$$W(n, n') = \tau\delta(m - n' - 1)E(n, m) + \tau\delta(m - n' + 1)E(n, m) . \tag{7.53}$$

The results of simulations are shown in Fig. 7.3, taken from (Chetverikov et al. 2011).

We see from the simulations the figure that any initial condition for the probability distribution converges to the same final distribution defined there by the given external potential. Following Boltzmann, the irreversibility is reflected by the behavior of the entropy. For a discrete set of states, the Shannon entropy and the Kullback entropy of a distribution $P = (p_1, p_2, \ldots, p_n)$ are given by

$$H(P) = -\sum_{i=1}^{s} p_i \log(p_i) \ , \qquad K\left(P, P'\right) = \sum_{i=1}^{s} p_i \log\left(\frac{p_i}{p_i'}\right) . \tag{7.54}$$

The Kullback entropy depends on two distributions. If they are identical, i.e., $P = P'$, then the Kullback entropy disappears. An important property of $K(P, P')$ is the positive definiteness $K(P, P') \geq 0$. The Kullback information can be interpreted as a kind of distance between the distributions P and P'. Now we make the choice $P' = P^{(0)}$, where $P^{(0)}$ is the stationary distribution which solves the homogeneous equation. This is a linear equation and the conditions for solvability are well known. In particular we know that the solution is unique.

Let us take first the equal distribution, this being the simplest example. Then we find

$$H\left(P^{(0)}\right) = \log s = H_{\max} \ , \tag{7.55}$$

which corresponds to the largest possible value of the Shannon entropy. For the Kullback entropy, we find

$$K\big(P, P^{(0)}\big) = \log s + \sum_{i=1}^{s} p_i \log p_i = H_{\max} - H > 0 \,.$$

This means that in the given case the Kullback entropy $K(P, P')$ is the difference between the maximal and the actual value of the Shannon entropy. For the case where the canonical distribution is the stationary distribution

$$p_i^{(0)} = \frac{1}{Q} \exp\left(-\beta E_i\right) = \frac{\exp\left(-\beta E_i\right)}{\sum\limits_{i=1}^{s} \exp\left(-\beta E_i\right)} \,, \tag{7.56}$$

we get

$$\begin{aligned} K\left(P, P^{(0)}\right) &= \sum_{i=1}^{s} p_i \log p_i + \sum_{i=1}^{s} \log Q + \beta \sum_{i=1}^{s} p_i E_i \\ &= \beta \langle E \rangle - H(P) + \log Q \end{aligned} \tag{7.57}$$

for the Kullback entropy. This gives

$$k_{\mathrm{B}} T K\big(P, P^{(0)}\big) = F - F_{\mathrm{eq}} > 0 \,. \tag{7.58}$$

We now show a theorem for the time evolution of the Kullback entropy.

7.3.2 H-Theorems

In order to check the time evolution of the function K, we differentiate with respect to time and use the Pauli equation for the evolution of the probabilities:

$$\frac{\mathrm{d}}{\mathrm{d}t}\Big(K\big(P, P^{(0)}\big)\Big) = \sum_{\substack{i=1 \\ j=1 \\ i \neq j}}^{s} W_{ij} p_j \log \frac{p_i p_j^{(0)}}{p_i^{(0)} p_j} \,. \tag{7.59}$$

Using the inequality

$$\log \frac{p_i p_j^{(0)}}{p_i^{(0)} p_j} \leq \frac{p_i p_j^{(0)}}{p_i^{(0)} p_j} - 1 \,, \tag{7.60}$$

we may transform (7.59) to an inequality

$$\frac{\mathrm{d}}{\mathrm{d}t}\left(K\left(P, P^{(0)}\right)\right) \leq \sum_{\substack{i=1\\j=1\\i\neq j}}^{s} W_{ij} p_j \left(\frac{p_i p_j^{(0)}}{p_i^{(0)} p_j} - 1\right) = 0\,. \qquad (7.61)$$

This leads us to the statement that the Kullback functional is either monotonically decreasing or constant. When the equal distribution is the stationary case, which corresponds to a microcanonical ensemble with $p_i^{(0)} = 1/s$, we find

$$-\dot{K}\left(P, P^{(0)}(1/s)\right) = \dot{H}(P) \leq 0\,. \qquad (7.62)$$

This means that the entropy H is monotonically increasing, and this is the analogue of Boltzmann's H-theorem for discrete systems. In the case of open systems in a heat bath, the theorem tells us that the free energy decreases monotonically up to the equilibrium value.

7.4 Fluctuation–Dissipation Relations

7.4.1 Classical Relations

Historically, the first *fluctuation–dissipation relation (FDR)* was given in 1905 by Einstein as a relation between the mean square displacement of a Brownian particle and the (spatial) diffusion coefficient D. Other early examples are the relations between the diffusion coefficient and velocity–velocity correlations found by Taylor in 1921 and extended by Kubo in 1957 to a relation between the conductivity σ of a system of electrons with density n and the velocity–velocity correlation function

$$D = \int_0^\infty \langle v(t)v(0)\rangle \mathrm{d}t\,, \qquad \sigma = \frac{ne^2}{k_\mathrm{B}T}\int_0^\infty \langle v(t)v(0)\rangle \mathrm{d}t\,. \qquad (7.63)$$

We explain Onsager's ideas for a specific example by considering an ensemble of variables x_i, $(i = 1, \ldots, n)$. We assume that the entropy is a function of these variables $S(x_i, (i = 1, \ldots, n))$. We assume that the equilibrium state corresponds to $x_1 = 0, \ldots, x_n = 0$ and that the deviation from the maximum entropy in thermodynamic equilibrium may be written in the form

$$S(x_1, \ldots, x_n) = S_\mathrm{max} - \frac{1}{2}\beta_{ij} x_i x_j\,. \qquad (7.64)$$

Here and in the following we automatically sum over repeated indices, often referred to as the Einstein convention.

Following Onsager's ideas described above, the thermodynamic forces and fluxes are given by the relations

$$X_i = -k_B \beta_{ij} x_j \ , \qquad J_i = -\dot{x}_i \ ; \tag{7.65}$$

The generalized linear Onsager ansatz and the conditions from the second law read

$$J_i = L_{ij} X_j \ , \qquad L_{ij} X_i X_j \geq 0 \ . \tag{7.66}$$

The inequality holds for any value of X_i and we have zero only for $X_i = 0$ for all $i = 1, \ldots, n$. This corresponds to the requirement of positive definiteness of the matrix L_{ij}. Inserting (7.65) into (7.66), we get

$$\dot{x}_i = -k_B L_{ij} X_j \ , \tag{7.67}$$

and introducing the matrix of relaxation coefficients of the linear processes near equilibrium states, we end up with

$$\dot{x}_i = -\lambda_{ij} x_j \ , \qquad \lambda_{ij} = k_B L_{ik} \beta_{kj} \ . \tag{7.68}$$

Since the matrix β_{ij} determines the dispersion of the stationary fluctuations, we have once again found a close relation between fluctuations and dissipation, i.e., a fluctuation–dissipation relation for a matrix of fluctuating and relaxing variables. The physical meaning of Onsager's assumption may be interpreted as follows. Onsager tacitly assumed that deviations from equilibrium and fluctuations around the equilibrium satisfy the same kinetics.

In order to illustrate this procedure, we take Brownian motion as a simple example, i.e., the motion of a heavy particle in a viscous liquid. As is well known, this problem was first studied over a hundred years ago. For the velocity of the particle, we assume the linear equation of motion

$$\dot{v} = -\gamma_0 v \ . \tag{7.69}$$

For the stationary correlation function, we get

$$\frac{\mathrm{d}}{\mathrm{d}\tau} \langle v(t+\tau) v(t) \rangle = -\gamma_0 \langle v(t+\tau) v(t) \rangle \ , \qquad \langle v(t)^2 \rangle = \frac{k_B T}{m} \ , \tag{7.70}$$

where the second equation plays the role of the initial condition. Solving these equations we find the correlation function and the spectrum of Brownian particles

$$\langle v(t+\tau) v(t) \rangle = \frac{k_B T}{m} \exp(-\gamma |\tau|) \ , \qquad S_{vv}(\omega) = \frac{k_B T}{m} \frac{2\gamma_0}{\gamma_0^2 + \omega^2} \ . \tag{7.71}$$

Sometimes this frequency distribution is called a red spectrum, since the maximum intensity is at low ω. A richer spectral structure is obtained if the particle is also under the influence of a harmonic force. Then the dynamical equations read

$$\dot{x} = v \,, \qquad \dot{v} = -\gamma_0 v - \omega_0^2 x \,, \tag{7.72}$$

and the corresponding system of equations for the correlation functions is

$$\dot{C}_{xv}(\tau) = C_{vv}(\tau) \,, \qquad \dot{C}_{vv} = -\gamma_0 C_{vv} - \omega_0^2 C_{xv} \,. \tag{7.73}$$

Following the approach described above, we find the spectrum

$$S_{vv}(\omega) = \frac{k_{\mathrm{B}} T}{m} \frac{2\gamma_0}{\gamma_0^2 + (\omega - \omega_0^2/\omega)^2} \,. \tag{7.74}$$

This distribution has a peak at $\omega \approx \omega_0$, a so-called resonance.

Important questions concern the symmetry relations associated with the Onsager matrix L_{ij}, which relates the thermodynamic fluxes and forces close to equilibrium. Empirically it is found that the matrix is symmetrical with respect to exchange of indices, although in a few cases one observes antisymmetry:

$$L_{ji} = \pm L_{ij} \,. \tag{7.75}$$

This macroscopic property means that, if a force X_j induces the thermodynamic flux J_i, we will also find a flux J_j generated by the force X_i. This observation was already made in Onsager's paper. For the proof of this relation, we use properties of the correlation function, in particular, $C_{ij}(\tau) = C_{ji}(-\tau)$ (Ebeling and Sokolov 2005). These relations belong to the fundamentals of linear irreversible thermodynamics and are well confirmed by many experiments.

The approach presented so far is restricted to the dynamics of classical variables close to equilibrium. In the next section, we discuss extensions to quantum variables. More difficult is the extension to far-from-equilibrium situations. There exist many approaches to solve this problem. We mention the early work of Onsager and Machlup in 1953, Ginzburg and Landau in 1950. There is an advanced theory of nonlinear irreversible processes due to Stratonovich (1994).

7.4.2 *Quantum Fluctuation–Dissipation Relations*

The first generalization of a classical FDR to quantum variables was proposed by Nyquist in 1928. Nyquist considered the fluctuations in the current and voltage in a linear circuit, as had been detected by J.B. Johnson in an earlier paper. Nyquist gave a theoretical deduction based on the well known circuit relations

$$\frac{\mathrm{d}Q}{\mathrm{d}t} = J \,, \qquad L\frac{\mathrm{d}J}{\mathrm{d}t} + RJ + \frac{Q}{C} = V \,, \tag{7.76}$$

where Q is the charge on the capacitor C and J is the current in a circuit with inductivity L and resistivity R. In the classical case, on the basis of the equipartition theorem, Nyquist derived the relation

$$(VV)_\omega = 2Rk_\mathrm{B}T \, . \tag{7.77}$$

For the current–current spectrum, we have

$$(JJ)_\omega = \frac{2R}{|Y(\omega)|^2} k_\mathrm{B}T \, , \tag{7.78}$$

where the complex admittance is

$$Y(\omega) = R - \mathrm{i}\omega L + \frac{\mathrm{i}}{\omega C} \, . \tag{7.79}$$

The corresponding energy contained in the fluctuating magnetic field is

$$U = \frac{1}{2}\langle LJ^2 \rangle = \int \frac{\mathrm{d}\omega}{2\pi} \frac{2Rk_\mathrm{B}T}{|Y|^2} = \frac{k_\mathrm{B}T}{2} \, . \tag{7.80}$$

Further Nyquist concludes as follows:

> In what precedes the equipartition law has been assumed, assigning a total energy per degree of freedom of $k_\mathrm{B}T$. If the energy per degree of freedom be taken
>
> $$\frac{h\nu}{\exp(h\nu/k_\mathrm{B}T) - 1} \, ,$$
>
> the expression for the electromotive force becomes [now in our notation]
>
> $$(VV)_\omega = 2Rk_\mathrm{B}T \frac{\hbar\omega}{\exp(\hbar\omega/k_\mathrm{B}T) - 1} \, . \tag{7.81}$$

Evidently, Nyquist was not yet aware of the ground state contribution $\hbar\omega/2$. This term was included only later by Callen and Welton, in 1951. The final expression for the Nyquist–Callen–Welton relation for the current–current correlations reads

$$S_{JJ} = (JJ)_\omega = 2\mathrm{Re}\big[Y(\omega)\big]k_\mathrm{B}T\hbar\omega \coth\left(\frac{\hbar\omega}{2k_\mathrm{B}T}\right) \, . \tag{7.82}$$

7.5 Plasma Fluctuations and Kinetic Equations

7.5.1 *Quantum Correlations of the Electrical Field*

As explained in Chaps. 4–6, there are other fruitful views beside the standard view on plasmas as neutral systems of charged particles. We know from the point of view of electrodynamics that plasmas are electromagnetic systems consisting of electric and magnetic fields distributed in space between point particles. These fields are changing very fast, they are essentially stochastic, and they carry electrical energy. The total Coulomb energy is determined by the Coulomb field between the particles. We neglect the magnetic fields and relativistic corrections here.

The electrical fields are essentially classical and the local energy density is given by the square of the local field strength. The correlations of the field determine the energy density. The study of the fluctuating fields in the plasma and their correlations is a very fruitful non-traditional view which was worked out in great detail by Klimontovich, Pines, and Silin and Rukhadse (Pines 1961, Klimontovich 1964, 1967; Silin and Rukhadse 1961). We use here the methods developed by Klimontovich (1967, 1982, 1986). In many respects this approach is also reminiscent of the formalism used to describe collective excitations or plasma waves. The corresponding name *dielectric formalism* is sometimes used for the technique, since the dielectric function plays a primary role (Pines 1961, Pines and Nozieres 1967). An attractive feature of these methods is that the classical and the quantum case are treated in a very similar way. The physical reason for this similarity is of course that the electric field in a plasma is a classical object while its quantum dynamics is determined by the quantum dynamics of the particles. Note that, in a medium, the vacuum field or the corresponding potential is to be divided by the dielectric constant of the medium ε_{r}.

Following Klimontovich, we introduce a time- and space-dependent electric field and a time-, space-, and momentum-dependent local plasma density, along with their Fourier representations:

$$\delta\,\mathbf{E}(\mathbf{r},t) = \int \mathrm{d}\mathbf{k}\,\mathrm{d}\omega\,\exp(\mathrm{i}\omega t - \mathrm{i}\mathbf{k}\cdot\mathbf{r})\,\delta\,\mathbf{E}(\mathbf{k},\omega)\ , \tag{7.83}$$

$$\delta\,N(\mathbf{r},\mathbf{p},t) = \int \mathrm{d}\mathbf{k}\,\mathrm{d}\omega\,\exp(\mathrm{i}\omega t - \mathrm{i}\mathbf{k}\cdot\mathbf{r})\,\delta\,N(\mathbf{k},\mathbf{p},\omega)\ . \tag{7.84}$$

In this Fourier picture, the Poisson equation reads

$$\delta\,\mathbf{E}(\mathbf{k},\omega) = -\mathrm{i}\frac{\mathbf{k}}{k^2}4\pi e \int \mathrm{d}\mathbf{p}\ \delta\,N(\mathbf{k},\mathbf{p},\omega)\ . \tag{7.85}$$

Together with the Vlasov equation for the densities, which is a kind of analog of an RPA approximation, this leads classically to the dielectric function

$$\varepsilon(\omega, k) = 1 + \frac{4\pi e^2}{k^2} n \int \mathrm{d}\mathbf{p} \, \frac{\mathbf{k} \cdot \partial f(\mathbf{p})/\partial \mathbf{p}}{\omega - kv + \mathrm{i}\Delta} \,. \tag{7.86}$$

Here the integral stands for the dielectric response function and $\mathrm{i}\Delta$ provides a small imaginary increment. For the density of the field energy we get, in Fourier space,

$$(\delta\,\mathbf{E}\ \delta\,\mathbf{E})_{\omega,k} = -\mathrm{Im}\frac{1}{\varepsilon(\omega, \mathbf{k})} \,. \tag{7.87}$$

By integrating over the frequencies, we come back to the previous relations for the k-dependent dielectric function and the energy density (Klimontovich 1964, 1967). However, the advantage of the new formulae is that, up to small details, they are also valid in the quantum case. This is of course due to the fact that the electric field is essentially a classical quantity. Therefore most parts of the quantum derivation are fully analogous to the classical one, if Klimontovich's form of the Wigner representation is used.

Referring to Klimontovich's textbook (see Chap. 17 of Klimontovich 1982, 1986), the Fourier components of the electric field and the potential may once again be represented by

$$\delta\,E(\omega, k) = -\mathrm{i}k\,\delta\,\varphi(\omega, k) \,, \qquad \delta\,U(\omega, k) = e\,\delta\,\varphi(\omega, k) \,, \tag{7.88}$$

and the correlations assume the form

$$(\delta\,\mathbf{E}\ \delta\,\mathbf{E})_{\omega,k} = \mathrm{Im}\frac{1}{\varepsilon(\omega, \mathbf{k})}\frac{8\pi}{\omega} F_\omega \,, \tag{7.89}$$

which appears instead of (7.87), where

$$F_\omega = \frac{\hbar\omega}{2} + \frac{\hbar\omega}{\exp(\beta\hbar\omega) - 1} \,. \tag{7.90}$$

The dielectric function has the same form as in the classical case, but the dielectric response function is essentially modified by the quantum effects (see Klimontovich 1975, 1986; Kraeft et al. 1986; Kremp et al. 2005):

$$\varepsilon(\omega, k) = 1 - \frac{4\pi e^2}{k^2} \Pi(\omega, \mathbf{k}) \,, \tag{7.91}$$

$$\Pi(\omega, \mathbf{k}) = -(2s+1) \int \frac{\mathrm{d}\mathbf{p}'}{(2\pi)^3} \frac{f(p' + p/2) - f(p' - p/2)}{\hbar\omega + E(p', p)} \,, \tag{7.92}$$

where

$$E(p', p) = \frac{\hbar^2}{2m}\left[(p' - p/2)^2 - (p' + p/2)^2\right] . \tag{7.93}$$

The dielectric function is related to the charge density–charge density correlations, i.e., the dynamic structure factor (Ichimaru 1992, Adamjan et al. 1993). For an electron gas, the relation is

$$S(\mathbf{k}, \omega) = \frac{k^2 \hbar \, \mathrm{Im}\big[\varepsilon^{-1}(\mathbf{k}, \omega)\big]}{4\pi^2 n e^2 \big[1 - \exp(-\beta\hbar\omega)\big]} \ . \tag{7.94}$$

We may mention studies in which the dynamical structure factor has been calculated from molecular dynamics simulations and compared with theoretical expressions for the right-hand side (Adamjan et al. 1993; Ortner et al. 1997, 1999). These and other relations can already be considered as special forms of fluctuation–dissipation relations (FDR), which will now be discussed from a more general point of view.

7.5.2 *Kinetic Equations and Fluctuation–Dissipation Relations*

The idea of a particle density operator in coordinate/spin space may be extended to phase space. In his textbook on statistical physics (Klimontovich 1982, 1986) and in further papers, Klimontovich developed a method which generalized the Wigner method and his method of random densities in phase space on the basis of the second quantization method. The phase density operator, called the *Klimontovich* operator by Kubo, is defined by (Klimontovich 1982, 1986)

$$\hat{N}(\mathbf{r}, \mathbf{p}, t) = \frac{\hbar^3}{V} \int \mathrm{d}\mathbf{g} \, \Psi^*(\mathbf{r} - \hbar\mathbf{g}/2) \Psi(\mathbf{r} + \hbar\mathbf{g}/2) \exp(-\mathrm{i}\mathbf{g} \cdot \mathbf{p}) \ . \tag{7.95}$$

The dynamics of the Klimontovich operator is given by

$$\begin{aligned} &\frac{\partial \hat{N}(\mathbf{r}, \mathbf{p}, t)}{\partial t} + \frac{\mathbf{p}}{m} \cdot \frac{\partial \hat{N}(\mathbf{r}, \mathbf{p}, t)}{\partial \mathbf{r}} \\ &\quad = \frac{\mathrm{i}}{\hbar (2\pi)^3} \int \Big[\hat{U}(\mathbf{r} - \hbar\mathbf{g}/2) - \hat{U}(\mathbf{r} + \hbar\mathbf{g}/2)\Big] \\ &\qquad\qquad \times \hat{N}(\mathbf{r}, \mathbf{p}', t) \exp\big[\mathrm{i}\mathbf{g} \cdot (\mathbf{p} - \mathbf{p}')\big] \mathrm{d}\mathbf{g} \, \mathrm{d}\mathbf{p}' \ , \end{aligned} \tag{7.96}$$

where Klimontovich introduced the potential energy operator

$$\hat{U}(\mathbf{r}, t) = U_0(\mathbf{r}) + \int V(\mathbf{r}, \mathbf{r}') \hat{N}(\mathbf{r}', \mathbf{p}', t) \frac{\mathrm{d}\mathbf{r}' \, \mathrm{d}\mathbf{p}'}{(2\pi\hbar)^3} \ . \tag{7.97}$$

The Klimontovich equation for the particle density operator in phase space is in fully analogous with the Wigner equation. However, an essential difference is that the new equation is not an approximation, valid only for one-particle problems, but a field operator equation for systems of interacting particles. In Chap. 8, we will come back to this operator form of Klimontovich's kinetic equation.

The final form of the Nyquist–Callen–Welton relation for the current–current correlations is

$$S_{JJ} = (JJ)_\omega = 2\mathrm{Re}\big[Y(\omega)\big]k_\mathrm{B}T\left[\frac{\hbar\omega}{\exp(\hbar\omega/k_\mathrm{B}T)-1}+\frac{\hbar\omega}{2}\right] \tag{7.98}$$

$$= 2\mathrm{Re}\big[Y(\omega)\big]\hbar\omega\coth\left(\frac{\hbar\omega}{2k_\mathrm{B}T}\right). \tag{7.99}$$

Here we have followed the original paper by Nyquist (1928) in the more refined version of Klimontovich (1986, 1995, 1999, 2001). Introducing a frequency-dependent temperature

$$T_\omega = \hbar\omega\coth\left(\frac{\hbar\omega}{2k_\mathrm{B}T}\right), \tag{7.100}$$

we may formulate the quantum FDT according to Klimontovich in the form

$$(JJ)_\omega = 2\mathrm{Re}\big[Y(\omega)\big]k_\mathrm{B}T_\omega\,. \tag{7.101}$$

A general quantum version of the fluctuation–dissipation theorem was given by Callen and Welton (1951) and Kubo (1957). This is explained in detail in the books by Landau and Lifshits (Landau and Lifshits 1980, Lifshits and Pitaevsky 1981). For some internal vector quantity of the system **X** (in the above, we considered the current as an example), the Callen–Welton FDT reads

$$(\delta X_i\delta X_j)_\omega = \hbar\,\mathrm{Im}(\alpha_{ij})\coth\left(\frac{\hbar\omega}{2k_\mathrm{B}T}\right). \tag{7.102}$$

According to Ginzburg and Pitaevski, this form of the FDT is an exact fundamental relation of quantum statistics (Ginzburg and Pitaevskii 1987; Stratonovich 1994). Interestingly enough, there is a some controversy over this statement (Klimontovich 1987). Following Klimontovich, the Callen–Welton derivation does not take into account the (weak) irreversibility present in all real processes, and its validity is therefore limited. The controversy over this statement, between Klimontovich (1987) on the one hand and Lifshits and Pitaevskii (1987) on the other, focuses largely on the meaning of the frequency which appears in the relation. In the Callen–Welton and Kubo derivation, there is no deeper discussion of this problem. The frequency appears simply as the running frequency of the Fourier spectral decomposition. In Nyquist's and in Klimontovich's derivations, ω is the frequency of the circuit under discussion. If Nyquist and Klimontovich are right, the FDT is not universal, but to some extent specific to the oscillatory phenomena and to the 'eigenfrequencies' of the systems or circuits under study. Klimontovich argues that the two frequencies agree completely only under the condition that the resonances are infinitely narrow

(Klimontovich 1995, 1999, 2001). In other words the condition for the Callen–Welton formula to be exact is that damping effects should be infinitely small.

In our view, the whole problem remains at least in part unsolved and constitutes a deep experimental and mathematical problem. As we can see from the formulae due to Sochotzki and Plemelj discussed earlier, even a very small term arising from dissipative effects may dramatically change the analytical properties and values of the integrals. According to Klimontovich, not all existing derivations and formulations of the Callen–Welton relations properly account for the fact that small irreversible effects are present in all macroscopic processes. We consider this question to be so deep and fundamental that it cannot be decided on the basis of purely theoretical arguments. However, experimental facts which would provide a proof of one or the other version of the quantum FDT are still absent, as far as we know. Evidently, this is one of the important open questions in quantum statistics.

The theory of fluctuations in plasmas is now a large field of statistical physics, and for further details we refer to the specialized literature (Pines, 1961, Silin and Rukhadse 1961; Klimontovich, 1964, 1967, 1982, 1987, 1995, 1999, 2001; Sitenko 1982; Ichimaru 1992; Bonitz 2016).

References

S.V. Adamjan, I.M. Tkachenko, J.L. Munoz-Cobo, G. Verdu, Martin, Dynamic and static correlations in model Coulomb systems. Phys. Rev. E **48**, 2067–2072 (1993)

N. Ashcroft. N.D. Mermin, Solid State Physics (Holt, Rinehardt and Winston, Philadelphia 1976, Mir Moscow 1979)

R. Balescu, *Equilibrium And Nonequilibrium Statistical Mechanics* (Wiley, New York, 1975)

K. Binder, G. Cicotti (eds.), *The Monte Carlo and Molecular Dynamics of Condensed Matter Systems* (SIF, Bologna, 1996)

N.N. Bogoliubov, N.N Bogoliubov Jr, Introduction to Quantum Statistical Mechanics (Gordon and Breach, New York, 1991)

N.N. Bogoliubov, *Selected Works, vol* (II, Quantum and Classical Statistical Mechanics (Gordon and Breach, New York, 1991)

N.N. Bogoliubov, Collected Papers. vol. 1–12. (Fizmatlit, Moscow 2005–2009)

M. Bonitz (ed.), *Progress In Nonequilibrium Green's Functions* (World Scientific, Singapore, 2000)

M. Bonitz, *Quantum kinetic theory* (Springer, Berlin, 2016)

H.B. Böttger, V.V. Bryksin, *Hopping Conduction in Solids* (Akademie-Verlag, Berlin, 1985)

L. Brizhik, A.P. Chetverikov, W. Ebeling, G. Röpke, M.G. Velarde, Electron pairing and Coulomb repulsion in one-dimensional anharmonic lattices. Phys. Rev. B **85**, 245105 (2012)

H.B. Callen, T.A. Welton, Irreversibility and generalized noise. Phys. Rev. **83**, 34–40 (1951)

R. Car, M. Parrinello, Unified Approach for Molecular Dynamics and Density-Functional Theory. Phys. Rev. Lett. **55**, 2471–247 (1985)

P. Chetverikov, W. Ebeling, G. Röpke, M.G. Velarde, Anharmonic excitations, time correlations and electric conductivity. Contrib. Plasma Phys. **47**, 465–478 (2007)

A. Chetverikov, W. Ebeling, M.G. Velarde, Local electron distributions and diffusion in anharmonic lattices. Eur. Phys. J. B **70**, 217–227 (2009)

A.P. Chetverikov, W. Ebeling, G. Röpke, Hopping transport and stochastic dynamics of electrons in plasma layers. Contr. Plasma Phys. **51**, 814–829 (2011)

A.P. Chetverikov, W. Ebeling, M.G. Velarde, in Without bounds a scientific canvas of nonlinearity and complex dynamics, ed. by R.G. Rubio, et al., Toward a theory of degenerated solectrons in doped lattices (Springer, Berlin, 2013)

A.P. Chetverikov, W. Ebeling, M.G. Velarde, On the temperature dependence of fast electron transport in lattices. Eur. Phys. J. B **88**, 202 (2014)

A.P. Chetverikov, W. Ebeling, M.G. Velarde, Controlling fast electron transfer at the nano-scale by solitonic excitations along crystallographic axes. Eur. Phys. J. B **85**, 291 (2015)

A.S. Davydov, *Solitons In Molecular Systems* (Reidel, Dordrecht, 1991)

C. Dorso, S. Duarte, J. Randrup, Excited electron dynamics modeling of warm matter, Phys. Lett. B 188 287 (1987); 215, 611 (1988)

T. Dittrich, P. Hänggi, G. Ingold, B. Kramer, G. Schön, W. Zwerger, *Quantum Transport And Dissipation* (Wiley-VCH, Weinheim, 1998)

W. Ebeling, Bound state effects in quantum transport theory. Ann. Physik (Leipzig) **33**, 350–358 (1976)

W. Ebeling, G. Röpke, Conductance theory of nonideal plasmas. Ann. Physik (Leipzig) **36**, 429–432 (1979)

W. Ebeling V.E. Fortov et al. (eds.), Transport Properties of Dense Plasmas (Birkhäuser, Basel, Boston 1984)

W. Ebeling, Förster, Thermodynamic, kinetics and phase transitions in dense plasmas, In: Elementary processes in dense plasmas (S. Ichimaru, S. Ogata, eds.), Addison-Wesley Reading 1995

W. Ebeling, I. Sokolov, *Statistical Thermodynamics and Stochastic Theory of Nonequilibrium Systems* (World Scientific, Singapore, 2005)

W. Ebeling, A. Filinov, M. Bonitz, V. Filinov, T. Pohl, The method of effective potentials in the quantum-statistical theory of plasmas. J. Phys. A Math. Gen. **39**, 4309–4317 (2006)

W. Ebeling, E. Gudowska, I. Sokolov, On stochastic dynamics in physics: History and terminology. Acta Phys. Pol. **39**, 1003–1018 (2008)

R. Feistel, W. Ebeling, *Physics Of Self-organization And Evolution* (Wiley, Weinheim, 2011)

V.L. Ginzburg, L.D. Landau, Zh Eksp, Teor. Fiz. **20**, 1064 (1950)

V.L. Ginzburg, L.P. Pitaevskii, Quantum Nyquist formula, applicability ranges of the Callen-Welton formula. Sov. Phys. Uspekhi **30**, 168–171 (1987)

L.P. Gorkov, Microscopic derivation of the Ginzburg-Landau equations for superconductivity, Sov. Phys. JETP E, 1364 (1959), JETP 36, 1918 (1959)

J.-P. Hansen, I.R. McDonald, E.L. Pollock, Phys. Rev. A **11**, 1025 (1975)

J.-P. Hansen, I.K. McDonald, Phys. Rev. A **23**, 2041 (1981)

N.M. Hugenholtz, Derivation of the Boltzmann equation for a Fermi gas. J. Stat. Phys. **32**, 231–254 (1983)

S. Ichimaru, *Statistical Plasma Physics* (Addison-Wesley, Redwood City, 1992)

J.C. Inkson, *Many-body Theory Of Solids* (Plenum Press, New York, 1984)

Y.L. Klimontovich, *Statistical theory of nonequilibrium processes in plasmas, Izdat MGU, Moskva 1964* (Engl. transl. Pergamon, Oxford, 1967)

Y.L. Klimontovich, Kinetic theory of nonideal gases and nonideal plasmas(in Russ.), (Nauka, Moskva 1975, Engl. transl. Pergamon Press, Oxford 1982)

Y.L. Klimontovich, Statistical Physics(in Russ.) (Nauka, Moscow 1982, Engl. transl. Harwood 1986)

Y.L. Klimontovich, *Statistical Theory of Open Systems* (Kluwer, Amsterdam, 1987)

Y.L. Klimontovich, Statistical Theory of Open Systems (in Russian). vol. I, II, III (Janus, Moscow 1995, 1999, 2001)

V.P. Yu, Klimontovich, Silin, The spectra of systems of interacting particles and collective losses during passage of charged particles through matter. Soviet Physics Uspekhi 3, 84–114 (1960)

W. Yu, Klimontovich, Ebeling, Quantum kinetic equations of nonideal gases and nonideal plasma. Zh. eksp. teor. Fiz. 63, 905–917 (1972)

W.D. Kraeft, D. Kremp, W. Ebeling, G. Röpke, Quantum Statistics of Charged Particle Systems, (Akademie & Pergamon Press, Berlin & New York 1986)

D. Kremp, M. Schlanges, W.D. Kraeft, *Quantum statistics of nonideal plasmas* (Springer, Berlin, 2005)

R. Kubo, Statistical-mechanical theory of irreversible processes. I. General theory and simple applications to magnetic and conduction problems. J. Phys. Soc. Jpn. **12**, 570–586 (1957)

J.D. Landau, E.M. Lifshits, Statistical Physics (part I), Nauka Moskva 1976, German transl. Berlin 1979, Engl. transl. Statistical Physics, Part 1. vol. 5, 3rd ed. Butterworth Heinemann 1980

L.D. Landau, *Collected papers (Oxford: Pergamon Press, 1965)* (Nauka, Moscow, 1969)

L.D. Landau, E.M. Lifshits. Quantum mechanics: Nonrelativistic theory, Moscow Nauka 1976, Engl. translation Pergamon Press, Oxford 1977, German transl. Akademie Verlag Berlin 1977

A.S. Larkin, V.S. Filinov, V.E. Fortov, Path integral representation of the Wigner function. Contr. Plasma Phys. **56**(3–4), 197–213 (1916)

E.M. Lifshits, P. Pitaevskii, Physical Kinetics, Course of Theoretical Physics, vol. 10 (Pergamon. New York, 1981)

B. Militzer, E.L. Pollock, Phys. Rev. E **61**, 3470 (2000)

E. Montroll, J. Ward, Quantum statistics of interacting particles. Phys. Fluids **1**, 55 (1958)

W.J. Nellis, S.T. Weir, A.C. Mitchell, Phys. Rev. B **59**, 3434 (1999)

H. Nyquist, Thermal agitation of electric charge in conductors. Phys. Rev. **32**, 110–113 (1929)

J. Ortner, F. Schautz, W. Ebeling, Quasiclassical molecular dynamics simulations of the electron gas: Dynamic properties. Phys. Rev. E **56**, 4665 (1997)

J. Ortner, I. Valuev, W. Ebeling, Sermiclassical dynamics and time correlations in two-component plasmas. Contr. Plasma Phys. **39**, 311–321 (1999)

J. Ortner, I. Valuev and W. Ebeling, Microfield Distribution in TCP, Contr. Plasma Phys. 40, 555–685 (2000). W. Pauli, Über das H-theorem vom Anwachsen der Entropie vom Standpunkt der neuen Quantenmechanik. In: Probleme der Modernen Physik (P. Debye, ed.), pp. 30–45, Leipzig Hirzel 1928

D. Pines, The Many-body Problem. A Lecture Note (Benjamin, New York 1961), (Russ. transl., Moskva 1963)

D. Pines, P. Nozieres, *The theory of quantum liquids(Benjamin, New York, Amsterdam 1966)* (Russ. Transl. Mir, Moskva, 1967)

R. Redmer, G. Röpke, Progress in the theory of dense strongly coupled plasmas. Contr. Plasma Phys. **50**, 970–985 (2010)

H. Reinholz, G. Röpke, S. Rosmej, R. Redmer, Conductivity of warm dense matter including electron-electron collisions. Phys. Rev. E **91**, 043105 (2015)

G. Röpke, W. Ebeling, W.D. Kraeft, Quantum-statistical conductance theory of nonideal plasmas by use of the force-force correlation function method. Physica A **101**, 243–254 (1980)

S.P. Sadykova, W. Ebeling, I.M. Tkachenko, Eur. Phys. J. D **61**, 117–130 (2011)

V.P. Silin, A.A. Rukhadse, Electromagnetic Properties of plasmas and plasma-like matter(in Russ.). (Atomisdat, Moskva 1961)

L. Spitzer, *Physics of fully ionized plasmas (Wiley, New York, 1961)* (Russ. transl, Mir Moskva, 1965)

O.G. Sitenko, *Fluctuations And Nonlinear Wave Interactions In Plasmas* (Pergamon Press, Oxford, 1982)

R.L. Stratonovich, *Nonlinear Nonequilibrium Thermodynamics I, Linear and Nonlinear Fluctuation-Dissipation Theorems* (Springer, Heidelberg, 1994)

W. Thirring, Lehrbuch der Mathematischen Physik. 4 Quantenmechanik großer Systeme Springer 1980

R.C. Tolman, *The Principles Of Statistical Mechanics* (University Press, Oxford, 1938)

V.M. Zamalin, G.E. Norman, V.S. Filinov, *The Monte Carlo Method in Statistical Thermodynamics* (Nauka, Moscow, 1977). (in Russian)
J. Von Neumann, *Mathematische Grundlagen der Quantenmechanik* (Springer, Berlin, 1932)
V.M. Zamalin, G.E. Norman, V.S. Filinov, Monte Carlo method in statistical mechanics (in Russ.), (Nauka, Moscow, 1977)
D.N Zubarev, V. Morozov, G. Röpke, Statistical Mechanics of Nonequilibrium Processes (Wiley VCH, Weinheim 1996, 1997)

Chapter 8
Hopping Kinetics, Quantum Dynamics and Transport

In this chapter we begin by examining simple models of electron hopping between bound states located on sites or atoms. In particular, we investigate electron hopping on classical dynamic lattices. Then we discuss models of molecular dynamics of electrons based on effective potential interactions, including models of Wigner dynamics with coordinate- and momentum-dependent interactions in phase space. Finally, we study several molecular dynamics models based on wave packet dynamics. The powerful methods of quantum Monte Carlo simulations will be treated in Chaps. 9 and 10.

8.1 Electron Hopping Kinetics

8.1.1 Hopping Dynamics of Electrons in Tight-Binding Models

The mechanism of hopping dynamics is in some sense opposite to the mechanism of standard plasma transport. In the usual plasma kinetics, and in particular in conductivity processes, the unperturbed state is the free motion of the electrons, while perturbations are acts of scattering. In the case of hopping kinetics, the unperturbed state is the rest of the electrons in one of the bound states, while the perturbation is the transition to another state by the hopping process.

The first problem we have to consider is the hopping of an electron between bound states of atoms. Assuming that the occupation number N_k of the bound state k is the relevant variable, we may assume the following set of kinetic equations, similar to those used in reaction kinetics (Ebeling et al. 1992; Leonhardt et al. 1993):

$$\frac{\mathrm{d}N_k}{\mathrm{d}t} = \sum_s (W_{ks}N_s - W_{sk}N_k) , \tag{8.1}$$

W. Ebeling et al., *Quantum Statistics of Dense Gases and Nonideal Plasmas*, Springer Series in Plasma Science and Technology, DOI 10.1007/978-3-319-66637-2_8

with appropriate expressions for the collision rate coefficients. In this way we may also study the time dependence of the occupation of individual levels, e.g., in hydrogen, helium, or carbon plasmas, as presented for instance in (Ebeling et al. 1992). In the work of Leonhardt et al. (1993), the dependence of ionization and recombination on the densities has been systematically studied. Instead of the occupation number itself, we may investigate occupation probabilities. This leads us to a Pauli-type dynamics, and allows the study of more complex occupation problems, like abundances in helium plasmas or polyethylene plasmas made from $(CH_2)_n$ (Ebeling et al. 1996, Förster et al. 1998). Another example of Pauli dynamics will be studied in a later next section.

Here we study first the dynamics of N noninteracting electrons interacting with a classical dynamical lattice in a rather simple approximation, the so-called tight-binding approximation. This is a standard model for the description of electrons in solid and polymer states (Ashcroft and Mermin 1979; Böttger and Bryksin 1985; Heeger et al. 1988). There exist quantum mechanical and quantum stochastic versions of the dynamics. We will assume the following hopping type tight-binding Hamiltonian for the electronic subsystem (Böttger and Bryksin 1985, Kraeft et al. 1986, Hennig et al. 2007):

$$H_\mathrm{e} = \sum_{jn} E_{jn} c_{jn}^+ c_{jn} + \sum_{jj'n} t_{jj'n} c_{j'n}^+ c_{jn} + \sum_{jnn'} U_{jnn} n_{jn} n_{jn'} \,. \tag{8.2}$$

Here j denotes the number of the atom and n the quantum number of the atomic state. This is a rather complex quantum number which denotes, not only the position of the atom to which the electron belongs, but also the internal quantum state. In our adiabatic approach, the atomic positions are assumed to be fixed at R_j. In a simplified version, we may simplify this expression by assuming that the index n uniquely defines the state and that transitions are restricted to nearest neighbors. Assuming further that there is only one atomic state per ion, we get the following standard form of the tight-binding model for 1D lattices (Hennig et al. 2007; Brizhik et al. 2012):

$$H_\mathrm{e} = \sum_n \left[E_n c_n^+ c_n - V_{nn-1}(c_n^+ c_{n-1} + c_n c_{n-1}^+) \right] . \tag{8.3}$$

With a further simplification, assuming that the energy levels E_n are constant, i.e., they do not depend on n, we can remove the first term by a transformation. Assuming further that the interactions depend exponentially on the distance between the ions, we get the transition matrix (Hennig et al. 2007)

$$V_{n,n-1} = V_0 \exp\left[-\alpha(R_n - R_{n-1}) \right] ; \tag{8.4}$$

This leads to equations of motion for the coefficients $c_n(t)$ which are coupled to the equations of motion for the scatterers (ions) which will be given in the next section for the case of a Morse lattice. The electron dynamics is described by differential equations for the expansion coefficients of the electronic wave function:

$$\begin{aligned} i\hbar\frac{dc_n}{dt} &= \big[\epsilon_0 - \chi(q_{n+1} - qn - 1)\big]c_n \\ &\quad -\tau\Big\{\exp\big[-\alpha(q_{n+1} - q_n)\big]c_{n+1} + \exp\big[-\alpha(q_n - q_{n-1})\big]c_{n-1}\Big\}. \end{aligned} \tag{8.5}$$

The equation for the (classical) lattice reads

$$\begin{aligned} \frac{d^2q_n}{dt^2} &= \Big\{1 - \exp\big[-(q_{n+1} - q_n)\big]\Big\}\exp\big[-(q_{n+1} - q_n)\big] \\ &\quad -\Big\{1 - \exp\big[-(q_n - q_{n-1})\big]\Big\}\exp\big[-(q_n - q_{n-1})\big] \\ &\quad -\alpha V\Big\{(c_{n+1}^*c_n + c_{n+1}c_n^*)\exp\big[-\alpha(q_{n+1} - q_n)\big]\Big\} \\ &\quad -\alpha V\Big\{(c_n^*c_{n-1} + c_nc_{n-1}^*)\exp\big[-\alpha(q_n - q_{n-1})\big]\Big\}. \end{aligned} \tag{8.6}$$

The adiabaticity parameter $\tau = V/(\hbar\omega_0)$, appearing on the right-hand side of (8.5) determines the separation between the time scales of the (fast) electronic and (slow) mechanical (acoustic) processes. For finite temperatures, we add a Langevin source to the equations for the lattice (8.6). To illustrate this, consider the values $\tau = 10$, $V = 0.1$, and $\alpha = 1.75$, which are relevant for ET in hydrogen bonded chains of biomolecules (Davydov 1979), while for conducting polymers we study the range of parameters $\alpha = 0.5$–1.75, $V = 0.1$–0.8 (Hennig et al. 2007; Chetverikov et al. 2007, 2011, 2012). This is the tight-binding Schrödinger equation for the complex numbers c_n. Tight-binding models have been studied in many papers (Hennig et al. 2007, Chetverikov et al. 2007, 2011, 2014).

Figure 8.1 shows an example of hopping kinetics on a potential landscape with two energy wells, similar to the one used in Chap. 7 for a Pauli-type simulation. An example of the same dynamics between two potential wells but at twice the distance is shown in Fig. 8.2. The temperature is so low that the lattice dynamics is almost irrelevant here.

The evolution of the density is quasiperiodic, which corresponds to the microscopic reversibility of the underlying Schrödinger equations. The comparison between the Pauli dynamics studied in Chap. 7 and Figs. 8.1 and 8.2 reveals the fundamental difference between quantum mechanical and quantum stochastic descriptions. The quantum kinetic equations, as exemplified here by the Pauli equation, describe irreversible processes induced in the Pauli case by the dissipative coupling to an heat bath, but the Schrödinger equation is reversible in time.

Of great interest for applications, in particular in solid state physics, is the case in which the sites where the electrons are located correspond to atoms which may change their positions $R_n(t)$ in time. In this case the wave functions are very much concentrated on the sites. In the simplest case, the electrons are just narrow wave packets around the sites and we have only one quantum state per site. Then, instead

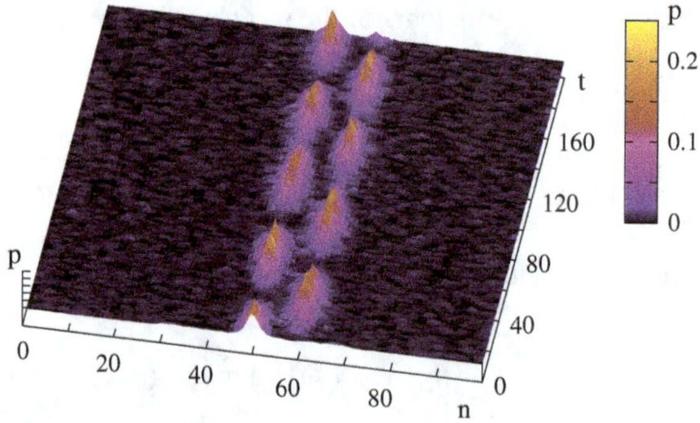

Fig. 8.1 Tight-binding Schrödinger dynamics of the electron density on a one-dimensional lattice (100 sites) at the temperature $T = 0.01$, with two energy wells at the small distance of 10 sites. Initially, we assume a delta-like electron distribution in the left well. The dynamics illustrates the quasi-reversible character, including oscillating coherent motions. $V_0 \hbar = 2$

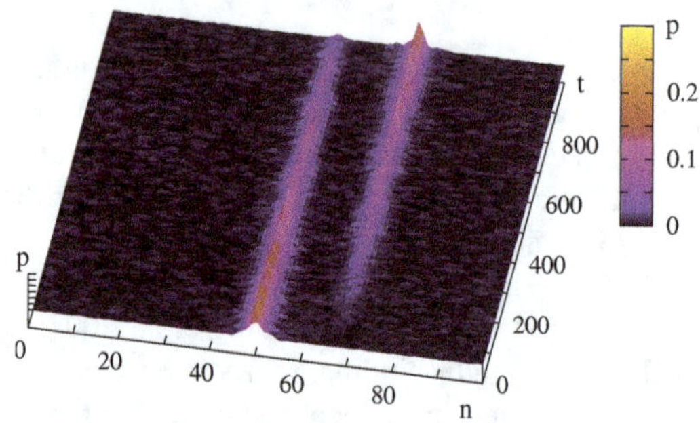

Fig. 8.2 Dynamics of the electron density corresponding to the TBA in a one-dimensional lattice at the temperature $T = 0.01$, with two energy wells at twice the distance, viz., 20 sites. Once again the electrons are located initially in the left well. The dynamics is quasi-reversible, exhibiting slow oscillating coherent motions. $V_0 \hbar = 2$

of a wave function or a state vector, we have just one complex number c_n per site, with N sites altogether, and standard differential equations for N complex numbers.

The TBA is well known from solid state physics (Ashcroft and Mermin 1979). We use here a version based on a Hamiltonian proposed by Davydov and independently by Su, Schrieffer and Heeger (Böttger and Bryksin 1985), and consider lattices which have their own dynamics. This Hamiltonian is of mixed quantum–classical type. For example, we investigate a one-dimensional lattice chain of molecules coupled by Morse forces, in which one excess electron has been injected (Chetverikov et al. 2007, 2012). Our model Hamiltonian for charge transport in the system was already discussed above. The equations of motion derived from the Hamiltonian were also given before.

In the two-dimensional case the equations of motion are more complicated. Assuming the same model for forces and friction as in the previous part and using complex coordinates $Z_n = x_n + \mathrm{i}y_n$, where x_n and y_n are Cartesian coordinates for the n th particle, we get the model (Chetverikov et al. 2012, 2014)

$$\frac{\mathrm{d}^2 Z_n}{\mathrm{d}t^2} = \sum_k \left[F_{nk}^{\mathrm{M}}(Z_{nk}) + F_{nk}^{\mathrm{P}}(Z_{nk})\right] z_{nk} - 2\alpha' \sum_{n'} \exp\left(-\alpha'|R_n - R_{n'}|\right)\sqrt{p_n p_{n'}}\ , \tag{8.7}$$

where the index n identifies one particle among the N particles of the atomic ensemble. We note that the dynamical equations which we use in 2D are different from the 1D equations, since due to phase averaging, only the real-valued variables $p_n = c_n c_n^*$ appear, rather than all the individual complex variables c_n.

In the 1D case, the electron dynamics was described by (8.5) for the expansion coefficients of the electronic wave function. In the 2D case, we have to modify the electron–ion interaction. A standard assumption is that this interaction is described by a pseudo-potential of polarization type. Several variants are in use. For convenience, we use a rather smooth potential (Chetverikov et al. 2011)

$$V_{\mathrm{ek}}(r) = -\frac{U_{\mathrm{e}} h^4}{(r^2 + h^2)^2}\ . \tag{8.8}$$

Here h is a characteristic distance and U_{e} the maximal polarization energy. Thermal lattice excitations moving across the lattice can significantly influence the local dynamics of the electrons and possibly even capture the light electrons. Several studies have considered the influence of solitonic excitations (Chetverikov et al. 2007, 2011). In particular, the influence on the conductivity of electron plasmas has been investigated (Chetverikov et al. 2014). Due to the exponential dependence of the transitions on the actual lattice distances, the lattice dynamics will strongly influence the electron dynamics. Thermal lattice excitations moving across the lattice can strongly influence the local dynamics of the electrons and possibly even capture the light electrons. (Chetverikov et al. 2011, 2014). An important result is that the capture of electrons by soliton-like excitations may lead to an increase in the conductivity. This is due to an increase in the long-range correlations in the dynamics of the electrons and a corresponding decrease in the collision frequency in the plasma.

We will now consider a stochastic model of tight-binding dynamics.

8.1.2 *Pauli Hopping Dynamics of Tight-Binding Electrons*

Let us consider a d-dimensional lattice ($d = 1, 2, 3$) and imbedded electrons which may carry electrical currents. The system consists of N classical ions and N electrons with periodic boundary conditions (ions are placed in 1D lattice of length L). For *heavy* ions, we assume that they obey the classical Langevin dynamics of lattice units with mass m and include a phenomenological damping γ_i. The ion particles are described by coordinates $\mathbf{r}_j(t)$ and velocities $\mathbf{v}_j(t)$, $j = 1, \ldots, N$. The Hamiltonian consists as usual of the classical atomic Hamilton function H_a, and an electronic contribution H_e in the TBA. The atomic part has the standard form, assuming that the atoms repel each other with strong forces. We approximate the potential of the forces between two atoms by the Morse potential (see Chap. 1). The dynamics studied above, and also in papers like (Hennig et al. 2007, Chetverikov et al. 2007, 2011, 2014), is based on the Schrödinger equation for the free electrons in the tight-binding approximation coupled to the Langevin equation for the classical lattice particles.

A serious problem with this approach is the very long relaxation time of the electrons due to the large differences between the time scales of the electrons and the lattice particles. This leads to some difficulties in extensive simulations. In the standard theory of electronic transport, this problem is solved by Boltzmann-type descriptions or by Fokker–Planck-type equations, which introduce an irreversible behavior. For tight-binding systems, the electronic states have a discrete character which allows a description in terms of discrete Markov chains and Pauli equations. In this way we move from the *reversible* Schrödinger equation for the tight-binding model to an *irreversible* Pauli master equation, as introduced in Sect. 8.1.1.

Using the Pauli perturbation approach, we get with some simplifications for the transition probabilities in the tight-binding model:

$$W(n, n') = \frac{V_0}{\hbar} \exp\left(- 2\alpha |q_n - q_{n'}|\right) E(n, n', \beta) \,. \tag{8.9}$$

Here the temperature-dependent thermal factors $E(n, n', \beta)$ are not symmetric with respect to the arguments and are subject to the condition of detailed balance. There are several variants for this even function E which have already been discussed above.

In order to use the master equation for computer simulations of electron hopping processes, we used the Monte Carlo procedure described above. This system provides a rather fast and therefore useful tool for simulating electron–lattice dynamics in thermal systems. Several computer simulations have been based on this approach (Chetverikov et al. 2011, 2012, 2015). Due to the way we treat electron relaxation effects, there are differences between the methodology using the coupled Schrödinger equation and Langevin equation system and the one using Pauli's approach, although they are minor differences at least for small and moderate values of the adiabaticity $\tau \sim 1$. For large τ, electron relaxation in the heat bath is very fast and the distribution approaches more and more a fractal distribution (Fig. 8.3).

An essential remark is that the description using the Pauli equation contains less information than the Schrödinger equation based on the TBA, since all phases are lost and only the probabilities p_n appear in the dynamical equations.

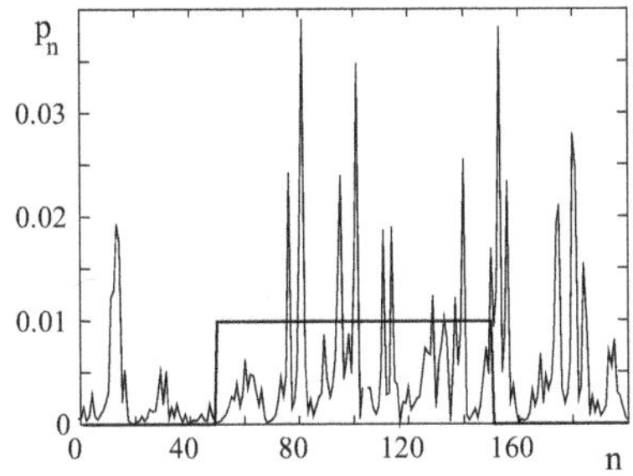

Fig. 8.3 Time evolution of the electron probability distribution according to Pauli's equation. We study a simulation for the temperature $T = 0.5$ (in units of $2D$). An initially rectangular distribution tends to spread irreversibly along the lattice. The initial rectangular distribution becomes localized around a local peak, thus illustrating the corresponding nonlinear lattice compressions (solitons)

8.2 Time Correlations and Linear Response

8.2.1 Time Correlations in the Tight-Binding Approximation

We begin by studying the velocity–velocity correlation function. The time correlation function of the electron is the essential quantity which in fact determines most transport properties, including:

- conductivity,
- diffusion and mean square displacement,
- reaction rates.

In order to give some examples let us recall the expression for the diffusion coefficient and the static conductivity. In some cases, it is most useful to estimate the transport properties from the mean square displacements which are easily observable by simulations (Chetverikov et al. 2014). We note the relations

$$D \approx \langle v; v \rangle_0 \,, \qquad \sigma = \frac{e^2}{k_{\mathrm{B}} T} \langle v; v \rangle_0 \,, \tag{8.10}$$

and the mean square displacement is

$$\langle \delta x(t)^2 \rangle \approx \langle v; v \rangle_0 t \,. \tag{8.11}$$

Since most transport properties are determined by the correlation time or more generally by the integral over the velocity correlation function, the calculation of this quantity is a question of primary importance.

In order to develop the statistical theory, we assume that the electrons move independently according to the Hamiltonian described above and introduce the position operator $\mathbf{R} = \sum_{N_e} \mathbf{r_j}$ of the N_e electrons imbedded in the lattice (Kraeft et al. 1986).

We then define the quantum-statistical time correlations of the velocities by $\langle v; v\rangle_0$ according to the linear transport theory (Kraeft et al. 1986; Chetverikov et al. 2007, 2011, 2014). In the framework of quantum statistics, the average is defined by

$$\langle A; B\rangle_\omega = \int_{0^\infty} \mathrm{d}\tau \exp\big[-(\eta + \mathrm{i}\omega)\tau\big]\frac{1}{\beta}\int_0^\beta \mathrm{Tr}\big[\rho_0 B(\tau - \mathrm{i}\hbar\lambda)A\big] , \tag{8.12}$$

with $\eta \to +0$. Following (Kraeft et al. 1986) we proceed as follows. We start from

$$\langle \dot{\mathbf{R}}\rangle = \mathrm{Tr}(\rho\dot{\mathbf{R}}) , \qquad \mathbf{R} = \sum_n \mathbf{R}_\mathbf{n} c_n^+ c_n , \tag{8.13}$$

where $\mathbf{R}$ is the position operator in the hopping form. For the derivative, we get

$$\dot{\mathbf{R}} = \frac{\mathrm{i}}{\hbar}[H, \mathbf{R}] = \frac{\mathrm{i}}{\hbar}\sum_{nn'} V_{nn'}(\mathbf{R}_n - \mathbf{R}_{n'})c_{n'}^* c_n . \tag{8.14}$$

We assume now that the positions of the lattice particles are fixed for a moment, so that we only have to follow the fast electron dynamics. Then using the standard procedure (see Kubo 1957; Böttger and Bryksin 1985; Kraeft et al. 1986, Zubarev et al. 1997), the velocity correlation function in perturbation theory is given by the expression (Chetverikov et al. 2007)

$$\langle v; v\rangle_0 = \frac{1}{2\hbar V}\sum_{n,n'} \big(R_n - R_{n'}\big)^2 V_{n',n} p_{n'}(1 - p_n) . \tag{8.15}$$

Here the term $1 - p_n$ takes into account the fact that the transition from state n' to state n can occur only if state n' is free (Pauli exclusion principle). Taking into account only nearest-neighbor transitions $n' = n \pm 1$ and neglecting Pauli effects, we get the final expression for the time-dependent velocity correlation function:

$$\langle v; v\rangle_0 = \frac{1}{\hbar N}\sum_n (\sigma + q_{n+1} - q_n)^2 |V_{n+1,n}|^2 . \tag{8.16}$$

The time correlation function depends slowly on the time through the displacements. We may show that the dynamics leads in general to an increase in the time correlations, and hence to an increase in transport (Chetverikov et al. 2014, 2015).

8.2.2 Linear Response Theory

We now discuss the statistical theory of conductivity, considering for simplicity the TBA. The coupling of the electrons to the external field $\mathbf{E}$ is described by

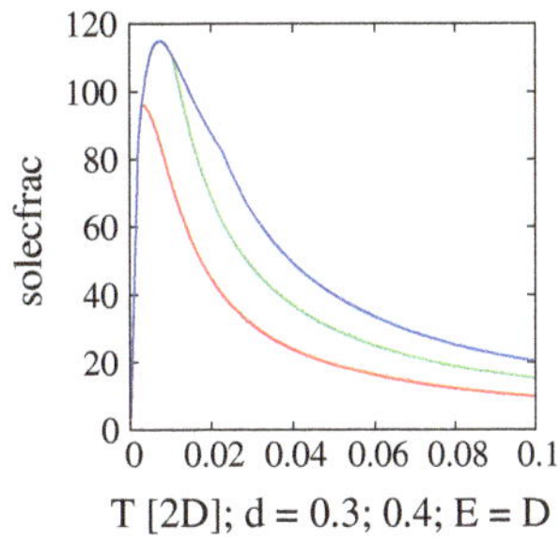

Fig. 8.4 Average enhancement of diffusion/conductance by nonlinear effects in a Morse lattice with 20 percent (*red*), 30 percent (*green*), and 40 percent (*blue*) doping. We see once again a typical maximum over the temperature

$$H_E = -e\mathbf{E} \cdot \mathbf{R} , \tag{8.17}$$

where the position operator of the electron is $\mathbf{R} = \sum \mathbf{r_j}$. Then, in the framework of quantum statistics, we find

$$\sigma = \frac{e^2}{k_\mathrm{B}T}\langle v; v\rangle_0 = \frac{1}{k_\mathrm{B}T}\langle j; j\rangle_0 , \tag{8.18}$$

where in the general case the average is defined by ($\eta \to +0$)

$$\langle A; B\rangle_\omega = \int_{0^\infty} \mathrm{d}\tau \exp\left[- (\eta + \mathrm{i}\omega)\tau\right]\frac{1}{\beta}\int_0^\beta \mathrm{Tr}\left[\rho_0 B(\tau - \mathrm{i}\hbar\lambda)A\right] . \tag{8.19}$$

One very important result from simulations (Chetverikov et al. 2007, 2011, 2014) is that we may consider solitonic excitations to lead to an increase in the conductivity in a certain temperature range. This is due to an increase in the long-range correlations in the dynamics of the electrons and a corresponding decrease in the collision frequency in the plasma. As an illustration, Fig. 8.4 shows the results of calculations by (Chetverikov et al. 2014) for the doping dependence of transport due to electrons. Here the doping is the number of charges in relation to the number of lattice sites. This is of some interest for applications, since we see that transport properties may be optimized.

For further development of transport theory, we follow (Böttger and Bryksin 1985; Chetverikov et al. 2007, 2011). We start from

$$\langle \mathbf{j}\rangle = \mathrm{Tr}(\rho\mathbf{j}) = \frac{e}{V}\mathrm{Tr}(\rho\dot{\mathbf{R}}) = \sigma\mathbf{E} . \tag{8.20}$$

We represent the position operator in the hopping form

$$\mathbf{R} = \sum_n \mathbf{R_n} c_n^+ c_n , \tag{8.21}$$

and find for the derivative

$$\dot{\mathbf{R}} = \frac{\mathrm{i}}{\hbar}[H, \mathbf{R}] = \frac{\mathrm{i}}{\hbar}\sum_{nn'} t_{nn'}(\mathbf{R}_n - \mathbf{R}_{n'})c^*_{n'}c_n \,. \tag{8.22}$$

In a first approach, we neglect the dynamic disorder (solitonic excitations) in the lattice. Then by introducing the derivative in (8.20), we find (Kraeft et al. 1986; Chetverikov et al. 2011) the main contribution to the current to be

$$\mathbf{j}_1 = \frac{\pi\beta e^2}{D\hbar V}\sum_{nn'} |t_{nn'}|^2(\mathbf{R}_n - \mathbf{R}_{n'})E(\mathbf{R}_n - \mathbf{R}_{n'})f_n(1 - f_n)\delta(E_n - E_{n'}) \,. \tag{8.23}$$

Correspondingly we get the main contribution to the conductivity in the form

$$\sigma = \frac{\pi\beta e^2}{D\hbar V}\sum_{nn'} |t_{nn'}|^2(\mathbf{R}_n - \mathbf{R}_{n'})^2 f_n(1 - f_n)\delta(E_n - E_{n'}) \,. \tag{8.24}$$

We call this the direct conductivity induced by the external field. However, this expression only gives the correct conductivity for regular lattices. In nonuniform lattices, we have a second contribution taking into account the nonuniformity (Böttger and Bryksin 1985).

We introduce the dynamic effects, applying the theory to a hopping-type Hamiltonian. We consider the transition matrix $t_{nn'}(R'_n, R_n) = t_{nn'}(r, R_n)$, where $\mathbf{r} = \mathbf{R}'_n - \mathbf{R}_n$, and introduce its Fourier transform

$$t_{nn'}(q; R_n) = \int \mathrm{d}r \; \exp(\mathrm{i}qr)t_{nn'}(r + R_n, R_n(t)) = \exp\left[\mathrm{i}qR_n(t)\right]t(q, n) \,. \tag{8.25}$$

We then introduce the *dynamic* Fourier transform of the structure factor, viz.,

$$S(q, \omega) = \int \mathrm{d}t \exp(\mathrm{i}\omega t)S(q, t) \,. \tag{8.26}$$

In this way, using perturbation theory, we find

$$\sigma = \frac{\pi\beta e^2}{D\hbar V}\sum_{nn'}\int \mathrm{d}\mathbf{q} \mid t_{nn'}(q)|^2 q^2 f_n(1 - f_n) \times S\left(\mathbf{q}, \frac{1}{\hbar}(E_{n'} - E_n)\right)\frac{\exp\left[\beta(E'_n - E_n)\right] - 1}{\beta(E'_n - E_n)} \,. \tag{8.27}$$

We see that we need the dynamical structure factor $S(\mathbf{q}, \omega)$ to calculate the conductivity. The dynamical structure factor may be calculated by molecular dynamics (Ortner et al. 1997; Ebeling et al. 2002, Karakhtanov et al. 2011). The function

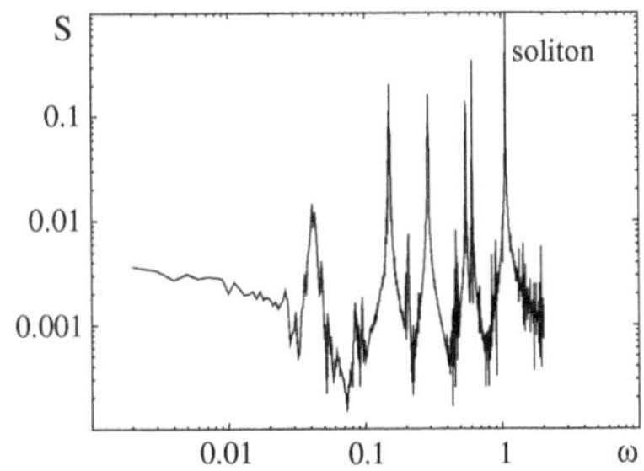

Fig. 8.5 Dynamic structure factor $S(k, \omega)$ of an anharmonic lattice with exponential repulsion, calculated for $N = 10$ with periodic boundary conditions as a function of ω (Chetverikov et al. 2002). We see a pronounced soliton peak near the frequency ω_{sol} corresponding to twice the velocity of the first (longest) phonon, and several other peaks corresponding to (lower) combination frequencies

$S(q, \omega)$ for classical systems of particles is also known from theoretical estimates by several authors like Adamjan and Tkachenko (for references, see Ortner et al. 1999, 2000; Sadykova et al. 2011). We may also mention experimental studies based on investigations of neutron scattering. However, at present, there seems to be no experimental data available which are precise enough to be used for conductivity calculations including the full (q, ω) spectrum.

For several 1D systems, features of the dynamical structure factor are available for several models of intermolecular forces (Ebeling, Schimansky–Geier and Romanovsky 2002). In order to give an example illustrating the general structure, we have computed the *dynamical* structure factor for a Morse lattice with $N = 10$ units, imposing periodic boundary conditions. The result is shown in Fig. 8.5. The effective temperature was chosen so that it corresponds to the transition region from the ordered (solid-like) lattice to (gas-like) disorder. We see a pronounced soliton peak at ω_{sol}, near to the frequency corresponding to twice the velocity of the first (longest) phonon, and further peaks at several (lower) combination frequencies. Note, however, that our knowledge of the dynamical structure factors for two- and three-dimensional conductors is quite limited (Ortner et al. 1997, 1999, 2000, Karakhtanov et al. 2011). For the extension of the linear response theory to a more general Hamiltonian, see Zubarev et al. (1997).

8.3 Molecular Dynamics with Effective Potentials

8.3.1 Simple Models of Effective Interactions

The first and simplest way to include quantum effects in classical molecular dynamics were simulations with Newton's equations and effective potentials. In these simulations, essentially based on classical dynamics, the quantum effects were included only through effective interactions. These first pioneering simulations by Hansen and Mc Donald (1981) used the simple Kramers–Kelbg–Deutsch potential

$$V_{ab}(r) = \frac{e_a e_b}{r}\big[1 - \exp(-\alpha r)\big]\,. \tag{8.28}$$

The properties of this potential were described in detail in Chaps. 3 and 4 (Hansen and Mc Donald 1981).

For the case of one-dimensional systems, a model of an electron–ion system has been simulated by molecular dynamics in (Chetverikov et al. 2007). The simulation was based on Langevin-type electron dynamics, where the electrons move on a one-dimensional ion lattice (Chetverikov et al. 2007). The electron–ion interactions were described by the effective potential (Pohl et al. 2002)

$$V_i(x - x_i) = -\frac{e e_i}{\sqrt{(x - x_i)^2 + h^2}}\,. \tag{8.29}$$

This model is not obtained by perturbation theory. It has a finite value at short range and can describe several atomic properties like bound states and polarization effects to a good approximation (Pohl et al. 2002). This will be explained in Sect. 8.5. One of the interesting results is the dynamical structure factor for a nonlinear lattice, which is needed for the calculation of transport properties, as shown in previous section. Looking at the formula (8.29), we see that any cluster of atoms generates a potential hole in which electron density might be concentrated. Furthermore, any displacement of the atoms changes the polarization energy. The electron will try to follow these changes. This is the basic effect leading to the solectron formation. The essential point which is relevant for physical effects are the running local compressions which generate a complex potential landscape. The rather deep potential wells moving (right to left or left to right) along the ion lattice can indeed strongly influence the local dynamics of the electrons, and may hold light electrons.

According to the Kubo theory, the velocity–velocity correlations are connected to the conductivity. In our case, this theory provides the conductivity σ and the resistivity ρ:

$$\sigma = \frac{ne^2}{k_B T}\int_0^\infty \langle v(\tau)v(0)\rangle\,, \qquad \rho = 1/\sigma\,. \tag{8.30}$$

A more advanced method of quantum molecular dynamics was proposed by Car and Parrinello in 1985. The Car–Parinello method is based on a kind of Born–Oppenheimer dynamics for the nuclei and wave equations which include the electrons in a single state. The idea uses the fact that electron motion is much faster than nuclear motion, so the largest possible time step is the one which allows us to integrate the electron equations of motion. In contrast, the dynamics of the nuclei can be integrated on a longer time scale given by nuclear motion. The new method can cut down the computational expenses of quantum molecular dynamics. The basic idea of the Car–

Parrinello approach is to map the two-component quantum/classical problem onto a two-component semi-classical problem with two separate energy scales. The Car–Parinello method has seen many applications, in particular to quantum chemistry and solid state physics.

8.3.2 Molecular Dynamics with Kelbg-Type Potentials

Here we study once again the more interesting 3D case of the free component in two-component plasmas. We have already mentioned MD simulations using the Kelbg–Deutsch potential, carried out by Hansen and McDonald (Hansen and Mc Donald 1981). Essential improvements were obtained by replacing the Kelbg–Deutsch potential by the full Kelbg potential (Ebeling et al. 1999). In this work, the thermodynamic functions, the dynamical structure factor, and the conductivity were calculated for the free component in mass-symmetric electron–ion plasmas. Subsequent, even more realistic simulations were based on better modifications of the Kelbg potential, as demonstrated in (Ortner et al. 1999, 2000; Sadykova et al. 2009) for general electron–ion plasmas.

The improved Kelbg potential contains corrections for small distances and uses the ansatz

$$U_{ab}(r) = \frac{e_a e_b}{r}\left[F(r/\lambda_{ab}) - k_\mathrm{B}T\tilde{A}_{ab}(\xi_{ab})\exp(-r^2/\lambda_{ab}^2)\right], \tag{8.31}$$

where

$$F(x) = 1 - \exp(-x^2) + \sqrt{\pi}x\left[1 - \mathrm{erf}(x)\right] \tag{8.32}$$

is the Kelbg function and the second term in (8.31) is a correction improving the behaviour at small distances. Note that, for high temperatures $T/I > 0.3$, the differences between the Kelbg potential corresponding to $\tilde{A}_{ab} = 0$ and the corrected Kelbg potential including $\tilde{A}_{ab}$ are very small. For lower values of T, the correction expressed by $\tilde{A}_{ab}$ becomes relevant and provides a way to model the free charges in a plasma even in the presence of bound state particles (atoms). The correction functions $\tilde{A}_{ab}$ were discussed in detail in (Ortner et al. 1999, 2000; Sadykova et al. 2009, 2011). Ortner et al. (1999) used the improved Kelbg potential to calculate the velocity–velocity and force–force autocorrelation functions of two-component plasmas, using MD simulations in comparison with the analytical theory:

$$C(t) = \frac{\langle \mathbf{v}(t)\mathbf{v}(0)\rangle}{\langle \mathbf{v}^2\rangle}, \qquad D(t) = \frac{\langle \mathbf{F}(t)\mathbf{F}(0)\rangle}{\langle \mathbf{F}^2\rangle}. \tag{8.33}$$

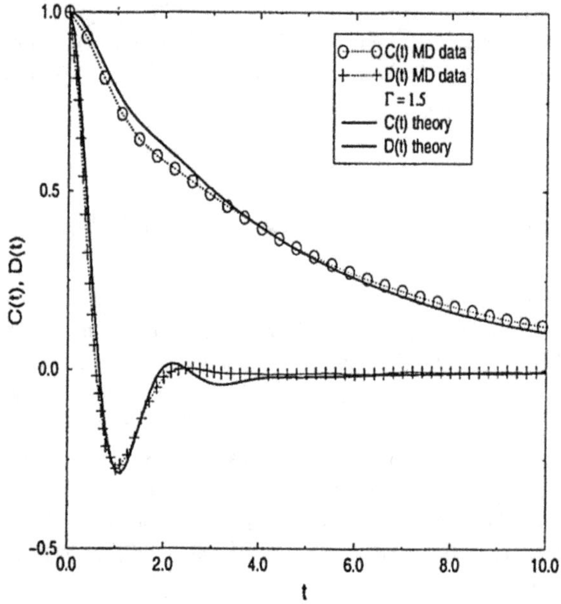

Fig. 8.6 Velocity–velocity and force–force correlations for a two-component plasma, calculated using an improved Kelbg potential from MD simulations for $\Gamma = 1.5$ in comparison with a good fit by Nevanlinna-type approximations (Ortner et al. 1999)

Recall the connection with the diffusion coefficient given by

$$D = \frac{k_{\mathrm{B}} T}{m_i} \int_0^\infty \mathrm{d}t \, \langle v(t); v(0) \rangle \, . \tag{8.34}$$

The correlation functions were calculated by Ortner et al. (1999) using a model based on an improved Kelbg effective potential (see Fig. 8.6), comparing with Nevanlinna-type approximations (Ortner and Tkachenko 1992). In this approach the correlation functions are represented as a linear combination of three exponentials, viz.,

$$C(t) = \sum_{i=1}^{3} C_i \exp(\mathrm{i}\Omega_i t) \, , \qquad D(t) = \sum_{i=1}^{3} C_i (\Omega_i^2/\omega_1^2) \exp(\mathrm{i}\Omega_i t) \, . \tag{8.35}$$

The frequencies Ω_i are the poles of the complex Nevanlinna expressions (Ortner and Tkachenko 1992; Ortner et al. 1999). They are defined as the solutions of some cubic equation:

$$\Omega(\Omega^2 - \omega_2^2) + \mathrm{i}h(\Omega^2 - \omega_1^2) = 0 \, . \tag{8.36}$$

The coefficients C_i are also given by these frequencies and h is a static correlation function. The constructed autocorrelation functions satisfy the following conditions:

- The exact short time behavior for the velocity autocorrelation function is reproduced to orders t^2 and t^4.

- The short time behavior of the force autocorrelation function is reproduced to order t^2.
- The long time behavior of the velocity autocorrelation function generates the exact diffusion constant.
- The correct connection is satisfied between the velocity and force autocorrelation functions.

The Nevanlinna approach, invented by Adamyan and coworkers and further developed by Ortner and Tkachenko (1992), is surprisingly good, providing in particular an excellent fit for our numerical results, as shown in Fig. 8.6.

8.4 Wigner Dynamics with Momentum-Dependent Potentials

8.4.1 Momentum-Dependent Interactions

Simulations with purely coordinate-dependent effective potentials have the great disadvantage that the velocity distributions are necessarily Maxwellian. This is wrong for problems including degeneracy. We introduce here interactions in the momentum space in order to overcome this shortcoming with the Maxwellian, and to find proper Fermi distributions. In the last few decades, several momentum-dependent interactions have been developed which take into account the fact that two electrons repel each other due to Pauli effects (Dorso et al. 1988; Ebeling and Schautz 1997; Ortner et al. 1997; Klimontovich 2001). We recall a hypothesis developed by several authors (Dorso 1987, 1988, Klimontovich 2001; Taruna et al. 2008), according to which we may incorporate the quantum effects into an effective Hamiltonian $H_{\text{eff}}(\mathbf{p}, \mathbf{x})$, generating an effective quasi-classical Hamiltonian and a corresponding 'Wigner dynamics' operating with momentum-dependent interactions (Dorso et al. 1988, Klimontovich 1997, 2001, Ortner et al. 1997).

In order to introduce the Wigner formalism, we start with the Hamilton operator, density operator, and von Neumann equation in the usual form

$$
\mathrm{i}\hbar\frac{\partial\hat{\rho}}{\partial t} = \left[\hat{H}, \hat{\rho}\right], \qquad \hat{H} = \frac{\hat{p}^2}{2m} + U(x) .
$$

In the matrix representation, we get

$$
\begin{aligned}
\mathrm{i}\hbar\frac{\partial\hat{\rho}(x, x', t)}{\partial t} &= \left\langle x|\hat{H}\hat{\rho}x'\right\rangle - \left\langle x|\hat{\rho}\hat{H}x'\right\rangle \\
&= -\frac{\hbar^2}{2m}\left(\frac{\mathrm{d}^2}{\mathrm{d}x^2} - \frac{\mathrm{d}^2}{\mathrm{d}x'^2}\right)\hat{\rho}(x, x', t) \\
&\qquad + \left[U(x) - U(x')\right]\hat{\rho}(x, x', t) .
\end{aligned} \tag{8.37}
$$

Following Wigner, in the 1D case, we define a function depending on both coordinates and momenta:

$$\rho(x, p, t) = \int \frac{\mathrm{d}x'}{h} \hat{\rho}(x, x', t) \mathrm{e}^{-\mathrm{i}p(x-x')/\hbar} . \tag{8.38}$$

This is an analogue of the classical probability on phase space. However, the Wigner function is only a quasi-probability, since $\rho(x, p)$ may not always be positive and real. Integrating the Wigner functions over the momentum, the diagonal element of the density matrix is found to be

$$\rho(x, t) = \int \mathrm{d}p\, \rho(x, p, t) = \hat{\rho}(x, x' = x, t) . \tag{8.39}$$

All operators with the structure

$$\hat{A} \to \hat{A}(x, \hat{p}) , \qquad \langle \hat{A}(x, \hat{p}) \rangle = A(x, p) , \tag{8.40}$$

may be calculated in 'classical way' using the Wigner function $\rho(x, p, t)$:

$$\langle \hat{A} \rangle = \frac{2\pi}{\hbar} \int \mathrm{d}x \int \mathrm{d}p\, \rho(x, p, t) A(x, p) . \tag{8.41}$$

We may easily show that, up to terms of order $O(h)$, the Wigner function satisfies the classical Liouville equation. All relations given here may be extended to the three-dimensional case by the definition

$$\rho(\mathbf{x}, \mathbf{p}, t) = \int \frac{\mathrm{d}\mathbf{x}'}{h^3} \rho(\mathbf{x}, \mathbf{p}, t) \exp\left[-\frac{\mathrm{i}}{\hbar} p(x - x') \right] . \tag{8.42}$$

Generalizing to N particles (see Klimontovich 1982, 1984), the N-particle Wigner function may also be expressed in terms of the N-particle density matrix in a non-diagonal coordinate representation:

$$\rho_N(\mathbf{q}, \mathbf{p}, t) = \frac{\hbar^{3N}}{V^N} \int \mathrm{d}\mathbf{g}\, \rho_N\big(\mathbf{q} + \hbar\mathbf{g}/2, \mathbf{p} - \hbar\mathbf{g}/2, t\big) \exp(-\mathrm{i}\mathbf{g} \cdot \mathbf{p}) . \tag{8.43}$$

For $N = 1$, we denote the Wigner function by $\rho(\mathbf{x}, \mathbf{p}, t)$, bearing in mind that we should restrict the class of densities to the real functions, considering a kind of 'Husimi density'. Then, in the space of real functions, we may write a hypothetical 'effective Liouville equation' in the form

$$\frac{\partial \rho(\mathbf{x}, \mathbf{p}, t)}{\partial t} + \frac{\partial H_{\mathrm{eff}}}{\partial p} \frac{\partial \rho(\mathbf{x}, \mathbf{p}, t)}{\partial \mathbf{x}} - \frac{\partial \rho(\mathbf{x}, \mathbf{p}, t)}{\partial \mathbf{p}} \frac{\partial H_{\mathrm{eff}}(p, x)}{\partial x} = 0 . \tag{8.44}$$

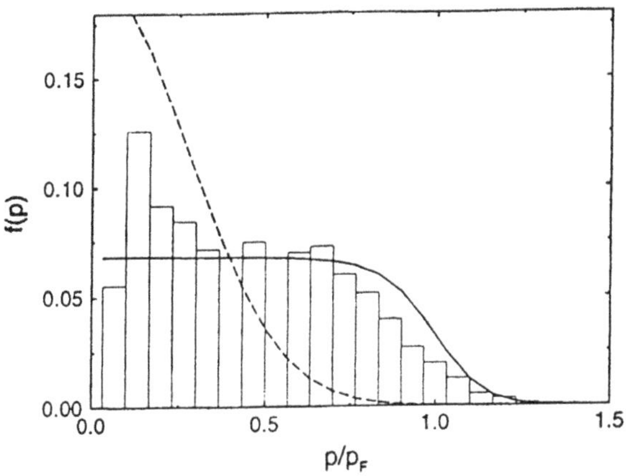

Fig. 8.7 Momentum distribution of an electron gas, calculated from MD simulations (Ebeling and Schautz 1997) and based on Wigner dynamics with momentum-dependent interactions ($T = 325$ K $= 0.1 T_F$, $V_0 = 0.26$ eV, $r_0 = 7.1$ Å). The *dashed curve* shows the Maxwell distribution for comparison

In this context we note also the Wigner trajectory method described by Grabowski et al. (2013). Here we use a different idea and develop simulations based on Hamilton-like equations with a model Hamiltonian which contains, beside the standard kinetic energy, non-traditional terms depending on the relative momenta of pairs of particles. Adapting the effective Hamiltonian as closely as possible to the exact Wigner Hamiltonian, which is a nonlocal integral operator, we arrive at a local dynamics in real phase space which we call Wigner-type dynamics. Assuming a real Wigner function and a real Hamiltonian, we arrive at useful approximations. This formalism cannot be exact, since, as mentioned, the true Wigner function is sometimes not a positive real function.

To give an example let us now discuss the quasi-classical dynamics of free charges in this model. The approach we are using was developed from the concept of minimum uncertainty wave packets by Wilets, Kirschbaum, Dorso, and others (see Dorso et al. 1987, 1988, Ebeling and Schautz 1997). The quantum mechanical effects corresponding to the Pauli and Heisenberg principles are modeled by constraints on the Hamiltonian. To derive the effective interactions, we use the concept of minimum uncertainty wave packets (Dorso et al. 1987, 1988, Klimontovich 1997, Ebeling and Schautz 1997; Ortner et al. 1997).

The wave function corresponding to a minimum uncertainty wave packet (also denoted as a coherent state) are Gaussians:

$$\psi_0(\mathbf{x}, t) = \text{const.} \times \exp\left[-\frac{(\mathbf{x}-\mathbf{r})^2}{2r_0^2} + \mathrm{i}\frac{(\mathbf{x}-\mathbf{r})\cdot\mathbf{p}}{\hbar} \right]. \tag{8.45}$$

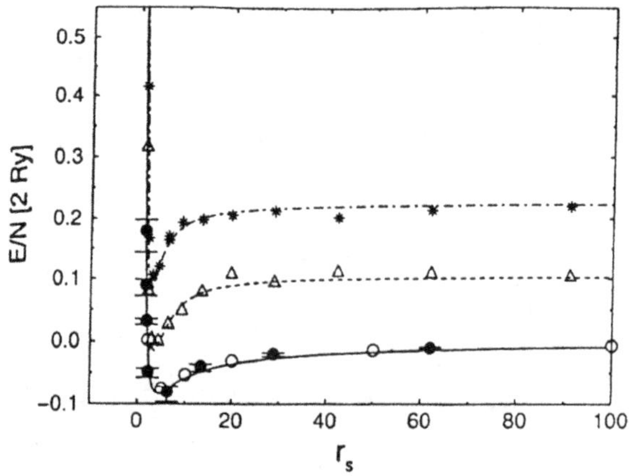

Fig. 8.8 Dependence on the parameter r_s of the energy per particle of an electron gas, calculated from MD simulations (Ebeling and Schautz 1997) using the model of velocity-dependent interactions for $T = 150$ K (*top curve*), $T = 22\,000$ K (*middle curve*), and $T = 47\,000$ K (*bottom curve*). *Lines* are from a Padé approximation (see Chap. 6) and *open circles* are from quantum MC calculations (Ceperley and Alder 1980). We note a good overall agreement between these approaches

Here r_0 is the mean dispersion of the wave function, which is a free parameter in this concept. The effective Hamiltonian is derived by averaging the original Hamilton operator with respect to the test wave functions (Dorso et al. 1987, 1988, Ebeling et al. 1997):

$$H(\mathbf{p}, \mathbf{q}; \hbar) = \int \mathbf{dx}\, \psi_0^*(\mathbf{x}) \hat{H} \psi_0(\mathbf{x}) \,. \tag{8.46}$$

From this expression one derives the following effective (quasi-classical) Hamiltonian:

$$H(\mathbf{p}, \mathbf{q}; \hbar) = \sum_i \frac{p_i^2}{2m_i} + \sum_{i<j} \delta_{ij} V_{\mathrm{P}}\left(\frac{p_{ij}}{p_0}, \frac{r_{ij}}{R_0}\right) + \sum_{i<j} \frac{e_i e_j}{r} \mathrm{erf}\left(\frac{r_{ij}}{r_0\sqrt{2}}\right) . \tag{8.47}$$

Here V_{P} is special potential describing interactions in the momentum space.

8.4.2 *Dorso Potential and Simulations of Plasmas*

A simple approximation for the momentum-dependent part of the interactions, which includes the essential effects and is suitable for simulations, is represented by the Dorso potential (Dorso et al. 1987, 1988)

$$V_{\mathrm{P}}\left(\frac{p_{ij}}{p_0}, \frac{r_{ij}}{r_0}\right) = \frac{\hbar^2}{4mr_0^2} \exp\left(-\frac{\Delta^2}{2}\right), \qquad \Delta^2 = \frac{p^2}{p_0^2} + \frac{r^2}{r_0^2}. \tag{8.48}$$

This is a two-body interaction depending on the phase-space distance corresponding to the minimum uncertainty condition $r_0 p_0 = \hbar$. In the effective potentials, r_0 is the width of the wave packets which we consider as a fixed free parameter. As in the Kelbg potential, for r_0, we take the thermal de Broglie wavelength, which is $r_0 = \sqrt{\pi}\Lambda_{\mathrm{e}}/2$ for electrons.

This was applied to the free Fermi gas and the electron gas, calculating the momentum distributions and the energy per particle, in (Ebeling et al. 1997). The results show reasonable momentum distributions corresponding to Fermi distributions, and also good results for the energy (see Fig. 8.8) and the dynamical structure factor (see Figs. 8.7 and 8.8 and Ortner et al. 1997). This means that, with respect to the momentum distributions and other dynamical properties, this method is an improvement on Kelbg-type potentials. On the other hand, the Dorso potential we used is only a first approximation and needs further development to higher approximations. Furthermore, we note that, regarding the mean energy, the QMD results are less precise than recent ab initio QMD results for the electron gas (Groth, Dornheim and Bonitz 2016, Dornheim et al. 2016). On the other hand, the QMD method provides additional dynamical information, programming is straightforward, and computation does not require exception numerical effort.

Having observed that the Wigner dynamics is a useful tool for the treatment of dense plasmas (see also Filinov et al. 2006), let us discuss briefly the question of a deeper statistical foundation and systematic development. The easiest quantum-statistical way to formulate an effective Wigner Hamiltonian was given by Klimontovich (1982, 2001). This approach was based on the kinetic equation for the Klimontovich operator given in Sect. 7.5.2 and the corresponding potential energy operator. According to Klimontovich the generalized form for the momentum-dependent operator of the interaction potential reads, including exchange effects,

$$\hat{U}_a(\mathbf{r}, \mathbf{p}) = \sum_b \int \frac{\mathrm{d}\mathbf{r}'\mathrm{d}\mathbf{p}'}{(2\pi\hbar)^3} \Bigg[V_{ab}(\mathbf{r} - \mathbf{r}') \pm \frac{\delta_{ab}}{2s_a + 1} \tilde{V}_{ab}\left(\frac{|\mathbf{p} - \mathbf{p}'|}{\hbar}\right) \delta(\mathbf{r} - \mathbf{r}') \Bigg] \hat{N}_b(\mathbf{r}', \mathbf{p}', t), \tag{8.49}$$

where $\hat{N}_b$ denotes the Klimontovich operator of the particles in the neighborhood in the Wigner representation. The upper sign + stands for Bose particles and the lower sign − stands for Fermi particles. Assuming that the operator $\hat{N}_b$ is delta-like, for the case of two identical particles with interactions having a Fourier transform $\tilde{V}_{12}(\mathbf{k})$, Klimontovich derived the effective Wigner Hamiltonian (Klimontovich 1982; 2001)

$$H_{\text{eff}}(\mathbf{r}_1, \mathbf{p}_1, \mathbf{r}_2, \mathbf{p}_2) = \frac{\mathbf{p}_1{}^2}{2m_1} + \frac{\mathbf{p}_2{}^2}{2m_2} + V(\mathbf{r}_{12}) \pm \frac{1}{(2\pi)^3} \frac{\delta(1,2)}{2s_a + 1} \tilde{V}_{12}\left(\frac{\mathbf{p}_{12}}{\hbar}\right) \delta(\mathbf{r}_{12}) . \tag{8.50}$$

Once again the $+$ applies to Bose pairs and the $-$ to Fermi pairs. For simulations, the delta functions have to be smeared out, e.g., by a Gaussian. Note that similar expressions play a role in the Landau theory of Fermi liquids. With the more general assumption that the surrounding b particles have a Gaussian distribution, viz.,

$$\hat{N}_b(\mathbf{r}', \mathbf{p}') \approx \frac{\hbar^2}{r_0^2 p_0^2} \exp\left(-\frac{\mathbf{r}'^2}{2r_0^2} - \frac{\mathbf{p}'^2}{2p_0^2} \right) , \tag{8.51}$$

we obtain once again the Dorso potential with some corrections in the momentum-dependent part. This shows that the Dorso potential we have used so far in our calculations is basically correct, but needs further development.

Note that other representations of Wigner dynamics, which are more generally based on path integrals, were derived by Larkin et al. (2014, 2016). This approach, which allows a relativistic generalization, will be explained in detail in Chap. 9.

8.5 Wave Packet Molecular Dynamics

A direct solution of the Schrödinger equations, which are partial differential equations for many-particle dense subsystems, seems to be a hopeless task, so we need approximations. One of them is certainly the TBA which we have already explained, and another is the density functional method (Parr and Yang 1989; Redmer et al. 2010; Holst et al. 2011; Karasiev et al. 2012). Density functional theory has its conceptual roots in the Thomas–Fermi model. It was put on a firm theoretical footing by the Hohenberg–Kohn theorems. The remarkable fact is that, not only is the Bohr theory an exact theory, but the Thomas–Fermi theory has the property of being exact, at least in the asymptotic sense as $N_e \to \infty$, as shown by Lieb, Simon and Thirring (see Thirring 1980).

The Thomas–Fermi theory is just a special case of a class of more general theories, now called density functional t heories. This is a trend in plasma theory which is increasing in importance (Redmer 2010). The general density functional approach is based on an important theorem due to Kohn, Sham and Hohenberg, which says that the energy may be expressed by some functional of the electron density, viz.,

$$E = E\big[n_e(\mathbf{r})\big] , \quad n_e(\mathbf{r}) = N_e \sum_{\sigma_i} \int d\mathbf{r}_2 \ldots d\mathbf{r}_{N_e} \left|\psi(\mathbf{r}, \mathbf{r}_2, \ldots \mathbf{r}_N)\right|^2 , \tag{8.52}$$

where ψ is the N-electron wave function. The theorem due to Kohn, Sham and Hohenberg (see Thirring 1980) states that the E functional is unique if the exact energy is considered. The exact form of the functional is of course not known. However, one is more interested in finding lower bounds and good approximate functionals. The density functional approach is now developing into a tool of great practical relevance for simulations of Coulomb systems. Since we are not experts in this field, we shall not try to explain the details here, so the reader is referred to other books (Parr and Yang 1989; Redmer et al. 2010). Instead we will introduce wave packet molecular dynamics (WPMD) as another new branch for the computer simulation of quantum systems.

8.5.1 Wave Packet Dynamics of Nearly Free Electrons

This method is based on the dynamics of proper wave functions, which are approximated by Gaussian wave packets with time-dependent width and location. The ion dynamics is classical and the electronic wave function is modeled by spherical Gaussian wave functions which are described by time-dependent parameters. The time dependence of the parameters is derived from the time-dependent variational principle (TDVP), while ions are usually treated classically with standard MD. This method can simulate equilibrium and nonequilibrium dynamics and provides reasonable approximations to the many-body time dependent Schrödinger equation. Wave packet molecular dynamics (WPMD) is also easy to implement in an MD code. The Gaussian ansatz was first used to simulate scattering between simple atoms by Heller (1975). It was later adopted by Feldmeier et al. (1995) to solve problems of nuclear structure and reactions. Klakow, Toepffer, and Reinhard were the first to apply WPMD to plasmas (Klakow et al. 1994, Jakob et al. 2007).

The Gaussian ansatz seems reasonable to describe free or nearly free electrons, but electrons simulated by WPMD show divergent width parameters leading to wave packet spreading (Klakow et al. 1994, Morozov and Valuev 2009, 2012, Grabowski et al. 2013; Valuev and Morozov 2015; Lavrinenko et al. 2016). Electrons thus spread and there is no mechanism to localize nearby nuclei, producing a nearly constant electron background at large times. Several methods were developed to avoid this spreading. Knaup et al. (2002, 2003) introduced an ad hoc fix to wave packet spreading by adding a harmonic constraint to the wave packet. Other groups imposed a position-dependent phase factor with constant nonzero coefficient on the variational ansatz. These approaches include an arbitrary fixed parameter that determines the width of the harmonic well acting on the WP width (Ebeling et al. 1996, 1997, 2002, Pohl et al. 2002, Jakob et al. 2007, Morozov and Valuev 2009, 2012). The impact of multiparticle properties such as antisymmetrization on wave packet spreading has not yet been well studied.

The basic idea of WPMD using effective potentials is to condense all quantum properties of the particles into some constant parameters, while their dynamics is tracked using classical equations of motion. Generally, the various parameters depend on the macroscopic state of the system, in such a way that every particle produces the same effective potential. While this concept is very successful in describing certain equilibrium aspects of strongly coupled plasmas, it is not well suited to studying properties such as velocity distributions, velocity correlations, transport properties, or nonequilibrium processes.

On the other hand, we have seen above that there is a strong analogy between the effective potentials and the resulting interactions between Gaussian wave packets. In fact the time evolution of such wave packets (WPs) can be obtained from classical equations of motion by employing the variational formulation of the Schrödinger equation. This concept of a wave packet molecular dynamics (WPMD), originally developed by Heller (1975), has been successfully applied in a number of situations ranging from collision processes in nuclear physics (Feldmeier et al. 1995) and atomic strong-field dynamics (Pohl et al. 2002) to the description of dense plasmas (Knaup et al. 2003, Jakob et al. 2007, Morozov and Valuev 2012, Grabowski et al. 2013).

Generally, one starts with a proper parametrization of the many-body wave function $\psi\left(\{\mathbf{r}_i\},\{q_j\}\right)$, where the dynamics of the parameters q_i is determined from the variational Dirac–Frenkel–McLachlan form of the Schrödinger equation:

$$\delta\int_{t_1}^{t_2}\left\langle\psi\left|\mathrm{i}\hbar\frac{\partial}{\partial t}-\hat{H}\right|\psi\right\rangle \mathrm{d}t=0\,. \tag{8.53}$$

After some algebra, we get the following equations of motion for the parameters:

$$\frac{\partial H}{\partial q_i}=\sum_j \mathcal{N}_{ij}\dot{q}_j\,,\qquad \mathcal{N}_{ij}=2\mathrm{Im}\left(\left\langle\frac{\partial\psi}{\partial j}\middle|\frac{\partial\psi}{\partial i}\right\rangle\right)\,, \tag{8.54}$$

where $H=\langle\psi|\,\hat{H}\,|\psi\rangle$ is the expectation value of the Hamiltonian operator.

In the simplest version we discuss the quasi-classical dynamics of the free charges in our model. The quantum-mechanical effects corresponding to the Pauli and Heisenberg principles are modeled by constraints on the Hamiltonian. To derive the effective interactions, we use the concept of minimum uncertainty wave packets. This means that the wave function corresponding to a minimum uncertainty wave packet (also referred to as a coherent state) is a Gaussian:

$$\psi_0(\mathbf{x},t)=C\exp\left[-\frac{(\mathbf{x}-\mathbf{r})^2}{2r_0^2}+\mathrm{i}\frac{\mathbf{p}\cdot(\mathbf{x}-\mathbf{r})}{\hbar}\right]\,. \tag{8.55}$$

Here r_0 is the mean dispersion of the wave function.

In a more advanced ansatz for the N-particle system, the width is considered as a variable parameter (Klakow et al. 1994) and the wave function is a product of single particle wave packets ϕ_i, where $\mathbf{r}_i$ and $\mathbf{p}_i$ denote the position and momentum of the center, while β_i and p_{β_i} denote the width and its corresponding momentum (Klakow et al. 1994). Further developments and extensive simulations were given by (Knaup et al. 2003).

As mentioned, electrons simulated by WPMD show divergent width parameters leading to wave packet spreading. Another way to avoid this is to introduce a phase term Ebeling et al. (2006). With this term we represent the single particle wave function ϕ_i by

$$\phi_i = C \exp\left[-\left(\frac{3}{4\beta} + \mathrm{i}p_\beta + \mathrm{i}\alpha\right)(\mathbf{x}_i - \mathbf{r}_i)^2 - \mathrm{i}\mathbf{p}_i\,(\mathbf{x}_i - \mathbf{r}_i)\right] .$$

Here $\mathbf{r}_i$ and $\mathbf{p}_i$ denote the position and momentum of the WP center, while β_i and p_{β_i} denote the WP width and its corresponding momentum. In contrast, to the standard ansatz (Knaup et al. 2003), we included an additional constant phase shift $\mathrm{i}\alpha(\mathbf{x}_i - \mathbf{r}_i)^2$, which helps us to treat several of the main problems. We note that exchange effects may be taken into account by using a properly symmetrized wave function (Knaup et al. 2003).

The main advantage of the new ansatz is that it results in Hamiltonian equations for the conjugate pairs $(\mathbf{r}_i, \mathbf{p}_i)$ and (β_i, p_{β_i}) with the effective Hamiltonian H. This ansatz with $\alpha = 0$ provides a semiclassical description of the particle dynamics, although the classical high-temperature limit, i.e., a strong narrowing of the individual WPs, cannot be recovered with such WPMD simulations. In fact, the divergence of the WP width not only leads to a divergent partition sum in parameter space, but also causes technical problems in numerical plasma simulations, even at finite temperatures. This is, in fact, not surprising as freely propagating WPs spread without bounds, while the transition to classical behavior results from localization effects due to decoherence caused by environmental influences.

Despite the ongoing discussion about how to extract thermodynamic properties from time averages of quantum mechanical time evolution in a restricted trail space, WPMD simulations have been successfully applied to describe statistical properties of strongly coupled plasmas (Klakow et al. 1994; Ebeling et al. 1996, 1997), as long as the problem of a diverging WP width has been avoided. Consequently, different ideas, such as fixing β at some value or assuming adiabatic dynamics for the WP width (see the next section), have been employed (Pohl et al. 2002; Knaup et al. 2003, Morozov et al. 2009, 2012). We may show that fixing the width by means of the thermal wavelength $\beta \propto \lambda^2$ leads us back to the previous approximation. Alternatively, we may multiply the trail state by the additional phase factor $\exp\left[-\mathrm{i}\alpha(\mathbf{x}_i - \mathbf{r}_i)^2\right]$ or again $\exp\left[-\mathrm{i}\alpha(\mathbf{x}_i - \mathbf{r}_i)^4\right]$ in order to model localization effects. It is important to note that the inclusion of this phase factor preserves the Hamiltonian character of the underlying equations of motion.

Effective Hamiltonian

Including this ansatz leads to the new effective Hamiltonian

$$H = \sum_i H_i + \sum_{i<j} V_{ij}\ , \quad H_i = \frac{p_i^2}{2m} + \frac{\hbar^2}{m}\left(\frac{9}{8\beta_i} + 2\beta_i p_{\beta i}^2 + 2\alpha^2\beta_i\right)\ , \tag{8.56}$$

$$V_{ij} = \frac{e^2}{r_{ij}}\,\mathrm{erf}\left[\sqrt{\frac{3}{2(\beta_i+\beta_j)}}\,r_{ij}\right]\ . \tag{8.57}$$

The effective interaction potential V_{ij} does not depend on the new phase factor α. Instead of fixing $\beta \propto \lambda^2$, which would lead us back to earlier approximations, we use here the full dynamics of $\beta(t)$. We notice that our phase factor α is closely related to the procedure proposed in (Knaup et al. 2003; Jakob et al. 2007; Valuev and Morosov 2015).

As illustrated in Fig. 8.9, the additional phase factor causes the width to describe closed trajectories, even in the case of free particles. Therefore the resulting parameter-space partition sum of free particles, viz.,

$$Z_\mathrm{p} = \frac{\pi^2}{\alpha}\Lambda^{-5}\exp\left(-\frac{3\hbar^2\alpha}{mk_\mathrm{B}T_\mathrm{p}}\right)\ , \tag{8.58}$$

is indeed finite, where $\Lambda = h/\sqrt{2\pi mk_\mathrm{B}T_\mathrm{p}}$ denotes the thermal wavelength and T_p is the temperature in the parameter space. Note that this temperature does not provide a unique measure of thermodynamic properties, as it depends on the particular choice of the trail wave function ψ. However, from the correct quantum mechanical entropy, one can find a relation between T_p and the physical system temperature T, resulting in an average WP width of

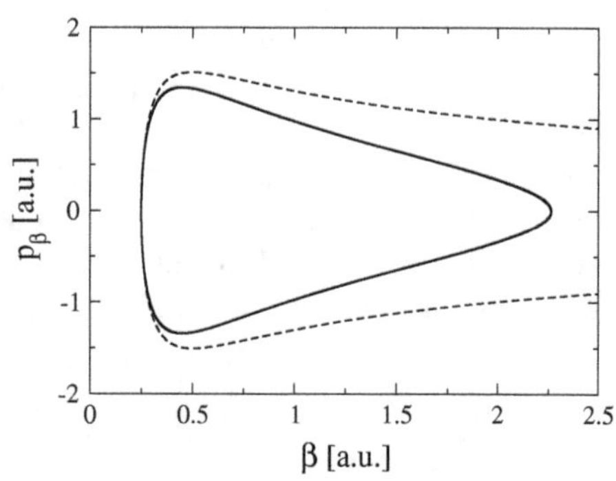

Fig. 8.9 Phase space trajectories of the WP width parameters, for $\alpha = 0$ (*dashed curve*) and $\alpha = 1$ a.u. (*continuous curve*)

$$\langle \beta \rangle = \frac{3}{4\alpha}\left(1 + \frac{mk_{\mathrm{B}}T}{3\hbar^2\alpha}\right) . \tag{8.59}$$

Comparing (8.57) with (8.47), one finds that $\langle \beta \rangle = 3\hbar^2/mk_{\mathrm{B}}T$, which finally leads to the desired result

$$\alpha = \frac{(3+\sqrt{89})}{40}\frac{mk_{\mathrm{B}}T}{\hbar^2} \approx 0.31\lambda^{-2} , \tag{8.60}$$

ensuring the correct high-temperature limit of the system in thermodynamic equilibrium.

Finally, we note that the problem of the diverging wave packets can be overcome directly by starting from a density matrix description rather than propagating the individual wave function of the system and doing the averaging afterwards. As shown by Militzer and Pollock (2000), a Gaussian ansatz for the equilibrium density matrix together with the corresponding variational principle leads to promising results for equilibrium properties of strongly coupled plasmas, and at the same time reproduces the known high temperature limit.

8.5.2 Bound Electrons in External Fields

The wave packet dynamics described above was successful for solving problems of free or weakly bound electrons. The situation is more complicated for strongly bound electrons moving in strong time-dependent fields, e.g., close to nuclei, and including bound to free transitions (Ebeling et al. 1996, 1997, 2000). We consider now a model study of the one-electron dynamics in internal and external fields (Pohl et al. 2002).

The Hamiltonian operator describes a one-electron atom in a periodic field and is given by

$$\widehat{H} = \frac{\widehat{p}}{2} - \frac{Z}{r} + V(\mathbf{r}, \mathbf{p}, t) , \qquad V(\mathbf{r}, t) = E_0 x \sin \omega t . \tag{8.61}$$

The laser field is treated in the dipole approximation, where only linearly polarized light is considered. For all further calculations we use atomic units, where $e = m_{\mathrm{e}} = \hbar = 1$. One possible semiclassical description of this problem is based on the approximations for the particle wave function we described above. They contain time-dependent parameters $q(t)$, which represent the coordinates in a generalized phase space. This technique has been widely applied in quantum molecular dynamics to study problems in plasma physics (Klakow et al. 1994; Ebeling et al. 1996) and nuclear physics (Feldmeier et al. 1995). The dynamics of the parameters is described by the time-dependent variational principle

$$\delta \int_{t_1}^{t_2} \left\langle \psi \left| \mathrm{i}\frac{\mathrm{d}}{\mathrm{d}t} - \widehat{H} \right| \psi \right\rangle \mathrm{d}t = 0 \,. \tag{8.62}$$

The equation of motion can be written in canonical form (Feldmeier et al. 1995), where the Hamiltonian function H is defined as the expectation value of the Hamilton operator $\widehat{H}$. As a trial wave function, we choose a Gaussian wave packet

$$\psi\left(\rho; r, p, \beta, p_\beta\right) = C \exp\left(-\left(\frac{3}{4\beta} + \mathrm{i}p_\beta\right)(\rho - r)^2 - \mathrm{i}p\,(\rho - r)\right) \,. \tag{8.63}$$

Here β and p_β parametrize the complex width of the distribution, while r and p are the average position and average momentum, respectively. With this choice of the wavefunction, we obtain the semiclassical Hamiltonian

$$H = \frac{p^2}{2} - \frac{Z}{r}\mathrm{erf}\left(r\sqrt{\frac{3}{2\beta}}\right) + 2p_\beta^2\beta + \frac{9}{8\beta} + E_0 x \sin \omega t \,. \tag{8.64}$$

The equations of motion obtained from this Hamiltonian are:

$$\begin{aligned}
\dot{x} &= p_x \,, \quad \dot{y} = p_y \,, \quad \dot{\beta} = 4p_\beta\beta \,, \\
\dot{p}_x &= -Z\frac{x}{r^3}\left[\mathrm{erf}\left(r\sqrt{\frac{3}{2\beta}}\right) - \sqrt{\frac{6}{\pi\beta}}r \exp\left(-\frac{3r^2}{2\beta}\right)\right] + E_0 \sin\left(\omega t\right) \,, \\
\dot{p}_y &= -Z\frac{y}{r^3}\left[\mathrm{erf}\left(r\sqrt{\frac{3}{2\beta}}\right) - \sqrt{\frac{6}{\pi\beta}}r \exp\left(-\frac{3r^2}{2\beta}\right)\right] \,, \\
\dot{p}_\beta &= \frac{9}{8\beta^2} - 2p_\beta^2 - Z\sqrt{\frac{2}{2\pi\beta^3}} \exp\left(-\frac{3r^2}{2\beta}\right) \,.
\end{aligned} \tag{8.65}$$

Using this form of dynamical equation, the width of the wave packet provides an additional degree of freedom. This fact leads to a divergent partition sum and to incorrect predictions for the thermal capacity, satisfying neither the classical nor the quantum limit. Thus, modifications of (8.65) are needed to solve these problems.

The simplest possibility is to fix the width at the value which gives the minimum ground state energy. Introducing a new parameter $r_0^2 = 2\beta/3$, this condition is satisfied for $r_0 = (3/4Z)\sqrt{\pi}$ with a ground state energy of $\varepsilon_0 = -0.424Z^2$, which is about 15% higher than the exact value provided by quantum mechanics. From the fixed width arises the problem that the ionization energy obtained from (8.64) does not agree with quantum mechanical predictions. The third summand in (8.64), the so called curvature energy, leads to an ionization energy of $\varepsilon_\mathrm{B} = 0.848Z^2$.

In order to avoid these difficulties we introduce a semiclassical adiabatic approximation (SAA), which is based on the assumption of a slowly varying width of the wave packet. From the condition $p_\beta = 0$, for given r and p, we find the following expression for r_0:

$$r_0 = \frac{3}{4Z}\sqrt{\pi}\exp\left(\frac{|r|^2}{r_0^2}\right). \tag{8.66}$$

This adiabatic approximation solves the problem of the incorrect ionization energy, as well as the problem of the additional degree of freedom. To get an explicit expression for the width of the wavepacket, we expand (8.66) in r_0 and obtain

$$r_0 = \frac{3}{8}\sqrt{\pi} + \frac{1}{8}\sqrt{9\pi + 64r^2}\,. \tag{8.67}$$

To test this result, we take the limit $Z \to 0$, which corresponds to a free electron in a linearly polarized laser field. By substituting the solution of (8.65) in (8.63), we obtain the Volkov wave function in the long wavelength approximation:

$$\psi = \exp\left(\mathrm{i}\mathbf{p}\cdot\mathbf{r} - \mathrm{i}\frac{p^2}{2}t - \mathrm{i}\frac{E_0^2}{4\omega^2}t + \mathrm{i}\frac{E_0 p_x}{\omega^2}\sin\omega t - \mathrm{i}\frac{E_0^2}{8\omega^3}\sin 2\omega t\right), \tag{8.68}$$

which is the exact quantum mechanical solution for the problem.

As mentioned above the SAA predicts an ionization energy which is smaller than the exact energy provided by exact quantum theory. With respect to this property, the SAA seems to be worse than the soft core potential describing atom–light interactions. However, the atomic potential should also give good predictions for the polarizability of the atom, defined by $\alpha = p/E$ where p is the dipole moment resulting from the static electric field E. Quantum mechanical perturbation theory predicts a value of $\alpha = 4.5$ a.u. for the hydrogen ground state. Using a harmonic expansion of the semiclassical potentials, the polarizability can be estimated by $\alpha = e^2/\omega_0^2$. The soft core potential fitted to the exact ground state energy yields $\alpha = 8$ a.u., which is about twice the exact value. Our SAA potential based on a Gaussian wave packet leads to $\alpha = 3.1$ a.u., which is only 30% off the exact value. Further calculations for hydrogen atoms with $Z = 1$ are given in (Pohl et al. 2002). The extension of this approach to many-atom systems is still an open task.

References

N. Ashcroft. N.D. Mermin, Solid State Physics (Holt, Rinehardt, and Winston, Philadelphia, 1976, Mir Moscow 1979)

H.B. Böttger, V.V. Bryksin, *Hopping Conduction in Solids* (Akademie-Verlag, Berlin, 1985)

L. Brizhik, A.P. Chetverikov, W. Ebeling, G. Röpke, M.G. Velarde, Phys. Rev. B **85**, 245105 (2012)

R. Car, M. Parrinello, Unified approach for molecular dynamics and density-functional theory. Phys. Rev. Lett. **55**, 241–247 (1985)

D.M. Ceperley, B. Alder, Phys. Rev. Lett **45**, 566 (1980)

A.P. Chetverikov, W. Ebeling, M. Jenssen, Yu. Romanovsky, Excitations on rings of molecules, in *Stochastic dynamics of reacting biomolecules*, ed. by W. Ebeling, et al. (World Scientific, Singapore, 2002)

P. Chetverikov, W. Ebeling, G. Röpke, M.G. Velarde, Anharmonic excitations, time correlations and electric conductivity. Contrib. Plasma Phys. **47**, 465–478 (2007)

A.P. Chetverikov, W. Ebeling, M.G. Velarde, Local electron distributions and diffusion in anharmonic lattices mediated by thermally excited solitons. Eur. Phys. J. B **70**, 217–227 (2011)

A.P. Chetverikov, W. Ebeling, G. Röpke, Hopping transport and stochastic dynamics in plasma layers. Contrib. Plasma Phys. **51**, 814–829 (2011)

A.P. Chetverikov, W. Ebeling, M.G. Velarde, Soliton-like excitations and solectrons in two-dimensional lattices. Eur. Phys. J. B **80**, 137–145 (2012)

A.P. Chetverikov, W. Ebeling, G. Röpke, M.G. Velarde, High electrical conductivity in nonlinear model lattice crystals mediated by thermal excitation of solectrons. Eur. Phys. J. B **87**, 153 (2014)

A.P. Chetverikov, W. Ebeling, M.G. Velarde, On the temperature dependence of fast electron transport in crystal lattices. Eur. Phys. J. B **88**, 202 (2015)

C. Dorso, S. Duarte, J. Randrup, Phys. Lett B **188**, 287 (1987)

T. Dornheim, S. Groth, A. Filinov, M. Bonitz, Permutation blocking path integral Monte Carlo: a highly efficient approach to the simulation of strongly degenerate non-ideal fermions. New J. Phys. **17**, 073017 (2015)

T. Dornheim, T. Schoof, S. Groth, A. Filinov, M. Bonitz, Permutation blocking path integral monte carlo approach to the uniform electron gas at finite temperature. J. Chem. Phys. **143**, 204101 (2015)

T. Dornheim, S. Groth, T. Schoof, C. Hann, M. Bonitz, Ab initio quantum Monte Carlo simulations of the uniform electron gas without fixed nodes: the unpolarized case. Phys. Rev. B **93**, 205134 (2016)

T. Dornheim, H. Thomsen, P. Ludwig, A. Filinov, M. Bonitz, Analyzing quantum correlations made simple. Contrib. Plasma Phys. **56**, 371 (2016)

T. Dornheim et al., Ab initio Quantum Monte Carlo simulation of the warm dense electron gas in the thermodynamic limit, Phys. Rev. Lett. **117**, 156403 (2016), arXiv: 1607.08076v2 [physics.plasm-ph]. 9 Sep 2016

C. Dorso, S. Duarte, J. Randrup, Excited electron dynamics modeling of warm matter, Phys. Lett. B **188**, 287 (1987); **215**, 611 (1988)

W. Ebeling, A. Filinov, M. Bonitz, V. Filinov, T. Pohl, The method of effective potentials in the quantum-statistical theory of plasmas. J. Phys. A: Math. Gen. **39**, 4309–4317 (2006)

W. Ebeling, V. Fortov, et al. (eds.), *Transport Properties of Dense Plasmas* (Birkhäuser, Boston, 1984)

W. Ebeling, I. Leike, U. Leonhardt, Bound states and ionization kinetics in dense plasmas, AIP Conf. Proc. (Atomic Processes in Plasmas, Portland ME 1991) **257**, 97–107 (1992)

W. Ebeling, A. Förster, V Yu. Podlipchuk, Quantum wave packets simulations of ionization processes in dense plasmas. Phys. Lett. A **218**, 297–303 (1996)

W. Ebeling, Bound state effects in quantum transport theory. Ann. Physik (Leipzig) **33**, 350–358 (1976)

W. Ebeling, G. Röpke, Conductance theory of nonideal plasmas. Ann. Physik (Leipzig) **36**, 429–432 (1979)

W. Ebeling, B. Militzer, F. Schautz, Quasiclassical theory and simulations of hydrogen quantum plasmas. Contrib. Plasma Phys. **37**, 137–148 (1997)

W. Ebeling, F. Schautz, Simulations of the quantum electron gas using momentum-dependent potentials. Phys. Rev. E **56**, 3498 (1997)

W. Ebeling, J. Ortner, Quasiclassical theory and simulations of strongly coupled plasmas. Phys. Scr. T **75**, 93–98 (1998)

W. Ebeling, G.E. Norman, A.A. Valuev, I. Valuev, Quasiclassical theory and molecular dynamics of two-component nonideal plasmas. Contr. Plasma Phys. **39**, 61 (1999)

W. Ebeling, T. Pohl, J.A. Holyst, Nonlinear effects and energy flow in the semiclassical dynamics of atomic electrons in strong laser fields. Laser Phys. **10**, 1069–1077 (2000)

W. Ebeling, T. Pohl, Quantum wave packet dynamics. in *Quantum limits on the second law*, ed. by D.P. Sheehan (AIP Conf. Proc., New York, 2002)

W. Ebeling, L. Schimansky-Geier, YuM Romanovsky (eds.), *Stochastic dynamics of interacting biomolecules* (World Scientific, Singapore, 2002)
W. Ebeling, A. Filinov, M. Bonitz, V. Filinov and T. Pohl, The method of effective potentials in the quantum-statistical theory of plasmas, J. Phys. A: Math. Gen. 39 , 4309–4317 (2006)
V. Filinov, M. Bonitz, A. Filinov, V. Golubnychiy, Wigner function quantum molecular dynamics, arXiv:cond-mat/0611560v1 [cond-mat.str-el]. 21 Nov 2006
H. Feldmeier, K. Bieler, J. Schnack, Nucl. Phys. A **596**, 493 (1995)
A. Förster, D. Beule, H. Conrads, W. Ebeling, Highly ionized carbon in capillary discharge plasma. Contr. Plasma Phys. **38**, 655–660 (1998)
A. Förster, T. Kahlbaum, W. Ebeling, Equation of state and phase diagram of fluid helium in the region of partial ionization. Laser Part. Beams **10**, 253–262 (1992)
P.E. Grabowski, A. Markmann, M.S. Murillo, C.A. Fichtl, D.F. Richards, V.S. Batista, F.R. Graziani, I.V. Morozov, I.A. Valuev, Wave packet spreading and localization in electron-nuclear scattering. Phys. Rev. E **87**, 063104 (2013)
J.-P. Hansen, I.R. McDonald, E.L. Pollock, Phys. Rev. A **11**, 1025 (1975)
S. Groth, T. Dornheim, M. Bonitz, Free energy of the uniform electron gas: testing analytical models against first principle results, arXiv:1611.05695v1 [physics.plasm-ph]. 17 Nov 2016
J.-P. Hansen, I.K. McDonald, Microscopic simulation of a strongly coupled hydrogen plasma. Phys. Rev. A **23**, 2041–2059 (1981)
A. Heeger, S. Kivelson, J.R. Schrieffer, W.P. Su, Rev. Mod. Phys. **60**, 781 (1988)
D. Hennig, M.G. Velarde, A.P. Chetverikov, W. Ebeling, Phys. Rev. E **76**, 046602 (2007)
E.J. Heller, J. Chem. Phys. **62**, 1544 (1975)
B. Holst, M. French, R. Redmer, Electronic transport coefficients from ab initio simulations and application to dense liquid hydrogen. Phys. Rev. B **83**, 235120 (2011)
B. Jakob, P.-G. Reinhard, C. Toepffer, G. Zwicknagel, Phys. Rev. E **76**, 036406 (2007)
G. Kalman (ed.), *Strongly Coupled Coulomb Systems* (Pergamon Press 1998)
V.S. Karakhtanov, R. Redmer, H. Reinholz, G. Röpke, Transport coefficients in dense plasmas including ion-ion structure factor. Contrib. Plasma Phys. **51**, 355–360 (2011)
V.V. Karasiev, T. Sjostrom, S.B. Trickey, Comparison of DFT and finite temperature HF approximation. Phys. Rev. E **86**, 056704 (2012)
G. Kelbg, Quantenstatistik der Gase mit Coulombwechselwirkung, Ann. Physik (Leipzig) **12**, 219–224, 354–360 (1963), **14**, 394–403 (1964)
D. Klakow, C. Toepffer, P.-G. Reinhard, Phys. Lett. A **192**(55) (1994), J. Chem. Phys. **101**(10766) (1994)
Y.L. Klimontovich, *Kinetic Theory Of Nonideal Gases And Nonideal Plasmas (Nauka, Moskva, 1975)* (Pergamon, Oxford, 1982)
Y.L. Klimontovich, *Statistical Physics* (in Russ.) (Nauka Moskva 1982), (Engl. transl., Harwood, 1984)
M. Knaup, P.-G. Reinhard, C. Toepffer, G. Zwicknagel, Wave packet molecular dynamics simulations of hydrogen at mbar pressures. Comput. Phys. Commun. **147**, 202–204 (2002)
M. Knaup, P.-G. Reinhard, C. Toepffer, G. Zwicknagel, J. of Physics A: Math. Gen. **36**, 6165 (2003)
Y.L. Klimontovich, *Statistical Theory of Open Systems*(in Russian). vol. I, II, III, (Janus, Moscow 1995, 1999, 2001)
W.D. Kraeft et al., *Quantum Statistics Of Charged Particle Systems* (Plenum, New York, 1986)
R. Kubo, Statistical-mechanical theory of irreversible processes. I. General theory and simple applications to magnetic and conduction problems. J. Phys. Soc. Jpn. **12**, 570–586 (1957)
A.S. Larkin, V.S. Filinov, Wigner's pseudo-particle relativistic dynamics in external potential field. Phys. Lett. A **378**, 1876–1882 (2014)
A.S. Larkin, V.S. Filinov, V.E. Fortov, Path integral representation of the Wigner function. Contr. Plasma Phys. **56**(3–4), 197–213 (2016)
Y.S. Lavrinenko, I.V. Morozov, A. Valuev, Reflecting boundary conditions for classical and quantum molecular dynamics. Contr. Plasma Phys. **56**, 448–458 (2016)

U. Leonhardt, W. Ebeling, Ionization and recombination coefficients of excited states in nonideal hydrogen plasmas. Physica A **192**, 249–261 (1993)

E.M. Lifshits, P. Pitaevskii, *in Physical Kinetics*, vol. 10 (Course of Theoretical Physics (Pergamon, New York, 1981)

W.R. Magro, D.M. Ceperley, C. Pierleoni, B. Bernu, Phys. Rev. Lett. **76**, 1240 (1996)

B. Militzer, E.L. Pollock, Phys. Rev. E **61**, 3470 (2000)

I.V. Morozov, I.A. Valuev, J. Phys. A: Math. Theor. **42**, (214044) (2009); Contributions Plasma Physics **52**, 140 (2012)

J. Ortner, I.M. Tkachenko, Phys. Rev. A **46**, 7882 (1992)

J. Ortner, F. Schautz, W. Ebeling, Quasiclassical molecular-dynamics simulations of the electron gas. Phys. Rev. E **56**, 4665–4670 (1997)

J. Ortner, F. Schautz, W. Ebeling, Investigations of the dynamic properties of the electron gas by quasi-classical simulations, in *Strongly coupled plasmas* ed. by G. Kalman et al. (Plenum Press, New York, 1998)

J. Ortner, I. Valuev, W. Ebeling, Semiclassical dynamics and time correlations in two-component plasmas. Contr. Plasma Phys. **39**, 311–321 (1999)

J. Ortner, I. Valuev, W. Ebeling, Electric microfield distribution in two-component plasmas. Contr. Plasma Phys. **40**, 555–685 (2000)

R.G. Parr, W. Yang, *Density-Functional Theory of Atoms and Molecules* (Oxford University Press, New York, 1989)

R. Redmer, B. Holst, F. Hensel (eds.), *Metal-to-Nonmetal Transitions* (Springer, Berlin, 2010)

H. Reinholz, G. Röpke, S. Rosmej, R. Redmer, Conductivity of warm dense matter including electron-electron collisions. Phys. Rev. E **91**, 043105 (2015)

G. Röpke, W. Ebeling, W.D. Kraeft, Quantum-statistical conductance theory of nonideal plasmas by use of the force-force correlation function method. Physica A **101**, 243–254 (1980)

T. Pohl, U. Feudel, W. Ebeling, Bifurcations of a semi-classical atom in a periodic field. Phys. Rev. E **65**, 046228 (2002)

S. Sadykova, W. Ebeling, I. Valuev, I. Sokolov, Contrib. Plasma Phys. **49**, 76 (2009)

S.P. Sadykova, W. Ebeling, I.M. Tkachenko, Eur. Phys. J. D **61**, 117–130 (2011)

V.P. Silin, A.A. Rukhadse, *Electromagnetic Properties of plasmas*(in Russ.) (Atomisdat, Moskva, 1961)

J. Taruna, J. Piekarewicz, M.A. Pèrez-Garcìa, Virtues and flaws of the Pauli potential, J. Phys. A: Math. Theor. **41**(3) (2008)

W. Thirring, Lehrbuch der Mathematischen Physik. 4 Quantenmechanik großer Systeme Springer 1980

I.A. Valuev, I.V. Morozov, Extension of the wave packet molecular dynamics. J. Phys. Conf. Series **653**, 012153 (2015)

D.N. Zubarev, V. Morozov, G. Röpke, *Statistical Mechanics of Nonequilibrium Processes*, vol. 1–2 (Wiley-VCH, Weinheim, 1997)

Chapter 9
Theoretical Approaches to Quantum Monte Carlo Methods

In this chapter we substantially extend the analysis of the previous chapters to two-component partially ionized Coulomb systems with positive and negative charges and a mass ratio M varying between one and two thousand. While low values of the mass ratio M are directly relevant for semiconductors, we do not consider any special solid state properties here, but focus on the fundamental aspects of Coulomb correlations in two-component plasmas which depend on M. In this way, we may cover the case of mass-symmetric plasmas on the one hand and the case of heavy-ion plasmas on the other as special cases.

9.1 Path Integral Approach to Coulomb Systems and the Task of Monte Carlo Methods

Over the last few decades significant progress has been made in theoretical studies of thermodynamic properties of strongly correlated fermions at non-zero temperatures, mainly due to the application of numerical simulations (see the review McMahon et al. 2012). The reason for this success is the possibility of an explicit representation of the partition function and density matrix in the form of the Wiener path integrals (Feynman and Hibbs 1965) and application of the Monte Carlo method for further calculations.

From a theoretical point of view, the complex processes of interest which involve strong correlations and quantum and spin effects are difficult to treat with analytical methods. Therefore, there has been considerable activity in the development of numerical treatments of two-component Coulomb systems (plasmas) (see, e.g., Bonitz et al. 1996; Egger et al. 1999; Klakow et al. 1994; Ebeling and Schautz 1997). An approach which is particularly well suited to describing equilibrium properties of two-component plasmas in the strong coupling and degeneracy regime is

W. Ebeling et al., *Quantum Statistics of Dense Gases and Nonideal Plasmas*, Springer Series in Plasma Science and Technology, DOI 10.1007/978-3-319-66637-2_9

the path integral quantum Monte Carlo (PIMC) method. Remarkable progress has been made in applying these techniques to Coulomb systems. For an overview, see, e.g., Kalman (1998), Kraeft and Schlanges (1996), Zamalin et al. (1977), Binder and Cicotti (1996), Berne et al. (1998).

The main difficulty for PIMC studies of Fermi systems arises due to anti-symmetrization of the density matrix (Feynman and Hibbs 1965). The result is that all thermodynamic quantities are presented as a sum of terms with alternating signs related to even and odd permutations. So the thermodynamic quantities are equal to the small difference between two large numbers, which are sums of positive and negative terms. This difficulty is known in the literature as the 'sign problem'. The PIMC calculations in this case are severely hampered and are only applicable for weak degeneracy, that is, at relatively high temperatures or low densities. Several strategies have been developed to overcome or at least reduce this difficulty (Egger et al. 1999; Ceperley 1991, 1992, 1995, Militzer and Pollock 2000; DuBois et al. 2014; Takahashi and Imada 1984; Filinov et al. 2004; Lyubartsev 2005; Suzuki 1990, 1995, Chin 1997, 2015, Schoof et al. 2011, 2015, Dornheim et al. 2015, 2016, Larkin et al. 2016, 2017).

The most widely known approach to extend the path integral Monte Carlo simulations to higher degeneracy is restricted PIMC (RPIMC) (Ceperley 1991, 1995, 1996), also known as the 'fixed node' approximation, where additional assumptions about the density operator are adopted which reduce the sum over permutations to even (positive) contributions alone. This requires knowledge of the nodes of the density matrix, and for interacting macroscopic systems, these are known only approximately (Militzer and Pollock 2000). This thereby introduces an uncontrollable systematic error. In addition, it has been shown analytically (Filinov 2014) (see Sect. 9.8) that RPIMC does not reproduce the exact density matrix in the limit of the ideal Fermi gas, whence the results become unreliable at increasing degeneracy. Analogous contradictions were obtained analytically many years ago in (Filinov 2001) from virial decomposition of the many fermion 'fixed-node' density matrix.

Approaches allowing the contribution of all permutations at once in the form of a determinant have occasionally been attempted by various groups (Imada 1984), but in general have not proven sufficiently precise or efficient for practical purposes. Further improvements have been made by combining determinants and modifying the measure of the path integral representation of the N-particle density operator in (Filinov et al. 2000–2004; Filinov et al. 2015). This allows for direct fermionic path integral Monte Carlo (DPIMC) simulations of dense plasmas over a broad range of temperatures and densities.

Some other original approaches are of interest here. The configuration path integral Monte Carlo (CPIMC) approach (Schoof et al. 2011, 2015) for degenerate correlated fermions with arbitrary pair interactions at finite temperatures is based on a representation of the N-particle density operator in a basis of (anti-)symmetrized N-particle states (configurations of occupation numbers). The main idea of this approach is to evaluate the path integral, not in configuration space, but in second quantization. This leads to path integrals in the occupation number representation, rather than paths in coordinate space, and allows one to treat arbitrary pair interactions in a continuous space. However, it turns out that the CPIMC method exhibits complementary behavior. This approach can be interpreted as a Monte Carlo simulation on a perturbation expansion around the ideal quantum system, which is why CPIMC works well for weak nonideality and strong degeneracy. Unfortunately, the physically most interesting region where both fermionic exchange and interactions are strong simultaneously remains out of reach.

Monte Carlo simulations at finite temperature over the entire fermion density range down to half the Fermi temperature have been carried out using the permutation blocking path integral Monte Carlo (PB-PIMC) approach (Dornheim et al. 2015, 2016). In order to simulate fermions in the canonical ensemble, a fourth-order approximation of the density matrix was combined, derived with a full antisymmetrization (by related determinants) on all time slices in discrete versions of the paths. It was demonstrated that the introduction of determinants effectively allows for the combination of the $N!$ configurations from standard PIMC into a single configuration weight, thereby reducing the complexity of the 'sign' problem. This has been used to obtain an accurate estimate of the ground state energy of degenerate, strongly nonideal electrons in a quantum dot (Chin 2015). Comparison of PB-PIMC results with the CPIMC method show excellent agreement, where results are available. On the contrary, results of CPIMC and PB-PIMC deviate at higher fermion degeneracy from analogous RPIMC results. Deviations are beyond the statistical errors and increase with density. In particular, it has been shown that RPIMC simulations for the uniform electron gas are limited to low densities ($r_s \approx 4$) and fail for higher densities. Treatment of interacting fermions in (Schoof et al. 2011, 2015; Dornheim et al. 2015, 2016; S. Groth 2016) have been carried out at high densities and results obtained for finite numbers of particles were extrapolated to the thermodynamic limit.

A completely different and very simple approach which can to a large extent overcome the 'sign problem' was suggested in the paper by Larkin et al. (2016, 2017), developing the new numerical version of the Wigner approach to quantum mechanics (Wigner 1932; Tatarskii 1983) to study the thermodynamic properties of degenerate systems of fermions (see Sect. 9.5). A new path integral representation of the quantum Wigner function in phase space was obtained for the canonical ensemble. An explicit analytical expression of the Wigner function was proposed to account for Fermi statistical effects using an effective pair pseudopotential. The derived pseudopotential depends on coordinates, momenta, and the degeneracy parameter of the fermions, and takes into account Pauli blocking of fermions in phase space. A new quantum Monte-Carlo method (WPIMC) has been proposed to calculate average values of arbitrary quantum operators in phase space. To test this approach, calculations of the momentum distribution function of the ideal system of Fermi particles have been carried out. According to WPIMC, the momentum distribution functions for degenerate ideal fermions are in good agreement with analytical Fermi distributions over a wide range of values of the momentum and degeneracy parameter. Comparing the resulting momentum distribution function of strongly coupled Coulomb systems of particles with Maxwell–Boltzmann and Fermi distributions shows the significant influence of the interparticle interaction.

Several approaches incorporating quantum effects into an effective quasi-classical Hamiltonian and corresponding Wigner dynamics to calculate quantum momentum distribution functions have been suggested in the last few decades (see Chap. 8 for details). These approaches operate with momentum-dependent interactions and take into account electron repulsion due to Pauli effects (Dorso et al. 1987; 1988; Ebeling and Schautz 1997; Ortner et al. 1997; Klimontovich 1995, 1999, 2001, 1999, 2001).

This chapter applies the DPIMC and WPIMC methods to mass-asymmetric electron–hole plasmas. In contrast to hydrogen (and similar) plasmas, here the heavy component (the holes) must also be treated quantum mechanically. A second extension is an improved treatment of exchange which allows consideration of higher densities and lower temperatures, as required to study the Mott effect and hole crystallization. Section 9.2 gives the theoretical background and contains a detailed derivation of the basic formulas used in the simulations. Most notably, expressions for thermodynamic quantities and the equation of state are presented. Moreover, the use of the effective quantum pair potential (Kelbg potential) in the simulations has been justified. Then an extensive numerical study of a two-component Coulomb systems is presented in Chap. 10. In particular the electron–hole configurations, the fraction of bound states, various (spin-dependent) pair distribution functions, charge structure factors, and also the energy and pressure are calculated for a wide range of coupling strengths, temperatures, and hole-to-electron mass ratio and densities.

9.2 Path Integral Representation of Thermodynamic Quantities

We consider a two-component plasma in equilibrium, with $N_e = N_h = N$ electrons and positive charges, e.g., holes, and described by the Hamiltonian $\hat{H} = \hat{K} + \hat{U}^c$, where $\hat{K}$ is the kinetic energy and $\hat{U}^c = \hat{U}^c_{hh} + \hat{U}^c_{ee} + \hat{U}^c_{eh}$ is the Coulomb interaction energy. In a canonical ensemble at temperature T and with volume V, the thermodynamic properties of this Coulomb system are fully described by the density operator $\hat{\rho} = e^{-\beta\hat{H}}/Z$ with the partition function

$$Z(N_e, N_h, V; \beta) = \frac{1}{N_e! N_h!} \sum_{\sigma} \int_V dq\, \rho(q, \sigma; \beta)\,, \tag{9.1}$$

where $\beta = 1/k_B T$, and $\rho(q, \sigma; \beta)$ denotes the diagonal matrix elements of the density operator at a given value σ of the total spin. Here $q = \{q_e, q_h\}$ and $\sigma = \{\sigma_e, \sigma_h\}$ are the spatial coordinates in units of thermal length and spin degrees of freedom of the electrons and holes, that is, $q_a = \{q_{1,a} \ldots q_{l,a} \ldots q_{N_a,a}\}$ and $\sigma_a = \{\sigma_{1,a} \ldots \sigma_{l,a} \ldots \sigma_{N_a,a}\}$, with $a = $ e, p.

Thermodynamic functions are defined by derivatives of the logarithm of the partition function with respect to the related thermodynamic variables. For example, for the pressure and internal energy we have

$$\beta p = \frac{\partial \ln Z}{\partial V} = \left[\frac{\alpha}{3V}\frac{\partial \ln Z}{\partial \alpha}\right]_{\alpha=1}, \tag{9.2}$$

$$\beta E = -\beta \frac{\partial \ln Z}{\partial \beta}, \tag{9.3}$$

where $\alpha = L/L_0$ is a length scaling parameter.

For low temperatures and high densities the exact density matrix of interacting quantum systems is not known, but it can be constructed using a path integral approach (Feynman and Hibbs 1965; Zamalin et al. 1977) based on the operator identity $\mathrm{e}^{-\beta\hat{H}} = \mathrm{e}^{-\Delta\beta\hat{H}}\mathrm{e}^{-\Delta\beta\hat{H}}\cdots\mathrm{e}^{-\Delta\beta\hat{H}}$, where $\Delta\beta = \beta/(n+1)$, which allows us to rewrite the integral in (9.1) in the form

$$\begin{aligned}
&\sum_\sigma \int \mathrm{d}q^{(0)} \rho(q^{(0)}, \sigma; \beta) \\
&= \int \mathrm{d}q^{(0)} \ldots \mathrm{d}q^{(n)} \rho^{(1)} \rho^{(2)} \cdots \rho^{(n)} \\
&\quad \times \sum_\sigma \sum_{P_\mathrm{e}} \sum_{P_\mathrm{h}} (\pm 1)^{\kappa_{P_\mathrm{e}}+\kappa_{P_\mathrm{h}}} \mathcal{S}(\sigma, \hat{P}_\mathrm{e}\hat{P}_\mathrm{h}\sigma'_a)\hat{P}_\mathrm{e}\hat{P}_\mathrm{h}\rho^{(n+1)}\big|_{q^{(n+1)}=q^{(0)},\sigma'=\sigma}\,.
\end{aligned} \tag{9.4}$$

Here the density matrix accounts for exchange effects through the permutation operators $\hat{P}_\mathrm{e}$ and $\hat{P}_\mathrm{h}$ acting on the electron and hole coordinates $q^{(n+1)}$ and spin projections σ'. The sum is over all permutations with parity κ_{P_e} and κ_{P_h}. In (9.4), the index $k = 1, \ldots, n+1$ labels the off-diagonal high-temperature density matrices

$$\rho^{(k)} \equiv \rho\big(q^{(k-1)}, q^{(k)}; \Delta\beta\big) = \big\langle q^{(k-1)}\big|\mathrm{e}^{-\Delta\beta\hat{H}}\big|q^{(k)}\big\rangle\,.$$

Accordingly, each particle is represented by a set of $n+1$ coordinates ('beads'). So the particle configuration is represented by a $3(N_\mathrm{e}+N_\mathrm{h})(n+1)$-dimensional vector $\tilde{q} \equiv \{q_{1,\mathrm{e}}^{(0)}, \ldots q_{1,\mathrm{e}}^{(n+1)}, q_{2,\mathrm{e}}^{(0)} \ldots q_{2,\mathrm{e}}^{(n+1)}, \ldots q_{N_\mathrm{e},\mathrm{e}}^{(n+1)}; q_{1,\mathrm{h}}^{(0)} \ldots q_{N_\mathrm{h},\mathrm{h}}^{(n+1)}\}$.

Representations of one (light) electron and one (heavy) hole are shown in Fig. 9.1 (Filinov et al. 2007). The circle around the electron beads symbolizes the main region that contributes to the partition function path integral. The size of this region is of the order of the thermal electron wavelength $\lambda_\mathrm{e}(T)$, while typical distances between electron beads are of the order of the electron wavelength taken at a temperature $(n+1)$ times higher. The same representation is valid for each hole, but due to the larger hole mass and smaller characteristic length scales, it is not shown.

To determine the energy in the path integral representation (9.4), each high temperature density matrix has to be differentiated in turn (Filinov 2001; Zamalin et al. 1977):

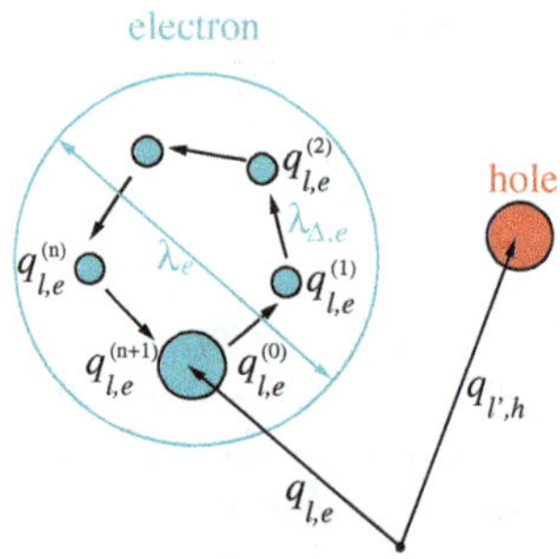

Fig. 9.1 Bead representation of electrons and holes. Here we have $\lambda_e^2 = 2\pi\hbar^2\beta/m_e$, $\lambda_{\Delta,e}^2 = 2\pi\hbar^2\Delta\beta/m_e$, $q_{l,e}^{(1)} = q_{l,e}^{(0)} + \lambda_{\Delta,e}\eta_{l,e}^{(1)}$, etc., and $\sigma = \sigma'$. The holes have a similar bead representation, but $\lambda_{\Delta,h}^2$ is m_h/m_e times smaller, so their bead distribution is not resolved in the figure (Filinov et al. 2007)

$$\beta E = -\frac{1}{Z}\int dq^{(0)}\dots dq^{(n)} \sum_{\sigma}\sum_{P_e}\sum_{P_h}(\pm 1)^{\kappa_{P_e}+\kappa_{P_h}} \mathcal{S}(\sigma, \hat{P}_e\hat{P}_h\sigma')$$
$$\times \sum_{k=1}^{n+1} \rho^{(1)}\cdots\rho^{(k-1)}\left(\beta\frac{\partial\rho^{(k)}}{\partial\beta}\right)\rho^{(k+1)}\cdots\rho^{(n)}\hat{P}_e\hat{P}_h\rho^{(n+1)}\Bigg|_{\substack{q^{(n+1)}=q^{(0)}\\ \sigma'=\sigma}}. \tag{9.5}$$

The matrix elements $\rho^{(k)}$ can be rewritten as

$$\begin{aligned}\rho^{(k)} &\equiv \langle q^{(k-1)}|e^{-\Delta\beta\hat{H}}|q^{(k)}\rangle \\ &= \int d\tilde{p}^{(k)}d\bar{p}^{(k)}\langle q^{(k-1)}|e^{-\Delta\beta\hat{U}^c}|\tilde{p}^{(k)}\rangle\langle\tilde{p}^{(k)}|e^{-\Delta\beta\hat{K}}|\bar{p}^{(k)}\rangle \\ &\qquad\times\langle\bar{p}^{(k)}|e^{-\Delta\beta^2[\hat{K},\hat{U}^c]/2}\dots|q^{(k)}\rangle,\end{aligned} \tag{9.6}$$

where $\tilde{p}^{(k)}(\bar{p}^{(k)})$ are conjugate variables to $q^{(k-1)}(q^{(k)})$. To evaluate the derivatives in (9.5), it is convenient to introduce dimensionless integration variables $\eta^{(k)} = (\eta_h^{(k)}, \eta_e^{(k)})$, where $\eta_a^{(k)} = \kappa_a(q_a^{(k)} - q_a^{(k-1)})$ for $k = 1, \dots, n$, and $\kappa_a^2 \equiv m_a/2\pi\hbar^2\Delta\beta = 1/\lambda_{\Delta,a}^2$. So differentiation of the density matrix now affects only the interaction terms:

$$\beta\frac{\partial\rho^{(k)}}{\partial\beta} = -\beta\frac{\partial[\Delta\beta\cdot U^c(X^{(k-1)})]}{\partial\beta}\rho^{(k)} + \beta\tilde{\rho}_\beta^{(k)}, \tag{9.7}$$

where

$$\tilde{\rho}_{\beta}^{(k)} = \int \mathrm{d}p^{(k)} \langle X^{(k-1)} | \mathrm{e}^{-\Delta\beta \hat{U}^{\mathrm{c}}} | p^{(k)} \rangle \exp\left[-\frac{\langle p^{(k)} | p^{(k)} \rangle}{4\pi(n+1)} \right] \times \left\langle p^{(k)} \left| \frac{\partial}{\partial\beta} \mathrm{e}^{-(\Delta\beta)^2 [\hat{K}, \hat{U}^{\mathrm{c}}]/2} \ldots \right| X^{(k)} \right\rangle , \quad (9.8)$$

with $p_a^{(k)} = \tilde{p}_a^{(k)} / \kappa_a \hbar$, $p(k) \equiv (p_{\mathrm{h}}^{(k)}, p_{\mathrm{e}}^{(k)})$, and use has been made of (9.6). Furthermore, $X^{(0)} \equiv (\kappa_{\mathrm{h}} q_{\mathrm{h}}^{(0)}, \kappa_{\mathrm{e}} q_{\mathrm{e}}^{(0)})$, $X^{(k)} \equiv (X_{\mathrm{h}}^{(k)}, X_{\mathrm{e}}^{(k)})$, where

$$X_a^{(k)} = \kappa_a q_a^{(0)} + \sum_{l=1}^{k} \eta_a^{(l)} ,$$

and k runs from 1 to n, with $X^{(n+1)} \equiv (\kappa_{\mathrm{h}} q_{\mathrm{h}}^{(n+1)}, \kappa_{\mathrm{e}} q_{\mathrm{e}}^{(n+1)}) = X^{(0)}$.

For $k = n + 1$, we have

$$\beta \frac{\partial}{\partial\beta} \hat{P}_{\mathrm{e}} \hat{P}_{\mathrm{h}} \rho^{(n+1)} = -\beta \frac{\partial \Delta\beta \cdot U^{\mathrm{c}}(X^{(n)})}{\partial\beta} \hat{P}_{\mathrm{e}} \hat{P}_{\mathrm{h}} \rho^{(n+1)} + \beta \hat{P}_{\mathrm{e}} \hat{P}_{\mathrm{h}} \tilde{\rho}_{\beta}^{(n+1)} . \quad (9.9)$$

Together with (9.5), we obtain for the energy

$$\beta E = \frac{3}{2}(N_{\mathrm{e}} + N_{\mathrm{h}}) - \frac{1}{Z} \int_V \mathrm{d}q^{(0)} \mathrm{d}\eta^{(1)} \ldots \mathrm{d}\eta^{(n)} \times \sum_{\sigma} \sum_{P_{\mathrm{e}}} \sum_{P_{\mathrm{h}}} (\pm 1)^{\kappa_{P_{\mathrm{e}}} + \kappa_{P_{\mathrm{h}}}} \mathcal{S}(\sigma, \hat{P}_{\mathrm{e}} \hat{P}_{\mathrm{h}} \sigma') \times \left\{ \sum_{k=1}^{n+1} \rho^{(1)} \cdots \rho^{(k-1)} \left[\beta \tilde{\rho}_{\beta}^{(k)} - \beta \frac{\partial \Delta\beta \cdot U^{\mathrm{c}}(X^{(k-1)})}{\partial\beta} \rho^{(k)} \right] \times \rho^{(k+1)} \cdots \rho^{(n)} \hat{P}_{\mathrm{e}} \hat{P}_{\mathrm{h}} \rho^{(n+1)} \right\} \Bigg|_{X^{(n+1)} = X^{(0)},\, \sigma' = \sigma} . \quad (9.10)$$

We must now find approximations for the high temperature density matrices $\rho^{(k)}$.

High Temperature Asymptotics for the Density Matrix

We now discuss approximations for the high temperature density matrix which can be used for PIMC simulations. It involves effective quantum pair potentials Φ_{ab}, which are approximated by the Kelbg potential (Kelbg 1964), as introduced and discussed in detail in Chap. 5. Details and further references can be found in (Filinov 2001).

Pair Approximation and Kelbg Potential

The N-particle high-temperature density matrix is expressed in terms of two-particle density matrices (higher order terms become negligible at sufficiently high temperature, i.e., for a large number n of time slices):

$$\rho_{ab}(q_{p,a}, q'_{p,a}, q_{t,b}, q'_{t,b}; \beta) = \frac{(m_a m_b)^{3/2}}{(2\pi\hbar\beta)^3} \exp\left[-\frac{m_a}{2\hbar^2\beta}(q_{p,a} - q'_{p,a})^2\right] \times \exp\left[-\frac{m_b}{2\hbar^2\beta}(q_{t,b} - q'_{t,b})^2\right] \exp(-\beta\Phi^{\mathrm{OD}}_{ab}) . \tag{9.11}$$

This results from factorization into kinetic and interaction parts, $\rho_{ab} \approx \rho_0^K \rho_{ab}^{U^c}$, which is exact in the classical case, i.e., at sufficiently high temperature. The error made at finite temperatures vanishes with the number of time slices as n^{-2} (Filinov 2001). The off-diagonal density matrix element (9.11) involves an effective pair interaction which is expressed approximately via its diagonal elements:

$$\Phi^{\mathrm{OD}}_{ab}(q_{p,a}, q'_{p,a}, q_{t,b}, q'_{t,b}; \beta) \approx \frac{1}{2}\left[\Phi_{ab}(q_{p,a} - q_{t,b}; \beta) + \Phi_{ab}(q'_{p,a} - q'_{t,b}; \beta)\right] ,$$

for which we use the Kelbg potential

$$\Phi_{ab}(x_{ab}; \beta) = \frac{e_a e_b}{\lambda_{ab} x_{ab}} \left\{1 - \mathrm{e}^{-x_{ab}^2} + \sqrt{\pi} x_{ab}\left[1 - \mathrm{erf}(x_{ab})\right]\right\} , \tag{9.12}$$

where

$$x_{ab} = |q_{p,a} - q_{t,b}|/\lambda_{ab} , \quad \lambda_{ab} = 2\pi\hbar^2\beta/m_{ab} , \quad 1/m_{ab} = 1/m_a + 1/m_b ,$$

and the error function is defined by

$$\mathrm{erf}(x) = \frac{2}{\sqrt{\pi}} \int_0^x \mathrm{d}t \, \mathrm{e}^{-t^2} .$$

Because quantum effects have been taken into account, the Kelbg potential is finite at zero distance. The validity of this approximation is restricted to temperatures substantially higher than the binding energy (Filinov 2004; Ebeling et al. 2006), and this puts a lower bound on the number of slices n. For a discussion of other effective potentials, we refer to (Klakow et al. 1994; Filinov et al. 2001, 2004, Ebeling et al. 2006).

We can conclude that, with the approximations (9.11), each of the high-temperature factors on the right-hand side of (9.4) carries an error of the order $1/(n+1)^2$, and we obtain the result

$$\rho^{(k)} = \mathrm{e}^{-\Delta\beta U(X^{(k-1)})} \rho_0^{(k)} + \mathcal{O}[(1/n+1)^2] , \tag{9.13}$$

where $\rho_0^{(k)}$ is the kinetic density matrix and U denotes the sum of all interaction energies, each consisting of the corresponding sum of pair interactions given by the Kelbg potentials:

$$U(X^{(k)}) = U_{\mathrm{hh}}(X_{\mathrm{h}}^{(k)}) + U_{\mathrm{ee}}(X_{\mathrm{e}}^{(k)}) + U_{\mathrm{eh}}(X_{\mathrm{h}}^{(k)}, X_{\mathrm{e}}^{(k)}) .$$

Estimator for the Total Energy

Let us discuss the final expressions following from (9.13). Note that in (9.10), special care has to be taken in performing the derivatives of the Coulomb potentials with respect to β. The products $\beta\partial\left[\Delta\beta \cdot U^{\mathrm{c}}(X^{(k-1)})\right]/\partial\beta$ have a singularity at zero interparticle distance which is integrable but leads to difficulties in the simulations. So we have to transform the e-e, h-h, and e-h contributions:

$$
\begin{aligned}
&\langle X^{(k-1)}|\mathrm{e}^{-\Delta\beta\hat{K}}|X^k\rangle\left[-\beta\frac{\partial}{\partial\beta}\left(\Delta\beta U^{\mathrm{c}}(X^{(k-1)})\right)\right] \\
&\approx \int_0^1 \mathrm{d}\alpha \int \mathrm{d}\tilde{X}^{(k-1)}\langle X^{(k-1)}|\mathrm{e}^{-\Delta\beta\alpha\hat{K}}|\tilde{X}^{(k-1)}\rangle\left[-\beta\frac{\partial}{\partial\beta}\left(\Delta\beta U^{\mathrm{c}}(\tilde{X}^{(k-1)})\right)\right] \\
&\quad\times\langle\tilde{X}^{(k-1)}|\mathrm{e}^{-\Delta\beta(1-\alpha)\hat{K}}|X^k\rangle + \mathcal{O}\left(1/(n+1)^2\right) \\
&\approx \langle X^{(k-1)}|\mathrm{e}^{-\Delta\beta\hat{K}}|X^k\rangle\left[-\beta\frac{\partial}{\partial\beta}\left(\Delta\beta U(X^{(k-1)})\right)\right] + \mathcal{O}\left(1/(n+1)^2\right) .
\end{aligned}
\tag{9.14}
$$

This means that, within the error $\mathcal{O}\left(n^{-2}\right)$, the sum of Coulomb potentials U^{c} can be replaced by the corresponding sum of Kelbg potentials U. This significantly improves the efficiency of the MC simulations. This result coincides with expressions which can be obtained if we first choose an approximation for the high-temperature density matrices $\rho^{(k)}$ using the Kelbg potential and then take the derivatives. Thus, the final result is

$$
\begin{aligned}
\beta E = &\frac{3}{2}(N_{\mathrm{e}} + N_{\mathrm{h}}) \\
&+ \frac{1}{Z}\sum_{s=0}^{N_{\mathrm{e}}}\sum_{k=0}^{N_{\mathrm{h}}}\int_V \mathrm{d}q^{(0)}\mathrm{d}\eta^{(1)}\ldots\mathrm{d}\eta^{(n)}\rho_{sk}\left(q^{(0)},\eta^{(1)},\ldots,\eta^{(n)},\beta\right) \\
&\times\left\{\sum_{l=0}^{n}\left[\sum_{p=1}^{N_{\mathrm{h}}}\sum_{t=1}^{N_{\mathrm{e}}}\Psi_{\mathrm{eh}}(|x_{pt}^l|) + \sum_{p<t}^{N_{\mathrm{h}}}\Psi_{\mathrm{ee}}(|r_{pt}^l|) + \sum_{p<t}^{N_{\mathrm{e}}}\Psi_{\mathrm{hh}}(|q_{pt}^l|)\right]\right. \\
&\quad + \sum_{l=1}^{n}\left[\sum_{p=1}^{N_{\mathrm{h}}}\sum_{t=1}^{N_{\mathrm{e}}}D(x_{pt}^l)\frac{\partial(\Delta\beta\Phi_{\mathrm{eh}})}{\partial|x_{pt}^l|} + \sum_{p<t}^{N_{\mathrm{e}}}C(r_{pt}^l)\frac{\partial(\Delta\beta\Phi_{\mathrm{ee}})}{\partial|r_{pt}^l|}\right. \\
&\qquad\qquad\left. + \sum_{p<t}^{N_{\mathrm{e}}}C(q_{pt}^l)\frac{\partial(\Delta\beta\Phi_{\mathrm{hh}})}{\partial|q_{pt}^l|}\right] \\
&\quad\left. - \frac{1}{\det\|\psi_{pt}^{n,0}\|_{sk}}\frac{\partial\det\|\psi_{pt}^{n,0}\|_{sk}}{\partial\beta}\right\},
\end{aligned}
\tag{9.15}
$$

with

$$C(r^l_{pt}) = \frac{\langle r^l_{pt}|y^l_{pt}\rangle}{2|r^l_{pt}|}, \quad C(q^l_{pt}) = \frac{\langle q^l_{pt}|\tilde{y}^l_{pt}\rangle}{2|q^l_{pt}|}, \quad D(x^l_{pt}) = \frac{\langle x^l_{pt}|y^l_p - \tilde{y}^l_t\rangle}{2|x^l_{pt}|},$$

and

$$\Psi_{ab}(x) \equiv \Delta\beta\partial[\beta'\Phi_{ab}(x,\beta')]/\partial\beta'|_{\beta'=\Delta\beta}.$$

Further, $\langle\ldots|\ldots\rangle$ denotes the scalar product, and q_{pt}, r_{pt}, and x_{pt} are differences between two coordinate vectors: $q_{pt} \equiv q_{p,\mathrm{h}} - q_{t,\mathrm{h}}$, $r_{pt} \equiv q_{p,\mathrm{e}} - q_{t,\mathrm{e}}$, $x_{pt} \equiv q_{p,\mathrm{e}} - q_{t,\mathrm{h}}$, $r^l_{pt} = r_{pt} + y^l_{pt}$, $q^l_{pt} = q_{pt} + \tilde{y}^l_{pt}$, $x^l_{pt} \equiv x_{pt} + y^l_p - \tilde{y}^l_t$, $y^l_{pt} \equiv y^l_p - y^l_t$, $\tilde{y}^l_{pt} \equiv \tilde{y}^l_p - \tilde{y}^l_t$, with $y^l_t = \Delta\lambda_\mathrm{e}\sum_{k=1}^{l}\eta^{(k)}_t$ and $\tilde{y}^l_p = \Delta\lambda_\mathrm{h}\sum_{k=1}^{l}\tilde{\eta}^{(k)}_p$.

The density matrices ρ_{sk} appearing in (9.15) are given by

$$\rho_{sk} = C^s_{N_\mathrm{e}} C^k_{N_\mathrm{h}} \mathrm{e}^{-\beta U} \prod_{l=1}^{n}\prod_{p=1}^{N_\mathrm{e}}\prod_{t=1}^{N_\mathrm{h}} \phi^l_p \tilde{\phi}^l_t \det\|\psi^{n,0}_{pt}\|_{sk}, \tag{9.16}$$

where

$$U = \frac{1}{n+1}\sum_{l=0}^{n}\Big[U_\mathrm{e}(X^{(l)}_\mathrm{e},\Delta\beta) + U_\mathrm{h}(X^{(l)}_\mathrm{h},\Delta\beta) + U_\mathrm{eh}(X^{(l)}_\mathrm{h}, X^{(l)}_\mathrm{e},\Delta\beta)\Big],$$

and $\phi^l_t \equiv \exp\big[-\pi|\eta^{(l)}_t|^2\big]$, $\tilde{\phi}^l_p \equiv \exp\big[-\pi|\tilde{\eta}^{(l)}_p|^2\big]$. Notice that the exchange problem is contained in the determinant related to a product of exchange matrices of electrons (index s) and holes (index k) where s (k) denotes the number of electrons and holes having the same spin projections [for more details, we refer to (Filinov 1975, 1976)]:

$$\begin{aligned}\|\psi^{n,0}_{pt}\|_{sk} = {} & \|\mathrm{e}^{-\frac{\pi}{\Delta\lambda^2_\mathrm{e}}|(r_a-r_b)+y^n_a|^2}\|_s \times \|\mathrm{e}^{-\frac{\pi}{\Delta\lambda^2_\mathrm{e}}|(r_a-r_b)+y^n_a|^2}\|_{N_\mathrm{e}-s} \\ & \times\|\mathrm{e}^{-\frac{\pi}{\Delta\lambda^2_\mathrm{h}}|(q_a-q_b)+\tilde{y}^n_a|^2}\|_k \times \|\mathrm{e}^{-\frac{\pi}{\Delta\lambda^2_\mathrm{h}}|(q_a-q_b)+\tilde{y}^n_a|^2}\|_{N_\mathrm{h}-k}.\end{aligned} \tag{9.17}$$

Estimator for the Equation of State

In a similar way, expressions can be derived for all other thermodynamic functions. Here, we only provide the corresponding result for the equation of state:

$$\begin{aligned}\frac{\beta pV}{N_\mathrm{e}+N_\mathrm{h}} = 1 - {} & \frac{1}{N_\mathrm{e}+N_\mathrm{h}}\frac{1}{3Z} \\ & \times\sum_{s=0}^{N_\mathrm{e}}\sum_{k=0}^{N_\mathrm{h}}\int \mathrm{d}q^{(0)}\mathrm{d}\eta^{(1)}\ldots\mathrm{d}\eta^{(n)}\rho_{sk}(q^{(0)},\eta^{(1)},\ldots,\eta^{(n)},\beta) \\ \times\Bigg\{ & \sum_{p=1}^{N_\mathrm{h}}\sum_{t=1}^{N_\mathrm{e}}|x_{pt}|\frac{\partial\Delta\beta\Phi_\mathrm{eh}}{\partial|x_{pt}|} + \sum_{p<t}^{N_\mathrm{h}}|q_{pt}|\frac{\partial\Delta\beta\Phi_\mathrm{hh}}{\partial|q_{pt}|} + \sum_{p<t}^{N_\mathrm{e}}|r_{pt}|\frac{\partial\Delta\beta\Phi_\mathrm{ee}}{\partial|r_{pt}|}\end{aligned}$$

$$+\sum_{l=1}^{n}\left[\sum_{p=1}^{N_\mathrm{h}}\sum_{t=1}^{N_\mathrm{e}} B(x_{pt}^l)\frac{\partial\Delta\beta\Phi_{\mathrm{eh}}}{\partial|x_{pt}^l|}+\sum_{p<t}^{N_\mathrm{e}} A(r_{pt}^l)\frac{\partial\Delta\beta\Phi_{\mathrm{ee}}}{\partial|r_{pt}^l|}+\sum_{p<t}^{N_\mathrm{h}} A(q_{pt}^l)\frac{\partial\Delta\beta\Phi_{\mathrm{hh}}}{\partial|q_{pt}^l|}\right]$$
$$-\frac{\alpha}{\det\|\psi_{pt}^{n,0}\|_{sk}}\frac{\partial\det\|\psi_{pt}^{n,0}\|_{sk}}{\partial\alpha}|_{\alpha=1}\Bigg\}\,, \tag{9.18}$$

with

$$B(x_{pt})=\frac{\langle x_{pt}^l|x_{pt}\rangle}{|x_{pt}^l|}\,,\quad A(r_{pt}^l)=\frac{\langle r_{pt}^l|r_{pt}\rangle}{|r_{pt}^l|}\,,\quad A(q_{pt}^l)=\frac{\langle q_{pt}^l|q_{pt}\rangle}{|q_{pt}^l|}\,.$$

In (9.15) and (9.18), the first terms on the right-hand side correspond to the classical ideal gas part. The ideal quantum contribution, in excess of the classical one, plus all correlation contributions are contained in the integral terms. The Coulomb correlation contributions arise from the terms with the Kelbg potentials Φ_{ab}, whereas the exchange contributions arise from the derivatives of the exchange matrix (last term).

The main advantage of the present expressions (9.15) and (9.18) for energy and pressure is that the explicit sum over permutations has been converted into the determinant which can be computed using standard linear algebra methods. Each of the sums in curly brackets in (9.15) and (9.18) is bounded, and the equations contain the important limit of a quantum plasma when the number of high-temperature factors increases ($n \to \infty$) (Filinov et al. 2001; Herman et al. 1982).

9.3 Monte Carlo Approach and the Metropolis–Hastings Algorithm

Now let us consider the general features of the numerical approach which can be used to calculate the above expressions for energy and pressure. As discussed in Chap. 1, the Monte Carlo method is often used to compute multidimensional definite integrals:

$$I=\frac{\int_\Omega P(\overline{\mathbf{x}})f(\overline{\mathbf{x}})\mathrm{d}\overline{\mathbf{x}}}{Q}\,, \tag{9.19}$$

where Ω is a subset of a multidimensional space R_m, and

$$Q=\int_\Omega P(\overline{\mathbf{x}})\mathrm{d}\overline{\mathbf{x}}\,. \tag{9.20}$$

In contrast to the deterministic approach which uses a regular grid to evaluate (Caflisch 1998), Monte Carlo integration evaluates the integrand at randomly chosen points (Weinzierl 2000). Each random set of points provides a different outcome and the final average outcome is an approximation of the correct value within appropriate error bars. This method is usually efficient for high-dimensional integrals (Press and Farrar 1998).

There are different ways to perform a Monte Carlo integration, such as uniform sampling, stratified sampling, and importance sampling. Importance sampling provides a widely used tool to perform Monte Carlo integration (Press and Farrar 1998). In importance sampling the random points $\overline{\mathbf{x}}$ can be drawn from any distribution $\overline{P}(\overline{\mathbf{x}})$. However, to decrease the variance of the average approximation of I, the distribution $\overline{P}(\overline{\mathbf{x}})$ has to be proportionate to $P(\overline{\mathbf{x}})$. This can make the sampling algorithm particularly useful, because it is often extremely difficult in practice to calculate the necessary normalization factor. From a given set of samples drawn from a distribution $P(\overline{\mathbf{x}})$, viz.,

$$\overline{\mathbf{x}}_1, \ldots, \overline{\mathbf{x}}_N \in \Omega, \tag{9.21}$$

the value of I can be approximated by the average outcome (Press and Farrar 1998)

$$I \approx I_N \equiv \frac{1}{N} \sum_{i=1}^{N} f(\overline{\mathbf{x}}_i) = \langle f \rangle \,. \tag{9.22}$$

This is because the law of large numbers ensures that

$$\lim_{N \to \infty} I_N = I \,. \tag{9.23}$$

This estimator is also valid for uniform sampling when $P(\overline{\mathbf{x}})$ is constant.

Given the estimate of I from I_N, the error bars of I_N can be estimated by the sample variance using the unbiased estimate of the variance:

$$\mathrm{Var}(f) \equiv \sigma_N^2 = \frac{1}{N-1} \sum_{i=1}^{N} \left[f(\overline{\mathbf{x}}_i) - \langle f \rangle \right]^2 , \tag{9.24}$$

which leads to

$$\mathrm{Var}(I_N) = \frac{1}{N^2} \sum_{i=1}^{N} \mathrm{Var}(f) = \frac{\mathrm{Var}(f)}{N} = \frac{\sigma_N^2}{N} \,. \tag{9.25}$$

As long as the sequence

$$\left\{ \sigma_1^2, \sigma_2^2, \sigma_3^2, \ldots \right\} \tag{9.26}$$

is bounded, this variance decreases asymptotically to zero as $1/N$. The estimate of the error of I_N is thus

$$\delta I_N \approx \sqrt{\mathrm{Var}(I_N)} = \frac{\sigma_N}{\sqrt{N}}\,, \tag{9.27}$$

which decreases as $1/\sqrt{N}$. This result does not depend on the number of dimensions of the integral, in contrast to most deterministic methods, which depend exponentially on the dimension (Lepage 1980). Note that random sampling may not uncover all the important features of the integrand and this may result in an underestimate of the error.

To finish this short review, let us note a useful generalization of (9.22), namely,

$$\langle f \rangle_w = \frac{\sum_{i=1}^{N} w(\overline{\mathbf{x}}_i) f(\overline{\mathbf{x}}_i)}{\sum_{i=1}^{N} w(\overline{\mathbf{x}}_i)}\,, \tag{9.28}$$

where $w(\overline{\mathbf{x}}_i)$ is an arbitrary weight function for the random quantity $\overline{\mathbf{x}}_i$ drawn from any distribution $P(\overline{\mathbf{x}})/\overline{Q}$, with $\overline{Q} = \int_\Omega P(\overline{\mathbf{x}})\mathrm{d}\overline{\mathbf{x}}$. This expression follows from

$$\langle f \rangle_w = \frac{\int_\Omega w(\overline{\mathbf{x}}) f(\overline{\mathbf{x}}) P(\overline{\mathbf{x}})\mathrm{d}\overline{\mathbf{x}}}{\overline{Q}} \frac{\overline{Q}}{Q}\,,$$

where $Q = \int_\Omega w(\overline{\mathbf{x}}) P(\overline{\mathbf{x}})\mathrm{d}\overline{\mathbf{x}}$ and

$$\frac{\int_\Omega w(\overline{\mathbf{x}}) f(\overline{\mathbf{x}}) P(\overline{\mathbf{x}})\mathrm{d}\overline{\mathbf{x}}}{\overline{Q}} \approx \frac{\sum_{i=1}^{N} w(\overline{\mathbf{x}}_i) f(\overline{\mathbf{x}}_i)}{N}\,,$$

with

$$Q/\overline{Q} \approx \frac{\sum_{i=1}^{N} w(\overline{\mathbf{x}}_i)}{N}\,.$$

If $w(\overline{\mathbf{x}}) \equiv 1$, then (9.28) coincides with (9.22).

Metropolis–Hastings Algorithm

The Metropolis–Hastings algorithm (Metropolis et al. 1953; Hastings 1970) can draw samples from any multi-dimensional probability distribution $P(x)$, provided one can compute the value of a function $\overline{P}(\overline{\mathbf{x}})$ which is proportional to the density of $P(x)$, as mentioned above. This sequence (a Markov chain) can be used to compute an integral such as an expected value by a Monte Carlo method (MC). The algorithm was first proposed in 1953 by Nicholas Metropolis, along with Arianna W. Rosenbluth, Marshall N. Rosenbluth, Augusta H. Teller, and Edward Teller, to calculate the equation of state in the canonical ensemble by computer. Further developments were made by W.K. Hastings, who extended it to the more general case in 1970 (Hastings 1970).

The Metropolis–Hastings algorithm generates a sequence of sample values with a distribution fitting the desired distribution $P(x)$. These random values are produced iteratively, with the distribution of the next sample depending only on the current sample value, making this sequence into a Markovian chain. Hence, at each Markovian step, the algorithm picks a candidate for the next sample value based on the current sample value. Then, with some probability, the candidate is accepted or rejected.

The algorithm uses a Markov process which asymptotically reaches a stationary distribution $\pi(x)$ coinciding with $P(x)$. A Markov process is uniquely defined by its transition probabilities $P(x \to x')$ between any two states x and x'. It has a unique stationary distribution π(x) when the following two conditions are met (Robert and Casella 2004):

1. **Existence of stationary distribution** π. A sufficient but not necessary condition is detailed balance, which requires that each transition $x \to x'$ is reversible: for every pair of states x, x', the probability of being in state x and transiting to state x' must be equal to the probability of being in state x' and transiting to state x :

$$\pi(x)P(x \to x') = \pi(x')P(x' \to x) \ . \tag{9.29}$$

2. **Uniqueness of stationary distribution** π. This is guaranteed by the ergodicity of the Markov process, which requires that every state must (i) be aperiodic, i.e., the system does not return to the same state at fixed intervals, and (ii) be positive recurrent, i.e., the expected number of steps for returning to the same state is finite.

The Metropolis–Hastings algorithm depends upon designing a Markov process (by constructing transition probabilities) which satisfies these two conditions, such that its stationary distribution $\pi(x)$ is equal to $P(x)$. The derivation of the algorithm starts with the condition of detailed balance:

$$P(x)P(x \to x') = P(x')P(x' \to x) \ , \tag{9.30}$$

which can be rewritten in the form

$$\frac{P(x \to x')}{P(x' \to x)} = \frac{P(x')}{P(x)} \ . \tag{9.31}$$

The idea is to separate the transition into two sub-steps: the proposal and acceptance/rejection. The proposal distribution $g(x \to x')$ is the conditional probability of proposing a state x' for given x, and the acceptance distribution $A(x \to x')$ the conditional probability of accepting the proposed state x'. The transition probability can be written as their product:

$$P(x \to x') = g(x \to x')A(x \to x') \ .$$

Inserting this relation in the previous equation, we have

$$\frac{A(x \to x')}{A(x' \to x)} = \frac{P(x')}{P(x)} \frac{g(x' \to x)}{g(x \to x')} . \tag{9.32}$$

Then one must choose an acceptance that satisfies detailed balance. One common choice is the Metropolis choice:

$$A(x \to x') = \min\left(1, \frac{P(x')}{P(x)} \frac{g(x' \to x)}{g(x \to x')}\right) . \tag{9.33}$$

This means that we always accept when the acceptance is bigger than 1 and we can accept or reject when the acceptance is smaller than 1.

The Metropolis–Hastings algorithm can formulated as follows:

1. To initialize, pick an initial state x at random.
2. Randomly pick a state x' according to $g(x \to x')$.
3. Accept the state according to the probability $A(x \to x')$. If it is not accepted, this means that $x' = x$, so there is no need to update anything. Otherwise, the system transits to x'.
4. Go to 2 until many M states are generated.
5. Save the state x, go to 2.

This algorithm means that the saved states are drawn from the distribution $P(x)$, while step 4 ensures that they are decorrelated. The value of M has to be at least of the order of the autocorrelation time of the Markov process (Newman and Barkema 1999).

It is important to note that, in a general problem, it is not clear which distribution $g(x \to x')$ one should use. It is a free parameter of the method which has to be adjusted to the particular problem at hand. This is usually done by calculating the acceptance rate, which is the fraction of proposed samples that is accepted during the last N samples. As has been shown theoretically, the ideal acceptance rate has to be in an interval of 23–50% (Roberts et al. 1997). The Markov chain starts from an arbitrary initial value x_0 and the algorithm is run for many iterations until this initial state is 'forgotten'. These samples, which are discarded, are known as 'burn-ins'. The remaining set of accepted values of x represents a sample from the distribution $P(x)$.

9.4 Path Integral Monte Carlo Procedure

Expressions for the energy (9.15) and pressure (9.18) are well suited for efficient numerical evaluation using Monte Carlo techniques if we take the states of the Markov chain $\overline{\mathbf{x}}$ to be $\{q^{(0)}, \eta^{(1)} \dots \eta^{(n)}; s, k, \}$ and the modulus of the density matrix $|\rho_{sk}(q^{(0)}, \eta^{(1)} \dots \eta^{(n)}, \beta)|$ to be the distribution $P(\overline{\mathbf{x}})$, with $\mathrm{sign}(\rho_{sk})$ the weight function $w(\rho_{sk}(q^{(0)}, \eta^{(1)} \dots \eta^{(n)}, \beta))$ (since the determinant in ρ_{sk} can take both positive or negative values) and make use of the expression (9.28) [see, e.g., Zamalin et al.

1977; Binder and Cicotti 1996; Bonitz and Semkat 2006]. Hence, in DPIMC calculations, the sign problem is reduced from the level of sign interference of the permutations to the level of sign interference of determinants.

To calculate the determinants and their ratios in the transition probability from one Markovian state to the next, the direct methods of linear algebra and the techniques of sparse matrices can be used. Thus, each Markovian step is equivalent to the $N!$ Markovian steps in the sum over permutations. This method has been tested for the example of an ideal Fermi gas (Zamalin et al. 1977). It was found that the method leads to agreement with the exact solution up to values of the degeneracy parameter $N_e\lambda_e^3/V \sim 10$–15, if the particle wavelength is smaller than the size of the Monte Carlo box. As shown in the calculations in (Zamalin et al. 1977; Filinov et al. 2001, 2007), this method works well enough for hydrogen and electron–hole plasmas.

In the Monte Carlo algorithm, to get the new state $\overline{\mathbf{x}}'$ of the Markov chain from the given state $\overline{\mathbf{x}}$, one can change s or k, or all coordinates $q_{t,\mathrm{e}}$ of a randomly chosen electron, or all coordinates $q_{p,\mathrm{h}}$ of a randomly chosen hole, or only individual electron $\eta_t^{(k)}$ or hole $\tilde{\eta}_p^{(k)}$ beads, until the calculated values converge.

The main obstacle for good convergence of calculated results is the very complicated structure of the configuration space (region of integration), divided by the determinant on isolated regions of positive and negative contributions to integrals. Interference of these contributions has to give a positive result, for example, for positive pair correlation functions. Although the convergence of calculated results is very slow due to this obstacle, it can be significantly improved by using the basic idea of Hasting's algorithm. One of the possible choices for the conditional probability $g(\overline{\mathbf{x}} \to \overline{\mathbf{x}}')$ of proposing a state $\overline{\mathbf{x}}'$ for given $\overline{\mathbf{x}}$ can be expressed in the form

$$g(\overline{\mathbf{x}} \to \overline{\mathbf{x}}') = \frac{\exp\left[\epsilon\varpi(\overline{\mathbf{x}}, \overline{\mathbf{x}}')\right]}{1 + \exp\left[2\epsilon\,\mathrm{sign}(\rho_{sk}(\overline{\mathbf{x}}'))\right]}, \tag{9.34}$$

where

$$\varpi(\overline{\mathbf{x}}, \overline{\mathbf{x}}') = \mathrm{sign}(\rho_{sk}(\overline{\mathbf{x}}'))\left[1 - \mathrm{sign}(\rho_{sk}(\overline{\mathbf{x}}'))\,\mathrm{sign}(\rho_{sk}(\overline{\mathbf{x}}))\right],$$

and $\epsilon > 0$ is a small free parameter ensuring good convergence. For all $\overline{\mathbf{x}}$ and $\overline{\mathbf{x}}'$, the function $\varpi(\overline{\mathbf{x}}, \overline{\mathbf{x}}')$ can take only two values 0 and $2\epsilon\,\mathrm{sign}(\rho_{sk}(\overline{\mathbf{x}}'))$, so the denominator is simply a normalization factor for the conditional probability.

Computer power usually allows one to consider up to several hundred electrons and holes with several tens of beads located in a small volume called the basic MC cell. Due to this limitation on the number of particles in the MC cell, we have to restrict the size of the basic MC cell for a given density. To reduce the influence of the finite small value of the volume and the finite number of particles, periodic boundary conditions (PBC) have to be put on this basic MC cell. The idea of the PBC is illustrated in Fig. 9.2 (Filinov et al. 2007). The yellow square (basic MC cell) in Fig. 9.2 has to be repeated periodically in the X, Y, and Z directions.

As mentioned before, the main contribution to the path integral representation of the partition function comes from configurations for which the typical size of

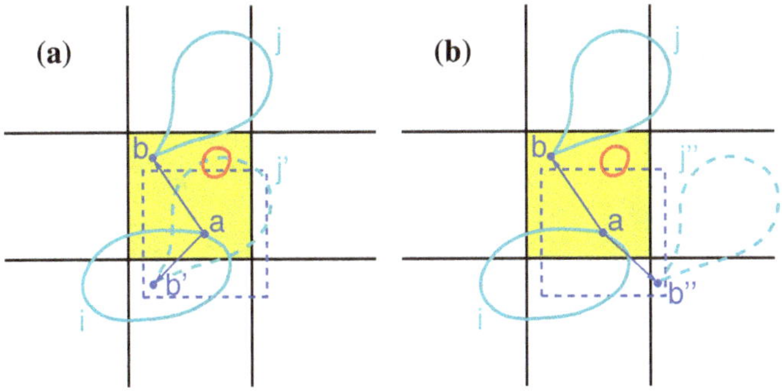

Fig. 9.2 Sketch of the boundary conditions used for the simulations. Electron and hole bead clouds are denoted by *solid green* and *red curves*, respectively. The *dashed line* shows the cell used to choose the image of particle b which is closest to particle a. For further explanation, see text (Filinov et al. 2007)

the clouds of electron (hole) beads is of the order of the thermal wavelength of the electrons (holes). In the case of a degenerate plasma the thermal wavelength and the typical size of the electron bead clouds may be of the same order or larger than the size of the basic MC cell. So beads of electrons belonging to the basic MC cell can penetrate into neighboring images of the main cell, just as electron beads from a neighboring cell can extend into the basic MC cell. In calculations, it can be assumed that an electron (hole) belongs to a certain cell if its physical coordinate $q^0_{t,a}$ $(t = 1, \ldots, N_a, a = \mathrm{e}, \mathrm{h})$ belongs to this cell. In Fig. 9.2, electron and hole bead clouds are indicated by solid green and red curves. Only a few periodic images of these particles are shown by dashed lines.

Let us consider the PBC for calculating the pressure and energy. For distances between beads with the same number k in the Kelbg potential and its derivatives [the first and second terms in curly brackets in (9.15) and (9.18)], we use the standard PBC shown symbolically in Fig. 9.2a. Hence, in the Kelbg potential and its derivatives, instead of the distance $\overline{ab}$ between beads with the same number k of electrons i and j, we take the smallest distance $\overline{ab'}$ to one of the electron images j'. The same applies to electron–hole and hole–hole distances.

Further, in calculations of the scalar products and derivatives of the Kelbg potential [terms C and D in (9.15) and terms A and B in (9.18)], the situation is more complicated due to the dependence of the scalar products on the angle between vectors to beads of the particle from the basic MC cell and its periodic images. In calculations, we first of all choose for a given particle i the nearest image j'', according to the distance between coordinates $q^0_{i,\mathrm{e}}$ and $q^0_{j'',\mathrm{e}}$ alone, as shown in Fig. 9.2b ($\overline{ab''}$ is the smallest distance for all ab). Then for this pair i, j'', we calculate all scalar product terms A, B, C, D and the related derivatives of the Kelbg potential. The same is done for electron–hole and hole–hole pairs.

For low degeneracy, particles belonging only to the basic Monte Carlo cell can be taken into account in the determinants of the exchange matrices. However, with increasing degeneracy $n\lambda^3$, the ratio of the particle thermal wavelength to the size of the Monte Carlo cell increases. When this ratio approaches unity, the exchange interaction between particles in the main MC cell and their images in the neighboring cells have to be included. Therefore, it is necessary to take into account the exchange interactions of electrons and holes from neighboring Monte Carlo cells, i.e., first from the $3^3 - 1$ nearest-neighbor cells, then from the $5^3 - 1$ next nearest-neighbor cells, and so on (Filinov 2004, 2007).

For simulations of dense electron–hole plasmas, both the particle number and the number of beads must be varied. It has been found that, for the densities and temperatures considered below, it is enough to simulate systems with particle numbers $N_e = N_h = 50, \ldots, 100$. Of course, the accuracy is significantly affected by the number of beads n. To preclude this dependence, the density matrices in the high-temperature decomposition have to be taken at temperatures above the Coulomb binding energy. In practice, a number of about $n = 20$ beads turn out to be sufficient. Furthermore, to simplify the computations, one can include only the dominant contribution in the sums over s and k (i.e., over the total electron and hole spin, respectively), which corresponds to $s = N_e/2$ electrons and $k = N_h/2$ holes having spin up and down, respectively. The contribution of the other terms is small and vanishes in the thermodynamic limit. We should emphasize that, for all the results presented below, the maximum statistical error is about 5%, which is sufficient for the physical analysis. This accuracy can be achieved at an acceptable cost of computer time. Of course, this error can be systematically reduced by increasing the length of the Monte Carlo run (Filinov 2004, 2007, Ebeling et al. 2006).

9.5 Path Integral Representation of the Wigner Function in Phase Space for the Canonical Ensemble

Here we discuss the new path integral representation of the quantum Wigner function in the phase space for the canonical ensemble (Larkin et al. 2016, 2017). An explicit analytical expression of the Wigner function has been proposed, accounting for Fermi statistical effects with an effective pair pseudopotential. The derived pseudopotential depends on coordinates, momenta, and the degeneracy parameter of fermions and takes into account Pauli blocking of fermions in phase space. A new quantum Monte Carlo method is put forward to calculate average values of arbitrary quantum operators in phase space. When calculated using the WPIMC method, the momentum distributions for degenerate ideal fermions are in good agreement with the analytical Fermi distribution over a wide range of values of the momentum and the degeneracy parameter. Comparing the resulting momentum distribution function for strongly coupled Coulomb systems of particles with the Maxwell–Boltzmann and Fermi distributions reveals the significant influence of interparticle interactions on the high energy asymptotics, which result in the appearance of quantum 'tails'.

Path Integrals and Antisymmerization

For one particle in the 1D case, the average value of an arbitrary quantum operator $\hat{A}$ can be written as its Weyl symbol $A(p, x)$, averaged over phase space with the Wigner function $W(p, x; \beta)$:

$$\langle \hat{A} \rangle = \int \frac{\mathrm{d}p\,\mathrm{d}x}{2\pi\hbar} A(p, x) W(p, x; \beta) , \tag{9.35}$$

where the Weyl symbol of the operator $\hat{A}$ is

$$A(p, x) = \int \mathrm{d}\xi\, \mathrm{e}^{-\mathrm{i}\langle \xi | p \rangle / \hbar} \langle x - \xi/2 | \hat{A} | x + \xi/2 \rangle . \tag{9.36}$$

Weyl symbols for standard operators like $\hat{p}$, $\hat{x}$, $\hat{p}^2$, $\hat{x}^2$, $\hat{H}$, $\hat{H}^2$, etc., are easily calculated directly from the definition (9.36). The Wigner function of a many-particle system in the canonical ensemble is defined as a Fourier transform of the off-diagonal matrix element of the density matrix operator in the coordinate representation:

$$W(p, x; \beta) = Z(\beta)^{-1} \int \mathrm{d}\xi\, \mathrm{e}^{\mathrm{i}\langle p | \xi \rangle / \hbar} \langle x - \xi/2 | \mathrm{e}^{-\beta \hat{H}} | x + \xi/2 \rangle . \tag{9.37}$$

Here $\rho = \exp(-\beta \hat{H})$ is the density operator of a quantum system of particles with Hamiltonian $\hat{H} = \hat{K} + \hat{U}$ equal to the sum of kinetic $\hat{K}$ and potential energy $\hat{U}$ operators, while $\beta = 1/k_{\mathrm{B}} T$.

The aim here is to obtain a new representation of the Wigner functions in the path integral form, allowing numerical simulations of strongly coupled quantum systems of particles in the canonical ensemble. Since the kinetic and potential energy operators in the Hamiltonian do not commute, there is in general no exact explicit analytical expression for the Wigner function. To overcome this difficulty, let us represent the Wigner function in a similar way to the path integral representation of the partition function discussed above.

As an example of Coulomb system of particles, we consider a 3D two-component asymmetric electron–hole plasma consisting of $N_{\mathrm{e}} = N_{\mathrm{h}} = N$ electrons and heavy holes in equilibrium (Filinov et al. 2007). The Hamiltonian of the system $\hat{H} = \hat{K} + \hat{U}^{\mathrm{c}}$ contains kinetic energy $\hat{K}$ and Coulomb interaction energy $\hat{U}^{\mathrm{c}} = \hat{U}^{\mathrm{c}}_{\mathrm{hh}} + \hat{U}^{\mathrm{c}}_{\mathrm{ee}} + \hat{U}^{\mathrm{c}}_{\mathrm{eh}}$ contributions. The thermodynamic properties in the canonical ensemble with given temperature T and fixed volume V are fully described by the diagonal elements of the density operator $\hat{\rho} = \mathrm{e}^{-\beta \hat{H}}/Z$ normalized by the partition function Z:

$$Z(N_{\mathrm{e}}, N_{\mathrm{h}}, V; \beta) = \frac{1}{N_{\mathrm{e}}! N_{\mathrm{h}}!} \sum_{\sigma} \int_V \mathrm{d}x\, \rho(x, \sigma; \beta) , \tag{9.38}$$

where $\rho(x, \sigma; \beta)$ denotes the diagonal matrix elements of the density operator $\hat{\rho}$. In (9.38), $x = \{x_{\mathrm{e}}, x_{\mathrm{h}}\}$ ($x_a = \{x_{1,a} \ldots x_{l,a} \ldots x_{N_a,a}\}$ with $a, b = \mathrm{e}, \mathrm{h}$ are the dimensionless coordinates in units of the thermal wavelength with related temperature $\lambda_a = \sqrt{2\pi\hbar^2\beta/m_a}$ and spin degrees of freedom $\sigma = \{\sigma_e, \sigma_h\}$ of the electrons and holes $\sigma_a = \{\sigma_{1,a} \dots \sigma_{t,a} \dots \sigma_{N_a,a}\}$ and $l, t = 1, \dots, N_a$.

Of course, the exact matrix elements of the density matrix of interacting quantum systems are not known (particularly for low temperatures and high densities), but they can be constructed using a path integral approach based on the operator identity $e^{-\beta\hat{H}} = e^{-\epsilon\hat{H}}e^{-\epsilon\hat{H}} \cdots e^{-\epsilon\hat{H}}$, where $\epsilon = \beta/M$, and this allows us to rewrite the integral in (9.38) in the form

$$\sum_\sigma \int dx^{(0)} \rho(x^{(0)}, \sigma; \beta) \tag{9.39}$$

$$= \int dx^{(0)} \dots dx^{(m)} \dots dx^{(M-1)} \rho^{(1)} \rho^{(2)} \cdots \rho^{(M-1)}$$

$$\times \sum_\sigma \sum_{P_e} \sum_{P_h} (\pm 1)^{\kappa_{P_e} + \kappa_{P_h}} \mathcal{S}(\sigma, \hat{P}_e \hat{P}_h \sigma'_a) \hat{P}_e \hat{P}_h \rho^{(M)} \Big|_{x^{(M)} = x^{(0)}, \sigma' = \sigma} .$$

The spin gives rise to the spin part of the density matrix $\mathcal{S}$, with exchange effects accounted for by the permutation operators $\hat{P}_e$ and $\hat{P}_h$ acting on the electron and hole coordinates $x^{(M)}$ and spin projections σ'. The sum is over all permutations with parity κ_{P_e} and κ_{P_h}. In (9.39), the index $m = 0, \dots, M-1$ labels the off-diagonal high-temperature density matrices $\rho^{(m)} \equiv \rho\left(x^{(m)}, x^{(m+1)}; \epsilon\right) = \langle x^{(m)} | e^{-\epsilon\hat{H}} | x^{(m+1)} \rangle$. With the error of order $1/M^2$ due to neglecting the commutator $\epsilon^2 [K, U^c]/2$, each high temperature factor can be represented in the form

$$\langle x^{(m)} | e^{-\epsilon\hat{H}} | x^{(m+1)} \rangle \approx \langle x^{(m)} | e^{-\epsilon\hat{U}^c} | x^{(m+1)} \rangle \rho_0^{(m)} ,$$

where $\rho_0^{(m)} = \langle x^{(m)} | e^{-\epsilon\hat{K}} | x^{(m+1)} \rangle$. In the limit $M \to \infty$, the error for the whole product ($\propto 1/M$) of high temperature factors is equal to zero, and we have an exact path integral representation of the partition function in which each particle is represented by a trajectory consisting of a set of M coordinates ('beads'). Hence, the whole particle configuration is represented by a $3(N_e + N_h)M$-dimensional vector

$$\tilde{x} \equiv \left\{ x_{1,e}^{(0)}, \dots, x_{1,e}^{(M-1)}, x_{2,e}^{(0)} \dots, x_{2,e}^{(M-1)}, \dots, x_{N_e,e}^{(M-1)}; x_{1,h}^{(0)} \dots, x_{N_h,h}^{(M-1)} \right\} .$$

Pair Approximation in High-Temperature Asymptotics for the Density Matrix

We now consider approximations for the high-temperature density matrix which can be used for efficient PIMC simulations. This involves the pseudopotential Φ_{ab} of the interparticle interaction and the effective pair quantum potentials v_{lt}^a allowing for Fermi statistical effects.

Kelbg Potential

Here we follow Sect. 9.2, where more details can be found. If we neglect terms of order $1/M^2$, the N-particle high-temperature density matrix $\rho^{(m)} = \langle x^{(m)}|\mathrm{e}^{-\epsilon\hat{H}}|x^{(m+1)}\rangle$ discussed above can be expressed in terms of two-particle density matrices (higher order terms become negligible at sufficiently high temperatures). This results from factorization into kinetic and interaction parts, $\rho_{ab} \approx \rho_0^K \rho_{ab}^U$. The off-diagonal density matrix element (9.11) involves an effective pair interaction which is expressed approximately via its diagonal elements:

$$\Phi_{ab}^{\mathrm{OD}}(x_{l,a}, x'_{l,a}, x_{t,b}, x'_{t,b}; \beta) \approx \frac{1}{2}\Big[\Phi_{ab}(x_{l,a} - x_{t,b}; \beta) + \Phi_{ab}(x'_{l,a} - x'_{t,b}; \beta)\Big],$$

for which we use the Kelbg potential Φ_{ab} (see Sect. 9.2). With these approximations, we obtain the result

$$\rho^{(m)} = \mathrm{e}^{-\epsilon U(x^{(m)}, x^{(m+1)})}\rho_0^{(m)} + \mathcal{O}[(1/M)^2], \tag{9.40}$$

where U denotes the sum of all interaction energies, each consisting of the corresponding sum of pair interactions given by Kelbg potentials:

$$\begin{aligned} U(x^{(m)}, x^{(m+1)}) = \frac{1}{2}\Big[& U_{\mathrm{hh}}(x_{\mathrm{h}}^{(m)}) + U_{\mathrm{ee}}(x_{\mathrm{e}}^{(m)}) + U_{\mathrm{eh}}(x_{\mathrm{h}}^{(m)}, x_{\mathrm{e}}^{(m)}) \\ & + U_{\mathrm{hh}}(x_{\mathrm{h}}^{(m+1)}) + U_{\mathrm{ee}}(x_{\mathrm{e}}^{(m+1)}) + U_{\mathrm{eh}}(x_{\mathrm{h}}^{(m+1)}, x_{\mathrm{e}}^{(m+1)})\Big]. \end{aligned}$$

The product of high temperature matrix elements $\rho^{(m)}$ in (9.39) gives an error proportional to $1/M$, and in the limit of large M, it represents the exact path integral representation of the partition function.

Exchange Effects in the Pair Approximation

To explain the basic ideas of our approach it is enough to consider the system of ideal electrons and holes, so here $\hat{U} \equiv 0$. The Hamiltonian of the system ($\hat{H} = \hat{K} = \hat{K}_{\mathrm{e}} + \hat{K}_{\mathrm{h}}$) consists of kinetic energy of electrons $\hat{K}_{\mathrm{e}}$ and holes $\hat{K}_{\mathrm{h}}$. As these operators commute, the representation of the density matrix (9.39) is exact for any finite number M. For our purpose, it is enough to consider the sum over permutations in the pair approximation at $M = 1$:

$$\begin{aligned} & \sum_{\sigma}\sum_{P_{\mathrm{e}}}\sum_{P_{\mathrm{h}}}(\pm 1)^{\kappa_{P_{\mathrm{e}}}+\kappa_{P_{\mathrm{h}}}} \mathcal{S}(\sigma, \hat{P}_{\mathrm{e}}\hat{P}_{\mathrm{h}}\sigma'_a)\, \hat{P}_{\mathrm{e}}\hat{P}_{\mathrm{h}}\rho\big|_{x^1=x^{(0)},\sigma'=\sigma} \\ & = \sum_{\sigma_{\mathrm{e}}}\sum_{P_{\mathrm{e}}}(\pm 1)^{\kappa_{P_{\mathrm{e}}}} \mathcal{S}(\sigma_{\mathrm{e}}, \hat{P}_{\mathrm{e}}\sigma'_{\mathrm{e}})\rho_{\mathrm{e}}\big|_{x_{\mathrm{e}}^1=x_{\mathrm{e}}^{(0)},\sigma'_{\mathrm{e}}=\sigma_{\mathrm{e}}} \\ & \qquad \times \sum_{\sigma_{\mathrm{h}}}\sum_{P_{\mathrm{h}}}(\pm 1)^{\kappa_{P_{\mathrm{h}}}} \mathcal{S}(\sigma_{\mathrm{h}}, \hat{P}_{\mathrm{h}}\sigma'_{\mathrm{h}})\rho_{\mathrm{h}}\big|_{x_{\mathrm{h}}^1=x_{\mathrm{h}}^{(0)},\sigma'_{\mathrm{h}}=\sigma_{\mathrm{h}}} \end{aligned}$$

$$
\begin{aligned}
&= \sum_{\sigma_e} \Big(1 - \sum_{l<t} f_{e,lt}^2 + \sum_{l,t,c} f_{e,lt} f_{e,lc} f_{e,tc} - \cdots \Big) \\
&\qquad \times \sum_{\sigma_h} \Big(1 - \sum_{l<t} f_{h,lt}^2 + \sum_{l,t,c} f_{h,lt} f_{h,lc} f_{h,tc} - \cdots \Big) \\
&\approx \sum_{\sigma_e} \prod_{l<t} (1 - f_{e,lt}^2) \sum_{\sigma_h} \prod_{l<t} (1 - f_{h,lt}^2) \\
&= \sum_{\sigma} \exp\Big(-\beta \sum_{l<t} \tilde{v}_{lt}^{e} \Big) \exp\Big(-\beta \sum_{l<t} \tilde{v}_{lt}^{h} \Big) , \qquad (9.41)
\end{aligned}
$$

where

$$
f_{a;lt} = \exp\left(-\frac{\pi |x_{l,a}^{(0)} - x_{t,a}^{(0)}|^2}{\lambda_a^2} \right)
$$

and

$$
\begin{aligned}
\tilde{v}_{lt}^{a} &= -kT \ln(1 - f_{a;lt} f_{a;tl}) \\
&= -kT \ln\left[1 - \delta_{\sigma_{l,a}\sigma_{t,a}} \exp\left(-\frac{2\pi |x_{l,a}^{(0)} - x_{t,a}^{(0)}|^2}{\lambda_a^2} \right) \right] \qquad (9.42)
\end{aligned}
$$

is the exchange potential (Huang 1963). This formula shows that the first corrections accounting for the antisymmetrization of the density matrix endow the particles with the pair exchange potential $\tilde{v}_{lt}^{a}$. Below, in order to take into account exchange effects in Wigner functions, we use an analogous pair potential depending on the phase space variables.

Antisymmetrized Wigner Function

The antisymmetrized Wigner function can be written in the form

$$
\begin{aligned}
W(p, x; \beta) &= \frac{1}{Z(\beta) N_e! N_h!} \sum_{\sigma} \sum_{P_e} \sum_{P_h} (\pm 1)^{\kappa_{P_e} + \kappa_{P_h}} \mathcal{S}(\sigma, \hat{P}_e \hat{P}_h \sigma')\big|_{\sigma'=\sigma} \\
&\times \int d\xi e^{i\langle \xi | p \rangle / \hbar^{6N}} \Big\langle x - \xi/2 \Big| \prod_{m=0}^{M-1} e^{-\epsilon \hat{U}_m^c} e^{-\epsilon \hat{K}_m} \Big| \hat{P}_e \hat{P}_h (x + \xi/2) \Big\rangle . \qquad (9.43)
\end{aligned}
$$

Now replacing the variables of integration $x^{(m)}$ for any given permutation $P_e P_h$:

$$
\begin{aligned}
x_e^{(m)} &= (P_e x_e - x_e) \frac{m}{M} + x_e + q_e^{(m)} - \frac{(M - m)\xi_e}{2M} + \frac{m P_e \xi_e}{2M} , \\
x_h^{(m)} &= (P_h x_h - x_h) \frac{m}{M} + x_h + q_h^{(m)} - \frac{(M - m)\xi_h}{2M} + \frac{m P_h \xi_h}{2M} , \qquad (9.44)
\end{aligned}
$$

we obtain

$$W(p,x;\beta) = \frac{C(M)}{Z(\beta)N_e!N_h!}\sum_{\sigma}\sum_{P_e}\sum_{P_h}(\pm 1)^{\kappa_{P_e}+\kappa_{P_h}}\mathcal{S}(\sigma,\hat{P}_e\hat{P}_h\sigma')\big|_{\sigma'=\sigma}$$
$$\times\int d\xi\int dq^{(1)}\ldots dq^{(M-1)}\exp\left\{-\pi\frac{\langle\xi|P_eP_h+E|\xi\rangle}{2M}+\mathrm{i}\langle\xi|p\rangle-\pi\frac{|P_eP_hx-x|^2}{M}\right.$$
$$-\sum_{m=0}^{M-1}\left[\pi|q^{(m)}-q^{(m+1)}|^2+\epsilon U\left((P_eP_hx-x)\frac{m}{M}+x+q^{(m)}\right.\right.$$
$$\left.\left.\left.-\frac{(M-m)\xi}{2M}+\frac{mP_eP_h\xi}{2M}\right)\right]\right\}, \tag{9.45}$$

where E is unit matrix, and the permutation matrix P_eP_h is equal to the unit matrix with appropriately transposed columns. Here and in the following, we shall take the momenta and coordinates to be dimensionless variables like $p_{l,a}\tilde{\lambda}_a/\hbar$ and $x_{l,a}/\tilde{\lambda}_a$, where $\tilde{\lambda}_a = \sqrt{2\pi\hbar\beta/m_aM}$. It will be shown that the constant $C(M)$ cancels out when we calculate the average values of operators.

When $\epsilon \to 0$, this multiple integral tends to an exact representation of the Wigner function $W(p,x;\beta)$ in the form of a path integral with a continuous dimensionless 'time' τ, which corresponds to m/M in the discrete case. In addition, the set of independent variables $q^{(m)}$ tends to a closed trajectory $q(\tau)$. This trajectory starts and ends at the point 0 when $\tau = 0$ and 1. Note that the integration here is related to the integration over the Wiener measure of all closed trajectories $q(\tau)$. In fact, a particle is represented by a trajectory with characteristic size of order $\lambda_a = \sqrt{2\pi\hbar\beta/m_a}$ in coordinate space. This is a manifestation of the uncertainty principle. The variable ξ is the integrand in the Fourier transform.

Harmonic and Linear Approximations for the Wigner Function

The expression (9.45) for the Wigner function is inconvenient for Monte Carlo simulations since, even for a free particle, for which $U(x) = 0$, it does not contain the explicit result of integration over ξ. In the general case, this integral cannot be calculated analytically. Exceptions are the linear and harmonic potentials.

To do this integration analytically and obtain an explicit expression for the Wigner function, let us take the approximation for the potential $U(x)$ given by the Taylor expansion up to the first or second order in the variables ξ:

$$U\left((P_eP_hx-x)\frac{m}{M}+x+q^{(m)}-\frac{(M-m)\xi}{2M}+\frac{mP_eP_h\xi}{2M}\right)$$
$$\approx U\left((P_eP_hx-x)\frac{m}{M}+x+q^{(m)}\right)$$
$$-\left\langle\frac{(M-m)\xi}{2M}-\frac{mP_eP_h\xi}{2M}\middle|\frac{\partial U((P_eP_hx-x)m/M+x+q^{(m)})}{\partial x}\right\rangle$$

$$+\frac{1}{2}\left\langle \frac{(M-m)\xi}{2M} - \frac{mP_eP_h\xi}{2M} \middle| \frac{\partial^2 U}{\partial x^2} \middle| \frac{(M-m)\xi}{2M} - \frac{mP_eP_h\xi}{2M} \right\rangle. \tag{9.46}$$

Here the second term is the scalar product of the vector related to the ξ combination with the multidimensional gradient of the pseudopotential energy, while the third term is a quadratic form with the matrix of the second derivatives.

This approximation for the Wigner function takes the form of a Gaussian integral and can be calculated analytically. Here, for simplicity, let us consider expressions accounting for the linear term in the expansion. It is straightforward to account for quadratic term. In the linear approximation, the Wigner function can be written in the form

$$\begin{aligned}
W(p,x;\beta) &= \frac{C(M)}{Z(\beta)N_e!N_h!} \\
&\times \sum_{\sigma}\sum_{P_e}\sum_{P_h}(\pm 1)^{\kappa_{P_e}+\kappa_{P_h}} \mathcal{S}(\sigma,\hat{P}_e\hat{P}_h\sigma')|_{\sigma'=\sigma} \int \mathrm{d}q^{(1)}\ldots\mathrm{d}q^{(M-1)} \\
&\times \exp\left\{-\sum_{m=0}^{M-1}\left[\pi|q^{(m)}-q^{(m+1)}|^2+\epsilon U\left((P_eP_hx-x)\frac{m}{M}+x+q^{(m)}\right)\right]\right\} \\
&\times \exp\left(-\pi\frac{|P_eP_hx-x|^2}{M}\right)\int \mathrm{d}\xi \exp\left\{-\pi\frac{\langle\xi|P_eP_h+E|\xi\rangle}{2M}\right. \\
&\left.+\left\langle\xi\middle|ip+\sum_{m=0}^{M-1}\left|\frac{(M-m)}{2M}E-\frac{m}{2M}P_eP_h\right|\frac{\partial[\epsilon U(\bar{x})]}{\partial\bar{x}}\right\rangle\Bigg|_{\bar{x}=(P_eP_hx-x)\frac{m}{M}+x+q^{(m)}}\right\}.
\end{aligned} \tag{9.47}$$

This expression for the Wigner function is obtained under the assumption that the potential energy U can be expanded with sufficient accuracy in a Taylor series in ξ. Let us discuss the validity of this assumption for one particle in the 1D case with the identity permutation. The Taylor expansion of the potential function contains powers of ξ multiplied by

$$\sum_{m=0}^{M-1}(m/M-1/2)^n\frac{\partial^n U(x+q^{(m)})}{n!\partial x^n}.$$

Using the mean value theorem, we can get a rough estimate of these sums as

$$\begin{aligned}
\sum_{m=0}^{M-1}(m/M-1/2)^n\frac{\partial^n U(x+q^{(m)})}{n!\partial x^n} &\approx \int_0^1 \mathrm{d}\tau\frac{1}{n!}\frac{\partial^n U(x+q^{(\tau)})}{\partial x^n}\left(\tau-\frac{1}{2}\right)^n \\
&\approx \frac{1}{n!}\frac{(1+(-1)^n)}{(n+1)2^{n+1}}\frac{\partial^n U(x+q^{(0)})}{\partial x^n},
\end{aligned} \tag{9.48}$$

where $q^{(0)} = q^{(M)} = 0$. We expect fast convergence of the Taylor series for potentials averaged along the trajectories $x + q^{(m)}$, with alternating sign for odd n and weight $(\tau - 1/2)^n$. The numerical value of this integral decreases rapidly: when $n = 2, 4, 6$ it equals 1/24, 1/1920 and 1/322560, respectively, and it is less for odd n. Thus we expect a negligible contribution from high order Taylor terms in the potential expansion. Numerical calculations for some arbitrary potentials in (Larkin et al. 2016, 2017) confirm this assumption.

As mentioned before, it is enough for our purposes [see (9.41)] to take into account pair permutations. In the degenerate system, the average distance between fermions is less than the thermal wavelength λ, and the trajectories in the path integrals (9.47) are strongly entangled. This is the reason why pair permutations cannot significantly affect the potential energy in (9.47), in comparison with the case of the identity permutation. So for any pair permutation, the potential energy in (9.47) can be represented as the energy related to the identity permutation (of order $N^2M/2$) and a small (in comparison with the identity permutation) difference (of order $2NM$):

$$\sum_{m=0}^{M-1} \epsilon U\left((P_e P_h x - x)\frac{m}{M} + x + q^{(m)}\right) - \sum_{m=0}^{M-1} \epsilon U\left(x + q^{(m)}\right). \tag{9.49}$$

So we take only the first terms of the perturbation series in the relative difference and obtain the following simplification:

$$\sum_{m=0}^{M-1} \epsilon U\left((P_e P_h x - x)\frac{m}{M} + x + q^{(m)}\right) \approx \sum_{m=0}^{M-1} \epsilon U\left(x + q^{(m)}\right).$$

A rigorous proof can be found by generalizing the Mayer expansion technique (the so-called algebraic approach) to nonideal systems of particles, as developed in (Zelener et al. 1981; Ruelle 1999). Now all permutations in (9.47) act only on variables x and ξ and can be taken out of the path integral.

Here we allow for the exchange effects in the pair approximation by using an effective pseudopotental like the one discussed above [see (9.41)], whence

$$\begin{aligned} W(p, x; \beta) \approx{} & \frac{C(M)}{Z(\beta) N_e! N_h!} \int \mathrm{d}q^{(1)} \ldots \mathrm{d}q^{(M-1)} \\ & \times \exp\left\{-\sum_{m=0}^{M-1}\left[\pi |q^{(m)} - q^{(m+1)}|^2 + \epsilon U(x + q^{(m)})\right]\right\} \\ & \times \exp\left\{\frac{M}{4\pi}\left| ip + \frac{\epsilon}{2}\sum_{m=0}^{M-1} \frac{(M - 2m)}{M} \frac{\partial U(x + q^{(m)})}{\partial x}\right|^2\right\} \\ & \times \sum_{\sigma_e}\left\{1 - \sum_{l<t} \delta_{\sigma_{l,e}\sigma_{t,e}} \exp\left(-2\pi \frac{|x_{l,e} - x_{t,e}|^2}{M}\right) \delta\left(\frac{(\tilde{p}_{l,e} - \tilde{p}_{t,e})\sqrt{M}}{2\pi}\right)\right\} \end{aligned}$$

$$\times \sum_{\sigma_{\rm h}} \left\{ 1 - \sum_{l<t} \delta_{\sigma_{l,\rm h}\sigma_{t,\rm h}} \exp\left(-2\pi \frac{|x_{l,\rm h} - x_{t,\rm h}|^2}{M} \right) \delta\left(\frac{(\tilde{p}_{l,\rm h} - \tilde{p}_{t,\rm h})\sqrt{M}}{2\pi} \right) \right\}, \tag{9.50}$$

where

$$\tilde{p}_{t,a} = p_{t,a} + \frac{\epsilon}{2} \sum_{m=0}^{M-1} \frac{\partial U(x + q^{(m)})}{\partial x_{t,a}} .$$

The main idea behind the derivation of (9.50) can be explained with the example of two electrons in 1D space. For two electrons, the sum over permutations consist of two terms related to the identity permutation (the matrix P is equal to the unit matrix E) and non-identical permutation (the matrix P is equal to the matrix E with transposed columns). To do the integration over ξ in (9.47), let us examine the eigenvalues of the matrix $P + E$. For the identity permutation, the eigenvalues are equal to each other and are equal to two, while the eigenvalues of the matrix $P + E$ related to non-identical permutation are equal to zero and two. Integration over ξ for the identity permutation is trivial, while for non-identical permutation the matrix $P + E$ before integration has to be represented in the form $P + E = ODO^{-1}$, where D is a diagonal matrix with zero and two as the diagonal elements. Here the matrix O and its inverse O^{-1} are

$$O = \begin{vmatrix} 1 & 1 \\ -1 & 1 \end{vmatrix}, \qquad O^{-1} = \frac{1}{2} \begin{vmatrix} 1 & -1 \\ 1 & 1 \end{vmatrix}.$$

Replacing the variables of integration by the relation $|\zeta, \eta\rangle = |O^{-1}|\xi\rangle$, one obtains the analogue of (9.50).

To obtain the final expression we have to approximate the delta function by the standard Gaussian exponent with small parameter α:

$$\begin{aligned} W(p, x; \beta) \approx{}& \frac{C(M)}{Z(\beta) N_{\rm e}! N_{\rm h}!} \int {\rm d}q^{(1)} \ldots {\rm d}q^{(M-1)} \\ &\times \exp\left\{ -\sum_{m=0}^{M-1} \left[\pi |q^{(m)} - q^{(m+1)}|^2 + \epsilon U(x + q^{(m)}) \right] \right\} \\ &\times \exp\left\{ \frac{M}{4\pi} \left| {\rm i}p + \frac{\epsilon}{2} \sum_{m=0}^{M-1} \frac{(M - 2m)}{M} \frac{\partial U(x + q^{(m)})}{\partial x} \right|^2 \right\} \\ &\times \sum_{\sigma} \exp\left(-\beta \sum_{l<t} v_{lt}^{\rm e} \right) \exp\left(-\beta \sum_{l<t} v_{lt}^{\rm h} \right), \end{aligned} \tag{9.51}$$

where

$$v_{lt}^{a} = -kT \ln\left[1 - \delta_{\sigma_{l,a}\sigma_{t,a}} \exp\left(-\frac{2\pi|x_{l,a} - x_{t,a}|^2}{M}\right) \times \frac{\sqrt{M}}{2\pi\alpha} \exp\left(-\pi \frac{|(\tilde{p}_{l,a} - \tilde{p}_{t,a})\sqrt{M}|^2}{(2\pi\alpha)^2}\right)\right], \tag{9.52}$$

and we imply analytic continuation for these functions v_{lt}^{a} on the complex plane. Note that the expression (9.51) explicitly contains a term relating to the classical Maxwell distribution. The difference is in the other terms, accounting for the influence of interaction on the momentum distribution function. To regularize the integration over the momenta in the limit of small α, p must be rescaled by a factor $\sqrt{M}/2\pi\alpha$ in the integration over the momenta in (9.35), so that one can use the simplified version of the effective pair exchange pseudopotential (π is included in small α^2):

$$v_{lt}^{a} \approx -kT \ln\left[1 - \delta_{\sigma_{l,a}\sigma_{t,a}} \exp\left(-\frac{2\pi|x_{l,a} - x_{t,a}|^2}{\lambda_a^2}\right) \times \exp\left(-\frac{|(\tilde{p}_{l,a} - \tilde{p}_{t,a})\lambda_a|^2}{(2\pi\hbar)^2\alpha^2}\right)\right]. \tag{9.53}$$

The momenta and coordinates here are written in natural units, with $\lambda_a^2 = 2\pi\hbar^2\beta/m_a$.

Average Values of Quantum Operators

To calculate the average values of quantum operators $\langle\hat{A}\rangle$, we develop a new PIMC method in the Wigner approach (WPIMC), using the basic ideas from the papers (Metropolis et al. 1953; Hastings 1970). To do this, we have to use path integrals in the discrete form of the multiple integral, like (9.51). As a result, we obtain final expressions for MC calculations in the form

$$\langle\hat{A}\rangle = \frac{\langle A(p,x) \cdot h(p,x,z_1,\ldots,z_{M-1})\rangle_w}{\langle h(p,x,z_1,\ldots,z_{M-1})\rangle_w}. \tag{9.54}$$

Here angle brackets $\langle g(p,x,z_1,\ldots,z_{M-1})\rangle_w$ on any function g of the variables $(p,x,z_1,\ldots,z_{M-1})$ denote the average

$$\langle g(p,x,z_1,\ldots,z_{M-1})\rangle_w = \int \mathrm{d}p\,\mathrm{d}x \int \mathrm{d}z_1 \ldots \mathrm{d}z_{M-1} \times g(p,x,z_1,\ldots,z_{M-1}) w(p,x,z_1,\ldots,z_{M-1}), \tag{9.55}$$

with positive weight $w(p,x,z_1,\ldots,z_{M-1})$ given by

$$
\begin{aligned}
&w(p, x, z_1, \ldots, z_{M-1}) \\
&\quad = \left| \cos\left[2\frac{M}{4\pi} \left\langle p \left| \frac{\epsilon}{2} \sum_{m=0}^{M-1} \frac{(M-2m)}{M} \frac{\partial U(x+q^{(m)})}{\partial x} \right. \right\rangle \right] \right| \\
&\quad \times \exp\left\{ -\frac{M}{4\pi} \left[|p|^2 + \left| \frac{\epsilon}{2} \sum_{m=0}^{M-1} \frac{(M-2m)}{M} \frac{\partial U(x+q^{(m)})}{\partial x} \right|^2 \right] \right\} \\
&\quad \times \exp\left\{ -\sum_{m=0}^{M-1} \left[\pi |q^{(m)} - q^{(m+1)}|^2 + \epsilon U(x+q^{(m)}) \right] \right\} \\
&\quad \times \sum_{\sigma} \exp\left(-\beta \sum_{l<t} v_{lt}^{\mathrm{e}} \right) \exp\left(-\beta \sum_{l<t} v_{lt}^{\mathrm{h}} \right),
\end{aligned}
\tag{9.56}
$$

and

$$
\begin{aligned}
&h(p, x, z_1, \ldots, z_{M-1}) \\
&\quad = \operatorname{sign}\left\{ \cos\left[2\frac{M}{4\pi} \left\langle p \left| \frac{\epsilon}{2} \sum_{m=0}^{M-1} \frac{(M-2m)}{M} \frac{\partial U(x+q^{(m)})}{\partial x} \right. \right\rangle \right] \right\}.
\end{aligned}
\tag{9.57}
$$

Note that denominator in (9.54) is equal to the numerator with $A(p, x) = 1$, so $C(M)$ in (9.51) cancels out.

To construct the Metropolis–Hasting algorithm, we introduce three types of WPIMC steps:

1. Variation of the momentum of some particle.
2. Variation of the overall position of some particle.
3. Variation of the trajectory representing some particle.

It is more convenient and reasonable to carry out calculations of the average values of quantum operators depending only on coordinates of particles in configuration space using the standard path integral Monte Carlo method (PIMC). In the context of the above approach, this can be done if we change the following functions:

$$
\begin{aligned}
\tilde{w}\left(x, q^{(1)}, \ldots, q^{(M-1)}\right) &= \exp\left\{ -\sum_{m=0}^{M-1} \left[\pi |q^{(m)} - q^{(m+1)}|^2 + \epsilon U(x+q^{(m)}) \right] \right\} \\
&\quad \times \sum_{\sigma} \exp(-\beta \sum_{l<t} \tilde{v}_{lt}^{\mathrm{e}}) \exp(-\beta \sum_{l<t} \tilde{v}_{lt}^{\mathrm{h}}),
\end{aligned}
\tag{9.58}
$$

$$
\tilde{h}(x, q^{(1)}, \ldots, q^{(M-1)}) \equiv 1,
$$

where $\tilde{v}_{lt}^{a}$ is defined by Eq. (9.52),

$$\left\langle g(x, q^{(1)}, \ldots, q^{(M-1)}) \right\rangle_{\tilde{w}} = \int \mathrm{d}x \int \mathrm{d}q^{(1)} \ldots \mathrm{d}q^{(M-1)} g(x, q^{(1)}, \ldots, q^{(M-1)}) \times \tilde{w}(x, q^{(1)}, \ldots, q^{(M-1)}) \,, \tag{9.59}$$

and

$$\langle \tilde{\hat{A}} \rangle = \frac{\left\langle \tilde{A}(x) \cdot \tilde{h}(x, q^{(1)}, \ldots, q^{(M-1)}) \right\rangle_{\tilde{w}}}{\langle \tilde{h}(x, q^{(1)}, \ldots, q^{(M-1)}) \rangle_{\tilde{w}}} \,. \tag{9.60}$$

The first thousand steps should be rejected during the calculation in order to 'forget' the initial configuration. Periodic boundary conditions have to be used to reduced finite number effects.

9.6 Phase Space Solution of the 'Sign' Problem in the Pair Approximation

We define the pair correlation function and momentum distribution function for holes ($a = h$) and electrons ($a = e$) by the following expressions:

$$w_a(|p|) = \frac{\left\langle \delta(|p_{1,a}| - |p|) \cdot h(p, x, q^{(1)}, \ldots, q^{(M-1)}) \right\rangle_{w}}{\left\langle h(p, x, q^{(1)}, \ldots, q^{(M-1)}) \right\rangle_{w}} \,,$$

$$g_{ab}(r) = \frac{\left\langle \delta(|x_{1,a} - x_{1,b}| - r) \cdot \tilde{h}(x, q^{(1)}, \ldots, q^{(M-1)}) \right\rangle_{\tilde{w}}}{\left\langle \tilde{h}(x, q^{(1)}, \ldots, q^{(M-1)}) \right\rangle_{\tilde{w}}} \,, \tag{9.61}$$

where δ is the delta function and a and b are the types of particles. The pair correlation functions g_{ab} give a probability density of finding a pair of particles of types a and b at a certain distance r from each other. They depend only on the coordinate difference due to the translational invariance of the system. In a noninteracting classical system, $g_{ab} \equiv 1$, whereas interactions and quantum statistics result in a redistribution of the particles.

The momentum distribution function $w_a(|p|)$ gives a probability density for a particle of type a to have momentum p. Due to the commutativity of the kinetic and potential energy operators, non-ideal classical systems of particles have a Maxwell distribution function (MD) proportional to $\exp[-(p\lambda_a)^2/4\pi\hbar^2]$, even at strong coupling. Quantum effects can affect the shape of the particle kinetic energy distribution function. Due to quantum statistics, ideal quantum systems of particles have Fermi or Bose momentum distribution functions, which are rather different from Maxwellian distributions. The interaction of a quantum particle with its surroundings restricts the volume of configuration space, and due to the uncertainty relation, this can also affect the shape of the momentum distribution function. In this section, WPIMC results are presented for both ideal and strongly coupled electron–hole plasmas with different hole masses.

Test Calculations for Ideal Electron–Hole Plasmas

To test the approach developed here, calculations of the path integral representation of the Wigner function in the form (9.54) have been carried out in (Larkin et al. 2016, 2017). To extend the region of applicability of the pair approximation, the small parameter α^2 was used as an adjustable function for the universal degeneracy parameter of ideal fermions $n\lambda^3$, namely $\alpha_a^2 = 0.00505 + 0.056n\lambda_a^3$. Calculations were carried out for two hundred particles, each represented by twenty beads. Results were obtained by averaging over one million particle configurations. To simplify calculations, we fixed the number of electrons and holes with the same spin projection equal to $N_\mathrm{e}/2$ and $N_\mathrm{h}/2$, respectively.

We begin by considering ideal fermions. For the ideal electron–hole plasma, Fig. 9.3 shows the momentum distributions $w_a(|p_a|)$, $(a = \mathrm{e}, \mathrm{h})$ and pair correlation functions $g_{ab}(|r|)$ $(a, b = \mathrm{e}, \mathrm{h})$ for electron degeneracy parameters equal to $n\lambda_\mathrm{e}^3 = 5.6$ $(T/E_\mathrm{F} = 0.208, k_\mathrm{F}\lambda_\mathrm{e} = 5.5)$ and $n\lambda_\mathrm{e}^3 = 10$ $(T/E_\mathrm{F} = 0.141, k_\mathrm{F}\lambda_\mathrm{e} = 6.66)$

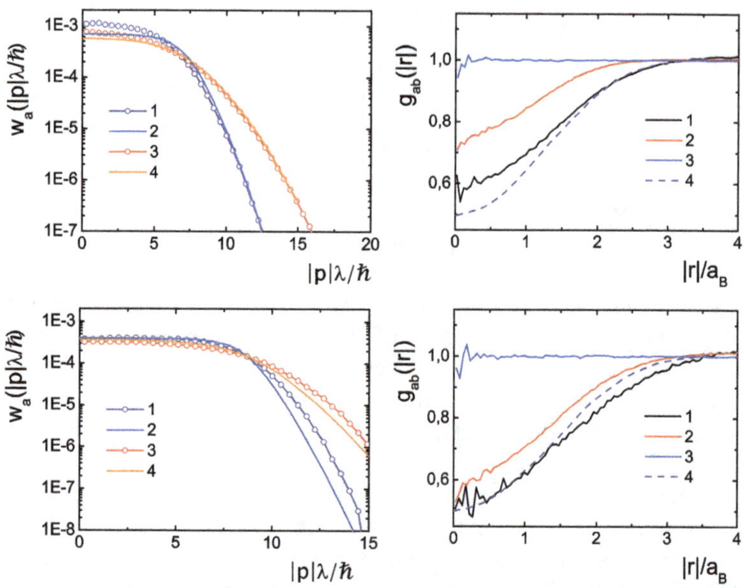

Fig. 9.3 Momentum distribution functions $w_a(|p|)$ (*left*) and pair correlation functions $g_{ab}(|r|)$ $(a, b = \mathrm{e}, \mathrm{h})$ (*right*) for an ideal electron–hole plasma. *Left*: *Curves* 1 and 3 show PIMC distributions $w_a(|p|)$ scaled by the ratio of the Planck constant to the electron thermal wavelength $(\hbar/\lambda_\mathrm{e})$, while *curves* 2 and 4 show the ideal Fermi distributions for electrons and holes that are twice as heavy, respectively. *Right*: *Curves* 1, 2, and 3 represent the PIMC electron–electron, hole–hole, and electron–hole correlation functions, respectively, scaled by the Bohr radius, while *curve* 4 shows the results of analytical approximations for electrons (Bosse et al. 2011)

(top and bottom rows, respectively). Note that, in these calculations, holes are twice as massive as electrons, so the corresponding degeneracy parameters are $2^{3/2}$ times smaller. The momentum distribution functions are normalized to unity. It follows from the analysis of Fig. 9.3 that the agreement between the PIMC calculations and the analytical Fermi distributions (FD) are good enough up to a value of the electron degeneracy parameter equal to $n\lambda_e^3 = 10\,(T/E_F = 0.141)$. Here, the integral characteristics such as the energy resulting from PIMC and the analytical distribution functions are practically equal.

In the right-hand column, we compare the pair correlation functions calculated here with analogous analytical approximations obtained in (Bosse et al. 2011) for electrons and find good agreement. At distances less than the thermal wavelength of the electrons or holes, the influence of Fermi repulsion is evident with increasing values of the degeneracy parameter (from top to bottom). At the same time the electron–hole pair correlation functions are identically equal to one, as there is no exchange interaction between the particles of different types.

The results presented here were obtained in the pair exchange approximation described above, i.e., using effective pair pseudopotentials (9.53). Figure 9.4 presents contour plots of pair exchange pseudopotentials for degeneracy parameter equal to 5.6. The momentum and coordinate axes are scaled by the electron thermal wavelength with the Planck constant and factor ten for the momentum. As before, holes are twice as massive as electrons. It follows from examination of Fig. 9.3 that the Pauli blocking in phase space accounted for by these exchange pseudopotentials provides agreement between the WPIMC calculations and analytical FD over a broad range of fermion degeneracies and fermion momenta, where decay of the distribution functions covers at least five orders of magnitude. It should be stressed that one of the reasons for the increasing discrepancy with degeneracy growth is the limitation of

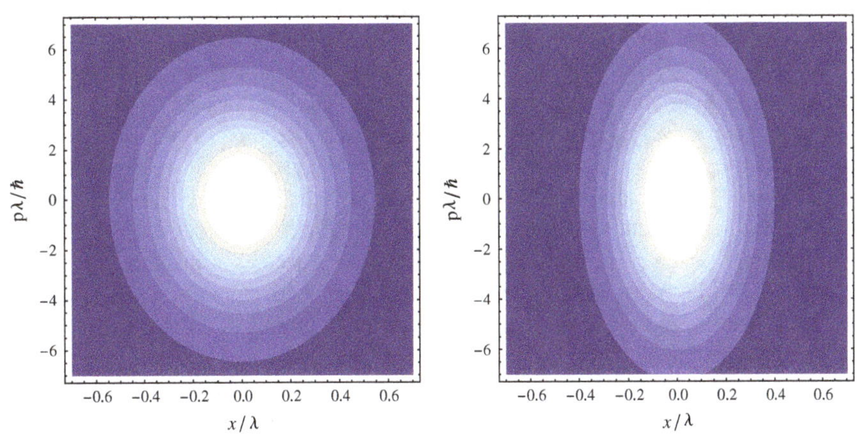

Fig. 9.4 Contour plots of the repulsive effective pair exchange pseudopotentials in phase space. *Left*: $m_e = 1$, *dark area* $\beta v_{ab}^{e} \approx 0$, *white area* $\beta v_{ab}^{e} \geq 1.9$. *Right*: $m_h = 2$, *dark area* $\beta v_{ab}^{h} \approx 0$, *white area* $\beta v_{ab}^{h} \geq 1.5$ (Larkin et al. 2016, 2017)

computing power available for doing calculations with several hundred particles in each Monte Carlo cell. When the degeneracy parameter is more than 10, the electron thermal wavelength is of the order of the Monte Carlo cell size, and the influence of the finite number of particles and periodic boundary conditions becomes significant, as attested in (Larkin et al. 2016, 2017).

9.7 Quantum 'Tails' in the Momentum Distribution Functions

The aim of this section is to show the influence of strong interparticle interactions on the quantum momentum distribution function $w_a(|p|)$. In quantum systems, the interparticle interaction can affect the Maxwellian shape of the particle kinetic energy distribution function, for the interaction of a particle with its surroundings restricts the volume of configuration space, and thus, due to the uncertainty relation, this results in an increase in the volume of the momentum space, i.e., a rise in the fraction of particles with higher momenta. On the other hand, Fermi statistical effects accounted for by Pauli blocking of fermions can also be important under the same conditions. Through these effects, the momentum distribution function may contain a power-law 'tail', even under conditions of thermodynamic equilibrium (Galitskii and Yakimets 1966; Kimball 1975; Starostin et al. 2002; Eletskii et al. 2005; Emelianov et al. 2011; Kochetov et al. 2016; Starostin and Petrushevich 2016).

As shown in these papers in the framework of certain models and perturbation theories, for any system of non-relativistic particles, the momentum distribution in the asymptotic region of large momenta p may be described by the sum of MD and quantum corrections proportional to const.$/p^8$ (P8) or MD and the product of const$/p^8$ and Maxwell distributions with an effective temperature that exceeds the temperature of the medium (P8MDEF) (Starostin and Petrushevich 2016).

WPIMC calculations for the electron and hole momentum distribution functions in an electron–hole plasma are presented in Fig. 9.5 by curves 1 and 4 for electrons and holes, respectively. Here the electron degeneracy parameter is equal to four and the hole masses are two (left) and five (right) times the electron mass. Figure 9.5 also shows several related analytical dependencies: FD (curves 2 and 5), MD (curves 3 and 6), P8 (curves 7 and 8), and P8MDEF (curves 9 and 10). Here, the constant in P8 and the effective temperature in P8MDEF have been used as adjustable parameters to fit the WPIMC momentum distribution functions in the asymptotic regions. All MD and FD were normalized to unity. It follows from examination of Fig. 9.5 that P8 may reliably fit WPIMC distributions in the intermediate region between FD and P8MDEF, while the P8MDEF fits WPIMC at large momenta in almost all the panels of Fig. 9.5.

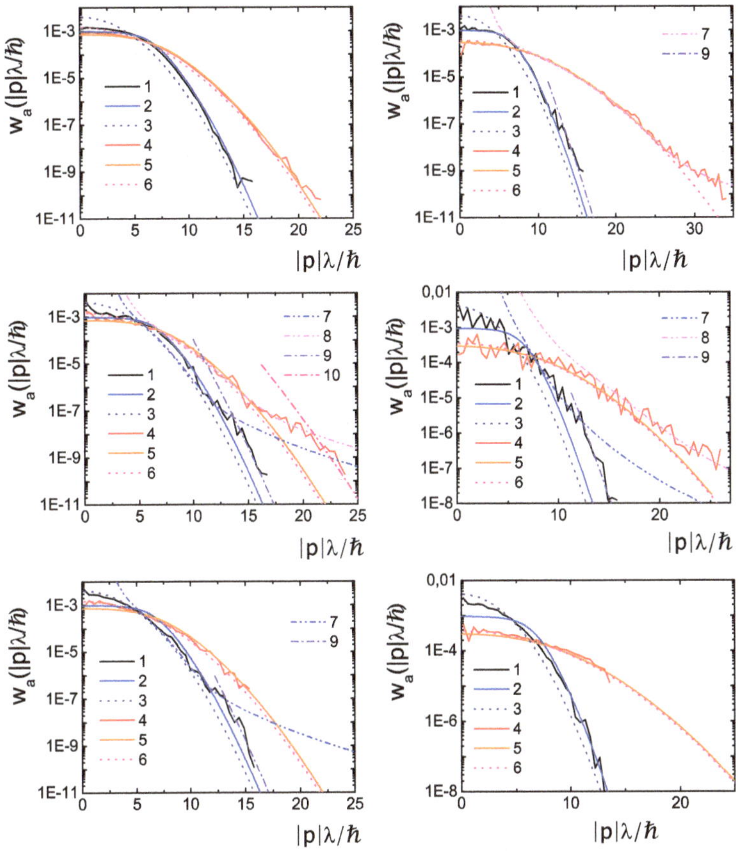

Fig. 9.5 Momentum distribution functions $w_a(|p|)$ for non-ideal electron–hole plasmas. Results for electrons and two and five times heavier holes are shown on the *left* and *right* plots, respectively. Electron momentum distributions are represented by *curves* 1, 2, 3, 7, and 9, while analogous results for holes are represented by *curves* 4, 5, 6, 8, and 10. A detailed description of the meaning of these curves is given in the text. The degeneracy of the electrons is equal to 4 and the plasma classical coupling parameter is equal to 1, 2, and 4 from top to bottom. Oscillations of the PIMC distribution functions indicate statistical errors (Larkin et al. 2017)

It should be stressed that each type of asymptotic fits the WPIMC momentum distribution function throughout a decay region of two or even more orders of magnitude. It is interesting to note the non-monotonic behavior of the discrepancy between the WPIMC distributions and FD or MD in the asymptotic regions with increasing values of the classical coupling parameter $\Gamma = \beta e^2/(3/4\pi n_a)^{1/3}$ from the top ($\Gamma = 1$) to bottom ($\Gamma = 4$) panels. Hence, for electrons and holes at low $\Gamma \approx 1$ and high enough $\Gamma \approx 4$, the WPIMC distributions approach the FD or MD better than at $\Gamma \approx 2$ or $\Gamma \approx 3$.

The physical meaning of this phenomenon is the following. At small $\Gamma \leq 1$, the influence of the interparticle interaction is negligible and, due to the exchange pseudopotentials $v_{lt}^a \approx 2$, the WPIMC momentum distribution functions approximate the related Fermi distributions. At large $\Gamma \geq 4$, the average interparticle interaction is more important than the weaker exchange interaction, and the WPIMC momentum distributions approach the MD, which are very close to the FD at large momenta. This means that the appearance of quantum 'tails' in the intermediate region of $\Gamma \approx 2$ or 3 may be explained by the simultaneous influence of Coulomb and interparticle exchange interactions.

Figure 9.6 shows the WPIMC distribution functions for an electron degeneracy parameter equal to two and $m_h = 2$ (left panel) and $m_h = 5$ (right panel). These

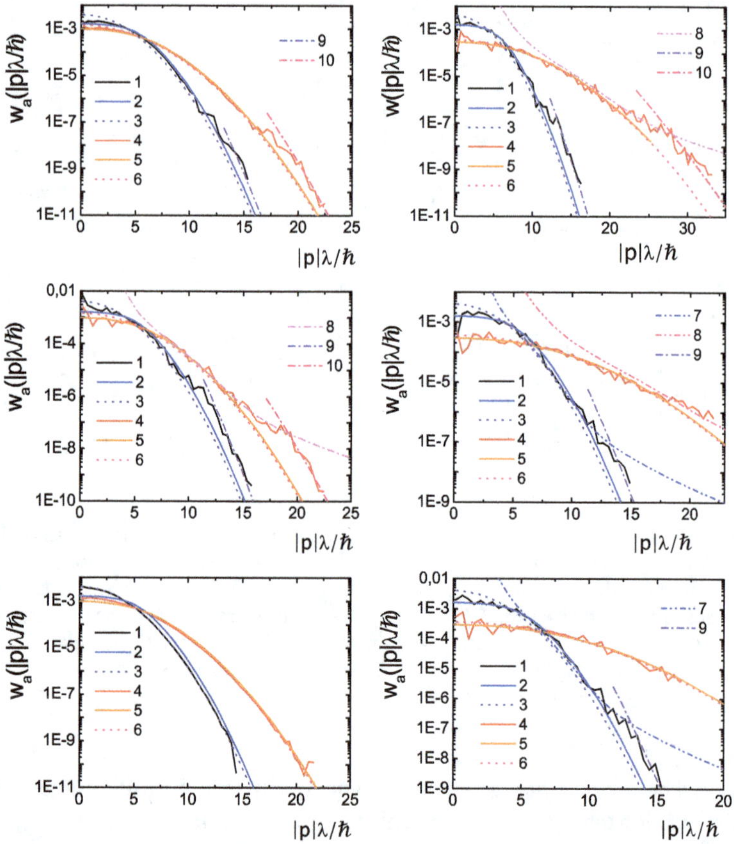

Fig. 9.6 Momentum distributions functions $w_a(|p|)$ for non-ideal electron–hole plasmas. Results for electrons and two and five times heavier holes are shown on the *left* and *right* plots, respectively. Electron momentum distributions are shown by *curves* 1, 2, 3, and 7, while analogous results for holes are shown by *curves* 4, 5, 6, 8, and 9. A detailed description of the meaning of these curves is given in the text. The degeneracy of the electrons is equal to 2 and the classical plasma coupling parameter is equal to 1.3, 2, and 2.6 from top to bottom. Oscillations of the WPIMC distribution functions indicate statistical errors (Larkin et al. 2017)

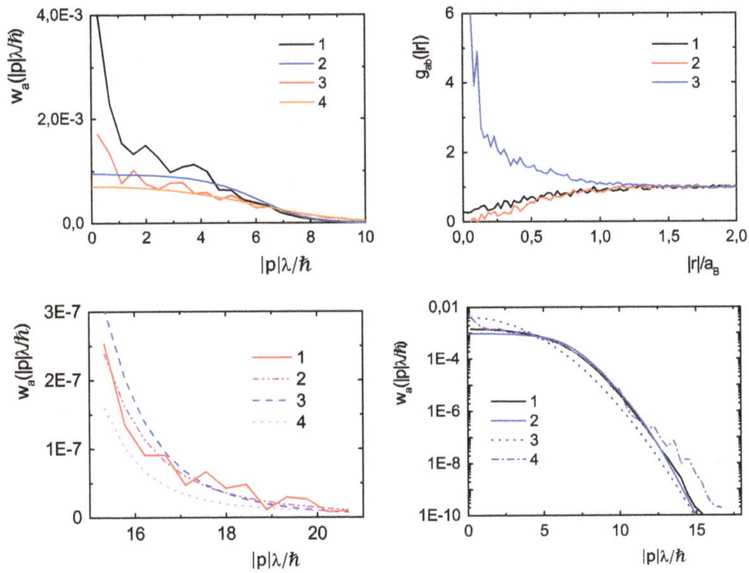

Fig. 9.7 Momentum and pair correlation functions for electrons and two times heavier holes. The degeneracy of the electrons is equal to 4, the classical plasma coupling parameter is equal to 2 (see Fig. 9.5). *Top left panel*: $w_a(|p|)$ at $p\lambda/\hbar \leq 10$. *Curves* 1, 2: WPIMC and FD for electrons. *Curves* 3, 4: WPIMC and FD for holes. *Top right panel*: $g_{ab}(|r|)$ $(a, b = \text{e}, \text{h})$. *Curves* 1, 2, 3 represent electron–electron, hole–hole, and electron–hole distribution functions, respectively. Small oscillations of WPIMC distribution functions indicate statistical error. *Bottom left panel*: const.$/p^n$ power-law approximations of the WPIMC momentum distribution in the intermediate region. *Curves* 1: WPIMC momentum distribution. *Curve* 2: $n = 8$. *Curve* 3: $n = 11$. *Curve* 4: $n = 5$. *Bottom right panel*: *Curve* 1: WPIMC for UEG. *Curve* 2: FD. *Curve* 3: MD. *Curve* 4: WPIMC for electron–hole plasma. Degeneracy parameter equal to 4 and classical coupling parameter Γ equal to two (the same as for Fig. 9.5) (Larkin et al. 2016, 2017)

results confirm even more clearly the basic features of the asymptotic behavior of the WPIMC distribution functions discussed above.

Figure 9.7 presents the WPIMC momentum and pair correlation functions. In the top left panel, for momenta $p\lambda/\hbar \leq 10$, the WPIMC distributions show systematic oscillations and exceed the FD for both electrons and holes. As can be seen from the right-hand panel, this is a consequence of the strong interparticle interaction resulting in partial ordering of the electrons and holes. Typical particle configurations are described by the pair correlation functions presented in the top right panel. This panel shows, as an example, the results of the calculation of pair correlation functions for a non-ideal electron–hole plasma. Now in contrast to an ideal plasma, the electron–hole pair correlation function increases at small interparticle distances due to the Coulomb attraction, while the electron–electron and hole–hole pair correlation functions decrease due to the Coulomb and Fermi repulsion. Coulomb electron–electron repulsion is less pronounced in comparison with hole–hole repulsion due to the fact that tunneling effects are taken into account, these being greater for lighter electrons.

Here the electron–electron Fermi repulsion is less pronounced in comparison with an ideal plasma, as the coupling parameter is strong enough (compare this function with the related g_{ee} of Fig. 9.3).

The bottom left panel of Fig. 9.7 shows the approximation of the WPIMC momentum distribution function in the intermediate region by several power-law dependencies for different values of the exponent. Examination of Fig. 9.7 clearly shows the validity of approximation described by the P8 dependence.

To stress the role of interparticle interaction and correlation effects, we have carried out calculations of the electron momentum distribution functions for the model of a uniform electron gas (UEG), this being the quantum mechanical model of interacting electrons in which positive charges are assumed to be uniformly distributed in space. The momentum distribution functions for the UEG model and the electron–hole plasma are shown in the bottom right panel of Fig. 9.7 for a degeneracy parameter equal to 4 and a classical coupling parameter equal to two (the same as for Fig. 9.5). A comparison of the results obtained here demonstrates the disappearance of the quantum 'tail' and very good agreement of the electron momentum distribution functions for UEG with the Fermi distribution.

9.8 Analytical Investigations of the 'Fixed-Node' Density Matrix

The basic background of the 'fixed-node' method (McMahon et al. 2012; Ceperley 1991, 1992; Militzer and Pollock 2000) is 'the path integral solution of the Bloch equation without minus signs' (Ceperley 1991, 1992). In this approach, integration over paths is restricted by the domain where the additional 'trial antisymmetric density matrix' is not negative. It is claimed that this restriction does not destroy the exact solution of the Bloch equation in the form of path integrals with standard antisymmetric initial conditions.

In the following section, we present an analytical proof that this restriction results in contradictions for explicit calculations of the density matrix, even for two ideal fermions. The analogous contradictions were obtained analytically many years ago from the virial decomposition of the many fermion fixed-node density matrix (Filinov 2001). The main result of this section is that the fixed-node method cannot reproduce even the well known ideal fermion density matrix and should be considered as an uncontrolled empirical approach for treating fermion thermodynamics.

Density Matrix in the Fixed-Node Method

Thermodynamic values of the fermion quantum system are defined by the appropriate derivatives of the logarithm of the partition function $Q_N = \mathrm{Tr}\{\rho\}$. Here $\rho = \exp(-\beta\hat{H})$ is the density matrix of a quantum system of particles with Hamiltonian $\hat{H} = \hat{K} + \hat{U}$ equal to the sum of kinetic $\hat{K}$ and potential energy $\hat{U}$ operators, while $\beta = 1/k_BT$. For our purposes, it is enough to consider a one-dimensional system of two ideal fermions. So $\hat{U} \equiv 0$, while the kinetic energy operator is the sum of two kinetic energy operators related to each particle $\hat{K} = \hat{K}_1 + \hat{K}_2$. The density matrix is the solution of the operator Bloch equation

$$\frac{\partial \hat{\rho}}{\partial \beta} = -\hat{K}\hat{\rho} \,, \tag{9.62}$$

with initial condition $\hat{\rho}|_{\beta=0} = \hat{1}$.

In the coordinate representation for fermions, this operator equation can be written in the form (Ceperley 1991, 1992)

$$\frac{\partial \rho_{\mathrm{F}}(R, R_0; \beta)}{\partial \beta} = -\hat{K}(R)\rho_{\mathrm{F}}(R, R_0; \beta) \,, \tag{9.63}$$

with the initial condition

$$\rho_{\mathrm{F}}(R, R_0; 0) = \frac{1}{N!}\sum_P (-1)^{\kappa_P} \delta(R - PR_0) \,, \tag{9.64}$$

where R is the set of coordinates of all particles and antisymmetry is put in by antisymmetrization. The sum runs over all permutations with parity κ_P acting on particle indices. Another possible coordinate representation of the fermion density matrix has the form

$$\rho_{\mathrm{F}}(R, R_0; \beta) = \frac{1}{N!}\sum_\alpha \exp(-\beta E_\alpha)\phi^*_\alpha(R)\phi_\alpha(R_0) \,, \tag{9.65}$$

where the sum is over the complete set of antisymmetric eigenfunctions $\phi_\alpha(R)$ of $\hat{H}$.

One popular exact coordinate representation of the operator $\hat{\rho}$ follows from the operator identity $\mathrm{e}^{-\beta\hat{K}} \equiv \mathrm{e}^{-\Delta\beta\hat{K}} \ldots \mathrm{e}^{-\Delta\beta\hat{K}} \ldots \mathrm{e}^{-\Delta\beta\hat{K}}$ for any integer fixed M (even of order unity). Here the right-hand side contains M identical factors with $\Delta\beta = \beta/M$. So one can represent the ideal density matrix exactly in the form of a finite-difference expression for the path integral:

$$\begin{aligned} \rho_{\mathrm{F}}(R_M, R_0; \beta) = \frac{1}{N!}\sum_P (-1)^{\kappa_P} \int \mathrm{d}R_1 \ldots \mathrm{d}R_{M-1} \\ \times \prod_{k=1}^{M-1} \rho(R_{k-1}, R_k; \Delta\beta)\rho(R_{M-1}, PR_M; \Delta\beta) \,, \end{aligned} \tag{9.66}$$

where N is the number of fermions. Here, for $N = 2$, the arguments in $\rho(R_{k-1}, R_k; \Delta\beta)$ are 2D vectors comprising the coordinates $X^{(1)}$ and $X^{(2)}$ of the first and second particles on 1D axes, and $\rho(R_{k-1}, R_k; \Delta\beta)$ are distinguishable particle density matrices. The density matrix is a function of the 2D space arguments R_M, R_0 in the

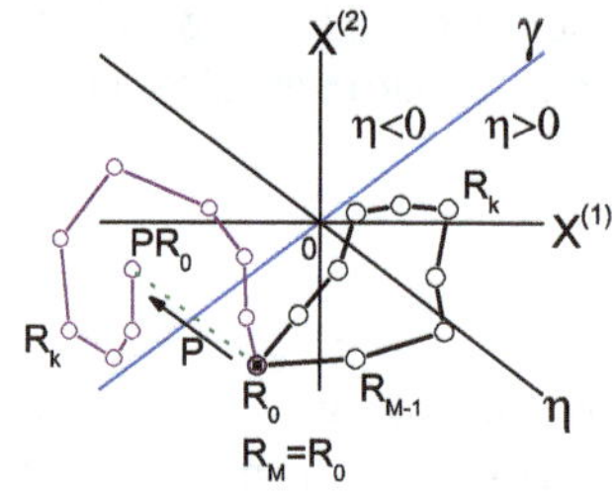

Fig. 9.8 Configuration space of two free fermions, showing the paths relating to the two possible permutations: identical (below the line γ) and non-identical (crossing the line γ) (Filinov 2014)

$(X^{(1)}, X^{(2)})$ plane and the reciprocal temperature β. Below we discuss the boundary conditions of the domain in the $(X^{(1)}, X^{(2)})$ plane and the initial conditions on β mentioned above for the parabolic equation (9.63).

The mathematical meaning of expression (9.66) for the density matrix of two particles is illustrated in Fig. 9.8 (Filinov 2014), where the vectors R_k are represented by circles (known as 'beads'), while the density matrices $\rho(R_{k-1}, R_k; \Delta\beta)$ are denoted by line segments. Sometimes, instead of the coordinates $\{X_k^{(1)}, X_k^{(2)}\}$, it is convenient to use $\{\gamma_k, \eta_k\}$ defined by

$$\gamma_k = 0.5\left[X_k^{(1)} + X_k^{(2)}\right], \qquad \eta_k = \left[X_k^{(1)} - X_k^{(2)}\right],$$

so that $R_k = \{X_k^{(1)}, X_k^{(2)}\} = \{\gamma_k, \eta_k\}$. The modulus of the Jacobian for the change of integration variables in (9.66) from the system $\{X_k^{(1)}, X_k^{(2)}\}$ to the system $\{\gamma_k, \eta_k\}$ is equal to unity. The action of a perturbation P is illustrated in Fig. 9.8 by the arrow carrying the letter P. For two fermions, the sum over permutations reduces to the sum of contributions of identical and non-identical permutations with opposite signs. The density matrix has the following general properties:

$$\rho_{\mathrm{F}}(R, R_0; \beta) = \rho_{\mathrm{F}}(R_0, R; \beta) = (-1)^{\kappa_P}\rho_{\mathrm{F}}(PR, R_0; \beta). \tag{9.67}$$

For further comparisons with the fixed-node density matrix, we recall the solution of the Bloch equation (9.63) with initial condition (9.64). The density matrices in (9.66) are well known for ideal particles and can be written in the form (Feynman and Hibbs 1965):

$$\begin{aligned}\rho(R_{k-1}, R_k; \Delta\beta) &= \frac{\exp\left(-\dfrac{\pi|R_k - R_{k-1}|^2}{\tilde{\lambda}^2}\right)}{\tilde{\lambda}^2} \\ &= \frac{\exp\left(-\dfrac{2\pi|\gamma_k - \gamma_{k-1}|^2}{\tilde{\lambda}^2}\right)\exp\left(-\dfrac{\pi|\eta_k - \eta_{k-1}|^2}{2\tilde{\lambda}^2}\right)}{\tilde{\lambda}^2},\end{aligned} \tag{9.68}$$

where $\tilde{\lambda}^2 = 2\pi\hbar^2\Delta\beta/m$ is the thermal wavelength related to $\Delta\beta$. The last factor in (9.66) for permutation P has the form

$$\rho(R_{M-1}, PR_M; \Delta\beta) = \frac{\exp\left(-\frac{\pi|PR_M - R_{M-1}|^2}{\tilde{\lambda}^2}\right)}{\tilde{\lambda}^2}$$
$$= \frac{\exp\left(-\frac{2\pi|P\gamma_M - \gamma_{M-1}|^2}{\tilde{\lambda}^2}\right)\exp\left(-\frac{\pi|P\eta_M - \eta_{M-1}|^2}{2\tilde{\lambda}^2}\right)}{\tilde{\lambda}^2}. \tag{9.69}$$

So the exact well known expression for two particle antisymmetrized density matrix looks like (Feynman and Hibbs 1965)

$$\rho_F(R_M, R_0; \beta) = \frac{1}{2}\int dR_1 \ldots dR_{M-1} \prod_{k=1}^{M-1} \frac{\exp\left(-\frac{\pi|R_k - R_{k-1}|^2}{\tilde{\lambda}^2}\right)}{\tilde{\lambda}^2}$$
$$\times \frac{\exp\left(-\frac{2\pi|\gamma_M - \gamma_{M-1}|^2}{\tilde{\lambda}^2}\right)}{\tilde{\lambda}}$$
$$\times \frac{\exp\left(-\frac{\pi|\eta_M - \eta_{M-1}|^2}{2\tilde{\lambda}^2}\right) - \exp\left(-\frac{\pi|P\eta_M - \eta_{M-1}|^2}{2\tilde{\lambda}^2}\right)}{\tilde{\lambda}}$$
$$\equiv \frac{1}{2}\frac{\exp\left(-\frac{2\pi|\gamma_M - \gamma_0|^2}{\lambda^2}\right)}{\lambda}$$
$$\times \frac{\exp\left(-\frac{\pi|\eta_M - \eta_0|^2}{2\lambda^2}\right) - \exp\left(-\frac{\pi|P\eta_M - \eta_0|^2}{2\lambda^2}\right)}{\lambda}, \tag{9.70}$$

with $\lambda^2 = 2\pi\hbar^2\beta/m$, $P\gamma_M = \gamma_M$, and $P\eta_M = -\eta_M$. If $\{\gamma_M, \eta_M\} = \{\gamma_0, \eta_0\}$, then $P\gamma_M = \gamma_0$ and $P\eta_M = -\eta_0$ (see Fig. 9.8).

The authors of the papers (Ceperley 1991, 1992) suggested the "fixed-node path integral solution of the Bloch equation without minus signs". They denote the second argument R_0 of the density matrix as the 'reference point' and "the set of points R_t for which there exists a continuous 'space–imaginary time' path with $\rho_F(R, R_0; t') > 0$ for $0 \leq t' \leq \beta$ the reach of R_0 or $\Gamma(R_0, \beta)$". For two ideal fermions, the reach can be obtained analytically (Ceperley 1991, 1992) and is shown in Fig. 9.8 (the half plane $\eta > 0$) for $\eta_0 > 0$. The reach does not depend on the temperature.

According to the papers (Ceperley 1991, 1992):

> It is a simple matter (see Appendix C) to show that the problematic initial condition (9.64) can be replaced by zero boundary conditions on the surface of the reach. It follows because the fermion density matrix is a unique solution to the Bloch equation (9.63) with the zero boundary condition.

However, the Bloch equation (9.63) is a linear equation and, with zero boundary conditions, has not only the trivial solution identically equal to zero, but also an infinite number of solutions distinguished at least by constant factors, while the Bloch equation (9.63) with initial condition (9.64) only really has a unique solution.

Let us consider the fixed-node approach to calculating the density matrices in the (γ, η) plane (see Fig. 9.8). The authors of the papers (Ceperley 1991, 1992) claim that, to obtain the fixed-node density matrix, "one simply restricts the paths in (9.66) to lie in $\Gamma(R_0, \beta)$". This means that restriction of integration over $R_1, \ldots, R_{M-1}$ in (9.66) by the reach has to give the exact solution of (9.63) with initial condition (9.64). The fallaciousness of this statement for $M = 2$, as well as for arbitrary integer M, is easily proved by explicit integration over variables $R_1, \ldots, R_{M-1}$ in the reach for two ideal fermions. The boundary surface of the 2D reach for two ideal fermions is exactly known, and according to (Ceperley 1991, 1992) is the line $\eta = 0$ in Fig. 9.8. So the density matrix with zero boundary conditions (when $\eta_M = 0$) in the whole configurational space (in both half planes $\eta > 0$ and $\eta < 0$, $\eta_M, \eta_0 \in \{X^{(1)}, X^{(2)}\}$), looks like

$$
\begin{aligned}
\rho_{\mathrm{F}}(R_M, R_0; \beta) = {} & \frac{C_1}{2} \int \mathrm{d}R_1 \ldots \mathrm{d}R_{M-1} \prod_{k=1}^{M-1} \frac{\exp\left(-\frac{\pi |R_k - R_{k-1}|^2}{\tilde{\lambda}^2}\right)}{\tilde{\lambda}^2} \\
& \times \frac{\exp\left(-\frac{2\pi |\gamma_M - \gamma_{M-1}|^2}{\tilde{\lambda}^2}\right)}{\tilde{\lambda}} \\
& \times \frac{\exp\left(-\frac{\pi |\eta_M - \eta_{M-1}|^2}{2\tilde{\lambda}^2}\right) - \exp\left(-\frac{\pi |P\eta_M - \eta_{M-1}|^2}{2\tilde{\lambda}^2}\right)}{\tilde{\lambda}} \\
& \times \theta(\eta_0)\theta(\eta_1) \cdots \theta(\eta_M) \\
& + \frac{C_2}{2} \int \mathrm{d}R_1 \ldots \mathrm{d}R_{M-1} \prod_{k=1}^{M-1} \frac{\exp\left(-\frac{\pi |R_k - R_{k-1}|^2}{\tilde{\lambda}^2}\right)}{\tilde{\lambda}^2} \\
& \times \frac{\exp\left(-\frac{2\pi |\gamma_M - \gamma_{M-1}|^2}{\tilde{\lambda}^2}\right)}{\tilde{\lambda}} \\
& \times \frac{\exp\left(-\frac{\pi |\eta_M - \eta_{M-1}|^2}{2\tilde{\lambda}^2}\right) - \exp\left(-\frac{\pi |P\eta_M - \eta_{M-1}|^2}{2\tilde{\lambda}^2}\right)}{\tilde{\lambda}} \\
& \times \theta(-\eta_0)\theta(-\eta_1) \cdots \theta(-\eta_M) \,,
\end{aligned}
\tag{9.71}
$$

where $\theta(\eta)$ is the theta function, equal to zero for $\eta < 0$ and unity for $\eta \geq 0$. The theta functions here restrict the domains of integration to the reach. Now assume that all 'basic statements' of the papers (Ceperley 1991, 1992) are correct. Then using the general properties of the density matrix mentioned above, we have to admit that integration over $R_2, \ldots, R_{M-1}$ in the reach gives the exact solution of (9.63) with initial condition (9.64). So the density matrix in the fixed-node method (9.71) can be transformed to the following integral over the last variable R_1 :

$$
\begin{aligned}
\rho_{\mathrm{F}}(R_M, R_0; \beta) &= \frac{1}{2}\int \mathrm{d}R_1 \frac{\exp\left(-\frac{\pi|R_1-R_0|^2}{\tilde{\lambda}^2}\right)}{\tilde{\lambda}^2} \frac{\exp\left(-\frac{2\pi|\gamma_M-\gamma_1|^2}{(M-1)\tilde{\lambda}^2}\right)}{\sqrt{(M-1)}\tilde{\lambda}} \\
&\times \left[C_1 \frac{\exp\left(-\frac{\pi|\eta_M-\eta_1|^2}{2(M-1)\tilde{\lambda}^2}\right) - \exp\left(-\frac{\pi|P\eta_M-\eta_1|^2}{2(M-1)\tilde{\lambda}^2}\right)}{\sqrt{(M-1)}\tilde{\lambda}} \theta(\eta_0)\theta(\eta_1)\theta(\eta_M) \right. \\
&\left. + C_2 \frac{\exp\left(-\frac{\pi|\eta_M-\eta_1|^2}{2(M-1)\tilde{\lambda}^2}\right) - \exp\left(-\frac{\pi|P\eta_M-\eta_1|^2}{2(M-1)\tilde{\lambda}^2}\right)}{\sqrt{(M-1)}\tilde{\lambda}} \theta(-\eta_0)\theta(-\eta_1)\theta(-\eta_M) \right] \\
&= \frac{1}{2} \frac{\exp\left(-\frac{2\pi|\gamma_M-\gamma_0|^2}{\lambda^2}\right)}{2\lambda} \left\{ C_1\theta(\eta_0)\theta(\eta_M) \right. \qquad (9.72) \\
&\times \left[\frac{\exp\left(-\frac{\pi|\eta_M-\eta_0|^2}{2\lambda^2}\right)\xi\big((M-1)\eta_0+\eta_M\big)}{\lambda} \right. \\
&\left. - \frac{\exp\left(\frac{\pi|P\eta_M-\eta_0|^2}{2\lambda^2}\right)\xi\big((M-1)\eta_0+P\eta_M\big)}{\lambda} \right] \\
&+ C_2\theta(-\eta_0)\theta(-\eta_M) \\
&\times \left[\frac{\exp\left(\frac{\pi|\eta_M-\eta_0|^2}{2\lambda^2}\right)\xi\big(-(M-1)\eta_0+\eta_M\big)}{\lambda} \right. \\
&\left.\left. - \frac{\exp\left(-\frac{\pi|P\eta_M-\eta_0|^2}{2\lambda^2}\right)\xi\big(-(M-1)\eta_0+P\eta_M\big)}{\lambda} \right] \right\},
\end{aligned}
$$

where

$$\xi(z) = \mathrm{erfc}\left[-\frac{\sqrt{\pi}z}{\sqrt{2M(M-1)}\tilde{\lambda}}\right].$$

Thus instead of the unique density matrix (9.70), we have an infinite number of fixed-node density matrices (9.72) (due to the two arbitrary constants C_1 and C_2) satisfying the 'zero boundary conditions' on the surface of the reach ($\eta_M = 0$). Moreover, this density matrix depends on the complementary error functions.

For a further detailed analysis of the function (9.72), we consider the limit $\Delta\beta \to 0$. Using the definition of the complementary error functions, one can transform (9.72) to the form

$$\begin{aligned}
\rho_{\mathrm{F}}(R_M, R_0; \beta) &= \frac{\exp\left(-\frac{2\pi|\gamma_M-\gamma_0|^2}{\lambda^2}\right)}{4\lambda} \\
&\times \left\{ C_1\theta(\eta_0)\theta(\eta_M)\frac{\left[\exp\left(-\frac{\pi|\eta_M-\eta_0|^2}{2\lambda^2}\right) - \exp\left(-\frac{\pi|P\eta_M-\eta_0|^2}{2\lambda^2}\right)\right]}{\lambda} \right. \\
&\qquad\qquad \times\left[1+\mathrm{sign}(\eta_0)\right] \\
&+C_2\theta(-\eta_0)\theta(-\eta_M) \\
&\left.\times\frac{\left[\exp\left(-\frac{\pi|\eta_M-\eta_0|^2}{2\lambda^2}\right) - \exp\left(-\frac{\pi|P\eta_M-\eta_0|^2}{2\lambda^2}\right)\right]\left[1-\mathrm{sign}(\eta_0)\right]}{\lambda}\right\} \\
&= \frac{\exp\left(-\frac{2\pi|\gamma_M-\gamma_0|^2}{\lambda^2}\right)}{4\lambda}\frac{\exp\left(-\frac{\pi|\eta_M-\eta_0|^2}{2\lambda^2}\right) - \exp\left(-\frac{\pi|P\eta_M-\eta_0|^2}{2\lambda^2}\right)}{\lambda} \\
&\quad\times\left\{C_1\theta(\eta_0)\theta(\eta_M) + C_2\theta(-\eta_0)\theta(-\eta_M)\right. \\
&\quad\left.+\left[C_1\theta(\eta_0)\theta(\eta_M) - C_2\theta(-\eta_0)\theta(-\eta_M)\right]\mathrm{sign}(\eta_0)\right\},
\end{aligned} \tag{9.73}$$

where

$$\begin{aligned}
\lim_{\Delta\beta\to 0}\mathrm{erfc}\left(-\sqrt{\pi}\frac{\pm(M-1)\eta_0+P\eta_M}{\sqrt{2M(M-1)}\tilde{\lambda}}\right) &= 1 - \lim_{\Delta\beta\to 0}\mathrm{erfc}\left(-\sqrt{\pi}\frac{\pm(M-1)\eta_0+P\eta_M}{\sqrt{2M(M-1)}\tilde{\lambda}}\right) \\
&= 1\pm\mathrm{sign}(\eta_0),
\end{aligned} \tag{9.74}$$

and erfc is the error function. Here sign(η_0) is equal to plus one for $\eta_0 > 0$ and minus one for $\eta_0 < 0$. In the limit $\Delta\beta \to 0$, the fixed-node density matrix (9.73) satisfies zero boundary conditions on the surface of the reach ($\eta_M = 0$) for two arbitrary constants C_1 and C_2.

To obtain a unique density matrix, we need to specify constants C_1 and C_2. The fixed-node density matrix (9.73) differs significantly from the exact expressions (9.70). For $C_1 = 1$ and $C_2 = 0$, we have a density matrix which coincides with the exact density matrix if both η_M and η_0 are positive. However, if η_M and η_0 have opposite signs the fixed-node density matrix is identically equal to zero and differs from the exact density matrix. Generally, the fixed-node density matrix (9.73) is identically equal to zero if η_0 and η_M lie in opposite half planes ($\eta > 0$ and $\eta < 0$) of the (γ, η) plane, while this is not the case for the exact density matrix (9.70). The fixed-node density matrix (9.73) is a non-analytical function, in contrast to the exact density matrix. Recall that all these expressions have been obtained on the assumption that all 'basic statements' in the papers (Ceperley 1991, 1992) are correct. All these contradictions mean that the integration in the reach cannot reproduce the exact solution of (9.63) with the initial condition (9.64), in spite of the 'basic statement' in (Ceperley 1991, 1992).

An alternative approach to Fermi systems without replacing the initial conditions by zero boundary conditions for the Bloch equation was discussed above and is known in the literature as the direct path integral Monte Carlo simulation (DPIMC) (Filinov 2001; Egger et al. 1999; Imada 1984; Filinov et al. 2000) In this approach, the sum over all permutations is represented identically as a determinant, which can be exactly calculated by the direct methods of linear algebra. The exactly calculated determinant can be positive or negative, but this approach significantly reduces the 'sign problem' discussed above and allows one to consider degenerate Fermi systems of particles. Comparing with the results of the DPIMC simulation, we find that the 'fixed-node method' describes the thermodynamic properties of strongly coupled fermions well only at weak degeneracy, when the main contribution to the partition function comes from the identity permutation (Filinov 2001; Filinov et al. 2000). The difference in the results obtained increases systematically with increasing degeneracy at high density and low temperatures (Filinov 2001; Filinov et al. 2000). The reason for this difference lies in the restriction by the 'reach' of integration over 'beads' in the fixed-node path integrals, this leading to an incorrect expression for the density matrix even for two ideal fermions. This restriction results in uncontrolled errors in calculations of thermodynamic quantities due to the incorrect description of statistical effects in the system of degenerate interacting and non-interacting fermions.

References

B.J. Berne, G. Ciccotti, D.F. Coker (eds.), *Classical and Quantum Dynamics of Condensed Phase* (World Scientific, Singapore, 1998)

K. Binder, G. Cicotti (eds.), *The Monte Carlo and Molecular Dynamics of Condensed Matter Systems* (SIF, Bologna, 1996)

M. Bonitz, D. Semkat (eds.), *Introduction to Computational Methods for Many Body Systems* (Rinton Press, Princeton, 2006)

M. Bonitz et al., J. Phys. B: Condensed. Matter **8**, 6057 (1996)

J. Bosse, K.N. Pathak, G.S. Singh, Phys. Rev. E **84**, 042101 (2011)

R.E. Caflisch, Acta Numerica **7**, 1 (1998)

D. Ceperley, J. Stat. Phys. **63**, 1237 (1991)

D. Ceperley, Phys. Rev. Let. **69**, 331 (1992)

D.M. Ceperley, Rev. Mod. Phys. **65**, 279 (1995)

D.M. Ceperley, in *The Monte Carlo and Molecular Dynamics of Condensed Matter Systems*, ed. by K. Binder, G. Cicotti (SIF, Bologna, 1996), pp. 447–482

S.A. Chin, Phys. Lett. A **226**, 344 (1997)

S.A. Chin, Phys. Rev. E **91**, 031301 (2015). (R)

T. Dornheim, S. Groth, A. Filinov, M. Bonitz, New J. Phys. **17**, 073017 (2015)

T. Dornheim, T. Schoof, S. Groth, A. Filinov, M. Bonitz, J. Chem. Phys. **143**, 204101 (2015)

T. Dornheim, S. Groth, T. Sjostrom, F.D. Malone, W.M.C. Foulkes, M. Bonitz, Phys. Rev. Lett. **117**, 156403 (2016)

C. Dorso, S. Duarte, J. Randrup, Phys. Lett B **188**, 287 (1987); **215**, 611 (1988)

C. Dorso, S. Duarte, J. Randrup, Excited electron dynamics modeling of warm matter, Phys. Lett. B **188**, 287 (1987); **215**, 611 (1988)

J.L. DuBois , B.J. Alder, E.W. Brown (2014), arXiv: 1409.3262

W. Ebeling, F. Schautz, Phys. Rev. E **56**, 3498 (1997)

W. Ebeling, A. Filinov, M. Bonitz, V. Filinov, T. Pohl, J. Phys. A Math. Gen. **39**, 4309 (2006)

R. Egger, W. Häusler, C.H. Mak, H. Grabert, Phys. Rev. Lett. **82**, 3320 (1999)

A.V. Eletskii, A.N. Starostin, M.D. Taran, Physics-Uspekhi **48**, 281 (2005)

A.V. Emelianov, A.V. Eremin, E.E. YuV Petrushevich, A.N. Sivkova, M.D. Starostin, V.E.Fortov Taran, ETP Lett. **94**(7), 530 (2011)

R.P. Feynman, A.R. Hibbs, *Quantum Mechanics and Path Integrals* (McGraw-Hill, New York, 1965)

A. Filinov, M. Bonitz, YuE Lozovik, J. Phys. A Math. Gen. **36**, 5957 (2003)

A. Filinov, V. Golubnychiy, M. Bonitz, W. Ebeling, J.W. Dufty, Phys. Rev. E **70**, 46411 (2004)

V. Filinov, J. Phys. A Math. Gen. **34**, 1665 (2001)

V. Filinov, P. Thomas, I. Varga, T. Meier, M. Bonitz, V. Fortov, S. Koch, Phys. Rev. B **65**, 165124 (2002)

V. Filinov, M. Bonitz, P. Levashov, V. Fortov, W. Ebeling, M. Schlanges, Contrib. Plasma Phys. **43**, 290 (2003)

V. Filinov, H. Fehske, M. Bonitz, V.E. Fortov, P.R. Levashov, Phys. Rev. E **75**, 36401 (2007)

V. Filinov, V. Fortov, M. Bonitz, Zh Moldabekov, Phys. Rev. E **91**, 033108 (2015)

V.S. Filinov, High Temp. **13**, 1065 (1975); **14**, 225 (1976)

V.S. Filinov, J. Phys. A **34**, 1665 (2001)

V.S. Filinov, High Temp. **52**, 615 (2014)

V.S. Filinov, M. Bonitz, V.E. Fortov, JETP Lett. **72**, 245 (2000)

V.S. Filinov, M. Bonitz, V.E. Fortov, JETP Lett. **72**, 245 (2000); Pisma v ZhETF, **72**, 361 (2000)

V.S. Filinov, V.E. Fortov, M. Bonitz, D. Kremp, Phys. Lett. A **274**, 228 (2000)

V.S. Filinov, P.R. Levashov, V.E. Fortov, M. Bonitz, in *Progress in Nonequilibrium Greens Functions*, ed. by M. Bonitz (World Scientific Publ, Singapore, 2000), p. 513

V.S. Filinov, V.E. Fortov, M. Bonitz, P.R. Levashov, JETP Lett. **74**, 384 2001); [Pisma v ZhETF **74**, 422 (2001)]

V.S. Filinov, M. Bonitz, W. Ebeling, V.E. Fortov, Plasma Phys. Control. Fusion **43**, 743 (2001)
V.S. Filinov, M. Bonitz, P.R. Levashov, V.E. Fortov, W. Ebeling, M. Schlanges, S.W. Koch, J. Phys. A Math. Gen. **36**, 6069 (2003)
V.S. Filinov, M. Bonitz, V.E. Fortov, W. Ebeling, P. Levashov, M. Schlanges, Contrib. Plasma Phys. **44**, 388 (2004)
V. Filinov, H. Fehske, M. Bonitz, V.E. Fortov, P.R. Levashov, Phys. Rev. E **75**, 36401 (2007)
V.M. Galitskii, V.V. Yakimets, Zh Eksp, Teor. Fiz. 51, 957 (1966). Sov. Phys. JETP **24**, 637 (1967)
S. Groth, T. Schoof, T. Dornheim, M. Bonitz, Phys. Rev. B **93**, 085102 (2016)
W.K. Hastings, Biometrika **57**(1), 97 (1970)
M.F. Herman, J. Chem. Phys. **76**, 29495 (1982). A similar energy estimator has been derived by Herman, but that result neglects the spin statistics and does not contain the correct noninteracting limit
K. Huang, *Statistical Mechanics* (Wiley, New York, 1963)
R. Imada, J. Phys. Soc. Jpn. **53**, 2861 (1984). However, we have found that this approach is not adequate for thermodynamic properties
G. Kalman (ed.), *Strongly Coupled Coulomb Systems* (Pergamon Press, New York, 1998)
G. Kelbg, Ann. Physik **12**, 219 (1963); **13**, 354; **14**, 394 (1964); W. Ebeling, H.J. Homann, G. Kelbg, Beitr. Plasmaphys. **7**, 233 (1967)
J.C. Kimball, J. Phys. A Math. Gen. **8**(9), 1513 (1975)
D. Klakow, C. Toepffer, P.-G. Reinhard, Phys. Lett. A 192, 55 (1994). J. Chem. Phys. **101**, 10766 (1994)
Yu.L. Klimontovich, Statistical Theory of Open Systems (in Russian), Vol. I (1995); II (1999); III (Janus, Moscow, 2001)
I.V. Kochetov, A.P. Napartovich, A.N. YuV Petrushevich, M.D.Taran Starostin, High Temp. **54**(4), 563 (2016)
W.D. Kraeft, M. Schlanges (eds.), in Proceedings of the International Conference on Strongly Coupled Plasmas, World Scientific, Singapore (1996)
G.P. Lepage, VEGAS: An Adaptive Multi-dimensional Integration Program, Cornell preprint CLNS 80–447 (1980)
A.S. Larkin, V.S. Filinov, V.E. Fortov, Contrib. Plasma Phys. **56**, 187 (2016)
A.S.Larkin, V.S. Filinov, V.E. Fortov (2017), arXiv:1702.04091 [physics.plasm-ph]
A.P. Lyubartsev, J. Phys. A Math. Gen. **38**, 6659 (2005)
J.M. McMahon, M.A. Morales, C. Pierleoni, D. Ceperley, Rev. Mod. Phys. **84**, 1607 (2012)
N. Metropolis, A. Rosenbluth, M. Rosenbluth, A. Teller, E. Teller, J. Chem. Phys. **21**, 1087 (1953)
B. Militzer, R. Pollock, Phys. Rev. E **61**, 3470 (2000)
M.E.J. Newman, G.T. Barkema, *Monte Carlo Methods in Statistical Physics* (Oxford University Press, USA, 1999)
J. Ortner, F. Schautz, W. Ebeling, Phys. Rev. E **56**, 4665–4670 (1997)
W.H. Press, G.R. Farrar, Comput. Phys. **4**, 190 (1998)
C. Robert, G. Casella, *Monte Carlo Statistical Methods* (Springer, New York, 2004)
G.O. Roberts, A. Gelman, W.R. Gilks, Ann. Appl. Prob. **7**(1), 110 (1997)
D. Ruelle, *Statistical Mechanics: Rigorous Results (World Scientific Publishing Co* (Pte. Ltd, Singapore, 1999)
T. Schoof, M. Bonitz, A.V. Filinov, D. Hochstuhl, J.W. Dufty, Contrib. Plasma Phys. **51**(8), 687 (2011)
T. Schoof, S. Groth, J. Vorberger, M. Bonitz, Phys. Rev. Lett. **115**, 130402 (2015)
T. Schoof, S. Groth, M. Bonitz, Contrib. Plasma Phys. **55**, 136 (2015)
M. Suzuki, Phys. Lett. A **146**, 319 (1990)
M. Suzuki, Phys. Lett. A **201**, 425 (1995)
A.N. Starostin, Yu.V. Petrushevich, Abstracts of Scientific-Coordination Workshop on Non-Ideal Plasma Physics, Moscow, Russia, 7–8 Dec 2016, http://www.ihed.ras.ru/npp2016/program/
A.N. Starostin, A.B. Mironov, N.L. Aleksandrov, N.J. Fischc, R.M. Kulsrudc, Physica A **305**, 287 (2002)

M. Takahashi, M. Imada, J. Phys. Soc. Jpn. **53**, 963–3765 (1984)
V.I. Tatarskii, Sov. Phys. Uspekhi **26**, 311 (1983)
S. Weinzierl, NIKHEF Theory Group (2000), arXiv:hep-ph/0006269v1
E.P. Wigner, Phys. Rev. **40**, 749 (1932)
V.M. Zamalin, G.E. Norman, V.S. Filinov, *The Monte Carlo Method in Statistical Thermodynamics* (Nauka, Moscow, 1977). (in Russian)
B.V. Zelener, G.E. Norman, V.S. Filinov, *Perturbation Theory and Pseudopotential in Statistical Thermodynamics* (Nauka, Moscow, 1981). (in Russian)

Chapter 10
Simulations of Gas-, Liquid-, and Crystal-Like States of Coulomb Systems

As example, we consider application of the theoretical approach discussed in the last chapter to treatment of a partially ionized strongly coupled dense electron-hole plasma. We take into account strong Coulomb correlation effects such as bound states (excitons, bi-excitons, clusters), their modification by the surrounding plasma, and their eventual breakup at high densities due to pressure ionization (Mott effect). Beyond the Mott density, we expect the possibility of hole crystallization if the hole mass is sufficiently large (Bonitz et al. 2005). We consider the results of direct path integral Monte Carlo simulations for a broad range of hole-to-electron mass ratios, temperatures, and densities. The density of the two-component plasma is characterized by the Brueckner parameters r_s, defined as the ratio of the mean distance between particles and the exciton Bohr radius. We discuss the spatial particle configurations, pair distribution functions, static structure factors, fractions of electrons and holes in bound states, internal energy, and equation of state.

10.1 Pair Distribution Functions, Bound States, Pressure, and Internal Energy for Electron–Hole Plasmas

Snapshots of the Electron–Hole Configuration

We first present some typical 'snapshots' of the electron–hole state in the simulation box for different particle densities. To make contact with the experiments on the Tm[Te,Se] system, the effective electron and hole masses were chosen as $m_e = 2.1$ and $m_h = 80$, respectively (in units of the free electron mass), and a relative dielectric constant $\varepsilon = 25$. Two temperatures ($T = 50$ K and $T = 200$ K) were taken below the exciton binding energy ($E_B/k_B = 517$ K) and within the measurement interval by (Wachter et al. 2004) (see Fig. 10.5).

Fig. 10.1 shows snapshots of the electron–hole configuration in the simulation box at different temperatures and densities. Each electron and hole were represented by twenty beads ($n = 20$), so for the upper panels, at $T = 50$ K, the high-temperature density matrices are taken at a temperature of 1 000 K. The spatial distribution of the beads of each quantum particle is proportional to its spatial probability distribution. Figure 10.1 shows that, if the hole mass is about forty times larger than the electron

W. Ebeling et al., *Quantum Statistics of Dense Gases and Nonideal Plasmas*, Springer Series in Plasma Science and Technology, DOI 10.1007/978-3-319-66637-2_10

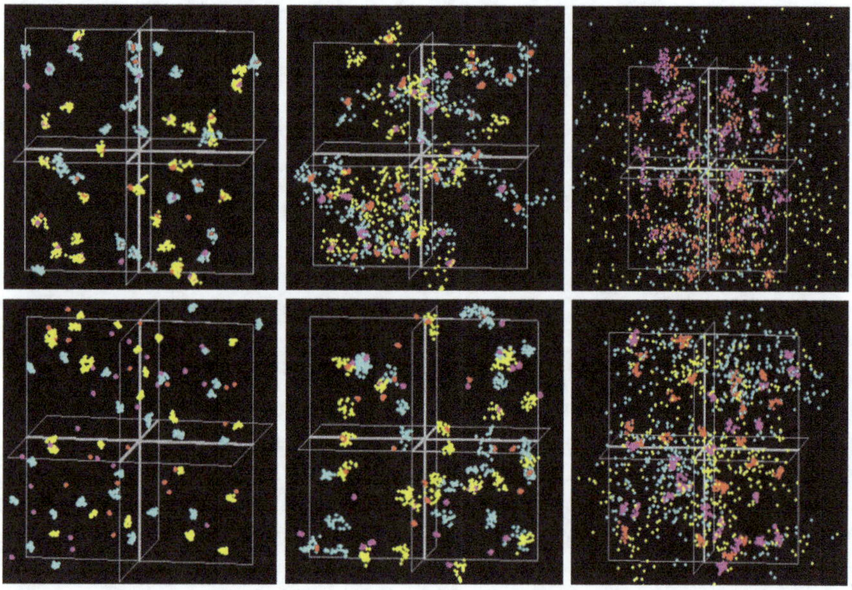

Fig. 10.1 Snapshots of particle configurations at three particle densities corresponding to $r_s = 10$ (*left*), $r_s = 4$ (*center*), and $r_s = 1$ (*right*). Spin up and spin down electrons (holes) are marked by *yellow and blue* (*red and pink*) clouds of dots, and the Monte Carlo cell is given by the *gray grid lines* (PBC were used). *Upper and lower panels* show results at low ($T = 50$ K) and high ($T = 200$ K) temperatures, respectively (Filinov et al. 2007)

mass, the typical size of the cloud of beads for electrons is several times larger than the one for the holes. Moreover, in the strongly correlated system, the extension of electron and hole 'bead' probability densities may differ from the de Broglie wavelength.

Note that, at low density ($r_s = 10$, left column) and temperature ($T = 50$ K), practically all holes are closely covered by electron beads. From a physical point of view this means that electrons and holes form bound states and the electron–hole plasma consists mainly of excitons. This statement will be supported below by the behavior of the pair distribution functions, exhibiting pronounced maxima at about one Bohr radius. Raising the temperature at fixed density leads to a (temperature-induced) ionization of the bound states and a substantial number of free electrons and holes in the simulations (the degree of ionization will be computed below).

At intermediate densities ($r_s = 4$, middle column) the formation of bi-excitons and many-particle clusters is observed at low temperatures ($T = 50$ K). So the electron–hole system is strongly inhomogeneous when the average distance d between particles is of the order of the electron wavelength. At $T = 200$ K, bi-excitons and many-particle clusters are absent due to thermal break up of exciton–exciton bound states.

When the particle density is high ($r_s = 1$), the electron wavelength becomes longer than the mean interparticle distance d and even longer than the size of the Monte Carlo cell, which is manifested by the broad extent of the electron bead clouds. For $r_s \leq 1.2$, excitons become unstable because two electrons bound to neighboring holes start to overlap, allowing for electron tunneling from one exciton to another. So bound states become unstable due to pressure ionization (Mott effect), and the system transforms into a strongly correlated electron–hole plasma. The structure of the hole presented by beads resembles a liquid state, since the hole wavelength is significantly shorter than the electron wavelength. If the hole mass exceeds a critical value, the holes may form a crystal (Bonitz et al. 2005).

Pair Distribution Functions and Structure Factors

We begin by discussing the behavior of the spin-averaged electron–electron, hole–hole, and electron–hole pair distribution functions:

$$g_{ab}(r) = \frac{N_e!N_h!}{Z} \sum_{\sigma} \int_{V} \mathrm{d}q \, \delta(r_{1,a} - q_{1,a})\delta(r_{2,b} - q_{2,b})\rho(q, \sigma; \beta) \, . \qquad (10.1)$$

Fig. 10.2 shows the three functions $g_{ab}(r)$ versus the interparticle distance $r = |r_{1,a} - r_{2,b}|$ for the same densities and temperatures as were used in the snapshots in Fig. 10.1. The fluctuations of g_{ee}, g_{hh}, and g_{eh} at small r reflect the magnitude of the statistical errors of simulation.

The pair distribution functions g_{ee} and g_{hh} are suppressed at small distances due to the Coulomb and Fermi (statistical) repulsions. The decay of the electron–electron correlations, however, is essentially different from that of the hole–hole correlations due to the large mass difference. The behavior of g_{ee} at small distances is defined mainly by electrons with opposite spin projections, as in this case where there is no Fermi repulsion and the main contribution to the repulsion comes from the effective quantum Kelbg potential, allowing for tunneling effects. This is also supported by the behavior of the spin-dependent pair distribution discussed later.

The sharp peak in g_{eh} at low densities ($r_s = 8$–10) is caused by excitons. This is confirmed by considering the function $r^2 g_{eh}(r)$, which has a pronounced maximum at about one exciton Bohr radius ($1a_B$), while the functions g_{ee} and g_{hh} exhibit no peak structure. So there is no indication of bound exciton–exciton formation (such as bi-excitons or even electron–hole droplets). With increasing temperature more and more excitons dissociate, and the maximum of $r^2 g_{eh}$ is reduced.

At low temperatures, varying the density over three orders of magnitude, the maximum of $r^2 g_{eh}$ is suppressed and finally vanishes. There is evidence of recombination into bi-excitons at about $r_s = 6$. Here the position of the maximum in $r^2 g_{eh}$ is shifted from 1 a_B to 2–3 a_B, indicating the increasing radii of the bound states. This is confirmed by the electron–electron and hole–hole pair correlation functions. For intermediate densities ($2 < r_s < 6$), g_{hh} and g_{ee} exhibit distinct peaks at larger r, pointing to formation of bi-excitons and many-particle clusters. In a rather dense electron–hole plasma ($r_s = 1$), the fraction of excitons is further reduced due to

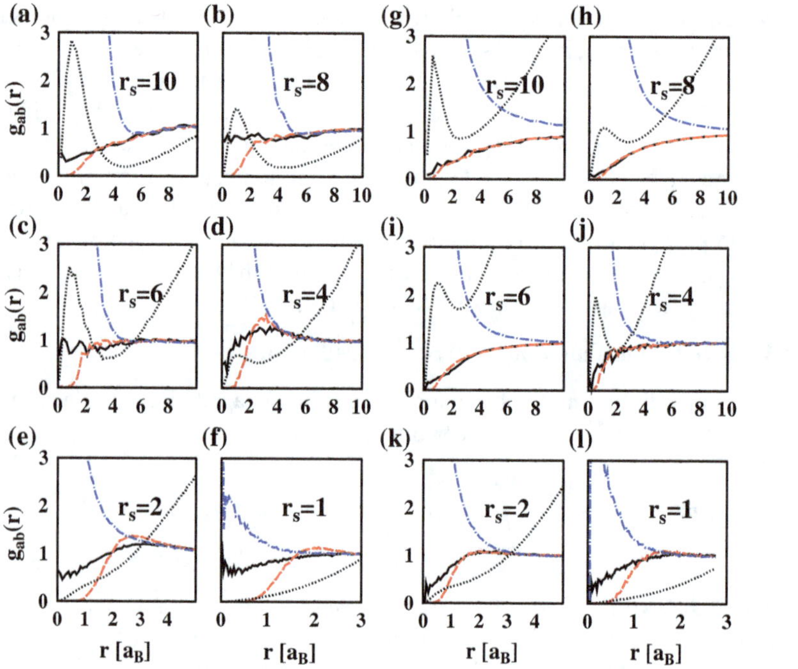

Fig. 10.2 Pair distribution functions g_{ee} (*black solid line*), g_{hh} (*red dashed line*), and g_{eh} (*blue dot-dashed line*) at $T = 50$ K (*two left columns*) and $T = 200$ K (*two right columns*). *Dotted lines* show $r^2 g_{eh}$, which is rescaled for better visibility of the exciton peak by: **a** and **b** factor $1/120$, **c** and **d** factor $1/30$, **e** and **f** $1/10$, **g** and **h** $1/30$, **i**–**l** $1/10$ (Filinov et al. 2007)

many-body effects (pressure ionization), and bi-excitons and many-particle clusters vanish. Now the short-ranged electron–hole correlations disappear and the density is close to the Mott density, prohibiting the formation of bound states.

Let us analyze the width of the peak of g_{eh} relative to the extent of the ground state wave function ($\sim 1 a_B$). At low densities ($r_s > 6$), the peak in g_{eh} is rather broad, indicating the population of excited states, while at high densities ($r_s \sim 1$), the width of the lowest peak in g_{eh} becomes significantly smaller. At the same time the hole–hole pair distribution function reveals an ordering of the holes into a fluid-like state.

Figure 10.3 shows the variation of the static charge structure factor, defined in reciprocal k-space by

$$S_{ab}(k) = \frac{\displaystyle\int_0^\infty \mathrm{d}r\, r^2\big[g_{ab}(r) - 1\big] \sin(kr)/kr}{\left|\displaystyle\int_0^\infty \mathrm{d}r\, r^2\big[g_{ab}(r) - 1\big]\right|}. \tag{10.2}$$

According to (10.2), positive (negative) values of $S_{ab}(k)$ indicate attraction (repulsion) (see also Fehske et al. 2005).

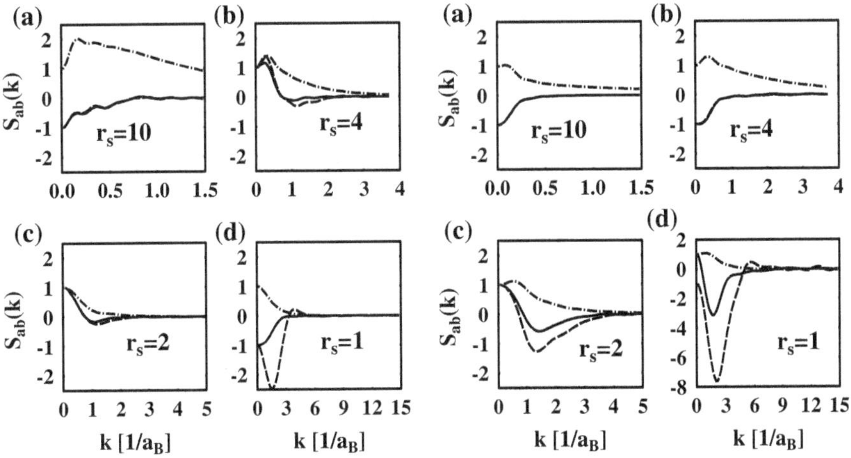

Fig. 10.3 Static structure factor S_{ee} (*black solid line*), S_{hh} (*red dashed line*), and S_{eh} (*blue dot-dashed line*) at $T = 50$ K (*two left columns*) and $T = 200$ K (*two right columns*) (Filinov et al. 2007)

At low densities ($r_s = 10$), the negative values of the electron–electron and hole–hole structure factors at small momenta (large distances) originate from the strong Coulomb repulsion of equally charged particles. The maximum of S_{eh}, as well as the modulations of S_{ee} and S_{hh} at finite k, are indicative of exciton formation. Excitons set a new length scale in the structure factor at about 0.2 ($1/2a_B$). These signatures are washed out at higher temperatures, where excitons break up. However, when the density is increased, bi-excitonic bound states occur at around $r_s = 4$, and then larger clusters of electrons and holes at $r_s \approx 2$, and this is reflected in the positive charge correlations between both electrons and holes. These correlations will be suppressed at high temperatures, but are more stable at higher densities (see right-hand columns of Fig. 10.3). At high densities ($r_s = 1$), the joint action of the Coulomb and Fermi repulsion, and even more importantly, the high mass ratio of electrons and holes, are responsible for the different behavior of the electron–electron and hole–hole pair correlation functions. This is observed particularly for low temperatures, when the electrons are perfectly delocalized but the holes still have fluid-like short-range correlations. The value of the normalization constant (denominator) in (10.2) is responsible for the different magnitudes of the hole–hole structure factor at $T = 50$ K and $T = 200$ K.

The electron–electron and hole–hole pair distributions and structure factors presented above were averaged over the spin degree of freedom of the particles. However, the DPIMC simulations allow inspection of the spin effects, which are important for high degeneracy ($n\lambda^3 \gg 1$). There may be qualitatively different behavior of $g_{aa}^{\uparrow\uparrow}$ and $g_{aa}^{\uparrow\downarrow}$ at small distances due to the influence of the Fermi statistics. This is confirmed by Fig. 10.4, which shows that the $g_{ee}^{\uparrow\uparrow}$ (Fig. 10.4b, d, and f) correlations are strongly suppressed compared to the $g_{ee}^{\uparrow\downarrow}$ (Fig. 10.4a, c, and e) correlations. We clearly see

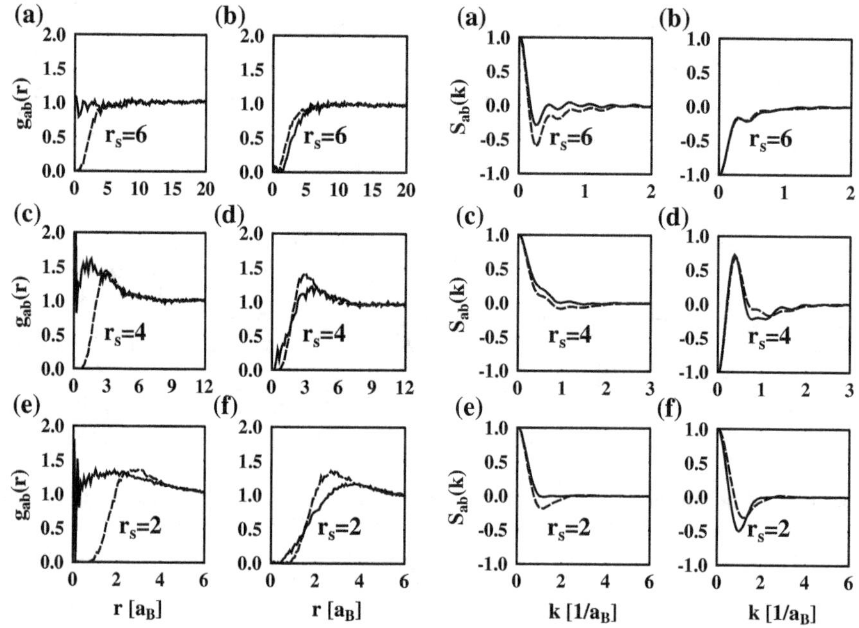

Fig. 10.4 Spin-resolved pair distribution functions $g_{ab}(r)$ and structure factor $S_{ab}(k)$ for electrons (*solid lines*) and holes (*dashed lines*) at $T = 50$ K. *Left* (*right*) columns correspond to antiparallel (parallel) spin projections of the particle pair (Filinov et al. 2007)

the hole exchange familiar from the ideal Fermi gas, but here it appears in a strongly interacting electron–hole plasma. In contrast, holes with masses forty times larger than the electrons behave more classically, and consequently the quantum statistical repulsion effect is much less pronounced. Here, decay of the hole–hole pair correlations for $r \to 0$ is mainly triggered by the (spin-independent) Coulomb interaction.

The snapshots depicted in Fig. 10.1 showed that, in the density regime $2 \leq r_s \leq 6$, the plasma contains a large fraction of bi-excitons. Here, Fig. 10.4 confirms this conclusion, since for $r_s = 4$, there is not much difference between $g_{hh}^{\uparrow\uparrow}$ and $g_{hh}^{\uparrow\downarrow}$. The two functions exhibit a similar peak at a hole–hole distance of about $3a_B$, while, in contrast, the electron behavior is strongly spin-dependent. There is a high probability of encountering two electrons with different spin projections between two holes at a distance $1a_B$–$2a_B$, as in bi-excitons or the hydrogen molecule. In addition, there are many-particle clusters which are also mainly built up from electrons with opposite spin projections.

The corresponding charge structure factors show that the difference between electrons and holes with respect to the spin correlations is most pronounced when bound states are formed. The differences are less important at very small and large k (corresponding to large and small distances, respectively). There is attraction (repulsion) for opposite (equal) spin projections of the two particles as $k \to 0$, except at high densities. Here, in the vicinity of the Mott transition, the difference between the structure factor for electrons and holes with the same and opposite spin projections is notably smaller, and the structure factors are close to their spin-averaged values presented in Fig. 10.3 for $T = 50$ K and $r_s = 2$.

Bound State Formation and Dissociation

As discussed above in the analysis of the snapshots and the pair distribution function, the mass-asymmetric electron–hole plasma contains free electrons and holes, excitons, bi-excitons, and many-particle clusters. This means that a varying fraction of electrons and holes are in bound states or in correlated scattering states. Strictly speaking, there is no clear subdivision into bound and free 'components', both contribute to the same pair distribution functions and it is rigorously impossible to obtain the 'fraction' of free and bound states, due to the mutual overlap of the electrons (and holes). Moreover, there is no rigorous criterion for excitonic or bi-excitonic bound states at high particle densities, due to the significant effect of the surrounding plasma on individual bound states.

Nevertheless, a rough estimate of the fraction of electron–hole bound states can be obtained by the following reasoning. Physically, the product $r^2 g_{\mathrm{eh}}(r)$ is the probability of finding an electron at a distance r from a hole. The product of r^2 and the corresponding quantum mechanical two-particle Slater sum, viz.,

$$\Sigma_{\mathrm{eh}} = 8\pi^{3/2}\lambda_{\mathrm{eh}}^3 \sum_{E_\alpha=E_0}^{\infty} |\Psi_\alpha(r)|^2 \exp(-\beta E_\alpha) , \tag{10.3}$$

has the same physical meaning, where E_α and $\Psi_\alpha(r)$ are the energy (without the center of mass energy) and the wave function of an electron–hole pair, respectively. Σ_{eh} is the diagonal part of the corresponding density matrix. In (10.3), the summation extends over all possible states α with contributions from the discrete (d) and continuous (c) parts of the spectrum:

$$\Sigma_{\mathrm{eh}} = \Sigma_{\mathrm{eh}}^{\mathrm{d}} + \Sigma_{\mathrm{eh}}^{\mathrm{c}} ,$$

where

$$\Sigma_{\mathrm{eh}}^{\mathrm{d}} = 8\pi^{3/2}\lambda_{\mathrm{eh}}^3 \sum_{E_\alpha=E_0}^{E'} |\Psi_\alpha(r)|^2 \exp(-\beta E_\alpha) \tag{10.4}$$

originates from the populated bound states between E_0 and E'. $\Sigma_{\mathrm{eh}}^{\mathrm{d}}$ can be calculated from the known wave functions $\Psi_\alpha(r)$. The contribution of the remaining (scattering) states is (Zamalin et al. 1977)

$$\Sigma_{\text{eh}}^{\text{c}} = \begin{cases} \left[1 - \text{erf}\left(\sqrt{\beta E' + \frac{\beta e^2}{r}}\right)\right] \text{e}^{-\beta e^2/r} + \frac{2}{\sqrt{\pi}}\left(\beta E' + \frac{\beta e^2}{r}\right), \\ \qquad\qquad\qquad\qquad \text{if } |\beta E'| < \frac{\beta e^2}{r}, \\ \text{e}^{-\beta e^2/r}, \quad \text{if } \beta E' > \frac{\beta e^2}{r}. \end{cases} \tag{10.5}$$

Here $\Sigma_{\text{eh}}^{\text{c}}(r, T)$ was derived using the quasi-classical approximation for the continuum and bound states with $E_\alpha > E'$. For low densities, it is reasonable to choose $E' \geq -1/\beta$ (Zamalin et al. 1977), while for high densities $E' \geq -\text{Ry}/r_{\text{s}}$ is appropriate, since in these quantum states, the electrons can be treated as free particles.

At lower temperatures than the electron–hole binding energy and shorter distances than several Bohr radii, the main contribution to the Slater sum comes from electron–hole bound states, while at larger distances free electron–hole (scattering) states give the dominant contribution. The product $r^2\Sigma_{\text{eh}}$ is sharply peaked around a_{B}, just like $r^2 g_{\text{eh}}(r)$ in Fig. 10.2, where the maximum is most pronounced at low temperatures. The height of the peak at about $1a_{\text{B}}$ increases with decreasing temperature. On the other hand, the peak is destroyed with increasing plasma density, due to the strong interaction between the particles. At high temperatures and densities, the quantity $r^2 g_{\text{eh}}(r)$ approaches the parabola r^2, as in an ideal plasma without bound states $g_{\text{eh}}(r) \equiv 1$. So the main contribution to the product $r^2\left[g_{\text{eh}}(r) - 1\right] > 0$ at short electron–hole distances, in the interval $0 \leq r \leq r^{\text{b}}$, comes from the bound states (to the fully bound plus continuum states). Then the fraction of electron–hole bound states can be roughly estimated from the ratio

$$\frac{N_{\text{eh}}^{\text{b}}}{N_{\text{eh}}^{\text{b}} + N_{\text{eh}}^{\text{c}}} = \frac{\displaystyle\int_0^{r^{\text{b}}} r^2\left[g_{\text{eh}}(r) - 1\right]\text{d}r}{\displaystyle\int_0^{r^{\text{b}}} r^2 g_{\text{eh}}(r)\text{d}r}, \tag{10.6}$$

where the denominator is the normalization constant and has the physical meaning of electrons being either in bound or free continuum states. An analogous ratio can be defined for the hole–hole bound-state fraction:

$$\frac{N_{\text{hh}}^{\text{b}}}{N_{\text{hh}}^{\text{b}} + N_{\text{hh}}^{\text{c}}} = \frac{\int_{r^{\text{b}'}}^{r^{\text{b}}} r^2\left[g_{\text{hh}}(r) - 1\right]\text{d}r}{\int_{r^{\text{b}'}}^{r_b} r^2 g_{\text{hh}}(r)\text{d}r} \tag{10.7}$$

with $[r^{\text{b}'}, r^{\text{b}}]$ being the interval where $r^2\left[g_{\text{hh}}(r) - 1\right]$ is positive.

Figure 10.5 presents the results of our calculations of the electron–hole bound state fraction according to (10.6) and (10.7). At low densities ($r_{\text{s}} > 6$) and low temperatures ($T < 100$ K), practically all electrons and holes are bound in excitons, i.e., bi-excitons and many-particle clusters are absent. Increasing the temperature to $T \geq 100$ K leads to a strong ionization of excitons, i.e., free particles dominate. In the intermediate density regime ($2 < r_{\text{s}} < 6$) and at low temperatures ($T < 100$ K), the highest fraction (up to about 20%) of hole–hole bound states (bi-excitons and many-particle clusters) is observed. At still higher temperatures ($T \sim 100$–300 K), the exciton fraction increases slightly for higher densities. Above the Mott density ($r_s \leq 1$), bound states do not exist due to the overlap of two electron wave functions from different excitons and strong screening for all temperatures.

The temperature dependence of the electron–hole bound state fraction at low densities ($r_s = 10$, in the excitonic regime) is presented in Fig. 10.5c). A larger dielectric constant ε leads to weakly bound excitons and shifts the ionization temperature to lower values. The same tendency is observed if the mass ratio m_h/m_e is lowered. Figure 10.5 shows that the ratio of temperatures corresponding to a fraction of electron–hole bound states of 50% is the same as the corresponding ratio of the reduced electron–hole masses $\mu = m_e m_h/(m_e + m_h)$ (keeping ε fixed). For example, a ratio of about 2 is observed for the data belonging to $m_h = 80$ and $m_h = 2.1$, and 1.5 for those belonging to $m_h = 80$ and $m_h = 4.2$. For an isolated exciton, the binding energy is proportional to the reduced electron–hole mass, while in this case, the excitons are embedded in an electron–hole plasma which influences the bound state spectrum, so recovery of the single-atom behavior is not a trivial result.

The electron–hole systems considered above with a mass anisotropy as high as 40 are indeed realized, e.g., in intermediate valence $Tm[Se_xTe_{1-x}]$ alloys (Wachter et al. 2004). The f-d-hybridization provides a narrow dispersive f-valence band and a large effective hole mass of the order of 50–100 (bare) electron masses. $TmSe_{0.45}Te_{0.55}$ is an indirect semiconductor with a gap of $E_\Delta = 130$ meV, where with $E_X^b \approx 50$–70 meV, an excitonic level has been observed below the bottom of the d-band. By applying pressure, the gap can be tuned and even closed, and it is speculated that, in the pressure region between 5 and 11 kbar, the material will become an excitonic insulator. The search for this has been running for a long time now (Halperin and Rice 1968). A necessary precondition is the existence of a high number density of (up to 10^{20}–10^{21} per cm^3) excitons of intermediate size (in order to avoid too strong an overlap of the excitonic bound states) (Wachter et al. 2004). Surprisingly, in $TmSe_{0.45}Te_{0.55}$, the excitonic phase is predicted to be stable at rather high temperatures (up to 200 K) (see the experimental phase diagram in Fig. 10.5).

Using the $Tm[Se_xTe_{1-x}]$ parameters in the considered simplified Coulomb model, the DPIMC simulations can, at least qualitatively, test experimental observations and expectations. Although DPIMC simulations do not exhibit a sharp transition from an excitonic to an ionized 'phase', a bound state fraction of 90% can be used (formally) as a condition for an excitonic phase. At low densities ($r_s = 10$), these bound states will be stable up to $T = 150$–250 K, depending on ε (see Fig. 10.5a). The high mass asymmetry between holes and electrons is crucial for these high temperatures. Further, we can give a rough estimate of the critical density where the transition from an insulator (electron–hole bound states) to a (semi-) metal takes place. As a qualitative criterion, we may assume that, in the metallic state, the bound state fraction should not exceed 5–10%. In addition, experiments indicate that the boundary should not depend much on the temperature (see the steep, high density branch of the phase boundary in the experiment, bottom part of Fig. 10.5). In DPIMC simulations (Fig. 10.5a), these conditions are indeed satisfied for $r_s \sim 1$, which corresponds to a

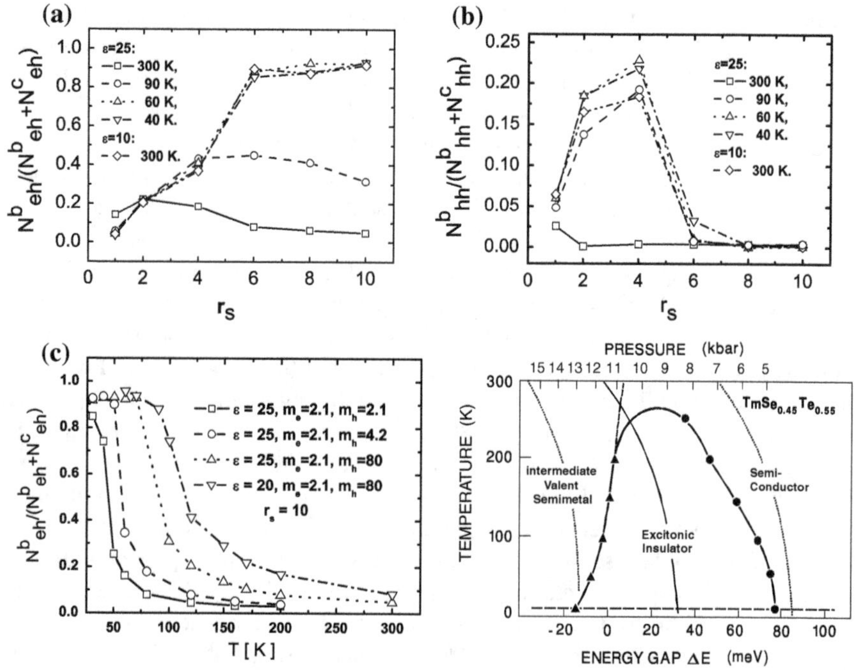

Fig. 10.5 Fraction of the electron–hole (**a**) and hole–hole bound states (**b**) versus the Brueckner parameter r_s for different temperatures and dielectric constants (indicated in the figures). (**c**) Temperature-induced breakup of electron–hole bound states (excitons) for different mass ratios and dielectric constants. *Bottom right* Excitonic phase diagram of $TmSe_{0.45}Te_{0.55}$ taken from (Wachter et al. 2004). Measured points are designated by symbols. 'Isobars' in the semi-conducting and semi-metallic phase are marked by *dotted lines*, whereas the 'isobar' entering the excitonic phase is denoted by a *continuous line*. The *lower abscissa* gives the corresponding energy gap ΔE (here negative values refer to the metallic state). For more details see Wachter et al. 2004, Filinov et al. 2007

particle density of the order of 10^{21} cm^{-3}, in reasonable agreement with the experimental estimates, and this density can be taken as the boundary between excitonic (insulating) and metal-like behavior (Mott transition).

According to the experimental prediction, there should be a very weak density dependence of the phase boundary near the maximum temperature of the excitonic phase. The same is observed in DPIMC simulations for the electron–hole bound state fraction for $r_s \sim 6$ (see Fig. 10.5a).

If the density is further reduced (at fixed temperature), the bound state fraction will decrease in the simulations as well, because of entropy ionization. If the average distance between electrons and holes becomes very large, the probability for recombination of ionized electrons and holes is negligible compared to the inverse process. The corresponding reduction of the bound state fraction for $r_s \gg 10$ is difficult to observe in DPIMC calculations, due to poor convergence in this regime. In this case the characteristic size of the exciton is much smaller than the average distance

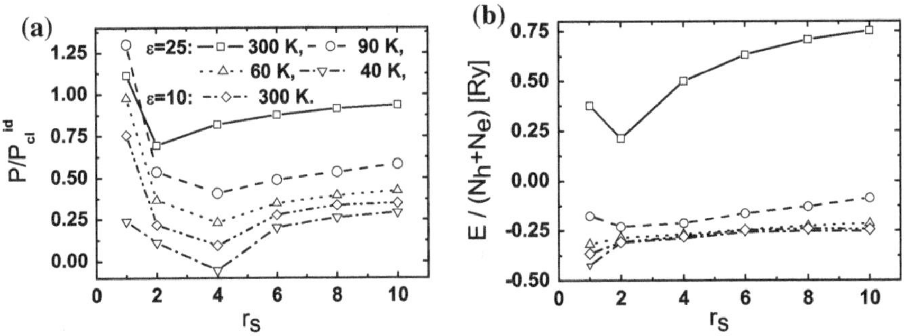

Fig. 10.6 Pressure and internal energy versus Brueckner parameter ($\sim n^{-1/3}$). Pressure is in units of the classical ideal pressure $P_{cl}^{id} = (n_e + n_h)k_B T$ (Filinov et al. 2007)

between particles, and the formation and decay of excitons are very rare events in the simulations, whence the statistical error becomes prohibitively large. Summarizing this analysis, the DPIMC simulations do indeed support the experimental prediction about an excitonic insulator phase in $TmSe_{0.45}Te_{0.55}$.

Pressure and Internal Energy

It is well-known that at high temperatures and low densities (weak interaction) both pressure and energy increase monotonically with density (as in the ideal case). The same happens due to the degeneracy of the electron (and possibly holes) at low temperatures and high densities (above the Mott density). Note that DPIMC plasma calculations yield the behavior of the ideal plasma asymptotically, not only at high densities (ideal mixture of degenerate electrons and holes), but also at low densities (ideal classical plasma). Figure 10.6 shows isotherms of a mass-asymmetric electron–hole plasma with different permittivities. The general trend is an increase in the pressure with increasing density ($r_s \leq 2$), which is most clearly seen for the highest temperature (300 K) and at very low densities (not shown).

At intermediate densities ($r_s \approx 2$–4), when the interaction effects increase, the pressure and energy of the nonideal electron–hole plasma deviate significantly from this trend. One can observe a non-monotonic behavior, in particular at low temperatures. The pressure and energy minima arise due to attractive Coulomb forces leading to formation of excitons and bi-excitons. Here, the pressure and energy are significantly lower than for an ideal plasma. A analysis of the low-temperature ($T = 50$ K) DPIMC results reveals that, for $2 \geq r_s \geq 4$, the homogeneous plasma state is unstable and forms clusters or droplets of higher density. These spatial inhomogeneities are clearly visible in the electron–hole configurations (‘snapshots’) and might be interpreted as an indication of a first order phase transition, as discussed by many authors (see Ebeling and Richert 1985, Beule et al. 1999, Schlanges et al. 1995, Filinov et al. 2001, and references therein). Let us emphasize that these effects have been observed in DPIMC simulations of electron–hole plasmas under conditions for which droplet formation is well established (Filinov et al. 2003a, b, c), and were observed experimentally three decades ago [Jeffries and Keldysh (eds.) 1988, Hensel et al. 1977, 2005]. This 'phase separation' does not exist above the Mott transition point, while in the intermediate density regime, there are large fluctuations of pressure and energy related to permanent formation and decay of clusters.

The present DPIMC simulations with small numbers of particles in the form of 'droplets' cannot adequately describe a macroscopic system in the two-phase region. This is because, in the macroscopic system, 'droplets' are expected to contain many orders of magnitude more particles. Small numbers of particles are the reason why the DPIMC pressure may become negative in a canonical simulation of a two-phase region. The clarification of this interesting issue will require more extensive simulations with substantially larger particle numbers and a transition to the grand canonical ensemble. Cluster formation has also been found in previous calculations for hydrogen plasmas (Filinov et al. 2001, Bonitz et al. 2003), and this supports the conclusion about the phase transition.

Discussion of Gas-like and Liquid-like States

The two-component Coulomb systems considered above with a mass ratio $M \sim 50$, which is intermediate between 'typical' semiconductors with $M \sim 2$–10 and 'conventional' plasmas such as hydrogen with $M \geq 1850$, behave in many respects like plasmas. The main difference is that the quantum and spin properties of the holes cannot be neglected. Such systems can be realized comparatively easily in intermediate valence semiconductors under pressure. From a plasma physics standpoint, these are very interesting materials because they can be used to investigate nontrivial high density phenomena, such as pressure ionization (Mott effect), the plasma phase transition, and the insulator–metal transition. These effects exist in 'conventional' plasmas only above a density of the order of $1\,\mathrm{g\,cm^{-3}}$, which can be achieved in laser or ion beam compression experiments and only for short periods of time. This makes measurements of the plasma properties very difficult. The same physical effects can be observed in the above-mentioned semiconductor materials under stationary conditions at densities and temperatures which are easily accessible in experiments. Qualitatively, the phase diagrams of dense plasmas and electron–hole plasmas are readily translated from one to another by rescaling the binding energy and the Bohr radius (Bonitz 2002, Bonitz et al. 2003).

For this reason, the phase diagram of Tm[Se,Te] shown in Fig. 10.5 is also of interest for dense partially ionized 'conventional' plasmas. In the central part of this phase diagram, there is a Mott transition and bound states. A theoretical treatment of these effects is very difficult, because many-body effects such as bound state formation, screening, and quantum effects have to be taken into account. While analytical methods fail for such complex systems, the path integral Monte Carlo approach is well suited.

Let us note that, above the Mott point, two-component plasmas with high mass anisotropy show interesting Coulomb correlation phenomena: with increasing density holes can undergo a phase transition to a Coulomb liquid and possibly to a Wigner crystal, these being embedded in a degenerate electron gas. However, crystal formation should be possible in a two-component plasma for mass ratios as low as 80 (Bonitz et al. 2005).

10.2 Thermodynamic Properties of Hydrogen Plasmas

Hydrogen plasma are attracting the interest of researchers in many fields, including plasmas, astrophysics, and solids [see G. Kalman (ed.) 1998, W.D. Kraeft and Schlanges (eds.) 1996 for an overview]. These overviews suggest that, in the temperature range Ha $\geq k_B T > 0.1$ Ry, we can expect the following general behavior of the internal energy per proton. At zero density, the energy is given by the ideal gas expression. With increasing density, the energy per proton becomes negative due to correlation effects (bound states, electron correlations, proton correlations). A minimum is formed, and then, at a density where the proton density is close to the cube of the reciprocal Bohr radius, the energy per proton turns to positive values and is determined by the ideal electron energy, increasing as $n^{2/3}$, corrected by correlation contributions of order $n^{1/3}$ which are determined by the Hartree term and by proton–proton coupling effects. In the following, it will be shown that this qualitative picture is supported by the results of DPIMC calculations.

The most interesting phenomena, such as metallic hydrogen (Da-Silva et al. 1997), the plasma phase transition, bound states, etc., occur in situations where the plasma is both strongly coupled and strongly degenerate. However, in this region, the thermodynamic properties of the plasma and its equation of state are only poorly known. Accounting for the strong Coulomb and quantum effects makes a theoretical treatment difficult because perturbative approaches, starting either from a weakly correlated plasma or from a weakly degenerate plasma, fail. Here it will be shown that, for many current problems in dense warm plasmas ($k_B T > 0.1$ Ry), the DPIMC simulations can in fact be carried out with acceptable efficiency, and interesting results have been obtained for the internal energy, equation of state, and pair distribution functions of partially ionized hydrogen over a wide range of coupling and degeneracy parameters, viz., $\Gamma = (4\pi n_e/3)^{1/3} e^2/4\pi\epsilon_0 k_B T$ and $\chi = n_e \lambda_e^3$, where λ_e is the electron thermal wavelength satisfying $\lambda_e^2 = 2\pi\hbar^2\beta/m_e$.

Ideal Plasmas. Test Calculations

To test the DPIMC procedure, ergodicity, and correct evaluation of the exchange contributions, the pressure and energy obtained by DPIMC can be compared with analogous values for a mixture of *ideal* electrons and protons for which thermodynamic quantities are known analytically (Kraeft et al. 1986). Figure 10.7 shows DPIMC results together with the theoretical curves. The agreement is clear, up to values of the degeneracy parameter χ as high as 10, and improves with increasing particle number (see inset of Fig. 10.7). This proves that DPIMC correctly samples the fermionic permutations.

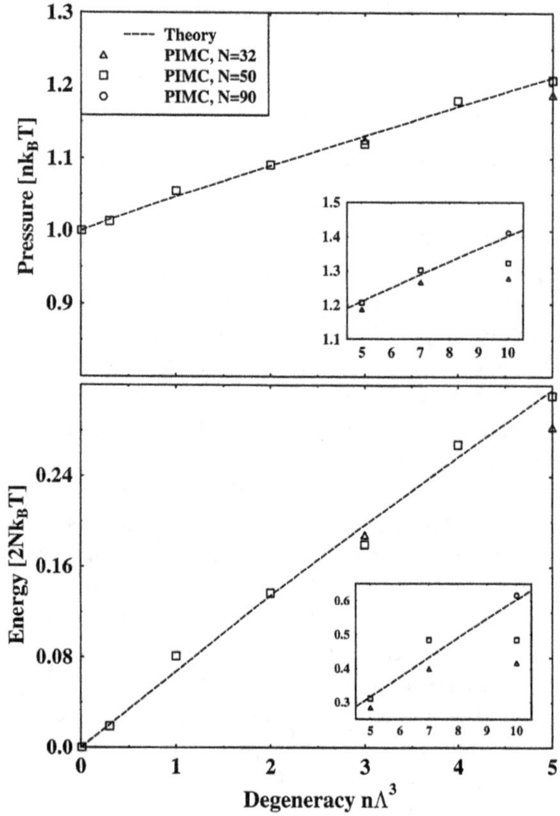

Fig. 10.7 Pressure and energy of an ideal plasma of degenerate electrons and classical protons. PIMC simulation results are shown for varying particle number: $N = 32$ (*triangles*), $N = 50$ (*squares*), and $N = 90$ (*circles*) are compared with the exact analytical result (*dashed line*) (Filinov et al. 2000a, b, c)

Nonideal Plasmas. Basic Physical States

Figure 10.8 and Table 10.1 present DPIMC results for nonideal hydrogen plasmas, in which the classical coupling parameter Γ is kept constant while the degeneracy parameter χ is varied. We see that, for weak coupling $\Gamma = 0.4$ and small degeneracy parameters $\chi < 0.5$, there is good agreement with analytical theories and QMC simulations without exchange (Riemann et al. 1996). However, as expected, with increasing χ and Γ, the deviations are growing rapidly. This figure also contains comparisons with restricted path integral results from Militzer et al. (Militzer and Ceperley 2000) (large triangles), corresponding to values of the coupling parameter in the range 0.17–1.35. Clearly, the agreement of the energies is very good. The deviations in the pressure are related to the fixed-node approximation used in (Militzer and Ceperley 2000).

Here the energy curves in Fig. 10.8 become almost parallel as the degeneracy increases. This indicates that, at high density and temperature, quantum and correlation effects are disentangled, i.e., increasing the coupling leads to an almost constant shift in the energy (Filinov et al. 2000b, c). In contrast, for $\Gamma > 0.6$, reduction of χ

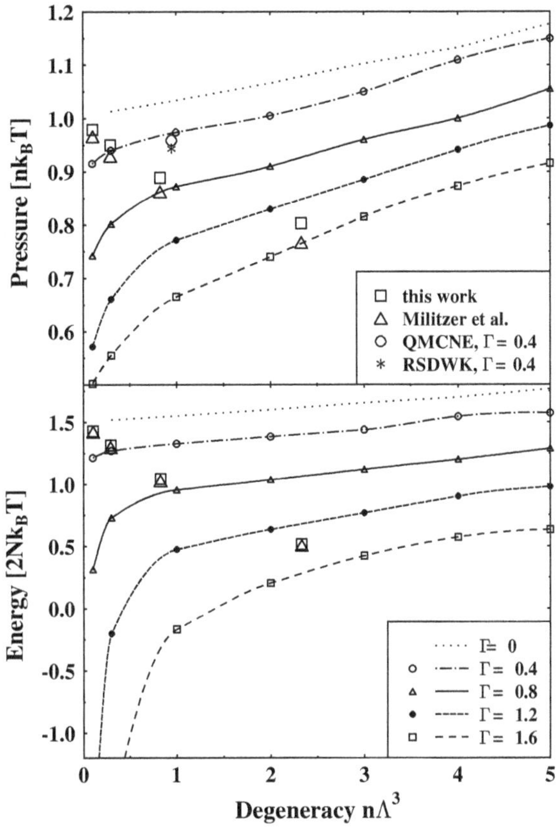

Fig. 10.8 Pressure (*upper*) and energy (*lower*) of a nonideal plasma of degenerate electrons and classical protons as a function of the quantum parameter χ. Curves correspond to different values of the coupling parameter Γ given in the inset of the lower figure. The *large circle* denotes quantum MC simulations without exchange (QMCNE) and the *large asterisk*, the weak coupling model of Riemann et al. (RSDWK) (Riemann et al. 1996). *Large triangles* are restricted PIMC results from Militzer et al. (Militzer and Ceperley 2000). These are compared with results (*large squares*) for four values of Γ. *Top to bottom*: $\Gamma = 0.169, 0.338, 0.672$, and 1.35 (Filinov et al. 2000a, b, c)

leads to a rapid decrease in the energy which is due to the formation of atoms and molecules, as will be shown below.

The main advantage of the DPIMC method is that it can be used to investigate dense plasmas in a variety of physical situations which are very difficult to describe reliably by other approaches. This includes partial ionization and dissociation, the Mott effect, and ionic ordering at high densities. Figure 10.9 shows the pair distribution functions for the four most interesting physical situations. Figure 10.9a clearly shows the existence of atoms and hydrogen molecules (note the peaks of the proton–proton and electron–electron pair distribution functions at a separation of about $1.4a_B$). At lower densities, the peak of the electron–proton function (multiplied by r^2) in DPIMC calculations appears at $r = 1a_B$ if no molecules are present. However, for the situation in Fig. 10.9a, the presence of molecules leads to a shift of the peak to greater distances.

Figure 10.9b shows that, with increasing temperature, atoms and molecules break up. This is clearly seen by the drastic lowering of the peaks of the pair distribution functions. At higher densities but at the same temperature, DPIMC calculations predict interesting physical phenomena. In Fig. 10.9b–d, one clearly sees increased

Table 10.1 Direct PIMC simulation results. *Upper line*: pressure $p(nk_BT)$. *Lower line*: internal energy $E(2Nk_BT)$. The statistical error of the data is below 1% (Filinov et al. 2000a, b, c)

$n\Lambda^3$	$\Gamma = 0.2$	$\Gamma = 0.4$	$\Gamma = 0.8$	$\Gamma = 1.0$	$\Gamma = 1.2$	$\Gamma = 1.6$
0.1	0.974	0.916	0.742	0.641	0.571	0.503
	1.401	1.215	0.315	−0.658	−1.978	−5.190
0.3	0.983	0.940	0.802	0.732	0.661	0.555
	1.425	1.271	0.731	0.333	−0.201	−1.680
1.0	1.017	0.974	0.872	0.825	0.771	0.665
	1.478	1.328	0.955	0.738	0.476	−0.165
2.0	1.048	1.005	0.910	0.878	0.830	0.740
	1.514	1.384	1.036	0.860	0.637	0.207
3.0	1.100	1.050	0.960	0.933	0.885	0.815
	1.604	1.438	1.120	1.005	0.769	0.424
4.0	1.145	1.100	1.000	0.975	0.941	0.873
	1.671	1.545	1.199	1.110	0.901	0.573
5.0	1.180	1.150	1.055	1.030	0.986	0.916
	1.706	1.573	1.287	1.191	0.978	0.635

ordering of protons from a partially ionized plasma behavior ($\Gamma \approx 1.2$ in Fig. 10.9b), through liquid-like ($\Gamma \approx 20$ in Fig. 10.9c), to solid-like ($\Gamma \approx 54$ in Fig. 10.9d). Note in particular the proton–proton pair distribution functions. For a further increase in density (Γ), crystallization of the ions is commonly expected.

We stress the qualitative change in the electron–electron function during the density increase from Fig. 10.9b–d: g_{ee} in Fig. 10.9b is typical for partially ionized plasmas, whereas a strong peak is observed at short distances in Fig. 10.9c. Further increase in the density leads to an almost uniform electron distribution in Fig. 10.9d. To better understand the electron behavior, Fig. 10.9c also shows the functions $r^2 g_{ee}$ and $r^2 g_{ii}$. The shoulders of these functions indicate that the most probable inter-electronic distance is almost twice as small as the average distance between two protons. A possible explanation for the behavior in Fig. 10.9c is *pairing of electrons* with opposite spin projections. Examination of the electronic bead distribution allows us to conclude that the 'extension' of the electrons is of the order of the inter-ionic distance, and that there is partial overlap of individual electrons. Under these conditions, pairing of the electrons minimizes the total energy of the system. This effect vanishes with increasing density due to the growing wave function overlap.

Figure 10.10 shows isotherms of pressure and internal energy at temperatures 10^4, 3×10^4, 5×10^4, and 10^5 K. As mentioned above, for an ideal plasma, the energy increases with density. The nonideal hydrogen plasma, on the other hand, shows a significant deviation from this trend, in full agreement with existing analytical predictions. Nevertheless, nonideal plasma data asymptotically approaches the standard results at low and high densities. At temperature $T = 10^4$ K, the ratio of the pressure to the ideal pressure approaches 0.5, indicating that electrons and protons

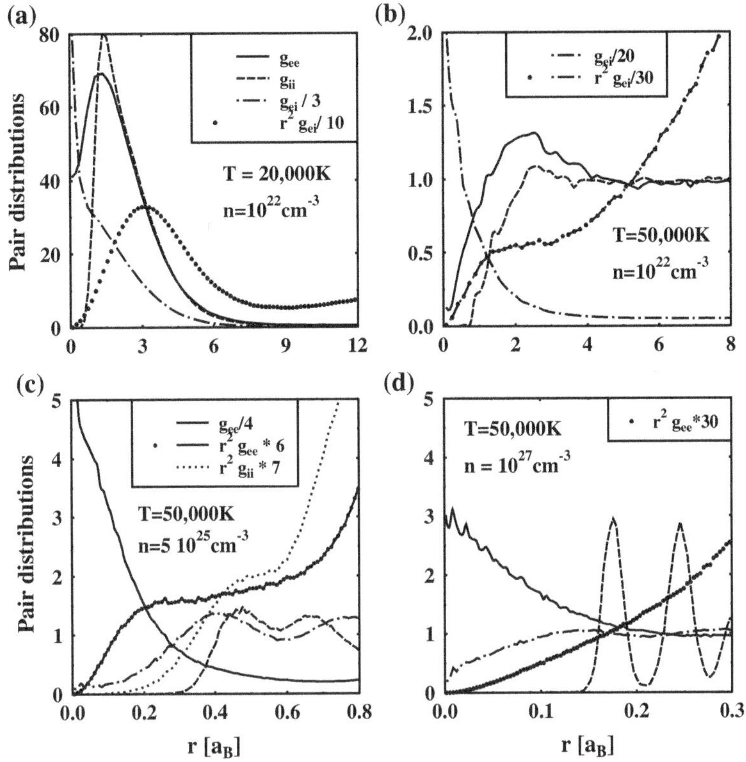

Fig. 10.9 Electron–electron (*continuous line*), ion–ion (*dashed line*), and electron–ion (*dash-dotted line*) pair distribution functions for $T = 20\,000$ K (**a**) and $T = 50\,000$ K (**b**–**d**) and varying density (see text in figures). The values for the coupling, degeneracy, and Brueckner parameters are (**a**) $\Gamma = 2.9$, $\chi = 1.46$, $r_s = 5.44$, (**b**) $\Gamma = 1.16$, $\chi = 0.37$, $r_s = 5.44$, (**c**) $\Gamma = 19.8$, $\chi = 1848$, $r_s = 0.318$, and (**d**) $\Gamma = 53.8$, $\chi = 37\,,000$, $r_s = 0.117$ (Filinov et al. 2000b, c)

form bound states (atoms). The formation of a pressure and energy minimum at intermediate densities (around $10^{22}\,\text{cm}^{-3}$) is due to the overall attractive Coulomb correlations, leading in particular to the formation of atoms and molecules. At higher densities, the energy and pressure increase rapidly, a consequence of electron degeneracy effects: increasing overlap of the electron wave functions leads to a breakup of bound states, giving rise to the Mott effect and an increase in the exchange energy. In the region of a possible plasma phase transition, the pressure becomes negative, indicating an instability. We observe large fluctuations in the pressure and energy (shown by vertical lines), related to formation and decay of many-particle clusters.

The DPIMC numerical results demonstrate the practical feasibility of simulations over a wide range of plasma conditions with comparatively low numbers of beads $n \leq 20$, up to temperatures of the order of 0.1 Ry. DPIMC simulations made it possible to analyze the region of partial ionization and dissociation in detail. Moreover, DPIMC results predict interesting new physical phenomena, such as phase transi-

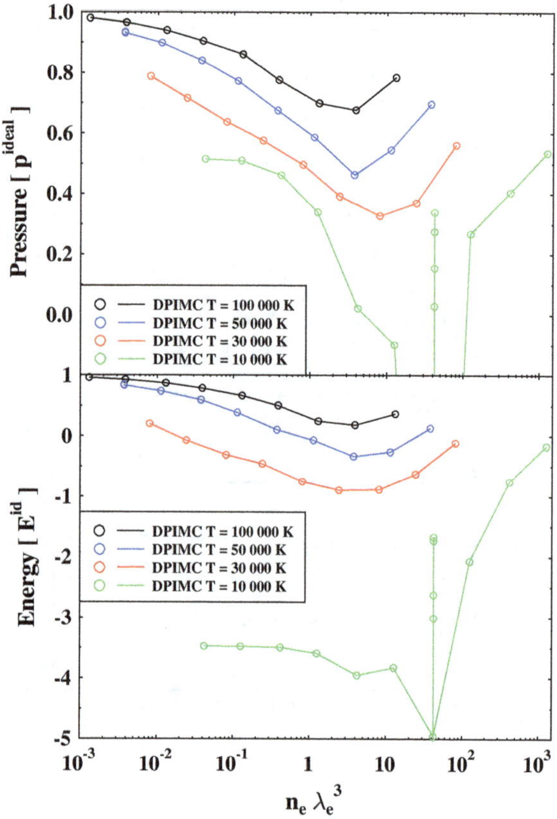

Fig. 10.10 Pressure and energy isotherms of dense hydrogen versus degeneracy parameter. λ_e is the electron thermal DeBroglie wavelength. For $T = 10\,000$ K, in the range $10 \le n_e \lambda_e^3 \le 100$, due to the formation of many particle clusters, strong fluctuations occur in the energy, whereas the simulation results for the pressure do not converge (Filinov et al. 2005)

tions or the pairing of electrons in high density hydrogen, which is related to the onset of proton ordering and vanishes at the transition to the crystal-like state.

Pressure and Energy

Here we compare the DPIMC simulation results with those from a sophisticated model taking into account many physical effects, such as dissociation and ionization, interactions between charged particles and neutral atoms and molecules, neutral–neutral interactions, high-pressure screening effects, excited electronic states of molecules, and a number of 'second-order' phenomena (Saumon et al. 1992, 1995). This model uses the chemical picture with classical statistics for molecules and ions and Fermi–Dirac statistics for the electrons. Owing to the complexity of the model (Saumon et al. 1992, 1995), the equations of state are presented in tabular form (Saumon et al. 1995), so the thermodynamic properties of pure hydrogen can be calculated by interpolation.

Figure 10.11 shows pressure and energy dependencies versus temperature along two isochors $n_e = 10^{20}$ and 10^{21} cm^{-3} of a hydrogen plasma. The agreement with the hydrogen EOS (Saumon et al. 1995) is good for the pressure at all temperatures

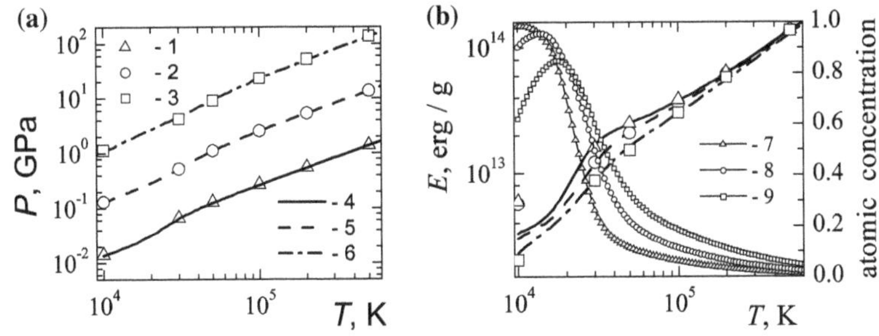

Fig. 10.11 DPIMC (symbols) and theoretical (lines) (Saumon et al. 1992, 1995) isochors for pressure (**a**) and energy (**b**) in hydrogen plasma versus temperature. Calculations: (1) isochor $n_e = 10^{20}$ cm^{-3}, (2) isochor $n_e = 10^{21}$ cm^{-3}. EOS calculations (Saumon et al. 1995): (3) isochor $n_e = 10^{20}$ cm^{-3}, (4) isochor $n_e = 10^{21}$ cm^{-3}, (5) and (6) relative number concentration of hydrogen atoms [(**b**), right axis] along the isochor $n_e = 10^{20}$ cm^{-3} and $n_e = 10^{21}$ cm^{-3}, respectively (Filinov et al. 2005)

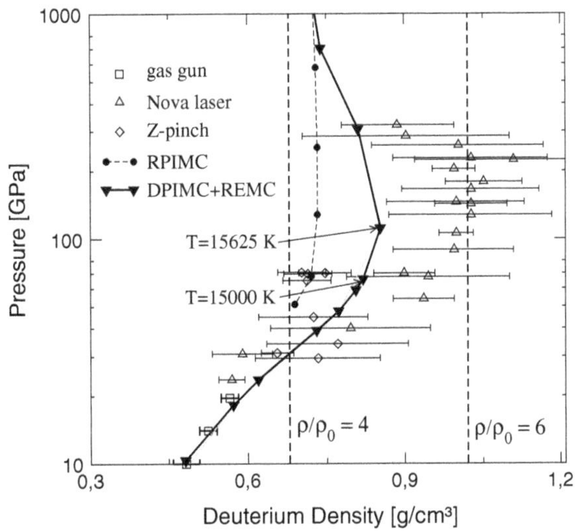

Fig. 10.12 Shock Hugoniot of deuterium. *Solid line and triangles*: combined Hugoniot (Bezkrovniy et al. 2004). *Dashed line and full points*: RPIMC calculations (Militzer and Ceperley 2000). Experimental data: gas gun, Nova laser, Z-pinch (Nellis et al. 1983, Collins et al. 1998, Knudson et al. 2001, Boriskovet al. 2005) (Filinov et al. 2005)

(Fig. 10.11a) and for the energy at temperatures higher than 3×10^4 K (Fig. 10.11b). The relative number concentrations of hydrogen atoms shown in Fig. 10.11b by dash-dotted and dotted lines indicate that hydrogen at $T = 10^4$ K and $n_e = 10^{20}$–10^{21} cm^{-3} consists mainly of atoms. At 10^4 K, in DPIMC calculations, the absolute energy value is higher than in the theoretical model, and this can be connected with the difficulties in finding a correct interpretation of bound states, in both the DPIMC and the chemical picture. Therefore this problem needs further investigation.

Comparison with Experimental Results

EOS high pressure experimental results are usually available in the form of a shock Hugoniot. The Hugoniot equation relates the state behind the front of the shock

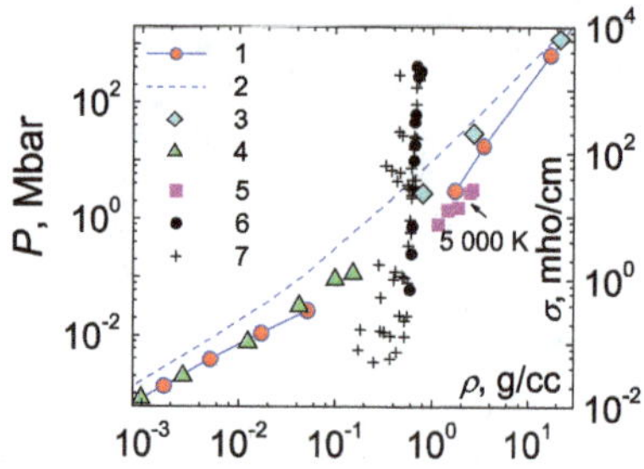

Fig. 10.13 Pressure (1–5) (*left axis*) and electrical conductivity (6,7) (*right axis*) for hydrogen at $T = 10^4$ K versus density. (1) DPIMC simulation, (2) ideal plasma, (3) DFT (Xu and Hansen 1998), (4) restricted PIMC computations (Militzer and Ceperley 2001). Experimental data: (5) break in the isentropic curve (Fortov et al. 2003, Mochalov et al. 2012, 2015), (6) electrical conductivity (William et al. 1996), (7) electrical conductivity (Ternovoi et al. Ternovoi et al. 1999a, b) (2005)

wave (namely, the density ρ, pressure P, and specific internal energy E) to the initial conditions ρ_0, p_0, and E_0. Figure 10.12 shows a theoretical Hugoniot, obtained by combining results for the EOS of the DPIMC technique with those of the reaction ensemble Monte Carlo (REMC) simulations performed in Rostock university. As has been shown by Bezkrovniy et al. (Bezkrovniy et al. 2004), REMC describes the low temperature region very well, and shows good agreement with the gas gun experiments by Nellis et al. (Nellis et al. 1983). The DPIMC simulation was performed at temperatures from 10^4 K and higher, over a wide range of particle densities. The influence of the plasma phase transition region allowed us to calculate the Hugoniot reliably only from 15 625 K. We then combined the lower part (up to 15 000 K) of the REMC Hugoniot and the DPIMC Hugoniot. We should stress here that these two methods are completely independent and no interpolation procedure is used. The final Hugoniot is plotted in Fig. 10.12 and shows a maximal compressibility of approximately 4.75, as compared to the initial normal condition deuterium density. This Hugoniot is located between the experimental values of Knudson et al. (Knudson et al. 2001) and Collins et al. (Collins et al. 1998), and shows higher compressibility of deuterium at pressures of about 1 Mbar than most of the other theories.

Metalization and Hydrogen Conductivity

Figure 10.13 presents the EOS obtained by different authors for temperature $T = 10^4$ K and experimental data on conductivity obtained by (Ternovoi et al. 1999a, b, Weir et al. 1996). For the pressure, blue diamonds are data obtained by Hansen and Xu in the framework of the density functional theory (DFT) (Xu and Hansen 1998). They stress that, at lower densities, they find large fluctuations in the pressure, indicating the existence of a phase transition. The data of Ceperley and Militzer (Militzer and Ceperley 2001) in the 'fixed-node approximation are represented by triangles. DPIMC calculations are represented by circles. The break in the EOS is related to the negative values of the pressure obtained in DPIMC. On the left side of this break, the homogeneous state of the plasma becomes unstable, giving rise to metallic-like many-particle electron–proton clusters. On the right side of this break, when these clusters overlapping, the plasma becomes uniform and stable. The *position of the break* on the quasi-isentrope obtained in experiments by Mochalov (Fortov et al. 2003, Mochalov et al. 2012, 2015) is shown by squares. This break is supposed to be connected with a plasma phase transition. The overlap of many particle clusters may result in a sharp increase in the conductivity, something that has been observed experimentally (Ternovoi et al. 1999a, b, Weir et al. 1996).

10.3 Thermodynamic Properties of Hydrogen–Helium Mixtures

Many astrophysical problems require knowledge of the thermodynamic properties of hydrogen and helium (Chabrier et al. 1992, Saumon et al. 1995, Gudkova and Zharkov 2000, Nellis 2000, Beule et al. 2000, Ebeling et al. 1991, 2007). To understand important physical effects in stellar evolution and structure, one should provide accurate modeling of the underlying physics, including peculiarities of the equation of state (EOS). In normal stars, where plasma is fully ionized and almost ideal, the EOS does not involve particular difficulties. However, investigation of the giant planets Jupiter and Saturn, and also brown dwarfs, demands thermodynamic information for hydrogen and helium in the range of temperatures $10^3 < T < 10^5$ K and mass densities $0.01 < \rho < 100$ g/cm^3. In this region, the complexity of an EOS calculation increases considerably when nonideal effects are compounded with chemical reactions associated with partial pressure dissociation and ionization equilibria (Saumon and Chabrier 1991, Saumon et al. 1992, Fortov et al. 2003, Norman and Starostin 1968a, b, Ebeling et al. 1976a, b). Moreover, the so-called plasma phase transition (PPT) has been predicted in this region (Norman and Starostin 1968a, b, Ebeling et al. 1976a, b).

A significant effort has been made to understand the behavior of dense fully ionized and partially ionized hydrogen and helium (see, for example, Fortov et al. 2003 and references therein). In these studies, the chemical picture is usually applied to calculate thermodynamic properties. This picture assumes that bound configurations, such as atoms and molecules, retain a definite identity and interact through pair potentials. So this model is valid only for weak interparticle interactions. However, at densities corresponding to pressure ionization, the electrons in bound configurations become delocalized and bound species lose their definiteness (Kraeft et al. 1986). Therefore there is great interest in DPIMC simulations, which avoid such approximations.

Here we present results of the DPIMC method for the thermodynamic properties of hydrogen–helium mixtures (Levashov et al. 2002, 2005, 2006). The results of these simulations are compared with the model based on the chemical picture (Chabrier et al. 1992, Saumon et al. 1995). The problem of the plasma phase transition in dense hydrogen–helium mixtures will be discussed, along with several theoretical and experimental predictions of this phenomenon.

Peculiarities of the DPIMC for Hydrogen–Helium Mixtures

Let us consider some peculiarities of the DPIMC for hydrogen–helium mixtures in the temperature range from 10^4 K to 2×10^5 K and electron densities from 10^{20} to $3 \times 10^{24}\,\mathrm{cm}^{-3}$. Under such conditions, electrons are degenerate, while protons and α particles can be treated as classical particles because of their relatively large masses. Thus for the case of an electro-neutral hydrogen–helium plasma with volume V, the partition function Z is given by

$$Z(N_\mathrm{e}, N_\mathrm{p}, N_\alpha, V, \beta) = \frac{1}{N_\mathrm{e}!N_\mathrm{p}!N_\alpha!} \sum_\sigma \int_V \mathrm{d}q_\mathrm{p}\, \mathrm{d}q_\alpha\, \mathrm{d}r\, \rho(q_\mathrm{p}, q_\alpha, r, \sigma; \beta)\,. \tag{10.8}$$

Here N_e, N_p, and N_α are the numbers of electrons, protons, and α particles, respectively, $\beta = 1/K_\mathrm{B}T$, T is the temperature, $q_\mathrm{p} \equiv \{\mathbf{q}_{\mathrm{p}1}, \mathbf{q}_{\mathrm{p}2}, \ldots, \mathbf{q}_{\mathrm{p}N_\mathrm{p}}\}$, $q_\alpha \equiv \{\mathbf{q}_{\alpha 1}, \mathbf{q}_{\alpha 2}, \ldots, \mathbf{q}_{\alpha N_\alpha}\}$ are the coordinates of the protons and α particles, respectively, $r \equiv \{\mathbf{r}_1, \mathbf{r}_2, \ldots, \mathbf{r}_{N_\mathrm{e}}\}$ are the coordinates of the electrons, and $\sigma \equiv \{\sigma_1, \sigma_2, \ldots, \sigma_{N_\mathrm{e}}\}$ are the spin variables of the electrons. The density matrix in (10.8) is expressed via a path integral:

$$\int\limits_V \mathrm{d}R^{(0)} \sum_\sigma \rho(R^{(0)}, \sigma; \beta) = \sum_P \sum_\sigma (-1)^{\kappa_P} \int\limits_V \mathrm{d}R^{(0)} \ldots \mathrm{d}R^{(n)} \tag{10.9}$$
$$\times \rho^{(1)} \rho^{(2)} \cdots \rho^{(n)} S(\sigma, \hat{P}\sigma') \hat{P} \rho^{(n+1)}\,,$$

where

$$\rho^{(i)} = \rho(R^{(i-1)}, R^{(i)}; \Delta\beta) \equiv \left\langle R^{(i-1)} \left| \mathrm{e}^{-\Delta\beta\hat{H}} \right| R^{(i)} \right\rangle\,, \quad \Delta\beta \equiv \frac{\beta}{n+1}\,,$$

with $\hat{P}$ the permutation operator, κ_P the parity of the permutation, S the spin matrix, $\hat{H} = \hat{K} + \hat{U}_\mathrm{C}$ the Hamiltonian of the system, $\hat{K}$ the kinetic energy, and $\hat{U}_\mathrm{C}$ the potential energy, consisting of the Coulomb interactions of the electrons, protons, and α particles:

$$\hat{U}_\mathrm{C} = \hat{U}_\mathrm{C}^\mathrm{p} + \hat{U}_\mathrm{C}^\mathrm{e} + \hat{U}_\mathrm{C}^\alpha + \hat{U}_\mathrm{C}^\mathrm{ep} + \hat{U}_\mathrm{C}^{\mathrm{e}\alpha} + \hat{U}_\mathrm{C}^{\mathrm{p}\alpha}\,.$$

We use the notation $R^{(i)} = (q_\mathrm{p}, q_\alpha, r_i)$, $i = 1, \ldots, n+1$, $R^{(0)} \equiv (q_\mathrm{p}, q_\alpha, r)$, $R^{(n+1)} \equiv R^{(0)}$, $\sigma' = \sigma$, for the particle coordinates. Thus electrons taking part in the simulation are represented by fermionic loops with n vertexes:

$$[R] \equiv [R^{(0)}; R^{(1)}; \ldots; R^{(n)}; R^{(n+1)}]\,.$$

Exchange effects for Fermi statistics are taken into account by the permutation operator $\hat{P}$ and a sum over permutations with parity κ_P. As we mentioned before, the expression (10.9) can be reduced to a form in which the sum over all permutations is replaced by the determinant of the exchange matrix $\psi_{ab}^{n,1}$. This technique allows us to improve simulation accuracy for strongly degenerate plasmas:

$$\sum_{\sigma} \rho(q_{\rm p}, q_{\alpha}, r, \sigma; \beta) = \frac{1}{\lambda_{\rm p}^{3N_{\rm p}} \lambda_{\alpha}^{3N_{\alpha}} \lambda_{\Delta}^{3N_{\rm e}}} \sum_{s=0}^{N_{\rm e}} \rho_s([R], \beta) ,$$

$$\rho_s([R], \beta) = \frac{C_{N_{\rm e}}^s}{2^{N_{\rm e}}} \exp\{ - \beta U([R], \beta)\} \prod_{l=1}^{n} \prod_{m=1}^{N_{\rm e}} \phi_{mm}^l \det \left| \psi_{ab}^{n,1} \right|_s . \tag{10.10}$$

Here $\lambda_p^2 = 2\pi\hbar^2\beta/m_{\rm p}$, $\lambda_\alpha^2 = 2\pi\hbar^2\beta/m_\alpha$, $\lambda_\Delta^2 = 2\pi\hbar^2\Delta\beta/m_{\rm e}$, and $m_{\rm p}$, m_α, and $m_{\rm e}$ are the masses of the proton, α particle, and electron, respectively. In (10.10),

$$U = U^{\rm pp} + U^{\alpha\alpha} + U^{\rm p\alpha} + \frac{1}{n+1} \sum_{l=1}^{n} \left(U_l^{\rm ee} + U_l^{\rm ep} + U_l^{\rm e\alpha} \right) ,$$

and $\phi_{mm}^l \equiv \exp\left[- \pi \left|\xi_m^{(l)}\right|^2 \right]$ are the functions generated from the kinetic energy density matrix, $\xi^{(1)}, \ldots, \xi^{(n)}$ are the dimensionless distances between neighboring vertices of fermionic loops which represent electrons $[R] \equiv [R^{(0)}; R^{(0)} + \lambda_\Delta \xi^{(1)}; R^{(0)} + \lambda_\Delta (\xi^{(1)} + \xi^{(2)}); \ldots]$. Elements of the exchange matrix $\psi_{ab}^{n,1}$ are defined by the expression:

$$\left\| \psi_{ab}^{n,1} \right\|_s \equiv \left\| \exp\left[-\frac{\pi}{\lambda_\Delta^2} \left| (r_a - r_b) + y_a^n \right|^2 \right] \right\|_s , \quad y_a^n = \lambda_\Delta \sum_{k=1}^{n} \xi_n^{(k)} .$$

The index s stands for the number of electrons with same spin projection.

As a high-temperature density matrix, one can use its asymptote in the limit $T \to \infty$. Every N-particle high-temperature density matrix is represented as a product of two-particle density matrices. For the two-particle density matrix there is an analytical solution of the Bloch equation by first-order perturbation theory (Ebeling et al. 1967):

$$\rho(\mathbf{r}_a, \mathbf{r}_a', \mathbf{r}_b, \mathbf{r}_b', \beta) = \frac{m_a m_b}{(2\pi\hbar^2\beta)^3} \exp\left[-\frac{m_a}{2\hbar^2\beta} (\mathbf{r}_a - \mathbf{r}_a')^2 \right] \times \exp\left[-\frac{m_b}{2\hbar^2\beta} (\mathbf{r}_b - \mathbf{r}_b')^2 \right] \exp\left(-\beta \Phi^{ab} \right) ,$$

where $\Phi^{ab}(\mathbf{r}_a, \mathbf{r}_a', \mathbf{r}_b, \mathbf{r}_b', \beta)$ is the nondiagonal effective two-particle pseudopotential:

$$\Phi^{ab}(\mathbf{r}_{ab}, \mathbf{r}_{ab}', \beta) = e_a e_b \int_0^1 \frac{{\rm d}\alpha}{d_{ab}(\alpha)} {\rm erf} \left(\frac{d_{ab}(\alpha)}{2\lambda_{ab}\sqrt{\alpha(1-\alpha)}} \right) . \tag{10.11}$$

Here $d_{ab}(\alpha) = |\alpha \mathbf{r}_{ab} + (1-\alpha)\mathbf{r}'_{ab}|$, $0 \le \alpha \le 1$, and

$$\mathrm{erf}(x) = \frac{2}{\sqrt{\pi}} \int_0^x \exp(-t^2)\mathrm{d}t$$

denotes the error function, $\lambda_{ab}^2 = \hbar^2\beta/2\mu_{ab}$, e_a and e_b are the charges of the particles, and m_a and m_b are the masses of the particles, with the reduced mass

$$\mu_{ab}^{-1} = m_a^{-1} + m_b^{-1} .$$

In the high-temperature limit, the two-particle nondiagonal effective potential can be approximated by a half-sum of diagonal pseudopotentials (10.11):

$$\Phi^{ab}(|\mathbf{r}_{ab}|, \Delta\beta) = \frac{e_a e_b}{\lambda_{ab} x_{ab}} \left\{ 1 - \exp(-x_{ab}^2) + \sqrt{\pi} x_{ab} \left[1 - \mathrm{erf}(x_{ab}) \right] \right\} , \qquad (10.12)$$

where $x_{ab} = |\mathbf{r}_{ab}|/\lambda_{ab}$. It is worth stressing that $\Phi^{ab}(|\mathbf{r}_{ab}|, \Delta\beta)$ tends to a finite value when $x_{ab} \to 0$ and to the Coulomb potential $e_a e_b / x_{ab}$ when $x_{ab} \to \infty$. It can be shown that the pseudopotential (10.12) coincides with an exact quantum potential at temperatures $T > 2 \times 10^5$ K (Filinov et al. 2003a, b, c). Thus, U^{aa} and U_l^{ab} in (10.10) are sums of the effective quantum pair interactions described by the Φ^{ab} between two charged plasma particles a and b.

All thermodynamic properties can be expressed through derivatives of the partition function. For example, the pressure and total energy are given by the formulas:

$$E = -\beta \frac{\partial \ln Z}{\partial \beta} , \qquad \beta P = \frac{\partial \ln Q}{\partial V} . \qquad (10.13)$$

The multiple integrals in (10.13) are calculated with the help of the standard Metropolis technique, in a cubic cell with periodic boundary conditions (Zamalin et al. 1977). The accuracy ϵ of the calculations depends on the number of factors n in (10.9), the temperature T, and the electron degeneracy parameter $\chi = n_e \lambda_e^3$, and it is given by the expression $\epsilon \sim (\beta \mathrm{Ry})^2 \chi/(n+1)$, where n_e is the particle density of electrons, $\lambda_e^2 = 2\pi\hbar^2\beta/m_e$, $\mathrm{Ry} = e^2/2a_B \approx 13.6$ eV. According to this estimate, in order to simulate a Coulomb system at temperature 10^4 K, it suffices to choose $n = 20$. The high temperature density matrix in (10.9) relates in this case to temperatures higher than Ry.

Pressure and Energy

Let us compare here the results of the DPIMC simulation with data on the hydrogen–helium mixture from (Saumon et al. 1992, 1995). As mentioned before, owing to the complexity of this model, the equations of state for hydrogen and helium are presented in tabular form, and the thermodynamic properties of pure hydrogen or helium can be calculated by interpolation. To obtain results for the hydrogen–helium mixtures, one must interpolate between the two pure equations of state. Using the

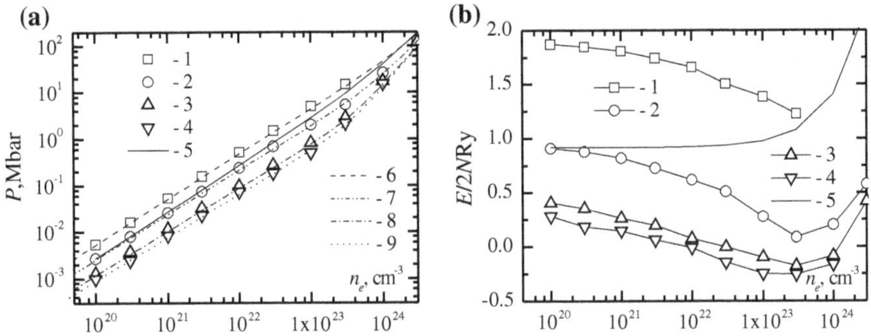

Fig. 10.14 Pressure (**a**) and energy (**b**) in a hydrogen–helium mixture with a mass concentration of helium $Y = 0.234$. Shown are isotherms calculated using the DPIMC method (*symbols* and *symbols with lines*) and related isotherms computed with the help of EOS (Saumon et al. 1992, 1995) (*lines*). DPIMC (Saumon et al. 1992, 1995) calculations: 1(6) 200 kK, 2(7) 100 kK, 3(8) 50 kK, 4(9) 40 kK. 5 at 100 kK is the isotherm for the ideal plasma (Levashov et al. 2005)

so-called 'linear mixing technique', the density $\rho(P, T)$ of the hydrogen–helium mixture can be approximated in terms of the mass fraction of helium Y at pressure P and temperature T:

$$\frac{1}{\rho(P,T)} = \frac{1-Y}{\rho^{\mathrm{H}}(P,T)} + \frac{Y}{\rho^{\mathrm{He}}(P,T)} \, .$$

During the mission of the Galileo spacecraft, the helium abundance in the atmosphere of Jupiter was determined as $Y = m_{\mathrm{He}}/(m_{\mathrm{He}} + m_{\mathrm{H}}) = 0.234$, close to the present-day protosolar value $Y = 0.275$. As the model of Jupiter is largely determined by its composition and EOS, it was interesting to simulate the thermodynamic properties of the mixture with such a composition in the region of pressure dissociation and ionization where traditional chemical models of plasma fail. So DPIMC calculations were carried out to determine the thermodynamic properties of a hydrogen–helium mixture with a composition corresponding to that of the outer layers of the Jovian atmosphere.

Calculations were carried out for temperatures $10^4 < T < 2 \times 10^5$ K and electron densities $10^{20} < n_{\mathrm{e}} < 3 \times 10^{24}$ cm^{-3}. The results of the comparison are shown in Figs. 10.14 and 10.15. The agreement between the DPIMC calculations and the model (Saumon et al. 1995) along the isotherms $T = 3 \times 10^4, 4 \times 10^4, 5 \times 10^4$, and 10^5 K is quite good, and gets better as the temperature increases. The smaller values of pressure on the DPIMC isotherm 3×10^4 K near the particle density value 10^{23} cm^{-3} can be explained by the strong influence of interaction and bound states in this region. These effects are taken into account only approximately in the model (Saumon et al. 1995). The formation of atoms and molecules is also the reason for the pressure and energy reduction along the 10^5 K isotherm, as compared with the isotherm for a non-interacting hydrogen–helium mixture.

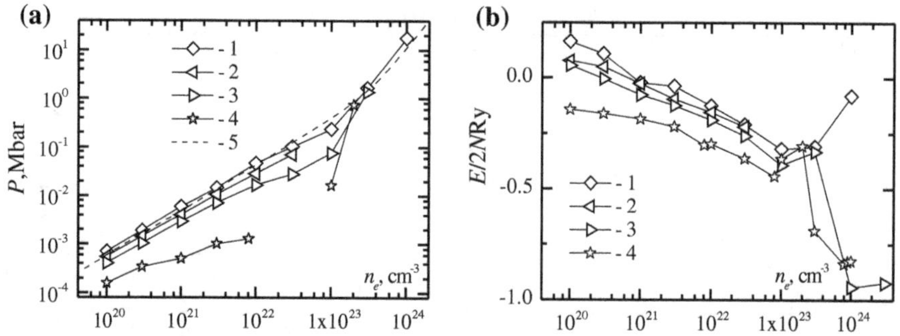

Fig. 10.15 Pressure (**a**) and energy (**b**) in a hydrogen–helium mixture with the mass concentration of helium $Y = 0.234$. Shown are isotherms calculated with the help of the DPIMC method (*symbols with lines*). DPIMC calculations: (1) 30 kK, (2) 25 kK, (3) 20 kK, (4) 10 kK. EOS calculations (Saumon et al. 1992, 1995) In (**a**), (5) 30 kK (Levashov et al. 2005)

Figure 10.15 shows the dependence of pressure and energy on the electron density along isotherms. At temperatures 30 and 25 kK, the isotherms have no peculiarities. However, along the isotherm $T = 2 \times 10^4$ K, there is a region for $n_e > 3 \times 10^{23}$ cm^{-3}, where the pressure fluctuates significantly and even becomes negative. Moreover, along the isotherms $T = 10^4$ K and $T = 1.5 \times 10^4$ K, there are two such regions $10^{22} < n_e < 10^{23}$ cm^{-3} and $n_e > 3 \times 10^{23}$ cm^{-3}.

Earlier, we discussed a similar effect for pure hydrogen at $T = 10^4$ K, in the region of pressure ionization. It was shown that, in the transition region, a number of large clusters (droplets) were formed (Filinov et al. 2001a, b, Levashov et al. 2001). In this pressure ionization region, the plasma phase transition was predicted by many authors (Norman and Starostin 1968a, b, Ebeling et al. 1976a, b, Kraeft et al. 1986, Saumon, and Chabrier 1989, Saumon and Chabrier 1991; Saumon and Chabrier 1992, Yan et al. 1991, Schlanges et al. 1995), and moreover, a sharp electrical conductivity rise was measured in (Ternovoi et al. 1999a, b). The instabilities in the DPIMC calculations indicate the existence of a phase transition in dense hydrogen.

In an electron–hole plasma of germanium semiconductor at low temperature, cluster formation and the phase transition was also found (Filinov 2003). This phase transition agrees well with the experimental phase diagram (Thomas et al. 1974). An important consequence of the presence of clusters (droplets) in a plasma is the decrease in the total energy of the system, as can be seen in Fig. 10.15.

The problem of the phase transition in a hydrogen–helium mixture is largely determined by the composition of the mixture Saumon and Chabrier (1992, Saumon et al. 1995, Schlanges et al. 1995). From shock-wave experiments, one can estimate the range of temperatures and densities in which there is a sharp rise in electrical conductivity. In quasi-isentropic compression, the transition from a low-conductivity state to a high-conductivity state for hydrogen occurs when $T \sim 3$–15 kK and $\rho =$ 0.4–0.7 g/cm^3 (Weir et al. 1996, Ternovoi et al. 1999a, b), whereas for helium it occurs when $T = 15$–40 kK and $\rho = 0.7$–1.25 g/cm^3 (Ternovoi et al. 2002).

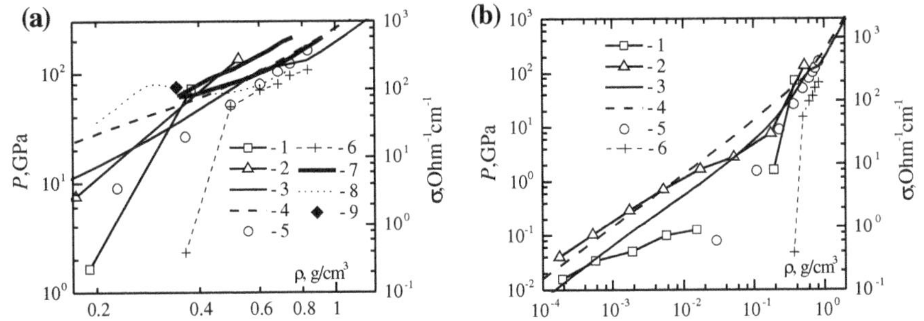

Fig. 10.16 DPIMC (*lines with squares and triangles*) and theoretical (Saumon et al. 1992, 1995) isotherms for pressure in a hydrogen–helium mixture with the mass concentration of helium $Y = 0.234$ versus density. DPIMC calculations: (1) 10 kK, (2) 20 kK. EOS calculations: (3) 10 kK, (4) 20 kK. Experiment (Ternovoi et al. 2004): (5) quasi-isentrope of hydrogen–helium mixture, $T \sim 5\,000$ K, (6) electrical conductivity of hydrogen–helium mixture along the quasi-isentrope (*right axis*), (7) phase boundaries of PPT in hydrogen (Saumon et al. 1992, 1995), (8) undercritical metastable isotherm $T = 1.2 \times 10^4$ K and $Y = 0.308$ (Schlanges et al. 1995), (9) critical point of PPT in hydrogen–helium mixture with $Y = 0.308$ (Levashov et al. 2005)

Another important point is determination of the critical point of the phase transition. According to the theoretical equations of state, the critical point of the phase transition in pure hydrogen is $T^{(1)}_{\mathrm{cr_H}} = 14.9$ kK, $P^{(1)}_{\mathrm{cr_H}} = 0.723$ Mbar (Schlanges et al. 1995) and $T^{(2)}_{\mathrm{cr_H}} = 15.3$ kK and $T^{(2)}_{\mathrm{cr_H}} = 0.61$ Mbar (Saumon et al. 1992, 1995). In pure helium, the critical point was found to be $T^{(1)}_{\mathrm{cr_{He}}} = 17$ kK, $P^{(1)}_{\mathrm{cr_{He}}} = 7.22$ Mbar (Schlanges et al. 1995). At the helium mass concentration $Y < 0.93$ and temperatures less than both critical temperatures, the properties of the hydrogen–helium mixture are determined for the main part by hydrogen, and there is only one PPT. At high values of $Y > 0.93$, both the hydrogen and the helium phase transition can occur. In DPIMC simulations, only one region of instability was found with negative pressure, at $T = 2 \times 10^4$ K, and two regions of instability at $T = 10^4$ K even for $Y = 0.234$. The results of the DPIMC simulation are shown in Fig. 10.16, together with experimental data and theoretical predictions.

Along the isotherm $T = 2 \times 10^4$ K, a region was found with poor convergence in the range of densities between 0.5 and 5 g/cm^3. Along the isotherms $T = 1.5 \cdot 10^4$ K and $T = 10^4$ K such region is even wider and begins from 0.38 g/cm^3. Surprisingly, there was another region where the pressure became negative, namely, in the range 0.015–0.19 g/cm^3. The physical reason for this phenomenon is connected with the formation of many particle clusters and requires further consideration. From Fig. 10.16, it is easy to see that other predictions of phase transitions in hydrogen or hydrogen–helium mixtures (Saumon et al. 1995, Schlanges et al. 1995) with low helium mass concentrations are located in the beginning of the region where the DPIMC simulation fails to converge. The sharp rise in the electrical conductivity of the hydrogen–helium mixture along the quasi-isentrope with initial state $T = 77.4$ K and $P = 8.1 \times 10^{-3}$ GPa is also observed experimentally in the range of densities 0.5–0.83 g/cm^3 (Ternovoi et al. 2004) (line with crosses in Fig. 10.16).

10.4 Coulomb Crystallization

In 1934 Wigner predicted the existence of a crystalline state of the electron gas at low densities—the electron Wigner crystal (Wigner 1934). Since then, there has been an active search for this phenomenon. Finally, crystallization of electrons was observed on the surface of helium droplets (Grimes and Adams 1979). The condition for there being a crystal in one-component plasmas (OCP) is that the mean Coulomb interaction energy $e^2/\bar{r}$ (where $\bar{r}$ denotes the mean interparticle distance) should exceed the mean kinetic energy (thermal energy $dk_{\rm B}T/2$ or Fermi energy $E_{\rm F}$ in classical or quantum plasmas, respectively) by a large factor $\Gamma^{\rm cr}$ which, in a classical OCP is given by 175 (137) in 3D (2D) (Grimes and Adams 1979, DeWitt and Slattery 1999). In a quantum OCP at zero temperature, the coupling strength is measured by the Brueckner parameter $r_{\rm s} \equiv \bar{r}/a_{\rm B}$, where $a_{\rm B}$ denotes the Bohr radius. Its critical value is $r_{\rm s}^{\rm cr} \approx 33$ (100/160) in 2D (3D for fermions/bosons) (Filinov et al. 2001, Ceperley and Alder 1980).

Coulomb crystallization has also been observed in neutral two-component plasmas (TCP), e.g., in colloidal and dusty plasmas (Thomas et al. 1994, Hayashi and Tachibana 1994). Besides classical TCP crystals, it is believed that in the interior of white dwarf stars and in the crust of neutron stars there exist crystals of fully ionized carbon, oxygen, and iron ions embedded in an extremely dense Fermi gas of electrons. It is of great interest to analyze the conditions for the existence of Coulomb crystals in a two-component plasma. Here we will concentrate on plasma-like systems which involve such charges as holes and electrons, as this allow us to consider Coulomb systems with different electron-to-hole mass ratios.

A Coulomb crystal of holes will survive in the presence of electrons only if the electrons and holes do not form bound states, as this would drastically reduce the correlation energy of the holes, thereby eventually reducing the coupling strength below the critical level necessary for crystallization. For qualitative estimates of TCP crystal parameters, let us consider the well known results for one-component plasmas (OCP). For the existence of a Coulomb crystal in the presence of a classical uniform gas of electrons, we require first of all that the heavy component should be able to form a classical OCP crystal, i.e., $\Gamma \geq \Gamma^{\rm cr}$ and, secondly, that the electrons should not destroy that crystal, e.g., through screening of the heavy particle interaction.

However, the main obstacle for the crystal turns out to be the formation of bound states (atoms, excitons with binding energy $E_{\rm B}$, etc.), because this reduces the hole–hole correlation energy. This crystal of heavy particles will survive in the presence of electrons only if the bound states are unstable, which occurs due to pressure ionization at high densities, above the Mott density, i.e., $r_{\rm se} \equiv \bar{r}/a_{\rm B} \leq r_{\rm s}^{\rm Mott} \approx 1{-}2$ ($a_{\rm B}$ is the effective Bohr radius). This has been confirmed by the first principle path integral Monte Carlo simulations since, according to the numerical data obtained for

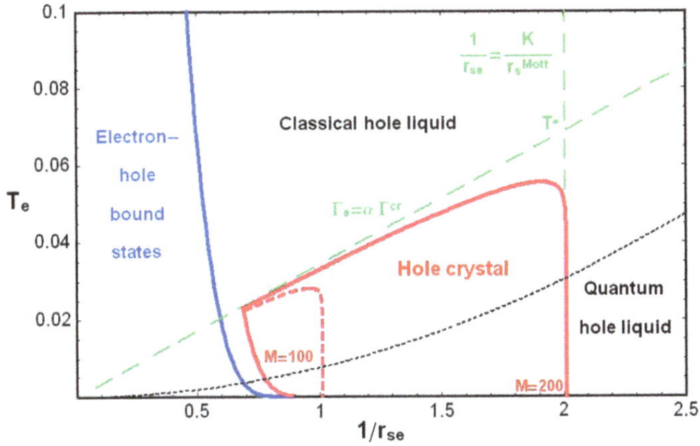

Fig. 10.17 Qualitative phase diagram of a two-component plasma in the plane of dimensionless electron temperature $T_e = 3k_B T/2E_B$ and density parameter $1/r_{se}$. The *blue line* indicates the boundary of the Coulomb bound state phase given by $r_s^{Mott}(T_e)$. The *red full (dashed) line* is the boundary of the hole crystal for $M = 200$ ($M = 100$), with the asymptotics given by $\Gamma \geq \Gamma^{cr}$ (*green dashed lines*) (Bonitz et al. 2005)

$r_{se} \leq 1.2$, the fraction of holes bounded in excitons and bi-excitons is less than 10% (see Fig. 10.17).

We now discuss the qualitative phase diagram of the TCP. Consider first the case of a hole crystal in semiconductors which is embedded in a dense Fermi gas of electrons (see Fig. 10.17). The holes behave classically above the black dotted line and quantum-mechanically below. This line is given by $n_h \Lambda_h^3 = 1$, where $\Lambda_h = h/\sqrt{2\pi m_h k_B T_h}$ is the hole thermal wavelength. The electron–hole bound state phase is on the left and contains excitons and bi-excitons (Bucher et al. 1991, 2004). Leaving this phase by crossing its boundary, given by the blue line $r_{se}(T_e) = r_s^{Mott}(T_e)$, the fraction of bound states rapidly vanishes in favor of unbound electron–hole pairs, with the holes exhibiting liquid-like behavior. Upon further compression (at low temperatures $T_e < T^*$ related to the condition $\Gamma > \Gamma^{cr}$), the hole liquid crystallizes, provided that the holes are classical particles with high enough mass (above the black dotted line) (Bonitz et al. 2005). At the density $1/r_{se} = 2$, quantum melting of the crystal is observed at zero temperature (vertical dashed green line). The entire hole Coulomb crystal phase for $M = 200$ is shown by the full red line in Fig. 10.17. When M is reduced, the crystal phase shrinks (see the red dashed line corresponding to $M = 100$), until it finally vanishes for $M = M^{cr}$.

Three-Dimensional Simulations

We now consider DPIMC simulations at high densities. There are two main factors that influence the DPIMC results. The first is quantum degeneracy. The practical limitations that must be imposed on the number of beads and particles necessarily makes DPIMC results less reliable for densities exceeding $r_s = 1$. The second factor is finite-size effects related to hole ordering.

Analyzing the electron–electron, electron–hole, and in particular hole–hole pair distribution functions, prominent features have been found relating to particle (hole) ordering phenomena (Bonitz et al. 2005, Filinov et al. 2005, Bonitz et al. 2006) at densities beyond the Mott point. With increasing hole density, the hole coupling parameter increases, leading to formation of a hole liquid, and eventually a hole lattice [provided the mass ratio between electrons and holes is sufficient large (Bonitz et al. 2005)], which is immersed in a delocalized sea of electrons. The simulation results for $r_s = 0.5$ and $T/E_B = 0.064$ are shown in Fig. 10.18, and clearly reveal the formation of a Wigner hole crystal. This effect has also been seen in previous simulations of ultradense hydrogen, where proton crystallization has been observed (Filinov et al. 2006).

Coulomb crystallization of the holes and quantum melting were directly verified by DPIMC simulation. The snapshots presented in Fig. 10.18 show the electron–hole states in the simulation box at fixed temperature and density, but for increasing values of $M = m_h/m_e$. In all figure parts, the electrons form a nearly homogeneous Fermi gas—individual electrons penetrate each other and extend far beyond the main simulation cell shown by the grid lines (to simulate a macroscopic system, this cell is periodically repeated in three space directions X, Y, Z). At the same time, the hole arrangement changes dramatically. At $M = 800$, holes are periodically ordered in space, at $M = 100$ and between $M = 100$ and 50 the hole crystal melts, at $M = 50$ the hole structure resembles a liquid, and at $M = 1$, the holes are in a gas-like state (similar to the electrons). The figure also clearly shows the mechanism of this quantum phase transition: with increasing M, the individual hole wave packets shrink continuously until, at $M = 800$, they collapse into a dot. The crystal melts when the decay length of the hole wave functions exceeds a critical size.

More subtle questions, such as the symmetry of the crystal and its energy cannot yet be answered conclusively (although the results indicate formation of a bcc lattice), because simulations are too small (only 50 electrons and 50 holes are presently feasible). Therefore, in order to obtain more accurate data, e.g., for the internal energy of a macroscopic two-component plasma at very high density, a significant increase in the simulation size is highly desirable.

The main results concern the relative distance fluctuations of holes shown in Fig. 10.19. According to Lindemann (Lindemann 1910), an increased amplitude of the vibrations of the particles around their equilibrium positions is the physical reason for the melting of the crystal. If the vibrational amplitude becomes so large that particles start to invade the space of their nearest neighbors, the melting process sets in. At the melting point, the root mean square vibration amplitude $\sqrt{\langle u^2 \rangle}$ exceeds a threshold value of about 10–28% of the nearest neighbor distance $\bar{r}$, depending on the type of pair interaction, dimensionality, and spin statistics (Bonitz et al. 2008).

In Fig. 10.19, the density and temperature are fixed in such a way that bound state formation is not possible, and the mass ratio from hydrogen to electron–hole plasmas is varied. At $M = 2\,000$, the distance fluctuations are small, and they remain almost unchanged when M is reduced. Around $M = 100$, a drastic increase is observed,

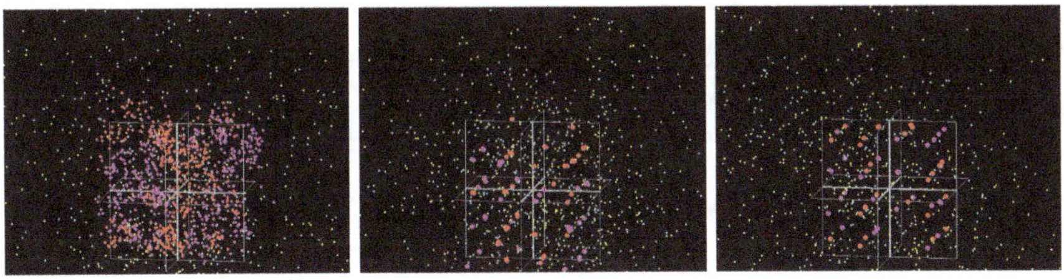

Fig. 10.18 Snapshots of a strongly correlated two-component Coulomb system of holes and electrons with mass ratio $M = m_h/m_e = 50$ (*left*), 100 (*center*), and 800 (*right*) at temperature $T/E_B = 0.064$ and $r_s = 0.5$. The color coding is the same as in Fig. 10.1 (Bonitz et al. 2005)

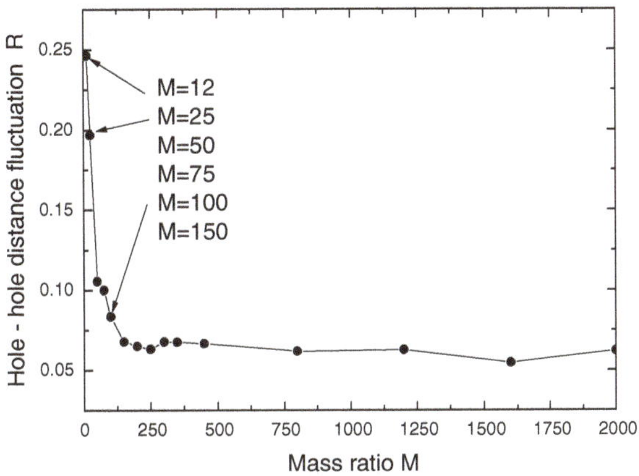

Fig. 10.19 Mean-square relative hole–hole distance fluctuations (normalized to the mean interparticle distance) as a function of the mass ratio M for $T_e = 0.096$ and $r_s = 0.63$ (Bonitz et al. 2006)

a clear indication of spatial delocalization of the holes. In fact, an analysis of the microscopic configuration in the simulation box and the pair distribution functions (Bonitz et al. 2005) clearly confirms this interpretation. At large M, the holes form a crystal which is embedded in a high-density electron gas. This crystal vanishes (melts) between $M = 100$ and $M = 50$. The predicted critical mass ratio is higher than in most conventional semiconductors. However, similar mass ratios have already been reported in intermediate valence semiconductors, such as Tm[Se Te] (Bucher et al. 2001, 2004). It has been possible to verify hole crystallization experimentally by means of neutron scattering.

The quantum crystal–liquid transition at fixed temperature and density ($r_{se} = 0.63$) is further supported by the behavior of the spin-averaged and spin-dependent pair distribution functions (PDFs) shown in Fig. 10.20a–d. The PDF g_{hh} indicates hole crystal formation with increasing hole mass. At the mass ratios, densities, and temperatures when hole ordering exists, the holes are still well localized, while the electrons are strongly degenerate and delocalized.

In the DPIMC simulations, the energy spectrum is not resolved, but a thermal average of all allowed energies is observed. At the same time, the simulations yield information about the spatial distribution of all the particles. As the PDFs g_{eh} and g_{ee}

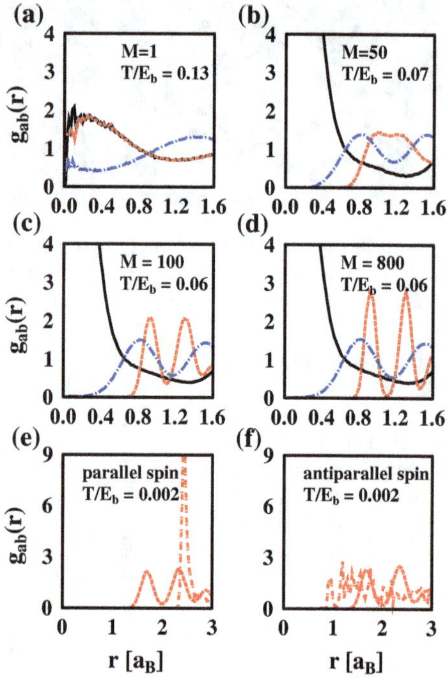

Fig. 10.20 Quantum liquid–crystal hole transition at $r_{se} = 0.63$, showing the PDFs g_{ab} averaged over spin projections in figures (**a**)–(**d**). *Solid lines* g_{ee}. *Dashed lines* g_{hh}. *Dash-dotted lines* g_{eh}. The PDFs g_{hh} for parallel (**e**) and antiparallel (**f**) spin projections are related to the transition of the hole crystal ($M = 400$, *dashed lines*) to the antiferromagnetic crystal structure ($M = 100$, *dash-dotted lines*) with decreasing hole mass at $r_{se} = 1.13$ (Filinov et al. 2005)

show, we observe an inhomogeneous space distribution of electrons. The physical evidence of this phenomenon can be understood from the behavior of g_{ee} and g_{eh} (Figs. 10.20b–d) and electron–hole snapshots in Fig. 10.21. The functions g_{ee} for ordered hole structures have sharp peaks at small distances between the electrons, while the g_{eh} are periodically oscillating functions. A peak in g_{ee} exists only if the holes are ordered. So the density of the highly degenerate nearly free electrons in a hole periodic and quasi-periodic potential field is nonuniform and has interference maxima (see Fig. 10.21) related to the variation of the modulus of the electron wave functions according to the Bloch theorem and the electron energy band structure. The quasi-lattice of negative charges related to these maxima lowers the crystal energy. So this quasi-lattice increases the critical temperature and reduces the critical value of Γ^{cr} for hole crystallization. As a consequence, at finite temperature, the crystal phase is more stable than predicted by the OCP plasma model, in which the electrons are assumed to form a strictly uniform background. For mass ratios smaller than 50 and temperatures $k_B T/E_B > 0.064$, the hole ordering is destroyed, and this peak in the electron–electron PDFs disappears. With decreasing hole mass, at some point, the hole wavelength becomes so long that holes can invade the space of their nearest neighbors with nonzero probability and disturb them, thereby initiating the melting process at $M = 75$–50. This crystallization threshold value of M in three-dimensional plasmas in semiconductors also agrees with the results of (Abrikosov 1978) ($M \sim 80$). Such values are feasible, e.g., in the intermediate valence system

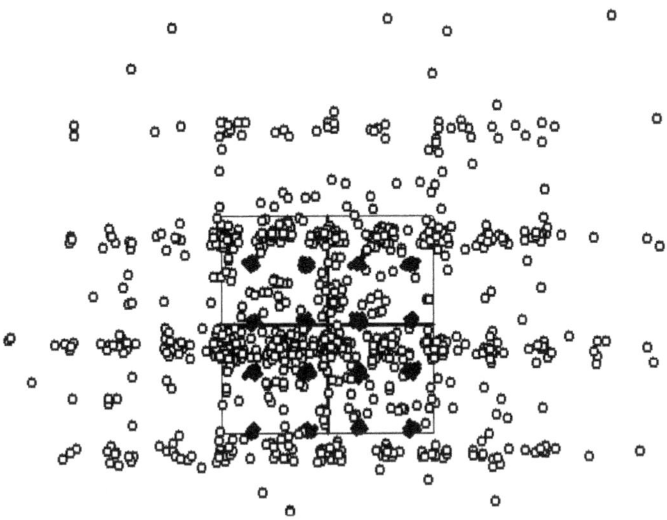

Fig. 10.21 Electron–hole snapshots for $k_B T/E_B = 0.002$, $r_{se} = 3$, $m_h = 800$, $m_e = 2.1$. *Empty (filled) circles* are electron (hole) beads

Tm[Se,Te] under pressure (Bucher et al. 1991, 2004). DPIMC calculations also show that quantum melting agrees with the Lindemann criterion.

Figure 10.20e and f show that, at fixed density and very low temperatures, the electron–hole system with decreasing hole mass can form an antiferromagnetic crystal-like structure, comprising two perfect hole lattices, each consisting of holes with the same spin projections. The physical reason for this phenomenon is the strong Fermi repulsion between particles with the same spin projection. Correlation between the lattices is defined only by an effective repulsive force, which is weaker at lower hole masses due to the tunneling by the holes.

Figure 10.22 is the electron–hole phase diagram obtained by DPIMC. In the bottom left corner, as in Fig. 10.17, electrons and holes are in bound states at low temperatures. This region is bounded by lines corresponding to the 50 and 10% fractions of electron–hole bound states. Estimates of the electron–hole bound states are given, for example, in (Fehske et al. 2005). With increasing density, but fixed temperature (moving rightwards), the bound states vanish and the Mott transition occurs. Red corresponds to the hole liquid phase and blue to the hole crystal. The blue line shows the hole liquid–crystal transition, while the green line corresponds to the transformation of the hole crystal to an antiferromagnetic crystal structure. As an example, the DPIMC calculations are shown by the colored triangles and circles.

The electron–hole phase in Fig. 10.22 agrees qualitatively with estimates from the phase diagram (Fig. 10.17) based on known data for the one-component plasma. The main temperature difference between the lines for the liquid–crystal transition in Figs. 10.22 and 10.17 (by a factor of order 1.5–2) is due to the greater stability of the TCP crystal discussed above. Periodic oscillations in the electron density lower

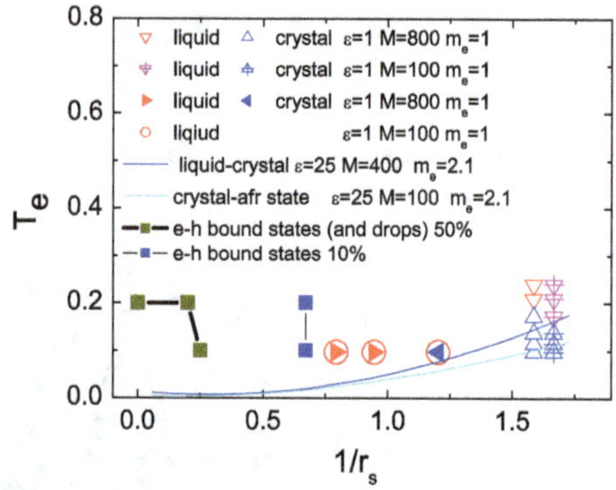

Fig. 10.22 DPIMC phase diagram of an electron–hole plasma. The DPIMC *triangles* directed *downward (upward)* and to the *right (left)* correspond to the liquid (crystal) structure. *Red*: Hole liquid phase. *Blue*: hole crystal

the total energy and the hole crystal can exist at higher temperature than in the OCP with fixed homogeneous negative background. The second difference between the phase diagrams concerns the transition from the hole crystal to the antiferromagnetic crystal structure.

Two-Dimensional Simulations

The strongly correlated two-dimensional Coulomb systems are the subject of intensive discussions (Bedanov et al. 1985, Stranburg 1988, Morf 1983, Bakker 1984, Abraham 1981, Lin et al. 2007, Zheng and Grieve 2006, Mak 2006, Keim et al. 2007). It is known that the competition between electrostatic and kinetic energy in a electron gas may be the reason for the unusual phase diagram of an electron system on a surface. The liquid state is stable when the kinetic energy dominates, while the electrostatically favored 'Wigner' triangular crystal is stable in the opposite case. In the region of interest, where there is strong competition between the two kinds of energy, there may exist different phases. Instead of the crystal-like long-range order in 2D systems, there may exist different phases with ordering characterized by different asymptotic behavior of translation and orientation symmetry. What is under discussion here is the type of phase transition and the related question of the existence of the anisotropic liquid (hexatic) phase intermediate between a crystal and an isotropic liquid. Note that the physical mechanism of melting can be influenced by interaction with the substrate and defects. All the possibilities mentioned for strongly correlated 2D charged particle systems make it necessary to use first principle numerical simulations.

Below, the density of the two-component plasma is characterized by the Brueckner parameter r_s defined as the ratio of the mean distance between particles $d = [1/\pi(n_e + n_h)]^{1/2}$, and the 3D Bohr exciton radius a_B, where n_e and n_h are the 2D electron and hole densities. The dimensionless temperature is taken as the ratio of the temperature and the 3D Rydberg constant, which includes the reduced effective mass and the dielectric constant.

According to the temperature decomposition of the density matrix, each electron and hole is represented by several tens of points ('beads'). The spatial distribution

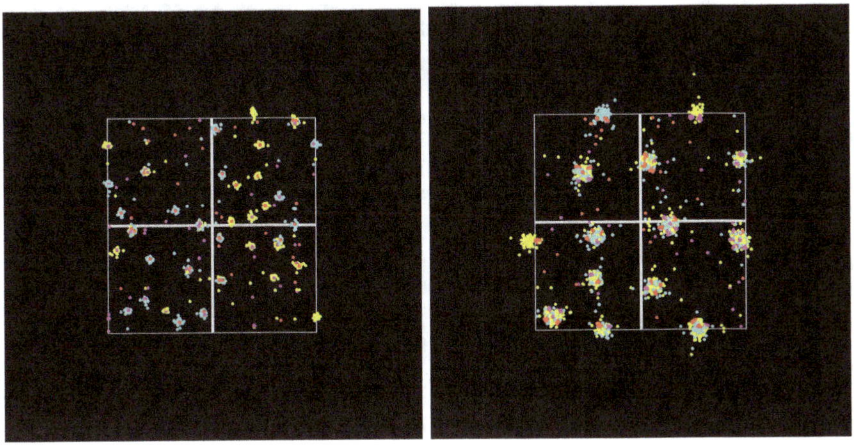

Fig. 10.23 Electron–hole snapshots at temperature $T/\mathrm{Ry} = 0.007$, hole-to-electron mass ratio $M = 800$, and densities such that the Brueckner parameter $r_s = 6$ (*left*) and $r_s = 2$ (*right*). *Red* and *pink points* represent beads of holes with different spin projections, while *yellow* and *green points* represent beads of electrons with opposite spins (Filinov et al. 2008)

of the beads of each quantum particle is proportional to its spatial probability distribution. Figure 10.23 shows that the typical size of the cloud of beads for electrons is several times larger than the one for the heavy holes. At low temperature and low ($r_s = 6$) and medium densities ($r_s = 2$), practically all holes are closely covered by electron beads. From a physical point of view this means that electrons and holes form bound states, i.e., excitons, bi-excitons and many-particle clusters. The existence of bound states is also supported by the behavior of the pair distribution functions, exhibiting pronounced maxima at distances of about half the Bohr radius and slightly greater distances (not shown here).

From Fig. 10.23, we see that a growth in the density results in an increase in the number of particles in clusters. Structural analysis of large many-particle clusters shows the hexagonal ordering of heavy holes inside liquid-like clusters. The growth in the number of particles in clusters with increasing density can be seen in the right-hand panel of Fig. 10.24. Here there are only two clusters: one is in the center of the Monte Carlo cell, while the other is divided into four parts (one in each corner) due to the periodic boundary conditions of the Monte Carlo cell. Here, besides the normal inner hexagonal structure, the holes due to the strong Coulomb repulsion form filament-like structures at the boundaries of clusters. These filament-like structures are a fully 2D topological effect, just as in the 3D case, the holes in the analogous many-particle clusters have normal liquid-like ordering.

If the particle density is high enough ($r_s = 0.25$), the electron wavelength becomes longer than the mean interparticle distance d and even longer than the size of the Monte Carlo cell, as can be seen from the broad extent of the electron bead clouds. For $r_s \leq 0.5$, clusters become unstable, because two electrons bound to neighboring holes start to overlap, allowing electron tunneling from one cluster to another (Mott effect).

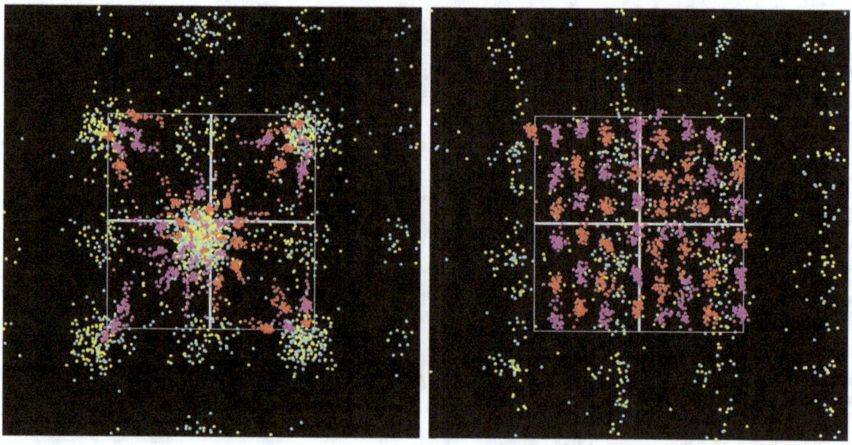

Fig. 10.24 Electron–hole snapshots at temperature $T/\text{Ry} = 0.007$, hole-to-electron mass ratio $M = 800$, and densities such that the Brueckner parameter $r_s = 0.5$ (*left*) and $r_s = 0.25$ (*right*) (Filinov et al. 2008)

Since the hole wavelength is significantly smaller than the electron wavelength, it may still be smaller than d, and the structure of the hole beads will resemble a liquid-like or crystal-like state. If the hole mass exceeds a critical value, the holes may even form a crystal-like structure (Bonitz et al. 2005) (right panel of Fig. 10.24). Here holes form crystal-like structures, while the electron density exhibits Bloch oscillations. At this very high density, the type of the hole 'crystal' is influenced by the boundary conditions of the Monte Carlo cell (finite-size effect). So a detailed analysis of the type of crystal-like structure should be carried out for a much greater number of particles in the Monte Carlo cell.

Proton Crystallization

About a decade ago, it was shown that, in a two-component (electrons and protons) dense hydrogen plasma at sufficiently low temperatures and high densities, the protons will spontaneously order to form a Coulomb crystal embedded in a highly degenerate electron gas (Filinov et al. 2005). A simplified description of this system is given by the one-component plasma (OCP) model, consisting of positive charges (protons, for example) in a uniform (rigid) neutralizing background of electrons. For the classical OCP, extensive studies have revealed the following regimes:

- A proton gas (plasma) state for values of the coupling parameter (CP) less than unity.
- A proton liquid state or proton glass state if CP exceeds unity.
- Finally, for still higher values of the CP (≥ 175 in three dimensions), a proton crystal state.

For a review see, e.g., (Ichimaru et al. 1987).

In order to localize structural transitions between different states of matter various criteria have been used in the literature, including many based on the analysis of the pair distribution functions (PDF). For example, a jump in the height of the first PDF peak was observed at the freezing point in a macroscopic 2D OCP [see Ott et al. (2011) and references therein]. Unfortunately, in simulations with a limited particle number—as is unavoidable in the quantum MC simulation of fermions—the height and shape of the peaks in the PDF are sensitive to the size of the Monte Carlo cell and the number of particles in it. Additional difficulties arise from the possible existence of metastable states. To overcome these difficulties, we may use the idea of Lindemann (Lindemann 1910) and analyze these fluctuations in the nearest-neighbor distances of the system for different states of matter.

In general, phase transitions between different states of matter and structural transitions within the solid phase are indicated by sharp jumps in the relative distance fluctuations between neighboring particles. Therefore, in this section, such jumps will be analyzed in order to localize structural transitions in a dense two-component hydrogen plasma. Here we discuss the phase diagram of a neutral two-component mass-asymmetric plasma. To this end, extensive DPIMC simulations have been carried out for a fully ionized macroscopic spatially homogeneous two-component electron–proton plasma in thermodynamic equilibrium, with determination of the temperature and density range of proton crystallization.

For a quantitative analysis of structural transitions in a dense hydrogen plasma, one has to determine the relative distance fluctuations of protons (RD-PP) $\sqrt{\langle u^2\rangle}/\bar{r}$. To get deeper physical insight into the behavior of the RD-PP, we consider the proton–proton pair distribution function (PP-PDF):

$$g_{\rm pp}(|r_{1,p}-r_{1,p}|)=\frac{N_{\rm e}!N_{\rm p}!}{Z}\sum_{\sigma}\int_{V}\mathrm{d}q\,\delta(r_{1,p}-q_{1,p})\delta(r_{2,p}-q_{2,p})\rho(q,\sigma;\beta)\ . \tag{10.14}$$

Figures. 10.25 and 10.26 give the results of DPIMC calculations for three characteristic temperatures at which we observe different phases. The top left panel of Fig. 10.25 illustrates the behavior of the RD-PP as a function of the mean interparticle distance. At high densities ($r_{\rm s}\leq 0.18$) and $T=20\,000$ K, RD-PP is practically constant. In this region, the PP-PDF is approximately equal to one, except for the correlation hole at very small distances, which is characteristic for a weakly coupled degenerate gas of protons (see top right panel of Fig. 10.25). When the density is decreased to $r_{\rm s}\approx 0.2$, a sudden jump occurs in the RD-PP. This is related to a transition to a perfect proton crystal, as can be concluded from the change in the PP-PDF. At $r_{\rm s}\approx 0.2$, $g_{\rm pp}$ exhibits well pronounced narrow peaks and dips (see the top right panel of Fig. 10.25). Thus, in the density interval from $r_{\rm s}=0.2$–0.23, protons are ordered in a crystal-like structure.

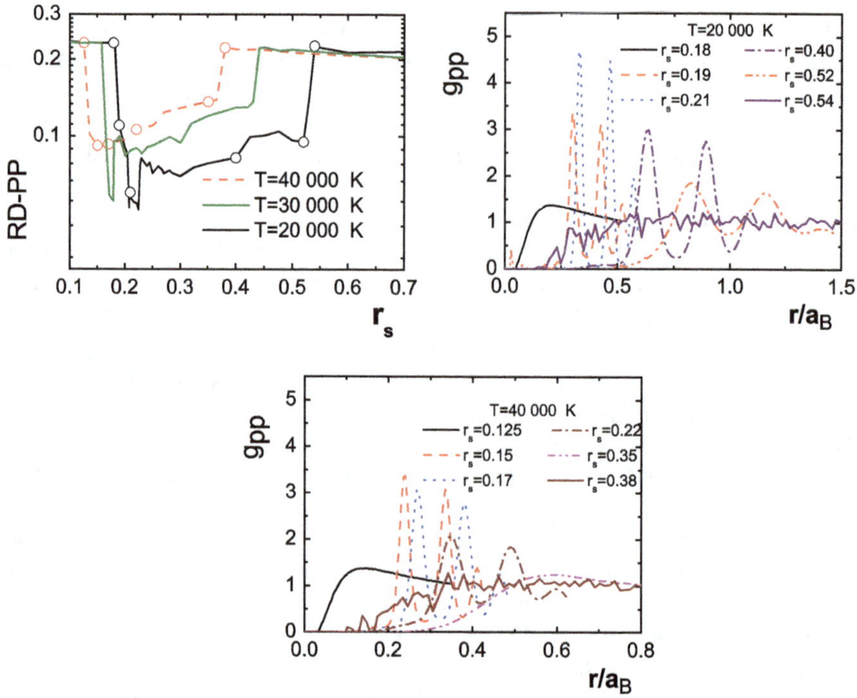

Fig. 10.25 *Top left* Relative proton–proton distance fluctuations for three temperatures. Sharp jumps signal structural transitions. For $T = 20\,000$ K: left jump, from a degenerate disordered phase of protons ($r_s \approx 0.18$) to an ordered phase ($r_s \approx 0.2$); middle jump, structural transition inside the ordered phase ($r_s \approx 0.225$–0.23); right jump, from a proton ordered phase ($r_s \approx 0.52$) to a disordered phase ($r_s \approx 0.53$). Similar behavior is observed for $T = 30\,000$ K. For $T = 40\,000$ K, the jump in the ordered phase is missing. *Top right* and *bottom* Proton–proton pair distribution functions for a fully ionized hydrogen plasma with $M = 2\,000$ for various particle densities at $T = 20\,000$ K and $40\,000$ K, respectively (Filinov et al. 2012)

A further density decrease ($r_s \geq 0.23$) results in a second sharp jump in the RD-PP, but with smaller magnitude. This jump is accompanied by a broadening and lowering of the peaks of the PP-PDF. As yet, the nature of the resulting state is not fully understood from the available data. A possible explanation is that the system is in another (metastable) crystalline state which vanishes only at $r_s = 0.52$, where a third large jump occurs. Another possibility is that we observe a phase transition to a liquid-like state that persists until $r_s = 0.52$.

As the density is further reduced to $r_s = 0.54$, the RD-PP exhibits another large jump, while the PP-PDF changes dramatically to a shape typical of a weakly coupled plasma. While the behavior of the RD-PP along the second isotherm $T = 30\,000$ K is very similar, the third isotherm, at $T = 40\,000$ K, does not exhibit the intermediate smaller jump, indicating that the ordered phase remains homogeneous throughout the density interval from $r_s = 0.15$ to 0.35.

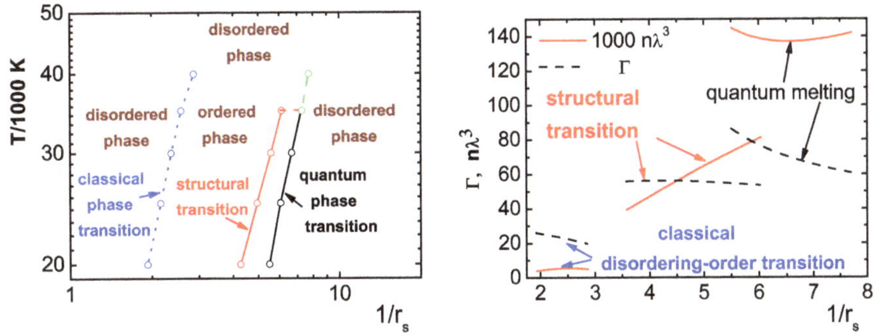

Fig. 10.26 *Left* Hydrogen phase diagram in the vicinity of proton crystallization. *Lines with circles* are DPIMC simulation results. The ordered phase with low values of the RD-PP is embedded in classical and quantum disordered phases, appearing to the *left of the blue line* and *right of the black line*, respectively. The *red line* indicates a structural transition inside the ordered phase. The ordered proton phase terminates in a triple point. *Right* Values of the classical coupling parameter Γ (*black dashed lines*) and quantum degeneracy parameter $n\lambda^3$ (*red full lines*) along the three lines along which the RD-PP has a jump (see left panel) (Filinov et al. 2012)

The DPIMC results for the PP-PDF and RD-PP are summarized in the phase diagram (see lines with circles in the left panel of Fig. 10.26). An increase in the density at fixed temperature first leads to a Mott transition from a neutral plasma (consisting of atoms and molecules) to a fully ionized disordered gas-like plasma. This transition is not shown in the phase diagram at the related $r_s \approx 1.5$–2. Upon further compression, the sharp jump in the RD-PP indicates a transition to a spatially ordered state (blue dotted line). The broad region of smooth variation of the RD-PP indicates that this state persists over a broad density interval. For temperatures below 40 000 K, a further increase in the density leads to a further reduction of the RD-PP, signaling another ordering transition of the protons. At a slightly higher density, the RD-PP jumps again, when the proton crystal melts into a proton quantum disordered phase as a consequence of overlapping proton wave functions. At temperatures above 40 000 K, the second jump in the RD-PP is missing. The characteristic values of the coupling parameter Γ and degeneracy parameter $n_p\lambda_p^3$ along the three phase boundaries are presented in the right-hand panel of Fig. 10.26. Here $\lambda_p^2 = 2\pi\hbar^2/mk_BT$ is the thermal proton de Broglie wavelength squared, and we find that Γ is of the order of 23, 60, and 80, respectively, whereas the proton degeneracy parameter at these phase boundaries is of the order of 0.01, 0.06, and 0.14, respectively.

To summarize, we should stress that we are presenting the first simulation results for the phase diagram of a two-component dense quantum plasma in the vicinity of a proton crystal. It should be emphasized that some of these results are still preliminary. In particular the interpretation of the ordered phase between the two large jumps of the RD-PP is not yet fully understood. Whether this whole density interval is characterized by a proton solid which undergoes several structural transitions at high densities or whether the proton crystal phase exists only in the narrow region of the lowest values of the RD-PP requires further analysis.

These findings differ substantially from earlier predictions based on the OCP model:

- In the classical part of the phase diagram, the crystal appears to be stabilized as compared to the OCP prediction. A nearly constant coupling parameter $\Gamma^{\rm cr}$ was found around 23 (60) at the disorder–order transition (at the structural transition inside the ordered phase).
- In the quantum part of the phase diagram, the crystal is destabilized compared to the OCP prediction of $r_{\rm s}^{\rm cr} = 100$, and it vanishes at lower densities. Moreover, the quantum melting temperature increases with density, in contrast to the OCP prediction.

We may thus conclude that the simplified OCP treatment of the liquid–solid transition in a two-component plasma has to be questioned. Apparently, the OCP assumption of a homogeneous rigid neutralizing background gives rise to substantial deviations of the critical parameters. More simulations are underway to check and generalize these predictions.

References

F.F. Abraham, Phys. Rep. **80**, 339 (1981)

A.A. Abrikosov, J. Less-Common Met. **62**, 451 (1978)

O. Arp, D. Block, A. Piel, A. Melzer, Phys. Rev. Lett. **93**, 165004 (2004)

A.F. Bakker, C. Bruin, H.J. Hilhorst, Phys. Rev. Lett. **52**, 449 (1984)

V.M. Bedanov, G.V. Gadijak, YuE Lozovik, JETF **88**, 1622 (1985)

D. Beule et al., Phys. Rev. B *59*, 14177 (1999); Contr. Plasma Phys. **39**, 21 (1999)

D. Beule, W. Ebeling, A. Förster, H. Juranek, R. Redmer, G. Röpke, J. de Phys. IV **10**(Pr5), 295 (2000)

V. Bezkrovniy, M. Schlanges, D. Kremp, W.-D. Kraeft, Phys. Rev. E **70**, 57401 (2004)

Bo-Jiun Lin, Li-Jen Chena, J. Chem. Phys. **126**, 034706 (2007)

M. Bonitz et al., J. Phys. A: Math. Gen. **36**, 5921 (2003)

M. Bonitz, V.S. Filinov, V.E. Fortov, P.R. Levashov, H. Fehske, Phys. Rev. Lett. **95**, 235006 (2005)

V. Filinov, H. Fehske, M. Bonitz, V.E. Fortov, P.R. Levashov, Phys. Rev. E **75**, 36401 (2007)

M. Bonitz et al., Phys. Plasmas **15**, 55704 (2008)

M. Bonitz, Phys. J. **7/8**, 69 (2002)

M. Bonitz, V.S. Filinov, V.E. Fortov, P.R. Levashov, H. Fehske, J. Phys. A: Math. Gen. **39**, 4717 (2006)

G.V. Boriskov, A.I. Bykov, R.I. Il'kaev, V.D. Selemir, G.V. Simakov, R.F. Trunin, V.D. Urlin, A.N. Shuikin, Phys. Rev. B **71**, 92104 (2005)

B. Bucher, P. Steiner, P. Wachter, Phys. Rev. Lett. **67**, 2717 (1991); P. Wachter, B. Bucher, J. Malar, Phys. Rev. B **69**, 945021 (2004); P. Wachter. Solid State Commun. **118**, 645 (2001)

G. Chabrier, D. Saumon, W. Hubbard, J. Lunine, Astrophys. J. **391**, 817 (1992)

D.M. Ceperley, B.J. Alder, Phys. Rev. Lett **45**, 566 (1980)

G. Collins, L. Da Silva, P. Celliers, D. Gold, M. Foord, R. Wallace, A. Ng, S. Weber, K. Budil, R. Cauble, Science **281**, 1178 (1998)
I.B. Da-Silva et al., Phys. Rev. Lett. **78**, 783 (1997)
H. DeWitt, W. Slattery, Contrib. Plasma Phys. **39**, 1 (1999)
F.J. Dyson, A. Lenard, J. Math. Phys. **8**, 423 (1967)
W. Ebeling, W. Richert, Phys. Lett. A **7108**, 80 (1985); phys. stat sol. (b) **128**, 167 (1985)
W. Ebeling, W.D. Kraeft, D. Kremp, *Theory Of Bound States And Ionization Equilibrium In Plasmas And Solids* (Akademie-Verlag, Berlin, 1976a)
W. Ebeling, W.D. Kraeft, D. Kremp, *Theory of Bound States and Ionization Equilibrium in Plasmas and Solids* (Akademie-Verlag, Berlin, 1976b)
W. Ebeling, H.J. Hoffmann, G. Kelbg, Contrib. Plasma Phys. **7**, 233 (1967)
W. Ebeling, A. Förster, V. Fortov, V. Gryaznov, A. Polishchuk, *Thermophysical Properties Of Hot Dense Plasmas*, Teubner Stuttgart 1991 (Russ. transl. Moskva, Ishevsk, 2007)
H. Fehske, V.S. Filinov, M. Bonitz, P. Levashov, V.E. Fortov, J. Phys.: Conf. Ser. **11**, 139 (2005)
H. Fehske, V.S. Filinov, M. Bonitz, V.E. Fortov, J. Phys.: Conf. Series **11**, 139 (2005)
H. Fehske, V.S. Filinov, M. Bonitz, V.E. Fortov, P.R. Levashov, J. Phys.: Conf. Ser. **11**, 139 (2005)
V.S. Filinov, V.E. Fortov, M. Bonitz, P.R. Levashov, JETP Lett. **72**, 245 (2000a)
V.S. Filinov, V.E. Fortov, M. Bonitz, P.R. Levashov, JETP Lett. **74**, 384 (2001); [Pis'ma v ZhETF **74**, 422 (2001)]
V. Filinov, M. Bonitz, P. Levashov, V. Fortov, W. Ebeling, M. Schlanges, Contrib. Plasma Phys. **43**, 290 (2003a)
V.S. Filinov, V.E. Fortov, M. Bonitz, D. Kremp, Phys. Lett. A **274**, 228 (2000b)
V.S. Filinov, M. Bonitz, V.E. Fortov, JETP Lett. **72**, 245 (2000c)
A.V. Filinov, M. Bonitz, W. Ebeling, J. Phys. A: Math. Gen. **36**, 5957 (2003b)
V.S. Filinov, V.E. Fortov, M. Bonitz, P.R. Levashov, JETP Lett. **74**, 384 (2001a)
V.S. Filinov, M. Bonitz, P.R. Levashov, V.E. Fortov, W. Ebeling, M. Schlanges, S.W. Koch, J. Phys. A.: Math. Gen. **36**, 6069 (2003c)
A.V. Filinov, M. Bonitz, YuE Lozovik, Phys. Rev. Lett. **86**, 3851 (2001b)
V.S. Filinov, M. Bonitz, V.E. Fortov, W. Ebeling, H. Fehske, D. Kremp, W.D. Kraeft, V. Bezkrovny, P. Levashov, J. Phys. A: Math. Gen. **38**, 1 (2005)
V.S. Filinov, H. Fehske, M. Bonitz, V.E. Fortov, P. Levashov, Phys. Lett. A **372**(31), 5208 (2008)
V.S. Filinov, M. Bonitz, H. Fehske, V.E. Fortov, P. Levashov, Contrib. Plasma Phys. **52**(3), 224 (2012)
V.E. Fortov, V.Y. Ternovoi, M.V. Zhernokletov, M.A. Mochalov, A.L. Mikhailov, A.S. Filimonov, A.A. Pyalling, V.B. Mintsev, V.K. Gryaznov, I.L. Iosilevskii, JETP **97**, 259 (2003)
C.C. Grimes, G. Adams, Phys. Rev. Lett. **42**, 795 (1979)
T.V. Gudkova, V.N. Zharkov, Planet. Space Sci. **47**, 671 (2000)
B.I. Halperin, T.M. Rice, Rev. Mod. Phys. **40**, 755 (1968)
Y. Hayashi, K. Tachibana, Jpn. J. Appl. Phys. **33**, L804 (1994)
S. Ichimaru, H. Iyeltomi, S. Tanaka, Phys. Rep. **149**, 91 (1987)
C.D. Jeffries, L.V. Keldysh (eds.) *Electron–hole droplets in semiconductors*, Nauka, Moscow 1988; J.C. Hensel, T.G. Phillips, G.A. Thomas, Solid State Phys. **32**, 88 (1977); **95**, 235006 (2005)
G. Kalman (ed.), *Strongly Coupled Coulomb Systems* (Pergamon Press, 1998)
P. Keim, G. Maret, H.H. von Grunberg, Phys. Rev. E **75**, 031402 (2007)
M. Knudson, D. Hanson, J. Bailey, C. Hall, J. Asay, W. Anderson, Phys. Rev. Lett. **87**, 225501 (2001)
W.D. Kraeft, M. Schlanges (eds.), *Proceedings of the International Conference on Strongly Coupled Plasmas* (World Scientific, Singapore 1996)
W.D. Kraeft, D. Kremp, W. Ebeling, G. Röpke, *Quantum Statistics of Charged Particle Systems* (Akademie-Verlag, Berlin, 1986)
W.D. Kraeft, D. Kremp, W. Ebeling, G. Röpke, *Quantum Statistics of Charged Particle Systems* (Akademie, Berlin, 1986)

P.R. Levashov, V.S. Filinov, V.E. Fortov, M. Bonitz, in *Shock Compression of Condensed Matter*, ed. by M.D. Furnish, N.N. Thardhani, Y. Horie (AIP, New York, 2002), p. 119
P.R. Levashov, V.S. Filinov, M. Bonitz, V.E. Fortov, J. Phys. A: Math. Gen. **39**, 4447 (2006)
P. Levashov, V.S. Filinov, M. Bonitz, V.E. Fortov, Contrib. Plasma Phys. **45**, 258 (2005)
M. Li, William, L. Johnson, W. A. Goddard III. Phys. Rev. B **54**, 12067 (1996)
S.Z. Lin, B. Zheng, S. Trimper, Phys. Rev. E **73**, 066106 (2006)
E.H. Lieb, W. Thirring, Phys. Rev. Lett. **31**, 111 (1975)
F. Lindemann, Z. Phys. **11**, 609 (1910)
P. Ludwig, S. Kosse, M. Bonitz, Phys. Rev. E **71**, 46403 (2005)
C.H. Mak, Phys. Rev. E **73**, 065104 (2006)
B. Militzer, D.M. Ceperley, Phys. Rev. Lett. **85**, 1890 (2000)
B. Militzer, D. Ceperley, Phys. Rev. E **63**, 66404 (2001)
M.A. Mochalov, R.I. Ilkaev, V.E. Fortov, A.L. Mikhailov, V.A. Arinin, A.O. Blikov, AYu. Baurin, V.A. Komrakov, V.A. Ogorodnikov, A.V. Ryzhkov, A.A. Yukhimchuk, Pis'ma v Zh. Eksper. Teoret. Fiz. **96**(3), 172 (2012)
M.A. Mochalov, R.I. Il'kaev, V.E. Fortov, A.L. Mikhailov, V.A. Arinin, A.O. Blikov, V.A. Komrakov, A.V. Ryzhkov, V.A. Ogorodnikov, A.A. Yukhimchuk, Pis'ma v Zh. Eksper. Teoret. Fiz. **101**(8), 575 (2015)
R.F. Morf, Helv. Phys. Acta **56**, 743 (1983)
W. Nellis, A. Mitchell, M. van Thiel, G. Devine, R. Trainor, J. Chem. Phys. **79**, 1480 (1983)
W.J. Nellis, Planet. Space Sci. **48**, 671 (2000)
G.E. Norman, A.N. Starostin, Teplofiz. Vys. Temp. **6**, 410 (1968a); **8**, 413 (1970), Sov. Phys. High Temp. **6**, 394 (1968a); **8**, 381 (1970)]
G.E. Norman, A.N. Starostin, Sov. Phys. High Temp. **6**, 410 (1968b)
G.E. Norman, IYu. Skobelev, V.V. Stegailov, Contrib. Plasma Phys. **51**, 411 (2011)
T. Ott, M. Stanley, M. Bonitz, Phys. Plasmas **18**, 63701 (2011)
J. Riemann, M. Schlanges, H.E. DeWitt, W.D. Kraeft, in *Proceedings of the International Conference on Strongly Coupled Plasmas*, ed. by W.D. Kraeft, M. Schlanges (World Scientific, Singapore, 1996), p. 82
D. Saumon, G. Chabrier, Phys. Rev. A **44**, 5122 (1991)
D. Saumon, G. Chabrier, H.M. Van Horn, Astrophys. J. Suppl. Ser. **99**, 713 (1995)
D. Saumon, G. Chabrier, Phys. Rev. A **46**, 2084 (1992)
D. Saumon, G. Chabrier, Phys. Rev. Lett. **62**, 2397 (1989)
M. Schlanges, M. Bonitz, A. Tschttschjan, Contrib. Plasma Phys. **35**, 109 (1995)
K.J. Stranburg, Rev. Mod. Phys. **60**, 161 (1988)
V. Ternovoi, A. Filimonov, V. Fortov et al., Phys. B **265**, 6 (1999a)
V.Y. Ternovoi, A.S. Filimonov, V.E. Fortov, S.V. Kvitov, D.N. Nikolaev, A.A. Pyalling, Phys. B **265**, 6 (1999b)
V.Y. Ternovoi, S.V. Kvitov, A.A. Pyalling, A.S. Filimonov, V.E. Fortov, JETP Lett. **79**, 8 (2004)
V.Y. Ternovoi, A.S. Filimonov, A.A. Pyalling, V.B. Mintsev, V.E. Fortov, in *Shock Compression of Condensed Matter*, ed. by M.D. Furnish, N.N. Thardhani, Y. Horie (AIP, New York, 2002), p. 107
H. Thomas et al., Phys. Rev. Lett. **73**, 652 (1994)
G.A. Thomas, T.-M. Rice, J.C. Hensel, Phys. Rev. Lett. **33**, 219 (1974)
S. Weir, A. Mitchell, W. Nellis, Phys. Rev. Lett. **76**, 1860 (1996)
E. Wigner, Phys. Rev. **46**, 1002 (1934). Wigner's original estimate for the lattice energy was later corrected, and we use an improved result. For a discussion of various estimates, see, e.g., Chap. V of G.D. Mahan in *Many-Particle Physics*, 2nd edn. (Plenum Press, 1990)
H. Xu, J.P. Hansen, Phys. Rev. E **57**, 211 (1998)

X. Yan, S. Tsai, S. Ichimaru, Phys. Rev. A **43**, 3057 (1991)
V.M. Zamalin, G.E. Norman, V.S. Filinov, *The Monte Carlo Method in Statistical Thermodynamics* (Nauka, Moscow, 1977). (in Russian)
B.V. Zelener, G.E. Norman, V.S. Filinov, *Perturbation Theory and Pseudopotential in Statistical Thermodynamics* (Nauka, Moscow, 1981). (in Russian)
X.H. Zheng, R. Grieve, Phys. Rev. B **73**, 064205 (2006)

Chapter 11
Applications to Matter with High Energy Density

The most fundamental approach to study relativistic quantum systems of many particles is currently a quantum field theory (QFT). In these theories, particles are usually considered as elementary excitations of the corresponding field. This approach automatically takes into account the creation and annihilation of particles, and treats interactions as exchange of virtual particles. The main working tool in QFT is the Feynman diagram technique (Schweber 1963), which represents a series generated in perturbation theory graphically, assuming that the coupling constants are small parameters. Feynmandiagrams are useless for strongly coupled systems of particles.

One of the most straightforward approaches that can be used to calculate the properties of strongly coupled nonideal systems is lattice QFT. However, calculations on lattices are very costly and require powerful supercomputers. An alternative approach to study relativistic systems is to make use of the path integral model of a system with a variable number of particles described by a Hamiltonian with relativistic kinetic energy operator $\sqrt{p^2c^2 + m^2c^4}$ and interacting through some effective pair potential. This method has been used to calculate thermodynamic and transport properties of quark–gluon plasmas (Filinov et al. 2012). This requires far fewer computer resources and could be applied to systems with large chemical potentials. Furthermore, an analogous approach has been used in applications to electromagnetic (Filinov et al. 2001) and electron–hole plasmas (Filinov et al. 2007).

11.1 Relativistic Partition Function of Newton–Wigner Particles in the Path Integral Representation

Here the relativistic spinless particle will be described by the Klein–Gordon equation. The thermodynamic properties of such a system can be expressed in the form of path integrals in an external potential field derived from the one-particle relativistic theory (Newton and Wigner 1949). This approach is equivalent to direct solution of the

W. Ebeling et al., *Quantum Statistics of Dense Gases and Nonideal Plasmas*, Springer Series in Plasma Science and Technology, DOI 10.1007/978-3-319-66637-2_11

relativistic wave equations, such as the Klein–Gordon equation. The concept of the Wiener path integral has been extended to the relativistic case. A new path integral Monte Carlo method has been developed for relativistic particles in an external potential field. This method was tested on a 1D model of the relativistic harmonic oscillator. The average energy and heat capacity of the oscillator were calculated over a broad range of temperature and 'frequency'. The limits of applicability of the available analytical approaches and results have been specified by comparison with Monte Carlo calculations. This newly developed path integral formalism can be directly extended to systems of many Newton–Wigner particles interacting with external fields and each other.

Klein–Gordon Equation

We consider a free spinless particle of mass m in 3D space. Then the well-known relativistic expression for the energy is $E^2 = \mathbf{p}^2c^2 + m^2c^4$. One can obtain the Klein–Gordon equation (KGE) for the free particle by formally substituting $E \to \mathrm{i}\hbar\partial/\partial t$, $\mathbf{p} \to -\mathrm{i}\hbar\nabla$. So in the coordinate representation (Schweber 1963):

$$\left(\frac{\hbar^2}{c^4}\frac{\partial^2}{\partial t^2} - \frac{\hbar^2}{c^2}\nabla^2 + m^2\right)\phi(\mathbf{x}, t) = 0 . \tag{11.1}$$

The solutions of this equation are plane waves $\phi(\mathbf{x}, t) = \mathrm{e}^{-\mathrm{i}(Et-\mathbf{px})/\hbar}$ with positive and negative energies $E = \pm\sqrt{\mathbf{p}^2c^2 + m^2c^4}$. However, these functions are also solutions of the differential equations

$$\mathrm{i}\hbar\frac{\partial}{\partial t}\phi(\mathbf{x}, t) = \pm\sqrt{m^2c^4 + c^2(-\mathrm{i}\hbar\nabla)^2}\,\phi(\mathbf{x}, t) , \tag{11.2}$$

the plus sign corresponding to plane waves with $E \geq 0$, and the minus sign to plane waves with $E \leq 0$.

We now proceed to a spinless particle with electric charge e in an external electromagnetic field, which is described by a 4D potential $A^\mu = (A^0, \mathbf{A})$ without time dependence. In this case, the KGE has the form (Schweber 1963):

$$\left\{\left[\mathrm{i}\hbar\frac{\partial}{\partial t} - eA^0(\mathbf{x})\right]^2 - c^2\left[-\mathrm{i}\hbar\nabla - \frac{e}{c}\mathbf{A}(\mathbf{x})\right]^2 - m^2c^4\right\}\phi(\mathbf{x}, t) = 0 . \tag{11.3}$$

Solutions with positive and negative energies satisfy differential equations which are obtained from (11.2) through generalization to include the electromagnetic field:

$$\left[\mathrm{i}\hbar\frac{\partial}{\partial t} - eA^0(\mathbf{x})\right]\phi(\mathbf{x}, t) = \pm\sqrt{m^2c^4 + c^2\left[-\mathrm{i}\hbar\nabla - \frac{e}{c}\mathbf{A}(\mathbf{x})\right]^2}\,\phi(\mathbf{x}, t) . \tag{11.4}$$

These equations are the relativistic analogue of the Schrödinger equation for a particle in an external electromagnetic field in the coordinate representation.

In quantum field theory, the positive-energy wave functions are interpreted as states of particles, while the states with negative energy are considered as antiparticles. The full solution of the KGE (11.3) is the superposition of wave functions for both particles and antiparticles.

Newton–Wigner Theory

Here we consider briefly the basic ideas of the one-particle relativistic theory suggested by Newton and Wigner in the 1940s (Newton and Wigner 1949). Although pair production is not taken into account in this theory, such an approach can be used for temperatures of the order of $0.1mc^2$, where pair generation will be negligible. At the same time, the relativistic corrections to quantum effects and dynamics may be important. So our main interest will be the so-called Newton–Wigner (pseudo)particle in an external potential field described by the Hamiltonian $\hat{H} = \omega(\hat{\mathbf{p}}) + V(\hat{\mathbf{x}})$, where $\omega(\mathbf{p}) = \sqrt{\mathbf{p}^2c^2 + m^2c^4}$ is the relativistic kinetic energy, while $\hat{\mathbf{p}}$ and $\hat{\mathbf{x}}$ are the momentum and position operators. The eigenvectors of the momentum operator $|\mathbf{p}\rangle$ have the Lorentz-invariant normalization and form a complete system (Rumer and Fet 1977):

$$\langle \mathbf{p}|\mathbf{p}'\rangle = (2\pi\hbar)^3 \frac{\omega(\mathbf{p})}{mc^2} \delta^{(3)}(\mathbf{p}-\mathbf{p}')\,, \quad 1 = \int \frac{\mathrm{d}^3\mathbf{p}}{(2\pi\hbar)^3} \frac{mc^2}{\omega(\mathbf{p})} |\mathbf{p}\rangle\langle\mathbf{p}|\,. \tag{11.5}$$

One can define the wave function for the state $|\mathbf{p}\rangle$ in the momentum representation by $\Psi(\mathbf{p}) = \langle\mathbf{p}|\Psi\rangle$, where we omit the time dependence for brevity. The scalar product $\langle\Phi|\Psi\rangle$ in the momentum representation has the form

$$\langle\Phi|\Psi\rangle = \int \frac{\mathrm{d}^3\mathbf{p}}{(2\pi\hbar)^3} \frac{mc^2}{\omega(\mathbf{p})} \Phi^*(\mathbf{p})\Psi(\mathbf{p})\,. \tag{11.6}$$

Note that $|\langle\Phi|\Psi\rangle|^2$ is proportional to the probability of transition from $|\Psi\rangle$ to $|\Phi\rangle$ and is Lorentz invariant (Rumer and Fet 1977).

The action of an arbitrary quantum operator $\hat{A}$ on wave functions is defined by $\hat{A}\Psi(\mathbf{p}) = \langle\mathbf{p}|\hat{A}|\Psi\rangle$. In particular, the action of the momentum operator $\hat{\mathbf{p}}$ in the momentum representation is simply multiplication by $\mathbf{p}$, while the coordinate operator $\hat{\mathbf{x}}$ looks like (Newton and Wigner 1949):

$$\hat{\mathbf{x}}\,\Phi(\mathbf{p}) = \left[\mathrm{i}\hbar\frac{\partial}{\partial\mathbf{p}} - \mathrm{i}\hbar\frac{\mathbf{p}c^2}{2\omega^2(\mathbf{p})} \right] \Phi(\mathbf{p})\,. \tag{11.7}$$

The quantum average of an arbitrary observable $\hat{A} = \hat{A}(\hat{\mathbf{p}}, \hat{\mathbf{x}})$ in the state $|\Psi\rangle$ can be calculated from

$$\langle\Psi|\hat{A}|\Psi\rangle = \int \frac{\mathrm{d}p}{2\pi\hbar} \frac{mc^2}{\omega(\mathbf{p})} \Psi^*(\mathbf{p}) A(\mathbf{p}, \hat{\mathbf{x}}) \Psi(\mathbf{p})\,. \tag{11.8}$$

The time evolution of the Newton–Wigner particle is governed by the differential equation

$$\mathrm{i}\hbar\frac{\partial}{\partial t}\Psi(\mathbf{p},t)=\hat{H}\Psi(\mathbf{p},t)\,. \tag{11.9}$$

Note that this equation can be obtained from the KGE (11.2) by inserting $V(\mathbf{x}) = eA_0(\mathbf{x})$ and $\mathbf{A(x)} = 0$, after the transition to the momentum representation for the positive-energy wave function $\phi(\mathbf{x}, t)$ (Schweber 1963):

$$\phi(\mathbf{x},t)=\frac{2\pi\hbar}{c\sqrt{2mc^2}}\int\frac{\mathrm{d}^4k}{(2\pi\hbar)^4} \times\delta\big([k^0-eA^0(\mathbf{x})]^2-\omega^2(\mathbf{p})/c^2\big)\theta(k^0)\mathrm{e}^{\mathrm{i}\mathbf{kx}/\hbar}\Phi(\mathbf{k},t)\,. \tag{11.10}$$

Partition Function of the Newton–Wigner Particle

The thermodynamic properties of a quantum system are defined by the partition function $Z = \mathrm{Tr}\big(\mathrm{e}^{-\beta\hat{H}}\big)$, where $\beta = 1/kT$ is the reciprocal temperature. For example, for the 1D case one can write a partition function for the Newton–Wigner particle, using the momentum eigenvectors, the completeness relation, the normalization (11.5), and the integral representation of the delta function:

$$Z=\int\mathrm{d}x\int\frac{\mathrm{d}p_N}{2\pi\hbar}\int\frac{\mathrm{d}p_0}{2\pi\hbar}\frac{mc^2}{\omega(p_0)}\langle p_N|\mathrm{e}^{-\beta\hat{H}}|p_0\rangle\mathrm{e}^{-\mathrm{i}(p_0-p_N)x/\hbar}\,. \tag{11.11}$$

The integral here is taken over the large finite 1D volume L. The partition function Z is thus a function of the variables β and L. Note that generalization to higher dimensions is straightforward.

To obtain an expression for Z in the form of a path integral, we use the usual manipulation for Feynman integrals [see, for example, (Zamalin et al. 1977; Marinov 1980)]. We divide the interval from 0 to β into a large number N of small parts with length $\epsilon = \beta/N$ and apply the completeness relation (11.5) $N - 1$ times. When $\epsilon \to 0$, the matrix elements have the form

$$\langle p_{n+1}|\mathrm{e}^{-\epsilon\hat{H}}|p_n\rangle=\left[\frac{\omega(p_n)}{mc^2}\right]^{1/2}\left[\frac{\omega(p_n)}{mc^2}\right]^{1/2} \times\int\mathrm{d}x_n\mathrm{e}^{-\mathrm{i}(p_{n+1}-p_n)x_n/\hbar-\epsilon\omega(p_n)-\epsilon V(x_n)}\,. \tag{11.12}$$

Substituting the matrix elements into the formula for Z, we obtain a multi-dimensional integral of the form

$$Z=\int_L\mathrm{d}x\prod_{k=1}^{N-1}\int\mathrm{d}x_k\prod_{n=1}^{N}\int\frac{\mathrm{d}p_n}{2\pi\hbar} \times\exp\left[\frac{\mathrm{i}}{\hbar}p_n(x_n-x_{n-1})-\epsilon\omega(p_n)-\epsilon V(x_n)\right]. \tag{11.13}$$

This result can be simplified by integrating over the momenta, and finally this expression for the partition function has the form of a path integral in the limit $N \to \infty$:

$$Z = \int \mathrm{d}x \prod_{k=1}^{N-1} \int \mathrm{d}x_k \tag{11.14}$$

$$\times \prod_{n=1}^{N} \frac{mcK_1\left(\frac{\beta mc^2}{N}\sqrt{1 + \frac{N^2}{\beta^2\hbar^2c^2}(x_n - x_{n-1})^2}\right)}{\pi\hbar\sqrt{1 + \frac{N^2}{\beta^2\hbar^2c^2}(x_n - x_{n-1})^2}} \mathrm{e}^{-\beta V(x_n)/N} ,$$

where $x_0 = x$, $x_N = x$ and $K_1(x)$ is the MacDonald function of the first order. This path integral can be write in a form similar to a Wiener integral:

$$Z = \int_L \mathrm{d}x \int \mathrm{d}_{\mathrm{RW}[0,x;\beta\hbar,x]}x(\tau) \exp\left[-\frac{1}{\hbar}\int_0^{\beta\hbar} \mathrm{d}\tau V\big(x(\tau)\big)\right] . \tag{11.15}$$

Here the measure d_{RW} is a relativistic generalization of the usual Wiener measure d_{W}.

Average Energy and Heat Capacity of the Newton–Wigner Particle

Thermodynamic quantities can be derived from the partition function Z. In particular, the average energy E and heat capacity at constant volume are obtained by differentiating Z with respect to $\beta = 1/kT$ (Zubarev 1977):

$$E = -\frac{1}{Z}\left(\frac{\partial Z}{\partial \beta}\right)_L , \quad C_L = k\beta^2\left[\frac{1}{Z}\left(\frac{\partial^2 Z}{\partial \beta^2}\right)_V - \left(\frac{1}{Z}\frac{\partial Z}{\partial \beta}\right)_V^2\right] . \tag{11.16}$$

However, direct differentiation of the path integral (11.14) leads to a numerically unstable expression, so one should transform the path integral to something similar to a Wiener integral (Gelfand and Yaglom 1956) to exclude derivatives of the measure. We make the change of variables $x = \eta L$, $x_k = \lambda\xi_k + \eta L$, where $\lambda = \lambda(\beta)$ is function we will choose below, and which has units of length. After such a change, the path integrals for the partition function and its derivatives can be written in the form

$$Z = (\lambda^{-1}L)\int_1 \mathrm{d}\eta\left(\prod_{k=1}^{N-1}\int \mathrm{d}\xi_k\right)\rho_K\rho_V ,$$

$$\frac{1}{Z}\left(\frac{\partial Z}{\partial \beta}\right)_L = -\lambda^{-1}\frac{\partial \lambda}{\partial \beta} + \frac{1}{Z}(\lambda^{-1}L)\int_1 \mathrm{d}\eta\left(\prod_{k=1}^{N-1}\int \mathrm{d}\xi_k\right)\rho_K\rho_V(\phi_K + \phi_V) ,$$

$$\frac{1}{Z}\left(\frac{\partial^2 Z}{\partial\beta^2}\right)_V = \left(\lambda^{-1}\frac{\partial\lambda}{\partial\beta}\right)^2 - \frac{\partial}{\partial\beta}\left(\lambda^{-1}\frac{\partial\lambda}{\partial\beta}\right) + \frac{1}{Z}(\lambda^{-1}L)\int_1 \mathrm{d}\eta\left(\prod_{k=1}^{N-1}\int \mathrm{d}\xi_k\right)$$
$$\times\rho_K\rho_V\left[(\sigma_K+\sigma_V)+(\phi_K+\phi_V)^2+(\phi_K+\phi_V)\left(2\lambda^{-1}\frac{\partial\lambda}{\partial\beta}-mc^2\right)\right]. \tag{11.17}$$

Here ρ_K is the factor which relates to the relativistic measure d_{RW}, while ρ_V relates to the potential part of the partition function:

$$\rho_K = \prod_{n=1}^{N}\frac{mc}{\pi\hbar}\lambda\frac{K_1\left(\frac{\beta mc^2}{N}\sqrt{1+\frac{\lambda^2N^2}{\beta^2\hbar^2c^2}(\xi_n-\xi_{n-1})^2}\right)}{\sqrt{1+\frac{\lambda^2N^2}{\beta^2\hbar^2c^2}(\xi_n-\xi_{n-1})^2}},$$

$$\rho_V = \prod_{n=1}^{N}\exp\left[-\frac{\beta}{N}V(\eta L+\lambda\xi_n)\right]. \tag{11.18}$$

The functions ϕ_K, ϕ_V, σ_K, and σ_V are the first and second derivatives of (11.18) with respect to β, respectively:

$$\phi_K = \left.\frac{\partial\rho_K}{\partial\beta}\right|_L, \quad \phi_V = \left.\frac{\partial\rho_V}{\partial\beta}\right|_L, \quad \sigma_K = \left.\frac{\partial^2\rho_K}{\partial\beta^2}\right|_L, \quad \sigma_V = \left.\frac{\partial^2\rho_V}{\partial\beta^2}\right|_L. \tag{11.19}$$

Explicit expressions for these functions are found by simple differentiation. However, they are rather cumbersome, so we shall not give them explicitly here.

We now turn to the choice of the function $\lambda(\beta)$. In the case of non-relativistic Wiener integrals, one uses the so-called thermal wavelength $\lambda_{\mathrm{therm}} = \sqrt{2\pi\hbar^2\beta/m}$, so in the formula (11.17), the term $\lambda^{-1}\partial\lambda/\partial\beta$ has to be equal to the average kinetic energy $kT/2$ of the free particle (Zamalin and Norman 1973). Therefore it is natural to select the function $\lambda(\beta)$ in such a way that it meets two requirements:

1. In the non-relativistic limit $\beta mc^2 \to \infty$, λ has to turn into a thermal wavelength λ_{therm}.
2. A quantity $\lambda^{-1}\partial\lambda/\partial\beta$ has to be equal to the average energy of the free relativistic Newton–Wigner particle at a temperature T.

When $\beta mc^2 \gg 1$, a partition function of the Newton–Wigner particle is reduced to a non-relativistic partition function Z_{nrel}, and its average energy turns into the energy E_{nrel} of the non-relativistic particle. So, taking into account the rest energy mc^2, we have:

$$Z \longrightarrow Z_{\mathrm{nrel}}\mathrm{e}^{-\beta mc^2}, \qquad E \longrightarrow E_{\mathrm{nrel}} + mc^2. \tag{11.20}$$

The partition function of the free Newton–Wigner particle in a volume L can be found from the definition $Z_0 = \mathrm{Tr}\left[\mathrm{e}^{-\beta\omega(\hat{p})}\right]$. This can be obtained using the completeness relation (11.5) for the momentum eigenvectors:

$$Z_0 = \frac{L}{2\pi\hbar}\int \frac{\mathrm{d}p}{2\pi\hbar}\frac{mc^2}{\omega(p)}\mathrm{e}^{-\beta\sqrt{p^2c^2+m^2c^4}} . \tag{11.21}$$

The integral can be expressed in terms of the MacDonald function (Gradstein and Ryzik 1963). Thus the partition function and average energy of the free relativistic particle are equal to

$$Z_0 = \frac{mcL}{2\pi^2\hbar^2}K_1(\beta mc^2) , \qquad E_0 = \frac{1}{\beta} + mc^2\frac{K_0(\beta mc^2)}{K_1(\beta mc^2)} . \tag{11.22}$$

It is easy to check that $\lambda(\beta)$ is of the form

$$\lambda = \frac{\pi\hbar}{mc}\frac{\mathrm{e}^{-\beta mc^2}}{K_1(\beta mc^2)} , \tag{11.23}$$

and satisfies requirements (1) and (2) above, so it is a relativistic generalization of the thermal wavelength.

Note an important detail in the expression (11.17). For a free particle, i.e., $V(x) = 0$, the functions ϕ_V and σ_V have to be equal to zero. But the functions ϕ_K and σ_K, arising from derivatives of the measure d_{RW} in the path integrals, look non-trivial. At the same time, the outer summands in (11.17) are already equal to thermodynamic quantities for the free Newton–Wigner particle (2). Therefore the integrals of ϕ_K and σ_K have to vanish. Of course, if an external field is applied, the vanishing of these integrals is not obvious. However, straight numerical calculations for different potentials $V(x)$ show that the integrals of ϕ_K and σ_K are negligible. One can thus keep only the 'potential' summands ϕ_V, σ_V in (11.17), and the measure in the path integral need not be differentiated.

Thermodynamic Behavior of the Relativistic Harmonic Oscillator

The Newton–Wigner particle in a quadratic potential is referred to as a relativistic harmonic oscillator (Larkin and Filinov 2014). Its Hamiltonian is $\hat{H} = \omega(p) + (m\omega^2/2)\hat{x}$, where the position operator is defined by the formula (11.7) and ω is a parameter of the oscillator, not to be confused with the kinetic energy $\omega(p)$. The thermodynamic properties of the oscillator are determined by two parameters: the temperature T and the 'frequency' ω (the volume L can be considered infinite due to the finite motion of the particle). Here we explore the domains of applicability of the available analytical approaches to the oscillator. At high temperatures, when both $kT \gg mc^2$ and $kT \gg \hbar\omega$, the oscillator is a relativistic system, and its quantum nature is irrelevant, so it can be described in terms of classical statistics. The average energy and heat capacity are

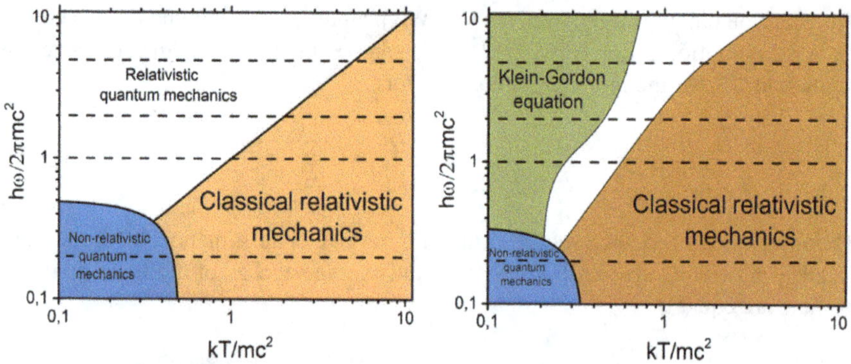

Fig. 11.1 *Left*: Qualitative bounds of applicability of the analytical approaches (11.24) and (11.26). *Right*: Bounds of applicability of analytical approaches corrected using analyses of Monte Carlo results: classical relativistic mechanics (11.24) *orange*, non-relativistic quantum mechanics *blue*, approximate relativistic quantum theory (11.26) *green* (Larkin and Filinov 2014)

$$
\begin{aligned}
E_{\mathrm{cl}} &= \frac{3}{2}\beta^{-1} + mc^2 \frac{K_0(\beta mc^2)}{K_1(\beta mc^2)}\,,\\
C_{L\mathrm{cl}} &= k\left\{1 + (\beta mc^2)^2 - \beta mc^2 \frac{K_0(\beta mc^2)}{K_1(\beta mc^2)} - \left[\beta mc^2 \frac{K_0(\beta mc^2)}{K_1(\beta mc^2)}\right]^2\right\}\,,
\end{aligned}
\tag{11.24}
$$

Note that in the high temperature limit $E_{\mathrm{cl}} \to 3kT/2$, $C_{L\mathrm{cl}} \to 2k$.

In contrast, when the temperature is relatively low ($kT \ll mc^2$), relativistic effects are negligible, so the oscillator can be studied using non-relativistic mechanics, either classical or quantum depending on the ratio between kT and $\hbar\omega$. The harmonic oscillator then behaves as depicted in Fig. 11.1, where the non-relativistic area is colored in blue, and classical relativistic mechanics is applicable in the orange area. In the remaining areas, one has to use relativistic quantum theory.

Due to the conformity between the Newton–Wigner theory and the KGE, one might expect equality between the states of the relativistic harmonic oscillator and the stationary positive-energy solutions of the equation

$$
\left[\left(\mathrm{i}\hbar\frac{\partial}{\partial t} - \frac{m\omega^2}{2}x^2\right)^2 - (-\mathrm{i}\hbar\nabla)^2 - m^2c^4\right]\phi(x,t) = 0\,. \tag{11.25}
$$

Explicit solutions of this equation are unknown, but one can use the approximations for energy levels at temperatures $kT \le mc^2$ (Harvey 1972):

$$
E_n = \pm mc^2\sqrt{1 + 2\left(n + \frac{1}{2}\right)\frac{\hbar\omega}{mc^2}}\,. \tag{11.26}
$$

A partition function can thus be calculated by direct summation over positive energies $Z(\beta) = \sum_n e^{-\beta E_n}$, and then the average energy and heat capacity can be obtained from derivatives of Z. The corresponding series converges fast, so practically speaking, one can take the sum over n from 0 to 10^5 with an accuracy of 0.01%. Comparing the results of the path integral and other calculations based on the available approximations, the limits of their applicability have been tested and are shown in Fig. 11.1. The figure shows the domains of applicability of the non-relativistic and classical theories, as well as the approximate solutions of the KGE (11.26). The corresponding areas are marked with different colors. No single analytical approach can be used in the uncoloured area, where the oscillator thermodynamics can only be studied using numerical calculations.

Energy and Heat Capacity of the Oscillator

Monte Carlo calculations were carried out for temperatures kT from $0.1mc^2$ to $10mc^2$ and for certain values of the 'vibrational quantum' $\hbar\omega$. Here we discuss the results and compare them with the predictions of classical mechanics (11.24) and approximate quantum theory (11.26). In particular, it will become clear how to define the plotted bounds of applicability in Fig. 11.1.

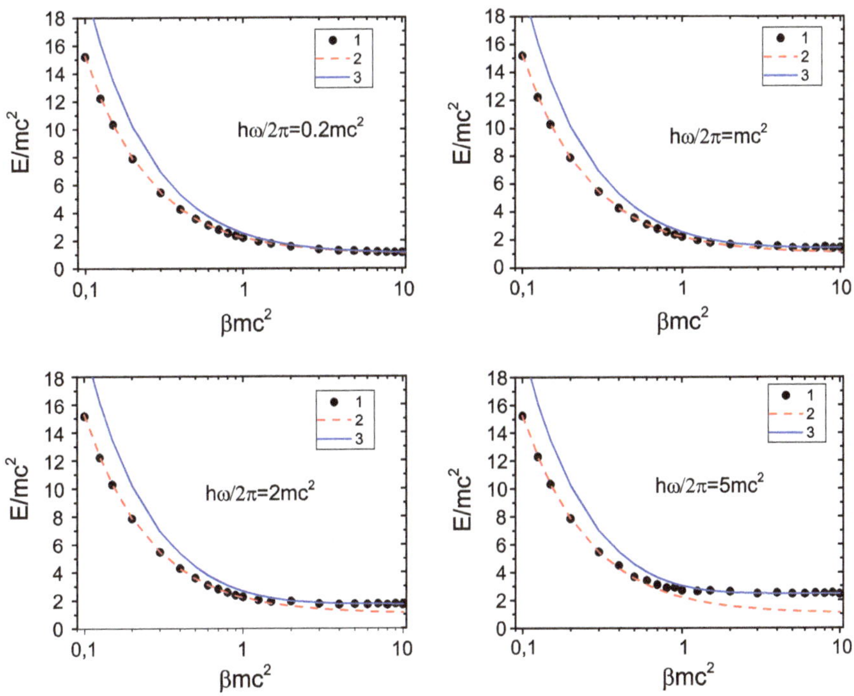

Fig. 11.2 Dependencies of the average energy on the reciprocal temperature for the harmonic oscillator for different values of $\hbar\omega$. *Top left*: $\hbar\omega = 0.2mc^2$. *Top right*: $\hbar\omega = mc^2$. *Bottom left*: $\hbar\omega = 2mc^2$. *Bottom right*: $\hbar\omega = 5mc^2$. (1) Monte Carlo (this work), (2) classical relativistic behavior (11.24), (3) theory based on approximate solution of the KGE (11.26) (Larkin and Filinov 2014)

Figure 11.2 shows the average energy for different values of the 'vibrational quantum' parameter $\hbar\omega$, viz., $0.2mc^2$, mc^2, $2mc^2$, and $5mc^2$. The numerical results are compared with the predictions of the analytical approaches (11.24) and (11.26). From Fig. 11.2, it follows that the oscillator can be considered as classical when the temperature is higher then $0.25mc^2$, $0.5mc^2$, mc^2, and $5mc^2$ for each value of $\hbar\omega$. The low temperature theory based on approximate solution of the KGE is justified for kT lower then $0.2mc^2$, $0.25mc^2$, $0.5mc^2$, and $0.5mc^2$, as was shown in Fig. 11.1.

We now consider the heat capacity of the relativistic oscillator. Note that the Nernst theorem demands that the heat capacity should be zero when $T \to 0$. However, one cannot avoid a statistical error in Monte Carlo calculations of derivatives of the partition function, which are multiplied by $\beta^2 = 1/(kT)^2$ according to (11.16). Therefore, when the temperature is low, the relative error becomes significant. This is the reason for presenting the dependence of $C_L/\beta^2 = Ck^2T^2$ on $\beta = 1/kT$, rather than the dependence of C_L. The results are shown in Fig. 11.3, where they are compared with the analytical approximations (11.24) and (11.26). The behavior of the oscillator is classical at high temperatures and can be described once again by the approximate solution of the KGE at relatively low temperatures. Note that $C_L \to 0$ when $T \to 0$, as required by the Nernst theorem.

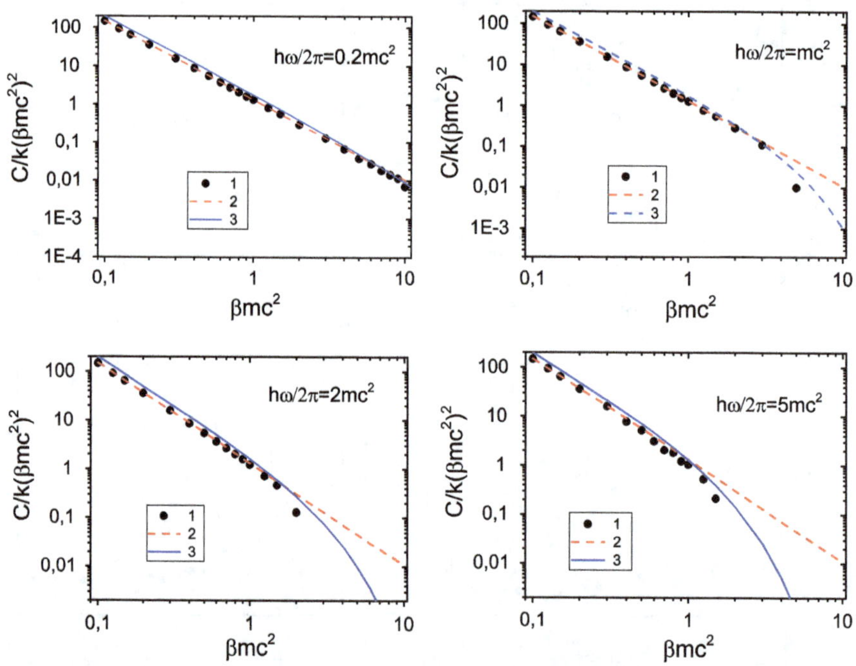

Fig. 11.3 Dependencies of C_L/β^2 on the reciprocal temperature for the harmonic oscillator for different values of $\hbar\omega$. *Top left*: $\hbar\omega = 0.2mc^2$. *Top right*: $\hbar\omega = mc^2$. *Bottom left*: $\hbar\omega = 2mc^2$. *Bottom right*: $\hbar\omega = 5mc^2$. (1) Monte Carlo (this work), (2) classical relativistic behavior (11.24), (3) theory based on approximate solution of the KGE (11.26) (Larkin and Filinov 2014)

11.2 Semiclassical SU(3) Partition Function of the Relativistic Quark–Gluon Plasma in the Color Path Integral Representation

Nowadays, studies of properties of the quark–gluon plasma (QGP) constitute one of the most important goals in high-energy nuclear physics. Experiments at the Relativistic Heavy-Ion Collider (RHIC) at Brookhaven National Laboratory (Shuryak 2009) and the Large Hadron Collider (LHC) at CERN (Schukraft 2011) have provided a wealth of data, from which one can obtain information on a number of features of the QGP. The most striking result (Shuryak 2009; Shen and Heinz 2012) is that deconfined quark–gluon matter behaves as an almost perfect fluid, rather than a perfect gas, as might have been expected from asymptotic freedom.

The most fundamental way to compute the properties of strongly interacting matter (Z. Fodor and Katz 2002, S. Borsanyiet al. 2010) is provided by lattice quantum chromodynamics (QCD). Although the equilibrium properties of the strongly interacting QGP are successfully computed using lattice QCD (Shuryak 2009, Borsanyi 2010, Cheng et al. 2010), these simulations are very time-consuming and not easy to interpret. Moreover, the physical interpretation of these very complicated computations requires application of various QCD-motivated, albeit schematic, models simulating various aspects of the full theory and allowing for a deeper physical understanding. Such models are also needed when lattice QCD fails, e.g., for large quark chemical potentials and out of equilibrium. It is therefore crucial to devise reliable and manageable theoretical tools for a quantitative description of non-Abelian QGP, both in and out of equilibrium.

A semi-classical approximation based on a quasiparticle picture has been introduced in (Litim and Manuel 1999, 2000, 2002; Hartmann et al. 2006). It is motivated by the expectation that the main features of non-Abelian plasmas can be understood in simple semi-classical terms without the difficulties inherent in a full quantum field-theoretical analysis. Independently, the same ideas were implemented in terms of molecular dynamics (MD) (Hofmann et al. 2000) and this approach was further developed in a series of studies (Gelman et al. 2006; Cho and Zahed 2009). The MD allows one to treat soft processes in the QGP which are not accessible by perturbative means.

The main goal of this section is to present and to test the color path-integral Monte-Carlo method (Filinov et al. 2009, 2011, 2012, 2013) against known lattice data (Fodor and Katz 2002) and to predict additional properties of the QGP, which are still inaccessible using lattice QCD. Here it will be shown that the DPIMC method is able to reproduce the QCD lattice equation of state at zero and non-zero baryon charge density, and also yields valuable insight into the internal structure of the QGP. This approach is based on a generalization of the Feynman path integral representation of quantum mechanics to high energy matter. Before deriving the main equations of the DPIMC approach, let us specify the simplifications and model parameters.

To begin with, let us summarize the main ideas of the path integral approach to calculations of the thermodynamic properties of the strongly correlated quark–gluon plasma. It has been established that hard modes (in terms of the *hard thermal loop* approximation (Pisarski 1989; Braaten and Pisarski 1990; Blaizot and Iancu 2002) behave like quasi-particles. Therefore, the masses of these quasiparticles should be deduced from nonperturbative calculations, taking into account hard field modes, e.g., they can be associated with pole masses deduced from lattice QCD calculations. At the same time, the soft quantum fields are characterized by very high occupation numbers per mode. Therefore, to leading order, they can be well approximated by soft classical fields. This is precisely the picture we are going to utilize: massive quantum quasiparticles (hard modes), interacting via classical color fields. The applicability of this approach was discussed in detail in (Litim and Manuel 1999, 2000, 2002; Gelman et al. 2006). The approach thereby developed differs from that of (Gelman et al. 2006) by a quantum treatment of quasiparticles, rather than the classical one, and additionally by a relativistic description of the kinetic energy, rather than the nonrelativistic approximation.

Table 11.1 compares the basic physical ingredients of an ('electromagnetic') electron–ion plasma model and a quark–gluon plasma model, since this provides the basis for realistic color DPIMC simulations. Although QCD was constructed by analogy with quantum electrodynamics, there exist fundamental differences. While Coulomb interacting charges are mapped onto fermions (or bosons) whose interactions are mediated by (usually weakly interacting) photons, the situation in QCD is different. Here also the field particles (gluons) providing the interaction between fermions (quarks and antiquarks) are, in general, strongly interacting. As a result, it is necessary to consider, not bare quarks, antiquarks, and gluons, but quasiparticles absorbing the 'hard modes', while the soft modes are incorporated into an effective color Coulomb interaction (Filinov et al. 2015).

Table 11.1 Comparison of an electromagnetic plasma and the semiclassical SU(3) quark–gluon plasma. T temperature, μ chemical potential, r_{ab} distance between particles a and b, QP quasiparticles, $Q_a \cdot Q_b$ scalar product of 8D vectors. [a]Only intrinsic, no orbital quantum numbers are listed. [b]Defined by the kinetic energy operator

	Electromagnetic plasma	Semiclassical SU(3) QGP
Basic particles or QP	Electrons, ions (i) or holes (h)	Quarks, antiquarks, gluons
Quantum numbers[a]	Spin	Spin, flavor, color
Renormalization	None (plasmas), QP in solids	QP
Masses	$m_\mathrm{e} \ll m_\mathrm{i}$ or QP masses $m_\mathrm{e} \sim m_\mathrm{h}$	Comparable QP masses $m_\alpha = m_\alpha(T, \mu)$
Charge	Fixed scalar electrical charge q_α	SU(3) Wong vector color charge variables Q_α
Coupling constant	Fixed value $\alpha = 1/137$	State and distance dependent $\alpha(T, \mu, r_{ab})$
Potential energy	Non-relativistic Coulomb potential $V_{ab} \sim q_a q_b / r_{ab}$	Non-relativistic color Coulomb potential $V_{ab} \sim \alpha(T, \mu, r_{ab}) Q_a \cdot Q_b / r_{ab}$
Kinetic energy	Non-relativistic	Relativistic
Path integral partition function[b]	Non-relativistic Gaussian measure	Relativistic Bessel and SU(3) group Haar measures

So the basic assumptions of the model presented in Table 11.1 are the following:

1. Quasiparticle masses (m) are of the same or higher order than the mean kinetic energy per particle. This assumption is based on the analysis of QCD lattice data (Petreczky et al. 2002, Liao and Shuryak 2006; Karsch and Kitazawa 2009). For instance, at zero net baryon density, it amounts to $m \approx T$, where T is a temperature.
2. In view of the first assumption, the interparticle interaction is dominated by a color–electric Coulomb potential. Magnetic effects are neglected as subleading terms.
3. Relying on the fact that the color representations are large, the color operators are replaced by their average values, i.e., by Wong's classical color vectors, which are eight-dimensional (8D) in SU(3), with quadratic and cubic Casimir conditions (Wong 1970).
4. We consider the 3-flavor quark model. For the sake of simplicity, we assume the masses of the up, down, and strange quarks to be equal. As for the gluon quasiparticles, we allow their mass to be different from (heavier than) the quark masses.

The nature of these approximations and their limitations were discussed in (Gelman et al. 2006). Thus, this model requires as input the following quantities as functions of the temperature (T) and the quark chemical potential (μ_q):

1. Quasiparticle masses, for quarks m_q and gluons m_g, and
2. The coupling constant g^2, or $\alpha_s = g^2/4\pi$.

Input quantities should be deduced from lattice QCD data or from an appropriate model simulating these data.

Semiclassical SU(3) QGP Model

Let us consider a multi-component QGP consisting of N color quasiparticles: N_g gluons, N_q quarks, and $\overline{N}_q$ antiquarks. In the following description of the QGP, we will use a system of units in which Planck's constant, the light velocity, and Boltzmann's constant are all equal to unity. The Hamiltonian of this system is $\hat{H} = \hat{K} + \hat{U}^C$, with the kinetic and color Coulomb interaction parts

$$\hat{K} = \sum_i \sqrt{\hat{\mathbf{p}}_i^2 + m_i^2(T, \mu_q)}\ , \quad \hat{U}^C = \frac{1}{2} \sum_{i \neq j} \frac{g^2(T, \mu_q)(Q_i \cdot Q_j)}{4\pi |\mathbf{r}_i - \mathbf{r}_j|}\ . \qquad (11.27)$$

Here the i and j summations run over the quark and gluon quasiparticles, $i, j = 1, \ldots, N$, with $N = N_q + \overline{N}_q + N_g$, where $N_q = N_u + N_d + N_s$ and $\overline{N}_q = \overline{N}_u + \overline{N}_d + \overline{N}_s$ are the total numbers of quarks and antiquarks of all flavours (up, down, and strange), respectively, The 3D vectors $\mathbf{r}_i$ are the quasiparticle spatial coordinates, the Q_i denote Wong's quasiparticle color variable, an 8D vector in the group $SU(3)$, and $(Q_i \cdot Q_j)$ denotes the scalar product of the color vectors. A nonrelativistic approximation is used for the potential energy, while we still keep the relativistic form for the kinetic energy, as the quasiparticle masses are not negligible compared with the temperature. The eigenvalue equation of this Hamiltonian is usually called the spinless Salpeter equation. It may be regarded as a well-defined approximation to the Bethe–Salpeter formalism (Salpeter and Bethe 1951) for the description of bound states in relativistic quantum field theories, obtained by assuming that all bound-state constituents interact instantaneously and propagate like free particles (Salpeter 1952). Among other things, it yields semi-relativistic descriptions of hadrons as bound states of quarks (Lucha et al. 1991, 1992).

In the classical approximation, this system is governed by Wong's equations of motion (Wong 1970)

$$\frac{\mathrm{d}\mathbf{p}_i(t)}{\mathrm{d}t} = \mathbf{F}_i(t) , \tag{11.28}$$

$$\frac{\mathrm{d}\mathbf{r}_i(t)}{\mathrm{d}t} = \mathbf{v}_i(t) , \tag{11.29}$$

$$\frac{\mathrm{d}Q_i^a(t)}{\mathrm{d}t} = \Upsilon_i^a(t) , \tag{11.30}$$

where $\mathbf{p}_i$ is the momentum of a quasiparticle, $\mathbf{v}_i = \mathbf{p}_i/\sqrt{\mathbf{p}_i^2 + m_i^2}$ is its velocity, $\mathbf{F}_i = -\partial U^{\mathrm{C}}/\partial \mathbf{r}_i$ is the color-electric force experienced by the quasiparticle, and

$$\Upsilon_i^a = \sum_{j\neq i}^{N} \sum_{b,c} \breve{f}_{abc} \frac{g^2 Q_i^b Q_j^c}{4\pi|\mathbf{r}_i - \mathbf{r}_j|} \tag{11.31}$$

is the driving force in the equation of motion for the color charge, with $\breve{f}_{abc}$ the structure constants of the group SU(3) and $a, b, c = 1, \ldots, 8$ (see the appendix in Sect. 11.4). Below we consider the spatial degrees of freedom quantum mechanically, but we continue to treat the color dynamics classically.

Thermodynamic properties in the grand canonical ensemble with given temperature (T), net quark number (μ_{q}) and strange (μ_{s}) chemical potentials, and fixed volume V are fully described by the grand partition function

$$\begin{aligned} &Z\left(\mu_{\mathrm{q}}, \mu_{\mathrm{s}}, \beta, V\right) \\ &= \sum_{\{N\}} \frac{\exp\left[\mu_{\mathrm{q}}(N_{\mathrm{q}} - \overline{N}_{\mathrm{q}})/T\right] \exp\left[\mu_{\mathrm{s}}(N_{\mathrm{s}} - \overline{N}_{\mathrm{s}})/T\right]}{N_{\mathrm{u}}!\, N_{\mathrm{d}}!\, N_{\mathrm{s}}!\, \overline{N}_{\mathrm{u}}!\, \overline{N}_{\mathrm{d}}!\, \overline{N}_{\mathrm{s}}!\, N_{\mathrm{g}}!} Z\left(\{N\}, V, \beta\right) , \end{aligned} \tag{11.32}$$

$$Z\left(\{N\}, V, \beta\right) = \sum_{\sigma} \int_V \mathrm{d}r\, \mathrm{d}\mu\, Q \rho(r, Q, \sigma; \{N\}; \beta) , \tag{11.33}$$

where $\{N\} = \left\{N_{\mathrm{u}}, N_{\mathrm{d}}, N_{\mathrm{s}}, \overline{N}_{\mathrm{u}}, \overline{N}_{\mathrm{d}}, \overline{N}_{\mathrm{s}}, N_{\mathrm{g}}\right\}$, and $\rho\big(r, Q, \sigma; \{N\}; \beta\big)$ denotes the diagonal matrix elements of the density matrix operator $\hat{\rho} = \exp(-\beta \hat{H})$ with

$\beta = 1/T$. Here r, σ, and Q denote the multi-dimensional vectors related to the spatial, spin, and color degrees of freedom, respectively, of all quarks, antiquarks, and gluons. The σ summation and spatial ($\mathrm{d}r \equiv \mathrm{d}^3r_1 \ldots \mathrm{d}^3r_N$) and color ($\mathrm{d}\mu Q \equiv \mathrm{d}\mu Q_1 \ldots \mathrm{d}\mu Q_N$) integrations run over all individual degrees of freedom of the particles, and $\mathrm{d}\mu Q_i$ denotes integration over the SU(3) group Haar measure (see the appendix in Sect. 11.4). The usual choice of the strange chemical potential is $\mu_\mathrm{s} = -\mu_\mathrm{q}$ (nonstrange matter), so that the total factor in front of $N_\mathrm{s} - \overline{N}_\mathrm{s}$ is zero. Therefore, below we omit μ_s from the list of variables. In (11.32), we explicitly wrote the sum over different quark flavors (u, d, and s). Below the sum over quark degrees of freedom is understood in the same way.

Since the masses and the coupling constant depend on the temperature and quark chemical potential, special care should be taken to preserve the thermodynamical consistency of this approach. To do this, thermodynamic functions such as pressure P, entropy S, baryon number N_B, and internal energy E, should be calculated from the corresponding derivatives of the logarithm of the partition function:

$$P = \frac{\partial(T \ln Z)}{\partial V} , \tag{11.34}$$

$$S = \frac{\partial(T \ln Z)}{\partial T} , \tag{11.35}$$

$$N_\mathrm{B} = \frac{1}{3}\frac{\partial(T \ln Z)}{\partial \mu_\mathrm{q}} , \tag{11.36}$$

$$E = -PV + TS + 3\mu_\mathrm{q} N_\mathrm{B} . \tag{11.37}$$

This is the conventional way of maintaining thermodynamical consistency in approaches of the Ginzburg–Landau type when they are applied in high-energy physics, e.g., in the PNJL model.

The exact density matrix $\rho = \mathrm{e}^{-\beta\hat{H}}$ of interacting quantum systems can be constructed using a path integral approach (Feynman and Hibbs 1965; Zamalin et al. 1977) based on the operator identity

$$\mathrm{e}^{-\beta\hat{H}} = \mathrm{e}^{-\Delta\beta\hat{H}}\mathrm{e}^{-\Delta\beta\hat{H}} \cdots \mathrm{e}^{-\Delta\beta\hat{H}} , \tag{11.38}$$

where the right-hand side contains $n+1$ identical factors with $\Delta\beta = \beta/(n+1)$. This allows us to rewrite the integral in (11.33) in the form

$$\begin{aligned}
&\sum_\sigma \int \mathrm{d}r^{(0)}\mathrm{d}\mu Q \rho(r^{(0)}, Q, \sigma; \{N\}; \beta) \\
&= \sum_\sigma \int \mathrm{d}\mu Q \mathrm{d}r^{(0)}\mathrm{d}r^{(1)} \ldots \mathrm{d}r^{(n+1)} \rho^{(1)}\rho^{(2)} \ldots \rho^{(n)} \\
&\times \sum_{P_\mathrm{q}} \sum_{P_{\bar{\mathrm{q}}}} \sum_{P_\mathrm{g}} (-1)^{\kappa_{P_\mathrm{q}} + \kappa_{P_{\bar{\mathrm{q}}}}} \mathcal{S}\big(\sigma^{(0)}, \hat{P}_\mathrm{q}\hat{P}_{\bar{\mathrm{q}}}\hat{P}_\mathrm{g}\sigma^{(n+1)}\big) \hat{P}_\mathrm{q}\hat{P}_{\bar{\mathrm{q}}}\hat{P}_\mathrm{g}\rho^{(n+1)}\big|_{\sigma^{(n+1)}=\sigma^{(0)}}
\end{aligned}$$

$$\times\, \delta\big(r^{(n+1)} - r^{(0)}\big)$$
$$\equiv \int \mathrm{d}\mu Q \int \mathrm{d}r^{(0)}\mathrm{d}r^{(1)} \ldots \mathrm{d}r^{(n+1)} R\big(r^{(0)}, r^{(1)}, \ldots, r^{(n+1)}; Q; \{N\}; \beta\big)$$
$$\times\, \delta\big(r^{(n+1)} - r^{(0)}\big)\,. \tag{11.39}$$

For notational convenience, we ascribe the superscript $^{(0)}$ to the original variables. Notice that the color charge Q is already a classical variable in the mixed (i.e., coordinate–momentum) representation (see the appendix in Sect. 11.4). Therefore, we do not build a chain of n different Q variables. The spin variable σ is the same in all $\rho^{(l)}$, except for $\rho^{(n+1)}$, where it is initially set as $\sigma^{(n+1)}$ and only changed to σ once the permutations have been carried out. The spin gives rise to the spin part of the density matrix ($\mathcal{S}$). To take into account the Fermi/Bose statistics of (anti)quarks/gluons, the density matrix must be antisymmetrized/symmetrized over the spatial, color, and spin variables. In the product of the $\rho^{(l)}$, it is enough to perform this antisymmetrization/symmetrization only in a single term (Huang 1963), since in fact the variables of any $\rho^{(l)}$ are related to the same set of quasiparticles.We choose it to be $\rho^{(n+1)}$. The (anti)symmetrization is done by the permutation operators $\hat{P}_{\mathrm{q}}$, $\hat{P}_{\bar{\mathrm{q}}}$, and $\hat{P}_{\mathrm{g}}$ acting on the related spatial $r^{(n+1)}$, spin $\sigma^{(n+1)}$, and color Q variables in $\rho^{(n+1)}$. The sum runs over all permutations, with parity factors $\kappa_{P_{\mathrm{q}}}$ and $\kappa_{P_{\bar{\mathrm{q}}}}$ corresponding to each permutation.

In (11.39),

$$\rho^{(l)} = \rho\left(r^{(l-1)}, r^{(l)}, Q; \{N\}; \Delta\beta\right) = \left\langle r^{(l-1)}, Q \left| \mathrm{e}^{-\Delta\beta\hat{H}} \right| r^{(l)}, Q \right\rangle \tag{11.40}$$

is an off-diagonal element of the density matrix. Accordingly, each quasiparticle is represented by a set of coordinates $\{r_i^{(0)}, \ldots, r_i^{(n)}\}$ ('beads') and an 8-dimensional color vector Q_i in the $SU(3)$ group. Thus, all the beads of each quasiparticle are characterized by the same spin projection, flavor, and color charge. Notice that the masses and coupling constant in each $\rho^{(l)}$ are the same as those for the original quasiparticles, i.e., these are still defined by the actual temperature T.

The main advantage of the decomposition (11.39) is that it allows us to use perturbation theory to obtain an approximation for the density matrices $\rho^{(l)}$, which is applicable due to the smallness of the artificially introduced factor $1/(n+1)$. This means that, in each $\rho^{(l)}$, the ratio

$$\frac{g^2(T, \mu_{\mathrm{q}})(Q_i \cdot Q_j)}{\left[4\pi |r_i^{(l)} - r_j^{(l)}| T(n+1)\right]}$$

can always be made much smaller than unity, and this in turn allows us to use perturbation theory with respect to the potential. Each factor in (11.39) should be calculated with an accuracy of order $1/(n+1)^{\theta}$, where $\theta > 1$, as in this case the error of the whole product in the limit of large n will be equal to zero. In the limit $(n+1) \to \infty$, ρ^{l} can be approximated by a product of two-particle density matrices $\rho_{ij}^{(l)}$ (Filinov et al. 2001; Feynman and Hibbs 1965; Zamalin et al. 1977). This approximation can be deduced from the operator expansion

$$\exp\left(-\Delta\beta\hat{H}\right) \approx \exp\left(-\frac{\Delta\beta}{2}\hat{K}\right)\exp\left(-\Delta\beta\hat{U}^{\mathrm{C}}\right)\exp\left(-\frac{\Delta\beta}{2}\hat{K}\right) \times\left\{\text{terms with }\left(\frac{\Delta\beta}{2}\right)^2\left[\hat{K},\hat{U}^{\mathrm{C}}\right]\ldots\right\}. \quad (11.41)$$

As a first approximation with error proportional to $1/(n+1)^2$, we can write

$$\exp\left(-\Delta\beta\hat{H}\right) \approx \exp\left(-\frac{\Delta\beta}{2}\hat{K}\right)\exp\left(-\Delta\beta\hat{U}^{\mathrm{C}}\right)\exp\left(-\frac{\Delta\beta}{2}\hat{K}\right), \quad (11.42)$$

or a simpler expression, neglecting the contribution of the commutator $\left[\hat{K},\hat{U}^{\mathrm{C}}\right]$, viz.,

$$\exp\left(-\Delta\beta\hat{H}\right) \approx \exp\left(-\Delta\beta\hat{K}\right)\exp\left(-\Delta\beta\hat{U}^{\mathrm{C}}\right). \quad (11.43)$$

It is very important that in both approximations the error of the whole product in (11.39) is proportional to $1/(n+1)$ and tends to zero in the limit $n\to\infty$. The second advantage of the decomposition of (11.39) is that it reduces the quantum multiparticle interaction to the pairwise sum of two-particle interactions described by two-particle classical density matrices in each $\rho^{(l)}$.

Thus, neglecting the commutator terms in (11.41), we arrive at the following expression for the density matrix of (11.40):

$$\rho^{(l)}_{\text{neglecting commutators}} = \rho_0^{(l)}\exp\left[-\Delta\beta U^{\mathrm{C}}\left(r^{(l)},Q\right)\right], \quad (11.44)$$

where $\rho_0^{(l)}$ is the corresponding density matrix of noninteracting particles. This approximation works well for potentials that are bounded below. However, the Coulomb potential can go to minus infinity and hence the result (11.44) diverges in this limit.

A more sophisticated treatment is required to avoid this divergence. All the calculations along this line can be rigorously performed for the two-particle density matrix $\rho^{[2]}(r,r',Q;\Delta\beta)$, where $r=\{\mathbf{r}_1,\mathbf{r}_2\}$, $r'=\{\mathbf{r}'_1,\mathbf{r}'_2\}$ and $Q=\{Q_1,Q_2\}$. Expanding the two-particle density matrix up to the second order in $1/(n+1)$, one arrives at the following result (Kelbg 1963):

$$\rho^{[2]}(r,r',Q;\Delta\beta) \approx \rho_0^{[2]}(r,r',Q;\Delta\beta) - \int_0^1 \mathrm{d}\tau\int \mathrm{d}^3\tilde{r}\,\frac{\Delta\beta g^2(Q_i\cdot Q_j)}{4\pi|\tilde{\mathbf{r}}|\Delta\lambda_{12}^2\sqrt{\tau(1-\tau)}} \times\exp\left[-\frac{\pi|\mathbf{r}_{12}-\tilde{\mathbf{r}}|^2}{\Delta\lambda_{12}^2(1-\tau)}\right]\exp\left[-\frac{\pi|\tilde{\mathbf{r}}-\mathbf{r}'_{12}|^2}{\Delta\lambda_{12}^2\tau}\right]$$

$$\approx \rho_0^{[2]} \exp\left[- \Delta\beta\Phi_{12}(\mathbf{r}_{12}, \mathbf{r'}_{12}, Q_1, Q_2)\right] , \tag{11.45}$$

where $\mathbf{r}_{12} = \mathbf{r}_1 - \mathbf{r}_2$, $\mathbf{r'}_{12} = \mathbf{r'}_1 - \mathbf{r'}_2$, $\Delta\lambda_{12} = \sqrt{2\pi\Delta\beta/m_{12}}$ is defined in terms of the reduced mass $m_{12} = m_1 m_2/(m_1 + m_2)$ of the pair of particles, and $\rho_0^{[2]}$ is the two-particle density matrix of noninteracting particles. In the end, (11.45) is presented in a form similar to (11.44), i.e., in terms of an off-diagonal two-particle effective quantum potential Φ_{12}, which is called a Kelbg potential (Kelbg 1963). Equation (11.45) is the definition of the color Kelbg potential. The diagonal part of the color Kelbg potential can be obtained analytically:

$$\Phi_{12}(\mathbf{r}_{12}, \mathbf{r}_{12}, Q_1, Q_2) \approx \frac{g^2\,(Q_i \cdot Q_j)}{4\pi\Delta\lambda_{12}x_{12}} \left\{1 - \mathrm{e}^{-(x_{12})^2} + \sqrt{\pi}x_{12}\left[1 - \mathrm{erf}(x_{12})\right]\right\} , \tag{11.46}$$

where $x_{12} = |\mathbf{r}_{12}|/\Delta\lambda_{12}$. Notice that the color Kelbg potential approaches the color Coulomb potential at distances greater than $\Delta\lambda_{12}$. What is of prime importance is that the color Kelbg potential is finite at zero distance, thus removing the classical divergences in a natural way and making any of the often applied artificial cut-offs, often applied obsolete [see, e.g., Gelman et al. 2006]. This color potential is a straightforward generalization of the corresponding potential of electromagnetic Coulomb plasmas (Filinov et al. 2004). The off-diagonal color Kelbg potential can be approximated by the diagonal ones by means of

$$\Phi_{12}(\mathbf{r}_{12}, \mathbf{r'}_{12}, Q_1, Q_2) \approx \frac{1}{2}\left[\Phi_{12}(\mathbf{r}_{12}, \mathbf{r}_{12}, Q_1, Q_2) + \Phi_{12}(\mathbf{r'}_{12}, \mathbf{r'}_{12}, Q_1, Q_2)\right] .$$

Following the experience gained with electromagnetic Coulomb plasmas, we apply the following widely used ansatz (Feynman and Hibbs 1965; Zamalin et al. 1977), which generalizes (11.45):

$$\rho^{(l)} = \rho_0^{(l)} \exp\left[-\Delta\beta\frac{1}{2} \sum_{i,j(i\neq j)}^{N} \Phi_{ij}\left(\mathbf{r}_i^{(l-1)} - \mathbf{r}_j^{(l-1)}, \mathbf{r}_i^{(l)} - \mathbf{r}_j^{(l)}, Q_i, Q_j\right)\right] . \tag{11.47}$$

We can now construct R of (11.39). The density matrix of noninteracting particles is known to be expressed in terms of determinants and permanents of single-particle density matrices in the standard way. These determinants and permanents take their origin from the (anti)symmetrization discussed in (11.39). Generalizing the electrodynamic plasma results (Filinov et al. 2001) to the quark–gluon plasma case, we write approximately

$$R\left(r^{(0)}, r^{(1)}, \ldots r^{(n+1)}; Q; \{N\}; \beta\right)$$

$$
\begin{aligned}
&= \exp(-\beta U) \sum_{\sigma} \left[\prod_{l=1}^{n} \prod_{i=1}^{N} \phi_{ii} \left(\mathbf{r}_i^{(l-1)}, \mathbf{r}_i^{(l)}, \Delta\beta \right) \right] \\
&\times \frac{\operatorname{per} \left\| \phi \left(r^{(n)}, r^{(0)}, \Delta\beta \right) \right\|_{N_{\mathrm{g}}}}{\Lambda_{\mathrm{g}}^{3(n+1)N_{\mathrm{g}}}(\Delta\beta)} \frac{\det \left\| \phi \left(r^{(n)}, r^{(0)}, \Delta\beta \right) \right\|_{N_{\mathrm{q}}}}{\Lambda_{\mathrm{q}}^{3(n+1)N_{\mathrm{q}}}(\Delta\beta)} \\
&\times \frac{\det \left\| \phi \left(r^{(n)}, r^{(0)}, \Delta\beta \right) \right\|_{\overline{N}_{\mathrm{q}}}}{\Lambda_{\overline{\mathrm{q}}}^{3(n+1)\overline{N}_{\mathrm{q}}}(\Delta\beta)} ,
\end{aligned} \tag{11.48}
$$

In (11.48), the effective total color interaction energy is

$$
U = \frac{1}{n+1} \sum_{l=1}^{n+1} \frac{1}{2} \sum_{i,j(i \neq j)}^{N} \Phi_{ij} \left(\mathbf{r}_i^{(l-1)} - \mathbf{r}_j^{(l-1)}, \mathbf{r}_i^{(l)} - \mathbf{r}_j^{(l)}, Q_i, Q_j \right) . \tag{11.49}
$$

Other quantities in (11.48) are defined as follows. To begin with,

$$
\Lambda_a^3(\beta) = \lambda_a^3 \sqrt{\pi/2(\beta m_a)^5} , \tag{11.50}
$$

where $\lambda_a = \sqrt{2\pi\beta/m_a}$ is a thermal wavelength of an a type quasiparticle, with $a = \mathrm{q}, \overline{\mathrm{q}}$, or g. The antisymmetrization and symmetrization are taken into account by the symbols 'det' and 'per', denoting the determinant and the permanent, respectively. Equation (11.48) is exact in the limit $n \to \infty$. Indeed, since each factor in (11.39) has an error of order $1/(n+1)^{\theta}$ with $\theta > 1$, the error in the whole product in the limit $n \to \infty$ equals zero. The matrix $\phi\left(r, r', \Delta\beta\right)$ is defined by its matrix elements

$$
\phi_{ij}\left(r, r', \Delta\beta\right) = \frac{K_2\left[z_{ij}\left(r, r', \Delta\beta\right)\right]}{\left[z_{ij}\left(r, r', \Delta\beta\right)\right]^2} \delta \left[\delta_{a_i,\mathrm{g}} + \left(\delta_{a_i,\mathrm{q}} + \delta_{a_i,\overline{\mathrm{q}}} \right) \delta_{f_i,f_j} \delta_{\sigma_i,\sigma_j} \right] , \tag{11.51}
$$

with

$$
z_{ij}\left(r, r', \Delta\beta\right) = \Delta\beta\, m_{\mathrm{i}} \sqrt{1 + \left| \mathbf{r}_i - \mathbf{r}'_j \right|^2 / \Delta\beta^2} \tag{11.52}
$$

defined in terms of the function K_2. These matrix elements are nonzero only for particles of the same type, i.e., $a_i = a_j$. Additional Kronecker symbols in the spin σ_i and flavor f_i indices of the particles are applicable only to quark and antiquark matrix elements. They prevent Pauli blocking for particles with different spins and flavors. The quantity ϕ describes the *relativistic measure* of trajectories in the color path integral. This measure is associated with the relativistic kinetic energy operator in (11.27). In the large mass limit, this measure coincides with the Gaussian one used in Feynman–Wiener path integrals. Due to the factors $\delta_{a_i,a_j}\, \delta_{f_i,f_j}$, the matrix ϕ has a block structure corresponding to different types of particles and different flavors of quarks and antiquarks. The subscripts N_a near the det and per operations refer precisely to the corresponding blocks, which in case of quarks and antiquarks are still subdivided into sub-blocks related to flavors.

The dominant contribution to the partition function comes from configurations in which the 'size' of the quasiparticle cloud of 'beads' is of the order of the Compton wavelength $\lambda_C = 1/m_{\rm i}$. Thus, this path integral representation takes into account the quantum uncertainty of the quasiparticle position. In the large mass limit, the spatial quasiparticle extension becomes much smaller than the average interparticle distance. This makes possible an analytical integration over the 'bead' positions using the method of steepest decent. As a result, the partition function is reduced to its classical limit involving point-like quasiparticles.

In fact, in Monte Carlo simulations, the pressure of the system is computed. To obtained the expression for the pressure, we change the variable in (11.34):

$$P = \frac{\partial T \ln Z}{\partial V} = T\left[\frac{\alpha}{3V}\frac{\partial \ln Z}{\partial \alpha}\right]_{\alpha=1}, \tag{11.53}$$

where $\alpha = L/L_0$ ($V = \alpha^3 L_0^3$) is the length scaling parameter introduced in physical quasiparticle coordinates. Details of the derivation of the final Monte Carlo pressure estimator and the final intricate formula for the path integral representation of the partition function are presented and discussed in (Zamalin et al. 1977; Filinov et al. 2001, 2007). Notice that only the maximal $\{N\}$ term in the sum of (11.32) is dominant in the thermodynamic limit of the box volume $V \to \infty$, corresponding to the canonical ensemble with $\{N\}$ particles. We arrive at the following expression for the Monte Carlo pressure estimator:

$$\begin{aligned}
\frac{P}{P_0} = 1 - \frac{1}{N}&\frac{\left[3Z(\{N\}, V, \beta)\right]^{-1}}{\Lambda_{\rm g}^{3(n+1)N_{\rm g}}(\Delta\beta)\Lambda_{\rm q}^{3(n+1)N_{\rm q}}(\Delta\beta)\Lambda_{\bar{\rm q}}^{3(n+1)\overline{N}_{\rm q}}(\Delta\beta)}\\
&\times \int {\rm d}\mu Q \int {\rm d}r^{(0)}{\rm d}r^{(1)}\ldots{\rm d}r^{(n+1)} R\left(r^{(0)}, r^{(1)}, \ldots, r^{(n+1)}; Q; \{N\}; \beta\right)\\
&\times\delta\left(r^{(n+1)} - r^{(0)}\right)\left\{\sum_{l=1}^{n}\sum_{i,j(i\neq j)}^{N}\frac{\left(\mathbf{r}_{ij}^{l}\cdot\mathbf{r}_{ij}^{0}\right)}{|\mathbf{r}_{ij}^{l}|}\frac{\partial(\beta U)}{\partial|\mathbf{r}_{ij}^{l}|}\right.\\
&-\left[\frac{\alpha}{\text{per}\left\|\phi\left(r^{(n)}, r^{(0)}, \Delta\beta\right)\right\|_{N_{\rm g}}\det\left\|\phi\left(r^{(n)}, r^{(0)}, \Delta\beta\right)\right\|_{N_{\rm q}}}\right]_{\alpha=1}\\
&\times\left[\frac{\alpha}{\det\left\|\phi\left(r^{(n)}, r^{(0)}, \Delta\beta\right)\right\|_{\overline{N}_{\rm q}}}\right]_{\alpha=1}\\
&\times\left[\frac{\partial\,\text{per}\left\|\phi\left(r^{(n)}, r^{(0)}, \Delta\beta\right)\right\|_{N_{\rm g}}\det\left\|\phi\left(r^{(n)}, r^{(0)}, \Delta\beta\right)\right\|_{N_{\rm q}}}{\partial\alpha}\right]_{\alpha=1}
\end{aligned}$$

$$\times \left[\frac{\partial \det \left\| \phi \left(r^{(n)}, r^{(0)}, \Delta\beta \right) \right\|_{\overline{N}_q}}{\partial \alpha} \right]_{\alpha=1} \Bigg\} , \tag{11.54}$$

where P_0 is the pressure of the ideal gas of quasiparticles, and $\mathbf{r}_{ij}^l = \mathbf{r}_i^l - \mathbf{r}_j^l$ is the distance between beads with a number l of quasiparticles with numbers i and j.

The structure of (11.54) is obvious. We have separated out the classical ideal gas part (first term). The ideal quantum part in excess of the classical one and the correlation contributions are contained in the integral term. The last term in the curly brackets in (11.54) is due to the explicit volume dependence of the exchange matrix. The main advantage of (11.54) is that the explicit sum over permutations has been converted into the determinant, which can be computed very efficiently using standard methods of linear algebra. Note that (11.54) contains the important limit of an ideal quantum plasma in a natural way.

Monte Carlo Simulations

For simulation of the thermodynamic properties of QGP, the Metropolis–Hasting algorithm has been used with the periodic boundary conditions discussed above, imposed on a cubic simulation box. The main idea of the simulations consists in constructing a Markov chain of different quasiparticle states in the color configuration space.

The computational procedure comprises two stages. In the first stage, a dominant, i.e., maximal, $\{N\}$-term in the sum of (11.32) is determined by calculations in the grand canonical ensemble. This term is indeed dominant in the thermodynamic limit of the box volume ($V \to \infty$). In the grand canonical ensemble, the quasiparticle numbers in the simulation box are varied, i.e., the consecutive states of the Markov chain can differ from each other by numbers of quarks, antiquarks, and gluons. Transitions between these states are the first type of Markovian elementary step. In the second type of elementary step, the coordinates of a single bead of a randomly chosen quasiparticle were changed, or this quasiparticle was shifted as a whole. The color variables are changed according to the SU(3) group Haar measure in the third type of Markovian elementary step. The Markov chain is generated until a full convergence of calculated values is achieved. This allows one to determine the average numbers of quarks, antiquarks, and gluons in the box at fixed temperature. Here, only densities of each species, i.e., the ratios of these average numbers with respect to the box volume, have physical meaning. Usually, after several million elementary steps, the average numbers of these quasiparticles become stable, and, for example, at the zero baryon chemical potential, the average number of quarks practically equals the average number of antiquarks. This equality can be considered as an inherent test of the consistency of the calculations.

In the second stage, the fixed number of quarks, antiquarks, and gluons is chosen to be equal to the obtained average values of quasiparticles, and calculations are performed in the canonical ensemble. Here only the second and third types of elementary step described above were used, and the pressure was calculated using (11.54). After several million elementary Markovian steps, the result for the pressure becomes stable.

An important difference with the case of electrodynamic plasmas is the use of the relativistic measure in the path integrals. This measure is associated with the relativistic kinetic energy operator, instead of the conventional Gaussian one arising from the non-relativistic kinetic energy operator in the Feynman–Wiener path integral.

The error in the Monte Carlo calculations of thermodynamic quantities related to the finite particle number (N) in the system with periodic boundary conditions is of order $1/N$ (Zamalin et al. 1977). However, too many particles, represented moreover by a large number of beads, would require too large computer resources. In practical DPIMC calculations, the total number of particles does not exceed $N = N_{\mathrm{q}} + N_{\bar{\mathrm{q}}} + N_{\mathrm{g}} = 126$ and the proper densities of species calculated in the first stage were adjusted by varying the total volume V of the box. The choice of particle and bead numbers is a compromise between acceptable accuracy and available computer resources. It has been checked that varying the number of beads from 15 up to 50 has practically no effect on the results. The number of beads for each particle was usually equal to $n = 20$.

As can be seen from Fig. 11.7, the degeneracy is moderate, i.e., the degeneracy parameter is of order of unity. This means that the well known sign problem in Monte Carlo simulations of Fermi particles is not severe here.

11.3 Thermodynamics of Quark–Gluon Plasmas

The parameters of this model should be deduced from QCD lattice data, but this task remains quite ambiguous. So analytical statements known for high temperatures have been extended to lower ones, and only one possible set of parameters was taken for the DPIMC simulations.

Parameters of the Model

The values of m_{g} and m_{q} for $T \gg T_{\mathrm{c}}$ can be taken in the form (Bellac 1996)

$$m_{\mathrm{g}}^2(\{\mu_{\mathrm{q}}\}, T) \approx \frac{1}{12}\left[(2N_{\mathrm{c}} + N_{\mathrm{f}})T^2 + \frac{3}{\pi^2} \sum_{q'=\mathrm{u,d,s}} \mu_{q'}^2\right] g^2(\{\mu_{\mathrm{q}}\}) , \quad (11.55)$$

$$m_{\mathrm{q}}^2(\{\mu_{\mathrm{q}}\}, T) \approx \frac{N_{\mathrm{g}}}{16N_{\mathrm{c}}}\left(T^2 + \frac{\mu_{\mathrm{q}}^2}{\pi^2}\right) g^2(\{\mu_{\mathrm{q}}\}) , \quad (11.56)$$

where N_{f} is the number of quark flavors that can be excited, $N_{\mathrm{c}} = 3$ for the SU(3) group, and g is the QCD running coupling constant, generally depending on T and all the μ_{q}. According to (11.55) and (11.56), the masses do not depend on T and μ_{q} separately, but on their combinations

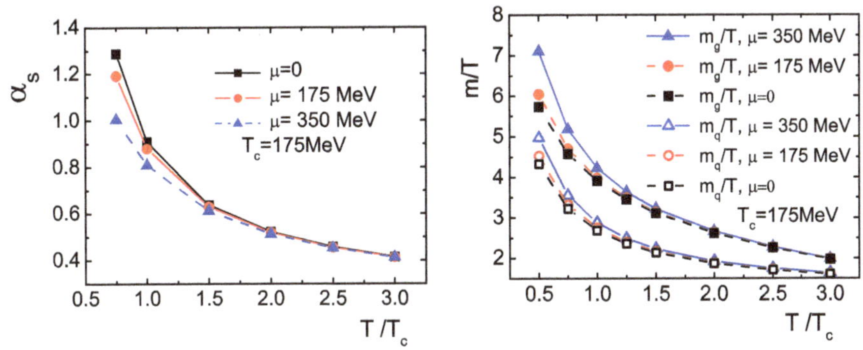

Fig. 11.4 Temperature dependencies of the model input quantities. *Left*: Coupling constant α_s. *Right*: Quasiparticle mass-to-temperature ratio ($m_q = m_{\bar{q}}$) (Filinov et al. 2013, 2015)

$$z_g = \left[T^2 + \frac{3}{\pi^2(2N_c + N_f)} \sum_{q'=u,d,s} \mu_{q'}^2 \right]^{1/2}, \qquad z_q = \left(T^2 + \frac{\mu_q^2}{\pi^2} \right)^{1/2},$$

respectively. It is also reasonable to assume that g^2 is a function of this single variable z_g. This choice is made because g^2 (like the gluons) is related to the whole system rather than to one specific quark flavor. Then we can use the same 'one-loop analytic coupling' (Shirkov and Solovtsov 1997, Prosperi et al. 2007)

$$\alpha_s(Q^2) = \frac{4\pi}{11 - 2N_f/3} \left[\frac{1}{\ln(Q^2/\Lambda_{QCD}^2)} + \frac{\Lambda_{QCD}^2}{\Lambda_{QCD}^2 - Q^2} \right], \tag{11.57}$$

and use $2\pi z_g$ for Q in our simulations. This coupling constant $\alpha_s(z_g) = g^2/(4\pi)$ is displayed in the left panel of Fig. 11.4. To obtain model input formulas for the masses m_g and m_q, we use the same related dependencies for thermodynamics and kinetics from our paper (Filinov et al. 2013), where we replace the temperature T by the expressions for z_g (z_q) written above. The final dependencies are presented in Fig. 11.4 (right).

QGP Equation of State

Figure 11.5 shows the equation of state (left panel) and the pressure difference compared with the limit of zero chemical potential (right panel) for different finite chemical potentials. As can be seen from Fig. 11.5, for small values of the chemical potential $\mu/T_c = 133/175 < 1$, the approach developed agrees rather well with available lattice QCD data based on a Taylor expansion (around $\mu = 0$), and we may thus hope that the results for larger values of the chemical potential (around $\mu = 175$ MeV) are reliable.

Another interesting thing is the behavior of the pressure versus the reciprocal density (specific volume, right panel in Fig. 11.6) and of the reciprocal density versus

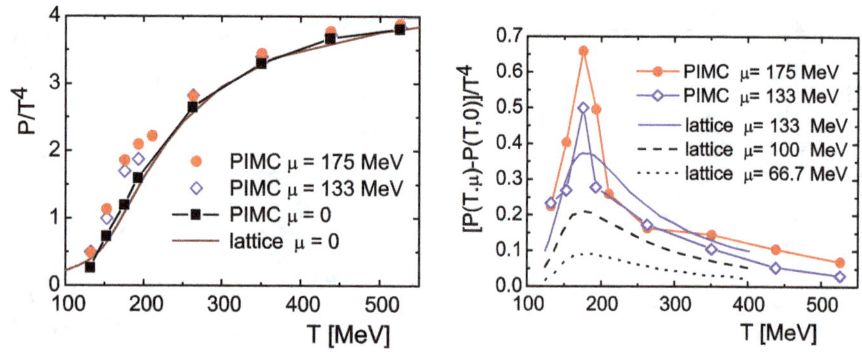

Fig. 11.5 *Left*: Equation of state of the QGP for different baryon chemical potentials. *Right*: Pressure ($\mu > 0$) in excess of the pressure at $\mu = 0$. PIMC results (*symbols*) are compared to lattice data (*lines*) from (Borsanyi et al. 2010, 2012) (Filinov et al. 2015)

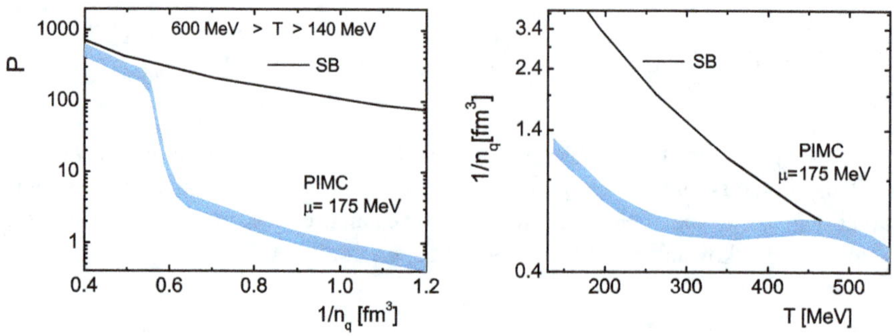

Fig. 11.6 *Left*: Stefan–Boltzmann (SB *black line*) limit and PIMC pressure (*blue*) versus reciprocal quark density for $\mu = 175$ MeV. *Right*: Reciprocal quark density versus temperature for $\mu = 175$ MeV. Comparison with the Stefan–Boltzmann (SB *black line*) limit and the present grand canonical color PIMC results (*blue*) (Filinov et al. 2015)

temperature (left panel). At high temperatures, the color PIMC results are close to the ideal Stefan–Boltzmann limit. However, at lower temperatures, the influence of interactions begins to grow and results in a sharp pressure decrease, whereas the quark density changes only weakly. Analogous behavior is observed in a dense electron–ion plasma and is connected with the formation of atomic and molecular bound states (Filinov et al. 2000, 2001). The same is observed in electron–hole plasmas in semiconductors, with the formation of excitons and bi-excitons (Filinov et al. 2007; Schleede et al. 2012). The physical origin of the analogous behavior in the present QGP is the strong quasiparticle interaction which may give rise to the formation of bound states (see below).

Figure 11.7 presents the entropy S/T^3 and trace anomaly $(\varepsilon - 3P)/T^4$ of the QGP. These quantities are calculated using (11.35)–(11.37). In order to avoid numerical noise, the derivative of a smooth interpolation between the DPIMC points was taken. Though agreement with the lattice data looks worse for the entropy and trace

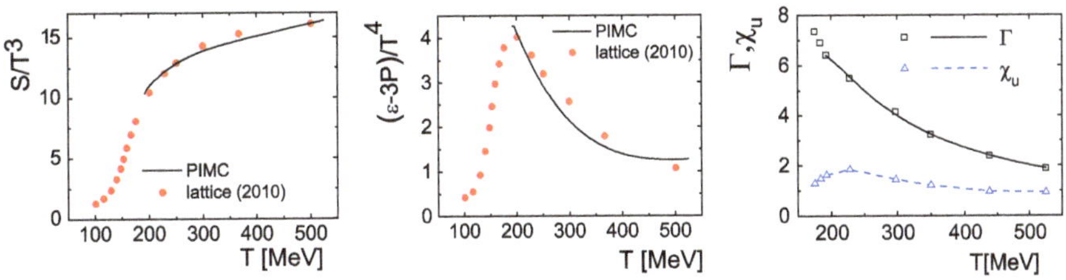

Fig. 11.7 *Left and Center*: Entropy and trace anomaly scaled with corresponding powers of temperature versus temperature from PIMC simulations. These are compared with lattice data from (Fodor and Katz 2002; Borsanyi et al. 2010). The *solid line* is a smooth interpolation between the PIMC points. *Right*: Quark degeneracy parameter χ_u and plasma coupling parameter Γ [see (11.58)] versus temperature (Filinov et al. 2011)

anomaly, in fact it is the same as for the pressure. Differentiation operations in (11.35) and (11.37) make the differences between the DPIMC results and the lattice data more pronounced.

To characterize physical conditions and the interplay of interaction and degeneracy in Fig. 11.7 let us consider a degeneracy parameter χ_u for up quarks and a plasma coupling parameter Γ given by

$$\chi_u = n_u \lambda_u^3 \,, \qquad \Gamma = \frac{\overline{q}_2 g^2}{4\pi r_s T} \,, \tag{11.58}$$

where the thermal wavelength λ_u is as defined previously [see the text after (11.50)], n_u is the density of u quarks, r_s is the Wigner–Seitz radius defined by $r_s^3 = 3/4\pi n$, n is the density of all quasiparticles (quarks, antiquarks, and gluons), and $\overline{q}_2$ is the quadratic Casimir value averaged over quarks, antiquarks, and gluons, for which $\overline{q}_2 = N_c^2 - 1$ is a good estimate. The plasma coupling parameter is a measure of the ratio of the average potential to the average kinetic energy, while the degeneracy parameter χ_u indicates whether a system is classical ($\chi_u \ll 1$) or quantum ($\chi_u \approx 1$). It turns out that Γ and χ_u are of order unity, which indicates that the QGP is a strongly coupled quantum ($\chi_u \approx 1$) liquid ($\Gamma \sim 1$) rather than a gas.

Pair Distribution Function and Bound States

We now consider the spatial arrangement of the quasiparticles in the QGP by studying a pair distribution function (PDF) $g_{ab}(R)$ defined by

$$g_{ab}\big(|\mathbf{R}_1 - \mathbf{R}_2|\big) = \left(\frac{V}{N}\right)^2 \sum_{\sigma} \sum_{i,j,i\neq j} \delta_{a_i,a}\delta_{a_j,b}\frac{1}{Z} \times \int \mathrm{d}r\, \mathrm{d}\mu Q\, \delta(\mathbf{R}_1 - \mathbf{r}_i)\delta(\mathbf{R}_2 - \mathbf{r}_j)\rho\big(r, Q, \sigma; \{N\}; \beta\big) \,, \tag{11.59}$$

where a_i and a_j are the particle types (q, $\overline{q}$, or g). The PDF gives the probability density for finding a pair of particles of types a and b at a certain distance

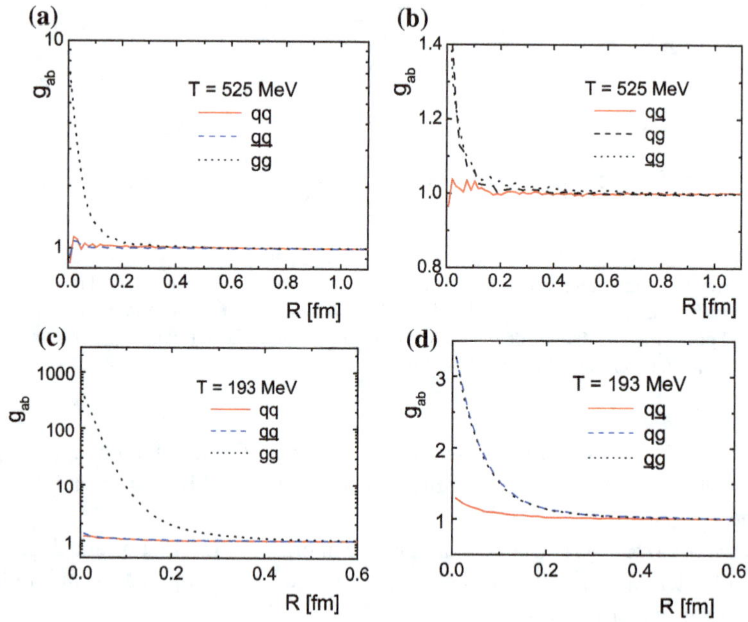

Fig. 11.8 Pair correlation functions of (**a**, **c**) identical and (**b**, **d**) different quasiparticles at temperatures $T = 525$ MeV (**a**, **b**) and $T = 193$ MeV (**c**, **d**) for $\mu = 0$ (Filinov et al. 2013)

$R = |\mathbf{R}_1 - \mathbf{R}_2|$ from each other. The PDF depends only on the coordinate difference because of the translational invariance of the system. In a non-interacting classical system, $g_{ab}(R) \equiv 1$, whereas interactions and quantum statistics result in a redistribution of the particles. At temperatures $T = 525$ MeV and $T = 193$ MeV, the PDF averaged over the quasiparticle spin, colors, and flavors are shown in Fig. 11.8.

At distances $R \approx 0.2$ or 0.3 fm, depending on the temperature, all PDFs are practically equal to unity (Fig. 11.8), like in an ideal gas, due to screening of the color Coulomb interaction. A drastic difference between the qq and gg PDFs (the $\overline{\text{qq}}$ and qq PDFs are identical) occurs at short distances. Here the gluon–gluon and gluon–quark PDFs increase monotonically when the distance goes to zero, while the qq and $\overline{\text{qq}}$ PDFs remain uncorrelated. One of the physical reasons for the difference between the PDFs is spatial quantum uncertainty and the different properties of Bose and Fermi statistics of the gluon and (anti)quark quasiparticles. Uncertainty in the particle localization is defined by the ratio T/m. Localization is better for heavier gluon quasiparticles. Fermi statistics results in effective quark–quark and aniquark–antiquark repulsion, while Bose statistics results in an effective gluon–gluon attraction. Oscillations in the PDF at very small distances $R \leq 0.1$ fm are related to the Monte Carlo statistical error, as the probability of quasiparticles being at short distances quickly decreases.

However, the qq and $\overline{\text{qq}}$ pair correlation functions reveal the almost total absence of Fermi repulsion. This happens because another physical reason comes into play.

The strong interaction between quasiparticles reduces the influence of degeneracy in the region of $\chi_u \sim 1$. This interaction is dominated by attraction at short distances. Indeed, the QGP lowers its total energy by minimizing the color Coulomb interaction energy via a spontaneous 'anti-ferromagnetic'-like ordering of color vectors, i.e., the color vectors of nearest neighbor quasiparticles become anti-parallel. A similar absence of Fermi repulsion was observed in hydrogen plasma at $\chi \sim 1$ (Zamalin et al. 1977). This short-distance attraction is stronger for gluon–gluon and gluon–(anti)quark pairs than for (anti)quark–(anti)quark pairs, because of the corresponding difference in the values of the quadratic Casimir invariants $\breve{q}_2$ (see the appendix in Sect. 11.4), which determine the maximal values of the effective color charge products $|Q_i \cdot Q_j|$ in the color Kelbg (Coulomb) potentials:

- For gluon–gluon pairs,
$$|Q_g \cdot Q_g|_{\max} = 24\,.$$
- For gluon–(anti)quark pairs,
$$|Q_g \cdot Q_q|_{\max} = |Q_g \cdot Q_{\bar{q}}|_{\max} \approx 10\,.$$
- For (anti)quark–(anti)quark pairs,
$$|Q_q \cdot Q_q|_{\max} = |Q_{\bar{q}} \cdot Q_{\bar{q}}|_{\max} = |Q_q \cdot Q_{\bar{q}}|_{\max} = 4\,.$$

Stronger gg attraction also enhances the correlation of the gluon–gluon pairs at short distances. At the same time, the short distance attraction is the only reason for the gluon–(anti)quark short distance correlation.

The short distance correlation implies formation of the gluon–gluon and gluon–(anti)quark clusters, which are uniformly distributed in space. In the case of the gluon–gluon clusters, we may even speak of gg bound states, i.e., glueballs, at zero and nonzero chemical potentials, as can be seen from Fig. 11.9. The maximum of

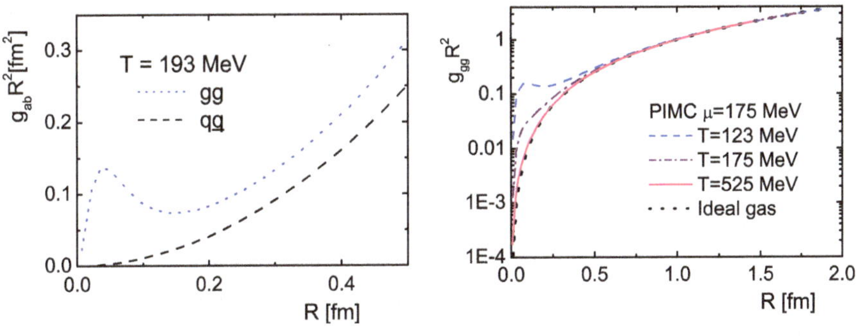

Fig. 11.9 *Left*: Gluon–gluon and quark–antiquark pair correlation functions multiplied by R^2 for $T = 193$ MeV and $\mu = 0$. *Right*: PIMC gluon–gluon quasiparticle pair distributions multiplied by R^2 for different temperatures and $\mu = 175$ MeV (Filinov et al. 2015)

$r^2 g(r)$ at small interparticle distances supports the interpretation in terms of gluon bound states (Filinov et al. 2007), and this explains the observed pressure and density behavior discussed above.

At the same time, the quark–quark and quark–gluon PDF do not exhibit such peaks and are close to an ideal PDF. The product $R^2 g_{ab}(R)$ is proportional (up to constant factor) to the probability of finding a pair of quasiparticles at a distance R from each other. As we know from consideration of hydrogen plasmas (Zamalin et al. 1977), a maximum in $R^2 g_{gg}(R)$ signals population of a bound state. For comparison, the quark–antiquark correlation function, i.e., $R^2 g_{q\overline{q}}(R)$, is also presented in Fig. 11.9. It demonstrates that there are no bound meson-like states. We can only speak about weak meson-like clustering at lower temperatures (see the short distance $q\overline{q}$ correlation at $T = 193$ MeV in Fig. 11.8. The possible existence of medium-modified meson-like bound states was actively discussed some time ago, e.g., in (Yukalov and Yukalova 1997) and later in (Shuryak and Zahed 2004; Brown et al. 2006), on the basis of results from lattice QCD calculations of spectral functions (Asakawa et al. 2001, Datta 2004). These results support (Koch et al. 2005) regarding the absence of $\mathrm{q}\overline{\mathrm{q}}$ bound states above the temperature of the phase transition. This finding is in contrast to our previous results on the SU(2) group (Filinov et al. 2001, 2009, 2012). In that case, well pronounced bound $\mathrm{q}\overline{\mathrm{q}}$ states were found just above the critical temperature, although they quickly dissolved with increasing temperature. This happens because the SU(3) plasma turns out to be essentially denser than the SU(2) plasma, and this is a consequence of the stronger effective attraction between constituents. As a result, possible bound states in the SU(3) plasma just melt. To check the relevance of all the trends discussed above, a more refined color-, flavor-, and spin-resolving analysis of the PDFs is necessary.

Discussion of the DPIMC Simulation of QGP Thermodynamic Properties

The results presented here have shown that color quantum Monte Carlo simulations based on the quasiparticle model of the QGP are able to reproduce the lattice equation of state at zero and finite baryon chemical potentials with realistic model parameters (quasiparticle masses and coupling constant), even near and below the critical temperature. They also yield valuable insights into the internal structure of the QGP.

A central question of the confinement problem in QCD is the nature of the chiral phase transition. The present color PIMC results indicate that, in the considered region of the baryonic chemical potentials, the transition from a phase dominated by colored degrees of freedom to a phase dominated by color-neutral degrees of freedom is an analytic 'cross-over'. This conclusion also agrees with lattice QCD calculations (Borsanyi et al. 2014).

In DPIMC simulations, the new relativistic path integral measure has been introduced and procedure has been developed for sampling color quasiparticle variables according to the SU(3) group Haar measure with appropriate Casimir conditions. Unfortunately, convergence of the DPIMC calculations becomes poor in the range of the expected phase transition, because the scheme suffers from jumps between stable and metastable states which turn out to be almost equally probable in this range. The DPIMC results indicate that the QGP exhibits *quantum liquid-like* (rather than gas-like) properties, up to the highest considered temperature of 525 MeV.

Short-distance correlations in the computed pair distribution functions of gluon–gluon and gluon–(anti)quark pairs indicate the formation of clusters. In the case of gluon–gluon clusters, one can even speak of gluon–gluon bound states, i.e., glueballs, at temperatures just above the phase transition.

The possible existence of medium-modified meson-like bound states was actively discussed some time ago (Shuryak and Zahed 2004, Brown et al. 2006). The DPIMC results support the conclusion of (Koch et al. 2005) regarding the absence of $q\overline{q}$ bound states above the temperature of the phase transition. This finding is in contrast to the previous DPIMC results on the SU(2) group (Filinov et al. 2009, 2011, 2012).In that case, well pronounced bound $q\overline{q}$ states were found just above the critical temperature, although they quickly dissolved with the temperature rise. This happens because the SU(3) plasma turns out to be essentially denser than the SU(2) plasma, and this in turn is a consequence of a stronger effective attraction between the constituents. As a result, possible meson-like bound states in the SU(3) plasma just melt.

The approach described here is highly sensitive to variations in the running coupling constant and quasiparticle masses. This allows for an accurate fit of our model to the lattice EOS by choosing optimal values of the gluon and quark quasiparticle masses.

11.4 Appendix. Integration with the SU(3) Group Haar Measure

In this appendix we consider details of integration with respect to the SU(3) Haar measure $\mathrm{d}\mu Q$ in (11.32). The measure for a single color charge in the case of the SU(3) group (Litim and Manuel 1999, 2000, 2002; Johnson 1989, Kelly et al. 1994) is defined by

$$\mathrm{d}\mu Q = \mathrm{d}^8 Q\, \delta(Q_a Q_a - \breve{q}_2)\delta(d_{abc} Q_a Q_b Q_c - \breve{q}_3)\ , \tag{11.60}$$

with summation over $a, b, c = 1, \ldots, 8$ and constants d_{abc} given in Table 11.2. For the group SU(N), the quadratic $\breve{q}_2$ and cubic $\breve{q}_3$ Casimirs are:

- $\breve{q}_2 = (N^2 - 1)C^2$ with $C_2 = N$ for gluons and $C_2 = 1/2$ for quarks and antiquarks,
- $\breve{q}_3 = 0$ for gluons and $\breve{q}_3 = (N^2 - 4)(N^2 - 1)/4$ for quarks, while $\breve{q}_3$ has opposite sign for antiquarks.

For random sampling of the Q variable in the Monte Carlo integration of (11.32), we must change to the related canonical Darboux variables for the SU(3) group. The set of canonical variables ϕ_α, π_α, $\alpha = 1, 2, 3$, is defined by the canonical Poisson bracket

Table 11.2 Nonzero constants $\breve{f}_{abc}$ and d_{abc} for the group SU(3)

$\breve{f}_{abc}$	$\breve{f}_{123}$	$\breve{f}_{147}$	$\breve{f}_{156}$	$\breve{f}_{246}$	$\breve{f}_{257}$	$\breve{f}_{345}$	$\breve{f}_{367}$	$\breve{f}_{458}$	$\breve{f}_{678}$
	1	$\frac{1}{2}$	$-\frac{1}{2}$	$\frac{1}{2}$	$\frac{1}{2}$	$\frac{1}{2}$	$-\frac{1}{2}$	$\frac{\sqrt{3}}{2}$	$\frac{\sqrt{3}}{2}$
d_{abc}	d_{118}	d_{146}	d_{157}	d_{228}	d_{247}	d_{256}	d_{338}	d_{344}	d_{355}
	$\frac{1}{\sqrt{3}}$	$\frac{1}{2}$	$\frac{1}{2}$	$\frac{1}{\sqrt{3}}$	$-\frac{1}{2}$	$\frac{1}{2}$	$\frac{1}{\sqrt{3}}$	$\frac{1}{2}$	$\frac{1}{2}$
d_{abc}	d_{366}	d_{377}	d_{448}	d_{558}	d_{668}	d_{778}	d_{888}		
	$-\frac{1}{2}$	$-\frac{1}{2}$	$-\frac{1}{2\sqrt{3}}$	$-\frac{1}{2\sqrt{3}}$	$-\frac{1}{2\sqrt{3}}$	$-\frac{1}{2\sqrt{3}}$	$-\frac{1}{\sqrt{3}}$		

$$\{A, B\}_{\text{PB}} = \frac{\partial A}{\partial r}\frac{\partial B}{\partial p} - \frac{\partial A}{\partial p}\frac{\partial B}{\partial r} + \frac{\partial A}{\partial \phi}\frac{\partial B}{\partial \pi} - \frac{\partial A}{\partial \pi}\frac{\partial B}{\partial \phi}, \tag{11.61}$$

where r and p are the conventional coordinate and momentum, respectively, satisfying

$$\{r_\alpha, p_\gamma\}_{\text{PB}} = \delta_{\alpha\gamma}, \qquad \{\phi_\alpha, \pi_\gamma\}_{\text{PB}} = \delta_{\alpha\gamma}. \tag{11.62}$$

The color variables Q_a form a representation of SU(3). In terms of canonical variables, their Poisson bracket reads

$$\{Q_a, Q_a\}_{\text{PB}} = \breve{f}_{abc} Q_c, \tag{11.63}$$

where $\breve{f}_{abc}$ are the structure constants of SU(3) given in Table 11.2.

The explicit transformations to canonical variables are given by (Litim and Manuel 1999, 2000, 2002; Johnson 1989; Kelly et al. 1994)

$$\begin{aligned}
Q_1 &= \pi_+\pi_- \cos\phi_1, \\
Q_2 &= \pi_+\pi_- \sin\phi_1, \\
Q_3 &= \pi_1, \\
Q_4 &= C_{++}\pi_+ A + C_{+-}\pi_- B, \\
Q_5 &= S_{++}\pi_+ A + S_{+-}\pi_- B, \\
Q_6 &= C_{-+}\pi_- A + C_{--}\pi_+ B, \\
Q_7 &= S_{-+}\pi_- A - S_{--}\pi_+ B, \\
Q_8 &= \pi_2,
\end{aligned} \tag{11.64}$$

in which we have used the definitions

$$\begin{aligned}
\pi_+ &= \sqrt{\pi_3 + \pi_1}, \\
\pi_- &= \sqrt{\pi_3 - \pi_1},
\end{aligned}$$

$$\begin{aligned} C_{\pm\pm} &= \cos\left[\frac{1}{2}\left(\pm\phi_1 + \sqrt{3}\phi_2 \pm \phi_3\right)\right] , \\ S_{\pm\pm} &= \sin\left[\frac{1}{2}\left(\pm\phi_1 + \sqrt{3}\phi_2 \pm \phi_3\right)\right] , \end{aligned} \tag{11.65}$$

and A and B are given by

$$\begin{aligned} A &= \frac{1}{2\pi_3}\sqrt{\left(\frac{J_1 - J_2}{3} + \pi_3 + \frac{\pi_2}{\sqrt{3}}\right)\left(\frac{J_1 + 2J_2}{3} + \pi_3 + \frac{\pi_2}{\sqrt{3}}\right)} \\ &\quad \times \sqrt{\left(\frac{2J_1 + J_2}{3} - \pi_3 - \frac{\pi_2}{\sqrt{3}}\right)} \\ B &= \frac{1}{2\pi_3}\sqrt{\left(\frac{J_2 - J_1}{3} + \pi_3 - \frac{\pi_2}{\sqrt{3}}\right)\left(\frac{J_1 + 2J_2}{3} - \pi_3 + \frac{\pi_2}{\sqrt{3}}\right)} \\ &\quad \times \sqrt{\left(\frac{2J_1 + J_2}{3} + \pi_3 - \frac{\pi_2}{\sqrt{3}}\right)} . \end{aligned} \tag{11.66}$$

In this expression the set, Q_1, Q_2, Q_3 forms an SU(2) subgroup with quadratic Casimir $Q_1^2 + Q_2^2 + Q_3^2 = \pi_3^2$. Note that two Casimirs depend only on J_1 and J_2. They can be computed using the values given in Table 11.2 as

$$\begin{aligned} Q_a Q_a &= \frac{1}{3}\left(J_1^2 + J_1 J_2 + J_2^2\right) , \\ d_{abc} Q_a Q_b Q_c &= \frac{1}{18}\left(J_1 - J_2\right)\left(J_1 + 2J_2\right)\left(2J_1 + J_2\right) . \end{aligned} \tag{11.67}$$

The phase space color measure for SU(3) given in (11.60) can be transformed to the new coordinates using (11.64) and evaluating the Jacobian

$$\left|\frac{\partial Q}{\partial(\phi, \pi)}\right| = \frac{\sqrt{3}}{48} J_1 J_2 (J_1 + J_2) . \tag{11.68}$$

Then the measure reads

$$\begin{aligned} \mathrm{d}\mu Q &= \mathrm{d}\phi_1 \mathrm{d}\phi_2 \mathrm{d}\phi_3 \mathrm{d}\pi_1 \mathrm{d}\pi_2 \mathrm{d}\pi_3 \mathrm{d}J_1 \mathrm{d}J_2 \frac{\sqrt{3}}{48} J_1 J_2 (J_1 + J_2) \\ &\quad \times \delta\left(\frac{1}{3}\left[(J_1)^2 + J_1 J_2 + (J_2)^2\right] - \breve{q}_2\right) \\ &\quad \times \delta\left(\frac{1}{18}(J_1 - J_2)(J_1 + 2J_2)(2J_1 + J_2) - \breve{q}_3\right) . \end{aligned} \tag{11.69}$$

Since the two Casimirs are independent, the δ-functions fix both J^1 and J^2. After integrating over J^1 and J^2, (11.69) gives a proper canonical volume element $\mathrm{d}\phi\,\mathrm{d}\pi$. Thus applying the Metropolis algorithm to the $\phi\pi$ variables, we can construct a Markovian chain in (ϕ, π) phase space and obtain random color variables Q for calculating the partition function according to the SU(3) group Haar measure with two Casimir conditions.

References

M. Asakawa, T. Hatsuda, Y. Nakahara, Prog. Part. Nucl. Phys. **46**, 459 (2001); Nucl. Phys. A **715**, 863 (2003) [Nucl. Phys. Proc. Suppl. **119**, 481 (2003)]; M. Asakawa, T. Hatsuda. Phys. Rev. Lett. **92**, 12001 (2004)

A. Bazavov et al., Phys. Rev. D **80**, 014504 (2009)

J. Blaizot, E. Iancu, Phys. Rep. **359**, 355 (2002)

M. Bonitz, D. Semkat (eds.), *Introduction to Computational Methods for Many Body Systems* (Rinton Press, Princeton, 2006)

S. Borsanyi, G. Endrodi, Z. Fodor, A. Jakovac, S.D. Katz, S. Krieg, C. Ratti, K.K. Szabo, JHEP **1011**, 77 (2010)

S. Borsanyi, G. Endrodi, Z. Fodor, S.D. Katz, S. Krieg, C. Ratti, K.K. Szabo, JHEP **08**, 53 (2012)

S. Borsanyi, Z. Fodor, C. Hoelbling, S.D. Katz, S. Krieg, K.K. Szabo, Phys. Lett. B **730**, 99 (2014)

E. Braaten, R.D. Pisarski, Nucl. Phys. B **337**, 569 (1990); **339**, 310 (1990)

G.E. Brown, B.A. Gelman, M. Rho, Phys. Rev. Lett. **96**, 132301 (2006)

M. Cheng et al., Phys. Rev. D **81**, 54504 (2010)

M. Cheng, Phys. Rev. D **81** (2010)

S. Datta, F. Karsch, P. Petreczky, I. Wetzorke, Phys. Rev. D **69**, 94507 (2004)

R.P. Feynman, A.R. Hibbs, *Quantum Mechanics and Path Integrals* (McGraw-Hill, New York, 1965)

V.S. Filinov, Yu.B Ivanov, M. Bonitz, V.E Fortov, P.R Levashov. Phys. Lett. A **376**, 1096 (2012)

V.S. Filinov, V.E. Fortov, M. Bonitz, P.R. Levashov, JETP Lett. **74**, 384 (2001)

V.S. Filinov, H. Fehske, M. Bonitz, V.E. Fortov, P.R. Levashov, Phys. Rev. E **75**, 036401 (2007)

V.S. Filinov, M. Bonitz, Y.B. Ivanov, E.-M. Ilgenfritz, V.E. Fortov, Contrib. Plasma Phys. **55**(2–3), 203 (2015)

V.S. Filinov, M. Bonitz, YuB Ivanov, E.-M. Ilgenfritz, V.E. Fortov, Plasma Phys. Control Fusion **57**, 44004 (2015)

V.S. Filinov, M. Bonitz, W. Ebeling, V.E. Fortov, Plasma Phys. Control Fusion **43**, 743 (2001)

A. Filinov, V. Golubnychiy, M. Bonitz et al., Phys. Rev. E **70**, 46411 (2004)

V.S. Filinov, M. Bonitz, W. Ebeling, V.E. Fortov, Plasma Phys. Control Fusion **43**, 743–759 (2001)

V.S. Filinov, V.E. YuB Ivanov, M. Fortov, P.R.Levashov Bonitz, Phys. Rev. C **87**, 35207 (2013)

V.S. Filinov, V.E. Fortov, M. Bonitz, D. Kremp, Phys. Lett. A **274**, 228 (2000)

V.S. Filinov, M. Bonitz, Y.B. Ivanov, V.V. Skokov, P.R. Levashov, V.E. Fortov, Contrib. Plasma Phys. **52**, 135 (2012)

V.S. Filinov, Y.B. Ivanov, M. Bonitz, P.R. Levashov, V.E. Fortov, Phys. Atom. Nucl. **74**, 1364 (2011)

V.S. Filinov, Y.B. Ivanov, M. Bonitz, P.R. Levashov, V.E. Fortov, Phys. Atom. Nucl. **75**, 693 (2012)

V.S. Filinov, M. Bonitz, Y.B. Ivanov, V.V. Skokov, P.R. Levashov, V.E. Fortov, Contrib. Plasma Phys. **49**, 536 (2009)

V.S. Filinov, M. Bonitz, Y.B. Ivanov, V.V. Skokov, P.R. Levashov, V.E. Fortov, Contrib. Plasma. Phys. **51**, 322–327 (2011)

Z. Fodor, S.D. Katz, Phys. Lett. B **534**, 87 (2002); Z. Fodor, S.D. Katz, arXiv:0908.3341 [hep-ph]

I.M. Gelfand, A.M. Yaglom, Uspekhi Mat. Nauk **11**, 77 (1956)

B.A. Gelman, E.V. Shuryak, I. Zahed, Phys. Rev. C **74**, 44908 (2006); ibid. **74**, 44909 (2006)

I.S. Gradstein, I.M. Ryzik, *Tables of Integrals, Sums, Series and Products, Moscow* (Gos. Izd. Phys.-Math, Lit, 1963)

P. Hartmann, Z. Donko, P. Levai, G.J. Kalman, J. Phys. A **42**, 214029 (2006); Nucl. Phys. A **774**, 881–884 (2006)

A.L. Harvey, Phys. Rev. D **6**, 1474 (1972)

M. Hofmann, M. Bleicher, S. Scherer, L. Neise, H. Stoecker, W. Greiner, Phys. Lett. B **478**, 161 (2000)

K. Huang, *Statistical Mechanics* (Wiley, New York, 1963)

K. Johnson, Ann. Phys. **192**, 104 (1989)

F. Karsch, M. Kitazawa, Phys. Rev. D **80**, 56001 (2009)

G. Kelbg, Quantenstatistik der Gase mit Coulombwechselwirkung. Ann. Physik (Leipzig) **12**, 219–224, 354–360 (1963)

P.F. Kelly, Q. Liu, C. Lucchesi, C. Manuel, Phys. Rev. Lett. **72**, 3461 (1994a)

P.F. Kelly, Q. Liu, C. Lucchesi, C. Manuel, Phys. Rev. D **50**, 4209 (1994b)

V. Koch, A. Majumder, J. Randrup, Phys. Rev. Lett. **95**, 182301 (2005)

A.S. Larkin, V.S. Filinov, Phys. Lett. A **378**, 1876 (2014)

M. Le Bellac, *Thermal Field Theory* (Cambridge University Press, Cambridge, 1996)

J. Liao, E.V. Shuryak, Phys. Rev. D **73**, 14509 (2006)

D.F. Litim, C. Manuel, Phys. Rev. Lett. **82**, 4981 (1999); Nucl. Phys. B **562**, 237 (1999); Phys. Rev. D **61**, 125004 (2000). Phys. Rep. **364**, 451 (2002)

W. Lucha, F.F. Schoberl, D. Gromes, Phys. Rep. **200**, 127 (1991)

W. Lucha, F.F. Schoberl, Int. J. Mod. Phys. A **7**, 6431 (1992)

M.S. Marinov, Phys. Rep. **60**, 1 (1980)

T.D. Newton, E.P. Wigner, Rev. Mod. Phys. **21**, 400 (1949)

P. Petreczky, F. Karsch, E. Laermann, S. Stickan, I. Wetzorke, Nucl. Phys. Proc. Suppl. **106**, 513 (2002)

R.D. Pisarski, Phys. Rev. Lett. **63**, 1129 (1989)

G.M. Prosperi, M. Raciti, C. Simolo, Prog. Part. Nucl. Phys. **58**, 387 (2007)

Y.B. Rumer, A.I. Fet, *Group Theory and Quantum Fields* (Nauka, Moscow, 1977)

E.E. Salpeter, H.A. Bethe, Phys. Rev. **84**, 1232 (1951)

E.E. Salpeter, Phys. Rev. **87**, 328 (1952)

J. Schleede, A. Filinov, M. Bonitz, H. Feshkse, Contrib. Plasma Phys. **52**, 819 (2012)

E.V. Shuryak, I. Zahed, Phys. Rev. C **70**, 21901 (2004). Phys. Rev. D **70**, 54507 (2004)

E. Shuryak, Prog. Part. Nucl. Phys. **62**, 48 (2009); ibid. **53**, 273 (2004)

E.V. Shuryak, Prog. Part. Nucl. **62**, 48 (2009)

J. Schukraft, (2011). arXiv:1112.0550, arXiv:hep-ex

S. Schweber, *An Introduction to Relativistic Quantum Field Theory* (Izd. Inostr. Lit, Moscow, 1963)

C. Shen, U. Heinz, Phys. Rev. C **85**, 54902 (2012)

D.V. Shirkov, I.L. Solovtsov, Phys. Rev. Lett. **79**, 1209 (1997); Nucl. Phys. Proc. Suppl. **64**, 106 (1998)

S.K. Wong, Nuovo Cimento A **65**, 689 (1970)

V.I. Yukalov, E.P. Yukalova, Physica A **243**, 382 (1997); Fiz. Elem. Chastits At. Yadra **28**, 89 (1997)

S. Cho, I. Zahed, Phys. Rev. C **79**, 44911 (2009); ibid. **80**, 14906 (2009); ibid. **82**, 54907 (2010); ibid. **82**, 64904 (2010); ibid. **82**, 44905 (2010); K. Dusling, I. Zahed. Nucl. Phys. A **833**, 172 (2010)

V.M. Zamalin, G.E. Norman, U.S.S.R. Comp, Math. Math. Phys. **13**, 169 (1973)

V.M. Zamalin, G.E. Norman, V.S. Filinov, *The Monte Carlo Method in Statistical Thermodynamics* (Nauka, Moscow, 1977). (in Russian)

B.V. Zelener, G.E. Norman, V.S. Filinov, *Perturbation Theory and Pseudopotential in Statistical Thermodynamics* (Nauka, Moscow, 1981). (in Russian)

D.N. Zubarev, *Nonequilibrium Statistical Thermodynamics* (Plenum Press, New York London, 1974)

Chapter 12
Transport Properties of Quark–Gluon Plasmas

Hydrodynamic simulations of relativistic heavy-ion collisions require knowledge, not only of the thermodynamic properties of the QGP, but also of the transport properties. While significant progress in calculations of the thermodynamic properties of QGP has been made in recent years, the transport properties are still poorly accessible using lattice QCD (Meyer 2007). It is therefore crucial to devise reliable and manageable theoretical tools for a quantitative description of non-Abelian QGP, both in and out of equilibrium. Unfortunately, the DPIMC method itself cannot directly predict transport properties. To simulate quantum QGP transport and thermodynamic properties in a unified approach, it is reasonable to combine the path integral and Wigner (in phase space) formulations of quantum mechanics. As will be shown below, to calculate kinetic coefficients according to the quantum Kubo formulas, the DPIMC method can be used to generate initial conditions (equilibrium quasiparticle configurations) for virtual dynamical Wigner trajectories in color phase space, describing the time evolution for space, momentum, and color variables. Correlation functions and kinetic coefficients are calculated as averages of Weyl's symbols for dynamic quantum operators along these trajectories. The basic ideas of this approach have been published in (Filinov et al. 2011). This method is applicable to systems with arbitrarily strong interactions. The self-diffusion coefficient and viscosity of the strongly coupled QGP have been calculated using this approach.

12.1 Wigner Approach to Quantum Dynamics

There are several different ways to formulate non-relativistic quantum mechanics. The first one is a standard method, making use of operators in Hilbert space. The second method uses continuous integrals (or path integrals) for quantization. These integrals were developed by Wiener (1923) and Feynman (Feynman and Hibbs 1965, but see also Kleinert 2004). This method is very useful in quantum thermodynamics and quantum field theory. The third way to formulate quantum mechanics was developed by Wigner (1932). It is based on a quasi-distribution function (called the

W. Ebeling et al., *Quantum Statistics of Dense Gases and Nonideal Plasmas*, Springer Series in Plasma Science and Technology, DOI 10.1007/978-3-319-66637-2_12

Wigner function) in phase space and Weyl's correspondence between ordinary c-number functions in phase space and quantum mechanical operators in Hilbert space (for an overview, see Tatarskii 1983).

The construction of relativistic quantum mechanics encountered many other well known difficulties, which finally led to the formulation of quantum field theory. However, before that, in their work (Newton and Wigner 1949), Newton and Wigner developed a much simpler theory, a relativistic particle theory that was nevertheless in conflict with quantum field theory in some points. In particular, this theory did not take into account the production or annihilation of particles. Notwithstanding, following the ideas of (Newton and Wigner 1949; Zav'jalov 2000) constructed the relativistic Wigner functions for free Wigner pseudo-particles, and further development of this formalism may be useful for treating relativistic quantum systems with strong interactions. Below we consider a generalization of this approach (Newton and Wigner 1949; Zav'jalov 2000) for Wigner pseudo-particles in an external potential field.

We shall derive a time-evolution equation for the relativistic Wigner function and discuss the Weyl correspondence between functions in phase space and quantum operators. As an example, we apply the developed formalism to one of the simplest relativistic systems, namely, the 1D relativistic quantum harmonic oscillator. However, due to the complexity of the problem even for this simplest system, only a numerical treatment is possible. So to consider its time evolution, a new approach combining Monte Carlo and molecular dynamics methods must be developed. We discuss the time evolution of the Wigner function, the momentum and position distributions, and the average values of the momentum and position, their dispersions, and the energy. To analyze the influence of relativistic dynamic effects, the results of these simulations will be compared with those for a 'semi-relativistic harmonic oscillator' (Larkin and Filinov 2013), where the initial non-relativistic expressions for the Wigner function are used.

Wigner Function in Non-relativistic Quantum Mechanics

Consider a spinless non-relativistic particle with mass m in three-dimensional space. The Hamiltonian for this particle has the form

$$\hat{H} = \frac{\hat{p}^2}{2m} + V(\hat{x}). \tag{12.1}$$

The usual approach to quantum mechanics uses wave functions, for example, in the coordinate representation $\Psi(x,t)$. The time evolution is described by the Schrödinger equation:

$$\mathrm{i}\hbar\Psi(x,t) = \hat{H}\Psi(x,t), \quad \Psi(x,0) = \psi(x), \tag{12.2}$$

where $\hat{x} = x$ and $\hat{p} = -\mathrm{i}\hbar\nabla$ are the position and momentum operators, respectively. This equation describing complex-valued functions is not always convenient for analytical or numerical considerations. The alternative approach, suggested by Wigner, uses real functions of quasi-probability in phase space, known as the Wigner functions (Wigner 1932; Tatarskii 1983):

$$W(p,x,t) = \frac{1}{(2\pi\hbar)^3}\int \Psi(x-\xi/2,t)\Psi^*(x+\xi/2,t)\mathrm{e}^{\mathrm{i}p\xi/\hbar}\mathrm{d}^3\xi. \qquad (12.3)$$

The time evolution of the quantum system in Wigner's formalism is described by the Wigner–Liouville equation, which is a generalization of the classical Liouville equation for the distribution function in phase space:

$$\frac{\partial W(p,x,t)}{\partial t} + \frac{p}{m}\frac{\partial W(p,x,t)}{\partial x} - \frac{\partial V(x)}{\partial x}\frac{\partial W(p,x,t)}{\partial p} = \int W(p-s,x,t)\omega(s,x)\mathrm{d}^3 s, \qquad (12.4)$$

with the initial condition $W(p,x,0) = W_0(p,x)$. Here the function $\omega(s,x)$ is defined by

$$\omega(s,x) = \frac{2}{(\pi\hbar^2)^3}\int V(x-x')\sin\frac{2sx'}{\hbar}\mathrm{d}^3x' - \frac{\partial V(x)}{\partial x}\frac{\mathrm{d}\delta(x)}{\mathrm{d}s}. \qquad (12.5)$$

The average value of the operator $\hat{A}$ in the quantum state $|\Psi\rangle$ can be written in Wigner's formalism as

$$\langle\Psi|\hat{A}|\Psi\rangle(t) = \int W(p,x,t)A(p,x)\mathrm{d}^3p\,\mathrm{d}^3x, \qquad (12.6)$$

similar to average values in thermodynamics. Here $A(p,x)$ is the Weyl symbol of the operator $\hat{A}$, defined by the expression

$$A(p,x) = \int \langle x-\xi/2|\hat{A}|x+\xi/2\rangle \mathrm{e}^{\mathrm{i}p\xi/\hbar}\mathrm{d}^3\xi. \qquad (12.7)$$

The Wigner function can be considered as a quantum generalization of the probabilistic distribution function in phase space. In the classical limit, the Eq. (12.4) reduces to the well known Liouville equation. Average values of operators can be calculated as average values of the Weyl symbols. However, this analogy with the probabilistic distribution function is not complete: the Wigner function may be both positive and negative.

Restrictions on the Initial Condition

To find solutions of (12.4), the initial function $W_0(x,p)$ has to be taken according to the definition (12.3). However, this definition imposes a certain restriction on the choice of possible functions in phase space. This is easily illustrated for the harmonic oscillator, i.e., a particle in a potential field of type $V(x) = V_0 + V_1x + V_2x^2$. In this case the Wigner–Liouville equation (12.4) reduces to the form

$$\frac{\partial W(x, p; t)}{\partial t} + \frac{p}{m}\frac{\partial W(x, p; t)}{\partial x} - V'_x(x)\frac{\partial W(x, p; t)}{\partial p} = 0, \tag{12.8}$$

and coincides with the classical Liouville equation. However, the solution of the quantum harmonic oscillator is totally different from the classical one. Consequently, the quantum and classical solutions of this task are distinguished from one another by the choice of initial condition $W_0(x, p)$. So we need an additional condition to choose the classical or quantum solution. For the density matrix of a pure state, this condition can be formulated as follows (Tatarskii 1983):

$$\frac{\partial^2}{\partial x_1 \partial x_2} \ln \rho(x_1, x_2; 0) = 0, \tag{12.9}$$

or in the Wigner representation,

$$\frac{\partial^2}{\partial x_1 \partial x_2} \ln \int_{-\infty}^{\infty} W_0\left(\frac{x_1 + x_2}{2}, p\right) \mathrm{e}^{\mathrm{i}p(x_1 - x_2)/\hbar} \mathrm{d}p = 0. \tag{12.10}$$

Another more complicated condition on the choice $W_0(x, p)$, which must be equivalent, can be found in Klimontovich (1956). One of the physical interpretations of this additional condition is connected with the requirement that the mean-square coordinate and momentum deviations should satisfy Heisenberg's uncertainty principle at time zero.

Non-relativistic Harmonic Oscillator

A non-relativistic harmonic oscillator with mass m and circular frequency ω has the Hamiltonian

$$\hat{H} = \frac{\hat{p}^2}{2m} + \frac{m\omega^2\hat{x}^2}{2}. \tag{12.11}$$

In the case of the harmonic potential $V(x) = m\omega^2x^2/2$, the Wigner–Liouville equation (12.4) has the simple form

$$\frac{\partial W(x, p; t)}{\partial t} + \frac{p}{m}\frac{\partial W(x, p; t)}{\partial x} - m\omega^2 x\frac{\partial W(x, p; t)}{\partial p} = 0. \tag{12.12}$$

As the initial condition for (12.12), we consider a coherent state of the harmonic oscillator, with wave function at $t = 0$ given by (Landau and Lifshitz 2008)

$$\Psi(x, 0) = \left(\frac{m\omega}{\pi\hbar}\right)^{1/4} \exp\left[\frac{\mathrm{i}\tilde{p}x}{\hbar} - \frac{m\omega(x - \tilde{x})^2}{2\hbar}\right] \exp\left(-\frac{\mathrm{i}\tilde{p}\tilde{x}}{2\hbar}\right).$$

Here $\tilde{p}$ and $\tilde{x}$ are the average values of the momentum and coordinate at the initial moment $t = 0$. According to the definition (12.3), the initial Wigner function is defined by

$$W_0(x, p) = \frac{1}{\pi\hbar} \exp\left[-\frac{m\omega(x-\tilde{x})^2}{\hbar} - \frac{(p-\tilde{p})^2}{\hbar m\omega}\right]. \tag{12.13}$$

The solution of (12.12) can be written in the form

$$\begin{aligned} W(x, p; t) &= \int \delta\big(p - \bar{p}(t; 0, p_0, x_0)\big)\delta\big(x - \bar{x}(t; 0, p_0, x_0)\big) W_0(x_0, p_0)\, \mathrm{d}p_0\, \mathrm{d}x_0 \\ &= \frac{1}{\sqrt{2\pi(\hbar/2m\omega)}} \exp\left\{-\frac{m\omega}{\hbar}\left[x - \left(\tilde{x}\cos\omega t + \frac{\tilde{p}}{m\omega}\sin\omega t\right)\right]^2\right\} \\ &\quad \times \exp\left\{-\frac{1}{\hbar m\omega}\left[p - \left(\tilde{p}\cos\omega t - m\omega\tilde{x}\sin\omega t\right)\right]^2\right\}, \end{aligned} \tag{12.14}$$

where $\bar{p}(t)$ and $\bar{x}(t)$ are virtual trajectories defined by the Hamilton function

$$H(p, x) = \frac{p^2}{2m} + \frac{m\omega^2 x^2}{2},$$

together with the related Hamilton equations and initial conditions $\bar{p}(0) = p_0$, $\bar{x}(0) = x_0$:

$$\bar{p}(t) = p_0 \cos\omega t - m\omega x_0 \sin\omega t, \qquad \bar{x}(t) = \frac{p_0}{m\omega}\sin\omega t + x_0\cos\omega t.$$

From (12.14), it follows that, in a coherent state of the harmonic oscillator, the average momentum and coordinate satisfy the classical law of motion

$$\langle p(t)\rangle = \tilde{p}\cos\omega t - m\omega\tilde{x}\sin\omega t, \qquad \langle x(t)\rangle = \tilde{x}\cos\omega t + \frac{\tilde{p}}{m\omega}\sin\omega t,$$

while the standard deviations of the momentum and coordinate are constant:

$$\langle \delta p^2\rangle = \frac{\hbar m\omega}{2}, \qquad \langle \delta x^2\rangle = \frac{\hbar}{2m\omega}.$$

Furthermore, Heisenberg's uncertainty formula assumes its minimum value:

$$\langle \delta p^2\rangle\langle \delta x^2\rangle = \hbar^2/4.$$

The average energy is constant, equal to

$$E = \frac{\hbar\omega}{2} + \frac{\tilde{p}^2}{2m} + \frac{m\omega^2\tilde{x}^2}{2}.$$

The more general case of composite states, slightly different from pure coherent states, is considered in Filinov et al. (1995).

Relativistic Wigner Function

We now consider a spinless relativistic particle with mass m in an external potential field $V(x)$. Generally speaking, this particle should be considered in the framework of quantum field theory, allowing for particle production and annihilation. The one-particle relativistic theory, introduced by Newton and Wigner in Newton and Wigner (1949), is not compatible with quantum field theory, but it may be used in practice in applications to relativistic systems of particles.

Basic Concepts and Formulas

We start with some concepts and results from Newton and Wigner (1949), considering a relativistic spinless particle in an external potential field with Hamiltonian

$$\hat{H} = \omega(\hat{p}) + V(\hat{x}), \tag{12.15}$$

where $\hat{p}$ and $\hat{x}$ are the momentum operator and the position operator, respectively (see below), and $\omega(\hat{p}) = \sqrt{\hat{p}^2c^2 + m^2c^4}$ is the well known expression for the kinetic energy. We introduce eigenstates $|p\rangle$ of the momentum operator:

$$\hat{p}|p'\rangle = p'|p'\rangle, \tag{12.16}$$

which has the relativistically invariant normalization (Rumer and Fet 1977):

$$\langle p|p'\rangle = (2\pi\hbar)^3 \left[\frac{\omega(p)}{mc^2}\frac{\omega(p')}{mc^2}\right]^{1/2} \delta(p - p'), \tag{12.17}$$

and satisfies the completeness condition

$$1 = \int \frac{\mathrm{d}^3 p}{(2\pi\hbar)^3}\frac{mc^2}{\omega(p)}|p\rangle\langle p|. \tag{12.18}$$

We define wave functions in the momentum representation $\Psi(p)$ (omitting the explicit dependence on time) of the state $|\Psi\rangle$:

$$\Psi(p) = \langle p|\Psi\rangle. \tag{12.19}$$

In particular, the wave function corresponding to the state with momentum p' has the form (12.17). The scalar product $\langle\Phi|\Psi\rangle$ can be written in terms of the wave functions in the form (Newton and Wigner 1949)

$$\langle\Phi|\Psi\rangle = \int \frac{\mathrm{d}^3 p}{(2\pi\hbar)^3}\frac{mc^2}{\omega(p)}\Phi^*(p)\Psi(p). \tag{12.20}$$

The square of the absolute value $|\langle\Phi|\Psi\rangle|^2$ is proportional to the transition probability from the state $|\Psi\rangle$ to the state $|\Phi\rangle$. This probability does not depend on the reference frame (Newton and Wigner 1949; Rumer and Fet 1977), and the scalar product of these states must be invariant under the Lorentz transformations. The definition (12.20) does satisfy the requirement of Lorentz invariance.

As usual, we define operator rules for wave functions (on the left, we mean an operator on functions, on the right, an operator on states):

$$\hat{A}\Psi(p) = \langle p|\hat{A}|\Psi\rangle, \tag{12.21}$$

where the momentum operator simply multiplies by the momentum, i.e., $\hat{p} = p$. The position operator has a more complicated form and differs from the non-relativistic case (Newton and Wigner 1949):

$$\hat{x}\Psi(p) = \left[\mathrm{i}\hbar\frac{\partial}{\partial p} - \mathrm{i}\hbar\frac{pc^2}{2\omega^2(p)} \right]\Psi(p). \tag{12.22}$$

The quantum average value of any operator $\hat{A} = A(\hat{p}, \hat{x})$ can be calculated in terms of the wave function by

$$\langle\Psi|\hat{A}|\Psi\rangle = \int \frac{\mathrm{d}^3 p}{(2\pi\hbar)^3}\frac{mc^2}{\omega(p)}\Psi^*(p)A(p,\hat{x})\Psi(p), \tag{12.23}$$

where the operator $\hat{x}$ is defined by (12.22). Note that if the operator is relativistically invariant, its average value is also relativistically invariant. Indeed, when the state $|\Psi\rangle$ transforms under a Lorentz transformation under the action of a unitary operator Λ, so that $|\Psi_\Lambda\rangle = \Lambda|\Psi\rangle$, this operator must commute with the relativistically invariant operator $\hat{A}$. Using commutativity of these operators and unitarity of Λ, one deduces that

$$\langle\Psi_\Lambda|\hat{A}|\Psi_\Lambda\rangle = \langle\Psi|\Lambda^{-1}\hat{A}\Lambda|\Psi\rangle = \langle\Psi|\hat{A}|\Psi\rangle.$$

The time evolution of the system can be described by the evolution operator (Newton and Wigner 1949)

$$\Psi(p,t) = \mathrm{e}^{-\mathrm{i}\hat{H}t/\hbar}\psi(p), \tag{12.24}$$

so the wave functions have to satisfy the Schrödinger equation:

$$\mathrm{i}\hbar\frac{\partial}{\partial t}\Psi(p,t) = \hat{H}\Psi(p,t), \qquad \Psi(p,0) = \psi(p). \tag{12.25}$$

In contrast to the scalar products and transition probabilities, this equation is not relativistically invariant, since the time plays an obvious role, and the Hamiltonian must transform like an energy under Lorentz transformations.

Following (Zav'jalov 2000), it is convenient to redefine the wave functions $\Psi(p,t)$ according to

$$\Psi'(p,t) = \left[\frac{mc^2}{\omega(p)}\right]^{1/2} \Psi(p,t). \tag{12.26}$$

The expression for the momentum operator remains unchanged, i.e., we still have $\hat{p}\Psi'(p,t) = p\Psi'(p,t)$, but the position operator for these functions has the simpler form

$$\hat{x}\Psi'(p) = \mathrm{i}\hbar\frac{\partial}{\partial p}\Psi'(p), \tag{12.27}$$

coinciding with the non-relativistic expression. The scalar product $\langle\Phi|\Psi\rangle$ is simpler in terms of Ψ' functions:

$$\langle\Phi|\Psi\rangle = \int \frac{\mathrm{d}^3 p}{(2\pi\hbar)^3}\Phi'^*(p)\Psi'(p), \tag{12.28}$$

with the usual integral measure $\mathrm{d}^3 p/(2\pi\hbar)^3$. An expression for the quantum average value of any operator $\hat{A} = A(\hat{p},\hat{x})$ also simplifies:

$$\langle\Psi|\hat{A}|\Psi\rangle = \int \frac{\mathrm{d}^3 p}{(2\pi\hbar)^3}\Psi'^*(p)A(p,\hat{x})\Psi'(p), \tag{12.29}$$

where the operator $\hat{x}$ is defined by (12.27). The time evolution of the system is described by the related Schrödinger equation, as above:

$$\mathrm{i}\hbar\frac{\partial}{\partial t}\Psi'(p,t) = \hat{H}\Psi'(p,t), \qquad \Psi'(p,0) = \psi'(p). \tag{12.30}$$

Wigner Function and Average Values of Operators

In Zav'jalov (2000), the definition of the Wigner function was generalized to free relativistic particles. This function satisfies a Lorentz invariant evolution equation which coincides nominally with the classical Liouville equation for free relativistic particles. This result is easily generalized to a particle in an external potential field. To do this, we define the Wigner function of such a system by

$$W(p,x,t) = \int \frac{\mathrm{d}^3 p_1 \mathrm{d}^3 p_2}{(2\pi\hbar)^6}\delta(p - p_1 \dotplus p_2)\mathrm{e}^{-\mathrm{i}(p_1 - p_2)x/\hbar}\Psi'^*(p_1,t)\Psi'(p_2,t), \tag{12.31}$$

where the wave functions $\Psi'(p_1,t)$ and $\Psi'(p_2,t)$ satisfy the Schrödinger equation (12.30). Here $p_1 \dotplus p_2$ is the sum on the so-called 'mass hyperboloid' $E^2 - p^2c^2 = m^2c^4$:

$$p_1 \dot{+} p_2 = \frac{mc^2(p_1 + p_2)}{\sqrt{2\left[m^2c^4 + \omega(p_1)\omega(p_2) - p_1 p_2 c^2\right]}}. \tag{12.32}$$

Note the following useful expression:

$$\omega(p_1 \dot{+} p_2) = \frac{mc^2\left[\omega(p_1) + \omega(p_2)\right]}{\sqrt{2\left[m^2c^4 + \omega(p_1)\omega(p_2) - p_1 p_2 c^2\right]}}. \tag{12.33}$$

Using the Fourier transform with respect to x in (12.31), it is easy to obtain the reverse transformation to the wave functions:

$$\Psi'^*(p_1, t)\Psi'(p_2, t) = (2\pi\hbar)^3 \, |J(p_1, p_2)| \int \mathrm{d}^3x \, W(p_1 \dot{+} p_2, x, t)\mathrm{e}^{\mathrm{i}(p_1 - p_2)x/\hbar}, \tag{12.34}$$

where $J(p_1, p_2) = \det |f_{ij}|$ is the Jacobian of the transformation from $p = p_1 \dot{+} p_2$ and $\rho = p_1 - p_2$ to p_1 and p_2:

$$f_{11} = \frac{\partial(p_1 \dot{+} p_2)}{\partial p_1}, \quad f_{12} = \frac{\partial(p_1 \dot{+} p_2)}{\partial p_2}, \quad f_{21} = \frac{\partial \rho}{\partial p_1}, \quad f_{22} = \frac{\partial \rho}{\partial p_1}.$$

We require an expression for the average value of any operator $\hat{A} = A(\hat{p}, \hat{x})$ in terms of the relativistic Wigner function (12.31). To obtain this, we use the completeness condition (12.18), definitions (12.19) and (12.26), and the expression (12.34):

$$\begin{aligned} \langle\Psi|\hat{A}|\Psi\rangle(t) = \int \frac{\mathrm{d}^3 p_1 \mathrm{d}^3 p_2}{(2\pi\hbar)^3} \mathrm{d}^3x \left[\frac{mc^2}{\omega(p_1)} \frac{mc^2}{\omega(p_2)}\right]^{1/2} \mathrm{e}^{\mathrm{i}(p_1 - p_2)x/\hbar} \\ \times W(p_1 \dot{+} p_2, x, t)\langle p_1|\hat{A}|p_2\rangle \, |J(p_1, p_2)| \,. \end{aligned} \tag{12.35}$$

After the change of variables $p = p_1 \dot{+} p_2$, $\rho = p_1 - p_2$, we have

$$\mathrm{d}^3 p \mathrm{d}^3 \rho = |J(p_1, p_2)| \mathrm{d}^3 p_1 \mathrm{d}^3 \rho.$$

The final formula for the quantum average value of the operator $\hat{A}$ is

$$\langle\Psi|\hat{A}|\Psi\rangle(t) = \int \mathrm{d}^3 p \, \mathrm{d}^3 x \, W(p, x, t) A(p, x), \tag{12.36}$$

where the function $A(p, x)$ is a generalization of the non-relativistic Weyl symbol of the operator $\hat{A}$ (12.7):

$$A(p, x) = \int \frac{\mathrm{d}^3 \rho}{(2\pi\hbar)^3} \left[\frac{mc^2}{\omega(p_1(p, \rho))} \frac{mc^2}{\omega(p_2(p, \rho))}\right]^{1/2} \mathrm{e}^{\mathrm{i}\rho x/\hbar}\langle p_1(p, \rho)|\hat{A}|p_2(p, \rho)\rangle. \tag{12.37}$$

It is easy to show that, in the particular cases, $\hat{A} = A(\hat{p})$ and $\hat{A} = A(\hat{x})$, the Weyl symbols of these operators are $A(p)$ and $A(x)$, using (12.37) and (12.22).

From the above, one can see that the relativistic Wigner function is something like a distribution function in the phase space. As in non-relativistic quantum mechanics, this function has real values, but may be negative.

Evolution of the Wigner Function

The time evolution of a non-relativistic particle in an external field is described by the Wigner–Liouville equation (12.4). To derive its relativistic analog, we differentiate relativistic Wigner function (12.31) with respect to time, taking into account the Schrödinger equation for wave functions (12.30):

$$\frac{\partial W(p,x,t)}{\partial t} = \int \frac{\mathrm{d}^3 p_1 \mathrm{d}^3 p_2}{(2\pi\hbar)^6} \delta(p - p_1 + p_2) \mathrm{e}^{-\mathrm{i}(p_1 - p_2)x/\hbar}$$
$$\times \frac{\mathrm{i}}{\hbar} \left\{ \left[\omega(p_1) - \omega(p_2) \right] + \left[V\left(\mathrm{i}\hbar \frac{\partial}{\partial p_1}\right) - V\left(\mathrm{i}\hbar \frac{\partial}{\partial p_2}\right) \right] \right\} \Psi'^*(p_1, t) \Psi'(p_2, t). \tag{12.38}$$

Using

$$\omega(p_1) - \omega(p_2) = \frac{(p_1 + p_2)c^2}{\omega(p_1 + p_2)}(p_1 - p_2), \tag{12.39}$$

we obtain the following expression for the first term in (12.38):

$$\int \frac{\mathrm{d}^3 p_1 \mathrm{d}^3 p_2}{(2\pi\hbar)^6} \delta(p - p_1 + p_2) \mathrm{e}^{-\mathrm{i}(p_1 - p_2)x/\hbar} \frac{\mathrm{i}}{\hbar} \left[\omega(p_1) - \omega(p_2) \right]$$
$$= -\frac{pc^2}{\omega(p)} \frac{W(p,x,t)}{\partial x}. \tag{12.40}$$

In fact, this result was already derived in Zav'jalov (2000). We then substitute the reverse formula (12.34) into the second term of (12.38) and add

$$\frac{\partial V(x)}{\partial x} \frac{W(p,x,t)}{\partial p}$$

to the left- and right-hand sides. Assembling this result and (12.40), we obtain an integro-differential equation of evolution:

$$\frac{\partial W(p,x,t)}{\partial t} + \frac{pc^2}{\omega(p)} \frac{W(p,x,t)}{\partial x} - \frac{\partial V(x)}{\partial x} \frac{W(p,x,t)}{\partial p} = \Phi(p,x,t). \tag{12.41}$$

Here the right-hand side has the operator form

$$\Phi(p,x,t) = -\frac{\partial V(x)}{\partial x}\frac{W(p,x,t)}{\partial p}$$
$$+\frac{\mathrm{i}}{\hbar}\int\frac{\mathrm{d}^3p_1\mathrm{d}^3p_2}{(2\pi\hbar)^3}\mathrm{d}^3y\,\delta(p-p_1\dot{+}p_2)\mathrm{e}^{-\mathrm{i}(p_1-p_2)x/\hbar}$$
$$\times\left[V\left(\mathrm{i}\hbar\frac{\partial}{\partial p_1}\right)-V\left(\mathrm{i}\hbar\frac{\partial}{\partial p_2}\right)\right]\mathrm{e}^{\mathrm{i}(p_1-p_2)y/\hbar}\,|J(p_1,p_2)|\,W(p_1\dot{+}p_2,y,t). \tag{12.42}$$

As mentioned, this is an integro-differential form of the evolution equation. Let us transform it into an integral form, as in the non-relativistic case (Tatarskii 1983). Without the right-hand side, Eq. (12.41) has quasi-characteristics which are solutions of the classical Hamilton equations for a particle in the potential field:

$$\frac{\mathrm{d}\bar{p}}{\mathrm{d}t} = -\frac{\partial V(x)}{\partial \bar{x}}, \qquad \frac{\mathrm{d}\bar{x}}{\mathrm{d}t} = \frac{\bar{p}c^2}{\omega(\bar{p})}. \tag{12.43}$$

We now define the dynamical Green function:

$$G(p,x,t;\,p',x',t') = \delta\big(p-\bar{p}(t;t',p',x',t')\big)\delta\big(x-\bar{x}(t;t',p',x',t')\big), \tag{12.44}$$

where $\bar{p}(t;\,p',x',t')$ and $\bar{x}(t;\,p',x',t')$ are solutions of these Hamilton equations under the initial conditions $\bar{p}(t)\big|_{t=t'} = p', \bar{x}(t)\big|_{t=t'} = x'$. These trajectories are called virtual trajectories. Then (12.41) with initial condition $W(p,x,t)\big|_{t=0} = W_0(p,x)$ can be represented in the integral form

$$W(p,x,t) = \int\mathrm{d}^3p_0\mathrm{d}^3x_0G(p,x,t;\,p_0,x_0,0)W_0(p_0,x_0)$$
$$+\int_0^t\mathrm{d}t'\int\mathrm{d}^3p'\mathrm{d}^3x'G(p,x,t;\,p',x',t')\Phi(p',x',t')\,, \tag{12.45}$$

where $\Phi(p',x',t')$ is defined by (12.42).

One can transform the right-hand side of this equation from an operator form to an integral one by getting rid of the position operators $\hat{x} = \mathrm{i}\hbar\partial/\partial p$. To do this, we consider the direct and reverse Fourier transform of the potential function $V(x)$:

$$V(x) = \int\frac{\mathrm{d}^3q}{(2\pi\hbar)^3}\tilde{V}(q)\mathrm{e}^{-\mathrm{i}qx/\hbar}\,, \qquad \tilde{V}(q) = \int\mathrm{d}^3x\,V(x)\mathrm{e}^{\mathrm{i}qx/\hbar}\,. \tag{12.46}$$

Note that the operator $e^{-iq\hat{x}/\hbar}$ is a translation in the momentum space (Zubarev 1974), i.e., $e^{-iq\hat{x}/\hbar} f(p) = f(p+q)$. So, using this result and the Fourier transform (12.46), we obtain the expression

$$V(i\hbar\partial/\partial p) f(p) = \int \frac{d^3 q}{(2\pi\hbar)^3} \tilde{V}(q) f(p+q) \,. \tag{12.47}$$

Using this formula in (12.42) and after simple calculations, we get the right-hand side of the evolution equation in integral form:

$$\begin{aligned} \Phi(p,x,t) = & \frac{1}{\hbar}\frac{\partial V(x)}{\partial x} \int \frac{d^3 q d^3 \xi}{(2\pi\hbar)^3} \xi \sin\left[\frac{(p-q)\xi}{\hbar}\right] W(q,x,t) \\ & -\frac{2}{\hbar} \int \frac{d^3 p_1 d^3 p_2}{(2\pi\hbar)^6} d^3 q \, d^3 y \, \delta(p - p_1 \dotplus p_2) \\ & \times \sin\left[\frac{qy}{\hbar} + \frac{(p_2 - p_1)(x-y)}{\hbar}\right] \tilde{V}(q) \, |J(p_1+q, p_2)| \\ & \times W\big((p_1+q) \dotplus p_2, y, t\big) \,. \end{aligned} \tag{12.48}$$

The solution of this integral equation can be presented in the form of an iterative series in which the first term describes the classical relativistic virtual trajectories. Subsequent terms are defined by the contribution of the virtual relativistic trajectories with momentum jumps, a consequence of the convolution structure of the right-hand side $\Phi(p,x,t)$. In the quasi-classical limit $\hbar \to 0$, the function $\Phi(p,x,t)$ tends to zero, so the Wigner function propagates along classical relativistic trajectories in phase space. For more detail, the reader is referred to Filinov et al. (1995), where non-relativistic particles are considered.

12.2 Relativistic Quantum Harmonic Oscillator

The simplest application of the developed formalism is the time evolution of the 1D relativistic quantum harmonic oscillator. Due to the complex structure of the evolution equation for the Wigner function, only a numerical treatment is possible, even for this simplest system. To solve the relativistic Wigner–Liouville equation, a combination of the Monte Carlo procedure and molecular dynamics method was used.

To analyze the influence of relativistic effects, the initial non-relativistic and relativistic Wigner functions related to the ground state of the harmonic oscillator were used, and then allowed to evolve according to the relativistic dynamics. It turns out that the relativistic treatment results in the appearance of new physical effects as compared with the non-relativistic case. One interesting feature is the complete change in the shape of the momentum and coordinate distribution functions, as well as the formation of 'unexpected' protuberances. To analyze the influence of relativistic effects on the average values of quantum operators, the time dependence of the average momentum and position, their dispersions, and the energy were calculated and compared with those for the non-relativistic dynamics. The resulting relativistic time dilation of the oscillator can be useful for considering the lifetime of particle bound states in traps.

Relativistic Wigner Function and Virtual Trajectories

The relativistic harmonic oscillator in one-dimensional space has the following Hamiltonian:

$$\hat{H} = \omega(\hat{p}) + \frac{m\omega^2}{2}\hat{x}^2 \,, \tag{12.49}$$

where the position operator $\hat{x}$ acts on wave functions $\Psi(p)$ according to the formula (12.22). [The reader should not confuse the circular frequency ω and the kinetic energy $\omega(p)$!] The relativistic Wigner function is defined by the 1D version of the formula (12.31):

$$W(p,x,t) = \int \frac{\mathrm{d}p_1 \mathrm{d}p_2}{(2\pi\hbar)^2} \delta(p - p_1 + p_2) \mathrm{e}^{-\mathrm{i}(p_1-p_2)x/\hbar} \Psi'^*(p_1,t)\Psi'(p_2,t) \,. \tag{12.50}$$

Let us take the function that corresponds to coherent states of the non-relativistic harmonic oscillator in the momentum representation at time $t = 0$:

$$\Psi(p) = \left(\frac{4\pi\hbar}{m\omega}\right)^{1/4} \exp\left[-\frac{(p-p_0)^2}{2\hbar m\omega} - \frac{\mathrm{i}px_0}{\hbar} + \frac{\mathrm{i}p_0x_0}{2\hbar}\right] . \tag{12.51}$$

Here p_0, x_0 are given free parameters, in fact average values of the momentum and position for this oscillator.

The related non-relativistic Wigner function (12.3) of this state has the form of a Gaussian exponent:

$$W(p,x,t)\big|_{t=0} = \frac{1}{\pi\hbar} \exp\left[-\frac{(p-p_0)^2}{\hbar m\omega} - \frac{(x-x_0)^2}{\hbar/m\omega}\right] , \tag{12.52}$$

so it is a Wigner function of the ground state of the non-relativistic harmonic oscillator, shifted in the phase space to p_0 and x_0. Although no analytic expression can be obtained for the relativistic Wigner function (12.50), one can calculate this function numerically.

Comparing the results obtained from these initial quantum non-relativistic and relativistic Wigner functions, one can study the appearance and influence of relativistic effects under different physical conditions. The virtual trajectories of the relativistic oscillator are described by the system of Hamilton equations

$$\frac{\mathrm{d}\bar{p}}{\mathrm{d}t} = -m\omega^2\bar{x} , \qquad \frac{\mathrm{d}\bar{x}}{\mathrm{d}t} = \frac{\bar{p}c^2}{\sqrt{\bar{p}^2c^2 + m^2c^4}} , \tag{12.53}$$

with initial conditions $\bar{p}(t)\big|_{t=0} = p_0$ and $\bar{x}(t)\big|_{t=0} = x_0$.

The time evolution of the relativistic oscillator is described by the one-dimensional version of the integral equation (12.45). The solution of this equation can be presented in the form of an iterative series. However, the problem of calculating all the terms in this series is a very complicated one, so let us consider only the first term of the series, allowing nevertheless for contributions from infinite powers of the Planck constant:

$$W(p, x, t) = \int \mathrm{d}p_0\, \mathrm{d}x_0 G(p, x, t; p_0, x_0, 0) W_0(p_0, \mathrm{d}x_0) . \tag{12.54}$$

Here $W_0(p, x)$ can be chosen as a quantum relativistic (12.50) or non-relativistic (12.52) Wigner function at the initial time $t = 0$.

Numerical Simulations

To consider the time evolution of the oscillator, the following numerical procedure was used. The initial Wigner function (12.31) was considered as a probabilistic distribution of points (p_0, x_0) in phase space. To sample these points, the Monte Carlo procedure was used. Each point (p_0, x_0) was considered as the origin of the virtual dynamic trajectory, described by Hamilton's equations (12.53). The relativistic molecular dynamics method was used to solve these equations. The distribution of virtual trajectories in phase space allows us to obtain the Wigner function $W(x, p; t)$ at any time t. The average value of any quantum operator can be obtained by calculating the time dependencies of the Weyl symbol for the operator along the virtual trajectories and averaging over the ensemble of all trajectories. For the momentum, coordinate, energy, and mean-square values of the momentum and coordinate, these averages were obtained at each time in the relevant time evolution interval.

In the dynamical simulations, 10^5 virtual trajectories were generated and the following implicit finite-difference scheme with centering (Potter 1973) may be used:

$$\begin{aligned} p_{m(i+1)} &= p_{m(i)} - \frac{\tau}{2}[x_{m(i+1)} + x_{m(i)}] , \\ x_{m(i+1)} &= x_{m(i)} + z\frac{\tau}{2}\left[\frac{p_{m(i+1)}}{\sqrt{z(z + p_{m(i+1)}{}^2)}} + \frac{p_{m(i)}}{\sqrt{z(z + p_{m(i)}{}^2)}}\right] . \end{aligned} \tag{12.55}$$

Here $\tau = 0.01$ is a time step. This system of algebraic equations was mainly solved by a method of simple iterations. For numerical calculations, the following system of units was used:

$$p = \frac{\hbar\omega}{c}\sqrt{z}p_m , \qquad x = \frac{c}{\omega}\frac{x_m}{\sqrt{z}} , \qquad t = \frac{c^2}{\hbar\omega}t_m ,$$

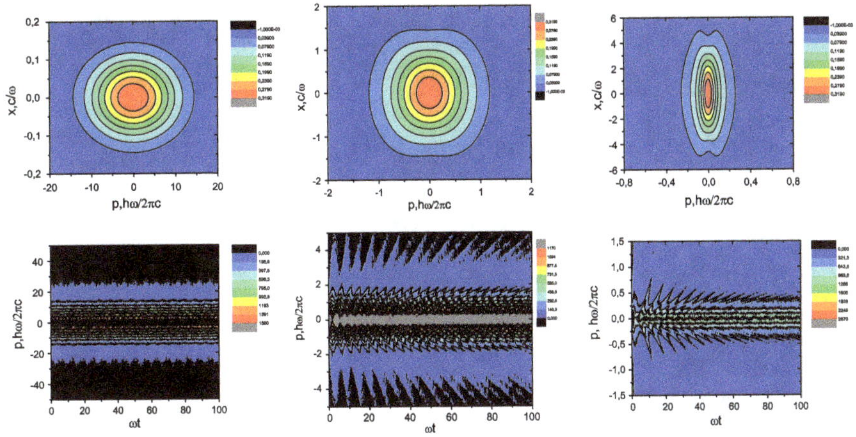

Fig. 12.1 Contour plots. *Upper row*: Wigner function at the initial time $t = 0$ for oscillators with $\hbar\omega = 0.01mc^2$ (*left*), $\hbar\omega = mc^2$ (*center*), $\hbar\omega = 10mc^2$ (*right*). *Lower row*: Time dependence of momentum distributions $W_{\rm d}(p, t)$. In all cases $p_0 = 0$, $x_0 = 0$ (Larkin and Filinov 2014)

where p_m, x_m, t_m are units in the scheme (12.55) for $\hbar = 1$, $c = 1$. However, all results below are written in the usual physical units. The physically dimensionless parameter $z = mc^2/\hbar\omega$ defines the degree of relativism of the oscillator. When $z \gg 1$, relativistic effects almost disappear.

Evolution of the Wigner Function

We now consider the differences between the initial relativistic Wigner function (12.50) and the non-relativistic expression (12.52). For simplicity, let us choose the initial values $x_0 = 0$ and $p_0 = 0$ [see (12.52)]. The initial Wigner function is represented in the upper row of Fig. 12.1 for three different values of the degree of relativism, i.e., the parameter $mc^2/\hbar\omega$ (in all these cases $p_0 = 0$, $x_0 = 0$). The upper left plot corresponds to an almost non-relativistic harmonic oscillator ($\hbar\omega = 0.01mc^2$). In this case, the Wigner function can be calculated from the formula (12.52). The upper central plot concerns the oscillator with $\hbar\omega = mc^2$, and the difference between relativistic and non-relativistic Wigner functions becomes significant. The function shrinks along the momentum axis and stretches along the position axis for large momentum values. This is especially clear in the case of the highly relativistic oscillator represented in the upper right plot ($\hbar\omega = 10mc^2$).

The time evolution of the momentum distribution $W_{\rm d}(p, t)$ is represented in the lower row of Fig. 12.1. This distribution is defined by the formula:

$$W_{\rm d}(p, t) = \int W(p, x, t)\,{\rm d}x \,. \tag{12.56}$$

Each plot in the lower row corresponds to evolution from the initial state represented above it. In the non-relativistic case (lower left contour), the distribution $W_{\rm d}(p, t)$ is

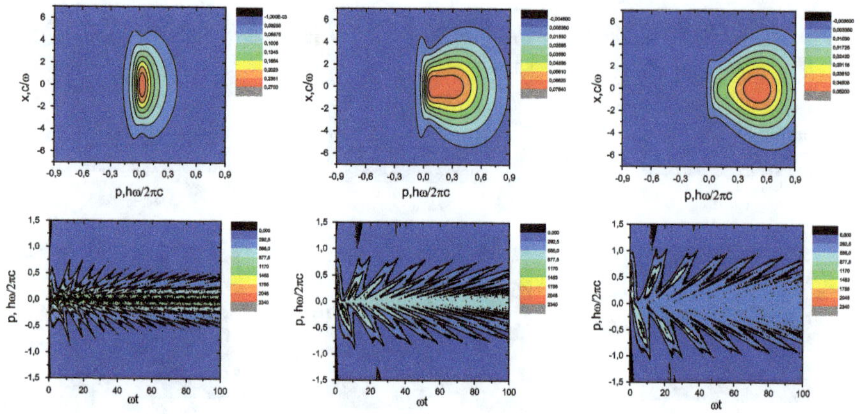

Fig. 12.2 Contour plots. *Upper row*: Wigner function at the initial time $t = 0$ for an oscillator with $\hbar\omega = 10mc^2$, $x_0 = 0$, and different values of p_0: $0.16\hbar\omega/c$ (*left*), $0.48\hbar\omega/c$ (*center*), $p\hbar\omega/c$ (*right*). *Lower row*: Time dependence of momentum distributions $W_{\rm d}(p, t)$ (Larkin and Filinov 2014)

almost constant, because the initial state (12.51) with $p_0 = 0, x_0 = 0$ is an eigenstate of the non-relativistic oscillator. As the parameter $\hbar\omega$ increases, relativistic effects become significant (lower center and right-hand plots). These effects cause broadening of the Wigner function in the phase space and the appearance of additional maxima of the distribution function at times $t \neq 0$. The last effect is caused by relativistic time dilation along the virtual trajectories with high energy, as described in detail in Larkin and Filinov (2013).

The initial Wigner functions for the highly relativistic oscillator, when $\hbar\omega = 10mc^2$, with $x_0 = 0$ and nonzero p_0 are shown in the upper row of Fig. 12.2. The upper left contour plot corresponds to $p_0 = 0.16\hbar\omega/c$, the upper center plot to $p_0 = 0.48\hbar\omega/c$, and the upper right to $p_0 = 0.64\hbar\omega/c$. In the relativistic case (12.50), the shape of the Wigner function depends heavily on the parameter p_0. This is an important difference with the non-relativistic Wigner function (12.52), which is simply shifted in the phase space by the values of x_0 and p_0.

The lower row of Fig. 12.2 shows the evolution of the momentum distribution, and each contour plot corresponds to the initial state depicted above it. Note the broadening of the distribution function once again. The evolution of the position distribution looks similar to the pattern for the momentum distribution, so we do not display the plots here.

The velocity distribution $W(v)$ related to (12.52) has an interesting shape, shown in Fig. 12.3 for oscillators with $z = 1$ (top) and $z = 0.1$ (bottom). Due to the transformation

$$v = \frac{pc}{\sqrt{p^2 + m^2c^2}} ,$$

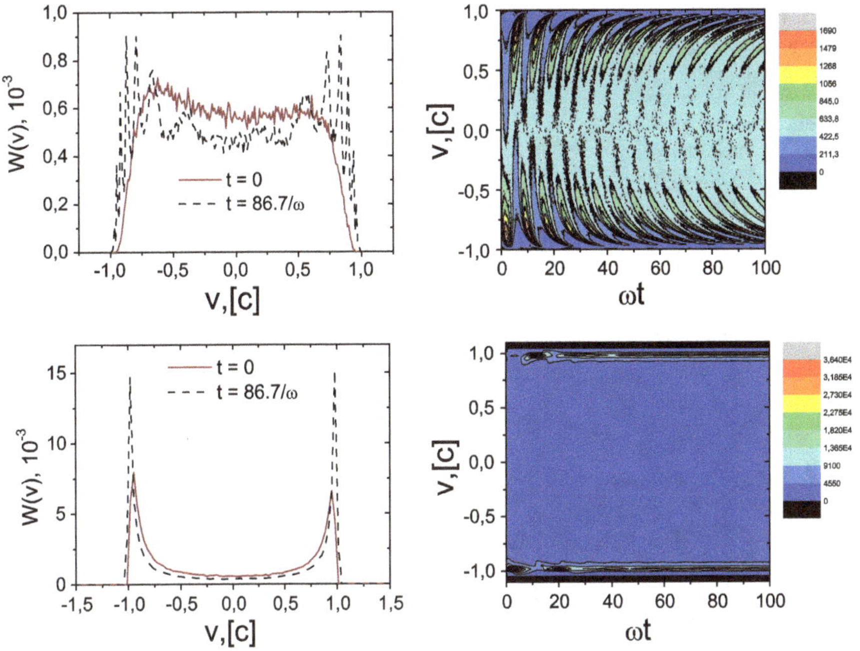

Fig. 12.3 Time evolution of the non-normalized velocity distributions $W(v)$ of an oscillator. *Top left*: $t = 0$ (*solid line*) and $\omega t = 86.7$ (*dotted line*) for $z = 1$. *Top right*: Contour plot for $z = 1$. *Bottom left*: $t = 0$ (*solid line*) and $\omega t = 86.7$ (*dotted line*) for $z = 0.1$. *Bottom right*: Contour plot for $z = 0.1$ (Larkin and Filinov 2013)

the positions of the maxima of the velocity distributions $W(v)$ at the initial time $t = 0$ do not coincide with the positions of the maxima of the momentum distribution at $p = 0$. Moreover, under the time evolution, the velocity distribution is restricted by the light speed, while the momentum distributions have no limits. The asymmetry of the velocity distribution at the initial time is the result of statistical errors, due to sampling exponentially rare events in the tails of the Gaussian momentum distributions. Note that not all the distributions are normalized to unity. The presented values show the number of virtual trajectories counted in the vicinity of each point.

Time Evolution of Quantum Average Values

We now consider the time dependence of the quantum average values of such physical quantities as energy, momentum, and position and their dispersions. Figure 12.4 shows the time dependence of the average energy. Here we see results for three different values of the relativism parameter, viz., $\hbar\omega = 0.01mc^2$, $\hbar\omega = mc^2$, and $\hbar\omega = 10mc^2$ (solid lines). The calculation for the non-relativistic expression (12.52) is represented by the dashed line for $\hbar\omega = 10mc^2$. Note that the non-relativistic expression for the Wigner function leads to an underestimate of the average energy of the oscillator.

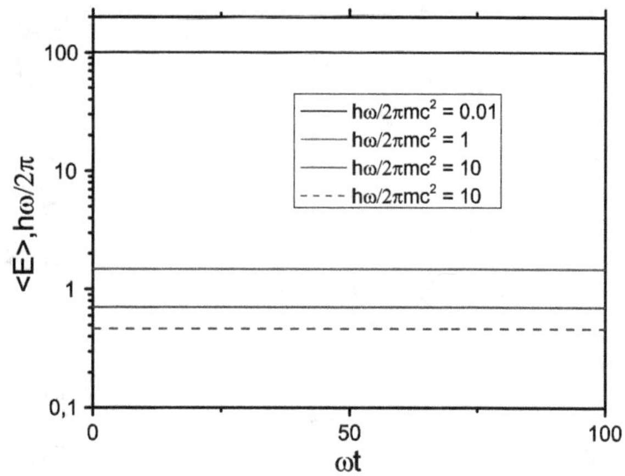

Fig. 12.4 Time dependence of the quantum average energy. *Solid lines* correspond to oscillators with different degrees of relativism. The initial Wigner function is calculated from (12.50) with $p_0 = 0, x_0 = 0$. The *dashed line* corresponds to a highly relativistic oscillator ($\hbar\omega = 10mc^2$) with non-relativistic initial Wigner function (12.52) (Larkin and Filinov 2013)

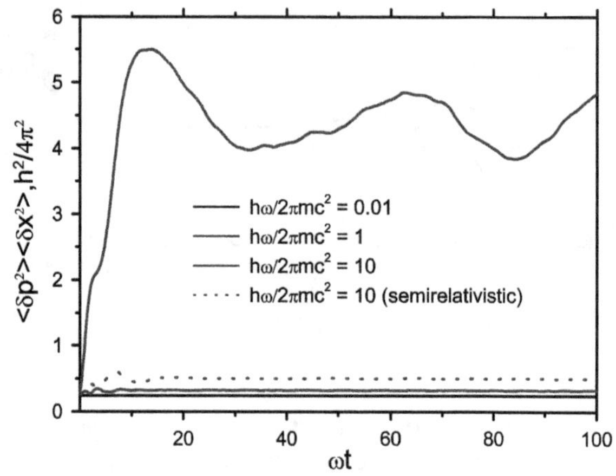

Fig. 12.5 Time dependence of the product of the momentum and position dispersions for oscillators with different degrees of relativism. The *dotted line* represents results for the oscillator with $\hbar\omega/mc^2 = 10$ and non-relativistic initial Wigner function (12.52) (Larkin and Filinov 2013)

The time evolution of the product of the momentum and position dispersions $\langle p^2 - \bar{p}^2\rangle\langle x^2 - \bar{x}^2\rangle$ is shown in Fig. 12.5 for the same values of the relativism parameter. This product must satisfy the Heisenberg principle

$$\langle \delta\, p^2\rangle\langle \delta\, x^2\rangle \geq \hbar^2/4\,, \tag{12.57}$$

and at the initial time $t = 0$, it has the minimum $\hbar^2/4$ [a property of the initial state (12.51)]. The initial state is the eigenstate of the non-relativistic oscillator, so this product is constant. In the relativistic case, this quantity increases quickly as result of the 'broadening' of the Wigner function in the phase space. Note that the non-relativistic initial Wigner function (12.52) broadens much less than with relativistic one (12.50).

The time dependencies of the average momentum and position are represented in Fig. 12.6 for $\hbar\omega = 10mc^2$. Solid lines correspond to the initial relativistic Wigner

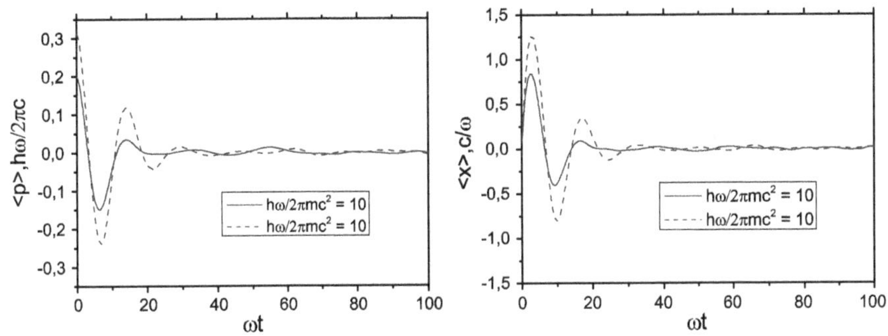

Fig. 12.6 Time dependencies of the average momentum (*left*) and average position (*right*) for an oscillator with $\hbar\omega = 10mc^2$. The *solid line* corresponds to a relativistic initial Wigner function, and the *dashed line* to a non-relativistic Wigner function (Larkin and Filinov 2013)

function (12.50), and dashed lines to the non-relativistic Wigner function (12.52). In both cases, $x_0 = 0$, $p_0 = 0.32\hbar\omega/c$. The damping of oscillations for the non-relativistic Wigner function is less, reflecting weaker broadening of the Wigner function in phase space. In addition, note that, for a non-relativistic initial Wigner function, the physical meanings of the parameters p_0 and x_0 are the initial average momentum and position, while this is not the case for the relativistic initial function, with $\langle p(0)\rangle = 0.2\hbar\omega/c$, $\langle x(0)\rangle = 0$.

Virtual Trajectories

To understand the physical reasons for the different properties of non-relativistic and relativistic harmonic oscillators, we consider the behavior of individual virtual trajectories having an initial delta distribution with fixed initial momentum and coordinate, viz.,

$$W_0(x_0, p_0; 0) = \delta(x - x_0)\delta(p - p_0) , \tag{12.58}$$

where (p_0, x_0) are the initial data for the virtual trajectories. Here, due to the restriction (12.9), the distribution W_0 cannot be considered as a physical Wigner function. However, this choice of the Wigner function can help in understanding peculiarities in the time behavior of the average values discussed above. The given choice of Wigner function (12.58) reduces the average over the ensemble of virtual trajectories to consideration of the contribution from only one virtual trajectory with fixed initial values x_0 and p_0, as in the classical case.

Note that in the relativistic case the virtual trajectories cannot be obtained in an explicit analytical form, as before. However, one can find the period of oscillation of the virtual trajectories as a function of the energy (Belendez et al. 2009). From (12.53), one obtains an equation for $p(t)$ and its first integral:

$$\frac{\mathrm{d}^2\bar{p}}{\mathrm{d}t^2} + m\omega^2 c\frac{\bar{p}}{\sqrt{\bar{p}^2 + m^2c^2}} = 0\,, \quad C_1 = \left(\frac{\mathrm{d}\bar{p}}{\mathrm{d}t}\right)^2 + 2m\omega^2\sqrt{\bar{p}^2c^2 + m^2c^4}\,, \tag{12.59}$$

where $C_1 = 2m\omega^2 E$ and the energy $E = \sqrt{p_0^2c^2 + m^2c^4} + m\omega^2 x_0^2/2$ is to be found from the initial conditions p_0 and x_0.

Integrating this equation (Ryzhik and Gradshteyn 1963), the period of oscillations depends on the energy of the trajectory according to

$$T(E) = \frac{4\sqrt{2}\sqrt{\frac{E}{mc^2}+1}}{\omega} E\left(\frac{E-mc^2}{E+mc^2}\right) - \frac{4\sqrt{2}}{\omega\sqrt{\frac{E}{mc^2}+1}} K\left(\frac{E-mc^2}{E+mc^2}\right) , \tag{12.60}$$

where $K(z)$ and $E(z)$ are elliptical integrals of the first and second kinds, viz.,

$$K(z) = \int_0^{\pi/2} \frac{d\phi}{\sqrt{1 - z\sin^2\phi}} , \quad E(z) = \int_0^{\pi/2} \sqrt{1 - z\sin^2\phi}\, d\phi . \tag{12.61}$$

This dependence is shown in the right-hand panel of Fig. 12.8.

Figure 12.7 shows the virtual trajectories as a function of time. For $z = 100$, the non-relativistic trajectories look like sinusoid curves with equal oscillation periods (trajectories 1 and 2 in the left panel). Note that the energy of the virtual trajectory is proportional to the square of x_0, and in the relativistic case (central and right-hand panels $z = 0.1$), the oscillation period increases sharply with increasing energy of the trajectory [see (12.60)]. The period of oscillations of $x(\omega t)$ for 4 is longer than for 3, due to the higher energy value. Moreover, 3 and 4 tend to become polygonal curves.

The velocity of the trajectory 4 (right-hand panel of Fig. 12.7) looks like a rectangular wave for high energy. This figure illustrates the fact that the virtual trajectory mainly has velocity v approaching the velocity of light c and only differs from this limit near the turning points. For the same z but lower energy, the velocity of trajectory 3 differs considerably from a rectangular wave.

The left-hand and central panels of Fig. 12.8 show the non-relativistic virtual trajectories versus time for $z = 100$ (lines 1, 2) and relativistic trajectories for $z = 0.1$ (lines 3, 4). The time dependencies of $p(\omega t)$ look almost like sinusoid curves. In the

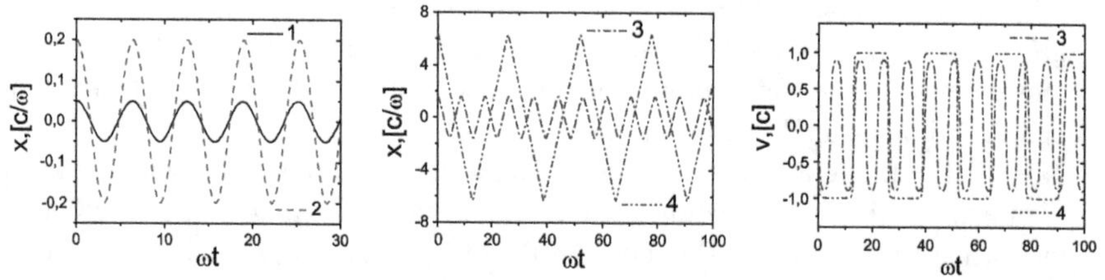

Fig. 12.7 Virtual trajectories for $p_0 = 0$. *Left*: Non-relativistic trajectories 1 (with $x_0 = 0.05c/\omega$) and 2 (with $x_0 = 0.2c/\omega$) for $z = 100$. *Centre and right*: Relativistic trajectories 3 (with $x_0 = 1.58c/\omega$) and 4 (with $x_0 = 6.3c/\omega$) $z = 0.1$ (Larkin and Filinov 2013)

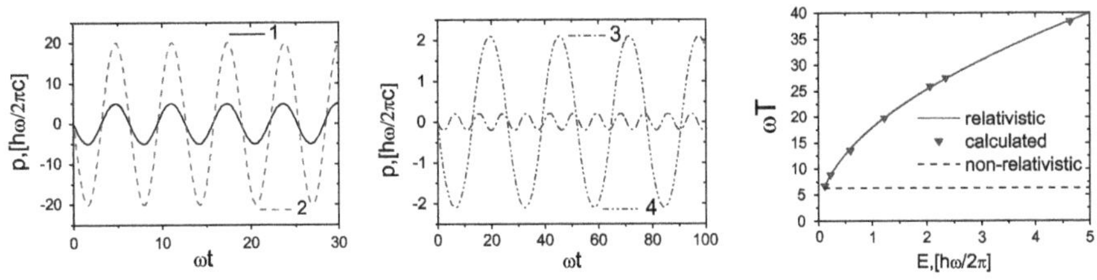

Fig. 12.8 Virtual trajectories for $p_0 = 0$. *Left*: Non-relativistic trajectories 1 and 2. *Center*: Relativistic trajectories 3 and 4. The oscillation period of the relativistic trajectory 4 is longer than that of 3 due to the higher value of the virtual energy. *Right*: Period of oscillation of the virtual trajectories versus energy for $z = 0.1$. *Solid line*: Eq. (12.60). *Triangles* are results of numerical simulations, the *dashed line* is the non-relativistic case (Larkin and Filinov 2013)

left-hand panel, period of the momentum oscillations for trajectories 1 and 2 is almost the same. In the central panel, the oscillation period is longer for the relativistic trajectory 4 with higher energy (a quantitative comparison will be given below).

Analytical and numerical calculations of the oscillation period are compared in the right-hand panel of the Fig. 12.8. The analytical dependence of the oscillation period on the energy according to the formula (12.60) is plotted in the solid curve for $z = 0.1$. Results of numerical calculations, represented by triangles, agree very well with the analytical dependence.

Time Dilation

One of the most important relativistic effects is time dilation, i.e., the proper time of a relativistic particle is slower than time in the non-relativistic (inertial) lab frame of reference. Let us consider this effect for a non-relativistic initial Wigner function of the harmonic oscillator. We begin by considering one virtual trajectory $p(t), x(t)$. The relationship between proper time and lab time is well known for the case when the particle moves with constant velocity, namely,

$$t' - t_0' = (t - t_0)\sqrt{1 - \frac{\dot{x}^2}{c^2}}\,, \tag{12.62}$$

where $t - t_0$ is the time interval between two events in the lab frame of reference, $t' - t_0'$ is the time interval between these two events in the rest frame of the particle, and $\dot{x}$ is the velocity of the particle. For the oscillator, however, the particle's velocity is not constant and (12.62) is correct only for infinitesimal time intervals:

$$\mathrm{d}t' = \mathrm{d}t\sqrt{1 - \frac{\dot{x}(t)^2}{c^2}}\,, \tag{12.63}$$

while for a finite time interval

$$t' - t_0' = \int_{t_0}^{t} \mathrm{d}t\sqrt{1 - \frac{\dot{x}(t)^2}{c^2}}\,. \tag{12.64}$$

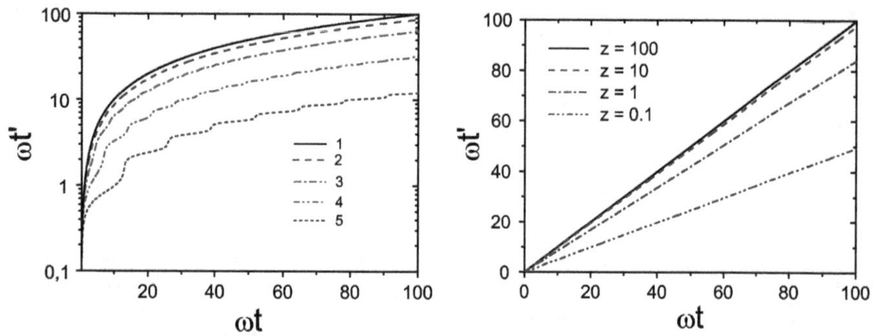

Fig. 12.9 Time dilation for $t_0 = 0$. *Left*: Time dilation for virtual trajectories with $z = 0.1$. Energies E_{tr} of the virtual trajectories: (1) $0.1\hbar\omega$ (resting particle), (2) $0.13\hbar\omega$, (3) $0.225\hbar\omega$, (4) $0.6\hbar\omega$, (5) $2.1\hbar\omega$. *Right*: Time dilation for semi-relativistic quantum oscillators (Larkin and Filinov 2013)

The time dependence of virtual trajectories with different energies is represented on a logarithmic scale in the left-hand panel of Fig. 12.9 for $z = 0.1$. Almost constant parts of the curves related to trajectories with high energy correspond to motion with velocity of the order of the speed of light c.

To obtain a physical result, one has to average the integral in (12.63) over all virtual trajectories. So for the quantum oscillator described by the Wigner function $W(x, p; t)$, the time dilation for a semi-relativistic oscillator can be calculated from

$$t' - t'_0 = \int_{t_0}^{t} \mathrm{d}t \int W(x, p; t) \frac{1}{\sqrt{1 + (p/mc)^2}} \mathrm{d}p \, \mathrm{d}x \, ,$$

which is related to the 'time dilation operator'

$$\mathrm{d}\hat{t}' = \mathrm{d}t \sqrt{1 - \frac{\hat{\dot{x}}(t)^2}{c^2}} \, . \tag{12.65}$$

In the numerical calculation, averaging was over all virtual trajectories for each time t. The results are shown in the right-hand panel of Fig. 12.9 for oscillators with $z = 100$, 10, 1, and 0.1.

12.3 Transport Properties of Quark–Gluon Plasmas

The DPIMC method itself cannot yield the transport properties of the QGP. One possible way to solve this task was suggested in Gelman et al. (2006). The authors of this paper used the classical color molecular dynamics (MD) simulation, allowing for quantum effects only phenomenologically via modification of a short-range potential. In contrast to these classical MD simulations (Gelman et al. 2006), we discuss here a more rigorous approach to study the transport properties of strongly coupled quark–gluon systems. This approach is based on a combination of the Feynman and Wigner formulations of quantum dynamics (Filinov et al. 2011). The basic ideas of this approach were discussed above, while here a generalization accounting for color degree of freedom will be presented. In particular, this approach allows one to calculate the self-diffusion coefficient and the viscosity of the strongly coupled QGP and compared them with available data.

Wigner Dynamics for Color Quantum Particles

Here we consider a multi-component QGP consisting of N color quasiparticles: N_{g} gluons, N_{q} quarks, and $\overline{N}_{\mathrm{q}}$ antiquarks. Henceforth, in order to describe the QGP, we shall use a system of units in which Planck's constant, the light velocity, and the Boltzman constant are all equal to unity. The Hamiltonian of this system is $\hat{H} = \hat{K} + \hat{U}^{\mathrm{C}}$ with the kinetic and color Coulomb interaction parts

$$\hat{K} = \sum_i \sqrt{\hat{\mathbf{p}}_i^2 + m_i^2(T, \mu_{\mathrm{q}})}\,, \quad \hat{U}^{\mathrm{C}} = \frac{1}{2}\sum_{i \neq j} \frac{g^2(T, \mu_{\mathrm{q}})(Q_i \cdot Q_j)}{4\pi|\mathbf{r}_i - \mathbf{r}_j|}\,. \tag{12.66}$$

Let us consider the possibility of using the Wigner formulation of quantum mechanics to treat QGP transport properties. To take into account the color degree of freedom of the QGP quasiparticles, one must derive the color Wigner–Liouville equation. As in the previous chapter, in the density matrix $\rho(r, r', Q, t)$ of the $3D$ N-particle system, the symbol Q denotes the color degrees of freedom of the quarks, antiquarks, and gluons. Here and above, the color charges Q are treated semi-classically, so only the diagonal of the density matrix with respect to colors will be considered. The Q variable includes both the canonical coordinate and momentum corresponding to color dynamics (see the appendix in Sect. 11.4). Therefore, the density matrix takes the form

$$\rho(r, r', Q, t) = \underline{\rho}(r, r', Q, t) \prod_i \delta\big(Q_i - Q_i(t)\big)\,, \tag{12.67}$$

where the product runs over all particles in the system and $Q_i(t)$ is a solution of the equation of motion for color (12.70). Here the quantity $\underline{\rho}(r, r', Q, t) = \Psi(r, Q, t)\Psi^*(r', Q, t)$ is a quantum part of the density matrix, with Ψ an eigenfunction of the Hamiltonian operator described by (12.66), and the Q are already c-numbers.

In the classic approximation, this system is governed by Wong's equations of motion (Wong 1970):

$$\frac{\mathrm{d}\mathbf{p}_i(t)}{\mathrm{d}t} = \mathbf{F}_i(t)\,, \tag{12.68}$$

$$\frac{\mathrm{d}\mathbf{r}_i(t)}{\mathrm{d}t} = \mathbf{v}_i(t)\,, \tag{12.69}$$

$$\frac{\mathrm{d}Q_i^a(t)}{\mathrm{d}t} = \Upsilon_i^a(t)\,. \tag{12.70}$$

Now the definition of the corresponding Wigner function (WF) reads

$$w\,(p,q,Q,t) = \frac{1}{(2\pi)^{3N}} \int \rho\left(q - \frac{\xi}{2}, q + \frac{\xi}{2}, Q, t\right) \mathrm{e}^{\mathrm{i}p\xi}\,\mathrm{d}\xi\,. \tag{12.71}$$

The quasiparticles are also characterized by spin and flavor, which we do not include explicitly in the list of quasiparticle degrees of freedom. Now the Wigner–Liouville equation (WLE) has the form (Filinov et al. 2011; Cho and Zahed 2010):

$$\frac{\partial w}{\partial t} + v\frac{\partial w}{\partial q} + F\frac{\partial w}{\partial p} + \Upsilon\frac{\partial w}{\partial Q} = \int \mathrm{d}s\, w\,(p-s,q,Q,t)\,\omega\,(s,q)\;, \tag{12.72}$$

where $v = \{\mathbf{v}_i\}$ is a $3N$-dimensional vector of velocities $\mathbf{v}_i = \mathbf{p}_i/\sqrt{\mathbf{p}_i^2 + m_i^2}$ of all the quasiparticles, $\mathbf{F}_i = -\partial U^{\mathrm{C}}/\partial \mathbf{r}_i$ is a set of color-electric forces experienced by all the quasiparticles [see (12.68)],

$$\Upsilon_i^a = \sum_{j\neq i}^{N} \sum_{b,c} \breve{f}_{abc} \frac{g^2 Q_i^b Q_j^c}{4\pi|\mathbf{r}_i - \mathbf{r}_j|} \tag{12.73}$$

is an $8N$-dimensional vector of driving forces in Wong's equation of motion for the color charge, and $\breve{f}_{abc}$ are structure constants of the group SU(3) with $a, b, c = 1, \ldots, 8$. So below we consider spatial degrees of freedom quantum mechanically, but the color dynamics still classically.

The classical part of the WLE (12.72), i.e., the left-hand side, can be easily derived, e.g., from Wong's equations of motion (12.68)–(12.70) for the color-charged particles (Cho and Zahed 2010). In particular, the term $\Upsilon\partial w/\partial Q$ naturally results from $(\mathrm{d}Q(t)/\mathrm{d}t)\partial w/\partial Q$ and Wong's equation (12.70).

Wigner Representation of Time Correlation Functions

In computations of transport properties, like viscosity, the starting point is the general Kubo expression for the canonical ensemble-averaged operator (Zubarev 1974):

$$\breve{C}_{BA}(t) = Z^{-1}\mathrm{Tr}\left(\mathrm{e}^{-\beta H}\hat{B}\mathrm{e}^{\mathrm{i}\hat{H}t}\hat{A}\mathrm{e}^{-\mathrm{i}\hat{H}t}\right)\,, \tag{12.74}$$

where $\hat{B}$ and $\hat{A}$ are quantum operators of dynamic quantities under consideration and $Z(N,V,T) = \mathrm{Tr}\big(\mathrm{e}^{-\beta\hat{H}}\big)$ is the canonical partition function. Frequently, a symmetric time-correlation function is also used (Doll et al. 1990):

$$C_{BA}(t) = Z^{-1}\mathrm{Tr}\left(\hat{B}\mathrm{e}^{\mathrm{i}\hat{H}t_c^*}\hat{A}\mathrm{e}^{-\mathrm{i}\hat{H}t_c}\right)\,, \tag{12.75}$$

where $t_c = t - i\beta/2$ is a complex-valued quantity including the reciprocal temperature $\beta = 1/T$. The Fourier transforms of $\breve{C}_{BA}(t)$ and $C_{BA}(t)$ are related by (Doll et al. 1990)

$$C_{BA}(\omega) = \exp\left(-\frac{\beta\omega}{2}\right)\breve{C}_{BA}(\omega) . \tag{12.76}$$

As a consequence, transport coefficients described by zero-frequency ($\omega = 0$) Fourier components can be obtained from the symmetric time-correlation functions, and this may offer certain computational advantages. This symmetric form is used below.

The Wigner representation of the time correlation function can be written in the form

$$C_{BA}(t) = (2\pi)^{-6N}\int \mathrm{d}\overline{pq\mu Q}\,\mathrm{d}\widetilde{pq\mu Q}\; B(\overline{pqQ})A(\widetilde{pqQ})\; W\left(\overline{pqQ};\widetilde{pqQ};t;\beta\right) , \tag{12.77}$$

where we have introduced a short-hand notation for phase space points in the $(6N + 8N)$-dimensional space, viz., $\overline{pqQ}$ and $\widetilde{pqQ}$, with p, q, and Q being the momenta, coordinates, and color variables, respectively, of all the particles in the system. Here $A(pqQ)$ denotes the Weyl symbol (Tatarskii 1983) of the operator $\hat{A}$:

$$A(\widetilde{pqQ}) = \int \mathrm{d}\widetilde{\xi}\exp(-\mathrm{i}\widetilde{p}\widetilde{\xi})\left\langle \widetilde{q} - \frac{\widetilde{\xi}}{2}, \widetilde{Q}\left|\hat{A}\right|\widetilde{q} + \frac{\widetilde{\xi}}{2}, \widetilde{Q}\right\rangle , \tag{12.78}$$

and similarly for the operator $\hat{B}$, while $W\left(\overline{pqQ};\widetilde{pqQ};t;\beta\right)$ is the spectral density expressed in the form

$$W\left(\overline{pqQ};\widetilde{pqQ};t;\beta\right) = Z^{-1}\sum_{\widetilde{\sigma},\overline{\sigma}}\int\int \mathrm{d}\overline{\xi}\,\mathrm{d}\widetilde{\xi}\,\mathrm{e}^{\mathrm{i}\overline{p}\overline{\xi}}\mathrm{e}^{\mathrm{i}\widetilde{p}\widetilde{\xi}}$$
$$\times\left\langle \overline{q} + \frac{\overline{\xi}}{2}, \overline{Q}\left|\mathrm{e}^{\mathrm{i}\hat{H}t_c^*}\right|\widetilde{q} - \frac{\widetilde{\xi}}{2}, \widetilde{Q}\right\rangle\left\langle \widetilde{q} + \frac{\widetilde{\xi}}{2}, \widetilde{Q}\left|\mathrm{e}^{-\mathrm{i}\hat{H}t_c}\right|\overline{q} - \frac{\overline{\xi}}{2}, \overline{Q}\right\rangle . \tag{12.79}$$

In (12.77), we tacitly assumed that the operators $\hat{A}$ and $\hat{B}$ do not depend on the spin variables. Therefore, summation over spins $\widetilde{\sigma}$ and $\overline{\sigma}$ can be safely moved to the definition of W. Here and below, we do not explicitly mention spin variables, if they are not essential. The time correlation function $C_{BA}(t)$ is a linear functional of the spectral density W. Thus, the problem of its treatment is reduced to considering the evolution of the spectral density.

It follows from (12.72) and Filinov (1996) that the following system of WL integro-differential equations describe the time evolution of the color spectral density W:

$$\frac{\partial W}{\partial t} + \overline{v}\frac{\partial W}{\partial \overline{q}} + F\frac{\partial W}{\partial \overline{p}} + \Upsilon\frac{\partial W}{\partial \overline{Q}} = \int \mathrm{d}s\, W\Big(\overline{p} - s, \overline{qQ}; \widetilde{pqQ}; t; \beta\Big)\omega\,(s, \overline{q},) \;, \tag{12.80}$$

$$-\frac{\partial W}{\partial t} + \widetilde{v}\frac{\partial W}{\partial \widetilde{q}} + F\frac{\partial W}{\partial \widetilde{p}} + \Upsilon\frac{\partial W}{\partial \widetilde{Q}} = \int \mathrm{d}s\, W\Big(\overline{pqQ}; \widetilde{p} - s, \widetilde{qQ}; t; \beta\Big)\omega\,(s, \widetilde{q}) \;, \tag{12.81}$$

where $\omega\,(s, q)$ is defined as in Chap. 11. These equations are derived in precisely the same way as (12.72), except that the Hamiltonian H appears here as the result of the differentiation of the exponent functions $\mathrm{e}^{\mathrm{i}\hat{H}t}$ and $\mathrm{e}^{-\mathrm{i}\hat{H}t}$ with respect to time. Note that, while (12.80) describes the evolution in the positive time direction, (12.81) specifies propagation in the reverse time direction. This happens because of the presence the direct time $\mathrm{e}^{-\mathrm{i}\hat{H}t}$ and reverse time $\mathrm{e}^{\mathrm{i}\hat{H}t}$ evolution operators in the definition of the time-correlation function (12.75).

Now using (12.80) and (12.81), we obtain an integral equation (Tatarskii 1983; Filinov et al. 1995; Ciccotti et al. 1999; Filinov 1996):

$$\begin{aligned} W\Big(\overline{pqQ}; \widetilde{pqQ}; t; \beta\Big) &= \int \mathrm{d}\overline{p_0 q_0 \mu Q_0}\, \mathrm{d}\widetilde{p_0 q_0 \mu Q_0}\, G\Big(\overline{pqQ}, \widetilde{pqQ}, t; \overline{p_0 q_0 Q_0}, \widetilde{p_0 q_0 Q_0}, 0\Big) \\ &\quad \times W(\overline{p_0 q_0 Q_0}; \widetilde{p_0 q_0 Q_0}; 0, \beta) \\ &\quad + \frac{1}{2}\int_0^t \mathrm{d}t' \int \mathrm{d}s \int \mathrm{d}\overline{p'q'\mu Q'}\, \mathrm{d}\widetilde{p'q'\mu Q'}\, G\Big(\overline{pqQ}, \widetilde{pqQ}, t; \overline{p'q'Q'}, \widetilde{p'q'Q'}, t'\Big) \\ &\quad \times \Big[W(\overline{p}' - s, \overline{q'Q'}; \widetilde{p'q'Q'}; t'; \beta)\omega(s, \overline{q}') \\ &\quad - W(\overline{p'q'Q'}; \widetilde{p}' - s, \widetilde{q'Q'}; t'; \beta)\omega(s, \widetilde{q}')\Big], \end{aligned} \tag{12.82}$$

with the Green function

$$\begin{aligned} &G\Big(\overline{pqQ}, \widetilde{pqQ}, t; \overline{p'q'Q'}, \widetilde{p'q'Q'}, t'\Big) \\ &= \delta\Big(\overline{p} - \overline{p}(t; \overline{p'q'Q'}, t')\Big)\, \delta\Big(\overline{q} - \overline{q}(t; \overline{p'q'Q'}, t')\Big)\, \delta\Big(\overline{Q} - \overline{Q}(t; \overline{p'q'Q'}, t')\Big) \\ &\times \delta\Big(\widetilde{p} - \widetilde{p}(t; \widetilde{p'q'Q'}, t')\Big)\, \delta\Big(\widetilde{q} - \widetilde{q}(t; \widetilde{p'q'Q'}, t')\Big)\, \delta\Big(\widetilde{Q} - \widetilde{Q}(t; \widetilde{p'q'Q'}, t')\Big). \end{aligned} \tag{12.83}$$

describing propagation of the spectral density along classical trajectories in the positive time direction:

$$\frac{d\overline{p}(t;\overline{p'q'Q'},t')}{dt} = \frac{1}{2}F(\overline{qQ_t}) , \quad \frac{d\overline{q}(t;\overline{p'q'Q'},t')}{dt} = \frac{1}{2}\overline{v}\big[\overline{p}(t;\overline{p'q'Q'},t')\big] ,$$
$$\frac{d\overline{Q}(t;\overline{p'q'Q'},t')}{dt} = \frac{1}{2}\Upsilon(\overline{qQ_t}) , \tag{12.84}$$

and in the reverse time direction

$$\frac{d\widetilde{p}(t;\widetilde{p'q'Q'},t')}{dt} = -\frac{1}{2}F(\widetilde{qQ_t}) , \quad \frac{d\widetilde{q}(t;\widetilde{p'q'Q'},t')}{dt} = -\frac{1}{2}\widetilde{v}\big[\widetilde{p}(t;\widetilde{p'q'Q'},t')\big] ,$$
$$\frac{d\widetilde{Q}(t;\widetilde{p'q'Q'},t')}{dt} = -\frac{1}{2}\Upsilon(\widetilde{qQ_t}) , \tag{12.85}$$

where $(\widetilde{qQ_t}) = \big[\widetilde{q}(t;\widetilde{p'q'Q'},t'),\widetilde{Q}(t;\widetilde{p'q'Q'},t')\big]$ and similarly for barred quantities. These equations of motion are supplemented by initial conditions at time $t = 0$:

$$\overline{p}(t;\overline{p_0q_0Q_0},0) = \overline{p_0} , \quad \overline{q}(t;\overline{p_0q_0Q_0},0) = \overline{q_0} , \quad \overline{Q}(t;\overline{p_0q_0Q_0},0) = \overline{Q_0} , \tag{12.86}$$

$$\widetilde{p}(t;\widetilde{p_0q_0Q_0},0) = \widetilde{p_0} , \quad \widetilde{q}(t;\widetilde{p_0q_0Q_0},0) = \widetilde{q_0} , \quad \widetilde{Q}(t;\widetilde{p_0q_0Q_0},0) = \widetilde{Q_0} , \tag{12.87}$$

and by initial conditions at time $t = t'$:

$$\overline{p}(t';\overline{p'q'Q'},t') = \overline{p}' , \quad \overline{q}(t';\overline{p'q'Q'},t') = \overline{q}' , \quad \overline{Q}(t';\overline{p'q'Q'},t') = \overline{Q}' , \tag{12.88}$$

$$\widetilde{p}(t';\widetilde{p'q'Q'},t') = \widetilde{p}' , \quad \widetilde{q}(t';\widetilde{p'q'Q'},t') = \widetilde{q}' , \quad \widetilde{Q}(t';\widetilde{p'q'Q'},t') = \widetilde{Q}' . \tag{12.89}$$

In fact, (12.84) are Wong's equations of motion but written for half-time $(t/2)$. Similarly, (12.85) are the half-time Wong equations of motion reversed in time. This happens because the time correlation is taken between instants in the past and the future, with the initial conditions fixed between these instants, i.e., at $t = 0$ the spectral density

$$W(\overline{pqQ_0};\widetilde{pqQ_0};t=0,\beta) = W_0(\overline{pqQ_0};\widetilde{pqQ_0};\beta)$$

is fixed as described in the next subsection. The right-hand sides of (12.84) and (12.85) include an interparticle interaction that can be arbitrarily strong.

The integral equation (12.82) can be solved in a form of iterative series with absolute convergence. Here only the first term of this iteration series will be taken into account:

$$W\left(\overline{pqQ};\widetilde{pqQ};t;\beta\right)\approx\int \mathrm{d}\overline{p_0q_0\mu Q_0}\,\mathrm{d}\widetilde{p_0q_0\mu Q_0}\\ \times G\left(\overline{pqQ},\widetilde{pqQ},t;\overline{p_0q_0Q_0},\widetilde{p_0q_0Q_0},0\right)W_0(\overline{pqQ_0};\widetilde{pqQ_0};\beta)\,. \tag{12.90}$$

Note that, if the initial $W_0(\overline{pqQ_0};\widetilde{pqQ_0};\beta)$ is chosen appropriately (Tatarskii 1983), i.e., so that it contains all powers of Planck's constant, then the first term of the iterative series, i.e., (12.90), describes propagation of an initial *quantum* spectral density along classical trajectories. Other (higher order) terms describe propagation of the initial spectral density along the analogous trajectories, but perturbed by momentum jumps resulting from the convolution structure of the integral term in (12.82). From the physical point of view, these jumps relate to quantum ($[p,q]$) uncertainty [for details, see the discussion in Filinov et al. (1995)]. As found in Filinov et al. (1995), Ciccotti et al. (1999), Filinov et al. (2002), the main contribution to WF comes from the trajectories without jumps, i.e., from the first term of the iterative series, if the motion takes place in a classically accessible region, which is the case here.

As we know, the classical limit for a multi-component Coulomb system does not exist, since the stability of Coulomb systems is only provided by quantum effects. This was the physical reason for adopting the color Kelbg potential in the partition function and (12.99) for the initial condition (see below). To take this quantum effect into account, we replace the color Coulomb potential U^{C} by the Kelbg potential Φ in the quantities F and Υ defining propagation of the spectral density along the classical trajectories [see (12.84) and (12.85)]. With this replacement, we are able to take into account certain higher-order quantum terms of the iterative series representing the solution of the integral equation (12.82). From a practical point of view, it allows us to avoid problems due to the singular character of the Coulomb potential.

Initial Conditions

The initial function W_0 is expressed in terms of the matrix elements of the density matrix. Accordingly to (12.79) for $t=0$, we have

$$W_0\left(\overline{pqQ};\widetilde{pqQ};\beta\right)=Z^{-1}\sum_{\widetilde{\sigma},\overline{\sigma}}\int \mathrm{d}\overline{\xi}\,\mathrm{d}\widetilde{\xi}\,\mathrm{e}^{\mathrm{i}\overline{p}\overline{\xi}}\mathrm{e}^{\mathrm{i}\widetilde{p}\widetilde{\xi}}\left\langle\overline{q}+\frac{\overline{\xi}}{2},\overline{Q}\left|\rho\left(\frac{\beta}{2}\right)\right|\widetilde{q}-\frac{\widetilde{\xi}}{2},\widetilde{Q}\right\rangle\\ \times\left\langle\widetilde{q}+\frac{\widetilde{\xi}}{2},\widetilde{Q}\left|\rho\left(\frac{\beta}{2}\right)\right|\overline{q}-\frac{\overline{\xi}}{2},\overline{Q}\right\rangle\delta(\overline{Q}-\widetilde{Q})\,\delta_{\widetilde{\sigma},\overline{\sigma}}\,. \tag{12.91}$$

Thus, the problem is reduced to calculating the matrix elements of the density matrix ρ, which is similar to what we did in Chap. 11 on thermodynamics. The difference is that we now need nondiagonal matrix elements, rather than diagonal ones.

We subdivide $\rho\,(\beta/2)$ into beads using the operator identity

$$\mathrm{e}^{-\beta\hat{H}/2} = \mathrm{e}^{-\Delta\beta'\hat{H}}\mathrm{e}^{-\Delta\beta'\hat{H}}\cdots\mathrm{e}^{-\Delta\beta'\hat{H}} ,$$

where the right-hand side contains $n+1$ identical factors with

$$\Delta\beta' = \beta/2(n+1) ,$$

so that

$$\left\langle \overline{q}+\frac{\overline{\xi}}{2}, Q \middle| \rho\left(\frac{\beta}{2}\right) \middle| \widetilde{q}-\frac{\widetilde{\xi}}{2}, Q \right\rangle \approx \int \mathrm{d}\overline{r}^{(1)}\ldots\mathrm{d}\overline{r}^{(n)}\,\overline{\rho}^{(1)}\,\overline{\rho}^{(2)}\,\overline{\rho}^{(3)}\cdots\overline{\rho}^{(n)}\overline{\rho}^{(n+1)} , \quad (12.92)$$

where $\overline{\rho}^{(l)}$ $(l=1,\ldots,n+1)$ are defined as in Chap. 11 with $\Delta\beta$ replaced by $\Delta\beta'$. The bar on $\overline{\rho}^{(l)}$ means that these quantities depend on barred variables $\overline{r}^{(l)}$. Similarly, after antisymmetrization, we obtain

$$\left\langle \widetilde{q}+\frac{\widetilde{\xi}}{2}, Q \middle| \rho\left(\frac{\beta}{2}\right) \middle| \overline{q}-\frac{\overline{\xi}}{2}, Q \right\rangle \approx \int \mathrm{d}\widetilde{r}^{(1)}\ldots\mathrm{d}\widetilde{r}^{(n+1)}\,\widetilde{\rho}^{(1)}\widetilde{\rho}^{(2)}\widetilde{\rho}^{(3)}\cdots\widetilde{\rho}^{(n)}$$
$$\times\left[\sum_{P_{\mathrm{q}}}\sum_{P_{\bar{q}}}\sum_{P_{\mathrm{g}}}(-1)^{\kappa_{P_{\mathrm{q}}}+\kappa_{P_{\bar{q}}}}\hat{P}_{\mathrm{q}}\hat{P}_{\bar{q}}\hat{P}_{\mathrm{g}}\widetilde{\rho}^{(n+1)}\right] , \quad (12.93)$$

with

$$\widetilde{r}^{(0)} = \widetilde{q}+\frac{\widetilde{\xi}}{2} , \qquad \widetilde{r}^{(n+1)} = \overline{q}-\frac{\overline{\xi}}{2} ,$$

where the functions $\widetilde{\rho}^{(l)}$ carrying a tilde depend on variables carrying a tilde. It is enough to perform the symmetrization–antisymmetrization in a single matrix element in (12.91).

In contrast to Chap. 11, we use here an alternative high temperature approximation. Notice that $\exp\left[-(\Delta\beta'/2)\hat{K}\right]$ is in fact a density matrix of the non-interacting system. This alternative symmetrized form results in the following approximation to the matrix elements:

$$\begin{aligned}\rho^{(l)} &= \rho\left(r^{(l-1)}, r^{(l)}, Q; \{N\}; \Delta\beta'/2\right)\\ &= \int \mathrm{d}q^{(l)}\,\rho_0\left(r^{(l-1)}, q^{(l)}, Q; \{N\}; \Delta\beta'/2\right)\,\rho_0\left(q^{(l)}, r^{(l)}, Q; \{N\}; \Delta\beta'/2\right)\\ &\quad\times\exp\left[-\Delta\beta'\frac{1}{2}\sum_{i,j(i\neq j)}^{N}\Phi_{ij}\left(\mathbf{q}_i^{(l)}-\mathbf{q}_j^{(l)}, \mathbf{q}_i^{(l)}-\mathbf{q}_j^{(l)}, Q_i, Q_j\right)\right] ,\end{aligned} \quad (12.94)$$

where Φ_{ij} is the diagonal part of the color Kelbg potential. This approximate form has the same accuracy as the one used in Chap. 11. At the same time, it allows us to perform Fourier transforms explicitly, since the ξ dependence now occurs only in the ρ_0 factors.

Thus, on the basis of the above approximation for $\rho^{(l)}$, we are able to evaluate the integrals over $\overline{\xi}$ and $\widetilde{\xi}$ explicitly:

$$\int \mathrm{d}\widetilde{\xi}\, \mathrm{e}^{\mathrm{i}\widetilde{p}\widetilde{\xi}} \overline{\rho}^{(n+1)} \widetilde{\rho}^{(1)} = \overline{\rho}^{(n+1)} \widetilde{\rho}^{(1)} \varphi\left(\widetilde{p}; \overline{r}^{(n)}, \widetilde{r}^{(1)}\right), \tag{12.95}$$

$$\int \mathrm{d}\overline{\xi}\, \mathrm{e}^{\mathrm{i}\overline{p}\overline{\xi}} \widetilde{\rho}^{(n+1)} \overline{\rho}^{(1)} = \widetilde{\rho}^{(n+1)} \overline{\rho}^{(1)} \varphi\left(\overline{p}; \widetilde{r}^{(n)}, \overline{r}^{(1)}\right), \tag{12.96}$$

where, on the right-hand side of these equations and below, the marginal coordinates already take the values

$$\overline{r}^{(0)} = \overline{q}\,, \qquad \overline{r}^{(n+1)} = \widetilde{q}\,, \qquad \widetilde{r}^{(0)} = \widetilde{q}\,, \qquad \widetilde{r}^{(n+1)} = \overline{q}\,,$$

and the complex-valued function φ is defined by

$$\varphi\left(p; r', r''\right) = \prod_{i=1}^{N} \left(2\lambda_i'^2\right)^{3/2} \exp\left[-\frac{1}{2\pi}\left(\mathbf{p}_i \lambda_i' + \mathrm{i}\pi \frac{\mathbf{r}_i' - \mathbf{r}_i''}{\lambda_i'}\right)^2\right], \tag{12.97}$$

with $\lambda_i' = \sqrt{\pi \Delta\beta'/m_i}$ being the i th particle thermal wavelength corresponding to the temperature $2/\Delta\beta'$.

Substituting these expressions into (12.91), we arrive at (Filinov 1996; Filinov et al. 2002; Feynman and Hibbs 1965)

$$\begin{aligned} &W_0(\overline{pqQ}; \widetilde{pqQ}; \beta) \\ &= \frac{1}{Z} \int \mathrm{d}\overline{r}^{(1)} \ldots \mathrm{d}\overline{r}^{(n)}\, \mathrm{d}\overline{q}^{(1)} \ldots \mathrm{d}\overline{q}^{(n+1)}\, \mathrm{d}\widetilde{r}^{(1)} \ldots \mathrm{d}\widetilde{r}^{(n)}\, \mathrm{d}\widetilde{q}^{(1)} \ldots \mathrm{d}\widetilde{q}^{(n+1)} \\ &\times \Psi\Big(\overline{pqQ}; \widetilde{pqQ}; \overline{r}^{(1)}, \ldots, \overline{r}^{(n)}; \overline{q}^{(1)}, \ldots, \overline{q}^{(n+1)}; \widetilde{r}^{(1)}, \ldots, \widetilde{r}^{(n)}; \\ &\qquad \widetilde{q}^{(1)}, \ldots, \widetilde{q}^{(n+1)}; \beta\Big), \end{aligned} \tag{12.98}$$

with

$$\begin{aligned} &\Psi\left(\overline{pqQ}; \widetilde{pqQ}; \overline{r}^{(1)}, \ldots, \overline{r}^{(n)}; \overline{q}^{(1)}, \ldots, \overline{q}^{(n+1)}; \widetilde{r}^{(1)}, \ldots, \widetilde{r}^{(n)}; \widetilde{q}^{(1)}, \ldots, \widetilde{q}^{(n+1)}; \beta\right) \\ &= \exp\left[-\beta\left(\overline{U} + \widetilde{U}\right)\right] \delta(\overline{Q} - \widetilde{Q}) \\ &\times \sum_{\widetilde{\sigma}, \overline{\sigma}} \left[\prod_{l=1}^{n+1} \rho_0\left(\overline{r}^{(l-1)}, \overline{q}^{(l)}, \overline{Q}; \{N\}; \Delta\beta'/2\right) \rho_0\left(\overline{q}^{(l)}, \overline{r}^{(l)}, \overline{Q}; \{N\}; \Delta\beta'/2\right)\right] \\ &\times \varphi\left(\widetilde{p}; \overline{r}^{(n)}, \widetilde{r}^{(1)}\right) \\ &\times \left[\prod_{l=1}^{n} \rho_0\left(\widetilde{r}^{(l-1)}, \widetilde{q}^{(l)}, \widetilde{Q}; \{N\}; \Delta\beta'/2\right) \rho_0\left(\widetilde{q}^{(l)}, \widetilde{r}^{(l)}, \widetilde{Q}; \{N\}; \Delta\beta'/2\right)\right] \\ &\times \varphi\left(\overline{p}; \widetilde{r}^{(n)}, \overline{r}^{(1)}\right) \end{aligned}$$

$$\times \sum_{P_q} \sum_{P_{\overline{q}}} \sum_{P_g} (-1)^{\kappa_{P_q}+\kappa_{P_{\overline{q}}}} \hat{P}_q \hat{P}_{\overline{q}} \hat{P}_g \rho_0 \left(\widetilde{q}^{(n+1)}, \widetilde{r}^{(n+1)}, \widetilde{Q}; \{N\}; \Delta\beta'/2 \right)$$
$$\times \delta_{\overline{\sigma},\widetilde{\sigma}} \, \mathcal{S}(\widetilde{\sigma}, \hat{P}_q \hat{P}_{\overline{q}} \hat{P}_g \widetilde{\sigma}')\big|_{\widetilde{\sigma}'=\widetilde{\sigma}} , \tag{12.99}$$

where

$$\overline{U} = \frac{1}{2} \frac{1}{n+1} \sum_{l=1}^{n+1} \sum_{i,j(i\neq j)}^{N} \Phi_{ij} \left(\overline{\mathbf{q}}_i^{(l)} - \overline{\mathbf{q}}_j^{(l)}, \overline{\mathbf{q}}_i^{(l)} - \overline{\mathbf{q}}_j^{(l)}, \overline{Q}_i, \overline{Q}_j \right) , \tag{12.100}$$

$$\widetilde{U} = \frac{1}{2} \frac{1}{n+1} \sum_{l=1}^{n+1} \sum_{i,j(i\neq j)}^{N} \Phi_{ij} \left(\widetilde{\mathbf{q}}_i^{(l)} - \widetilde{\mathbf{q}}_j^{(l)}, \widetilde{\mathbf{q}}_i^{(l)} - \widetilde{\mathbf{q}}_j^{(l)}, \widetilde{Q}_i, \widetilde{Q}_j \right) . \tag{12.101}$$

So we finally arrive at

$$\Psi \left(\overline{pqQ}; \widetilde{pqQ}; \overline{r}^{(1)}, \ldots, \overline{r}^{(n)}; \overline{q}^{(1)}, \ldots, \overline{q}^{(n+1)}; \widetilde{r}^{(1)}, \ldots, \widetilde{r}^{(n)}; \widetilde{q}^{(1)}, \ldots, \widetilde{q}^{(n+1)}; \beta \right)$$
$$= \exp\left[-\beta \left(\overline{U} + \widetilde{U} \right) \right] \delta(\overline{Q} - \widetilde{Q})$$
$$\times \sum_{\widetilde{\sigma},\overline{\sigma}} \left[\prod_{l=1}^{n+1} \left(\prod_{i=1}^{N} \phi_{ii} \left(\overline{r}^{(l-1)}, \overline{q}^{(l)}, \Delta\beta'/2 \right) \right) \left(\prod_{i=1}^{N} \phi_{ii} \left(\overline{q}^{(l)}, \overline{r}^{(l)}, \Delta\beta'/2 \right) \right) \right]$$
$$\times \varphi \left(\overline{p}; \overline{r}^{(n)}, \overline{r}^{(1)} \right)$$
$$\times \left[\prod_{l=1}^{n} \left(\prod_{i=1}^{N} \phi_{ii} \left(\widetilde{r}^{(l-1)}, \widetilde{q}^{(l)}, \Delta\beta'/2 \right) \right) \left(\prod_{i=1}^{N} \phi_{ii} \left(\widetilde{q}^{(l)}, \widetilde{r}^{(l)}, \Delta\beta'/2 \right) \right) \right]$$
$$\times \frac{\det \left\| \phi \left(\widetilde{q}^{(n+1)}, \widetilde{r}^{(n+1)}, \Delta\beta'/2 \right) \right\|_{N_q}}{\Lambda_q^{6(n+1)N_q} (\Delta\beta'/2)} \frac{\det \left\| \phi \left(\widetilde{q}^{(n+1)}, \widetilde{r}^{(n+1)}, \Delta\beta'/2 \right) \right\|_{\overline{N}_q}}{\Lambda_{\overline{q}}^{6(n+1)\overline{N}_q} (\Delta\beta'/2)}$$
$$\times \frac{\mathrm{per} \left\| \phi \left(\widetilde{q}^{(n+1)}, \widetilde{r}^{(n+1)}, \Delta\beta'/2 \right) \right\|_{N_g}}{\Lambda_g^{6(n+1)N_g} (\Delta\beta'/2)} \varphi \left(\overline{p}; \overline{r}^{(n)}, \overline{r}^{(1)} \right) \delta_{\overline{\sigma},\widetilde{\sigma}} , \tag{12.102}$$

with

$$\overline{r}^{(0)} = \overline{q} , \qquad \overline{r}^{(n+1)} = \widetilde{q} , \qquad \widetilde{r}^{(0)} = \widetilde{q} , \qquad \widetilde{r}^{(n+1)} = \overline{q} .$$

In the limit $n \to \infty$, this expression gives the product of matrix elements in (12.91) exactly in the form of path integrals multiplied by a limiting expression of the φ functions. According to the Lebesque–Dirac delta theorem (see the appendix at the end of this section), in the limit $n \to \infty$, the $\overline{\rho}^{()} \widetilde{\rho}^{()} \varphi$ products in the integral (12.98) are equivalent to the following real-valued expressions:

$$\overline{\rho}^{(n+1)} \widetilde{\rho}^{(1)} \varphi \left(\widetilde{p}; \widetilde{r}^{(n)}, \widetilde{r}^{(1)} \right) \overset{n\to\infty}{\longrightarrow} \overline{\rho}^{(n+1)} \widetilde{\rho}^{(1)} \prod_{i=1}^{N} \left[\left(2\lambda_i^2 \right)^{3/2} \exp\left(-\frac{\widetilde{\mathbf{p}}_i^2 \lambda_i^2}{2\pi} \right) \right] \left(\frac{\lambda_i'}{\pi} \right)^3$$

$$\times \delta\left(\overline{\mathbf{r}}_i^{(n)} - \widetilde{\mathbf{r}}_i^{(1)}\right), \qquad (12.103)$$

$$\widetilde{\rho}^{(n+1)}\overline{\rho}^{(1)}\varphi\left(\overline{p}; \widetilde{r}^{(n)}, \overline{r}^{(1)}\right) \stackrel{n\to\infty}{\longrightarrow} \overline{\rho}^{(n+1)}\widetilde{\rho}^{(1)} \prod_{i=1}^{N}\left[\left(2\lambda_i^2\right)^{3/2}\exp\left(-\frac{\overline{\mathbf{p}}_i^2\lambda_i^2}{2\pi}\right)\right]\left(\frac{\lambda_i'}{\pi}\right)^3$$

$$\times \delta\left(\widetilde{\mathbf{r}}_i^{(n)} - \overline{\mathbf{r}}_i^{(1)}\right), \qquad (12.104)$$

where $\lambda_i = \sqrt{\pi\beta/2m_i}$. Analytical integration over the delta function simplifies the final path integral used for further computation of the real-valued W_0 by means of the Monte Carlo method.

Transport Coefficients

An important aspect of the strongly coupled QGP is its transport properties, which differ significantly from those expected for weakly coupled plasmas. The approach developed here on the basis of the Wigner formulation of quantum mechanics (Wigner dynamics WD) is used to calculate the QGP self-diffusion constant and shear viscosity, as these quantities can be compared with the corresponding values deduced from analysis of experimental data on heavy-ion collisions and also the predictions of lattice QCD computations. More precisely, a summary of shear viscosity deduced from analysis of experimental elliptic flow and an extensive review of theoretical works on viscosity are presented in Khvorostukhin et al. (2010), Bass (2011), while the heavy-quark diffusion constant is available from experimental analysis (Gossiaux et al. 2012, He et al. 2013) and QCD lattice computations (Ding et al. 2012; Banerjee et al. 2012). The WD self-diffusion and heavy-quark diffusion constants are comparable in order of magnitude.

A natural way to obtain these transport coefficients is to use the quantum Green–Kubo relations. These give the transport coefficients in terms of integrals of equilibrium time-dependent correlation functions. According to (12.77), the self-diffusion constant D is the integral of the velocity autocorrelation function

$$D = \lim_{t\to\infty} D(t)\,, \quad D(t) = \frac{1}{3}\int_0^t \mathrm{d}\tau \left\langle v(0)\cdot v(\tau)\right\rangle, \qquad (12.105)$$

$$\left\langle v(0)\cdot v(\tau)\right\rangle = (2\pi)^{-6N}\int \mathrm{d}\overline{pq\mu Q}\,\mathrm{d}\widetilde{pq\mu Q}$$

$$\times W\left(\overline{pqQ}; \widetilde{pqQ}; \tau; \beta\right) v\left(\overline{p}(\tau)\right)\cdot v\left(\widetilde{p}(\tau)\right),$$

where the product of the 3-velocities is

$$v\left(\overline{p}(\tau)\right)\cdot v\left(\widetilde{p}(\tau)\right) = \frac{1}{N}\sum_{i=1}^{N}\frac{\overline{\mathbf{p}}_i(\tau)\cdot\widetilde{\mathbf{p}}_i(\tau)}{\sqrt{\overline{\mathbf{p}}_i^2(\tau)+m_i^2}\sqrt{\widetilde{\mathbf{p}}_i^2(\tau)+m_i^2}}\,. \qquad (12.106)$$

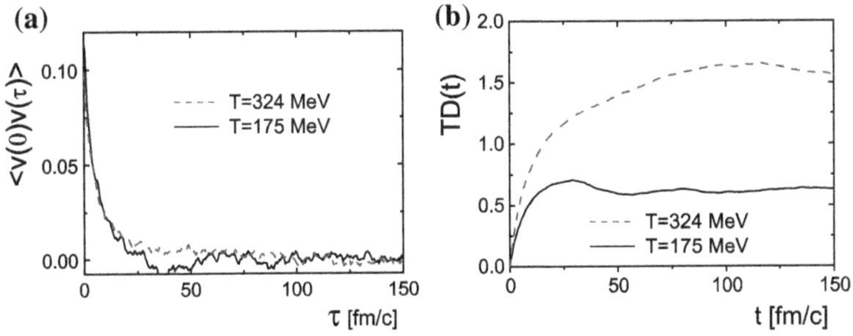

Fig. 12.10 Velocity autocorrelation function **a** and temperature-scaled self-diffusion function $TD(t)$ **b** versus time for two temperatures (Filinov et al. 2013)

where the spectral density $W(\overline{pqQ}; \widetilde{pqQ}; \tau; \beta)$ is given by (12.90), while trajectories in positive and reverse time directions, shown with a bar or a tilde, are defined by (12.84) and (12.85), respectively. Figure 12.10 shows examples of the velocity–velocity autocorrelation and its antiderivative functions. The self-diffusion constant is a limiting value of the related antiderivative function at $t \to \infty$.

The autocorrelation functions are calculated in the canonical ensemble and involve a combination of Monte Carlo sampling of initial conditions $\overline{pqQ_0}$ and $\widetilde{pqQ_0}$ for the trajectories and solution of the system of dynamic equations (12.84) and (12.85). The initial conditions $\overline{pqQ_0}$ and $\widetilde{pqQ_0}$ for the trajectories are sampled by the Monte Carlo method using the probability distribution $W_0(\overline{pqQ_0}; \widetilde{pqQ_0}; \beta)$. The autocorrelator (12.106) is calculated as a function of time along the trajectories (12.84) and (12.85), which are themselves computed by means of a numerical scheme for solution of a system of ordinary first order differential equations. It is reasonable to use an explicit numerical scheme with automatically adapted time step. To check the correctness of the calculations, the three integrals of motion, viz., the energy and the quadratic and cubic Casimirs, were considered. Their variations in WD calculations amount to less than 1–2%. Several thousand generated trajectories are usually required for convergence of the antiderivative of the autocorrelation function up to an accuracy of 5–10%. The convergence is fast enough because the autocorrelation function involves averaging (i.e., summation) over all quasiparticles.

Analogously, the Green–Kubo relation for the shear viscosity is the integral of the autocorrelation function of the stress–energy tensor:

$$\eta = \lim_{t\to\infty} \eta(t)\,, \quad \eta(t) = \frac{1}{VT}\int_0^t \mathrm{d}\tau \langle \sigma_{xy}(0)\sigma_{xy}(\tau)\rangle$$
$$\langle \sigma_{xy}(0)\sigma_{xy}(\tau)\rangle = (2\pi)^{-6N} \int \mathrm{d}\overline{pq\mu Q}\, \mathrm{d}\widetilde{pq\mu Q}\, W\left(\overline{pqQ}; \widetilde{pqQ}; \tau; \beta\right)$$
$$\times \sigma_{xy}\left(\overline{pqQ(\tau)}\right) \sigma_{xy}\left(\widetilde{pqQ(\tau)}\right)\,, \tag{12.107}$$

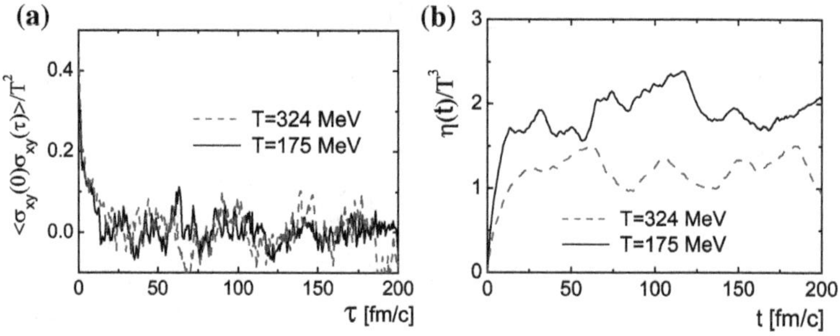

Fig. 12.11 Temperature-scaled stress–energy tensor autocorrelation function **a** and shear-viscosity function $\eta(t)/T^3$ **b** versus time for two temperatures (Filinov et al. 2013)

where the off-diagonal stress–energy tensor is

$$\sigma_{xy}\big(pqQ(\tau)\big) = \sum_{i=1}^{N} \frac{p_{i,x}(\tau)\, p_{i,y}(\tau)}{\sqrt{\mathbf{p}_i^2(\tau) + m_i^2}} - \frac{1}{2} \sum_{i \neq j}^{N} q_{ij,x}(\tau) \frac{\partial U\,(qQ)}{\partial q_{ij,y}}(\tau)\,, \tag{12.108}$$

with $\mathbf{q}_{ij} = \mathbf{q}_i - \mathbf{q}_j$, and U is the sum of the color Kelbg potentials defined by (12.100) and (12.101) with $n = 0$. Examples of the stress–energy-tensor autocorrelation and its antiderivative function are presented in Fig. 12.11. The shear viscosity is defined by the limiting value of the related antiderivative function at $t \to \infty$.

The self-diffusion constant and the viscosity-to-entropy ratio are presented in Fig. 12.12 as a function of temperature. The entropy density s is taken from results of calculations presented in Chap. 11. WD results are presented in the form of bands. The width of the bands represents the theoretical uncertainty associated with oscillations of the antiderivative functions at large times (see Figs. 12.10 and 12.11). Slowly decaying oscillations of the time correlation functions are inherent in liquid-like systems of strongly interacting particles, in contrast to exponentially decaying oscillations in gas-like systems. As known from hydrogen plasmas, these oscillations arise because of quasi-closed chaotic orbits, and are caused by strong interparticle interactions. In liquids, these oscillations decay according to a power law, rather than exponentially. Therefore, extremely long (in time) trajectories are required for more accurate calculations of the diffusion constant and viscosity. Due to CPU time limitations, these computations were stopped before the decay of oscillations was complete.

Unfortunately, the self-diffusion constant is unavailable from other calculations. Therefore, the WD results were compared with the heavy-quark diffusion constant, anticipating that these should be of the same order of magnitude. The heavy-quark diffusion constant is available from recent QCD lattice computations (Ding et al. 2012; Banerjee et al. 2012), presented in Fig. 12.12a. WD results (labeled 'Wigner

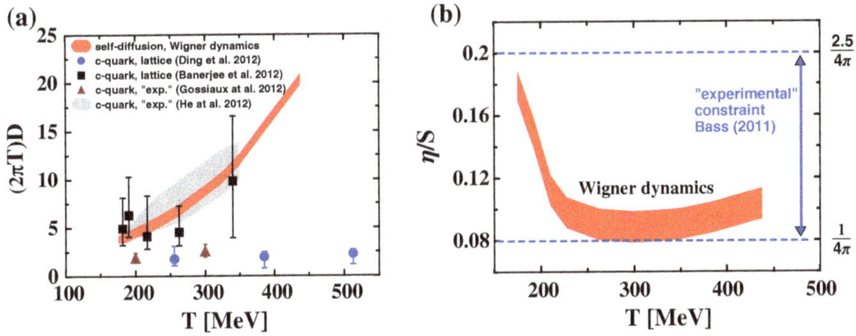

Fig. 12.12 **a** The self-diffusion constant [self-diffusion, Wigner dynamics] as a function of temperature, compared with heavy-quark diffusion constants predicted by QCD lattice computations [c quark, lattice (Ding et al. 2012)] and (Banerjee et al. 2012) [c-quark, lattice (Banerjee et al. 2012)], and deduced from analysis of experimental data (Gossiaux et al. 2012) [c-quark, "exp." (Gossiaux et al. 2012)] and (He et al. 2013) [c-quark, "exp." (He et al. 2013)]. **b** The ratio of shear viscosity to entropy density as a function of temperature (*shaded area* marked 'Wigner dynamics'). *Horizontal dashed lines* indicate the range of constraint on the viscosity-to-entropy ratio deduced from numerous hydrodynamical simulations of heavy-ion experimental data, as summarized in Bass (2011), Filinov et al. (2013)

dynamics') agree well with lattice data from Banerjee et al. (2012), while they essentially overestimate those from Ding et al. (2012).

The heavy-quark diffusion constant is also available from analysis of experiments on heavy-quark quenching in ultrarelativistic heavy ion collisions at RHIC. For comparison, the two studies (Gossiaux et al. 2012; He et al. 2013) were taken. Here the results are also rather diverse. Estimates from (He et al. 2013) compare well with the WD result and lattice data from Banerjee et al. (2012), while the estimate from Gossiaux et al. (2012), He et al. 2013 is considerably lower and conforms better to lattice results from Ding et al. (2012).

WD results on the shear viscosity are presented in Fig. 12.12b. The WD results on the shear viscosity are in the range of those deduced from the analysis of the experimental elliptic flow in ultrarelativistic heavy ion collisions at RHIC, as summarized in (Bass 2011), i.e., in terms of the viscosity-to-entropy ratio, $1/4\pi \leq \eta/S < 2.5/4\pi$, in the temperature range from 170 to 440 MeV. Lattice data on the shear viscosity in the realistic case of the SU(3) group are not available, but an extensive review of theoretical work on viscosity in QCD-motivated models can be found in Khvorostukhin et al. (2010). The minimum of the viscosity-to-entropy ratio is reached at a temperature ($\approx$300 MeV), above the expected phase transition rather than at the phase transition point, as commonly expected. This minimum turns out to be very shallow. The value of the viscosity-to-entropy ratio at the minimum is very close (from above) to the lower bound of $\eta/S = 1/4\pi$ for this quantity (Kovtun et al. 2005). As the temperature decreases, i.e., towards the hadronic phase, the viscosity rises rapidly.

12.4 Appendix. Lebesque–Dirac Delta Theorem

Let f be a summable function of a real argument such that $\int f(x)\mathrm{d}x = I_0$. Then $Mf(M(x-x')) \to I_0\delta(x-x')$ when $M \to \infty$.

Proof Let φ be any test function. Then

$$\begin{aligned}\lim_{M\to\infty}\int Mf\big(M(x-x')\big)\varphi(x)\mathrm{d}x &= \lim_{M\to\infty}\int f(s)\varphi(s/M+x')\mathrm{d}s \\ &= \int f(s)\Big[\lim_{M\to\infty}\varphi(s/M+x')\Big]\mathrm{d}s \\ &= \varphi(x')\int f(s)\mathrm{d}s = I_0\varphi(x')\ .\end{aligned}$$

References

D. Banerjee, S. Datta, R. Gavai, P. Majumdar, Phys. Rev. D **85**, 14510 (2012)
S.A. Bass, Nucl. Phys. A **862–863**, 174 (2011)
A. Belendez, D.I. Mendez, M.L. Alvarez, C. Pascual, T. Belende, Int. J. Mod. Phys. B **23**, 521 (2009)
S. Cho, I. Zahed, Phys. Rev. C **82**, 054907 (2010)
G. Ciccotti, C. Pierleoni, F. Capuani, V. Filinov, Comp. Phys. Comm. **121–122**, 452 (1999)
H.T. Ding, A. Francis, O. Kaczmarek, F. Karsch, H. Satz, W. Soeldner, Phys. Rev. D **86**, 14509 (2012)
J.D. Doll, D.L. Freeman, T.L. Beck, Adv. Chem. Phys. **78**, 61–127 (1990)
R.P. Feynman, A.R. Hibbs, *Quantum Mechanics and Path Integrals* (McGraw-Hill, New York, 1965)
V.S. Filinov, Yu.B Ivanov, V.E. Fortov, M. Bonitz, P.R. Levashov, Phys. Rev. C **87**, 35207 (2013)
V.S. Filinov, M. Bonitz, Y.B. Ivanov, V.V. Skokov, P.R. Levashov, V.E. Fortov, Contrib. Plasma. Phys. **51**, 322 (2011)
V.S. Filinov, J. Mol. Phys. **88**, 1517–1529 (1996)
V. Filinov, Y. Medvedev, V. Kamskii, Mol. Phys. **85**(4), 711 (1995)
V. Filinov, P. Thomas, I. Varga, T. Meier, M. Bonitz, V. Fortov, S. Koch, Phys. Rev. B **65**, 165124 (2002)
B.A. Gelman, E.V. Shuryak, I. Zahed, Phys. Rev. C **74**, 44908 (2006); ibid. **74**, 44909 (2006)
P. Gossiaux, J. Aichelin, T. Gousset, Prog. Theor. Phys. Suppl. **193**, 110 (2012); P.B. Gossiaux, J. Aichelin, M. Bluhm, T. Gousset, M. Nahrgang, S. Vogel, K. Werner. Nucl. Phys. A **910–911**, 301 (2013)
M. He, R.J. Fries, R. Rapp, Phys. Rev. Lett. **110**, 112301 (2013); arXiv:1204.4442 [nucl-th]
A.S. Khvorostukhin, V.D. Toneev, D.N. Voskresensky, Nucl. Phys. A **845**, 106 (2010)
H. Kleinert, *Path Integrals in Quantum Mechanics, Statistics, Polymer Physics and Financial Markets* (World Scientific Publishing Co., Pte. Ltd., Singapore, 2004)
YuL. Klimontovich, Doklady Akademii Nauk SSSR **108**, 1033 (1956)
P. Kovtun, D.T. Son, A.O. Starinets, Phys. Rev. Lett. **94**, 111601 (2005)
L.D. Landau, E.M. Lifshitz, *Quantum Mechanics* (PhysMathLit, Moscow, 2008)
A.S. Larkin, V.S. Filinov, Phys. Lett. A **377**, 1171 (2013)
A.S. Larkin, V.S. Filinov, Phys. Lett. A **378**, 1876 (2014)
H.B. Meyer, Phys. Rev. D **76**, 101701 (2007)

T.D. Newton, E.P. Wigner, Rev. Mod. Phys. **21**, 3 (1949)
D.E. Potter, *Computational Physics* (Wiley, Chichester, 1973)
Y.B. Rumer, A.I. Fet, *Group Theory and Quantum Fields* (Nauka, Moscow, 1977)
I.M. Ryzhik, I.S. Gradshteyn, *Table of Integrals, Series, and Products* (PhysMathLit, Moscow, 1963)
V. Tatarskii, Sov. Phys. Uspekhi **26**, 311 (1983)
N. Wiener, J. Math. Phys. **2**, 132 (1923)
E. Wigner, Phys. Rev. Phys. Rev. **40**, 749 (1932); **46**, 1002 (1934)
S.K. Wong, Nuovo Cimento A **65**, 689 (1970)
O.I. Zav'jalov, Proc. Steklov Inst. Math. **228**, 126 (2000)
D.N. Zubarev, *Nonequilibrium Statistical Thermodynamics* (Plenum Press, New York, 1974)

Index

W. Ebeling et al., *Quantum Statistics of Dense Gases and Nonideal Plasmas*, Springer Series in Plasma Science and Technology, DOI 10.1007/978-3-319-66637-2

www.ingramcontent.com/pod-product-compliance
Lightning Source LLC
Chambersburg PA
CBHW071600010626
46009CB00003B/42
9783319666365